T0329421

Industrial Process Automation Systems

Design and Implementation

Industrial Process Automation Systems

Design and Implementation

Y. Jaganmohan Reddy

AMSTERDAM • BOSTON • HEIDELBERG • LONDON
NEW YORK • OXFORD • PARIS • SAN DIEGO
SAN FRANCISCO • SINGAPORE • SYDNEY • TOKYO

Butterworth-Heinemann is an Imprint of Elsevier

Butterworth-Heinemann is an imprint of Elsevier
The Boulevard, Langford Lane, Kidlington, Oxford OX5 1GB, UK
225 Wyman Street, Waltham, MA 02451, USA

British Library Cataloguing-in-Publication Data
A catalogue record for this book is available from the British Library

Library of Congress Cataloging-in-Publication Data
A catalog record for this book is available from the Library of Congress

ISBN: 978-0-12-800939-0

For information on all Butterworth-Heinemann publications
visit our website at http://store.elsevier.com/

Typeset by Thomson Digital

Printed and bound in The United States of America

14 15 16 17 18 10 9 8 7 6 5 4 3 2 1

Contents

CHAPTER

INDUSTRIAL AUTOMATION

1

1.1 INTRODUCTION

Industrial automation of a plant/process is the application of the process control and information systems. The world of automation has progressed at a rapid pace for the past four decades and the growth and maturity are driven by the progression in the technology, higher expectations from the users, and maturity of the industrial processing technologies. Industrial automation is a vast and diverse discipline that encompasses process, machinery, electronics, software, and information systems working together toward a common set of goals – increased production, improved quality, lower costs, and maximum flexibility.

But it's not easy. Increased productivity can lead to lapses in quality. Keeping costs down can lower productivity. Improving quality and repeatability often impacts flexibility. It's the ultimate balance of these four goals – productivity, quality, cost, and flexibility that allows a company to use automated manufacturing as a strategic competitive advantage in a global marketplace. This ultimate balance is difficult to achieve. However, in this case the journey is more important than the destination. Companies worldwide have achieved billions of dollars in quality and productivity improvements by automating their manufacturing processes effectively. A myriad of technical advances, faster computers, more reliable software, better networks, smarter devices, more advanced materials, and new enterprise solutions all contribute to manufacturing systems that are more powerful and agile than ever before. In short, automated manufacturing brings a whole host of advantages to the enterprise; some are incremental improvements, while others are necessary for survival. All things considered, it's not the manufacturer who demands automation. Instead, it's the manufacturer's customer, and even the customer's customer, who have forced most of the changes in how products are currently made. Consumer preferences for better products, more variety, lower costs, and "when I want it" convenience have driven the need for today's industrial automation. Here are some of the typical expectations from the users of the automation systems.

As discussed earlier, the end users of the systems are one of the major drivers for the maturity of the automation industry and their needs are managed by the fast-growing technologies in different time zones. Here are some of the key expectations from major end users of the automation systems. The automation system has to do the process control and demonstrate the excellence in the regulatory and discrete control. The system shall provide an extensive communication and scalable architectures. In addition to the above, the users expect the systems to provide the following:

- Life cycle excellence from the concept to optimization. The typical systems are supplied with some cost and as a user, it is important to consider the overall cost of the system from the time the purchase is initiated to the time the system is decommissioned. This includes the cost of the system; cost of the hardware; and cost of services, parts, and support.

- Single integration architecture needs to be optimum in terms of ease of integration and common database and open standards for intercommunication.
- Enterprise integration for the systems needs to be available for communication and data exchange with the management information systems.
- Cyber security protection for the systems due to the nature of the systems and their deployment in critical infrastructure. Automation systems are no more isolated from the information systems for various reasons. This ability brings vulnerability in the system and the automation system's supplier is expected to provide the systems that are safe from cyber threats.
- Application integration has to be closely coupled, but tightly integrated. The systems capabilities shall be such that the integration capabilities allow the users to have flexibility to have multiple systems interconnected and function as a single system: shop floor to top floor integration or sensor to boardroom integration.
- Productivity and profitability through technology and services in the complete life cycle, in terms of ease of engineering, multiple locations based engineering, ease of commissioning, ease of upgrade, and migration to the newer releases.
- Shortening delivery time and reducing time of start-up through the use of tools and technologies. This ability clearly becomes the differentiator among the competing suppliers.
- SMART service capabilities in terms of better diagnostics, predictive information, remote management and diagnostics, safe handling of the abnormal situations, and also different models of business of services such as local inventory and very fast dispatch of the service engineers.
- Value-added services for maximization in profit, means lower product costs, scalable systems, just-in-time service, lower inventory, and technology-based services.
- Least cost of ownership of the control systems.
- Mean time to repair (MTTR) has to be minimum that can be achieved by service center at plant.

The above led to continuous research and development from the suppliers for the automation systems to develop a product that are competitive and with latest technologies and can add value to the customers by solving the main points. The following are some of the results of successful automation:

- *Consistency*: Consumers want the same experience every time they buy a product, whether it's purchased in Arizona, Argentina, Austria, or Australia.
- *Reliability*: Today's ultraefficient factories can't afford a minute of unplanned downtime, with an idle factory costing thousands of dollars per day in lost revenues.
- *Lower costs*: Especially in mature markets where product differentiation is limited, minor variations in cost can cause a customer to switch brands. Making the product as cost-effective as possible without sacrificing quality is critical to overall profitability and financial health.
- *Flexibility:* The ability to quickly change a production line on the fly (from one flavor to another, one size to another, one model to another, and the like) is critical at a time when companies strive to reduce their finished goods inventories and respond quickly to customer demands.

The earliest "automated" systems consisted of an operator turning a switch on, which would supply power to an output – typically a motor. At some point, the operator would turn the switch off, reversing the effect and removing power. These were the light-switch days of automation.

Manufacturers soon advanced to relay panels, which featured a series of switches that could be activated to bring power to a number of outputs. Relay panels functioned like switches, but allowed for more complex and precise control of operations with multiple outputs. However, banks of relay panels generated a significant amount of heat, were difficult to wire and upgrade, were prone to failure, and occupied a lot of space. These deficiencies led to the invention of the programmable controller – an electronic device that essentially replaced banks of relays – now used in several forms in millions of today's automated operations. In parallel, single-loop and analog controllers were replaced by the distributed control systems (DCSs) used in the majority of contemporary process control applications.

These new solid-state devices offered greater reliability, required less maintenance, and had a longer life than their mechanical counterparts. The programming languages that control the behavior of programmable controls and DCSs could be modified without the need to disconnect or reroute a single wire. This resulted in considerable cost savings due to reduced commissioning time and wiring expense, as well as greater flexibility in installation and troubleshooting. At the dawn of programmable controllers and DCSs, plant-floor production was isolated from the rest of the enterprise operating autonomously and out of sight from the rest of the company. Those days are almost over as companies realize that to excel they must tap into, analyze, and exploit information located on the plant floor. Whether the challenge is faster time-to-market, improved process yield, nonstop operations, or a tighter supply chain, getting the right data at the right time is essential. To achieve this, many enterprises turn to contemporary automation controls and networking architectures.

Computer-based controls for manufacturing machinery, material-handling systems, and related equipment cost-effectively generate a wealth of information about productivity, product design, quality, and delivery. Today, automation is more important than ever as companies strive to fine tune their processes and capture revenue and loyalty from consumers. This chapter will break up the major categories of hardware and software that drive industrial automation; define the various layers of automation; detail how to plan, implement, integrate, and maintain a system; and look at what technologies and practices impact manufacturers. Industrial automation is a field of engineering on application of control systems and information technologies to improve the productivity of the process, to improve the energy efficiency, to improve the safety of equipment and personnel, and to reduce the variance in the product quality and hence improve the quality.

The terminology and nomenclature of the industrial automation systems differ based on the industry of the applications. The term for computer-integrated manufacturing (CIM) is used in the manufacturing industry context and plant wide control in a process industry context. The essential of both these terms is to interconnection of information and control systems throughout a plant in order to fully integrate the coordination and control of operations. The automation engineering spans from the sensing technologies of the physical plant variables to the networks, computing resources, display technologies, and database technologies.

Improved human operator productivity will be realized through the implementation of individual workstations, which proved the tools for decision-making as well as information that is timely, accurate, and comprehensible. Time lines of data will be assured through the interconnection of all workstations and information processing facilities with a high-speed, plant-wide LAN network and a global relational database.

The broad goal is to improve the overall process and business operations by obtaining the benefits that will come from a completely integrated plant information system. The continual growth of the linkage of the process operations data with product line, project, and business systems data will be supported. The system will make such data readily available, interactively in real time, to any employee with a need to know, at workstations scattered throughout the plant and, above all, easy to use. The resulting comprehensive plant information management system will be the key to long-range improvements to process control, product line management, plant management, and support of business strategies.

One of the major challenges in today's automation jobs is evaluating the suppliers. This challenge is more apparent in the recent days because the systems appear same across the suppliers. Here are some of the guidelines that can be considered in the selection process.

These guidelines helps to set out your organization's needs, understand how suppliers can meet them, and identify the right supplier for you. The 10 Cs are Competency, Capacity, Commitment, Control, Cash, Cost, Consistency, Culture, Clean, and Communication. Used as a checklist, the 10 Cs model can help to evaluate potential suppliers in several ways. First it helps to analyze different aspects of a supplier's business: examining all 10 elements of the checklist will give a broad understanding of the supplier's effectiveness and ability to deliver the system on time, on budget with quality while having a sustained relationship for the rest of the life cycle including engineering, installation, precommissioning, commissioning, operation, and services.

1.2 INNOVATORS

The industrial automation cannot be described without remembering the pioneering works of the various scientists, whose contributions helped these technologies to mature and become commercially viable over a period of time. Few of the pioneering scientist's contributions are listed below.

ALESSANDRO VOLTA (1745–1827)

Alessandro Volta, Italian physicist is known for his pioneering work in electricity (Figure 1.1). Volta was born in Como and educated in public schools there. In 1800, he developed the battery called Upper Volta, a pioneer of the electric battery, which produces a constant flow of electricity. The electrical unit known as the volt was named in his honor owing to his work in the field of electricity. A year later, he improved and popularized the electrophorus, a device that produced static electricity. His promotion of it was so extensive that he is often credited with its invention, even though a machine operating on the same principle was described in 1762 by the Swedish experimenter Johan Wilcke.

ANDRE MARIE AMPERE (1775–1836)

Andre Marie Ampere was a French physicist and mathematician who was generally regarded as one of the main founders of the science of classical electromagnetism, which he referred to as electrodynamics (Figure 1.2). The SI unit of the measurement of electric current, the ampere is named after him. Ampere showed that two parallel wires carrying electric currents attract or repel each other, depending on whether the currents flow in the same or opposite directions, respectively – this laid the foundation of electrodynamics. The most important of these was the principle that came to be known as Ampere's law, which states that the mutual action of two lengths of current-carrying wire is proportional to their lengths and to the intensities of their currents.

FIGURE 1.1 Alessandro Volta

FIGURE 1.2 Andre Marie Ampere

GEORG SIMON OHM (1789–1854)

Ohm was a German physicist and mathematician who conducted research using the electrochemical cells (Figure 1.3). Using equipment of his own, Ohm found a directly proportional relationship between the potential difference (voltage) applied across a conductor and the resultant electric current. This relationship is called the Ohm's law. In his work (1827), the galvanic circuit investigated mathematically gave complete theory of electricity. In this work, he stated that the electromotive force acting between the extremities of any part of a circuit is the product of the strength of the current and the resistance of that part of the circuit.

WERNER VON SIEMENS (1815–1892)

Siemens was a German inventor and industrialist (Figure 1.4). Siemens name has been adopted as the SI unit of electrical conductance, the Siemens. He was also the founder of the electrical and telecommunications company Siemens.

THOMAS ALVA EDISON (1847–1931)

He is an American inventor and businessman (Figure 1.5). He developed many devices that greatly influenced life around the world, including the phonograph, the motion picture camera, and a long-lasting, practical electric light bulb. He was one of the first inventors to apply the principles of mass production and large-scale teamwork to the process of invention, and because of that, he is often credited with the creation of the first industrial research laboratory.

Edison is the fourth most prolific inventor in history, holding 1093 U.S. patents in his name, as well as many patents in the United Kingdom, France, and Germany. More significant than the number of

FIGURE 1.3 Georg Simon Ohm

FIGURE 1.4 Werner Von Siemens

FIGURE 1.5 Thomas Alva Edison

Edison's patents are the impacts of his inventions because Edison not only invented things, his inventions established major new industries worldwide, notably, electric light and power utilities, sound recording, and motion pictures. Edison's inventions contributed to mass communication and, in particular, telecommunications.

Edison developed a system of electric power generation and distribution to homes, businesses, and factories – a crucial development in the modern industrialized world. His first power station was on Pearl Street in Manhattan, New York.

FIGURE 1.6 Michael Faraday

MICHAEL FARADAY (1791–1867)

He is an English scientist who contributed to the fields of electromagnetism and electrochemistry (Figure 1.6). His main discoveries include those of electromagnetic induction, diamagnetism, and electrolysis. He was one of the most influential scientists in history. It was by his research on the magnetic field around a conductor carrying a direct current that Faraday established the basis for the concept of the electromagnetic field in physics. Faraday also established that magnetism could affect rays of light and that there was an underlying relationship between the two phenomena. He similarly discovered the principle of electromagnetic induction, diamagnetism, and the laws of electrolysis. His inventions of electromagnetic rotary devices formed the foundation of electric motor technology, and it was largely due to his efforts that electricity became practical for use in technology.

JOHN BERDEEN (1908–1991)

He is an American physicist and electrical engineer, the only person to have won the Nobel Prize in Physics twice: first in 1956 with William Shockley and Walter Brattain for the invention of the transistor; and again in 1972 with Leon N. Cooper and John Robert Schrieffer for a fundamental theory of conventional superconductivity known as the BCS theory (Figure 1.7). The transistor revolutionized the electronics industry, allowing the Information Age to occur, and made possible the development of almost every modern electronic device, from telephones to computers to missiles. Bardeen's developments in superconductivity, which won him his second Nobel, are used in nuclear magnetic resonance (NMR) spectroscopy or its medical subtool magnetic resonance imaging (MRI).

WALTER H. BRATTAIN (1902–1987)

He is an American physicist at Bell Labs who along with John Bardeen and William Shockley, invented the transistor (Figure 1.8). They shared the 1956 Nobel Prize in Physics for their invention. He devoted much of his life to research on surface states. His early work was concerned with thermionic

FIGURE 1.7 John Berdeen

FIGURE 1.8 Walter H. Brattain

emission and adsorbed layers on tungsten. He continued on the field of rectification and photo-effects at semiconductor surfaces, beginning with a study of rectification at the surface of cuprous oxide. This was followed by similar studies of silicon.

WILLIAM SHOCKLEY (1910–1989)

He is an American physicist and inventor (Figure 1.9). Along with John Bardeen and Walter Houser Brattain, Shockley coinvented the transistor, for which all three were awarded the 1956 Nobel Prize in Physics. Shockley's attempts to commercialize a new transistor design in the 1950s and 1960s led to California's "Silicon Valley" becoming a hotbed of electronics innovation. In his later life, Shockley was a professor at Stanford and became a staunch advocate of eugenics.

JACK KILBY (1923–2005)

He is an American electrical engineer who took part (along with Robert Noyce) in the realization of the first integrated circuit while working at Texas Instruments (TI) in 1958 (Figure 1.10). He was awarded the Nobel Prize in Physics in 2000. He is also the inventor of the handheld calculator and the thermal printer.

ROBERT NOYCE (1927–1990)

Nicknamed "the Mayor of Silicon Valley," cofounded Fairchild Semiconductor in 1957 and Intel Corporation in 1968 (Figure 1.11). He is also credited with the invention of the integrated circuit or microchip, which fueled the personal computer revolution and gave Silicon Valley its name. Noyce's leadership in the field of computers and his invention in the field of electronics and physics led to the computer chip we know today.

FIGURE 1.9 William Shockley

FIGURE 1.10 Jack Kilby

FIGURE 1.11 Robert Noyce

GORDON E. MOORE (1929–TODAY)

Engineer and entrepreneur Gordon Moore earned degrees in Chemistry and Physics from CalTech, was hired by William Shockley's Shockley Semiconductor in 1956, and was one of the "traitorous eight" engineers who left Shockley in 1957 to form the pioneering electronics firm, Fairchild Semiconductor

FIGURE 1.12 Gordon E. Moore

(Figure 1.12). At Fairchild, he became one of the world's foremost experts on semiconductive materials, which have much lower resistance to the flow of electrical current in one direction than in the opposite direction, and are used to manufacture diodes, photovoltaic cells, and transistors. In 1968, with Robert Noyce, he cofounded Intel, which has since become the world's largest maker of semiconductor chips.

1.3 INDUSTRIAL REVOLUTIONS

Human's ability to use the machine power as a replacement for the human or animal effort can be treated as the first revolution of the industry. From the above perspective, steam engine can be treated so because of its ability to create a large amount of work with minimum human effort. This lead to a large need to automate these machines because the true value of these machines can be realized only if they can be controlled as needed. The second industrial revolution can be seen as electricity and its benefits.

The electrical power and its ability to transport the power from one place to another changed the industrial landscape and created a large growth and new possibilities for industry. Microelectronics created another revolution with the ability to miniaturize and use less power to create bigger things. Subsequent revolution can be seen after the invention of micro controllers. At present, world is experiencing the revolution created by the Internet and the way it can impact all the industries and common people. The wealth of information and ability to connect multiple things together changes the way industries operate and yield efficiencies.

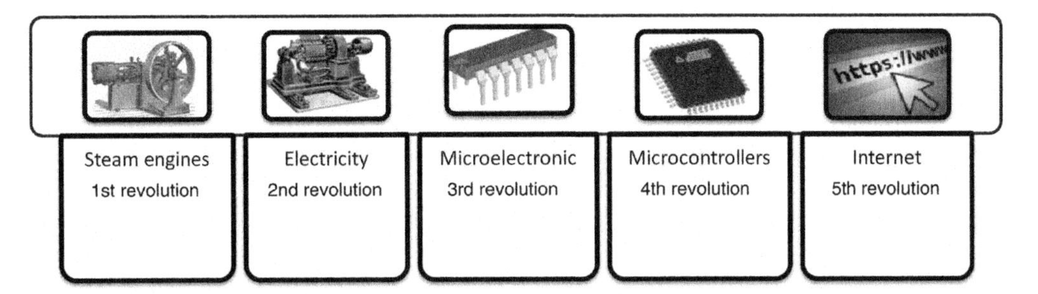

FIGURE 1.13

Evolution of automation from technology perspective

1.4 EVOLUTION OF AUTOMATION FROM NEEDS PERSPECTIVES

Initial automation needs in the industrial process operations were driven by the needs to automate the existing manual process and hence gain better yield and consistency in the product quality. The technologies available from the late 1970s fulfilled this need. Subsequently, the automation needs of the industry are driven by the standards and regulations for keeping up the environmental standards. In the 1990s the safety of the equipment and people became an additional requirement in the automation of the plant and led to the widespread usage of the safety systems and emergency shutdown and startup operations being automated. During the subsequent years in the 1990s the pressure on the process industries for improving the bottom line revenue led to the importance of having better efficiency in the plant operations. Industrial process automation systems play a bigger role in assisting the process operations for better efficiency (Figure 1.13).

1.5 EVOLUTION OF AUTOMATION FROM TECHNOLOGY PERSPECTIVES

First-generation automation systems were mostly used in the place where there is a need to replace human force by machines. The mechanization of the industries leads to automation using mechanical gears and fixtures. The control signals and mechanisms were mostly hydraulic. The pressure of the liquid in small volume is used to transmit the signals from one place to another and is subsequently manipulated for multiplication and division. However, due to the inherent nature of liquid compressibility and the energy required to create the signals, the pneumatic systems come into the place. The pneumatic signals were used to drive the sensors and valves and also for transporting the signals from one place to another.

The next generation sees that the electrical signals in the form of voltage were used as the technology of choice for the control systems. The fourth generation of the systems used the digital electronics as the technology of choice in the control systems. In the 1960s, the 4–20 mA analog interface was established as the de facto standard for instrumentation technology. As a result, the manufacturers of instrumentation equipment had a standard communication interface on which to base their products. Users had a choice of instruments and sensors, from a wide range of suppliers, which could be integrated into their control systems.

With the advent of microprocessors and the development of digital technology, the situation has changed. Most users appreciate the many advantages of digital instruments. These include more information being displayed on a single instrument, local and remote display, reliability, economy, self-tuning, and diagnostic capability. There is a gradual shift from analog to digital technology. There are a number of intelligent digital sensors, with digital communications and capability for most traditional applications. These include sensors for measuring temperature, pressure, levels, flow, mass (weight), density, power system parameters, and analytical sensors such as oxygen O_2, carbon dioxide CO_2, pH, dissolved oxygen, UV, IR, and GC. These new intelligent digital sensors are known as "smart" instrumentation.

The main features that define a "smart" instrument are intelligent, digital sensors, digital data communications capability, and ability to be multidropped with other devices in the system. There is also an emerging range of intelligent, communicating, digital devices that could be called "smart" actuators. Examples of these are devices such as variable-speed drives, soft starters, protection relays, and switchgear control with digital communication facilities. The microprocessor has had an enormous impact on instrumentation and control systems.

Historically, an instrument had a single dedicated function. Controllers were localized and, although commonly computerized, they were designed for a specific purpose. It has become apparent that a microprocessor, as a general-purpose device, can replace localized and highly site-specific controllers. Centralized microprocessors, which can analyze and display data as well as calculate and transmit control signals, are capable of achieving greater efficiency, productivity, and quality gains.

Currently, a microprocessor connected directly to sensors and a controller requires an interface card. This implements the hardware layer of the protocol stack and in conjunction with appropriate software, allows the microprocessor to communicate with other devices in the system. There are many instrumentation and control software and hardware packages; some are designed for particular proprietary systems and others are more general purpose. Interface hardware and software now available for microprocessors cover virtually all the communications requirements for instrumentation and control. As a microprocessor is relatively cheap, it can be upgraded as newer and faster models become available, thus improving the performance of the instrumentation and control system.

1.6 CHALLENGES THREE DECADES BACK

Automation systems have seen a large change in the last three decades. Due to limited communication systems, the communication between the users and the systems was limited. Due to the lack of real-time information, large inventories were planned. There were many islands of automation. The information flow from the plant floor to the executives is very limited leading to high energy costs of operations due to noncoordinated systems. The discrete and process automation is seen as two different distinct systems and information flow is slow or limited. The information on new products and new solutions available in the market is very limited. There is limited information on how other peer industries are performing and their best practices with the automation systems. All the above-mentioned problems have been resolved now due to technology and differently trained operators. The technology made more data available at an affordable price and communication revolution made the data available at any convenient location at a much faster rate. The overall plant operations have changed with the new technologies.

1.7 CURRENT CHALLENGES

Current challenges in the industry are fewer human resources, same or fewer assets and an increase in demand for more production, demands for less waste, more efficiency, better quality and improved tracking. Regulations and compliance aspects are increasing across all the domains of plant operations, especially at pharmaceuticals and petrochemicals, equipment types as distinct process and discrete is disappearing and is going toward a mix of both analog and digital. The goal to reduce energy utilization is another challenge to face. The challenges in automation and control are, demand for reduced development, engineering, and commissioning costs, and multidisciplinary control systems, namely discrete, motion, process, drives, and safety applications for improving machine uptime and reduction in machine down times, plant floor to enterprise connectivity, remote diagnostics, and web access.

1.8 TECHNOLOGY TRENDS

Overall, the industrial automation is undergoing technological changes and the major trends are moving from hardwired/proprietary systems to bus-based open systems. This is driven by the need of the users to have the lowest cost of ownership on the capital expenditure and also to have the ability and flexibility to upgrade and use multivendor products. The communication between the systems within the same plant operations is becoming an open, object-based, standard-based interface.

The general architecture of the automation systems is moving from a single large monolithic solution to a scalable and modular system. This approach is driven by the need to have the lowest capital investment in the beginning of the plant and grow with the operations. This approach also is driven by the needs such as lowest single point of failures in the system. The information collected from the automation system is moving from a repository of process information to more qualitative information supporting the business decisions. Various custom-built applications deployed on the automation systems provide valuable insight into the process operations such as inventory, and scheduling. The information so collected becomes or is becoming real-time archival instead of an archival that occurs on batches or some scheduled time of a day.

The smart field instruments became a source of information with process diagnostics and prognostics from a dumb process measuring instrument. This led to more information for the process operations on the maintenance strategies and shut-down planning. Similarly, the communication between various subsystems is moving toward wireless for the industrial environment from a point-to-point wired communications (Figure 1.14).

The automation engineers in the last three decades have experienced a wide change in the apparatus used for various automation applications. The different types of appliance used for different purposes are indicated below. The local indicators typically used to communicate the process variable to the field engineer or operator have used different display technologies. Initially all the local indicators used to be mechanical in nature and directly convert the process variable into some mechanical movement in a scale or a direct indication of the position of some specific level.

The same indicating instruments today come with LCD displays for the local operator and the display technologies have long life and the flexible orientation for ease of access. Similarly, the remote indicators are used to communicate the process variable at some other place than the measurement point and the signal

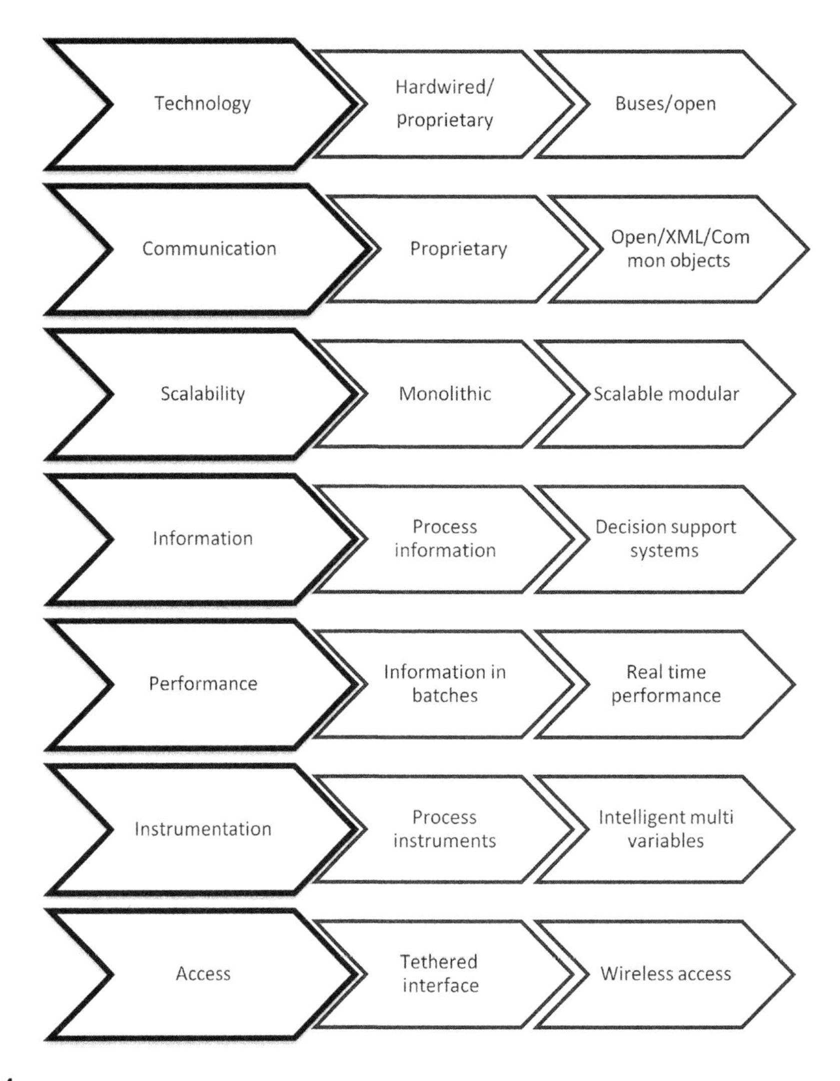

FIGURE 1.14

Technology trends

is carried out using pneumatic signals and later to analog electronics, digital electronics, and then single- and multiloop configurable microcontrollers and, subsequently, to control systems that are programmable.

Local Indication

1. Gauges
2. Direct indicating instruments
3. Transmitters
4. Pneumatics

Remote Indication

5. Analog electronics
6. Digital electronics
7. Microprocessors (configurable)
8. Bus communications (FF&PB)
9. DCS and programmable logic controllers (PLC) (programmable)

The recorders used for measuring and recording some important measurement were initially started as paper- and pen-based recorders with some mechanical motor moving the paper at a constant rate with respect to time. The technology subsequently changed to thermal charts for long duration of the recording papers. Now the paper recorders are upgraded to digital recorders with advanced communication capabilities, removable media for the storage and configurability using the open tools and operating systems.

Recording

1. Paper and pen
2. Thermal charts
3. Digital trending

Alarm systems has a journey from a local alarm indication from some sensors to a centralized audio visual annunciations in the control room to much more advanced alarm logs in computer-based systems that record the alarms, events, and sequence of events. These systems became much more advanced and an essential part of the process operations and as tools for the investigation on the plant operations.

Alarms and Events

1. Local
2. Remote audio visual annunciators
3. Alarm logs
4. Alarm management

The control of the process operations started with a local control instrument and subsequently centralized equipments in common panels. The control instruments became panel mounted versus its predecessors like field-mounted equipment. The evolution continued from local panels to plant-wise control rooms wherein all the plant operations got centralized in one control room. Subsequently, the operations became more centralized with central control rooms for the entire site operations for the supervision with the plant-wise control rooms still prevailing for the local operations. The field data and control remained assigned to process interface buildings and many such process interface buildings were connected to central control building using dual redundant fiber optic networks.

Control

1. Local instruments
2. Local panels
3. Plant wise control room
4. Centralized control room and PIB's

1.8.1 TRANSMISSION MEDIA & TECHNOLOGY

Signal transmission is one of the key enabler for the maturity of the automation and the world of automation witnessed many changes in this area. The research and technology in this area is primarily motivated for more information, which is faster at a low price. All the communication and transmission media technologies enabled the automation systems to create a central control room where the humans can comfortably sit in one place and know the plant information and control when needed at ease.

Pneumatic 3–15 psi

Signal transmission is carried pneumatically, that is, air in the tubes. Hence air compressors with oil and dirt-free air need to be designed. Signal range is 0.2–1 kg/cm^2 or 3–15 psi. Most of the control valve signals today still use the pneumatic signals due to the inherent advantages (Figure 1.15).

Electrical 4–20 mA

Electrical analog 4–20 mA: The 4–20 mA current loop is a very robust sensor signaling standard. Current loops are ideal for data transmission because of their inherent insensitivity to the electrical noise. In a 4–20 mA current loop, all the signaling current flows through all components; the same current flows even if the wire terminations are less than perfect.

All the components in the loop drop voltage due to the signaling current flowing through them. The signaling current is not affected by these voltage drops as long as the power supply voltage is greater than the sum of the voltage drops around the loop at the maximum signaling current of 20 mA. The 4–20 mA loops are available in two-wire and four-wire options. The four wires can be used for the instruments that require large power for the operation of the device. For example, a magnetic flow meter can be four-wired option, with two wires meant for carrying the power and the remaining two wires for carrying the signals. The general preferred choice is two wires, due to less cost and less complexity.

Electrical 4–20 mA with HART

Electrical Hybrid HART: 4–20 mA superimposed with DC: Smart (or intelligent) instrumentation protocols are designed for applications where actual data are collected from instruments, sensors, and actuators by digital communication techniques. These components are linked directly to DCSs or controllers. The highway addressable remote transducer (HART) protocol is a typical smart instrumentation fieldbus that can operate in a hybrid 4–20 mA digital fashion. At a basic level, most smart instruments provide core functions, such as control of range/zero/span adjustments, diagnostics to verify functionality, memory to store configuration, and status information (such as tag numbers and serial numbers). Accessing these functions allows major gains in the speed and efficiency of the

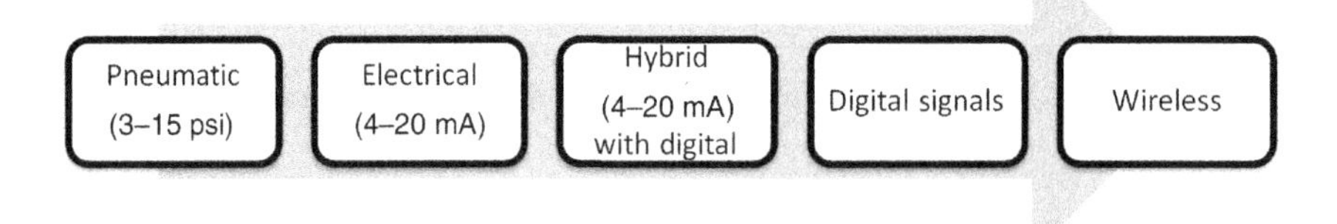

FIGURE 1.15

Evoluation of communication technology

installation and maintenance process. For example, the time-consuming 4–20 mA loop check phase can be achieved in minutes, and the device can be made ready for use in the process by zeroing and adjustment for any other controllable aspects such as the damping value.

FF and Profibus

Electrical Digital Busses (FF and Profibus DP): Foundation fieldbus (FF) takes full advantage of the emerging "smart"-field devices and modern digital communications technology allowing end user benefits, such as reduced wiring, communications of multiple process variables from a single instrument, advanced diagnostics, interoperability between devices of different manufacturers, enhanced field level control, reduced startup time, and simpler integration. The concept behind FF is to preserve the desirable features of the present 4–20 mA standard (such as a standardized interface to the communications link, bus power derived from the link and intrinsic safety options) while taking advantage of the new digital technologies. This will provide the features noted above because of the following: reduced wiring due to the multidrop capability, flexibility of supplier choices due to interoperability, reduced control room equipment due to distribution of control functions to the device level, increased data integrity and reliability due to the application of digital communications.

ProfiBus (PROcess FIeld BUS) is a widely accepted international networking standard, commonly found in process control and in large assembly and material handling machines. It supports single-cable wiring of multi-input sensor blocks, pneumatic valves, complex intelligent devices, smaller subnetworks (such as AS-i), and operator interfaces. ProfiBus is nearly universal in Europe and also popular in North America, South America, and parts of Africa and Asia. It is an open, vendor-independent standard. It adheres to the OSI model and ensures that devices from a variety of different vendors can communicate together easily and effectively.

Fiber Optic

Signals are transmitted by light rays using fiber optic cables. Fiber optic communication uses light signals guided through a fiber core. Fiber optic cables act as waveguides for light, with all the energy guided through the central core of the cable. The light is guided due to the presence of a lower refractive index cladding surrounding the central core. None of the energy in the signal is able to escape into the cladding and no energy is able to enter the core from any external sources. Therefore, the transmissions are not subject to electromagnetic interference. The core and the cladding will trap the light ray in the core, provided the light ray enters the core at an angle greater than the critical angle. The light ray will then travel through the core of the fiber, with minimal loss in power, by a series of total internal reflections.

Some of the advantages include capacity to transmit over large range distance – the glass purity of today's fiber, combined with improved system electronics, enables fiber to transmit digitized light signals well beyond 100 km without amplification. The other advantages are few transmission losses, low interference, high bandwidth potential, and easy to install because of its lightweight, small size, and flexibility.

Wireless

Signal transmission without wire. Conventionally control signals are transmitted from the field instruments to the control room via instrument cables. Instrument cables are meant for transmitting the signal from the field instruments and also to supply power to the filed instruments. In wireless technology, the control signal is transmitted without wires from the field to the control room and filed instruments are powered by battery with long life. Hence, the basic difference with wired systems is that instead of cables, data are transmitted through radio waves. From the above description, it is clear that the radio becomes a heart of the wireless communication devices. The performance of the wireless communicating devices depends on frequency and power.

Typical frequencies are 900 MHz and 2.4 GHz. Out of the two 2.4 GHz is license-free globally and 900 MHz is not allowed in some countries such as India as it interferes with licenses GSM 900 MHz spectrum. Ultimately the difference between these spectrums is data throughput and range.
The higher the frequency, the more the data are possible but less range. Many times the government regulations mandate the maximum power allowed. Standardized protocols available such as ISA100.11a and Wireless HART, which enable the field instruments from multiple vendors get connected and function with multiple host systems or control systems. The wireless systems get networked in a mesh form. Individual field devices can communicate with each other and their gateway and repeaters. Within the mesh network there are two approaches, meshed instrument network and meshed node network.

1.9 DEVICE CONNECTIVITY

Connectivity of the devices to the host systems has progressed significantly from a simple one to one connections to complex mesh networks (Figure 1.16).

One to One

Connection from the device to the host systems is generally connected in a one-to-one mode for all the years. The 4–20 mA signals are considered one of the easy examples of one-to-one connection. This type of connectivity is considered costly in terms of the cost of the cable and installation. However, the signal integrity is high and easy to troubleshoot and replace. The fault in one connection does not affect the others in the system. In a one-to-one connectivity, there shall be an equivalent connection in the host for each of the device (Figure 1.17).

Multiplexing

Multiplexing type of connectivity provides the signal integrity with less cable in the trunk. Multiple devices/sensors from the field shall be connected to one single multiplexer, which in turn carries a single cable to the host system. The multiplexer connects each device for a predefined time and carries the signal to the host system. The host system can demultiplex the signal and provides the data to upper

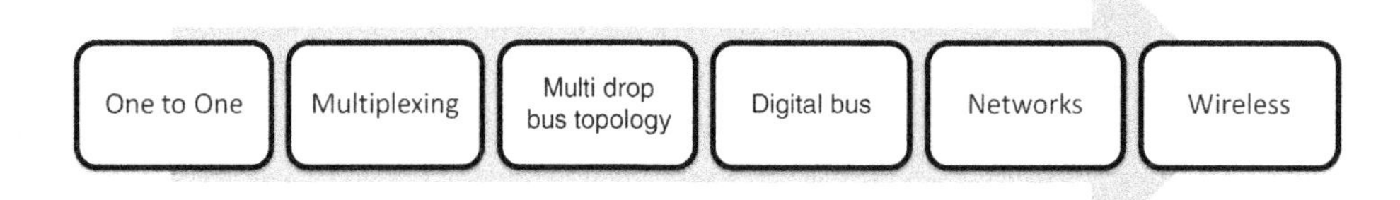

FIGURE 1.16

Evolution of connectivity

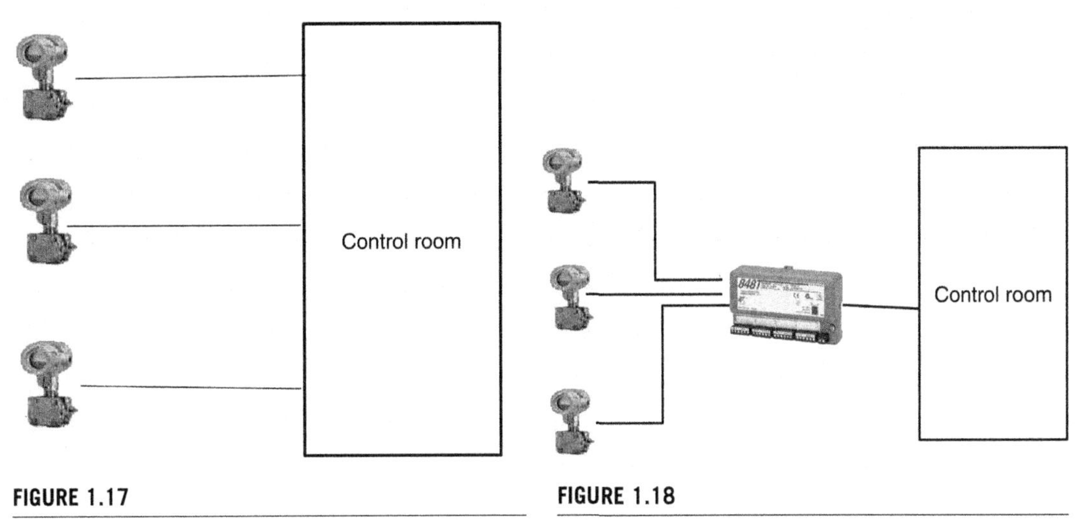

FIGURE 1.17

One to one connections

FIGURE 1.18

Multiplexed connection

layers. This type of connectivity also provides some level of integrity, less cost and less installation cost. However, the additional multiplexing device cost has to be considered. Similarly the host system shall be provided with the means to demultiplex the signals to extract the data and hence special I/O modules. The speed of the data transfer is limited by the multiplexing speed and hence is used in the places where the process variable changes slowly. For example, temperature in the tanks and large vessels changes slowly and may not need one-to-one connectivity and hence considered appropriate for the multiplexed connectivity (Figure 1.18).

Multidrop-Bus Topology

Multidrop type of connectivity is another type where in the devices shall be connected to a single cable across the devices and to the host system. The devices carry an address for each and hence each of the devices shall get some sort of allocated time for the transmission. The host system plays the role of arbitration and hence the signal integrity is maintained. The type of connectivity saves the cost of cabling and installation. However, the reliability of the installation needs to be maintained due to single point of failures on the cable side. Most of the serial communication devices are used in practice with the multidrop connectivity. The analog communication devices with HART communication provide the ability to multidrop; however, the industry acceptance is limited.

Digital Bus

Most advanced field communication protocols such as FF, Profibus, and Device net provide the digital bus for the communication (Figure 1.19).

In digital bus, the devices shall be connected to the common bus in the form of spurs from a main trunk and the trunk can be extended to a fixed length. The signal and power are carried in the same bus for the devices. Host system or any one of the devices acts as an arbitrator for the bus bandwidth allocation for each device. In the digital bus participating devices, the devices themselves acts like controllers with much advanced features embedded in them. The host system collects the processed data for the display and achieving purposes. The main advantage of the digital bus is that the signal

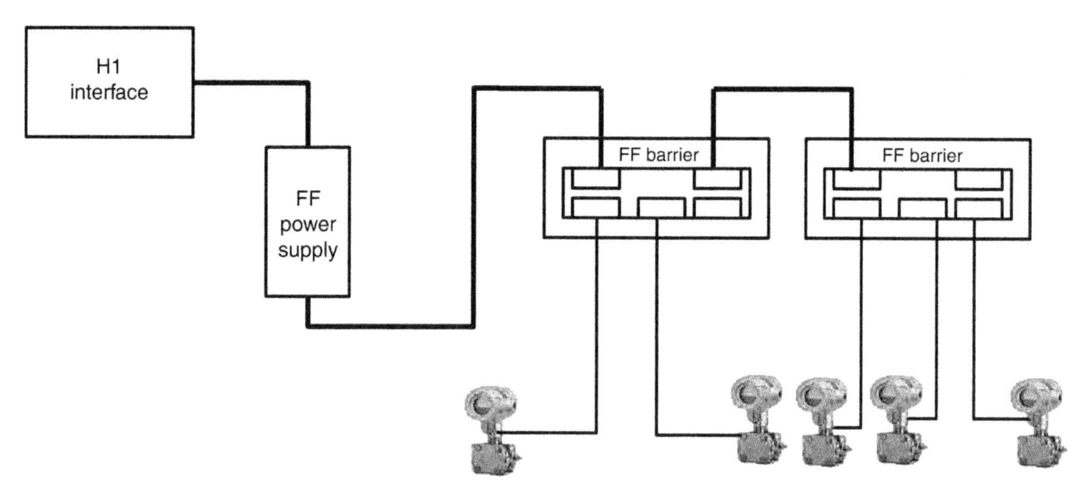

FIGURE 1.19

Bus connections

integrity is maintained at the same time the cost of installation and cost of the cables get reduced. Due to the advantages from all the sides, the digital busses are gaining acceptance from the users and hence, there has been a lot of improvement in the last decade expecting to continue the growth in the coming decade as well.

Network

Network-based connectivity is similar to the connections made in Ethernet, in which the devices are connected to a switching device near to the participating field instrument or the controllers. The switching device in turn connects to some other switching device or higher layer switches. The networking-based connectivity is gaining acceptance from the user base due to its simplicity and easy maintenance. Various technology options are available to increase the length or distance between the devices and increase the bandwidth. Special switches are available for installation in a plant environment. With the advances in the switching technologies, the speed of the packet communication is increased from 10 mbps to 1 gbps and hence, the latency is increased multifold. Redundancies are available in various forms and hence, the reliability of the networks has increased. The technology maturity brought the concept of networking security to protect the control systems from the cyber security perspective. A separate chapter is provided on the topic for detailed study (Figure 1.20).

Wireless

The most recent change in the process control systems and automation is the adaptation of the wireless communications (Figure 1.21). The industry is moving from the proprietary communications to an open, standards-based communication protocols. The infrastructure for the wireless devices is going to be standard-based and multiple types of the devices can participate in the network. The industry is moving toward using the wireless infrastructure for the process control instruments as well as the devices such as cameras and access control devices and laptops to participate in the network. The industry standards communication protocols such as HART are available in the wireless format like

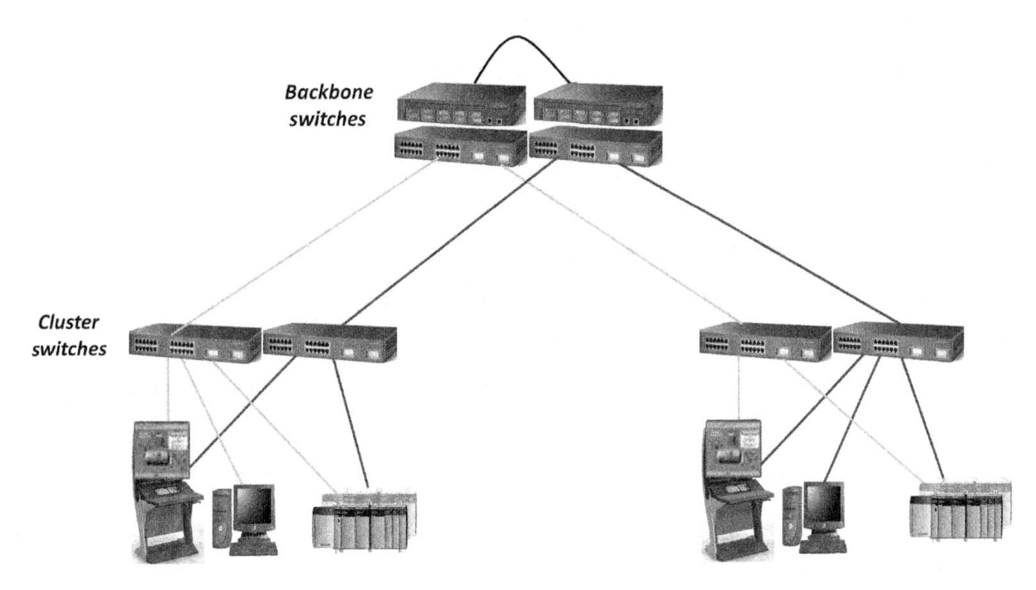

FIGURE 1.20

Networked connection

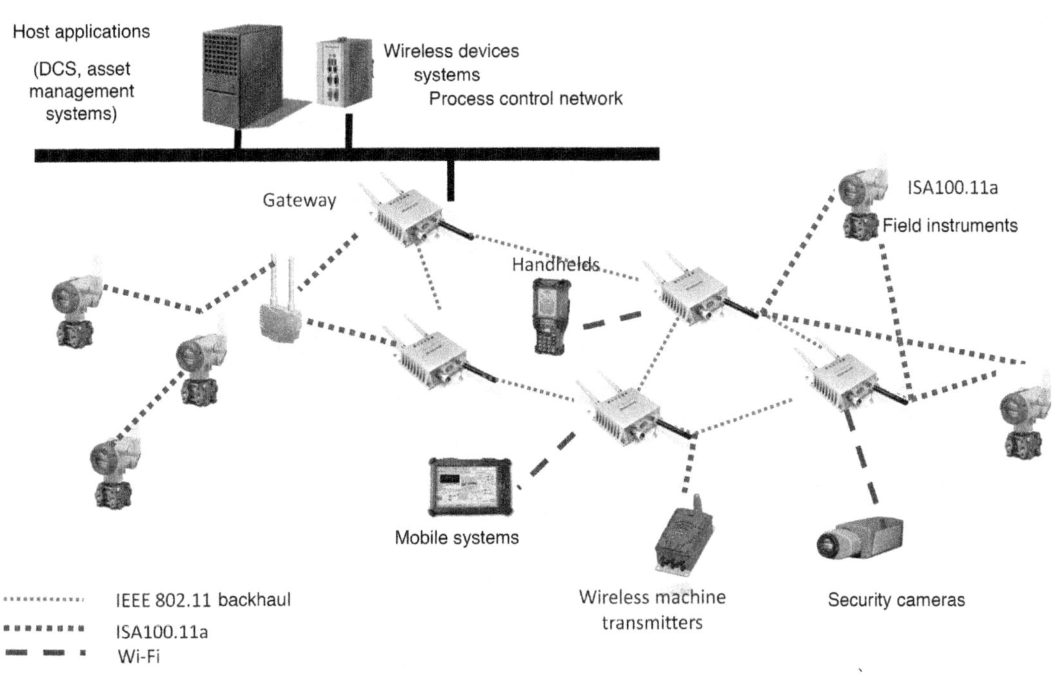

FIGURE 1.21

Mesh connection

ISA100.11a and Wireless HART so that the users of the system need less time to adopt the same in the new network environment. Currently wireless devices are not used much for closed-loop control as speed of communication and battery life are limited. Advancement in the next few years may make it fully available for total closed-loop control. A separate chapter is dedicated to details about the wireless communications with more information.

1.10 AUTOMATION SYSTEM CONTROLLERS

Controllers are seen as the heart of the automation system and in many cases, the cost of the overall system is governed by these controllers. Just like any other subsystem, controllers have changed over a period of time adopting various technologies at that point of time. The initial controllers were used to have mechanical control with gear mechanisms. The subsequent generations adopted pneumatic, hydraulic, and electronic- and microprocessor-based systems. The capacity of each controller has increased over a period of time, hence reducing the cost of automation per closed loop and making the systems more affordable. The increase in density of the loops in each controller has reduced not only the cost, but also the footprint and improved the ease of operations and speed of the control.

Mechanical Controllers

The control systems designed using the mechanical gear movements with springs and bellows can be considered as the mechanical controllers. The mechanical controllers are the first type of controllers in the modern world with the steam engine governor. The concept of the same can be seen in many applications wherein the movement of the gears and operation of the springs can bring the desired action of the process variable. However, due to mechanical movements and complex designs and also due to the relatively easy and advanced technology, use of modern electronics replaced the mechanical controllers in most of the places.

Pneumatic Controllers

A pneumatic controller is a mechanical device designed to measure temperature or pressure and transmit a corrective air signal to the final control element. A controller is a comparative device that receives an input signal from a measured process variable, compares this value with a predetermined control point value (set point), and determines the appropriate amount of output signal required by the final control element to provide corrective action within a closed control loop. The controller is made with nozzle and baffle systems and highly accurate mechanisms. There are many installations across the world using these technologies due to inherent nature of no electrical connections.

Hydraulic Controllers

A hydraulic controller is a device that uses a liquid control medium to provide an output signal that is a function of an input error signal. The error signal is the difference between a measured variable signal and a reference or set-point signal. Self-contained closed-loop hydraulic controllers continue to be used for certain types of process control problems, but as the use of the computer, with its electrical output, expands in process control applications, the electro hydraulic servo or proportional valve gains in usage. The combination adds the advantages of hydraulic control to the versatility of the computer. Also contributing to the expanding use of hydraulics is the steady improvement in fire-resistant fluids. Since the servo or proportional valve does not accept low-level digital input directly, a digital-to-analog

(D/A) converter and an amplifier are required. So where it can be used, a hydraulic controller that senses a controlled variable directly is preferred in the industrial environment for its easy maintainability.

Electronic Controllers

As it is true of other instrumentation and control, stand-alone controllers have derived numerous technical advancements from digital electric circuitry, microprocessors, modern displays, and very creative software. Contemporary stand-alone controllers differ markedly (perhaps even radically) from the devices available as recently as a decade or so ago. Instrument features blue vacuum fluorescent dot-matrix display (4 lines of 10 characters per line), and self-tuning based on one-shot calculation of the optimum PID values based on a "cold" start. Calculated PID values are stored and can be displayed or changed on demand. Both direct- and reverse-acting outputs can be self-tuned. Four PID settings can be stored for each control output. PID set selection may be programmed as part of a profile, initiated by process variable value, internal trigger, keystroke, or logic input signal. Controller uses a dynamic memory rather than fixed allocation of memory. User can assign memory as needed during configuration for best memory utilization.

Security by user ID number can be selected at any program level to prevent unauthorized access and program or data changes. Self-diagnostics are used on start-up and during normal operation by monitoring internal operations and power levels. Upon detection of an out-of-tolerance condition, controller shuts off output(s), activates alarm(s), and displays appropriate message. Controller can perform custom calculations using math programs. Device maintains a history file of event triggers as a diagnostic tool. Two process inputs and four outputs can be assigned to two separate control loops creating a dual-loop controller. Optional features include parallel printer output, logic I/O, digital communications, and PC interface.

DCS

A DCS is a technology in which the controllers and other subsystems are geographically and functionally distributed across the plant. The DCS has evolved to meet the specific needs of process applications such as pulp and paper, utility, power, refining, petrochemical, and chemical processing. DCSs are generally used in applications in which the proportion of analog to digital is higher than a 60:40 ratio, and/or the control functions performed are more sophisticated. A DCS typically consists of unit controllers that can handle multiple loops, multiplexer units to handle a large amount of I/O, operator and engineering interface workstations, a historian, communication gateways, and an advanced control function in dedicated proprietary controllers. All these are fully integrated and usually connected by means of a communication network. A DCS typically takes a hierarchical approach to control, with the majority of the intelligence housed in microprocessor-based controllers that can each handle 10–1000 inputs/outputs. The advent of fieldbus has resulted in the development of field-based DCS (and PLC) that facilitates both the movement of the control functions to the field and the use of additional measurements and diagnostics from smart instrumentation and control valves.

1.10.1 CONTROL LOGICS

Discrete control is one of the major subsystems in the industrial automation and needs of the manufacturing sectors are the primary driver for the growth and maturity. A dedicated chapter is provided to detail the technology behind these products. The discrete control and automation have started the journey with the sensors and relays wired to realize the logic. However, the flexibility expected by the

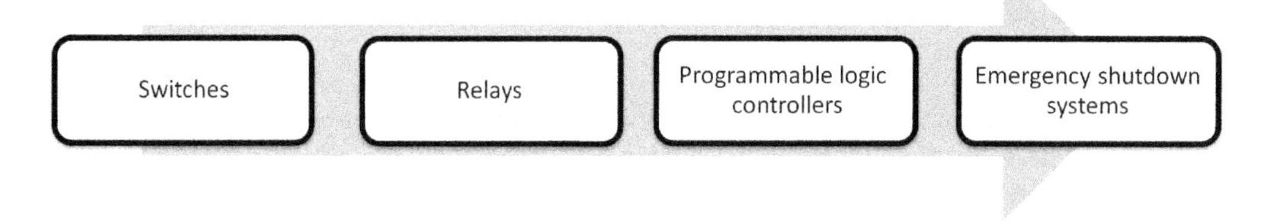

FIGURE 1.22

manufacturing industry brought forward the concepts such as PLC fuelled by the microprocessors and microcontrollers and matured as emergency shutdown systems (ESDs) by adopting networking and distributed processing technologies (Figure 1.22).

Switches

In the past, machine control could be achieved by wiring switches and relays physically together to perform logic functions. Hard-wired ladder diagrams were created for the circuit to work; current had to flow through each rung. The main drawback of this system was that a change in logic function required a physical rewire of the circuit. This involved maintenance personnel conducting a time-consuming job and often resulted in lost production time, as the machine would have to be switched off.

Relay

Relays serve as switching, timing, and multiplying mechanisms for input devices, such as push buttons, selector switches, and photoelectric sensors. Relays are fairly intuitive; however, as mechanical devices, they do not offer the same programming and troubleshooting flexibility found in modern programmable control systems. Relays are also known for taking up considerable amounts of space, requiring extensive wiring, and needing regular maintenance.

PLC

The first programmable controller, introduced in 1970, was developed in response to a demand from General Motors for a solid-state system that had the flexibility of a computer, yet could be programmed and maintained by plant engineers and technicians. These early programmable controllers took up less space than the relays, counters, timers, and other control components they replaced, and they offered much greater flexibility in terms of their reprogramming capability. The initial programming language, based on the ladder diagrams and electrical symbols commonly used by electricians, was key to industry acceptance of the programmable controller. Because programmable controllers can be programmed in relay ladder logic, it is relatively simple to convert electrical diagrams to the programmable controller program. This process involves defining the rules of operation for each control point, converting these rules to ladder logic, and identifying and labeling outputs (addressing). Today's work force has a mix of engineers – some of whom have been around for a while and are familiar with ladder logic as well as newer engineers more comfortable with computer-centric based programming and control. This has led to a mix of programming technologies that are applied based on user background and application need.

ESD

The ESD shall minimize the consequences of emergency situations, related to typically uncontrolled flooding, escape of hydrocarbons, or outbreak of fire in hydrocarbon-carrying areas or areas that may otherwise be hazardous. Traditionally risk analyses have concluded that the ESD system is in need of a high safety integrity level (SIL), typically SIL 2 or 3. Basically the system consists of field-mounted sensors, valves and trip relays, system logic for processing of incoming signals, alarm, and HMI units. The system is able to process input signals and activating outputs in accordance with the cause and effect charts defined for the installation. Shutdown of part systems and equipment is to isolate hydrocarbon inventories, isolate electrical equipment, prevent escalation of events, stop hydrocarbon flow, depressurize/blow down, emergency ventilation control, and close watertight doors and fire doors.

1.10.2 OBJECTIVES OF THE PLANT INFORMATION AND CONTROL SYSTEMS

In order to support the broad objectives of the plant, a more specific set of objectives is needed for the various technology systems projected to meet the long-range needs of the plant. These include the following:

1. Process control systems
2. Communications networks
3. Database management systems
4. Process optimization and process improvement systems
5. Decision support systems

Process Control Systems: Process control systems make computer-automated control available in all areas on all processes. In addition, the technology expanded the scope of conventional control to include the following supporting goals:

- Minimize the manual entry and recording of all measurements and operational decisions to minimize errors and expedite data acquisition.
- Simplify the conducting of economic and operational studies to permit quick analysis of unusual operating conditions.
- Increase the process and system engineer's productivity through readily accessible, efficient, and comprehensive analysis and design tools.
- Increase the scope and interactive access to history data to permit thorough analysis of process and operational problems.
- Expedite the process of system expansion and growth.

Communication Networks: Communication networks provide plant-wide information exchanges with appropriate interactive workstations and permit ready access to plant information by all users of data. The following are the supporting goals:

- Connectivity and interoperability between systems of different vendors must be provided for adaptability and ease of expansion.
- Integrity and security of data in transmission and access to databases must be assured for reliable plant operation.
- Delay or latency in transmission must be minimized with highest economic speed for timely analysis and decisions.

- Internetwork bridges, routers, and gateways must be supplied where needed to provide connectivity; making DCS as a single point of interaction between man and machine integrates ESD; machine monitoring systems, analyzer management systems, PLC's, automatic tank gauges, MOV control systems, and fiscal metering systems.
- Voice, data, and video image transmission must be integrated where needed to provide consistency of information.

Database Management Systems: Database management systems should be global in nature and must interconnect, interrelate, and integrate all department and area databases of the plant, including corporate, business, research, and marketing strategies as well as plant operations and production control. The following supporting goals must be included:

- Industry standard relational database structures and systems must be employed to permit easy integration.
- Ease of access through a user-friendly, ad hoc query language must be supported to permit timely analysis of plant operations problems.
- Integrity of temporal data must be maintained via high-speed network access rather than large-scale collection and copying.
- Security of the data must be maintained while providing access to all users with a need.
- Support of plant-wide information gathering for formulation of management decisions with simultaneous access of a single user program or person to multiple databases as the system grows.
- During data integration, control data are separated by demilitarized zone (DMZ) for security reasons. Data are provided outside of DMZ for use by other users such as technology groups, and technical services.

Process Optimization and Improvement: Process optimization must permit the expansion of efforts in simulation, optimization, and scheduling of process operations, including the following supporting goals:

- Support of execution of process analysis and modeling tools from all levels to extend their use throughout the plant.
- Support of effective management of materials with timely and comprehensive real-time inventory, and demand and supply data.
- Provide for dynamic acquisition of energy and material balance information to support the optimization of their utilization and reduction of overall costs.
- Support access to process modeling systems from throughout the plant to permit more thorough analysis.
- Expand total processing power available throughout the plant.

Decision Support Systems: Decision support tools must be provided to assist people in accessing, manipulating, analyzing, displaying, and documenting data. Included in this group are the following:

- A broad and flexible databases management system for comprehensive versatile problem analysis.
- A user-friendly, multiple access, database query method, or language to permit rapid access to plant-wide data.
- All-purpose report generators, capable of combining text, graphics, data table, calculations, and formatting to permit effective presentations of process conditions and problems.
- Structured data analysis (spread sheets) to afford extensive extrapolations of plant data to determine plant operating conditions.

- Statistical analysis packages to determine operational and demand characteristics (trends).
- Support for long-range decision making with market and business simulation systems.
- A network management system is required for management of large networks using several layers of switches.

1.11 THE GENERIC DUTIES OF AN AUTOMATION SYSTEM IN HIERARCHICAL FORM

The automation of any such industrial plant manages the plant's information systems to ensure that necessary information is collected and used whenever it can enhance the plants operation – true information systems technology in its broadest sense.

Another major factor should also be called to our attention here. It has been repeatedly shown that one of the major benefits of the use of digital computer systems in the automation of industrial plants has been in the role of a control system enforcer. In this mode, one of the control computers main tasks is to continually ensure that the control system equipment is actually carrying out the job that it was designed to do in keeping the units of the plant's production system operating at some best (near optimal) level. That is, to be sure that those in the continuous process plant, for instance, the controllers have not been set on manual and that the optimal set points are being maintained. Likewise, it is the task of dynamic control to ensure that the plant's production schedule is carried out, that is, to enforce the task set by the production scheduling function.

Often the tasks carried out by these control systems have been the ones that a skilled and attentive operator could have readily done himself. The difference is the degree of attentiveness to the task at hand that can be achieved over the long run.

As stated earlier, all of this must be factored into the design and operation of the control system that will operate the plant, including the requirements for maximum productivity and minimum raw material and energy usage. As the overall requirements, both energy and productivity based, become more complex, more sophisticated and capable control systems are necessary.

1.12 FUNCTIONAL REQUIREMENTS OF AN INTEGRATED INFORMATION AND AUTOMATION SYSTEMS: A GENERIC LIST

1. An extensive system for the automatic monitoring of a large number of different plant variables operating over a very wide range of process operations and of process dynamic behavior. Such monitoring will detect and compensate for current or impending plant emergencies or production problems.
2. Development of a large number of quite complex, usually nonlinear relationships for the translation of some of the above plant variable values into control correction commands.
3. Transmission of these control correction commands to another very large set of widely scattered actuation mechanisms of various types.
4. Improvement of all aspects of the manufacturing operations of the plant by guiding them toward likely optima of the appropriate economic or operational criteria. Results may be applied to the control correction commands of item 2 above and/or to the plant scheduling functions of the item 8 given below.

5. Reconfiguration of the plant production system and/or of the control system as necessary and possible to ensure the applicable production and/or control system for the manufacturing situation at hand.
6. Keeping plant personnel, both operating and management, aware of the current status of the plant and of each of its process and their production including suggestion for alternate actions, where necessary.
7. Reduction of the plant operational and production data and product quality data to form a historical database for reference by plant engineering, other staff functions, and marketing.
8. Adjusting the plant's production schedule and product mix to match its customer's needs, as expressed by the new order stream being continually received, while achieving high plant productivity and the lowest practical production costs. This function must also provide for appropriate plant preventive or corrective maintenance functions.
9. Determination of and provision for appropriate inventory and use levels for raw materials, energy, spares, goods in process, and products to maintain desired production and economic for the plant.
10. Assuring the overall availability of the control system for carrying out its assigned tasks through the appropriate combination of fault detection and fault tolerance, redundancy, and fail-safe techniques.
11. Maintaining interfaces with the external entities that interact with the plant production system such as corporate management, marketing, accounting, corporate engineering, external transportation, suppliers and vendors, purchasing, customers, and contractors.

An overall plant automation system must provide the following:

1. An effective dynamic control of each operating unit of the plant to ensure that it is operating at its maximum efficiency of production capability, product quality, and/or energy and materials utilization based upon the production level set by the scheduling and supervisory functions listed below. This thus becomes the control enforcement component of the system. This control reacts directly to compensate for any emergencies, which may occur in it's own unit.
2. A supervisory and coordinating system that determines and sets the local production level of all units working together between inventory locations in order to continually improve (i.e., optimize) their operation. This system ensures that no unit is exceeding the general area level of production and thus using excess raw materials or energy. This system also responds to emergencies in any of the units under its control in cooperation with those units' dynamic control system to shutdown or systematically reduce the output in these and related units as necessary to compensate for the emergency. In addition, this system is responsible for efficient reduction of plant operational data from the dynamic control units, described just above, to ensure its availability for use by any plant entity requiring access to it as well as its use for the historical database of the plant.
3. An overall production control system capable of carrying out the scheduling functions for the plant from customer orders or management decisions so as to produce the required products for these orders at the best (near optimum) combination of customer service and the use of time, energy, inventory, manpower, and raw materials suitably expressed as cost functions.
4. A method of assuring the overall reliability and availability of the total control system through fault detection, fault tolerance, redundancy, uninterruptable power supplies, maintenance planning, and other applicable techniques built into the system's specification and operation.

Because of the ever-widening scope of authority of each of the first three requirements in turn, they effectively become the distinct and separate levels of a superimposed control structure, one on top of the other. Also in view of the amount of information that must be passed back and forth among the above four "tasks" of control, a distributed computational capability organized in a hierarchical fashion would seem to be the logical structure for the required control system. This must be true of any plant regardless of the industry involved.

As noted, hierarchical arrangement of the element of a distributed, computer-based control system seems to be an ideal arrangement for carrying out the automation of the industrial plant just described. Figure 1.23, Figure 1.24 shows one possible form of the distributed, hierarchical computer control system for overall plant automation.

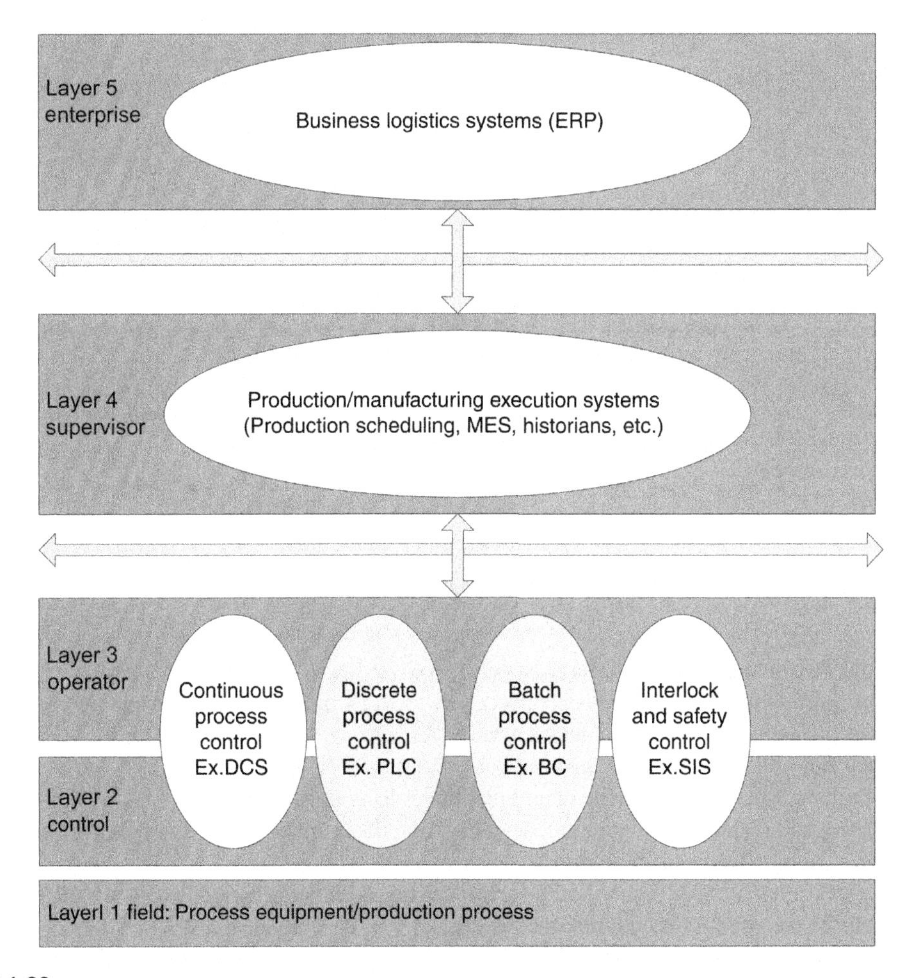

FIGURE 1.23

Layered functions of automation systems

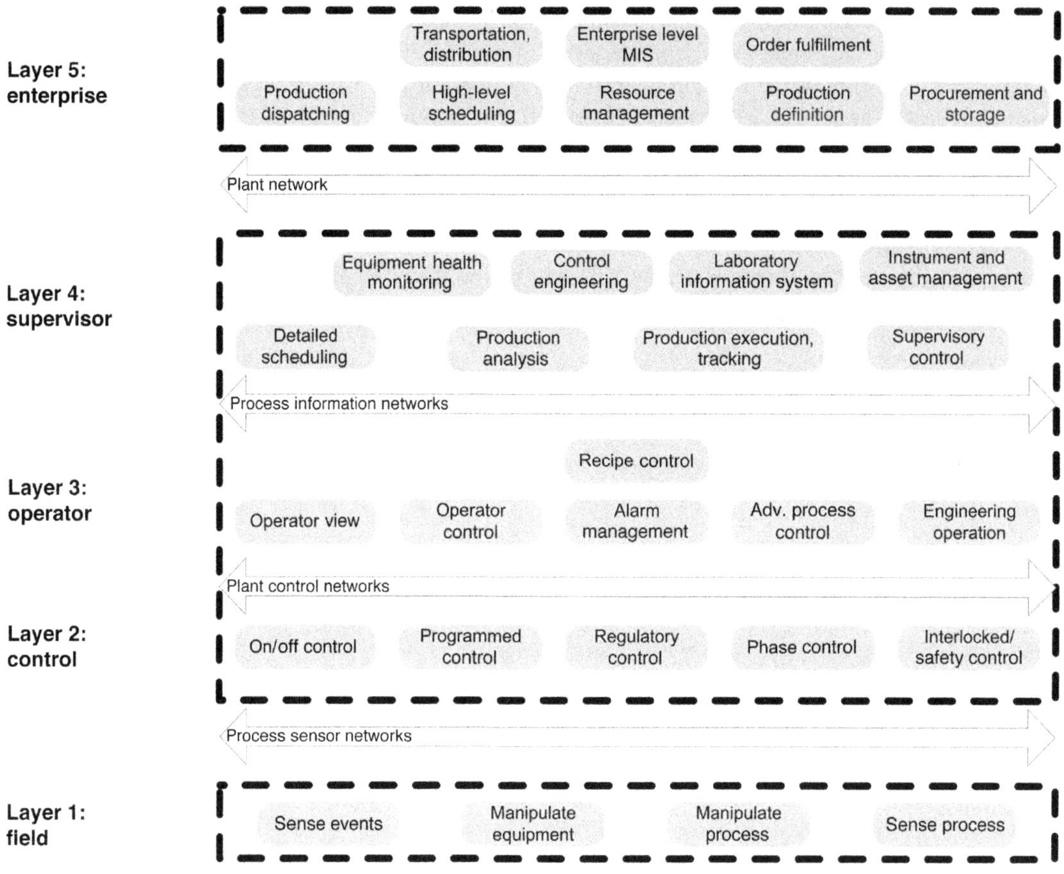

Layer 5: enterprise
Transportation, distribution
Enterprise level MIS
Order fulfillment
Production dispatching
High-level scheduling
Resource management
Production definition
Procurement and storage
Plant network
Layer 4: supervisor
Equipment health monitoring
Control engineering
Laboratory information system
Instrument and asset management
Detailed scheduling
Production analysis
Production execution, tracking
Supervisory control
Process information networks
Layer 3: operator
Recipe control
Operator view
Operator control
Alarm management
Adv. process control
Engineering operation
Plant control networks
Layer 2: control
On/off control
Programmed control
Regulatory control
Phase control
Interlocked/ safety control
Process sensor networks
Layer 1: field
Sense events
Manipulate equipment
Manipulate process
Sense process

FIGURE 1.24

Functional distribution in layers

Functional Requirements for Control Level (Level 1): Level 1 defines the activities involved in sensing and manipulating the physical processes. Level 1 element is the sensors and actuators attached to the control functions in automation systems.

I. Control Enforcement
   - Maintain direct control of the plant units under its cognizance.
   - Detect and respond to any emergency condition that may exist in these plant units.

II. System Coordination and Reporting
   - Collect information on unit production, raw material, and energy use and transmit to higher levels.
   - Service the operator's man/machine interface.

Functional Requirements for the Supervisory Level (Level 2): Level 2 defines the activities of monitoring and controlling the physical processes and in automated systems this includes equipment control and equipment monitoring. Level 2 automation and control systems have real-time responses measured in subseconds and are typically implemented on PLC, DCSs, and open control systems (OCSs).

I. Control Enforcement
- Respond to any emergency condition that may exist in its region of plant cognizance.
- Optimize the operation of units under its control within limits of established production schedule. Carry out all established process operational schemes or operating practices in connection with these processes.

II. System Coordination and Operational Data Reporting
- Collect and maintain data queues of production, inventory, and raw material, spare parts and energy usage for the units under its control.
- Maintain communications with higher and lower levels.
- Service the man/machine interfaces for the units involved.

III. Reliability Assurance
- Perform diagnostic on itself and lower-level machines.
- Update all standby systems.

Functional Requirements of the Area Level (Level 3): Level 3 defines the activities that coordinate production resources to produce the desired end products. It includes workflow "control" and procedural "control" through recipe execution. Level 3 typically operates on time frames of days, shifts, hours, minutes, and seconds. Level 3 functions also include maintenance functions, quality assurance, and laboratory functions, and inventory movement functions, and are collectively called manufacturing operations management. Level 3 functions directly related to production are usually automated using manufacturing execution systems (MES).

I. Production Engineering
- Establish the immediate production schedule for its own area including maintenance, transportation, and other production-related needs.
- Optimize locally the costs for its individual production area while carrying out the production schedule established by the production control computer system (level 4A) (i.e., minimize energy usage or maximize production, for example).
- Modify production schedules along with level 4A to compensate for plant production interruptions, which may occur in its area of responsibility.

II. System Coordination and Operational Data Reporting
- Make area production reports including variable manufacturing costs.
- Use and maintain area practice files.
- Collect and maintain area data queues for production, inventory, manpower, raw materials, spare parts, and energy usage.
- Maintain communication with higher and lower levels of the hierarchy.
- Operations data collection and off-line analysis as required by engineering functions including statistical quality analysis and control functions.
- Service the man/machine interface for the area.
- Carry out needed personnel functions, such as
  - **a.** Work period statistics (time and task)
  - **b.** Vacation schedules
  - **c.** Work force schedules

    d. Union line of progression
    e. In-house training and personnel qualification.

III. Reliability Assurance

- Diagnostics of self and lower-level functions.

Required Tasks of the Intra Company Communications Control Systems (Level 4B): Level 4 defines business-related activities that manage a manufacturing organization. Manufacturing-related activities include establishing the basic plant schedule (such as material use, delivery, and shipping), determining inventory levels, logistics "control," and material inventory "control" (making sure materials are delivered on time to the right place for production). Level 4 is called business planning and logistics. Level 4 typically operates frames of months, weeks, and days. Enterprise resource planning (ERP) logistics systems are used to automate level 4 functions.

I. System Coordination and Reporting

- Maintain interfaces with
    a. Plant and company management
    b. Sales and shipping personnel
    c. Accounting, personnel, and purchasing departments
    d. Production scheduling level (level 4A)
- Supply production and status information as needed to
    a. Plant and company management
    b. Sales and shipping personnel
    c. Accounting, personnel, and purchasing departments
    d. This information will be supplied in the form of
        1. Regular production and status report
        2. Online enquiries
- Supply order status information as needed to sales personnel.

II. Reliability Assurance

- Perform self check and diagnostic check on itself.

Functional Requirements of the Production Scheduling and Operational Management Level (Level 4A)

I. Production Scheduling

- Establish basic production schedule.
- Modify the production schedule for all units per order stream received energy constraints, power demand levels, and maintenance requirements.
- Develop optimum preventive maintenance and production unit renovation schedule in coordination with required production schedule.
- Department the optimum inventory levels of raw materials, energy sources, spare parts, and goods in process at each storage point. The criteria to be used will be the trade-off between customer service (i.e., short delivery time) and the capital cost of the inventory itself, as well as the trade-offs in operating costs versus costs of carrying inventory level. This function will also include the necessary material requirements planning (MRP) and spare parts procurements to satisfy the production schedule planned. (This is an off-line function.)
- Modify production schedule as necessary whenever major production interruptions occur in downstream units, where such interruptions will affect prior or succeeding units.

II. Plant Coordination and Operational Data Reporting

- Collect and maintain raw material and spare parts use and availability of inventory, and provide data for purchasing for raw material and spare parts order entry, and for transfer to accounting.
- Collect and maintain overall energy use and availability of inventory, and provide data for purchasing for energy source order entry, and for transfer to accounting.
- Collect and maintain overall goods in process and production inventory files.
- Collect and maintain the quality control file.
- Collect and maintain machinery and equipment use and life history files necessary for preventive and predictive maintenance planning.
- Collect and maintain manpower use data for transmittal to personnel and accounting departments.
- Maintain interfaces with management interface-level function and with area-level systems.

III. Reliability Assurance

- Run self-check and diagnostic routines on self and lower-level machines.

Notes: There are no control functions as such required at this level. This level is for the production scheduling and overall plant data functions.

1.13 CONCEPTUAL/FUNCTIONAL TOPOLOGY OF AN AUTOMATION SYSTEM

In the previous section various levels of an automation system and the primary duties performed at each level are discussed. In this section, we shall discuss the conceptual topology of a system in relation to the reference model. We already know that there are five layers/levels of an automation system. The physical realization of such a system will be discussed now (Figure 1.25).

The conceptual topology diagram and functional areas are shown in Figure 1.25. The level 1 of the reference model can be mapped to the transmitters, control valves, solenoid valves, and limit switches. The level 1 device functionally is meant to sense process, manipulate process, sense events, and manipulate equipment. The level 1 device communicates with the level 2 system using a communication mechanism. The communication mechanisms differs from system to system and they starts from a base current based 4–20 mA to a digital communication technology.

The level 2 contains the controllers such as DCS, PLC, safety systems, batch controllers, single loop controllers. The functional responsibility of the level 2 controllers is on/off control, programmed control, continuous control, phase control, interlock and safety control. The level 2 controllers communicate with the level 2 systems using the plant communication network. The level 2 systems constitute HMI systems, and SCADA systems. Functionally the level 2 systems are responsible for alarm management, operator visibility, operator control, supervisory control, and recipe control.

Levels 2 and 3 are connected by a network called plant information network and systems from the various vendors are interfaced using the standards-based interfaces. The level 3 systems contain MES and LIMS and are functionally expected to perform detailed scheduling, production execution, and production analysis.

Levels 3 and 4 are connected using a plant network and standards-based interfaces. The level 4 devices contain ERP and SAP and are functionally expected to provide management information systems and decision support system.

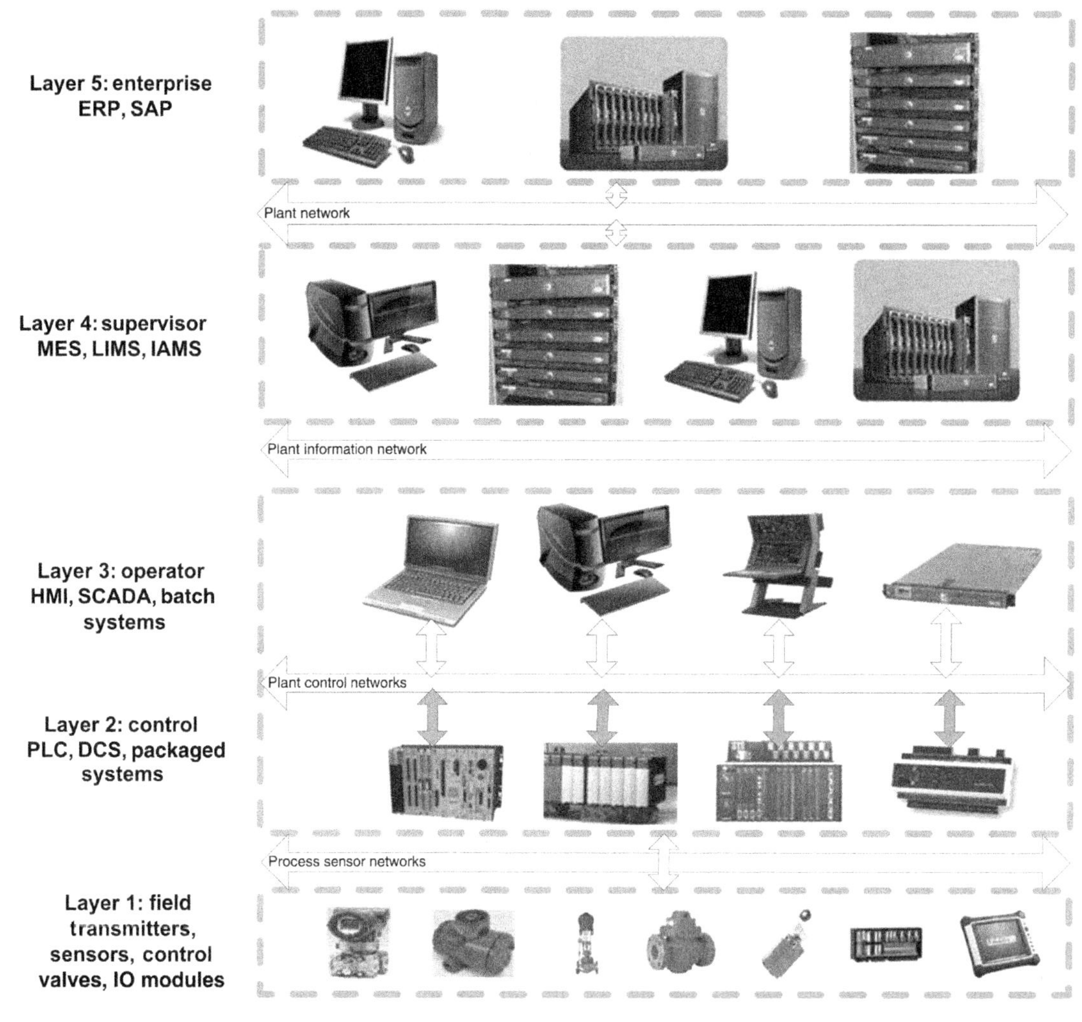

FIGURE 1.25

Systemic view of layered architecture

1.13.1 PHYSICAL ARCHITECTURE

The physical architecture of an automation system is shown in Figure 1.26. The systems shown are only tentative. It is not necessary that all the systems will be the same. The diagram can be used as a reference.

The network starts with the level 1 itself for the sensor to communicate. The level 2 layer has a communication network and is connected to the above systems which will be a network router. The network router will provide a separate LAN for the process communication network and interconnects the different systems that are secure, separate, and logically connected together.

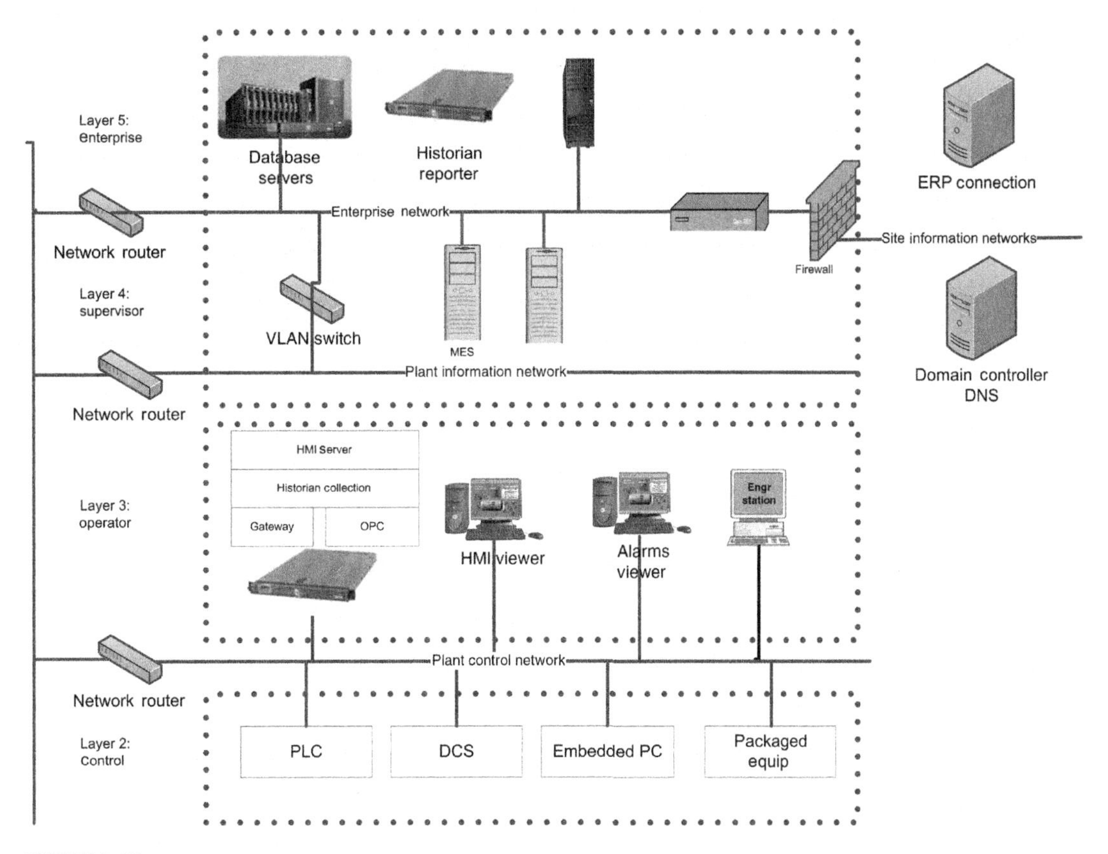

FIGURE 1.26

Physical view of layered automation systems

The level 3 is connected to the level 4 separated by a firewall and a DNS server. Some of the systems/nodes that are accessible from both the levels will be placed in DMZ. The typical systems placed in the DMZ are wireless access points, plant history access, web access, and data access from plant-wide LAN.

FURTHER READINGS

Anderson, N.A., 1998. Instrumentation for Process Measurement and Control, Third ed. CRC Press.
ANSI/ISA-5.1-1984 (R1992) – Instrumentation Symbols and Identification.
www.wikipedia.com
ANSI/ISA 95.
Enterprise – control systems integration – ISA.
Boyes, W. (Ed.), 2003. Instrumentation Reference Book. Third ed. Butterworth-Heinemann.
Coggan, D.A. (Ed.), 2005. Fundamentals of Industrial Control. Second ed. ISA, Practical Guides for Measurement and Control Series.

Goettsche, L.D. (Ed.), 2005. Maintenance of Instruments and Systems. Second ed. ISA, Practical Guides for Measurement and Control Series.
Hashemian, H.M., 2005. Sensor Performance and Reliability. ISA.
Hughes, T.A., 2002. Measurement and Control Basics, Third ed. ISA.
ISA, 2003. The Automation, Systems, and Instrumentation Dictionary, Fourth ed. ISA.
Lipták, B.G., Editor-in-Chief, 2003.Instrument Engineer's Handbook: Process Measurement and Analysis, Volume I, fourth ed. ISA & CRC Press.
ANSI/ISA-50.1-1982 (R2002) – Compatibility of Analog Signals for Electronic Industrial Process Instruments.

CHAPTER

THE PROGRAMMABLE LOGIC CONTROLLER

2

2.1 INTRODUCTION TO THE PROGRAMMABLE LOGIC CONTROLLER

A Programmable Logic Controller (PLC) is treated as the first building block of the automation systems. A PLC is a special form of microprocessor-based controller that uses a programmable memory to store instructions and implement functions such as logic, sequencing, timing, counting, and arithmetic to control machines and processes, as illustrated in Figure 2.1, and is designed to be operated by engineers with perhaps a limited knowledge of computers and computing languages. Computer programmers are not needed to set up or change the programs. PLC designers preprogram it so that the control program can be entered using a simple, rather intuitive, form of language. The term logic is used because programming is primarily concerned with implementing logic and switching operations, for example, if A or B occurs switch on C and if A and B occurs switch on D. Input devices, for example, sensors such as switches, and output devices in the system being controlled, for example, motors, valves, and so on, are connected to the PLC. The operator enters a sequence of instructions, that is, a program, into the memory of the PLC. The controller then monitors the inputs and outputs according to this program and carries out the programmed control rules.

PLCs have the advantage of enabling the use of the same basic controller with a wide range of control systems. The control system and the rules to be used can be easily modified by an operator by simply keying in a different set of instructions. This results in a flexible, cost-effective, system without rewiring, which can be used with control systems that vary widely in their nature and complexity. PLCs are similar to computers, but whereas computers are optimized for calculation

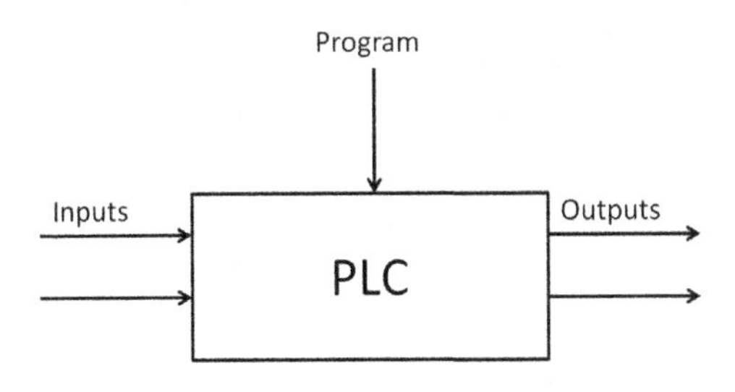

FIGURE 2.1

A programmable logic controller

and display tasks, PLCs are optimized for control tasks and the industrial environment. Thus, PLCs are

- Rugged and designed to withstand vibrations, temperature, humidity, and noise.
- Have interfacing for inputs and outputs already inside the controller.
- Easily programmed and have an easily understandable programming language that is primarily concerned with logic and switching operations.

The first PLC was developed in 1969. They are now widely used and extend from small self-contained units for use with perhaps 20 digital inputs/outputs to modular systems, which can be used for large numbers of inputs/outputs, handle digital or analog inputs/outputs, and also carry out proportional–integral–derivative control modes.

2.2 HARDWARE

2.2.1 FUNCTIONAL COMPONENTS OF A PLC

Typically a PLC system has the basic functional components of a processor unit, memory, power supply unit, input/output (I/O) interface section, communications interface, and the programming device. Figure 2.2 shows the basic arrangement of a PLC.

- The processor unit or central processing unit (CPU) contains the microprocessor that interprets the input signals and carries out the control actions, according to the program stored in its memory, communicating the decisions as action signals to the outputs.
- The power supply unit is needed to convert the mains a.c. voltage to the low d.c. voltage (5 V) necessary for the processor and the circuits in the input and output interface modules.
- The programming device or a personal computer loaded with programming software is used to enter the program into the memory of the processor. The program is developed in the device and then transferred to the memory unit of the PLC.

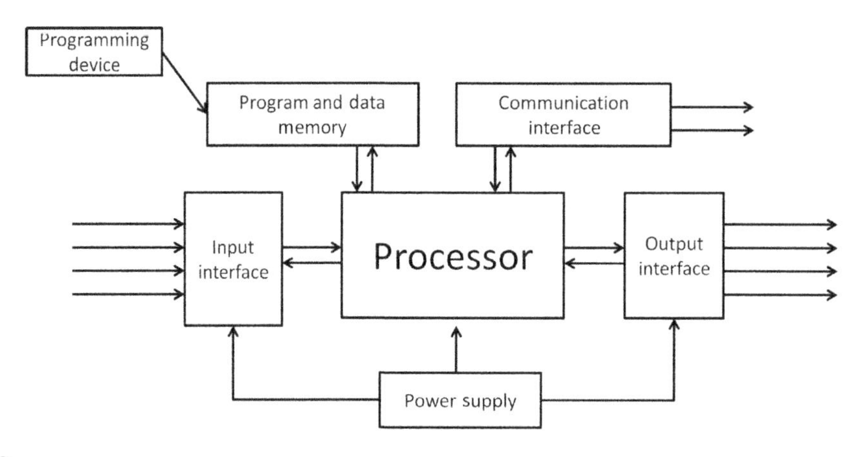

FIGURE 2.2

The PLC system

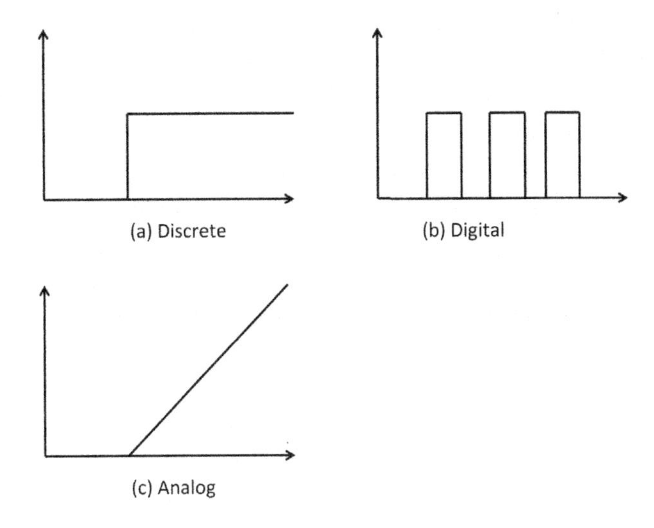

FIGURE 2.3

Signals. (a) discrete, (b) digital, and (c) analog

- The memory unit is where the program is stored to be used for the control actions to be exercised by the microprocessor and data stored from the input for processing and for the output for outputting.
- The input and output sections are where the processor receives information from external devices and communicates information to external devices. The inputs might thus be from switches, with the automatic drill, or other sensors such as photoelectric cells, as in the counter mechanism, temperature sensors, or flow sensors. The outputs might be to motor starter coils, solenoid valves, and so on.
- Input and output devices can be classified as giving signals, which are discrete, digital, or analog, as illustrated in Figure 2.3. The signals of devices giving discrete or digital signals are either OFF or ON. Thus, a switch is a device giving a discrete signal, either no voltage or a voltage. Digital devices can be considered to be essentially discrete devices that give a sequence of ON–OFF signals. Analog devices give signals whose size is proportional to the size of the variable being monitored. For example, a temperature sensor may give a voltage proportional to the temperature.
- The communications interface is used to receive and transmit data on communication networks from or to other remote PLCs. It is concerned with such actions as device verification, data acquisition, synchronization between user applications, and connection management. Refer to Figure 2.4 for the communication link being a backbone across the various components of a PLC.

2.3 INTERNAL ARCHITECTURE

The basic internal architecture of a PLC is illustrated in Figure 2.5. It consists of a CPU containing the system microprocessor, memory, and I/O circuitry. The CPU controls and processes all the operations within the PLC. It is supplied with a clock with a frequency of typically between 1 and 8 MHz. This frequency determines the operating speed of the PLC and provides the timing and synchronization for all elements

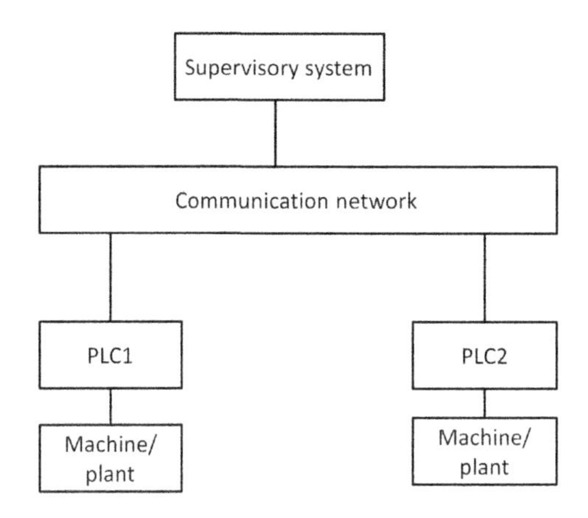

FIGURE 2.4

Basic communications model

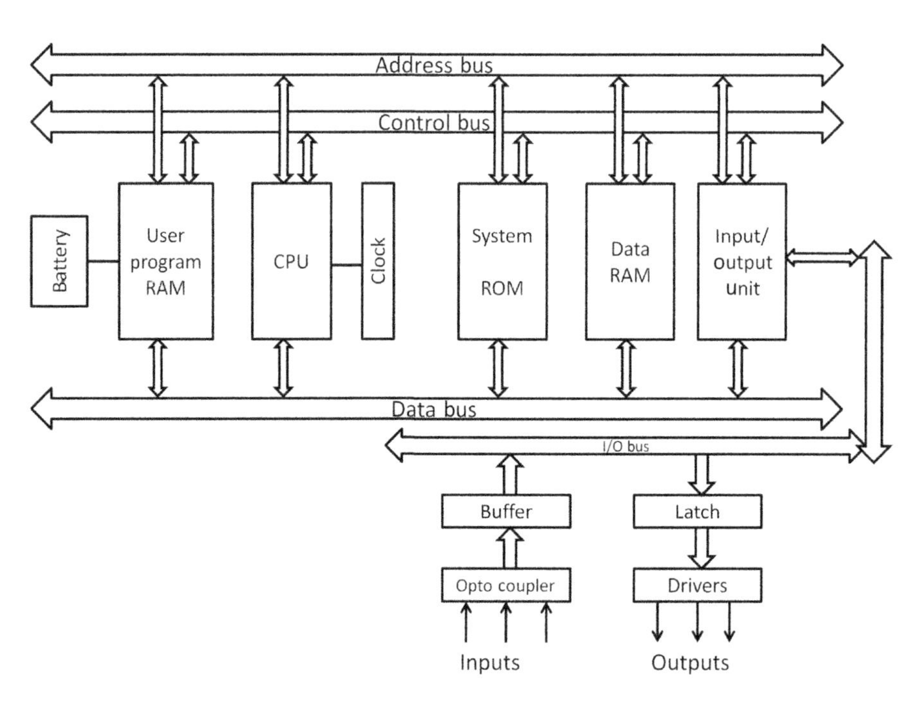

FIGURE 2.5

Architecture of a PLC

in the system. The information within the PLC is carried by digital signals. The internal paths along which digital signals flow are called buses. In the physical sense, a bus is just a number of PLCs with conductors along which electrical signals can flow. It might be tracks on a printed circuit board or wires in a ribbon cable. The CPU uses the data bus for sending data between the constituent elements, the address bus to send the addresses of locations for accessing stored data, and the control bus for signals relating to internal control actions. The system bus is used for communications between the I/O ports and the I/O unit.

2.3.1 SOURCING AND SINKING

The terms sourcing and sinking are used to describe the way in which d.c. devices are connected to a PLC. With sourcing, using the conventional current flow direction as from positive to negative, an input device receives current from the input module, that is, the input module is the source of the current (see Figure 2.6a). If the current flows from the output module to an output load then the output module is referred to as sourcing (Figure 2.6b). With sinking, using the conventional current flow direction as from positive to negative, an input device supplies current to the input module, that is, the input module is the sink for the current (Figure 2.7a). If the current flows to the output module from an output load then the output module is referred to as sinking (Figure 2.7b).

There are two common types of mechanical design for PLC systems: a single unit and the modular/rack type. The single unit type (or, as sometimes termed, brick) is commonly used for small programmable controllers and is supplied as an integral compact package complete with power supply, processor, memory, and I/O units. Typically such a PLC might have 6, 8, 12, or 24 inputs; 4, 8, or 16 outputs; and a memory, which can store some 300–1000 instructions.

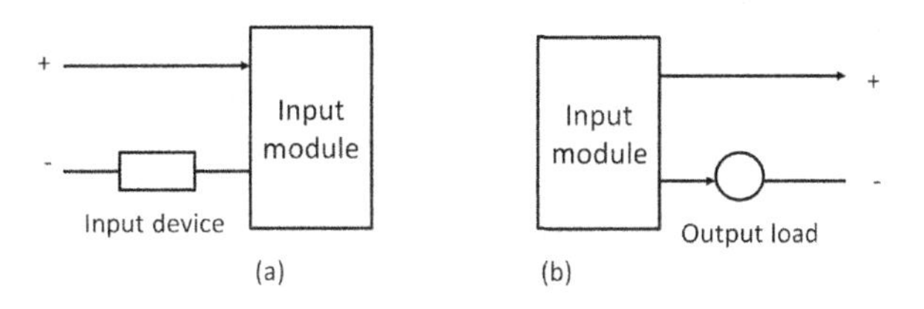

FIGURE 2.6

Sourcing. (a) input (b) output

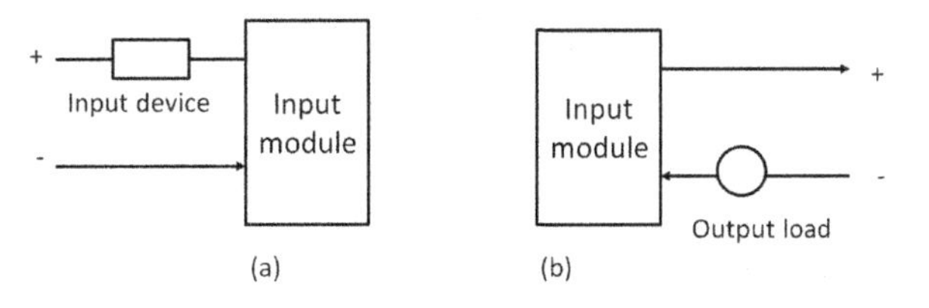

FIGURE 2.7

Sinking. (a) input (b) output

Systems with larger numbers of inputs and outputs are likely to be modular and designed to fit in racks. The modular type consists of separate modules for power supply, processor, and so on, which are often mounted on rails within a metal cabinet. The rack type can be used for all sizes of programmable controllers and has various functional units packaged in individual modules, which can be plugged into sockets in a base rack. The user decides the combination of modules required for a particular purpose and the appropriate ones are then plugged into the rack. Thus, it is comparatively easy to expand the number of I/O connections by just adding more I/O modules or to expand the memory by adding more memory units.

2.3.2 PROGRAMMING PLCS

PLCs can be programmed using a hand-held device, a desktop console, or a computer. The program is designed on the programming device and transferred to the memory unit of the PLC.

1. Hand-held programming devices normally contain enough memory to allow the unit to retain programs while being carried from one place to another.
2. Desktop consoles are likely to have a visual display unit with a full keyboard and screen display.
3. Personal computers are widely configured as program development workstations.

Some PLCs only require the computer to have appropriate software; others require special communication cards to interface with the PLC. A major advantage of using a computer is that the program can be stored on the hard disk or a CD and network shares and copies easily made.

2.4 I/O DEVICES

This section includes a brief description of typical input and output devices used with PLCs. The input devices include digital and analog devices such as mechanical switches for position detection, proximity switches, photoelectric switches, encoders, temperature and pressure switches, potentiometers, linear variable differential transformers, strain gauges, thermistors, thermotransistors, and thermocouples. Output devices considered include relays, contactors, solenoid valves, and motors.

2.4.1 INPUT DEVICES

The term sensor is used for an input device that provides a usable output in response to a specified physical input. For example, a thermocouple is a sensor that converts a temperature difference into an electrical output. The term transducer is generally used for a device that converts a signal from one form to a different physical form. Thus, sensors are often transducers. Other devices can also be transducers, for example, a motor that converts an electrical input into rotation. Sensors that give digital/discrete, that is, ON–OFF, outputs can be easily connected to the input ports of PLCs. Sensors that give analog signals have to be converted to digital signals before inputting them to PLC ports. The following are some of the more common terms used to define the performance of sensors.

- Accuracy is the extent to which the value indicated by a measurement system or element might be wrong. For example, a temperature sensor might have an accuracy of ±0.1 °C. The error of a measurement is the difference between the result of the measurement and the true value of the quantity being measured.

- The range of variable of system is the limits between which the input can vary. For example, a resistance temperature sensor might be quoted as having a range of −200 to +800 °C.
- When the input value to a sensor changes, it takes some time to reach and settle down to the steady-state value. The response time is the time that elapses after the input to a system or element is abruptly increased from zero to a constant value up to the point at which the system or element gives an output corresponding to some specified percentage, for example, 65%, of the value of the input. The rise time is the time taken for the output to rise to some specified percentage of the steady-state output. Often the rise time refers to the time taken for the output to rise from 10% of the steady-state value to 60% or 65% of the steady-state value. The settling time is the time taken for the output to settle to within some percentage, for example, 2% of the steady-state value.
- The sensitivity indicates the change in output of an instrument system or system element in response to a change in the quantity being measured by a given amount, that is, the ratio output/input. For example, a thermocouple might have a sensitivity of 20 V/ °C and so give an output of 20 V for each 1 °C change in temperature.
- The stability of a system is its ability to give the same output when used to measure a constant input over a period of time. The term drift is often used to describe the change in output that occurs over time. The drift may be expressed as a percentage of the full range output. The term zero drift is used for the changes that occur in output when there is zero input.
- Repeatability refers to the ability of a measurement system to give the same value for repeated measurements of the same value of a variable. Common cause of lack of repeatability is random fluctuations in the environment, for example, changes in temperature and humidity. The error arising from repeatability is usually expressed as a percentage of the full range output. For example, a pressure sensor might be quoted as having a repeatability of ±0.1% of full range. Thus, with a range of 20 kPa this would be an error of ±20 Pa.
- The reliability of a measurement system, or element in such a system, is defined as being the probability of its operation to an agreed level of performance, for a specified period, subject to specified environmental conditions. The agreed level of performance might be that the measurement system gives a particular accuracy. The following are examples of some of the commonly used PLC input devices and sensors.

2.4.1.1 Mechanical switches

A mechanical switch generates an ON–OFF signal or signals as a result of some mechanical input causing the switch to open or close. Such a switch might be used to indicate the presence of a workpiece on a machining table, the workpiece pressing against the switch and so closing it. The absence of the workpiece is indicated by the switch being open and its presence by it being closed. Thus, with the arrangement shown in Figure 2.8a, the input signals to a single input channel of the PLC are thus the logic levels:

Workpiece not present 0
Workpiece present 1
The 1 level might correspond to a 24 V d.c. input, thc 0 to a 0 V input.

With the arrangement shown in Figure 2.8b, when the switch is open the supply voltage is applied to the PLC input, and when the switch is closed the input voltage drops to a low value. The logic levels are thus:

Workpiece not present 1
Workpiece present 0

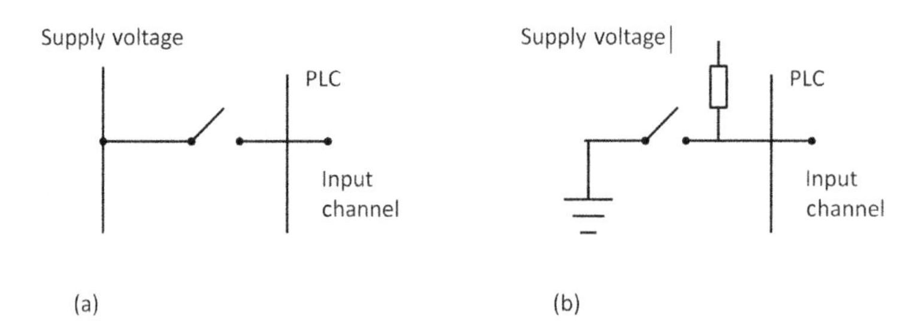

FIGURE 2.8

Switch sensors. (a) switch input to single input (b) switch input to supply voltage

Switches are available with *normally open (NO) or normally closed (NC)* contacts or can be configured as either by choice of the relevant contacts. An NO switch has its contacts open in the absence of a mechanical input and the mechanical input is used to close the switch. An NC switch has its contacts closed in the absence of a mechanical input and the mechanical input is used to open the switch.

2.4.1.2 Proximity switches

Proximity switches are used to detect the presence of an item without making contact with it. There are many types of proximity switches, some being only suitable for metallic objects. The eddy current type of proximity switch has a coil that is energized by a constant alternating current and produces a constant alternating magnetic field. When a metallic object is close to it, eddy currents are induced in it. The magnetic field due to these eddy currents induces an e.m.f. back in the coil, leading to a change in the voltage amplitude needed to maintain the constant coil current. The voltage amplitude is thus a measure of the proximity of metallic objects. The voltage can be used to activate an electronic switch circuit, basically a transistor that has its output switched from low to high by the voltage change, and so give an ON–OFF device. Such objects can be typically detected over a range of about 0.5–20 mm.

2.4.1.3 Photoelectric sensors and switches

Photoelectric switch devices operate as either transmissive types where the object being detected breaks a beam of light, usually infrared radiation, and stops it from reaching the detector, as illustrated in Figure 2.9a or reflective type where the object being detected reflects a beam of light onto the detector (Figure 2.9b). In both cases, the radiation emitter is generally a light-emitting diode (LED). The radiation detector might be a phototransistor, often a pair of transistors, known as a Darlington pair. The Darlington pair increases the sensitivity. Depending on the circuit used, the output can be made to switch to either high or low when light strikes the transistor. Figure 2.9c shows a U-shaped form where the object breaks the light beam. Another possibility is a photodiode. Depending on the circuit used, the output can be made to switch to either high or low when light strikes the diode.

Light is converted to a current, voltage, or resistance change in all these sensors. If the output is to be used as a measure of the intensity of the light, rather than just the presence or absence of some object in the light path, the signal is amplified and then converted from analog to digital by an analog-to-digital converter. Alternately, a light-to-frequency converter is used, the light then being converted to a sequence of pulses with the frequency of the pulses being a measure of the light intensity.

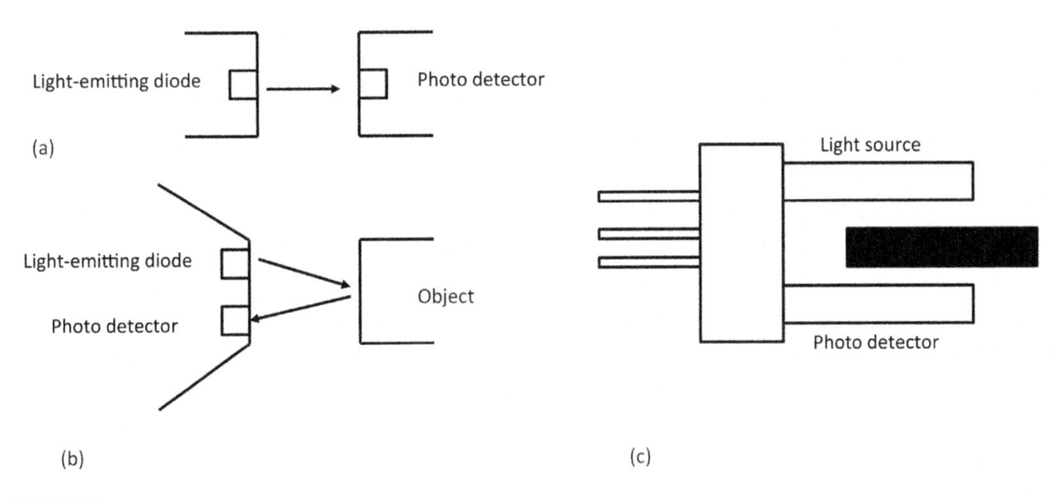

FIGURE 2.9

Photoelectric sensors. (a) light detection (b) photo reflection (c) u shaped object

2.4.1.4 Encoders

Encoders are devices that provide a digital output as a result of angular or linear displacement. An increment encoder detects changes in angular or linear displacement from some datum position, while an absolute encoder gives the actual angular or linear position. A beam of light, from perhaps a LED, passes through slots in a disc and is detected by a light sensor, for example, a photodiode or phototransistor. When the disc rotates, the light beam is alternately transmitted and stopped. Therefore, a pulsed output is produced from the light sensor. The number of pulses is proportional to the angle through which the disc has rotated, the resolution being proportional to the number of slots on a disc.

2.4.1.5 Temperature sensors

Bimetal element is a simple form of temperature sensor used to provide an ON–OFF signal when a particular temperature is reached. This consists of two strips of different metals, for example, brass and iron, bonded together, as illustrated in Figure 2.10. Since the two metals have different coefficients of

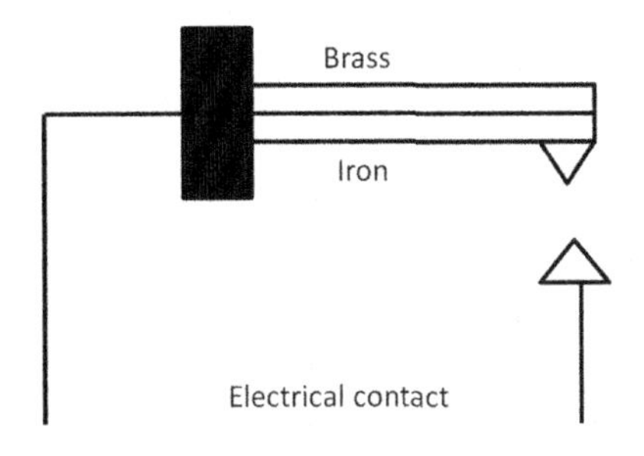

FIGURE 2.10

Bimetallic strip

expansion, increase in the temperature of the bimetal strip leads to the strip curving, to enable one of the metals to expand more than the other. The higher expansion metal is on the outside of the curve. As the strip cools, the bending effect is reversed. This movement of the strip can be used to make or break electrical contacts and hence, at some particular temperature, give an ON–OFF current in an electrical circuit. The device is not very accurate but is commonly used in domestic central heating thermostats.

2.4.1.6 Resistive temperature detector

Another form of temperature sensor is the resistive temperature detector (RTD), based on the principle that the electrical resistance of metals or semiconductors changes with temperature. In the case of a metal, the ones most commonly used are platinum, nickel, or nickel alloys, the resistance of which varies in a linear manner with temperature over a wide range of temperatures, although the actual change in resistance per degree is fairly small. Semiconductors, such as thermistors, show very large nonlinear changes in resistance with temperature. Such detectors can be used as one arm of a Wheatstone bridge and the output of the bridge taken as a measure of the temperature, as illustrated in Figure 2.11a. Another possibility is to use a potential divider circuit with the change in resistance of the thermistor changing the voltage drop across a resistor as in Figure 2.11b. The output from either type of circuit is an analog signal that is a measure of the temperature.

2.4.1.7 Thermodiodes and thermotransistors

Thermodiodes and thermotransistors are also used as temperature sensors since the rate at which electrons and holes diffuse across semiconductor junctions is affected by temperature. Integrated circuits are available that combine such a temperature-sensitive element with the relevant circuitry to give an output voltage related to temperature. Another commonly used temperature sensor is the thermocouple. The thermocouple essentially consists of two dissimilar wires A and B forming a junction, as illustrated in Figure 2.12. When the junction is heated so that it is at a higher temperature than the other junctions in the circuit, which remain at a constant cold temperature, an e.m.f. is produced that is related to the hot junction temperature. The voltage produced by a thermocouple is small and needs amplification

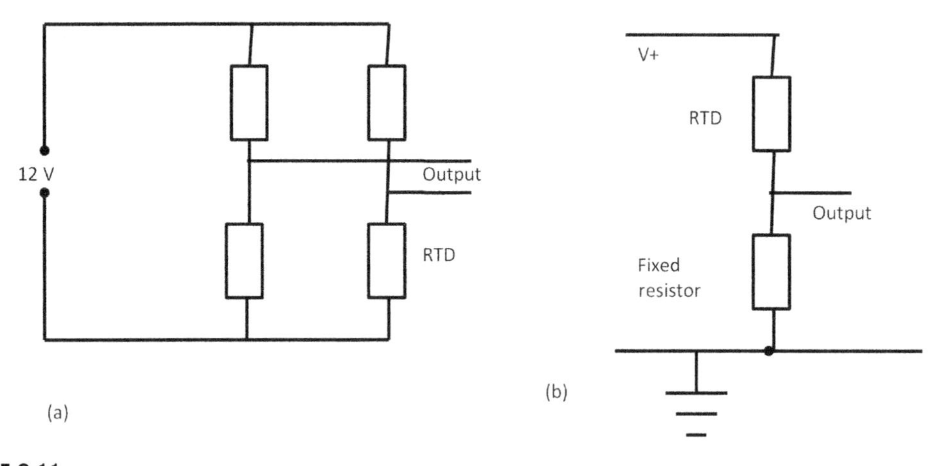

FIGURE 2.11

(a) Wheatstone bridge and (b) potential divider circuits

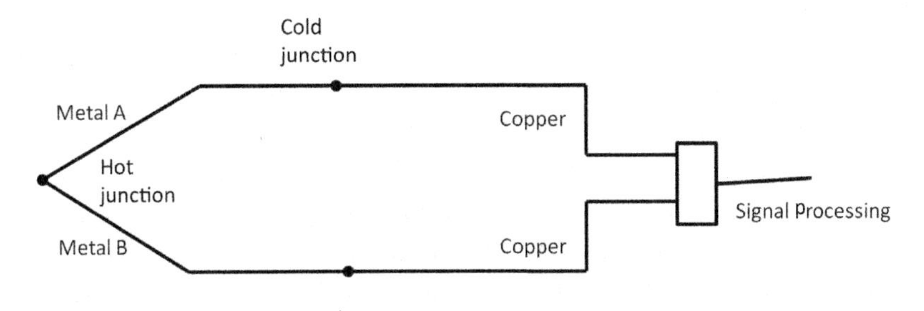

FIGURE 2.12

Thermocouple

before it can be fed to the analog channel input of a PLC. There is also circuitry required to compensate for the temperature of the cold junction since its temperature affects the value of the e.m.f. given by the hot junction. The amplification and compensation, together with filters to reduce the effect of interference from the 50 Hz mains supply, are often combined in a signal-processing unit.

2.4.1.8 Pressure sensors

Diaphragm and bellows type are commonly used pressure sensors that give responses related to the pressure. The diaphragm type consists of a thin disc of metal or plastic, secured round its edges. When there is a pressure difference between the two sides of the diaphragm, the center of the diaphragm deflects. The amount of deflection is related to the pressure difference. This deflection may be detected by strain gauges attached to the diaphragm, by a change in capacitance between it and a parallel fixed plate or by using the deflection to squeeze a piezoelectric crystal. When a piezoelectric crystal is squeezed, there is a relative displacement of positive and negative charges within the crystal and the outer surfaces of the crystal become charged. Hence, a potential difference appears across it. Other versions with one side of the diaphragm open to the atmosphere are used to measure gauge pressure; others allow pressures to be applied to both sides of the diaphragm and are used to measure differential pressures.

2.4.1.9 Output devices

The output ports of a PLC are of the relay type or optoisolator with transistor or triac types depending on the devices connected to them, which are to be switched ON or OFF. Generally, the digital signal from an output channel of a PLC is used to control an actuator, which in turn controls some process. The term actuator is used for the device that transforms the electrical signal into some more powerful action, which further results in the control of the process. The following are some of the examples.

Relay

Solenoids form the basis of a number of output control actuators. When a current passes through a solenoid, a magnetic field is produced that can attract ferrous metal components in its vicinity. One example of such an actuator is the relay, the term contactor being used when large currents are involved. When the output from the PLC is switched ON, the solenoid magnetic field is produced and pulls on the contacts thereby closing a switch or switches, as illustrated in Figure 2.13. The result is that much larger currents can be switched ON. Therefore, the relay might be used to switch on the current to a motor.

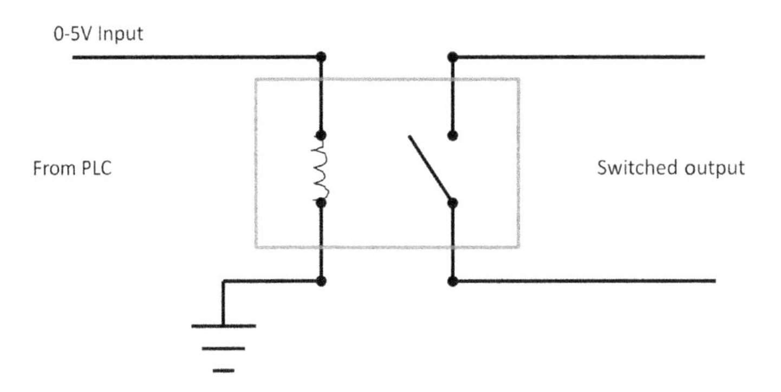

FIGURE 2.13

Relay used an output device

2.4.1.10 Directional control valves

Another example of the use of a solenoid as an actuator is a solenoid operated valve. The valve may be used to control the directions of flow of pressurized air or oil and used to operate other devices such as a piston moving in a cylinder.

2.4.1.11 Motors

A d.c. motor has coils of wire mounted in slots on a cylinder of ferromagnetic material, this being termed the armature. The armature is mounted on bearings and is free to rotate. It is mounted in the magnetic field produced by permanent magnets or current passing through coils of wire, these being termed the field coils. When a current passes through the armature coil, forces act on the coil leading to rotation. Brushes and a commutator are used to reverse the current through the coil every half rotation to keep the coil rotating. The speed of rotation can be controlled by changing the size of the current to the armature coil. However, because fixed voltage supplies are generally used as the input to the coils, the required variable current is often obtained by an electronic circuit. This can control the average value of the voltage, and hence current, by varying the time for which the constant d.c. voltage is switched ON. Pulse width modulation (PWM) is used to control the average d.c. voltage applied to the armature. A PLC might thus control the speed of rotation of a motor by controlling the electronic circuit used to control the width of the voltage pulses.

2.5 I/O PROCESSING

Signal conditioning plays an important role to achieve signal at a particular level when there are many inputs or outputs between the PLC controller and the outside world and also to isolate it from possible electrical hazards such as high voltages. Let us attempt to understand the forms of typical I/O modules, in an installation where sensors are some distance from the PLC processing, and there are many inputs or outputs.

The following input signals from sensors and the outputs are required for actuating devices:

1. Analog, a signal whose size is related to the size of the quantity being sensed.
2. Discrete, essentially just an ON–OFF signal.
3. Digital, a sequence of pulses.

The I/O units of PLCs are designed to enable a range of input signals to be changed into 5 V digital signals and also to ensure the availability of a range of outputs to drive external devices. It is this in-built facility of enabling the handling of a range of inputs and outputs that makes PLCs so easy to use. The following is a brief indication of the basic circuits used for input and output units. In the case of rack instruments they are mounted on cards that can be plugged into the racks and so the I/O characteristics of the PLC can thus be changed by changing the cards. A single box form of PLC has I/O units incorporated by the manufacturer.

2.5.1 OUTPUT UNITS

When a PLC output unit provides the current for the output device, as illustrated in Figure 2.14a, it is said to be sourcing, and when the output device provides the current to the output unit, it is said to be sinking, as illustrated in Figure 2.14b. Often, sinking input units are used for interfacing with electronic equipment and sourcing output units for interfacing with solenoids.

The output units can be relay, transistor, or triac.

2.5.2 REMOTE CONNECTIONS

When many inputs or outputs are located in considerable distance from the PLC, while it would be possible to run cables from each such device to the PLC a more economic solution is to use I/O modules in the vicinity of the inputs and outputs. This would need the use of just a single core cable to connect each, over the long distances, to the PLC instead of the multicore cable that would be needed without such distant I/O modules (Figure 2.15).

In some situations a number of PLCs may be linked together with a master PLC unit sending and receiving I/O data from the other units (Figure 2.15). The distant PLCs do not contain the control program since all the control processing is carried out by the master PLC. The cables used for communicating data between remote I/O modules and a central PLC, remote PLCs, and the master PLC are typically twisted-pair cabling, often routed through grounded steel conduit in order to reduce electrical "noise." Coaxial cable enables higher data rates to be transmitted and does not require the shielding of

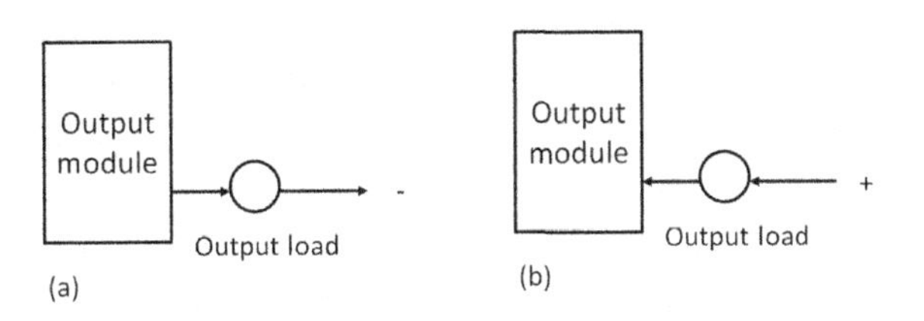

FIGURE 2.14

Output unit (a) sourcing and (b) sinking

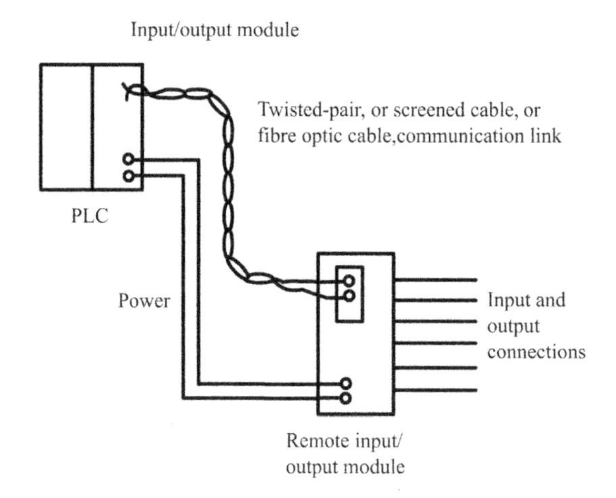

FIGURE 2.15

Units located some distance from the PLC

steel conduit. Fiberoptic cabling has the advantage of resistance to noise, small size, and flexibility and is now becoming more widely used.

2.5.3 SERIAL AND PARALLEL COMMUNICATIONS

Transmission of data one bit at a time, as illustrated in Figure 2.16a, is called serial communication. Therefore, if an 8-bit word is to be transmitted, the eight bits are transmitted one at a time in sequence along a cable. This means that a data word has to be separated into its constituent bits for transmission and then reassembled into the word when received. Transmission of all the constituent bits of a word simultaneously along parallel cables, as illustrated in Figure 2.16b, is called parallel communication. This allows data to be transmitted over short distances at high speeds.

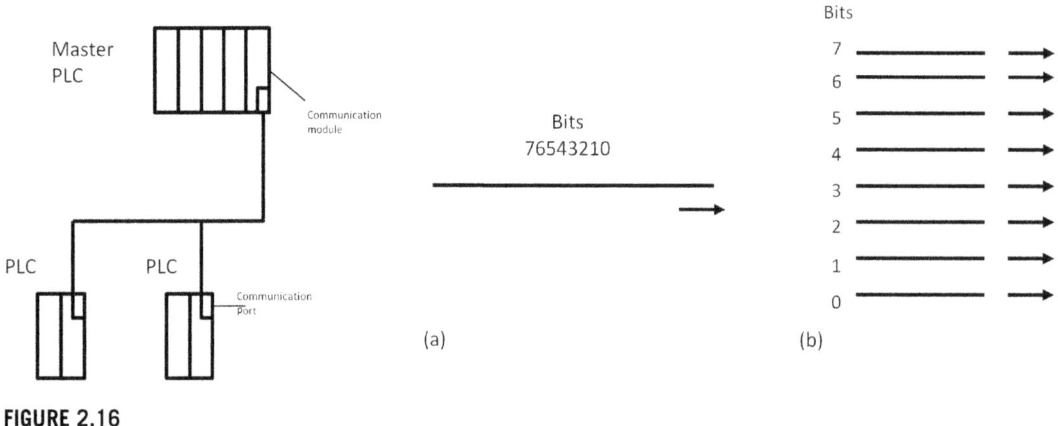

FIGURE 2.16

(a) Serial communication and (b) parallel communication

Serial communication is used for transmitting data over long distances, as it is much cheaper to run a single core cable over a long distance than the multicore cables needed for parallel communication. Serial communication is used for the connection between a computer, when used as a programming terminal, and a PLC. Parallel communication is used when connecting laboratory instruments to the system. However, internally, PLCs work, for speed, with parallel communications. Thus, circuits called universal asynchronous receivers–transmitters (UARTS) have to be used at I/O ports to converts serial communications signals to parallel. The serial communication details are available in subsequent chapters.

2.5.4 DISTRIBUTED SYSTEMS

Often PLCs figure in an entire hierarchy of communications, as illustrated in Figure 2.17. Thus, at the lowest level we have input and output devices such as sensor and motors interfaced through I/O with the next level. The next level involves controllers such as small PLCs or small computers, linked through a network with the next level of larger PLCs and computers exercising local area control. These in turn may be part of a network involved with a large mainframe company computer controlling all.

Systems that can control and monitor industrial processes are being increasingly used. This involves control and the gathering of data. The term SCADA, which stands for supervisory control and data acquisition system, is widely used for such a system.

2.5.5 I/O ADDRESSES

The PLC has to be able to identify each particular input and output. It does this by allocating addresses to each input and output. With a small PLC this is likely to be just a number, prefixed by a letter to indicate whether it is an input or an output. Thus for the Mitsubishi PLC, we might have inputs with addresses X400, X401, X402, etc., and outputs with addresses Y430, Y431, Y432, etc., the X indicating

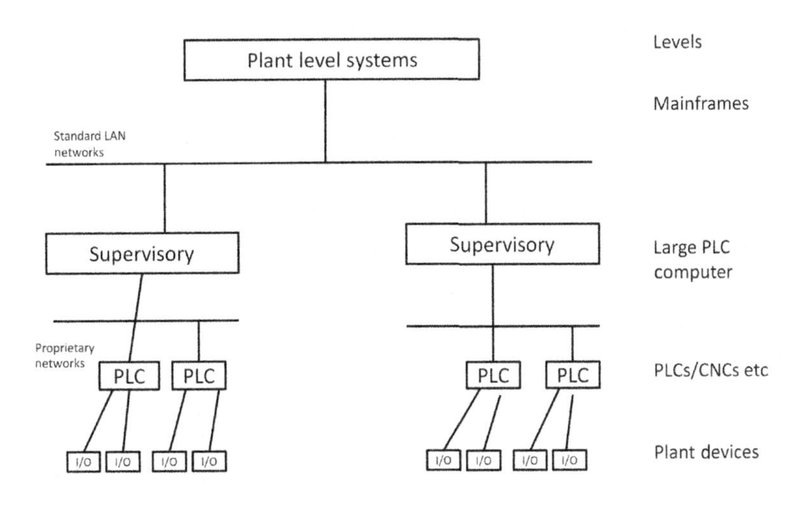

FIGURE 2.17

Hierarchical architecture

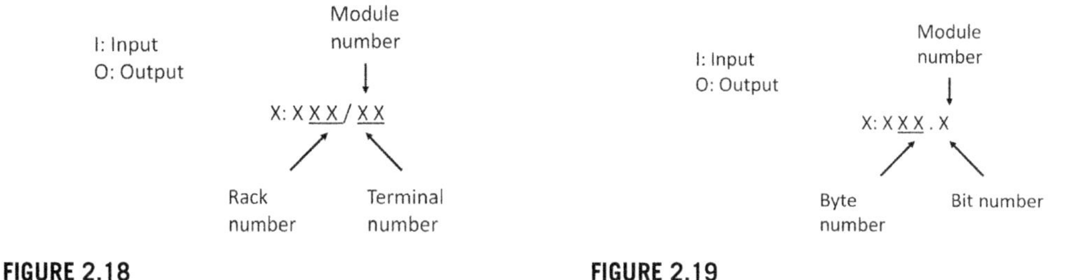

FIGURE 2.18

Allen-Bradley PLC-5 addressing

FIGURE 2.19

Siemens SIMATIC S5 addressing

an input and the Y an output. Toshiba uses a similar system. With larger PLCs having several racks of input and output channels, the racks are numbered. With the Allen-Bradley PLC-5, the rack containing the processor is given the number 0 and the addresses of the other racks are numbered 1, 2, 3, etc. according to how set-up switches are organized. Each rack can have a number of modules and each one deals with a number of inputs and/or outputs. Thus, addresses can be of the form shown in the following. For example, we might have an input with address I:012/03. This indicates an input, rack 01, module 2, and terminal 03 (Figure 2.18).

With the Siemens SIMATIC S5, the inputs and outputs are arranged in groups of eight. Each group of eight is termed a byte and each input or output within a group of eight is termed a bit. The inputs and outputs thus have their addresses in terms of the byte and bit numbers, effectively giving a module number followed by a terminal number, a full stop (.) separating the two numbers. Figure 2.19 is an illustration of the Siemens system. Thus, I0.1 is an input at bit 1 in byte 0, and Q2.0 is an output at bit 0 in byte 2.

2.6 LADDER AND FUNCTION BLOCK PROGRAMMING

Programs for microprocessor-based systems have to be loaded in machine code, this being a sequence of binary code numbers to represent the program instructions. However, assembly language based on the use of mnemonics can be used, for example, LD is used to indicate the operation required to load the data that follows the LD, and a computer program called an assembler is used to translate the mnemonics into machine code. Programming can be made even easier by the use of high-level languages, for example, C, BASIC, PASCAL, FORTRAN, and COBOL. These use prepackaged functions, represented by simple words or symbols descriptive of the function concerned. For example, with C language the symbol & is used for the logic AND operation. However, the use of these methods to write programs requires some skill in programming and PLCs are intended to be used by engineers without any great knowledge of programming. This led to the development of ladder programming. This is a means of writing programs that can then be converted into machine code by some software for use by the PLC microprocessor. This method of writing programs was adopted by most PLC manufacturers; however, each developed their own versions and so an international standard has been adopted for ladder programming and the methods used for programming PLCs. The standard, published in 1993, is IEC 1131-3 (International Electrotechnical Commission). The IEC 1131-3 programming languages are ladder diagrams (LAD), instruction list (IL), sequential function charts (SFCs), structured text (ST), and function block diagrams (FBDs). Let us attempt to understand the basic techniques involved in developing ladder and function block programs to represent basic switching operations, involving the logic functions of AND, OR, Exclusive OR, NAND, and NOR, and latching.

2.6.1 LADDER DIAGRAMS

As an introduction to LAD, consider the simple wiring diagram for an electrical circuit in Figure 2.20a. The diagram shows the circuit for switching ON or OFF an electric motor. We can redraw this diagram in a different way, using two vertical lines to represent the input power rails and stringing the rest of the circuit between them. Figure 2.20b shows the result. Both circuits have the switch in series with the motor and supplied with electrical power when the switch is closed. The circuit shown in Figure 2.20b is termed a LAD.

With such a diagram the power supply for the circuits is always shown as two vertical lines with the rest of the circuit as horizontal lines. The power lines, or rails as they are often termed, are like the vertical sides of a ladder with the horizontal circuit lines like the rungs of the ladder. The horizontal rungs show only the control portion of the circuit; in Figure 2.21 it is just the switch in series with the motor. Circuit diagrams often show the relative physical location of the circuit components and how they are actually wired. LADs do not show the actual physical locations and

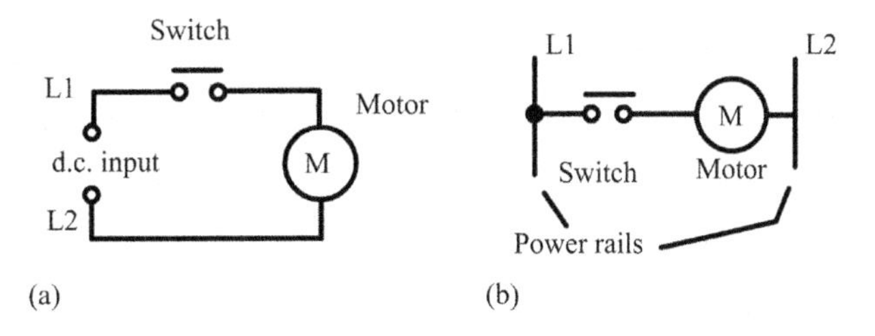

FIGURE 2.20

Ways of drawing the same electrical circuit (a) electrical connection (b) ladder representation

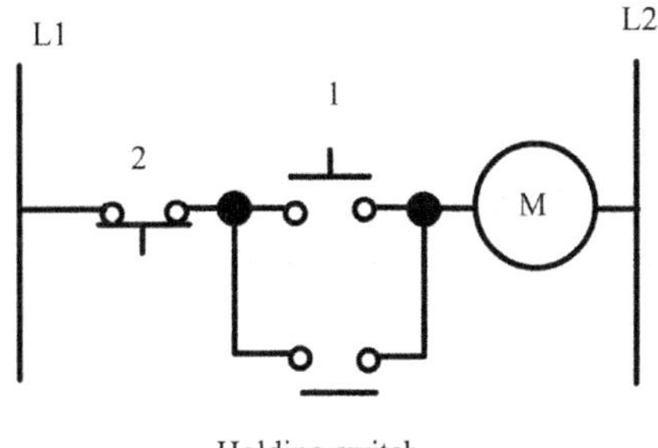

FIGURE 2.21

Stop-start switch

focus on clearly showing how the control is exercised. Figure 2.21 depicts an example of a LAD for a circuit used to start and stop a motor using push buttons. In the normal state, push button 1 is open and push button 2 closed. When button 1 is pressed, the motor circuit is completed and the motor starts. Also, the holding contacts wired in parallel with the motor close and remain closed as long as the motor is running. Thus, when the push button 1 is released, the holding contacts maintain the circuit and hence the power to the motor. To stop the motor, button 2 is pressed. This disconnects the power to the motor and the holding contacts open. Thus when push button 2 is released, there is still no power to the motor. Thus, we have a motor that is started by pressing button 1 and stopped by pressing button 2.

2.6.2 PLC LADDER PROGRAMMING

A very commonly used method of programming PLCs is based on the use of LAD. Writing a program is then equivalent to drawing a switching circuit. The LAD consists of two vertical lines representing the power rails. Circuits are connected as horizontal lines, that is, the rungs of the ladder, between these two verticals. The following conventions are used when drawing a LAD:

- The vertical lines of the diagram represent the power rails between which circuits are connected. The power flow is assumed to be from the left-hand vertical across a rung.
- Each rung on the ladder defines one operation in the control process.
- A LAD is read from left to right and from top to bottom. Figure 2.22 depicts the scanning motion employed by the PLC. The top rung is read from left to right. Then the second rung down is read from left to right and so on. When the PLC is in its run mode, it goes through the entire ladder program to the end, the end rung of the program being clearly denoted, and then promptly resumes at the start. This procedure of going through all the rungs of the program is termed a cycle. The end rung might be indicated by a block with the word END or RET for return, since the program promptly returns to its beginning.

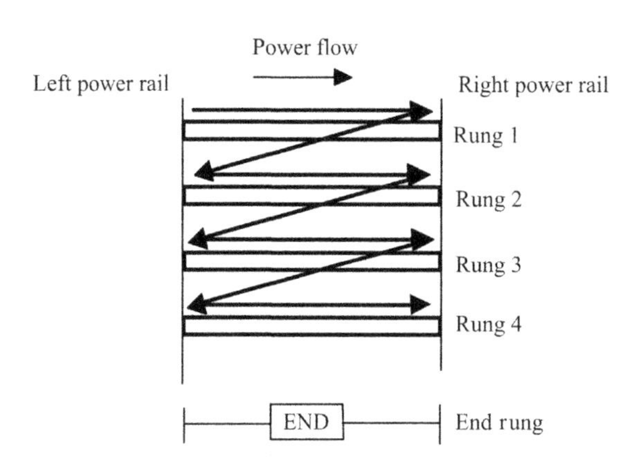

FIGURE 2.22

Scanning the ladder program

- Each rung must start with an input or inputs and must end with at least one output. The term input is used for a control action, such as closing the contacts of a switch, used as an input to the PLC. The term output is used for a device connected to the output of a PLC, for example, a motor.
- Electrical devices are shown in their normal condition. Thus, a switch that is normally open until some object closes it is shown as open on the LAD. A switch that is normally closed is shown closed.
- A particular device can appear in more than one rung of a ladder. For example, we might have a relay that switches on one or more devices. The same letters and/or numbers are used to label the device in each situation.
- The inputs and outputs are all identified by their addresses. The notation used for addressing varies, depending on the PLC manufacturer. This is the address of the input or output in the memory of the PLC. Figure 2.23 shows standard IEC 1131-3 symbols used for input and output devices. Some slight variations occur between the symbols when used in semigraphic form and when in full graphic. Note that inputs are represented by different symbols representing normally open or normally closed contacts. The action of the input is equivalent to opening or closing a switch. Output coils are represented by just one form of symbol (Figure 2.24).

To illustrate the drawing of the rung of a LAD, consider a situation where the energizing of an output device, for example, a motor, depends on a normally open start switch being activated by being closed. The input is thus the switch and the output the motor. Figure 2.24a depicts the LAD.

Starting with the input, we have the normally open symbol | | for the input contacts. There are no other input devices and the line terminates with the output, denoted by the symbol ( ). When the switch is closed, that is, there is an input, the output of the motor is activated. Only as long as there is an

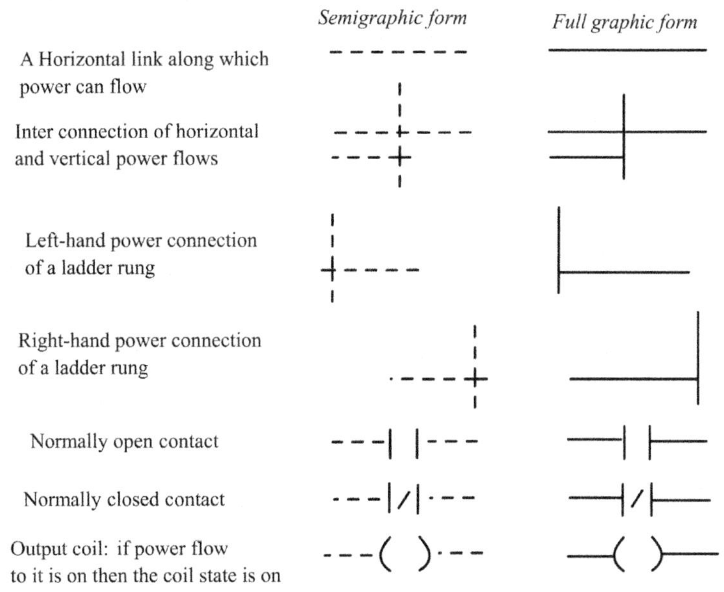

	Semigraphic form	*Full graphic form*
A Horizontal link along which power can flow	-------	
Inter connection of horizontal and vertical power flows		
Left-hand power connection of a ladder rung		
Right-hand power connection of a ladder rung		
Normally open contact	---| |---	
Normally closed contact	---|/|---	
Output coil: if power flow to it is on then the coil state is on	---( )---	

FIGURE 2.23

Basic symbols

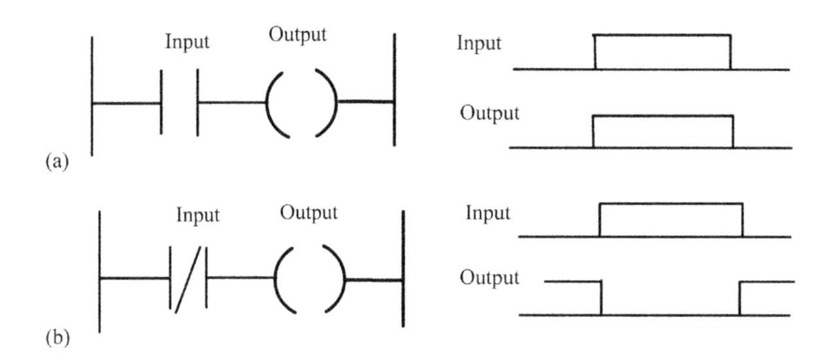

FIGURE 2.24

A ladder rung (a) Ladder Rung (b) Signal

input to the contacts there is an output. If there had been a normally closed switch |/| with the output (Figure 2.24b), then there would have been an output until that switch was opened. Only while there is no input to the contacts is there an output.

In drawing LAD the names of the associated variable or addresses of each element are appended to its symbol. Figure 2.25 shows how the LAD of Figure 2.24a would appear using (a) Mitsubishi, (b) Siemens, (c) Allen-Bradley, and (d) Telemecanique notations for the addresses. Thus Figure 2.25a indicates that this rung of the ladder program has an input from address X400 and an output to address Y430. When wiring up the inputs and outputs to the PLC, the relevant ones must be connected to the input and output terminals with these addresses.

There are many control situations requiring actions to be initiated when a certain combination of conditions is realized. Thus, for an automatic drilling machine, there might be the condition that the drill motor is to be activated when the limit switches are activated that indicate the presence of the workpiece and the drill position as being at the surface of the workpiece. Such a situation involves the AND logic function, condition A and condition B both to be realized for an output to occur. Let us understand such logic functions.

Input X400 Output Y430 (a)

Input Output (b)

Input I:001/01 Output O:010/01 (c)

Input I0,0 Output O0,0 (d)

FIGURE 2.25

Notation: (a) Mitsubishi, (b) Siemens, (c) Allen-Bradley, and (d) Telmecanique

2.6.2.1 AND

Figure 2.26a shows a situation where an output is not energized unless two, normally open, switches are both closed. Both switch A and switch B have to be closed, which gives an AND logic situation. We can think of this as representing a control system with two inputs A and B (Figure 2.26b). Only when A and B are both on is there an output. Thus, if we use 1 to indicate an on signal and 0 to represent an off signal, then for there to be a 1 output we must have A and B both 1. Such an operation is said to be controlled by a logic gate and the relationship between the inputs to a logic gate and the outputs is tabulated in a form known as a truth table. Thus, for the AND gate we have the following.

An example of an AND gate is an interlock control system for a machine tool so that it can only be operated when the safety guard is in position and the power switched ON. Figure 2.27a shows an AND gate system on a LAD. The LAD starts with | |, a normally open set of contacts labeled input A, to represent switch A and in series with it | |, another normally open set of contacts labeled input B, to represent switch B. The line then terminates with O to represent the output. For there to be an output, both input A and input B have to occur, that is, input A and input B contacts have to be closed. In general, on a LAD contacts in a horizontal rung, that is, contacts in series, represent the logical AND operations.

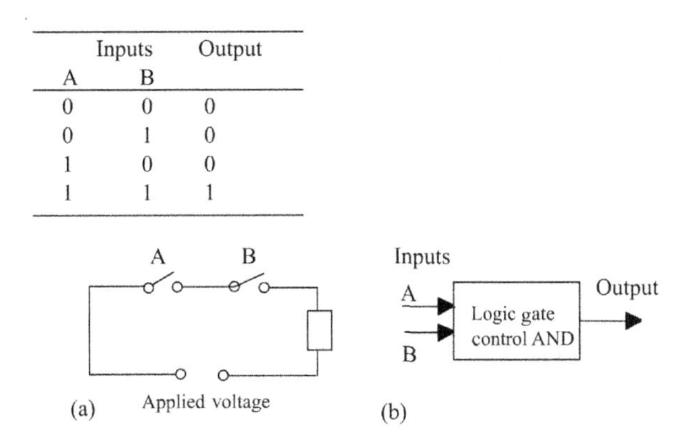

Inputs		Output
A	B	
0	0	0
0	1	0
1	0	0
1	1	1

FIGURE 2.26

(a) AND circuit and (b) AND logic gate

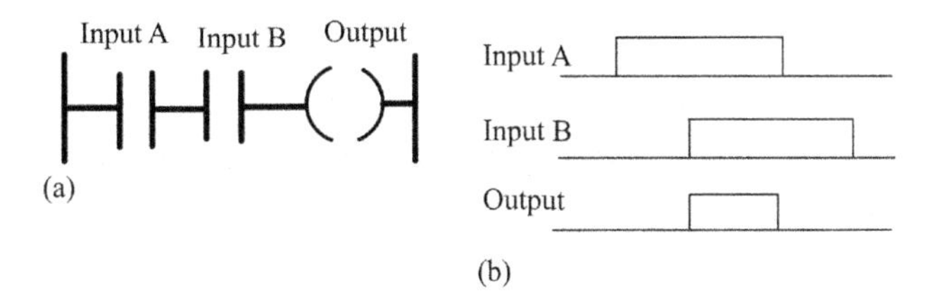

FIGURE 2.27

AND gate with a ladder diagram rung. (a) Ladder diagram (b) Signal Diagram

2.6.2.2 OR gate

Figure 2.28a shows an electrical circuit where an output is energized when switch A or B, both normally open, are closed. This describes an OR logic gate (Figure 2.28b) in that input A or input B must be on for there to be an output. The truth table is:

Alternative paths provided by vertical paths from the main rung of a LAD, that is, paths in parallel, represent logical OR operations (Figure 2.29).

An example of an OR gate control system is a conveyor belt transporting bottled products to packaging where a deflector plate is activated to deflect bottles into a reject bin if either the weight is not within certain tolerances or there is no cap on the bottle.

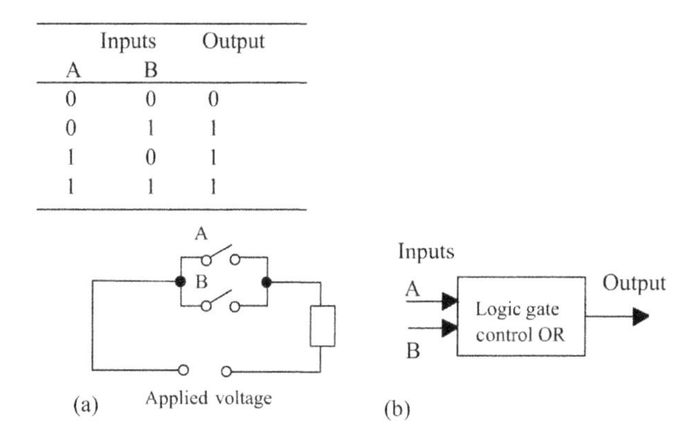

Inputs		Output
A	B	
0	0	0
0	1	1
1	0	1
1	1	1

FIGURE 2.28

(a) OR electrical circuit and (b) OR logic gate

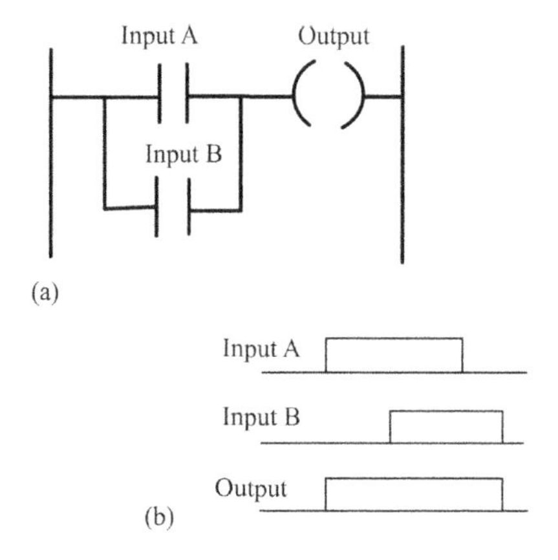

FIGURE 2.29

OR gate. (a) Ladder diagram (b) signal diagram

2.6.2.3 Latching

There are often situations where it is necessary to hold an output energized, even when the input ceases. A simple example of such a situation is a motor started by pressing a push button switch. Although the switch contacts do not remain closed, the motor is required to continue running until a stop push button switch is pressed. The term latch circuit is used for the circuit used to carry out such an operation. It is a self-maintaining circuit in that, after being energized, it maintains that state until another input is received. An example of a latch circuit is shown in Figure 2.30. When the input A contacts close, there is an output. However, when there is an output, another set of contacts associated with the output closes. These contacts form an OR logic gate system with the input contacts. Thus, even if the input A opens, the circuit still maintains the output energized. The only way to release the output is by operating the normally closed contact B.

As an example of the application of a latching circuit, consider a motor controlled by stop and start push button switches and for which one signal light must be illuminated when the power is applied to the motor and another when it is not applied. Figure 2.31 shows the LAD with Mitsubishi notation for the addresses.

X401 is closed when the program is started. When X400 is momentarily closed, Y430 is energized and its contacts close. This results in latching and also the switching off of Y431 and the switching on of Y432. Note that the stop contacts X401 are shown as being programmed as open. If the Stop switch used is normally closed, then X401 receives a start-up signal to Close. This gives a safer operation than programming X401 as normally Closed. To switch the motor off, X401 is pressed and opens. Y430 contacts open in the top rung and third rung, but close in the second rung. Thus, Y431 comes on and Y432 off. Latching is widely used with start-ups so that the initial switch on of an application becomes latched on.

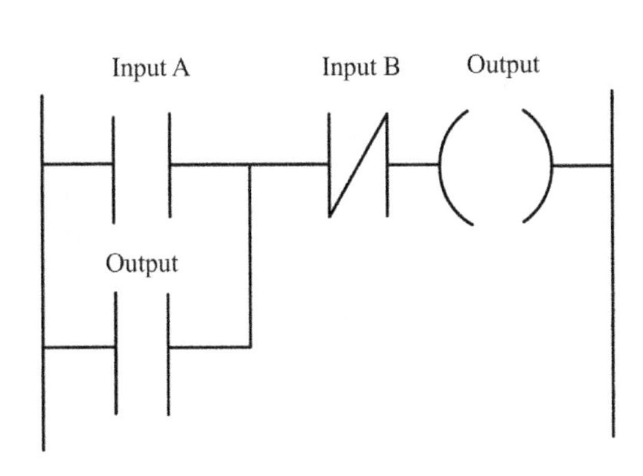

FIGURE 2.30

Latched circuit

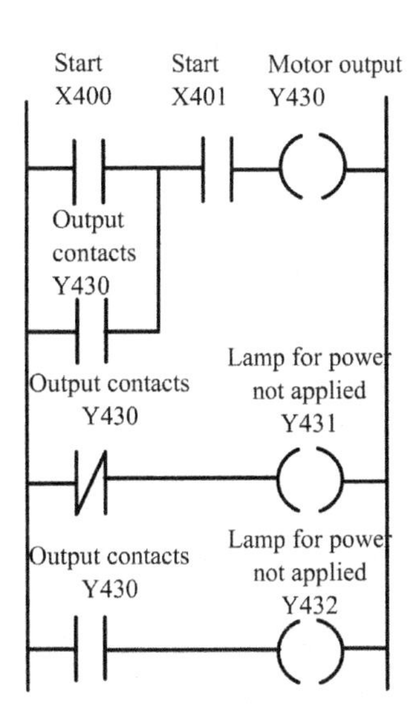

FIGURE 2.31

Motor on–off, with signal lamps, ladder diagram.

2.6.3 MULTIPLE OUTPUTS

With LAD, there can be more than one output connected to a contact. Figure 2.32 shows a ladder program with two output coils. When the input contacts close both the coils give outputs. For the ladder rung shown in Figure 2.33, output A occurs when input A occurs. Output B only occurs when both input A and input B occur.

2.6.4 ENTERING PROGRAMS

Each horizontal rung on the ladder represents an instruction in the program to be used by the PLC. The entire ladder represents the complete program. There are several methods used for keying in the program into a programming terminal. Whatever method is used to enter the program into a programming terminal or computer, the output to the memory of the PLC has to be in a form that can be handled by the microprocessor. This is termed machine language and is just binary code, for example, 0010100001110001.

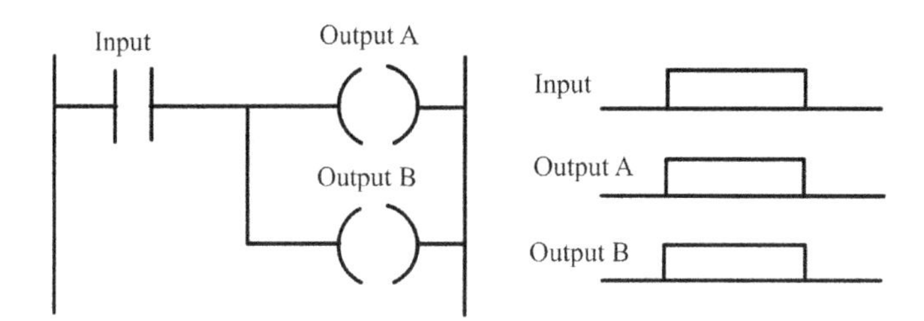

FIGURE 2.32

Ladder rung with two outputs

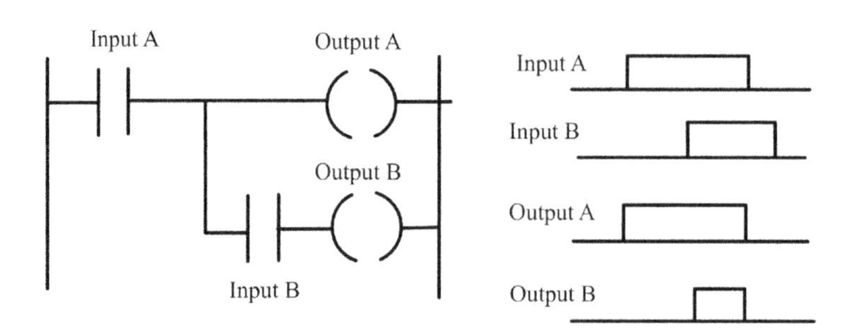

FIGURE 2.33

Ladder rung with two inputs and two outputs

2.7 FUNCTION BLOCKS

The term FBD is used for PLC programs described in terms of graphical blocks. It is described as being a graphical language for depicting signal and data flows through blocks, these being reusable software elements. A function block is a program instruction unit that, when executed, yields one or more output values. Thus, a block is represented as in Figure 2.34 with the function name written in the box.

A function block is depicted as a rectangular block with inputs entering from the left and outputs exiting from the right. The function block type name is shown in the block, for example, AND, with the name of the function block in the system shown above it, Timer1. Names of function block inputs are shown within the block at the appropriate input and output points. Cross diagram connectors are used to indicate where graphical lines would be difficult to draw without cluttering up or complicating a diagram and show where an output at one point is used as an input at another.

Function blocks can have standard functions, such as those of the logic gates or counters or times, or have functions defined by the user, for example, a block to obtain an average value of inputs.

2.7.1 LOGIC GATES

Programs are often concerned with logic gates. Two forms of standard circuit symbols are used for logic gates, one having originated in the United States and the other being an international standard form (IEEE/ANSI), which uses a rectangle with the logic function written inside it. The 1 in a box indicates that there is an output when the input is 1. The OR function is given by ≥ 1; this is because there is an output if an input is greater than or equal to 1. A negated input is represented by a small circle on the input, and a negative output by a small circle on the output as in Figure 2.35. In FBD diagrams the notation used in the IEEE/ANSI form is often encountered.

Figure 2.36 shows the effect of such functional blocks in PLC programs.

To illustrate the form of such a diagram and its relationship to the LAD, Figure 2.37 shows an OR gate. When A or B inputs are 1 then there is an output.

Consider the development of a FBD and LAD for an application in which a pump is required to be activated and pump liquid into a tank when the start switch is closed, the level of liquid in the tank is below the required level, and there is liquid in the reservoir from which it is to be pumped. In such a situation, we need an AND logic situation between the start switch input and a sensor input, which is on when the liquid in the tank is below the required level. We might have a switch that is on until the liquid is at the required level. These two elements are then in an AND logic situation with a switch indicating that there is liquid in the reservoir. The FBD, and the equivalent LAD, is depicted in the Figure 2.38 as follows (Figure 2.39).

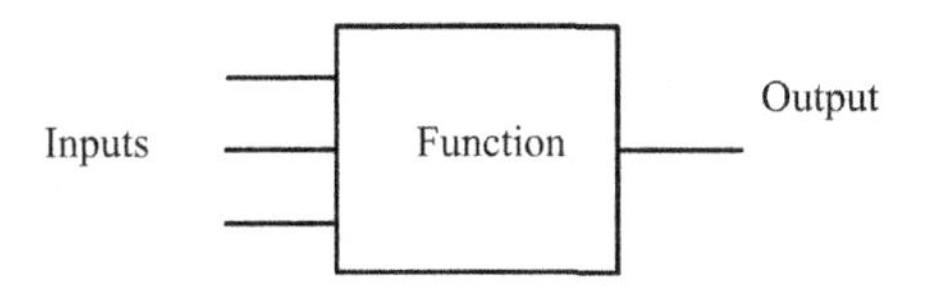

FIGURE 2.34

Functional block

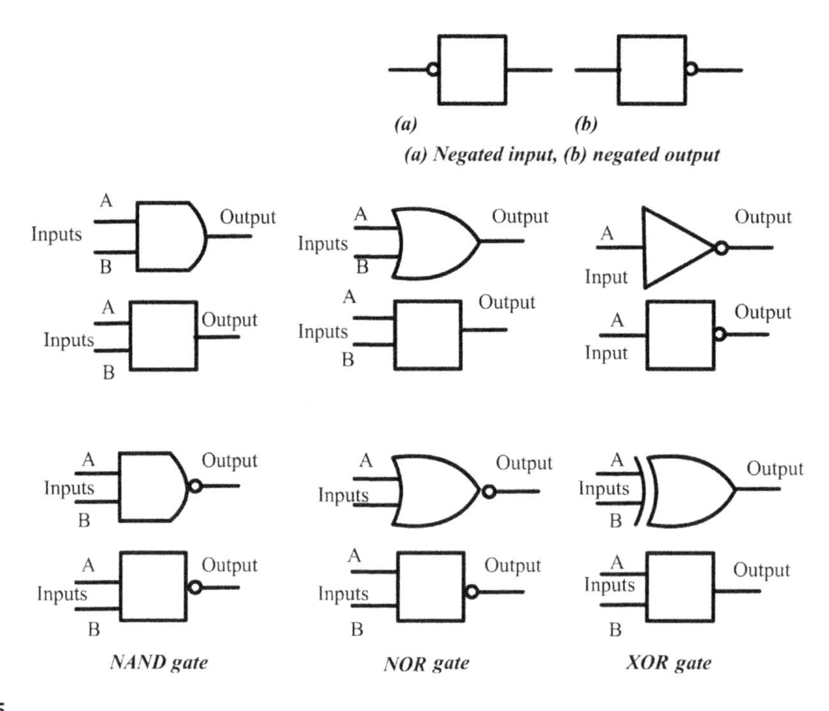

FIGURE 2.35

Logic gate symbols

2.7.2 PROGRAM EXAMPLES

The following tasks are intended to illustrate the application of the programming techniques discussed this far. A signal lamp is required to be switched ON if a pump is running and the pressure is satisfactory, or if the lamp test switch is closed. For the inputs from the pump and the pressure sensors we have an AND logic situation since both are required if there is to be an output from the lamp. We, however, have an OR logic situation with the test switch in that it is required to give an output of lamp on regardless of whether there is a signal from the AND system. The FBD and the LAD are thus of the form shown in Figure 2.40. Note that with the LAD, we tell the PLC when it has reached the end of the program by the use of the END or RET instruction.

As another example, consider a valve to be operated to lift a load when a pump is running and either the lift switch is operated or a switch operated indicating that the load has not already been lifted and is at the bottom of its lift channel. We have an OR situation for the two switches and an AND situation involving the two switches and the pump. Figure 2.41 depicts possible program.

As another example, consider a system where there has to be no output when any one of four sensors gives an output, otherwise there is to be an output. One way we could write a program for this is for each sensor to have contacts that are normally closed so there is an output. When there is an input to the sensor, the contacts open and the output stops. We have an AND logic situation. Figure 2.42 shows the FBD and the LAD of a system that might be used.

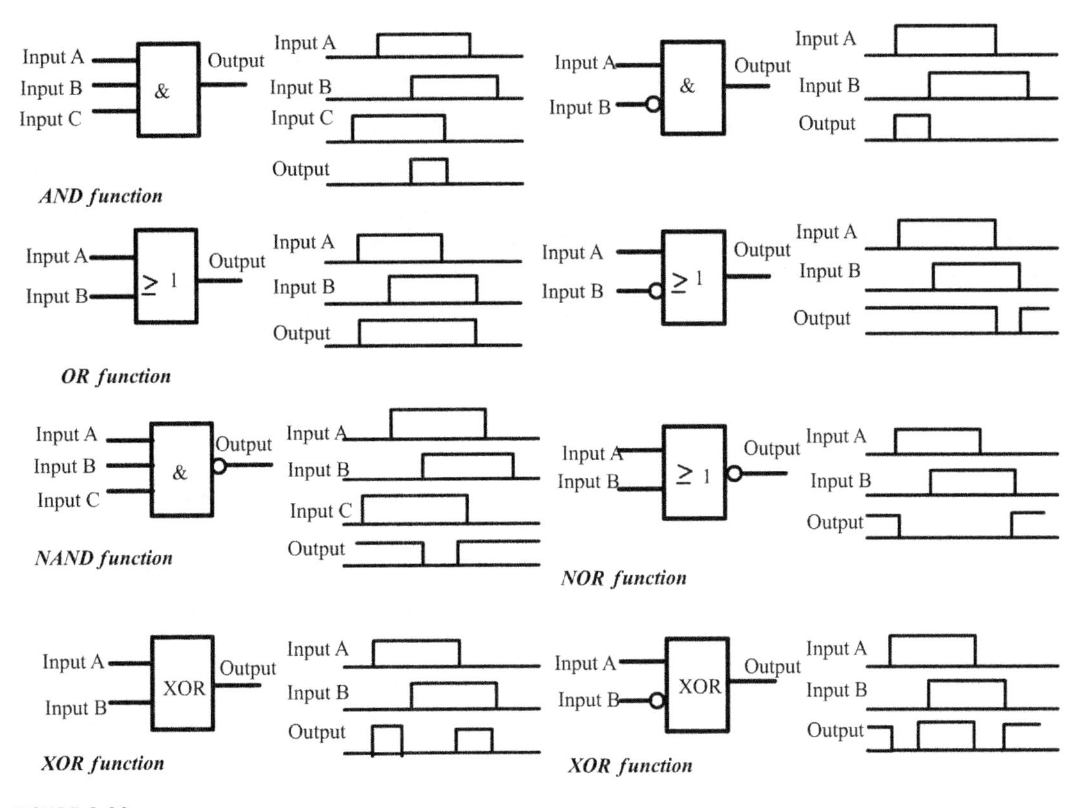

FIGURE 2.36

Function blocks

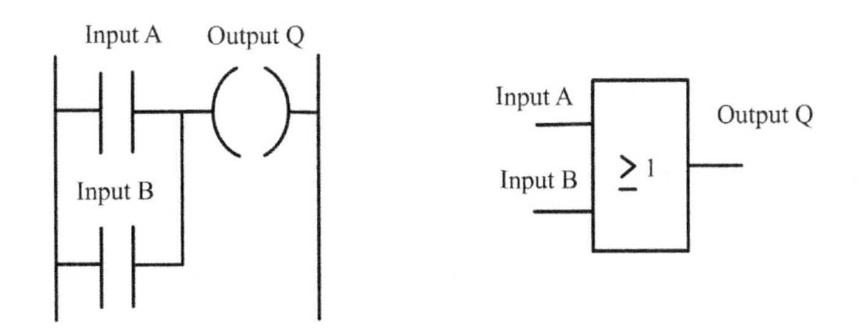

FIGURE 2.37

Ladder diagram and equivalent functional block diagram

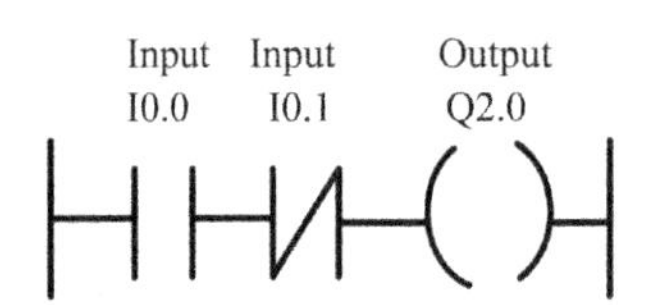

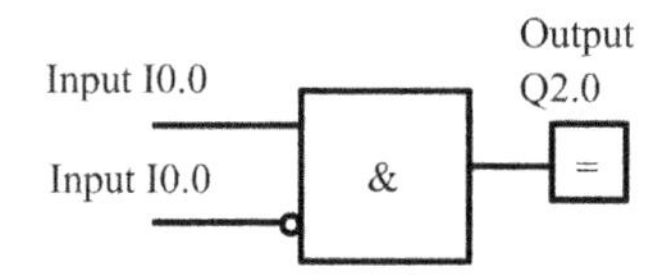

FIGURE 2.38

Ladder diagram and equivalent function block diagram

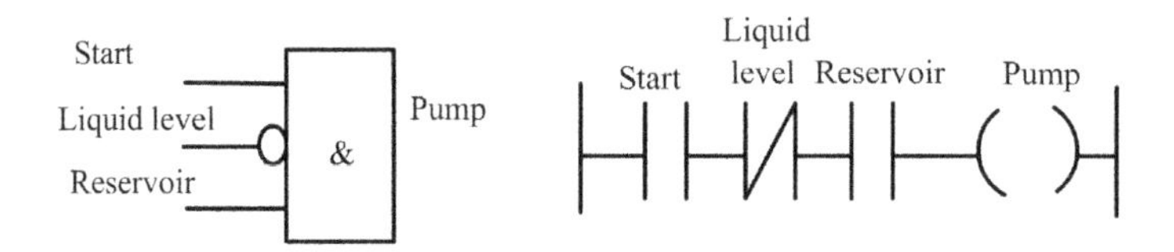

FIGURE 2.39

Pump application

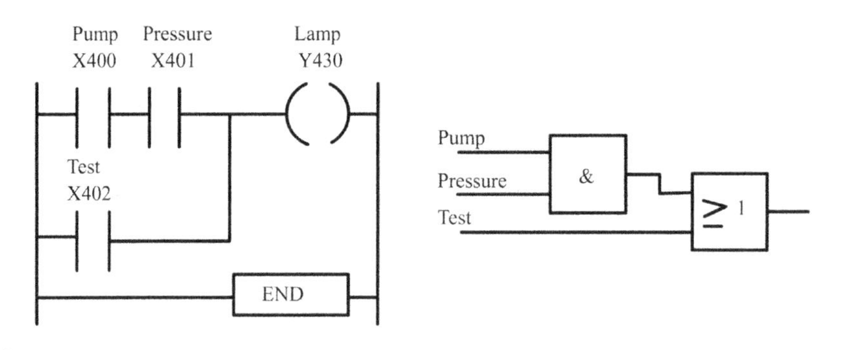

FIGURE 2.40

Signal lamp task

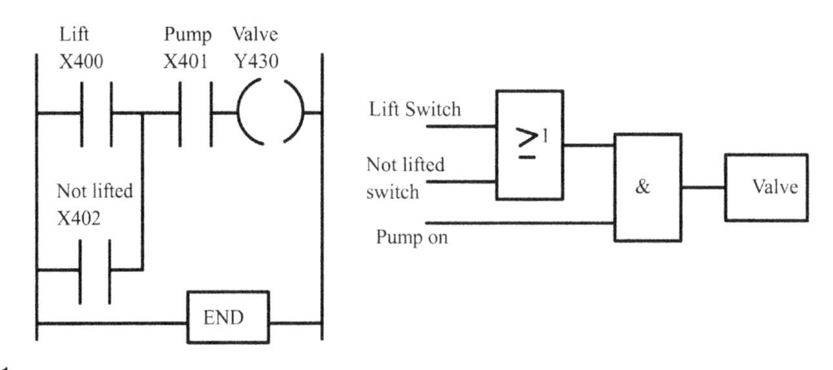

FIGURE 2.41

Valve operation program

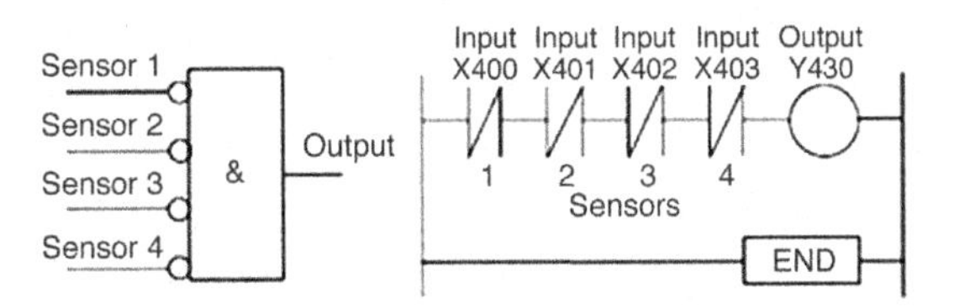

FIGURE 2.42

Output switched off by any of four sensors being activated

2.8 IL, SFC, AND ST PROGRAMMING METHODS

Let us understand some of the other IEC 1131-3 programming languages, that is, ILs, SFCs, and ST in this section.

2.8.1 INSTRUCTION LISTS

The programming method, of entering a ladder program using text, is IL. Instruction list gives programs that consist of a series of instructions, each instruction being on a new line. An instruction consists of an operator followed by one of more operands, that is, the subjects of the operator. In terms of LAD an operator may be regarded as a ladder element. Each instruction may either use or change the value stored in a memory register. Mnemonic codes are used, each code corresponding to an operator/ladder element. The codes used differ to some extent from manufacturer to manufacturer, although a standard IEC 1131-3 has been proposed and is being widely adopted. Table 2.1 shows some of the codes used by manufacturers, and the proposed standard, for instructions used in this section.

Table 2.1 Instruction code mnemonics

IEC 1131-3 Operators	Mitsubishi	OMRON	Siemens/ Telemecanique	Operation	Ladder Diagram
LD	LD	LD	A	Load operand into result register	Start a rung with open contacts
LDN	LDI	LD NOT	AN	Load negative operant into result register	Start a rung with closed contacts
AND	AND	AND	A	Boolean AND	A series element with open contacts
ANDN	ANI	AND NOT	AN	Boolean AND with negative operand	A series element with closed contacts
OR	OR	OR	O	Boolean OR	A parallel element with open contacts
ORN	ORI	OR NOT	ON	Boolean OR with negative operand	A parallel element with closed contacts
ST	OUT	OUT	=	Store result register into operand	An output

As an illustration of the use of IEC 113-1 operators, consider the following:

```
LD    A     (*Load A*)
AND   B     (*AND B*)
ST    Q     (*Store result in Q, i.e. output to Q*)
```

In the first line of the program, LD is the operator, A the operand, and the words at the ends of program lines and in brackets and preceded and followed by * are comments added to explain what the operation is and are not part of the program operation instructions to the PLC. LD A is thus the instruction to load A into the memory register. It can later be called on for further operations. The next line of the program has the Boolean operation AND performed with A and B. The last line has the result stored in Q, that is, outputted to Q. Labels can be used to identify various entry points to a program and they are useful for jumps in programs, and these precede the instruction and are separated from it by a colon. Thus, we might have: PUMP_OK: LD C (* Load C*).

With there being the instruction earlier in the program to jump to, PUMP_OK if a particular condition occurs.

With the IEC 113-1 operators, an N after it is used to negate its value.

For example, if we have:

```
ANDN B (*AND NOT B*)
LD A (*Load A*)
```

Thus, the ANDN operator inverts the value of ladder contacts and ANDs the result.

Consider an example of an OR gate. Figure 2.43 shows the gate with Mitsubishi notation.

The instruction for the rung start with an open contact is LD X400. The next item is the parallel OR set of contacts X401. Thus, the next instruction is OR X401. The last step is the output, hence OUT Y430. The IL would therefore be:

```
LD    X400
OR    X401
OUT   Y430
```

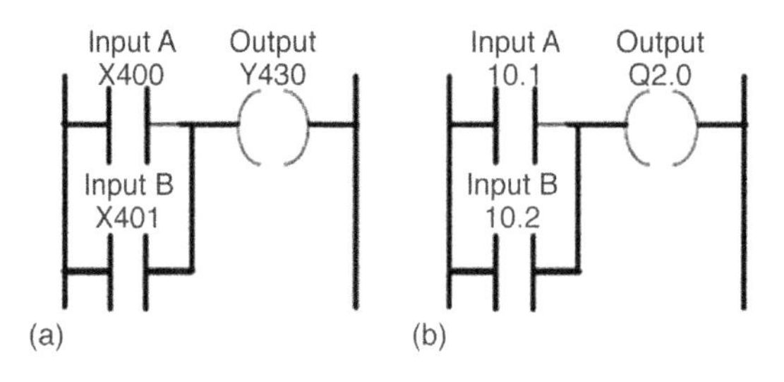

FIGURE 2.43

OR gate (a) Mitsubishi and (b) Siemens

2.8.2 SEQUENTIAL FUNCTION CHARTS

A traffic lamp sequence could be represented as a sequence of functions or states such as red light state and green light state and the inputs and outputs to each state. Figure 2.44 illustrates this. State 0 has an input that is triggered after the green light has been on for 1 min and an output of red light on. State 1 has an input that is triggered after the red light has been on for 1 min and an output of green light is on.

The term SFC is used for a pictorial representation of a system's operation to show the sequence of events involved in its operation. SFC charts have the following elements:

- The operation is described by a number of separate sequentially connected states or steps represented by rectangular boxes, each representing a particular state of the system being controlled. The initial step in a program is represented differently than the other steps, as shown in its following representation (Figure 2.45).
- Each connecting line between states has a horizontal bar representing the transition condition that must be met before the system can move from one state to the next. Two steps can never be directly connected; they must always be separated by a transition. Two transitions can never directly follow from one to another; they must always be separated by a step.
- When the transfer conditions to the next state are met, the next state or step in the program occurs. The process thus continues from one state to the next until the complete machine cycle is completed.
- Outputs/actions at any state are represented by horizontally linked boxes and occur when that state has been realized. As an illustration, Figure 2.46 shows part of an SFC and its equivalent LAD.

To illustrate the principles of SFC, consider the situation with, say, part of the washing cycle of a domestic washing machine where the drum is to be filled with water. When full a heater has to be switched ON and remain on until the temperature reaches the required level. Then the drum is to be rotated for a specified time. We have a sequence of states that can be represented in the manner shown in Figure 2.47.

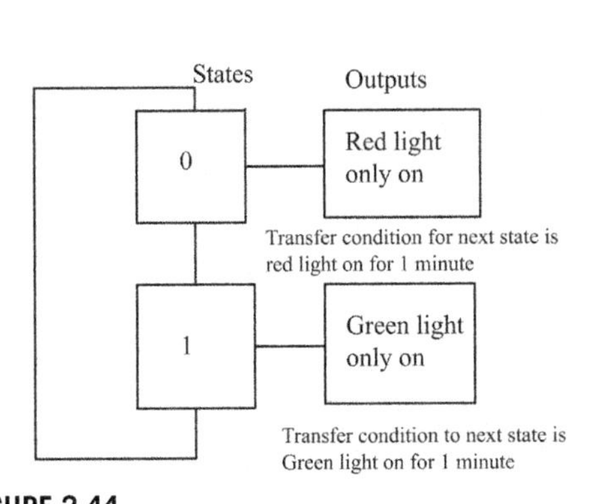

FIGURE 2.44

Sequence for traffic lights

Start
Initial step in which the system is held ready to start
Transition condition
State/Step
Output
Transition condition
etc.
Stop
Final step

FIGURE 2.45

A state and its transition

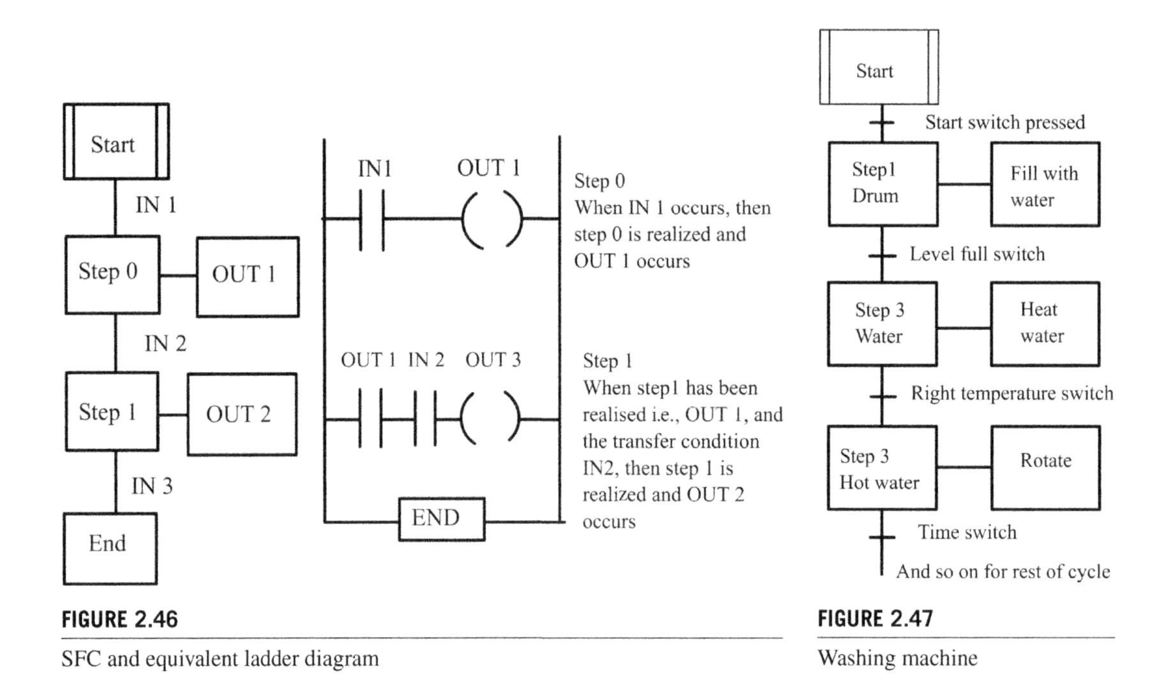

FIGURE 2.46

SFC and equivalent ladder diagram

FIGURE 2.47

Washing machine

2.8.2.1 Branching and convergence

Selective branching allows for different states to be realized depending on the transfer condition that occurs, as illustrated in Figure 2.48.

Parallel branching, represented by a pair of horizontal lines, allows for two or more different states to be realized and proceed simultaneously, as illustrated in Figure 2.49.

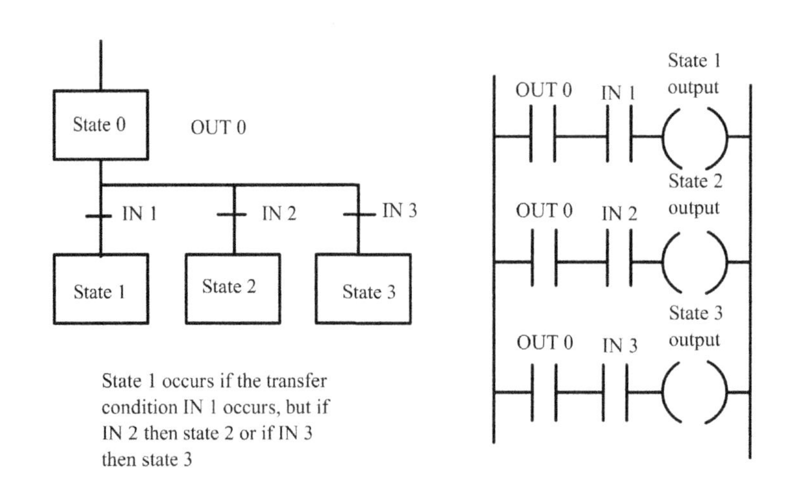

FIGURE 2.48

Selective branching: the state that follows State 0 will depend on whether transition IN1, IN2, or IN3 occur

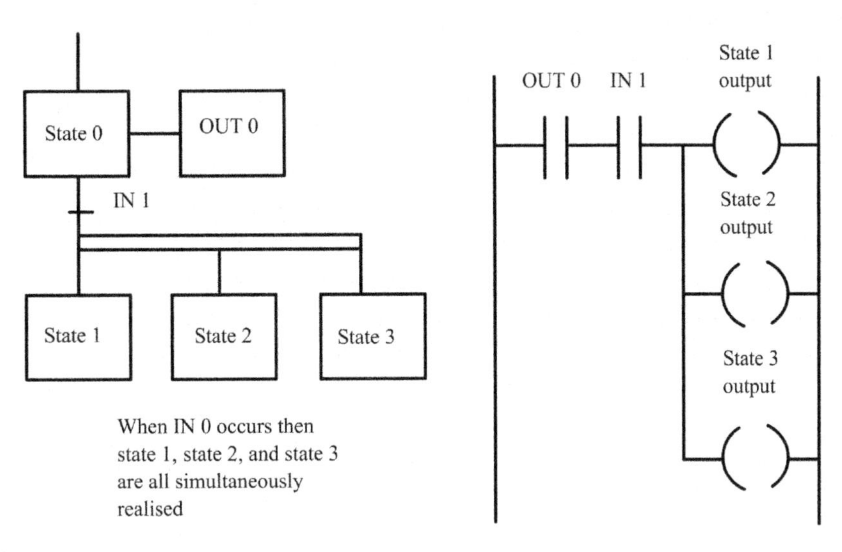

FIGURE 2.49

Parallel branching states 1, 2, and 3 occur simultaneously when transition IN1 occurs

Figures 2.50 and 2.51 show how convergence is represented by an SFC. The sequence is allowed to go from state 2 to state 4 if IN 4 occurs or from state 3 to state 4 if IN 5 occurs. In Figure 2.51, the sequence can go simultaneously from both state 2 and state 3 to state 4 if IN 4 occurs.

2.8.2.2 Actions

With states, there is an action or actions that have to be performed. Such actions, like the outputs in the above example, are depicted as rectangular boxes attached to the state. The behavior of the action can be represented using a LAD, FBD, IL, or ST. Therefore, where a LAD is used, the behavior of the action is represented by the LAD being enclosed within the action box. The action is then activated when there is a power flow into the action box, as illustrated in Figure 2.52.

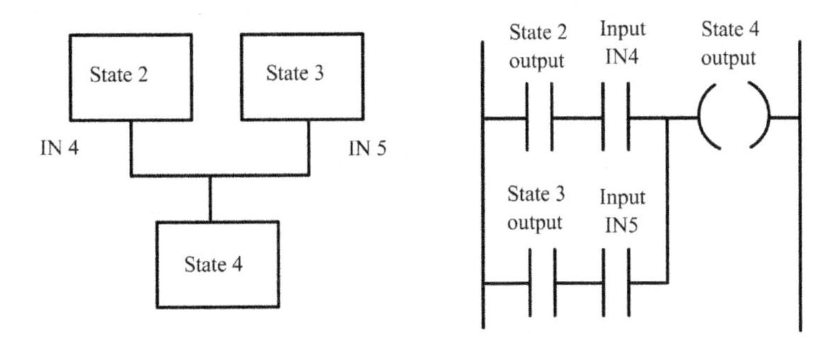

FIGURE 2.50

Convergence: state 4 follows when either IN4 or IN5 occur

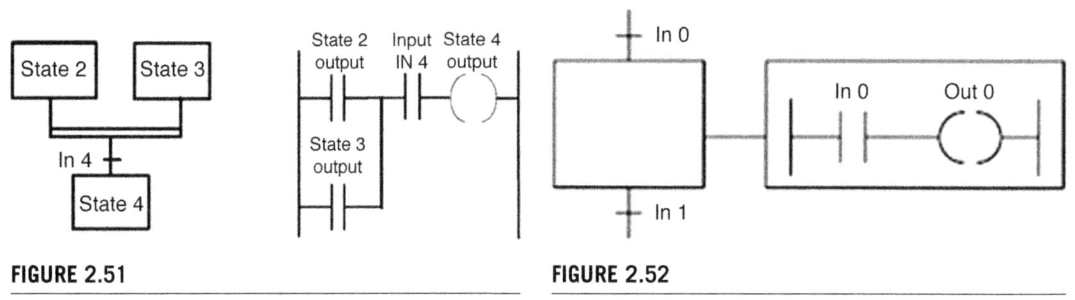

FIGURE 2.51

Simultaneous convergence: When IN4 occurs State 4 follows from either State 2 or 3

FIGURE 2.52

Action represented by a ladder diagram

Action boxes are generally preceded by qualifiers to specify the conditions to exist for the action. In the absence of a qualifier or the qualifier N, the action is not stored and executed continually while the associate state is active. The qualifier P is used for a pulse action that executes only once when a step is activated. The qualifier D is used for a time delayed action that only starts after a specified period. The qualifier L is used for a time-limited action that terminates after a specified period.

2.8.3 STRUCTURED TEXT

Structured text is a programming language that strongly resembles the programming language PASCAL. Programs are written as a series of statements separated by semicolons. The statements use predefined statements and subroutines to change variables, these being defined values, internally stored values, or inputs and outputs. Assignment statements are used to indicate how the value of a variable must be changed, for example,

Light := SwitchA;

is used to indicate that a light is to have its "value" changed, that is, switched ON or off, when switch A changes its "value," that is, is on or off. The general format of an assignment statement is:

X := Y;

where Y represents an expression that produces a new value for the variable X. Another example is:

Light :- SwitchA OR SwitchB;

to indicate that the light is switched ON by either switch A or switch B. Some of the operators, like the OR in the above statement, that are used in ST programs and their relative precedence when an expression is being evaluated are given in Table 2.2. For example,

InputA = 6

InputB = 4

InputC = 2

OutputQ := InputA/3 + InputB/(3 - InputC)

Table 2.2 Structured text operators

Operation	Description	Procedure
(...)	Parenthesized (bracketed) expression	Highest
Function(...)	List of parameters of a function	
**	Raising to a power	
−, NOT	Negation, Boolean NOT	
*, /, MOD	Multiplication, division, modulus operation	
+, −	Addition, subtraction	
<, >, <=, >=	Less than, greater than, less than or equal to, greater than or equal to, inequality	
=, <>	Equality, inequality	
AND, &	Boolean AND	
XOR	Boolean XOR	
OR	Boolean OR	Lowest

has (3 − InputC) evaluated before its value is used as a divisor so the second part of the OutputQ statement is 4/(3 − 2) = 4. Division has precedence over addition and so the first part of the statement is evaluated before the addition, that is, 6/3.

So we have for OutputQ the value 2 + 1 = 3.

ST is not case sensitive; thus, lower case or capital letters can be used necessarily to aid clarity. Likewise spaces and indenting lines are not necessary but can be used to aid clarity. All the identities of directly represented variables start with the % character and followed by one or two letter code to identify whether the memory location is associated with inputs, outputs, or internal memory and whether it is bits, bytes, or words, for example, %IX100 (*Input memory bit 100*).

```
%ID200 (*Input memory word 200*)
%QX100 (*Output memory bit 100*)
```

The first letter is I for input memory location, Q for output memory location, and M for internal memory. The second letter is X for bit, B for byte (8 bit), W for word (16 bits), D for double word (32 bits), and L for long word (64 bits). AT is used to fix the memory location for a variable. Thus, we have:

```
Input1 AT %IX100; (*Input1 is located at input memory bit 100*)
```

2.8.3.1 Conditional statements

IF ... THEN ... ELSE are used when selected statements are to be executed only on meeting certain conditions. For example,

```
IF (Limit_switch1 AND Workpiece_Present) THEN
Gate1 :- Open;
Gate2 :- Close;
ELSE
Gate1 :- Close;
Gate2 :- Open;
End_IF;
```

Note that the end of the IF statement has to be indicated. Another example, using PLC addresses, is

```
IF (I:000/00 = 1) THEN
O:001/00 :- 1;
ELSE
O:000/01 = 0;
End_IF;
```

So if there is an input to I:000/00 to make it 1 then output O:001/00 is 1, otherwise it is 0.

CASE is used to give the condition that selected statements are to be executed if a particular integer value occurs else some other selected statements. For example, for temperature control we might have:

```
CASE (Temperature) OF 0 ... 40 ;
Furnace_switch :- On; 40 ... 100
Furnace_switch :- Off;
ELSE
Furnace_switch :- Off;
End_CASE;
```

Note, as with all conditional statements, the end of the CASE statement has to be indicated.

2.8.3.2 Iteration statements

Iteration statements are used where it is necessary to repeat one or more statements several times, depending on the state of some variable. The FOR ... DO iteration statement allows a set of statements to be repeated depending on the value of the iteration integer variable. For example, FOR Input :- 10 to 0 BY -1

```
DO
Output :- Input;
End_FOR;
```

has the output decreasing by 1 each time the input, drops from 10 to 0,

decreases by 1.

WHILE ... DO allows one or more statements to be executed while a particular Boolean expression remains true, for example,

```
OutputQ :- 0;
WHILE InputA AND InputB
DO
OutputQ =: OutputQ + 1;
End_WHILE;
```

REPEAT ... UNTIL allows one or more statements to be executed and repeated while a particular Boolean expression remains true.

```
OutputQ :- 0
REPEAT
OutputQ = OutputQ + 1;
UNTIL (Input1 = Off) OR (OutputQ > 5)
End_REPEAT;
```

FURTHER READINGS

IEC 61131-5 Ed. 1.0 en: 2000. Part 5 Communications.
IEC 61131-7 Ed. 1.0 b: 2000. Part 7 Fuzzy Control Programming.
IEC 61131-1 Ed. 2.0 en: 2003. Part 1 General Information.
IEC 61131-2 Ed. 2.0 en: 2003. Part 2 Equipment Requirements and Tests.
IEC 61131-3 Ed. 2.0 en: 2003. Part 3 Programming Languages.
IEC/TR 61131-8 Ed. 2.0 en: 2003. Part 8 Guidelines for the Application and Implementation of Programming Languages.
The Automation, Systems, and Instrumentation Dictionary, fourth ed. ISA, 2003.

IEC/TR 61131-4 Ed. 2.0 en: 2004. Part 4 User Guidelines.

IEC 61131 – Programmable Controllers.

Coggan, D.A. (Ed.), 2005. Fundamentals of Industrial Control. Second ed. ISA, Practical Guides for Measurement and Control Series.

Hughes, T.A., 2004. Programmable Controllers, Fourth ed. ISA.

Lewis, R.W., 1998. Programming Industrial Control Systems Using IEC 1131-3, Revised ed. IEEE.

Liptak, B.G., Editor-in-Chief, 2003. Instrument Engineer's Handbook: Volume 1 – Process Measurement and Analysis, fourth ed. CRC Press and ISA, 2003.

John, K.-H., Tiegelkamp, M., 2001. IEC 61131-3: Programming Industrial Automation Systems – Concepts and Programming Languages, Requirements for Programming Systems, Aids to Decision-Making Tools. Springer.

CHAPTER

DISTRIBUTED CONTROL SYSTEM

3

3.1 INTRODUCTION

Automatic control typically involves the transmission of signals or commands/information across the different layers of system, and calculation of control actions as a result of decision-making. The term DCS stands for distributed control system. They used to be referred to as distributed digital control systems (DDCS) earlier, implying that all DCS are digital control systems. They use digital encoding and transmission of process information and commands. DCS are deployed today not only for all advanced control strategies but also for the low-level control loops. The instrumentation used to implement automatic process control has gone through an evolutionary process and is still evolving today. In the beginning, plants used local, large-case pneumatic controllers; these later became miniaturized and centralized onto control panels and consoles. Their appearance changed very little when analog electronic instruments were introduced. The first applications of process control computers resulted in a mix of the traditional analog and the newer direct digital control (DDC) equipment located in the same control room. This mix of equipment was not only cumbersome but also rather inflexible because the changing of control configurations necessitated changes in the routing of wires. This arrangement gave way in the 1970s to the distributed control system (DCS).

DCS controllers are distributed geographically and functionally across the plant and they communicate among themselves and with operator terminals, supervisor terminals to carry out all necessary control functions for a large plant/process. The scope of control is limited to the part of the plant it is distributed in. DCS is most suited for a plant involving a large number of continuous control loops, special control functions, process variables, and alarms. Most of the DCS architectures are generally similar in the way they are designed and laid out. Operator consoles are connected to controllers housed in control cubicles through a digital, fast, high-integrity communications system. The control is distributed across by the powerful and secure communication system. The process inputs are connected to the controllers directly or through IO bus systems such as Profibus, Foundation FieldBus, and so on. Some systems also use proprietary field bus systems. The DCS offered many advantages over its predecessors. For starters, the DCS distributed major control functions, such as controllers, I/O, operator stations, historians, and configuration stations onto different boxes. The key system functions were designed to be redundant. As such the DCS tended to support redundant data highways, redundant controllers, redundant I/O and I/O networks, and in some cases redundant fault-tolerant workstations. In such configurations, if any part of the DCS fails the plant can continue to operate. Much of this change has been driven by the ever-increasing performance/price ratio of the associated hardware. The evolution of communication technology and of the supporting components has dramatically altered the fundamental structure of the control system. Communication technology such as Ethernet and TCP/UDP/IP combined with standards such as OPC allowed third-party applications to be integrated into the control system. Also, the general acceptance of object-oriented design, software component design, and supporting tools for implementation has facilitated the development of better user interfaces and the implementation of reusable software.

With advancing technologies, DCS have rapidly expanded their capabilities in terms of features, functions, performance and size. The DCSs available today can perform very advanced control functions, along with powerful recording, totalizing, mathematical calculations, and decision-making functions. The DCS can also be tailored to carry out special functions, which can be designed by the user. An essential feature of modern-day DCS is the integration with ERP and IT systems through exchange of various pieces of information.

To understand DCS, it is a good idea to review the evolution of control systems. This includes hardware elements, system implementation philosophies, and the drivers behind this evolution. This will help in understanding how process control, information flow, and decision-making have evolved over the years.

3.2 EVOLUTION OF TRADITIONAL CONTROL SYSTEMS

3.2.2.1 Pneumatic control

Earliest implementations of automatic control systems involved pneumatic transmission of signals. They used compressed air as the medium for signal transmission and actuation. Actual control commands were computed using elements such as springs and bellows. Plants used local, pneumatic controllers, which were large mechanical structures. These later became miniaturized and centralized onto control panels and consoles.

A pneumatic controller has high margin for safety, and since it is explosion proof, it could be used in hazardous environments. However, they have slow response and are susceptible to interference.

The common industry standard pneumatic signal range is 3–15 psig where 3 psig corresponds to the lower-range value (LRV) and the 15 psig corresponds to the upper-range value (URV).

3.2.2.2 Electronic analog control

Over time, electronic analog control was introduced. Electrical signals were used as the mode of transmission in this implementation. As in pneumatic control, computation devices are mechanical. Electrical signals to pressure signals converter (E/P transducers) and pressure to electrical (P/E transducers) are used to transmit signals to enable coexistence of pneumatic and electrical signals. Disadvantage with analog signals is the susceptible to contamination from stray fields, leading to degradation in signal quality over long distances.

The most common standard electrical signal is the 4–20 mA current signals. With this signal, a transmitter sends a small current through a set of wires. The current signal is a kind of gauge in which 4 mA represents the lowest possible measurement, or zero, and 20 mA represents the highest possible measurement.

3.2.2.3 Digital control

In this mode, the transmission medium is still an electrical signal, but the signals are transmitted in binary form. Digital signals are discrete levels or values that are combined in specific ways to represent process variables and also carry diagnostic information. The methodology used to combine the digital signals is referred to as protocol. Manufacturers may use either an open or a proprietary digital protocol. Open protocols are those that anyone who is developing a control device can use. Proprietary protocols are owned by specific companies and may be used only with their permission. Open digital protocols include the HART® (highway addressable remote transducer) protocol, FOUNDATION™ FieldBus, Profibus, DeviceNet, and the Modbus® protocol.

Digital signals are far less sensitive to noise. In digital signaling, we look for two levels of signals, and the magnitude of the signals is expressed as a combination of 1 and 0, corresponding to the magnitude expressed as a binary number. Therefore, the impact of noise is reduced compared to an analog signal. The computational devices are digital computers embedded processors with real time operating systems. Digital computers are more flexible because they are programmable. They are more versatile because there is virtually no limitation to the complexity of the computations it can carry out. The limitation is on the computing power of the computer, how many computations it can perform in a given unit of time. Moreover, it is possible to carry out computation with a single computing device, or with a network of such devices.

Many field sensors naturally produce analog voltage or current signals. Transducers that convert analog signals to digital signals (A/D) and those that convert them back to analog signals (D/A) are used as interface between the analog and digital elements of the modern control system. With the development of digital implementation systems, on which DCS are based, it is possible to implement many sophisticated control strategies at very high speeds.

3.2.2.4 Modes of computer control

Computer control is usually carried out in the following modes:

1. Direct digital control (DDC)
2. Supervisory control
3. Hierarchical computer control system.

3.2.2.5 Direct digital control

In DDC, a digital computer computes control signals that directly operate the control devices. A single computer digitally performs signal processing, indication, and control functions and therefore the name "direct digital control." Initially computers were very large and housed in large buildings with substantial environment controls. As the electronics evolved, and medium scale integration (MSI) and large-scale integration (LSI) integrated circuits (ICs) became available, powerful and relatively small mini computers became a reality. These mini computers were first deployed to realize DDC (Figure 3.1).

3.2.2.5.1 Disadvantages of DDC

The following are some of the disadvantages of a DDC:

- Centralized control – a single central processor is used to perform several tasks necessary for control:
  - IO scanning
  - Data logging

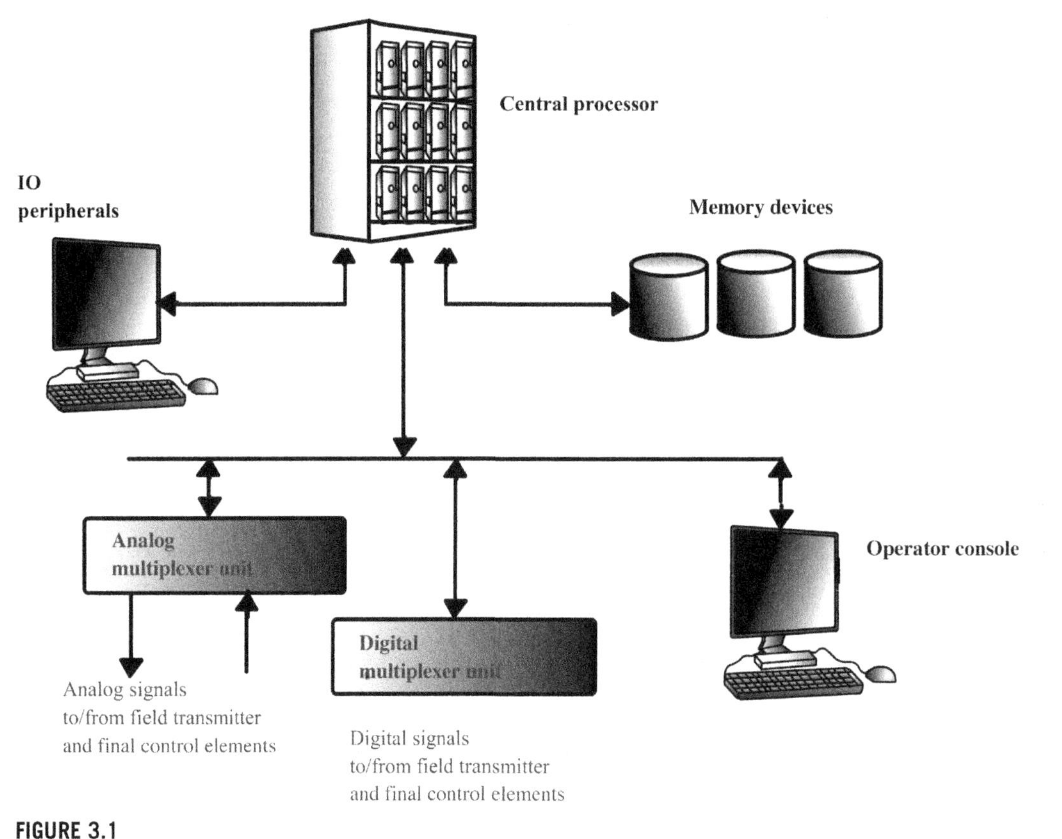

FIGURE 3.1

DDC architecture

- Control execution
- Alarm generation
- Database update
- Serving operator display updates
- Periodic and on-demand report generation
- Serve peripherals such as printers and recorders
- Process optimization

In addition, DDC systems support software systems (compilers) configuration and engineering functions.

- CPU memory requirements and management were challenges.
- Poor system reliability and performance – single failure is catastrophic to plant control, with no provision for redundancy.
- Costly and complex – difficulty in troubleshooting.
- Creating customized software for advanced process control and optimization was cumbersome.

3.2.3 SUPERVISORY CONTROL

In supervisory control, a digital computer generates signals that are used as reference (set-point) values for conventional analog controllers. This is described in the block diagram (Figure 3.2). Measurements are transmitted to computer and control signals are transmitted from the computer to control valves at specific time intervals known as sampling time. In a supervisory control system, analog control subsystem and panel instrumentation are used for controlling but are interfaced to the supervisory control computer through an interfacing hardware. The supervisory control computer provides the facility for monitoring process (which provides the process mimics on the video display units for operators with features such as alarm handling, data storage, and real-time values).

3.2.3.1 Advantages of supervisory control

The following are the advantages of supervisory control:

- With supervisory control architecture, primary loop control is returned to analog controllers, while the supervisory control computer monitors the process and adjusts the set points. The computer is relieved of intense computational tasks and therefore could be utilized for process optimization and plant management functions, which are less time critical and computation intensive.
- Analog controllers added to DDC computer system enhance the overall reliability.
- Control algorithms, which provide the set point to the analog control subsystem, accommodate higher complexity.

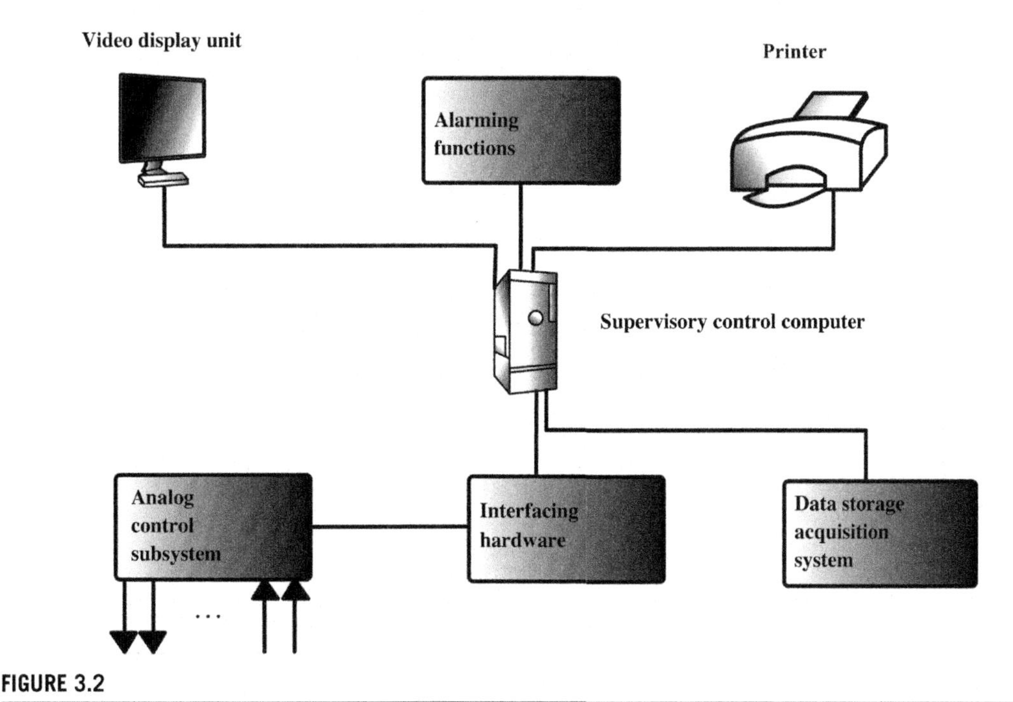

FIGURE 3.2

Architecture of Supervisory control

3.2.3.2 Disadvantages of supervisory control

The following are some of the disadvantages of supervisory control:

- Extensive wiring is required between the analog controllers and the supervisory computer and also between other instrumentation and the computer system.
- Interfacing equipment from multiple vendors (interfacing one vendor's computer system to another vendor's analog instrumentation) is difficult.
- Supervisory control is costlier than DDC.

All vendors who later started offering DCS offered both DDC and supervisory control options. In the earlier versions, both versions coexisted with the DCS. Many of the current DCS control technologies had their origins in DDC and supervisory control systems. However, DDC is rarely used in current industrial automation scenarios.

3.2.4 HIERARCHICAL COMPUTER CONTROL SYSTEM

A hierarchical system is a network of process and/or information management computer systems integrated to serve common functions such as management and control of large plants and geographically distributed systems such as pipeline networks. (Figure 3.3)

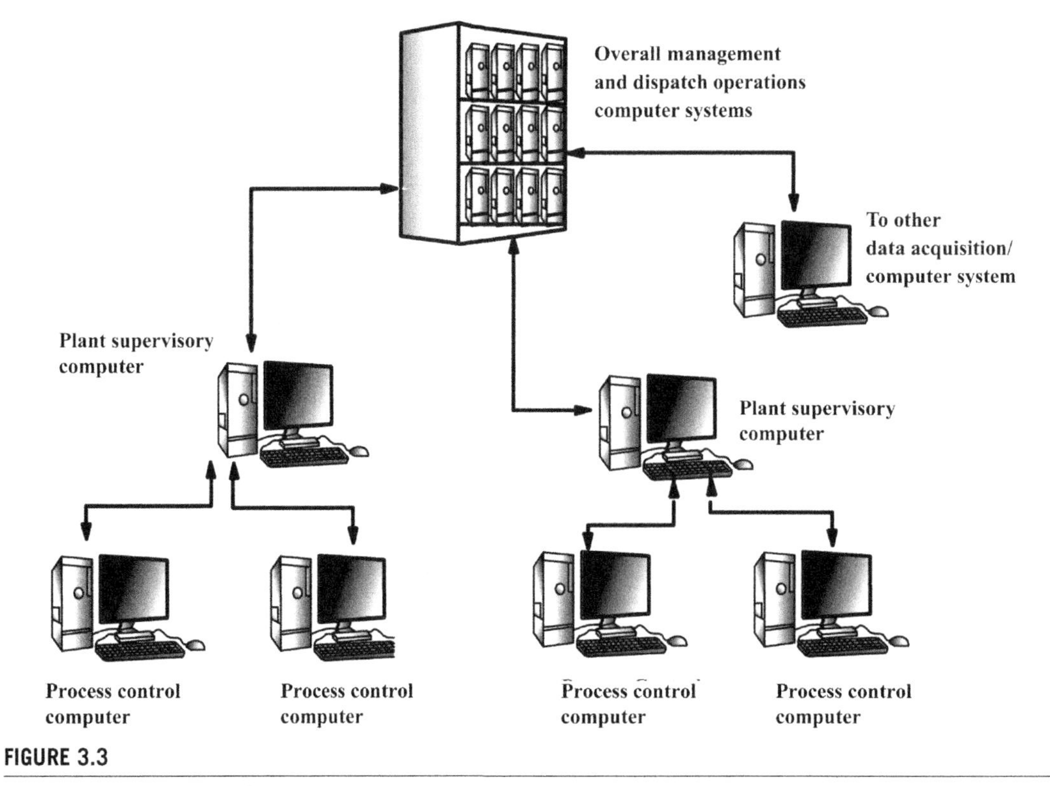

FIGURE 3.3

Architecture of hierarchical control

- Information is passed up and down between primary level of process monitoring and control, through supervisory levels, to decision making/top management level.
- The computer network architecture usually parallels the organizational structure of a company itself.
- With hierarchical systems, primary computers provide direct control of process and they could be a combination of DDC, supervisory control, or microcontroller-based controllers.

3.3 DISTRIBUTED CONTROL SYSTEMS

With the advent of microcontrollers, individual controllers became powerful. They could execute more number of control algorithms and more complex algorithms. They could also control a larger set of control steps. It became easy to move the intelligence involved in controls to lower levels and improve the signal processing in transmitters. Powerful microcontrollers also enabled the design of faster networks. Together the concept of DCSs could be turned into a reality.

3.3.1 PROGRAMMABLE LOGIC CONTROLLERS

Programmable logic controller (PLC) was the first manifestation of a distributed controller. A PLC is specialized for process control of noncontinuous systems such as batch processes or discrete manufacturing systems that encompass equipment or control elements that operate discontinuously. A PLC is programmed to execute the desired Boolean logic and implement sequencing of operations. Therefore, a PLC is used in many instances where interlocks are required. For example, a flow control loop must not be actuated unless a pump responsible for the flow has been turned on. Similarly, during startup or shutdown of continuous processes, elements must be correctly sequenced; for example, upstream flows and levels must be established before downstream pumps can be turned on. The PLC implements a sequence of logical decisions to implement a control.

The inputs to the PLC are generally a set of relay contacts representing the state of various process elements. Various operator inputs are also provided. The outputs are generally a set of relays energized (activated) by the PLC that can turn a pump on or off, activate lights on a display panel, operate solenoid valve, and so on.

Although PLCs were initially conceived to implement simple binary logic, they started using powerful microcontrollers and could therefore handle comparatively complex functions such as proportional, integral, and derivative (PID) controls. Currently, PLCs can handle thousands of digital I/O and hundreds of analog I/O and continuous PID control. However, PLCs lack the flexibility for expansion and reconfiguration. The choice of operator interfaces to PLC systems is also limited. However, PLCs continue to breach their limits and are overcoming several limitations and are today positioned for many complex control tasks.

PLCs are not typically applied in traditional continuous process plants. However, for operations such as sequencing and interlocks, the speed and power of PLCs can be used very effectively. Where sophisticated process control strategies are needed, PLCs are a cost-effective alternative to DCSs in the places where there is a large number of discrete operations and small number of analog control. PLC architectures have always focused on flexible and fast local control. Recent advancements in PLC technology have added process control features. When PLCs and HMI software packages are integrated, the result looks a lot like a DCS.

PLCs and DCSs can be combined by design in a hybrid system where PLCs are connected through a link to a controller forming part of a larger DCS, or are connected directly to network of the DCS.

3.3.2 DISTRIBUTED CONTROL SYSTEMS

A DCS is defined as a system comprising of functionally and physically separate automatic process controllers, process monitoring and data logging equipment all of which are interconnected through a fast, digital network. This ensures sharing of relevant information for optimum control of the plant. In large-scale manufacturing or process plants, there are hundreds of control loops to be monitored and controlled. For such large processes, the commercial DCS is normally the control system of choice. Figure 3.3.1 is the most common architecture of the distributed control systems in the world today.

The hardware and software of the DCS are quite flexible and easy to modify and configure. They are capable of handling a large number of loops. Modern DCSs are equipped with optional software elements for optimization, and various controls based on process models. They also come with tools for defining high-performance models.

As signified by the term "distributed," DCS architecture enables distribution of the controllers and the operator input elements through the network, called process control network, which connects the different parts. Elements closest to the process transmit and receive raw data between them and the local computers while those farther away from the process exchange mostly processed data at lesser frequencies but for a wider set of consumers of the data. All data exchanged such as the presentation information for the multidisplays on various operator control panels and historical data to and from archival storage have to pass through the data highway. The data highway is, therefore, the backbone of the DCS system. A supervisory computer is normally at the top of the hierarchy and is responsible

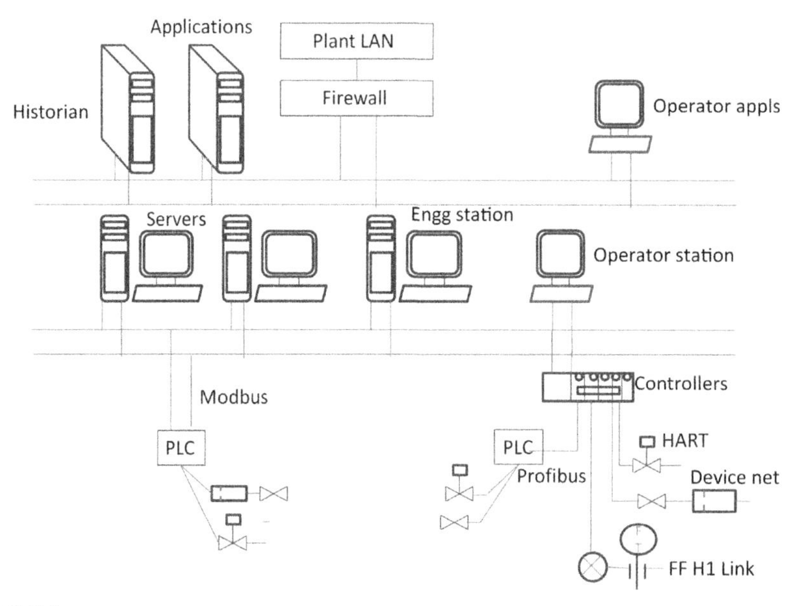

FIGURE 3.3.1

Architecture of a simple DCS

for performing many higher-level functions. These could include optimization of the process operation over varying time horizons (days, weeks, or months), carrying out special control procedures such as plant startup or product grade transitions, and calculating the data required to track the economic performance of the plant.

In the common DCS architecture, the microcomputers attached to the process are known as front-end computers. They are usually less sophisticated equipment employed for low-level functions. One such example is an IO module. Typically an IO module acquires process data from the measuring devices and converts them to standard engineering units. The results are then passed upward to the controllers that are responsible for more complex operations. Processed data is then passed on to computers higher in the hierarchy, which can be programmed to perform more advanced calculations. DCSs are not limited to continuous control applications; they are also capable of carrying out many of the functions of a PLC, in the same way that the PLCs have evolved to perform several tasks traditionally done by the DCS. A range of functions are designed into the controllers, such that the entire general plant control operation functions, whatever the type, could be carried out by the DCS. These functions include continuous control, cyclic control, logic control, motor control, and batch control. One key difference is in the processing speed of sequence functions. A PLC's scan time (the time taken to scan the inputs) is in 10 s of milliseconds, while the same for a DCS could be in 100 s of milliseconds or seconds (typical fastest speed 0.25 s). In most process control applications this speed of execution of the DCS is quite adequate. If higher speeds are necessary in selected operations, PLCs can be interfaced directly to most DCS using standard components, which make the PLC transparent to the operator. However, it calls for using different engineering tools for the DCS and the PLC. On some DCS the need for higher speeds is addressed through the use of a high-speed controller and IO cards and they provide the speed of execution necessary for special applications. These controllers come with a cost premium and are used judiciously to ensure cost-effectiveness of the solution.

Therefore, the standard elements of functionality available with DCS today are sufficient to provide an integrated control system capable of controlling most processes, either by themselves or through integration with other controllers. They are also capable of providing necessary information for the overall management of the production facility.

Traditionally, DCSs have been more expensive to purchase than a PLC-based system and many processing plants had lower demands in terms of production rates, yield, waste, safety and regulatory compliance than they are experiencing today. A PLC-based system offered a lower capital investment and from a functional point of view was "good enough." Demands on manufacturing companies have risen – and the purchase price of the DCS has come down. Yet, DCSs do have advantages over PLC systems.

The DCS architecture has always been focused on distributing control on a network so that operators can monitor and interact with the entire scope of the plant and the classic DCS originated from an overall system approach. Coordination, synchronization, and integrity of process data over a high-performance and deterministic network are at the core of the DCS architecture.

The major advantages of functional distribution of hardware and software characteristic of DCS are:

- Flexibility in system design
- Ease of expansion
- Reliability
- Ease of maintenance.

Local control can be maintained even if central components fail or are degraded functionally to a substantial extent. We then say the plant is operating as a set of "islands of automation." The DCS architecture is such that complete loss of the data highway does not cause complete loss of system capability. Often local units can continue operation with no significant loss of function over moderate or extended periods of time. This greatly enhances the availability and reliability of the system.

The control network is the most important component of DCS. Suppliers ensure this through comprehensive maximum topology testing and subject the network to high levels of message volume in test labs to ensure reliable network performance in demanding environments. Most of the DCSs are provided with redundant industrial Ethernet networking technology utilizing inexpensive off-the-shelf components to provide a high-availability solution. Industrial Ethernet continuously monitors the process control network (PCN) by providing network diagnostics that are tracked and reported as a part of the DCS.

Control performance is another area where DCSs have great advantages. Good process control is built on reliable and repeatable execution of the control strategy. While the PLC runs "as fast as it can" the process controller favors repeatability. That means, the control strategy runs on fixed clock cycles – running faster or running slower are not tolerated. Other system services are also designed to give priority to solving the controller configuration. Most DCSs come with function blocks (FBs) with a complete set of parameter-based functions, using which the user can develop and fine-tune control strategies without designing control functions. All necessary functions are available and documented as configurable selections. The application engineer simply assembles the blocks into the desired control configuration with a minimum of effort. A self-documenting, programming-free controller configuration makes the DCS architecture efficient to engineer and troubleshoot.

Moreover, the DCS network is versatile and allows different modes of control implementation such as manual/auto/supervisory/computer operation for each local control loop. In the manual mode, the operator manipulates the final control element directly. In the auto mode, the final control element is manipulated automatically through a low-level controller usually a PID block executed in controller. The set point for this control loop is provided by the operator. In the supervisory mode, an advanced digital controller is placed on top of the low-level controller, which sets the set point for the low-level controller. The set point for the advanced controller can be set either by the operator or can be the outcome of steady-state optimization.

DCS vendors also supply the control building tools, a data historian, trend tools, alarm management, asset management, back up and archive, OPC servers, remote maintenance servers, web servers, documentation servers, network management server, business integration software, and graphics needed to run a plant as a single package that can be easily deployed on the DCS. The capabilities of DCS architecture allow all of the control applications to load correctly, are guaranteed to be the correct version, and are tested to work together. This becomes very significant when DCSs stay deployed for longer periods of time and are expanded to meet changing plant requirements.

On account of the systems approach to DCS design, the software elements can be integrated to share a single data model, no matter where a data element resides, it can be used by any element of the architecture and that particular data element need not be duplicated. That is a significant advantage given the integrated nature of a typical industrial automation system. Refer to Figure 3.3.2 for the DCS components from the figure, it can be derived that a DCS is a combination of controllers, I/Os, networks, operation consoles and engineering consoles.

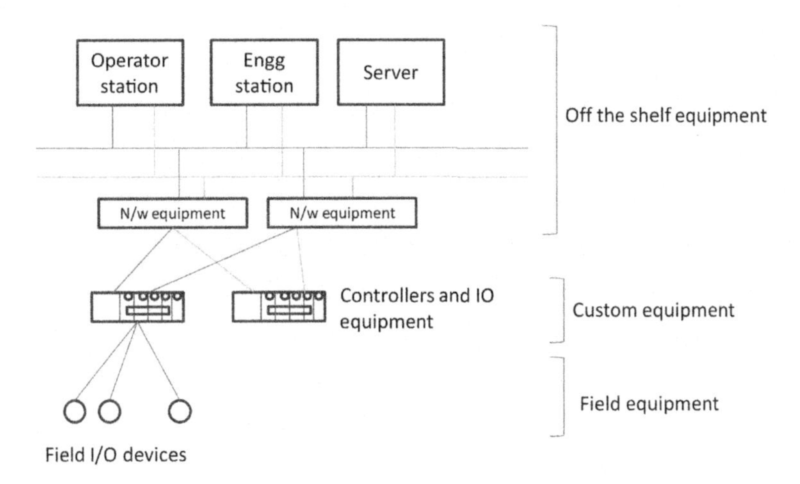

FIGURE 3.3.2

Functional components of a DCS

The field Instrumentation also plays a role in terms of the architecture of the DCS due to the evolution of various protocols.

3.3.3 DCS DESIGN CONSIDERATIONS

Any modern DCS is expected to meet the following requirements from plant operations and maintenance perspective. These form the considerations during the design of the DCS:

- High Reliability: Any control or hardware component failure could have devastating consequences leading to loss of life and property. Therefore, reliability of the control system is a life-critical requirement. During design, development and manufacturing, all electronic components of the DCS are normally subjected to extensive periods of cycling at temperatures exceeding the extremes listed in equipment specifications. This process weeds out the components most likely to fail. Suppliers provide redundancy in their design, and as a redundant component in the architecture because reliability is still probable. Power supplies, data highways, traffic directors, and controller electronics are important single points of failure in the system and are considered as candidates for having redundancy. It is essential to have automatic transfer between redundant parts, so that if one fails the other takes over without disturbance of the operation or output. At the same time, there must be some form of alarm to alert the operator to draw his attention to the fact that a failure has occurred.
- High Availability: High availability is as important as reliability. Defining availability as the ratio of mean time between failure (MTBF) to mean time between failure plus mean time to repair (MTBF + MTTR), it is clear that a system is most available when it is very reliable (high MTBF) and can be quickly repaired (low MTTR). Since distributed control equipment is highly modular and contains many printed circuit cards, time to repair can be very short if sufficient spare parts are available and the components can be quickly brought into service and the necessary software updated online without affecting the control functions in any way.

- Low Cost: If we consider together the initial purchase price, cost of initial implementation, and the cost of making subsequent changes to the system over time, the DCS can be much less expensive. The total project costs include the expenses required to build a working solution that accomplishes the long-term goal of effective process control. Maintenance costs and costs of changes to accommodate growth of operations over time are key factors to consider while estimating the cost of the system. This is called life cycle cost. This life cycle cost turns out to be lower in case of DCS compared to providing comparable level of functionality using PLCs because the built-in functions and inherent integration capabilities available in a DCS enable implementation and maintenance of a more effective system with reduced labor, plant life cycle cost while avoiding degradation in functionality over time.
- High Alarm Management Capability: DCS systems must be capable of intelligent alarm management to aid in abnormal situation management. Alarm management is necessary in a manufacturing process environment using a control system, such as a DCS or a system of PLCs. Such a system may have hundreds of individual alarms that up until very recently have probably been configured with limited consideration of the interdependencies among all alarms in the system. Humans can pay attention to a limited number of stimuli and messages at a time. Therefore, we need a way to ensure that alarms are presented at a rate that can be assimilated and acted upon by a human operator, particularly when the plant is in a disturbed state or in an emergency condition. Alarms also need to be capable of directing the operator's attention to the most important problem that he or she needs to act upon methodically, using a prioritization scheme. The capabilities of alarm management in the system must go beyond the basic level of utilizing multiple alarm priority levels. It is to be noted that alarm priority itself is often dynamic. Likewise, disabling an alarm based on unit association or suppressing audible annunciation based on priority do not meet operational requirements in a complex plant. We need dynamic, selective alarm annunciation in such cases. DCSs are designed with an alarm management system that dynamically filters the process alarms based on the current plant operation and conditions so that only the currently significant alarms are annunciated.
- Scalability: A small SCADA/PLC system is easy to design and configure. As the system grows bigger, the effort involved to properly engineer and configure the system grows in a nonlinear fashion. It also increases the risks of errors creeping into the process of engineering. It is easy to design and implement a single loop PID controller in a SCADA/PLC system and it can be done quickly. However, to design and implement the base layer control for a refinery using a SCADA/PLC system can be a daunting endeavor. One of the chief considerations in the design of the engineering tools for a DCS is that engineering time for system expansion and other changes must be considerably less. Features such as batch updates, replication of application programs with suitable substitution are provided in the engineering tools.
- Distributed Systems: A DCS has to share real-time data across a network, despite the fact that the components are geographically distributed. The need for a seamless transfer of control signals among the distributed controllers, the supervisory controllers, operator workstations, plant computers, and so on can never be overstated. This makes networking a major component in the DCS architecture. The objectives of a networking topology in industrial automation systems include the following.
    - Enable wide distribution of the components
    - Connectivity to different machines and nodes
    - Reliable data gathering and sharing

- Redundant communication medium
- Deterministic transmission and receipt of data
- Sufficient speed to match the plant requirements

3.3.4 HIERARCHY OF PLANT OPERATIONS

Plant operations could be classified into three control zones:

- PCN (plant/process control network) layer where the process control operations and data transfer occurs.
- DMZ (demilitarized) layer – where servers such as OPC, remote maintenance, web server can be operated.
- PIN (plant information network) layer – for plant and office personnel access.

Field devices are the first level that comes in the hierarchy, intended to communicate the parameters in field such as flow, level, pressure/temperature/proximity/analysis, and so on. They are inputs. The outputs are actuators, valves, and motors, which are final control elements controlling the elements such as liquid/gas flow.

They communicate over multiple protocols such as HART/Foundation FieldBus/Profibus, etc. They also support the traditional 4–20-mA signal. The adoption of the IEC1158-2 Fieldbus standard by major DCS manufacturers ushered in the next generation of control and automation products and systems. Based on this standard, fieldbus capability can be integrated into a DCS system to provide:

- Advanced function added to field instruments
- Expanded view for the operator
- Better device diagnostics
- Reduced wiring and installation costs
- Reduced I/O equipment by one half or more
- Increased information flow to enable automation of engineering, maintenance, and support functions
- Lower CAPEX and OPEX

At the control level, the signals from the transmitters in the field are processed to generate commands to the actuators. The usual equipment is a PLC or process control system (PCS). Based on the signals received, valves are opened and closed or pumps and motors stopped and started. PCS is generally a customary controller offered by DCS vendors. All DCS vendors have a customized protocol to communicate with their subsystems and the controllers also support interfacing with multiple vendors through commonly supported protocols such as Modbus.

Supervisory platforms are intended to monitor downstream control and translate the control into a user viewable format, by creating the engineering design and then loading into controllers to view it. The vital components of a supervisory system are as follows:

- Engineering stations
- Operator stations
- Application stations

Engineering station has the software that has libraries used for analog/digital control. Using the library, control strategies are written to design logic to run the devices in the field.

This logic flow can be validated from an operator station that mimics the process in the field. It is not uncommon to use proprietary protocols until this layer to regain the hold on control that is supported by each vendor.

A system has to be flexible enough to support popular and open protocols available in market because they are bound to deal with most of the popular protocols and solutions available in the market once they are in the field. This adds to a point where a solution that is offered leaves the existing working controls of multiple vendors undisturbed.

Rate of control is an important aspect at the control level because controls are very time critical and often needs less latency at this level. In a layer above this, users may have to access data from other DCS in which case technologies such as OPC are used. This data access is not very critical for control; therefore, even if the latency is higher than at the control level, it can be tolerated.

Plant diagnostics and integration are often referred as applications. Plant diagnostics offers a smarter way of viewing things that are happening in the plant. For instance, plant diagnostics could include an application that consolidates all the alarms that are happening in the lower layers, to warn the user about a trend in the alarms that could lead to a potential process failure. It may include a historian just dedicated to collect the entire history over the plant and maintain it to derive various conclusions for the data that is stored.

Integration refers to communicating with third party devices over technologies such as OPC, a common example is; data from one DCS vendor to another is transferred over OPC. The distinction between integration and control level is with respect to criticality; delay in transferring and receiving data cannot be tolerated at the supervisory layer whereas at the control layer it is tolerated to certain extent. And data available at the control layer is filtered and made noncritical, when compared to supervisory systems layer, for security reasons. The control layer that normally runs over IT local area network (LAN) has higher risk of getting hacked. Hence, noncritical controls or purely supervisory data such as history is transferred through OPC.

The enterprise level is where process information flows into the office management world, for example, to aid ordering and billing via service access point (SAP) or production planning. This is the world of supply chain management that collates and evaluates such information as the quantities of raw materials in store, the amount of water being used every day, the energy used in heating and refrigeration, the quantities of product being processed, and the amount that can be sold within the next day, week, or month.

3.4 FUNCTIONAL COMPONENTS OF DCS

DCSs are made of several components namely:

- Input/Output Subsystems
- Controller Subsystem
- Networks (PCN, PIN)
- Plant Information Network
- Operation and control Subsystems
- Gateways
- Engineering/configuration subsystem

Majority of the field devices even today are hardwired to the controllers. They can also be connected via analog, digital, or combined analog/digital buses. The devices, such as valves, valve positioners, switches, and transmitters or direct sensors, are located in the field. Smart field devices can communicate on an IO buses while performing control calculations, alarming functions, and other control functions locally. Control strategies that reside in field-mounted controllers send their signals over the hardwired cables or over the field buses to the final control elements, which they control.

IO subsystems perform a variety of functions such as analog-to-digital conversion, conversion to engineering values, limit checking, quality tagging, and so on. They also enable assignment of addresses to the field signals for use by the controllers.

Controllers subsystems execute the control logic once in a fixed interval of time, which constitutes the controller cycle time. During the engineering process, the tools ensure that the given package of control logic can be executed in a deterministic way within the cycle time of the controller.

The controllers consist of a base firmware that enables the controllers to communicate on the network, exchange data, and commands among themselves or with the operator and supervisory interfaces. The application programs are plant specific and are engineered for every plant where the DCS is being implemented. The engineering tools enable the creation and validation of the application program. The application programs are downloaded into the controllers after the validation process and are typically stored and executed from a designed separate memory area in the controller.

Information from the field devices and the controller are transmitted to the operator workstations, application workstations, data historians, report generators, centralized databases, and so on over the plant control network. These workstations run applications that enable the operator to perform a variety of operations. An operator may interact with the system through the displays and keyboards to change control parameters, view the current state of the process, or access and acknowledge alarms that are generated by field devices and controllers. The operator can also change the control software executing in the controllers and in some of the field devices. Some systems also support process simulation for training personnel or for testing the process control software.

Information from the operator workstations is passed on to the other workstations, which execute a variety of higher-level applications. The operator workstations and the supervisory workstations exchange data through the PIN. The PIN could also be part of the enterprise network, although there are substantial safeguards to prevent unauthorized access to the networks in the realm of the control systems. Figure 3.3.3 provides the integrated view of a typical closed loop as seen from different sub systems of a DCS.

Gateways enable two-way data transfer. Gateways can be made available on any of the networks to enable data exchange from other control systems that are used in the DCS either to aid control decisions or to store for historical references. They could alternatively be used to share data from the DCS with other plant/business systems.

DCS are always sold as packages because the parts function together as a system. Since the components of the system communicate over a shared data highway, no change is required to the wiring when the process and its control logic are modified. However, standards have enabled interoperability between components from different vendors. DCS users are no longer tied to a single vendor.

3.4.1 FIELD COMMUNICATION

Digital communications technology reduces wiring and improves end-to-end signal accuracy and integrity in modern digital plants. Digital technology enables new innovative and more powerful devices, wider

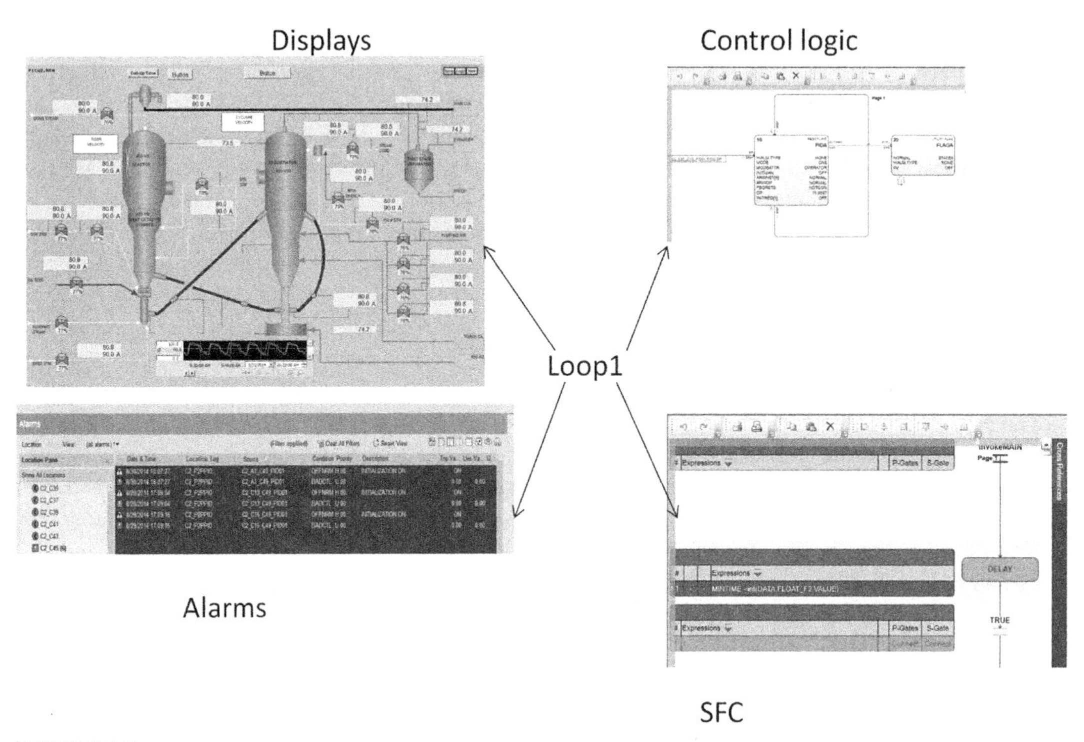

FIGURE 3.3.3

Data integrity of a DCS

measurement range, elimination of range mismatch, and access to more information. Overall, use of digital technology can reduce automation project costs significantly by in addition to operational improvement.

One of the greatest advantages of digital communications over analog is the ability to communicate vast amounts of data over a limited number of data channels. Using digital communications and multidrop Fieldbus wiring instead of conventional wiring also has many advantages including reduction in cable and connections as many devices connect to a single bus. Other benefits include better accuracy for control loops as no precision is lost in D/A and A/D conversion, and higher integrity as distortions can be detected using 8- or 16-bit error checking. Two-wire devices get more current, allowing delivery of new and faster diagnostics over the bus, enabling plants to adopt a predictive maintenance program. Further, digital values may be transferred in engineering units, allowing transmitters to be used over their full range and eliminating range mismatch. Access to more information is also a key to intelligent device management.

There exist many different digital communication standards designed to interconnect industrial instruments and various IO modules that understand such protocols. Some common and popular digital communication standards and their description are as follows:

- HART
- Modbus
- FOUNDATION FieldBus

- Profibus
- AS-I
- CANbus
- ControlNET
- DeviceNet

HART: The HART technology is managed by the HART Communication Foundation (HCF) headquartered in Austin, Texas, USA. HART was designed specifically for use in process control instrumentation such as temperature, pressure, level, flow, conductivity, density, concentration, resistivity, dissolved oxygen, oxygen transmitters as well as final control elements such as control valve positioners. Because of its importance for process control, HART is supported in all modern digital automation systems, in leading SIS logic solvers, and in device management software part of asset management suites. Additionally, a handheld communicator is available to work on HART devices in the field. HART is a hybrid, superimposing digital communications on top of 4–20 mA signals, and is usually used in a point-to-point scheme, multidrop in some applications. A separate chapter is included to discuss more on HART protocol.

Modbus/RTU: Modbus/RTU is one of the bus technologies managed by the Modbus organization headquartered in North Grafton, Massachusetts, USA. Modbus/RTU has been adopted in a very wide range of distributed peripherals such as conventional I/O blocks, flow computers, remote terminal units (RTU), and weighing scales. Final control elements such as a.c. and d.c. drives are also available. Because of its simplicity, Modbus/RTU is supported in all digital automation systems, DCS, and most PLCs. For this reason Modbus/RTU is often used to integrate package unit controllers to the main control system. Refer to the separate chapter to learn more on Mod-bus.

Foundation FieldBus H1: FOUNDATION FieldBus H1 is one of the bus technologies managed by the FieldBus Foundation organization headquartered in Austin, Texas, USA. FOUNDATION FieldBus H1 was designed specifically for use in process control instrumentation for measuring temperature, pressure, level, flow, pH/ORP, conductivity, density, concentration, resistivity, dissolved oxygen, and oxygen transmitters as well as machinery health monitors. Final control elements such as control valve positioners, electric actuators, discrete switches, on/off valves, and signal converters are also available. Because of its importance for process control, FOUNDATION FieldBus H1 is supported in all modern digital automation systems and in device management software residing in asset management suites. Additionally, a handheld communicator is available to work on Fieldbus devices in the field. Refer to the separate chapter to learn more on FF.

PROFIBUS: PROFIBUS is one of the bus technologies managed by the Profibus International (PI) organization headquartered in Karlsruhe, Germany. PROFIBUS was designed specifically for Distributed Peripherals (DP) such as conventional I/O blocks and weighing scales. Final control elements such as drives, motor starters, circuit breakers, and solenoid valve manifolds are also available.

DeviceNet: DeviceNet is one of the bus technologies managed by the Open DeviceNet Vendor Association headquartered in Ann Arbor, Michigan, USA. A wide range of products are available using DeviceNet including conventional I/O blocks, inductive and optical switches, encoders and resolves, barcode readers and RFID, final control elements such as electric and pneumatic actuators, and valves, a.c. and d.c. drives, motor starters, and solenoid valve manifolds.

Finally since different areas of automation and different levels of the control system hierarchy have different communication needs, many different Fieldbus technologies exist. All types of devices are not available with all the different protocol options, and therefore it is necessary to use more than one protocol in control systems. For example, transmitters and valves will communicate using FOUNDATION FieldBus because the bus must be synchronized for good PID control. Electric drives will use PROFIBUS DP because of the higher speed possible at short distances, although DeviceNet is also an option. Discrete I/O may use either DeviceNet or AS-I. Modbus/RTU is used when integrating real-time control and interlock signals from OEM-packaged units to the main control system. The control system must integrate these buses, requiring that the digital system to have the interface cards for direct connection of these buses; using gateways or multiplexers is costly, time-consuming, error prone, and less reliable. The engineering station and engineering software must support the different protocols being integrated, as a fully-featured engineering tool will eliminate the need for special applications software for each protocol, which would be too difficult to manage.

3.4.2 I/O SUBSYSTEMS

DCS operation and control depends on how the physical measurements are made and transferred to digital control systems. Sensor or transducer measures the primary parameters in the field such as temperature/pressure or position/presence of an object. Based on the nature of parameter an analog or digital is generated. This signal is fed to the controllers in the DCS. A signal cannot be directly fed to the control system; it must be converted or processed before being sent to the control system. The signals are generally processed in three different ways:

1. analog signals directly being sent to the control system
2. analog-to-digital or vice versa conversion is done and sent to systems using digital instruments
3. by dealing with the signals purely in digital form as digital-to-digital input and output.

Traditional modules are the ones that operate on 4–20 mA or 0–10 V or 1–5 V. These are the standard inputs signals and often used across multiple systems. They do not have better diagnostics when compared to many systems available today. However, they are still preferred mechanisms in terms of data reliability and cost effectiveness.

Typical features of I/O modules are as follows:

- Isolated or nonisolated grounding on a per-point/channel or per-board basis
- Level of fusing protection on a per-point/channel, per-circuit, or per-board basis
- Accuracy and linearity of the sampling frequency
- Protection from external disturbance such as electromotive force and transients
- Fault option for output modules
- Overload, short circuit, and surge protection
- Impedance matching with field devices
- Loop feedback sensing or read-back option
- Back initialization feature to avoid bumps
- Manual override of loop control
- Criticality to indicate if the board fails, an indication of what else will be affected.

Coming to physical form factor, I/O modules can be broadly categorized into two types as rack/chassis-mounted and rail-mounted type and depending on how they interface with field devices they can again be broadly categorized into two types. In one they would normally have a terminal block in the face of the module to which field device wiring can be done or the IO module would have a separate unit called field termination, which would be connected to IO module using a prefabricated cable and field device would terminate at Field Termination. But in both the cases, direct termination at module level or at FTA level is not practical and a marshalling panel is used in between.

I/Os are broadly classified into the following categories:

- Analog inputs
- Analog outputs
- Digital inputs
- Digital outputs
- Pulse inputs
- FF inputs
- FF outputs

3.4.2.1 Analog inputs

Analog inputs are classified into

- Traditional modules
- Special modules

The wiring is differently done for generating a 4–20 mA from that of voltage signal. The most popular design for precision analog measurements ranging from 4 to 20 mA in current mode or from 1 to 5 V in voltage mode is 16 bits of resolution. Generally a single input is provided for each output channel. It is important to provide a suitable instrumentation ground for these single-ended input circuits. Each analog channel has built-in current protection circuitry, such that each channel open circuits itself before any circuit damage will occur.

Special modules are used for identifying the inputs from certain modules, for example RTD inputs and thermocouple inputs, which have to be processed specially due to very nature of the inputs. Some recent intelligent systems offer the wiring provisions in such a way that the wiring can be done on the same field termination for both RTD and thermocouple inputs with certain settings therefore saving on the hardware cost. With direct connection of the RTD input module, the temperature data (°C) can be converted into a digital value to be processed to certain number of digits after the point as a digital value and this decides the precision of reading. The RTD input module has Pt100, Jpt100, or cable burn-out function at their every channel. The RTD input module detects the out-of-range temperature that is input by Pt100 or JPt100. If a part of the connected RTD or cable is disconnected, the out-of-range voltage is input by the internal open wire circuit and the connection or disconnection is detected.

RTDs have different shield connections because the RTDs are electrically insulated from the process vessel; therefore, cut the shield wire off at the process end and connect the shield wires at the field termination. The Field Termination internal shield bus then connects to the local master reference ground bus bar.

On the software configuration it is capable of

- Analog-to digital conversion
- PV characterization
- Open wire detection
- Range checking and PV filtering
- PV source selection

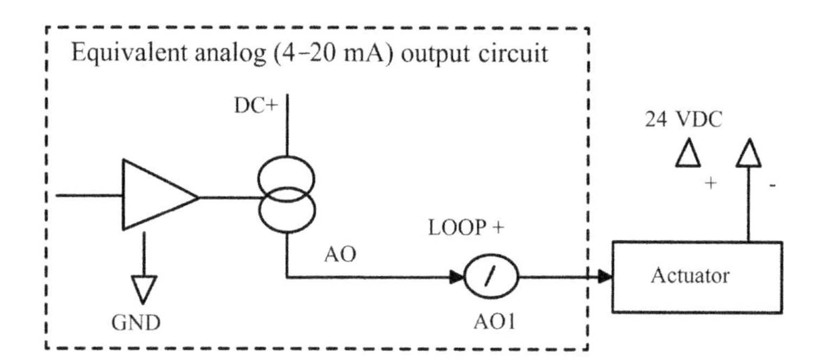

Figure 3.4

Analog output

The type of analog input needed by the I/O module is based on the type of field sensor that is providing the input to the channel and the characterization options selected by the user. Certain channels are generally used for control points or data acquisition points.

The signal received from the field is characterized based on the entries user provides for the type of sensor, input direction, and so on. The input signal is first converted to a raw PV signal whose units can be percentage, ratio, millivolts, microvolts, or milliohms depending on the sensor type.

3.4.3 ANALOG OUTPUTS

Analog outputs generate current output with certain resolution at each channel (Figure 3.4). A single input is provided for each output channel. It is essential to provide a suitable instrumentation ground for these single-ended input circuits. Software configuration of analog output channel converts the output value (OP) to a 4–20 mA output signal for operating final control elements such as valves and actuators in the field. The OP parameter value can be controlled from the control that was declared or by the operator to override the control.

To convert the OP value to a 4–20 mA signal, the AO channel performs:

- direct/reverse output function
- nonlinear output characterization

User specifies whether the output of the data point is direct acting (where 4 mA = 0%, and 20 mA = 100%) or reverse acting (where 4 mA = 100%, and 20 mA = 0%). The default mode can be anything based on the provision given by the vendor. In order to process the signal for linearization it has to be verified in which direction the output is acting.

3.4.4 DIGITAL INPUTS

An example of a digital input circuit is illustrated in Figure 3.5.

An input count feature uses input registers to accumulate the positive translations of each input. Positive d.c. voltage must be applied to an input to indicate an ON. All channels are referenced to a common return, which is connected to the negative pole (ground) of the d.c. power source. One wire

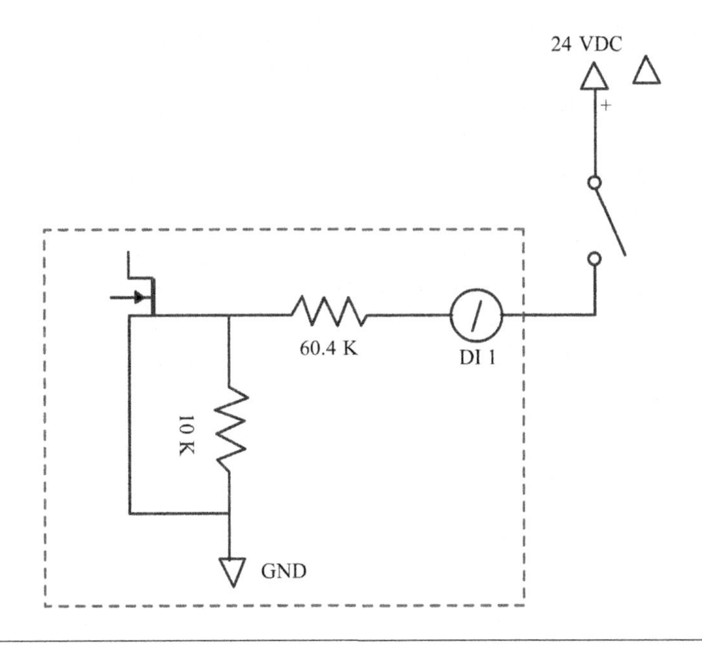

FIGURE 3.5

Digital Input

from each sourcing field input should be bussed together and connected to the terminal. The connection becomes vice versa in case of a sinking input.

The DI channel block can belong to DI or DISOE. This digital input channel converts a digital PVRAW signal received from the field to a PV that can be used by other data points in the DCS. The soft configuration has the capability to perform the following functions:

- Bad PV
  - When the PV source is being changed or when the module is made inactive in the software.
  - The PV source is auto and the PV is not being updated because the channel is inactive or the module is idle or there is a channel soft failure.

The digital input channel can also be configured as a status input or a latched input.

- Status Digital Input Channel: For this digital input type, the PV value represents the state of the raw input signal after the direct/reverse conversion is performed. The current PV state is available as an input to logic blocks.
- Latched Digital Input Channel: To capture the occurrence of momentary digital inputs, such as from push buttons, the digital input channel is configured as a latched input. Configuring the channel as latched is accomplished by a setting made to latched. Although this option varies across vendors, its primary purpose is to detect latched inputs. When configured as a latched input channel, an input pulse for a minimum time is detected and latched for the specified amount of time. This ensures that any control FB that needs to monitor this input executes at least once during the time that the signal is latched on. When the inputs are latched it generally will not support SOE. Open wire capability is same as that of analog inputs.

- Sequence of Events (SOE) Mode: The purpose of SOE is to keep track of events in case of faster running process such as turbine and generate the events with proper timestamp in order to detect the sequence of operations in case of a failure. When the digital input is set to SOE, the input channel operates in SOE mode. In this mode, the input is sampled every millisecond and PV is processed at certain time and SOE functionality is available in this mode.

3.4.5 DIGITAL OUTPUTS

The digital output channel provides a digital output to the field, based on the origin of the input and the configured parameters.

A wide category of digital output points are as follows:

- pulse-width modulated (PWM) output,
- status output (SO),
- pulse-on output (PULSEON), and
- pulse-off output (OFFPULSE).

The output type is selected based on the user application. The PWM type is used in combination with analog control mostly regulatory controls to provide true proportional control. The status and pulsing output types are for digital outputs that are connected to digital control blocks. Actual output action also depends on the configuration of the digital control point.

3.4.5.1 Status output (SO) type

The SO type can be controlled from a digital control block output or a logic block output.

3.4.5.1.1 Pulse-width modulated (PWM) output type

The PWM output type can receive its input from analog controls through a user-configured output connection. The length of the pulse is derived from the regulatory blocks, which are upstream to the digital outputs. Because OP is in percent, the percent value becomes the percent on time for the pulse whose period can be modified while defining the block.

The pulse on-time for direct and reverse acting outputs is calculated as follows:

For direct action:

$$\text{PulseOnTime} = \frac{\text{OP\%} \times \text{PERIOD}}{100}$$

For reverse action:

$$\text{PulseOnTime} = \frac{100 - \text{OP\%} \times \text{PERIOD}}{100}$$

If the value of OP is less than 0%, it is clamped to 0%; an OP with a value greater than 100% is clamped to 100%.

3.4.5.2 On-pulse and off-pulse output type

The on-pulse and off-pulse output types can be controlled from a digital control block output, a logic block output as determined by the parameter connection. Pulsed operation (pulse on or pulse off) can be obtained by linking the output connections to the determined parameters in the block. In case of on-pulse dominant conditions the output is on for the on state parameters and at the end of the pulse time it is set to off and vice versa is true in case of off-pulse dominant conditions.

3.4.6 PULSE INPUT MODULES

Wiring diagram of a pulse input module is as follows. It is extremely helpful while dealing with faster reading sensors such as encoders they sense the pulses and provide the inputs to channel in form of number of pulses. Based on the number of pulses, the position of the conveyor/item as sensed by the encoder is determined by the pulse input channel. A sample-wiring diagram of pulse input is as follows (Figure 3.6):

3.4.7 WIRING CONSIDERATIONS

3.4.7.1 Installation considerations

Installing a DCS is the process of placing the components, environmental conditioning, power distribution, and wiring in physical locations.

But there are many preconditions that must be considered while installing the hardware. Some critical and common ones are listed as follows but there may be many depending on the type of hardware.

- Electrostatic Discharge: The hardware may be sensitive to electrostatic discharge, which can cause internal damage and affect normal operation of the hardware. Hence, some guidelines such as below may have to be considered when handling the hardware.
  - Touch a grounded object to discharge potential static.
  - Wear an approved grounding wrist strap.
  - Do not touch connectors or pins on component boards.

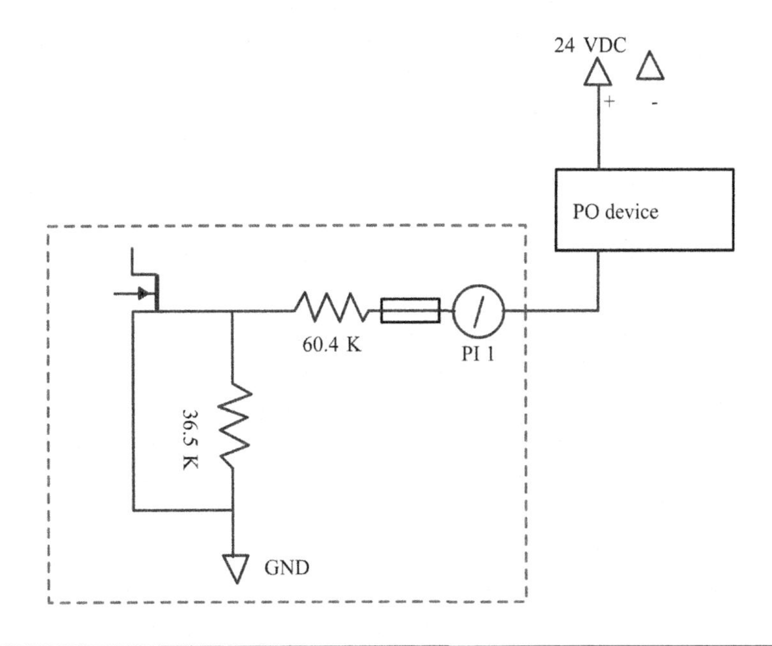

Figure 3.6

Pulse Input

    - Do not touch circuit components inside the equipment.
    - Use a static-safe workstation, if available.
    - Store the equipment in appropriate static-safe packaging when not in use.
- Removal and Insertion Under Power (RIUP): This mainly depends on the design of the hardware. Some hardware support RIUP and some may not support. Hence, it is very essential to read the manual to clearly understand its implications. And depending on the location this may again vary. Therefore, although hardware may support RIUP in normal locations, it may not support the same in hazardous locations. When you insert or remove the module while backplane power is on, an electrical arc can occur. This could cause an explosion in hazardous location installations. Therefore, ensure that power is disconnected or the area is nonhazardous before proceeding. Repeated electrical arcing causes excessive wear to contacts on both the module and its mating connector. Worn contacts may create electrical resistance that can affect module operation.

3.4.7.2 Grounding considerations

A proper grounding provides protection against fire and electric shock hazards. So one has to ensure a proper wire connection from the ground terminal of the power supply on the rack or rail to the ground screw attached to the rack- or rail-mounted hardware.

- Identification of hardware to be installed: Identification of the individual hardware to be installed such as power supply, various other hardware such as communication module, controller, I/O modules, etc. and writing such as power supply connections and field connections has to be made. It is recommended to refer the related manual for the installation procedure before starting installation.

  Once the above conditions are checked, then one can proceed with the identification of location and enclosure or cabinet where the actual hardware must be mounted. This enclosure/cabinet could be a panel that will have all the hardware within it or it could be mounted on the wall itself depending on the type of the environment that has to be mounted and protection required from the environment.

3.4.7.3 Wiring and marshalling

Wiring mainly involves identification of type of wires and termination to be used for power supply wiring, field wiring to the actual field devices, and wiring involved between marshalling to termination assembly of the IO module and how they are actually wired. This marshalling cabinet function is to interface the incoming field cable (which is normally a multipair cable) and the I/O card connection. The existence of a marshalling cabinet for a large number of I/O ensures smooth maintenance by operation personnel. Cross-wiring function is another important interface function of the marshalling cabinet. Cross wiring is always necessary since the incoming field signal and the channel quantity of the I/O card are always different. For example, let us say we have a 24-pair incoming field cable that carry 20 field analog signal. This field signal is split at least into two analogs 16 channel I/O card. The first 16 I/O are in the first 16 channel I/O card and the other 4 are in the next 16 channel I/O card. To have this split wiring (cross wiring) in the system cabinet is not a good practice from the operation and maintenance point of view. Thus, the marshalling cabinet is needed.

Another cause of the cross wiring is mixed I/O signals coming from the field. The incoming multipair cable can have mixed analog input (AI) and analog output (AO) signal in the same multipair cable. This mixed signal is split in the marshalling cabinet so that the AI signal is terminated in the dedicated AI termination board and so do the AO signal.

In some application such as voting application for safety system (i.e., 2003 voting) there is a requirement to have the I/O signal allocated in the different I/O card so that the common fault cause can be avoided. This application needs a cross wiring in the marshalling cabinet.

In a typical application, the marshalling cabinet has the following cable routing.

1. The multipair field cables enter the marshalling cabinet through the bottom part of the cabinet.
2. Then each wire of the incoming multipair field cable is terminated in the surge protection devices or surge arrester. If there is no requirement of such surge protection devices, each wire is terminated in the terminal block.
3. From the surge protection devices, in non-IS application, there is a cross wiring that matches the field signal and the I/O address assignment in the termination board. If the system is IS, then before cross wiring, each wire connection is routed first to the IS barrier.
4. The dedicated system cable that has a "plug and play" connection is then routed from the termination board to the I/O card in the system cabinet.
5. Some termination board may require a dedicated d.c. power supply that can be taken from the d.c. power supply in the marshalling cabinet.

3.4.7.4 Barriers

The barriers, installed in a safe area, provide the separation between the intrinsically safe devices in the field and the nonintrinsically safe devices (DCSs, PLCs, recorders, indicators, etc.) in the control room. The barriers must also be certified and be suitable for being connected to field-mounted apparatus, which are installed in a particular hazardous area and exposed to a particular group of gases and temperature class.

An intrinsically safe barrier has terminals for connecting field and control room wiring, which minimizes the number of panel terminals. Also, the barrier system provides a sharp line of demarcation between the hazardous and safe area. The barrier is a completely passive device that requires no power source and passes a 4–20 mA d.c. signal at a nominal 24 V d.c. power rating with virtually no degradation (less than 0.1%).

A barrier contains wire-wound resistors to limit current and redundant Zener diodes to limit the voltage. A fuse is in series with the resistors. A fuse alone cannot do the job because if a high voltage were placed on the input terminals with only a fuse and resistors present (without diode), an explosion could occur in the hazardous area before the fuse could blow. A fuse can pass sufficient energy to cause the explosion before it opens. It takes only a few microseconds, under the right conditions, for an explosion to occur. Without some method for limiting voltage, a fault voltage that appears on the barrier input terminals, when combined with increased current, could be passed along to the hazardous area.

3.5 DIAGNOSTICS IN IOs

IO modules have power-on self-test (POST) diagnostics during power on process. POST is the initial set of diagnostic tests performed by the IO module when it is powered on. Tests that fail are relayed to the user via the use of POST codes, beep codes, or on-screen POST error messages immediately after the IO module is powered on. Similarly, during runtime operations also an IO module keeps running internal diagnostics to detect faults and reports the same as and when it occurs. Modules also try to rectify the problems on its own and take necessary actions needed such as switchover to redundant module thinking the fault may be on its own channel and may not be present in its redundant partner.

3.5.1 PROTECTION METHODS, OPTICAL ISOLATORS, RELAY

Voltage, current, or temperature measurements through IO modules are an integral part of industrial and process control applications. Often these applications involve environments with hazardous voltages, transient signals, common-mode voltages, and fluctuating ground potentials capable of damaging measurement systems or affecting the measurements. To overcome these challenges, systems designed for industrial applications such as IO modules make use of some protection methods. This type of protection implemented within the IO module is called galvanic isolation.

3.5.1.1 Galvanic isolation

Galvanic isolation is the principle of isolating functional sections of electrical systems to prevent current flow; no direct conduction path is permitted. Energy or information can still be exchanged between the sections by other means, such as by optical, capacitance, induction or electromagnetic waves, or acoustic or mechanical means.

Galvanic isolation is used where two or more electric circuits must communicate, but their grounds may be at different potentials. It is an effective method of breaking ground loops by preventing unwanted current from flowing between two units sharing a ground conductor. Galvanic isolation is also used for safety, preventing accidental current from reaching ground through a person's body.

A transformer is the most widespread example of galvanic isolation.

Some different forms of galvanic isolation methods are as follows:

- Optical Isolation: LEDs produce light when a voltage is applied across them. Optical isolation uses a light-emitting diode (LED) along with a photodetector device to transmit signals across an isolation barrier using light as the method of data translation. A photodetector receives the light transmitted by the LED and converts it back to the original signal.

  Optical isolation is one of the most commonly used methods for isolation. Immunity to electrical and magnetic noise is one of the benefits of using optical isolation. Transmission speed, which is restricted by the LED switching speed, high-power dissipation, and LED wear, is however a great disadvantage.
- Capacitive Isolation: Capacitive isolation is based on an electric field that changes with the level of charge on a capacitor plate. This charge is detected across an isolation barrier and is proportional to the level of the measured signal.

  One advantage of capacitive isolation is its immunity to magnetic noise. Compared to optical isolation, capacitive isolation can support faster data transmission rates because there are no LEDs that need to be switched. Capacitive coupling involves the use of electric fields for data transmission, making it susceptible to interference from external electric fields.
- Inductive Isolation: In the early 1800s, Hans Oersted, a Danish physicist, discovered that current through a coil of wire produces a magnetic field. It was later discovered that current can be induced in a second coil by placing it in close vicinity of the changing magnetic field from the first coil. The voltage and current induced in the second coil depend on the rate of current change through the first. This principle is called mutual induction and forms the basis of inductive isolation.

  Inductive isolation uses a pair of coils separated by a layer of insulation. Insulation prevents any physical signal transmission. Signals can be transmitted by varying current flowing through one of the coils, which causes a similar current to be induced in the second coil across the insulation barrier. Inductive isolation can provide high-speed transmission similar to capacitive techniques. Because inductive coupling involves the use of magnetic fields for data transmission, it can be susceptible to interference from external magnetic fields.

3.5.1.2 Calibration

The ISA Instrument Calibration Series defines calibration as "Determination of the experimental relationship between the quantity being measured and the output of the device which measures it; where the quantity measured is obtained through a recognized standard of measurement."

Sufficient data must be collected to calculate the IO modules operating errors to check if the IO needs calibration. This is typically accomplished by performing a multiple point test procedure that includes the following steps.

- IO module should not be in the running condition. It has to be either in idle or program mode depending on the manufacturer.
- Enable or enter calibration mode in the IO module.
- An input representing the minimum IO module signal or 0% should be applied. So if the module has a 4–20 mA range, then 4 mA should be applied. The output should be adjusted to be correct if it is not. Adjustment normally involves a reset or shorting the jumpers so that minimum or 0% is set for minimum signal.
- Next the maximum signal or 100% is applied which in this case could be 20 mA. The range is then again adjusted (either by resetting or shorting jumpers) to give the required output if it is not 20 mA so that 20 mA output is set for 100% of the signal.
- Also procedure can be repeated for intermediate values to get better results such as 25%, 50%, and 75% instead of limiting to 0% and 100%.

3.6 CONTROLLERS

Controllers are an integral part of a DCS – the heart of the system. Physically a controller can be mounted in a rack and placed on a rail. Racks are often caged structures where in different hardware modules are placed in the respective slots in the cage as shown in Figure 3.7.

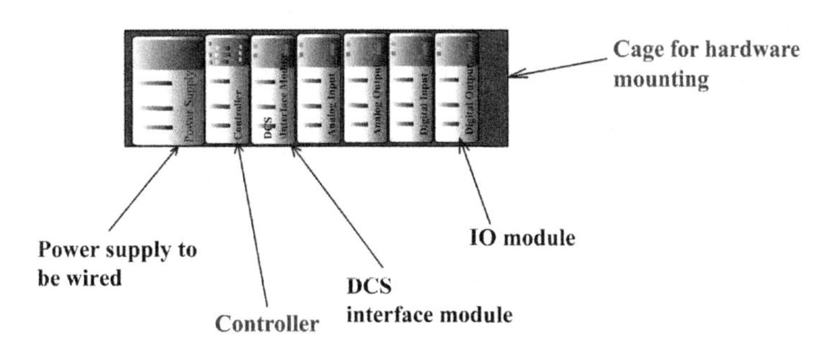

FIGURE 3.7

Sample controller

Innovation in human engineering and ergonomics, have given way to new designs that make controllers cage free, easy to mount, and more isolated. Some solutions offer controllers and its related I/O modules to be mounted vertically in dedicated cabinets.

Both methods have their own advantages and disadvantages. Communication in the cage method happens, though backplane for some of the IOs, thereby reducing the wiring costs. In solutions where mounting is vertical and distributed, heat dissipation is better. This ensures longer life for the hardware. But wiring has to be done for each and every sub system of the hardware. However, wiring costs can be optimized by following proper designs. Providing the power supply through bus bars where the controller is mounted, instead of powering the controllers or IO modules separately through wires, reduces the chances of human errors while wiring, which generally leads to malfunctioning introduced by loose contacts while wiring.

Controllers are available for the commonly supported ranges as 230 V/110 V/24 V.

IOLink connections communicate with the IO modules from controller. Since I/O modules are of the same family as controller they have certain proprietary protocols to communicate. The connection from controller goes to IO module over an IOBus and the rate of data transfer depends upon the type of the system; it may range from few kilobits per second to megabits per second.

IO modules that are analog/digital/pulse outputs/inputs can be communicated over IOBus. Apart from the data communication, diagnostic information pertaining to I/O modules such as alarms/events, etc. are also transferred through this link.

Redundancy is an important aspect of DCS design. With respect to controller, the redundancy is maintained at all levels, namely IOBus level, IO module level, controller level, power supply level.

IOBus redundancy is designed to maintain the connection from the IO module to the controller, when one of the links breaks. Any of the links could break especially when the connections are daisy chained. To avoid losses a redundant link is readily available to take over the control without any loss, in such a case. The typical switchover time is a few milliseconds – small enough for any glitch in the control to go undetected except for the alarms that are populated by the module.

Common connection methodologies for IOs are as follows:

Depending on the type of IO module, communication protocol, and the way they are connected to controller, there exist various topologies that vary from vendor to vendor. There can be two types of topologies on chassis or rack-mounted type, based on the controller and IO modules position or slot within the same rack or different rack. In all cases, modules communicate with each other within the chassis through a common backplane circuit board. Figure 3.8 illustrates a nonredundant combination, and Figure 3.9 illustrates a redundant combination.

Power supply
Comm. module
Controller
Analog input module
Analog output module
Digital input module
Digital output module

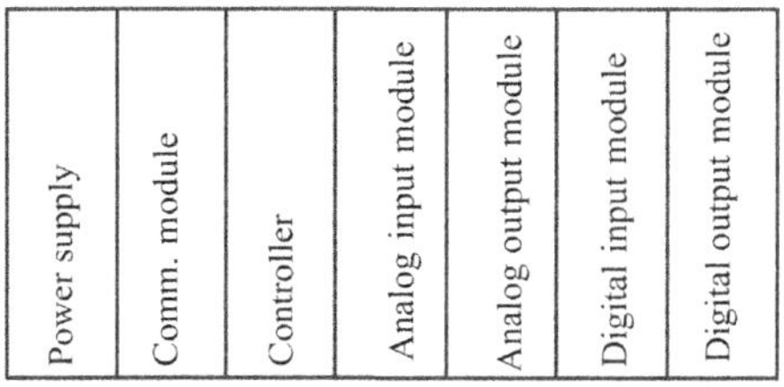

FIGURE 3.8

Rack mounted non redundant controller

FIGURE 3.9

Rack mounted redundant controller

Figures 3.10 and 3.11 illustrate nonredundant and redundant types of topology based on position of controllers and IO in different racks with IO rack connected to controller through a ribbon cable.

The primary and secondary chassis remains synchronized through the redundancy modules, which are in turn connected to each other through a redundancy cable, in a redundant combination. It is

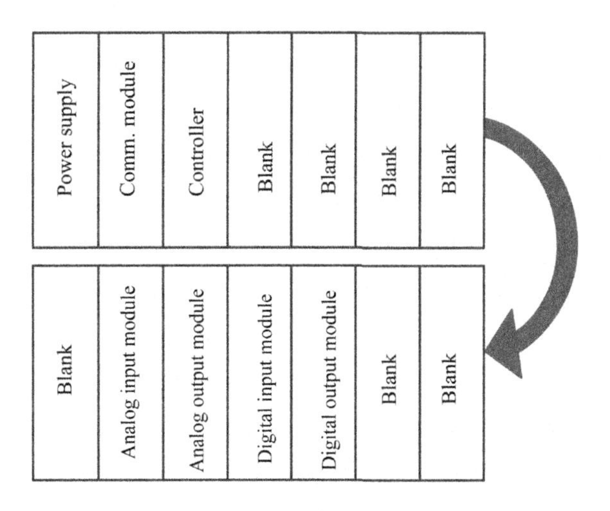

FIGURE 3.10

Rack mounted redundant IO Chassis

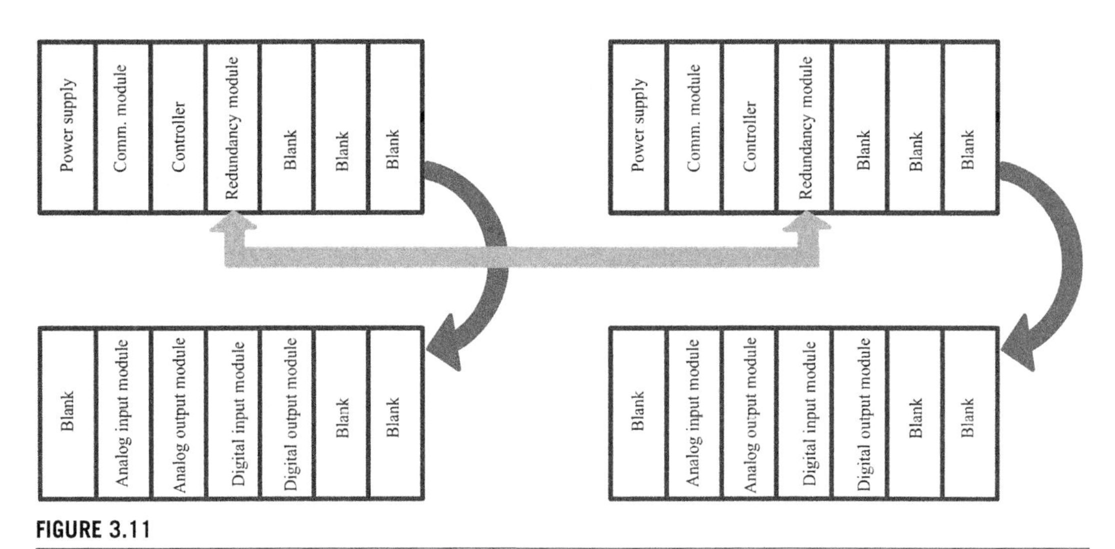

FIGURE 3.11

Rack mounted redundant controller and IO

essential for the position or slot of all the modules in both primary and secondary chassis to match some times for the system to be synchronized.

The topology for rail-mounted controller and IO modules slightly differs in two combinations. In one combination, IO modules are mounted on a rail normally on a base and these bases are connected to another base for communication and a gateway module is used for communication with controller modules which in turn sits on a rack as shown in Figure 3.12. In the second combination, controller and IO modules all sit on a rail and are connected to each other through a cabling system in a daisy chain manner as illustrated in Figure 3.13.

Controller-level redundancy helps in maintaining control in case of abnormal situation occurring in the system with the potential of causing loss of control or view. It supports a hot-swap, enabling the

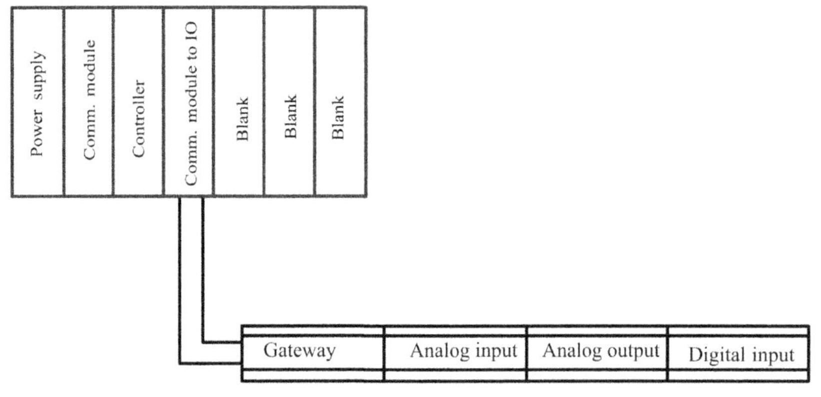

FIGURE 3.12

Remote IO Bus

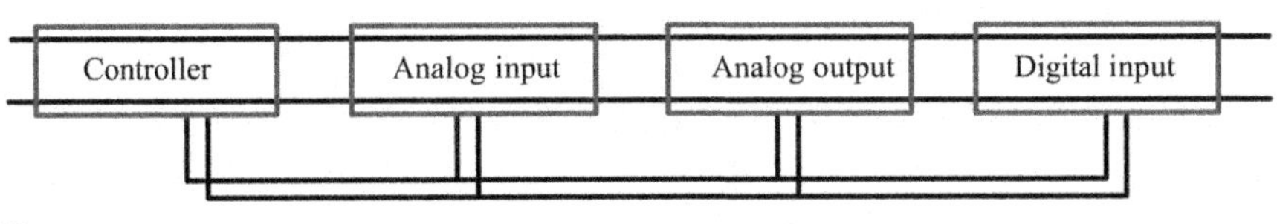

FIGURE 3.13

Bus connected IOs

change of a controller module without having to power down the system. Modules can also autoconfigured after a hot swap.

It is essential to have a proper backup of the components of the controller in a healthy state to maintain redundancy. Otherwise, in case of failure the new primary may provide reduced functionality, which violates the whole purpose of redundancy. The controller should have healthy communication cables, IObus cables, and redundancy cable. Redundancy is lost if any of these are impaired and switchover takes place onto the backup controller, which is error free. For example, in case of faulty communication, the switchover to back up takes place, which ensures control is retained.

Where there is a redundancy cable, one is primary and the other is backup and they are monitored periodically for sustaining redundancy. Time synch is used for precise applications such as SOE. SOEs are helpful when fast-running processes such as turbine may trip and user has to analyze the sequence of operations that lead to failure. However to generate the events, the source of time for IO module is taken from controller. The controller is equipped to take time from the domain controller through communication cables. In case of need for more precision in timestamps, direct time synch connection to Global Positioning System (GPS) Time Server is made. System Designers decide what needs to be used for their application.

Communication cables are used for communicating to workstations/supervisory nodes/peer controllers at the speeds of 10/100 Mbps through a cat5/5e cable over proprietary protocols or open protocols. Although in certain controllers multiple connectors are shown as RJ-45 kind of connectors, their application is different and clearly stated on the controller for better user understanding. Communications and configurations are explained in greater detail in the coming sections.

Controller architecture can be classified as follows:

- CPU
- RAM
- Communication interface

CPU is the heart of controller; the entire electronics that help in control is designed around this; often a powerful processor from manufacturers such as Motorola, TI, Xilinx is used. RAM is another vital part wherein the live execution data is present in the RAM and retrieved at a faster rate during data access control mechanisms. Controller must have the capability to communicate as follows:

- Communication with peer
- Communication with interfaces
- Communication with system

Any sort of communication that the controller should establish involves the following configurations:

- IP address
- Speed of communication

- Duplex settings
- Station settings

Communication with supervisory networks is enabled using an IO address, which is set manually through a mechanical switch by providing a unique ID. Alternately, a software is used to generate IP addresses on the basis of Mac addresses using technologies such as Dynamic Host Configuration Protocol (DHCP). In the absence of a provision on the controller to configure IP address, the serial communication is converted to TCP/IP through terminal server to ensure faster transmission. Along with IP address, speed setting and duplex are very important for communicating across networks. The vendor provides these settings recommendation. In case of a mismatch in settings and the recommendations, it will result in malfunctioning.

Station settings are extremely useful in case of profibus and serial communication protocols. As per the protocol, design stations are the unique identifiers for the master devices to communicate with slaves. Therefore, each node that has a data to provide is designated a unique station number. Compliance to other settings pertaining to speed is also applicable.

Peer communication refers to communicate with a controller of the same category. A PLC or a DCS controller communicating with the same PLC or DCS over specified protocols is a good example of communication with peers. The idea behind such communication is to maintain perfect control of the process. A cascade control of PID is an example of such a control. If a loop is running a cascade control and if the primary block is in area 1 and final control element is in area 2 along with secondary PID then it is always preferred to have a peer control to handle failure scenarios such as valve failure through features such as back-initialization, and these solutions are more reliable as they are provided by one manufacturer. However, there are exceptions, where the controller supports popular protocols such as foundation FieldBus/HART/Profibus; any manufacturer of a controller supporting these standards irrespective of the vendor achieves the same level of control on the process.

Difference arises with reference to latency. When a controller communicates directly with the controller of its own category over a proprietary protocol, the response is generally faster. When it communicates with the devices supporting aforementioned popular standards, the communication is routed through a gateway.

Communication with interfaces is a very vital part as it helps the user understand what is happening around the field and within the controller. Communication with interfaces is broadly classified into two categories:

- Communication with operator consoles
- Communication with engineering consoles

Communication with operator console is aimed at providing the user a feel of process that is running in the plant; it could be a display of a boiler control, distillation column, or conveyor. The controller runs an RTOS where the displays are generally html based. Therefore, a proper interface driver is called for to enable display-oriented connections.

Communication with engineering stations enables basic configuration of the controller. Basic configuration of controller refers to configuration necessary to establish communication in an intended environment. Engineering station is also helpful for upgrades, when certain performance-related issues are noticed during controller run time. Controller enhancements or software related defects are updated into software and flashed to the controller for improving performance and making the control more robust. It can either be flashed through an interface from the engineering consoles or load port is used to connect the controller with a PC through a dedicated port from where the firmware is flashed. This is normally the case when the fixes on controller are not from the DCS vendor but from the manufacturer of the ICs used in the controllers or there are third party systems in the DCS calling for updates.

A library of an engineering console contains multiple controller options. These can be multiple generations of controllers from the same vendor or controllers supported by a vendor due to their popularity in the market. Once the controller is selected, it must be configured based on its attributes for communication, for example, the controller may be intended to run on serial communication with a certain baud rate. Similarly a controller may run with 100 Mbps/10 Mbps and few may run their proprietary protocols; therefore, defining this configuration is the key to successful communication in the targeted environment. Once these attributes are defined, a download/load is performed. This step helps in validating that the configuration made in the engineering station reaches out to the controller in the field. If the physical wiring and the communication attributes are correct, it returns a message to the user stating that the operation succeeded and the health of the communication and loaded attributes can be verified to confirm that the controller is operative.

Communication with ij involves execution within itself and its IOs. A controller has certain operations to be performed to execute in its environment first. If a loop is considered, the flow of data would be from AI-> PID-> AO AI, and AO may be part of the controller if it has IO modules connected or it can be from any of the devices that support protocols such as Profibus/foundation FieldBus.

In the above case, if we eliminate AI communication (because it would receive the information anyway) the controller processes the data as per the parameters given by the data acquisition and PID. This blocks execution happens purely within the controller and the data processed are again given back to self or to the specific devices running the supported protocols.

Apart from this, the controller provides more diagnostics to user in two ways:

- Physical health representation
- Soft health representation

The LED indicator displays the equipment's physical health. The green or red light can be enabled corresponding to the status of the equipment. The LED status and corresponding color indication and operational condition.

Soft health includes majority of the information that engineers need to ensure that controller is behaving as expected. This information is available in the engineering station and some of it is found in the operator stations also. This contains information such as statistics of controller, for example, the time since controller power up, the controllers operating temperature, the rate at which the controller is communicating with peers, and so on.

3.6.1 CONTROL TECHNOLOGY

Control technology primarily consists of the libraries, which is a vital part of control technology as it defines what a system is capable of controlling. Primarily library can be classified into three categories:

- Standard Control Library – Standard library offers digital and analog controls with a variety of FBs. The interfacing and programming adheres to IEC61131-3 standards.

- Standards Communication Library – Standards library consists of blocks that are needed to adhere to certain standards such as S88/Profibus/HART, and so on.
- User-Defined Library – User-defined library allows users to create blocks and customize the control for an application.

There are many possible options for blocks in the library. Following are some of the common process scenarios and the solutions for them. Each method uses different control strategy. Valves play an important role in analog controls because they are the final control element to regulate the process. Therefore, let us consider analog controls in that area.

3.7 WORKSTATIONS

3.7.1 OPERATOR CONSOLES

Operator console are the most important and integral part of a DCS. Major part of a desktop-based operator interface is the consoles, which provides options to view the process, system diagnostics, alarms, reports, and so on. The console enables users to view the plant and control it to a certain extent. Operators can change modes/set points of the loops and control them. These operator consoles are normally located in the central control room or closer to the process in the shop floor. Operator consoles have provision to display process graphics, and process and system alarms. Trends can be viewed to understand how the process is varying. Data from trends can be historized and restored at a later point of time to understand the process responses.

3.7.1.1 Physical properties of the operator console

Operator consoles can be classified as follows:

- Customized operator interfaces
- Generic operator interfaces

Customized operator interfaces consist of components provided by the DCS provider. For example, a DCS provider provides a key board and mouse that have customized keys or track ball movements to navigate across the display pages in an operator interface. Dedicated keys are also provided for acknowledging alarms, changing access levels.

Dedicated monitors are provided to view the display with a provision to combine multiple monitors in a single console for better view of process.

Generic operator interface consists of components that are available in market and support is received from popular hardware manufacturers such as DELL/IBM/HP. However, the recommendation for the configuration including hardware such as hard disk, RAM, graphic cards, monitors, and network cards along with their supported slots is given by the DCS provider.

Currently, operator interfaces are more inclined toward the usage of generic hardware as it lessens the burden of maintenance and is easy to operate. Operator interfaces are generally desktop grade machines. Embedded operator panels are also popular for industrial controls, which are of limited capability and are discussed later. However, the flip side of maintenance is with frequent replacement of hardware and the shortened life cycle of the operator interfaces; therefore, a solution is emerging in terms of virtualizing the machines. Operator interfaces are virtualized and thin clients are connected to the virtualized machines.

Virtual machines are server-grade machines that are more powerful than desktop machines. Based on the capacity, a virtual server can host tens of operator interfaces. Only problem with virtual operator interface is that if the virtual server fails then multiple operator interfaces fails in one shot. Hence, system has to have very less MTBF. Another alternative from the design perspective is to distribute the operator interfaces related to one process across multiple virtual servers to avoid the loss of view of process.

3.7.1.2 Software

Software used for operator interfaces has its base on operating system. The operating systems supported for desktops vary from UNIX to XP/WIN7. The proprietary software from DCS vendors is installed on top of these operating systems. DCS manufacturers provide the recommendations for installing the operating interfaces on specific desktop machines.

Proprietary software primarily contains the following components; although they are generic software components, their usage methodology leads to the evolution of proprietary software:

- SQL
- HTML
- XML
- Ethernet
- Scripting techniques such as VB
- Generic communication protocols such as Modbus

Proprietary components that evolve through above components are as follows:

- Database for storage and retrieval of information
- Stations for users to navigate across the displays and alarms
- Scripting to perform specific task at given point of time/situation
- Trends, reports

3.7.1.3 Database

Per definition database is a set of data held in a structured format available for access to multiple systems. A database on an operator interface is used for data access in the station; a station is a page where user can view the displays, alarms, trends, and so on. SQL, Access, and Oracle are the most popular software on which databases are built.

Most of the DCS manufacturers follow server/client-based approach for control systems. SQL and oracle are used for heavy applications such as servers and historians, which are heart of a DCS system at various levels. Access and SQL-CE (compact edition) are used for lighter applications such as operator interfaces and purely SCADA applications.

SQL resides on the server having information of critical control data. The data connects between lighter databases such as SQL/access database residing on the operator interfaces. A part of the server data is continually synchronized with operator interfaces databases associated with the server. This data contains the information related to control. This method is bound by one of the principles of DCS design, that is, redundancy. Data synchronization happens across all operator interfaces; in case the server loses communication with controllers the connection is maintained by the operator interfaces. Therefore, even in the absence of servers user can largely control the loops.

Only difference an absence of server makes is with respect to the engineering changes that one can do. In the absence of a server, a user may not be able to change a report configuration or change the logic to work in a different manner. However, control is not interrupted.

Access databases are lighter version of databases that can be used to hold a specific data and pass on the information to SQL. Applications such as SCADA configuration are created in access database and the data is passed across controllers for SCADA applications. Current trend in process control industry is shifting toward more secured applications. In this context multiple databases of different combinations, for example, access to SQL makes it vulnerable to security risks and therefore not a preferred solution. Therefore, the design focus is shifting toward having a better-secured single database component that does multiple tasks.

Such a design does not impact the redundancy or availability because it is still available in all nodes in the form of a lighter edition of database.

Displays are primary source of information for operators. They are designed with a back-end technology of HTML or Visual studio components. Displays designed with the help of HTML technology enable the customization of the displays for respective application. Sometimes the displays can be browser based.

Alarms and trends also have HTML components for presenting the data; when it comes to scripts VB is the most preferred scripting method. Scripting enables a way of expanding the capability of traditionally available features in the configuration. For example, if a user wants to raise an alarm purely based on the various inputs received from SCADA without intervening with the control configuration then scripting is the preferred method. In short, scripting enables customization of the application.

Reports can take multiple paths; the back-end technology that it runs could be HTML but the interaction does not end there. Reports can be generated in the form of a spreadsheet with technologies such as Excel, or it can be in HTML format for browser readable method. It can also be in the form of XML for other applications to read data and present in better graphics for user.

Considering the embedded operator panels, Windows compact edition is a preferred version for an operator panel in an embedded environment. Operator panel developed on compact edition does not have equivalent capacity as that of a desktop-based operating interface. The panel is confined to control a specific area that has a provision to view alarms, trends, mimic diagrams. However, the capacity is not as good as desktop machines.

Desktop machines need more maintenance when it comes to security. Although the software's in the DCS makes it proprietary, one has to remember that the base is still an operating system platform such as XP and WIN7 and are prone to risks such as remote code execution. System's maintenance has to be taken up periodically with the imminent cyber security threats. Therefore, security concerns are handled at two levels: one is platform security and another is control system security. Although there are some overlaps, there is a fine line of difference between both. A compromise in platform security can lead to compromise in control system security. But a compromise in control system security can be purely due to proprietary software design also. The differences between desktop-based and embedded-based interfaces are explained in Table 3.1.

Operator interface has a primary component called console, from where user can get a view of multiple aspects of the process that is getting controlled. Following are the features of a station, which are explained in greater detail in subsequent sections:

- Displays
- History
- Trends

Table 3.1 Differences between desktop-based and embedded-based interfaces

HMI Panels	Operator Interfaces (Desktop-Based)
Compact in size, can mount in a marshalling cabinet	Need dedicated space, can mount in central control room or close to process with proper protection
Rugged – can withstand very high temperatures, humidity, dust	Relatively less rugged when compared to HMI panels
Embedded OS consumes less resources	Desktop OS consumes more resources
Used for dedicated applications such as controlling-specific area/process in a plant. Area of control is limited.	Has access to entire plant unless restricted, can control all loops for which it was given access
Less secure by design	More secure by design
Less maintenance for sustaining security	More maintenance for sustaining security
Often supports open protocols for communication such as Modbus	Dedicated to a specific DCS vendor hence will communicate over proprietary protocols only
Powerful diagnostics are not available by default. Need to choose among the variants to achieve this.	By default powerful diagnostics are available as this is the integral part of DCS, only restricted features for security reason.
Modifying the control after mounting is not easy because it needs dedicated connectors, software, and machines to do that	Control can be modified with ease because the machine is well equipped

- Alarms
- Reports
- Supervisory control and configuration
- Security

3.8 FUNCTIONAL FEATURES OF DCS

The major architectural components of any DCS comprise of the following (refer to the Figure 3.3.5):

- System configuration
- Communications
- Control
- Alarms and events
- Diagnostics
- Redundancy
- Historical data
- Security
- Integration

3.8.1 SYSTEM CONFIGURATION/PROGRAMMING

Every DCS controller is a computer, although not endowed with all the peripherals of a computer system, images of which one tends to conjure up when the word "computer" is mentioned. The controllers therefore need instructions to execute the control actions. Two distinct terms must be distinguished here – programming and configuration. Every controller comes with inbuilt firmware. In addition, an application program is downloaded into another partition of the controller memory. In some cases, manufacturers provide the application programs also and a way to set certain parameters to make the generic logic work in the particular plant. For example, a DCS manufacturer may provide an application program to control a set of three compressors. The same controller might be sold by the same vendor with another application program say for control of a set of pumps. At the field, it is just a matter of defining the minimum and maximum pressures, the number and type of relay contacts to be operated, and the measurement range for field signals. This process is called configuration. Consider the case where the DCS vendor provides the controller with the application program memory in a blank state. DCS users can write custom application programs to cater to specific logic; this process is termed as programming.

Engineering tools enable configuration programming of controllers, depending on the applications. The actual features might be controlled by licensing mechanisms. The engineering tools also hide the complexities of programming the microcontrollers (which have their own specific instruction sets) by providing a common programming language with suitable user interfaces. Therefore, application and process engineers describe the control logic mostly graphically, which are translated into the instruction set of the microcontrollers. Typically, the control strategies are made up of interconnected FBs, sequential function charts (SFC), and equipment and unit representations, which perform functions within the control scheme based on inputs. The FBs also provide outputs to other FBs and/or physical I/O within the control scheme. The set of FBs are invariably provided by the DCS vendor as libraries. This eliminates the need for any special software language to program the microprocessors used in the system. In addition, the users can create custom FBs by combining the predefined FBs to more complex library of functions. The vendor pretested FBs enable a DCS to be applied to any plant very quickly, cutting down on the debugging necessary on programmed software. The engineering tools invariably follow ISO standards on the types of FBs, the representation of the blocks and the accepted form of interconnection. The programmed control schemes are stored as files/control programs in a configuration database. The engineering tool is used to download these strategies via the control network to distributed controllers, consoles, and devices. Refer to Figure 3.3.4 for an engineering view.

In addition to software configuration, some hardware configuration is also solicited, like setting the addresses of the IO modules and controllers. Sometimes depending upon the type of connections, different links might have to be connected to adopt the same card to suit different field interfaces. While the hardware configuration needs the actual hardware in the field, software programming can be accomplished remotely and downloaded to the controllers at the plant.

The configuration/engineering application also allows a designer to create or change operator interfaces, such as plant schematics and process control diagrams viewed on the operator displays through a viewing application. These diagrams displayed on the screen enable the operator to change settings within the PCS.

Many DCS can also be reconfigured without need to take the system off-line which enables plant modification, and so on with minimum down-time, provided the necessary precautions are taken to ensure integrity of the running process. This capability is particularly useful in applications where the process is continuous and shutdown maintenance periods are short or limited.

3.8.2 COMMUNICATIONS

DCS systems vary in size from very small to very large, depending on the size and complexity of the plant being controlled. Today's systems are enabled with integrated web services for plant integration

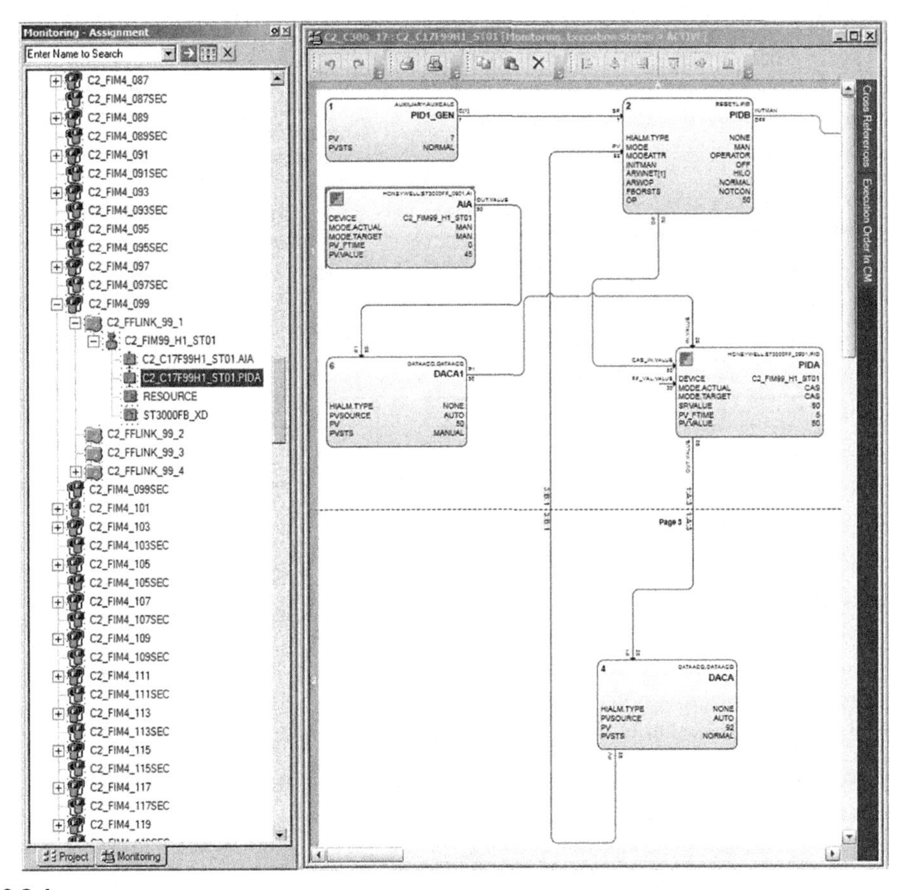

FIGURE 3.3.4

Loop configuration in DCS

while supporting a variety of open standards, such as OPC, for communicating with data sources external to the system.

The communication infrastructure comprising the control network supports:

- Connections between different subsystems in the system
- Unsolicited communications for real-time data changes in the process
- Synchronous and asynchronous read/writes
- Configuration downloads to nodes, controllers, and devices
- Autosensing of workstations, controllers, IO cards, devices
- Diagnostics of system components and control strategies
- Online upgrades of system in operation
- Hot/warm/cold restart of control strategy from backup
- Secure and unsecured access to information in the system
- Alarms and events generated by process and system
- Device alerts generated by devices and equipment in the system

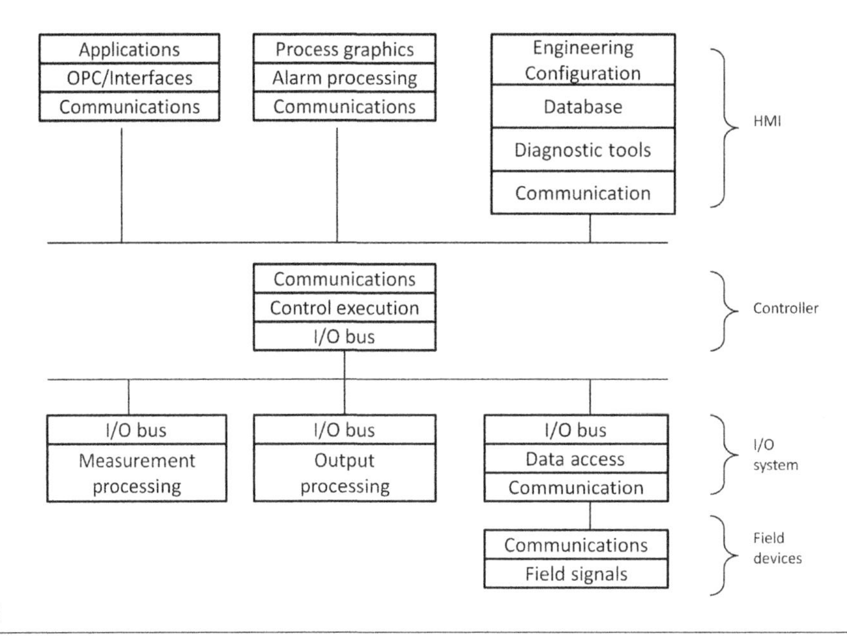

FIGURE 3.3.5

Functional architecture of DCS

- Time synchronization across nodes, devices, and I/O
- Deterministic communication of plant data so that data exchange is guaranteed in the system

The data highway is the communication medium that allows a DCS to permit distribution of the controlling function through a large plant area.

Depending on the speed of transmission, bandwidth supported, and physical characteristics of the medium of the data highway, the highway length could vary. However, data highways are designed as segments with suitable bridges/extenders connecting segments so that the length of a segment of data highway is not a limiting factor. The most popular physical medium is Ethernet CAT5 cable. However, several suppliers still offer communication over twisted and shielded coax cables. Several modern DCS also have implemented the data highway using fiber optic cables. Some of them also have successfully incorporated wireless exchange of data instead of a physical medium such as the data highway.

Optic fiber cables are used most commonly for point-to-point connection between switches and hubs. Optical fiber is attractive for use as a data highway medium because it eliminates problems of electromagnetic and radiofrequency interference, ground loops, and common mode voltages. It is safe in explosive or flammable environments. It can carry more information than copper conductors. It is inert to most chemicals, and is lighter and easier to handle than coaxial cable. However, special equipment and skilled labor are needed to terminate and connect optical fibers.

3.8.3 CONTROL

The DCS is connected to field sensors and actuators and uses set-point control to control the process in the plant. The most common example is a set-point control loop consisting of a pressure sensor, controller, and control valve. Pressure or flow measurements are transmitted to the controller, usually through transmitted and signal conditioning I/O cards. When the measured variable reaches a certain point, the controller instructs a valve or actuation device in the field to open or close until the fluidic flow process reaches the desired set-point. Large oil refineries have many thousands of I/O points and employ very large DCSs. This is a typical example where processes are not limited to fluidic flow through pipes. DCS controls can also include things such as paper machines and their associated quality controls, variable speed drives, and motor control centers. DCS are widely used in control of cement kilns, mining operations, and ore-processing facilities, among others.

A typical DCS consists of functionally and/or geographically distributed digital controllers capable of executing several (actual numbers depend upon the model of the controller and the complexity of the individual loops) regulatory control loops in one controller. The I/O devices can be either colocated with the controller or located remotely via a field network. Today's controllers have extensive computational capabilities and, in addition to PID control, can generally perform logic and sequential control. Modern DCSs also support neural networks and fuzzy applications.

3.8.4 ALARMS AND EVENTS

A critical part of the DCS is the integrated alarms and events processing subsystem. The engineering software is used to configure to get notified of significant system states. This enables monitoring the system states and acknowledging them. Priorities can also be associated with the events so as to monitor the events in the plant. Events represent significant changes in state for which some action is potentially required. An active state indicates that the condition that caused the event still exists. When an operator has seen the message and acknowledges the same, the event enters the acknowledged state.

In most DCS event types can also be defined. The event type specifies the message to be displayed to an operator for the various alarm states and the associated attributes whose value should be captured when an event of this type occurs. Event priorities can also be defined. An event priority type defines the priority of an event for each of its possible states.

Many DCS systems also support device and equipment alerts. Like process alarms, alerts can be assigned priority, can be acknowledged, and convey information related to the condition that caused them. Unlike process alarms, however, these alerts are generated by the DCS hardware or devices and equipment external to the DCS. Alarms and alerts are presented to the operators in alarm banners and summaries. Operators use these specialized interfaces, to quickly observe and respond to conditions. They then typically navigate to a specific display to view additional details and take appropriate actions. Operators can also suppress and filter alarms. Alarm suppression is typically used to temporarily remove alarms from the system for which some condition exists that the operator knows about (e.g., a piece of equipment has been shut down or is under maintenance). Alarm filtering provides a way for the operator to view collections of alarms and efficiently manage alarms when there is a flood of alarm messages and to suppress several alarms which result as a consequence of a basic alarm condition.

Alarms are the most vital part of a system. In facts alarms are a subset of alerts. Alerts can be broadly classified into three categories:

- Alarms
- Events
- Messages

Alarms, events, and messages attract user attention with high priority when they appear in the system. Alarms are classified based on the priority; priorities can be urgent, high, and low. They should have a provision of audio and display annunciation.

The difference between alarms and annunciation is that alarms appear in the system by default and the user has to do some engineering to dedicate an annunciation panel for the same. Annunciating panel comprises of the following:

- Hooter
- Push button switches with lights for visual indication.

Hooter provides the audio alerts for alarms. Push button switches with visual indication provide the option of acknowledging the alarms for the panel, on seeing what the indication is against.

Alarms that are of much lesser priority can be configured as an event; events are dedicated for certain operations as per system design. For example, a user logging in with a certain privilege is logged as an event or if a new batch for manufacturing is started it is generated as an event; this is as per system configuration and user has no control over it. Alternately, noncritical process or equipment malfunction generating annoying alarms can be marked as only to be reported as events. Also, there is an advanced feature available in case of nuisance alarms. These alarms can be hidden for certain period to avoid unnecessary attention. The groups of alarms that are a result of equipment malfunction are known to the operators; certain permissions enable alarm-hiding features to keep these alarms out of regular list of alarms.

Alarms are mainly classified into two categories:

- Process alarms
- Diagnostic alarms

Process alarms represent a malfunctioning in the control loop. When a process is upset it could be because of multiple reasons; when a valve goes back it will initialize the PID controlling it, resulting in a mode initialization; under this condition a process alarm is seen. This kind of process upsets are reported under process alarms. PV in the following example means process variable, when there is an upset in process that is represented as HiHi (High High)/Hi alarms/Lo alarm/LoLo (Low Low alarm). This means that the process variable is varying between extreme limits on the high limit and the low limit; therefore, if the flow rate is more than the high limit and HiHi limit mentioned for the loop then that alarm is seen; same applies for LoLo limits as well (Figure 3.14).

For example, if the cable connecting to an operator console is disconnected then the cable bad alarm is shown. This is the diagnostic representation of equipment health and reported under separate category of diagnostic alarms. The following is a sample illustration of diagnostic alarms (Figure 3.15).

Observe that the last alarm in the list shows a cable fault being reported. Every alarm generates an event whereas event does not generate an alarm. An event is representative of multiple happenings in the system. The following is a representation of events (Figure 3.16).

There is no representation of priority for events, like in alarms because they are intended for informative purpose. The events logging is more than alarms because numerous operations are logged periodically for the purpose of providing information to users on demand, to understand what is happening in the system and when no alarms are present.

Messages are another feature that is key to advising certain actions to users when needed. Messages are also present in the station along with alarms but in a different page. An alarm is defined on the basis of abnormal condition of process. A message is used to make user aware of certain operation

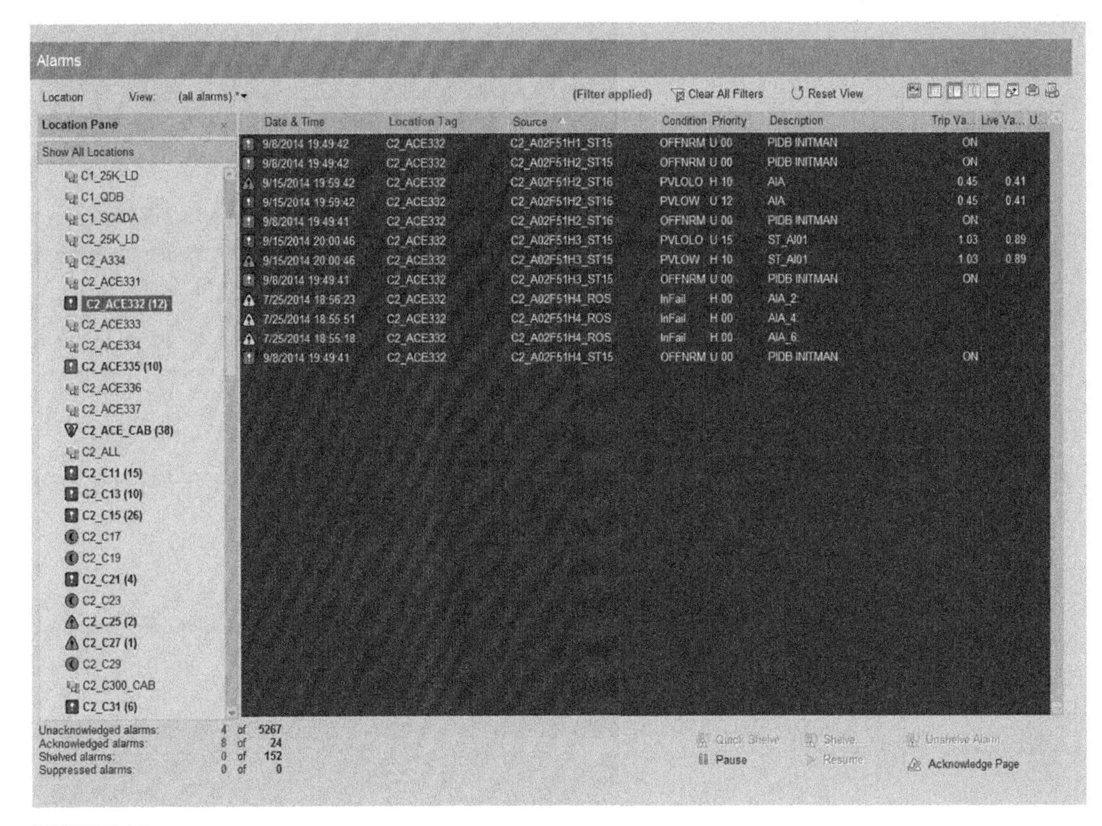

FIGURE 3.14

Process alarm display

in the plant for and it can recommend the action to be taken. Since these messages can be customized they can be clearly maintained in a way equipment is communicating with user. For example, if a valve got damaged then the upstream block such as PID does not function in the regular mode and generates an alarm. Although alarm is useful, a user can also be passed on a message saying that valve in X area malfunctioned. Please check for spares in y department.

However, there is a flipside to the alarms, events, and messages. Like history it is important to maintain the record of events; in fact when a process glitch is seen in trend it is events that help users understand what went wrong with the process. Hence, they play a very vital role in diagnosis. But again maintenance of event records comes at the cost of disk space. Therefore, if there are too many events, alarms, and messages over a period, the system is likely to get overloaded. Hence, users are advised to limit their rate of alerts by the DCS manufacturer.

3.8.4.1 Engineering guidelines for alarms

- Ensure that displays are available to view disabled and inhibited alarms. Operators need to be aware of specific abnormal plant conditions that are not annunciated because the alarm states are disabled or inhibited/hidden. Operators have easy access to a list of alarms that are hidden without

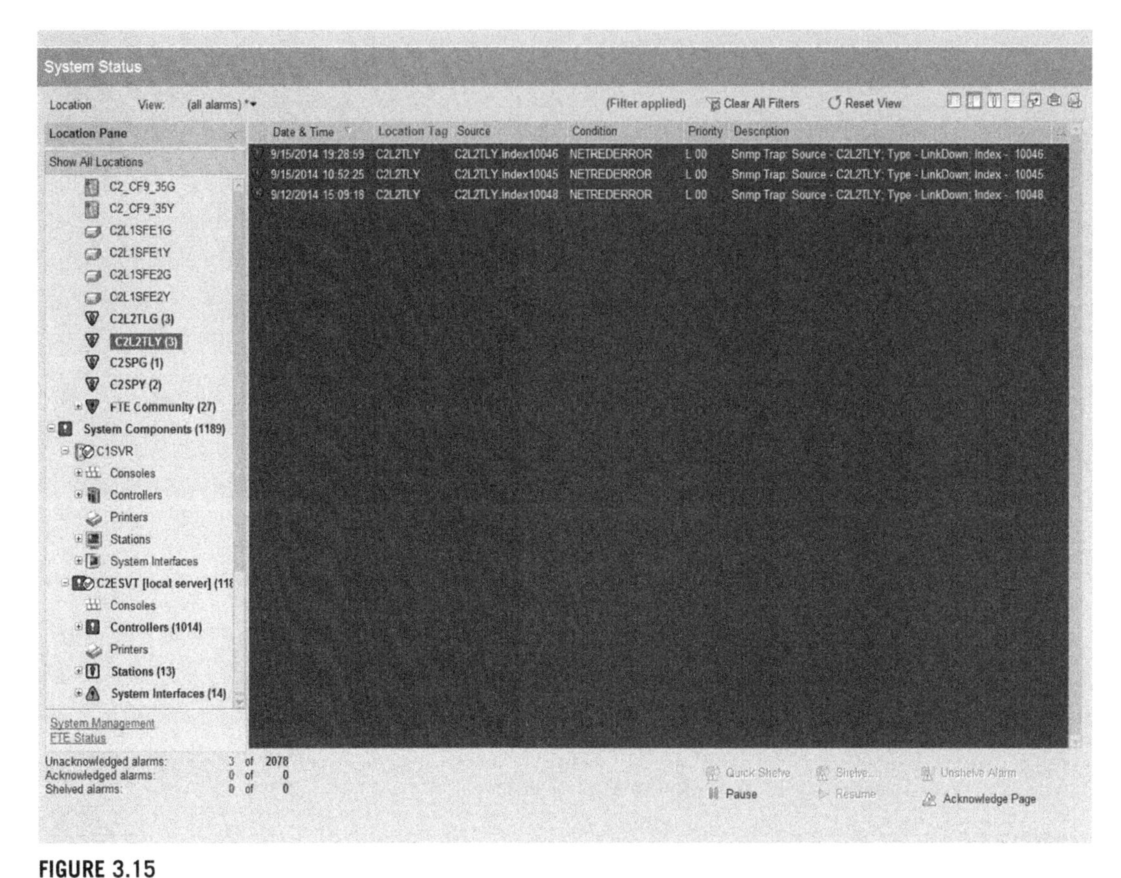

FIGURE 3.15

Diagnostic alarm display

a printout. If operators can see a list of all these alarms in a single view, they can easily perceive the abnormal situation. It is advisable to ensure that this list is actually reviewed periodically.

- Use a formal process to define and maintain alarm configuration. A poorly defined alarm configuration scheme may confuse and interfere with the operator's ability to handle significant disturbances. Hence, other alarm annunciation practices provide less value if a formal process does not ensure that the alarms are meaningful and appropriate to their annunciation.
- Use appropriate integration of DCS and hardwired panel alarms. The alarm presentation in the DCS is consistent with the presentation in a hardwired panel system. If the implementation duplicates alarms in both locations, the operator does not have to acknowledge alarms in both locations.
- Provide information on alarm configuration settings that deviate from the designed values. A mechanism to review deviations in the alarm configuration settings periodically ensures that operators are appropriately aware of temporary changes in the alarm configuration and reduce the possibility of oversight in returning the values to their designed settings in a timely fashion.
- Provide access to alarm rationalization information. Quick online access to this kind of information, especially for infrequent or uncommon alarms, improves both the speed and accuracy of operator responses.

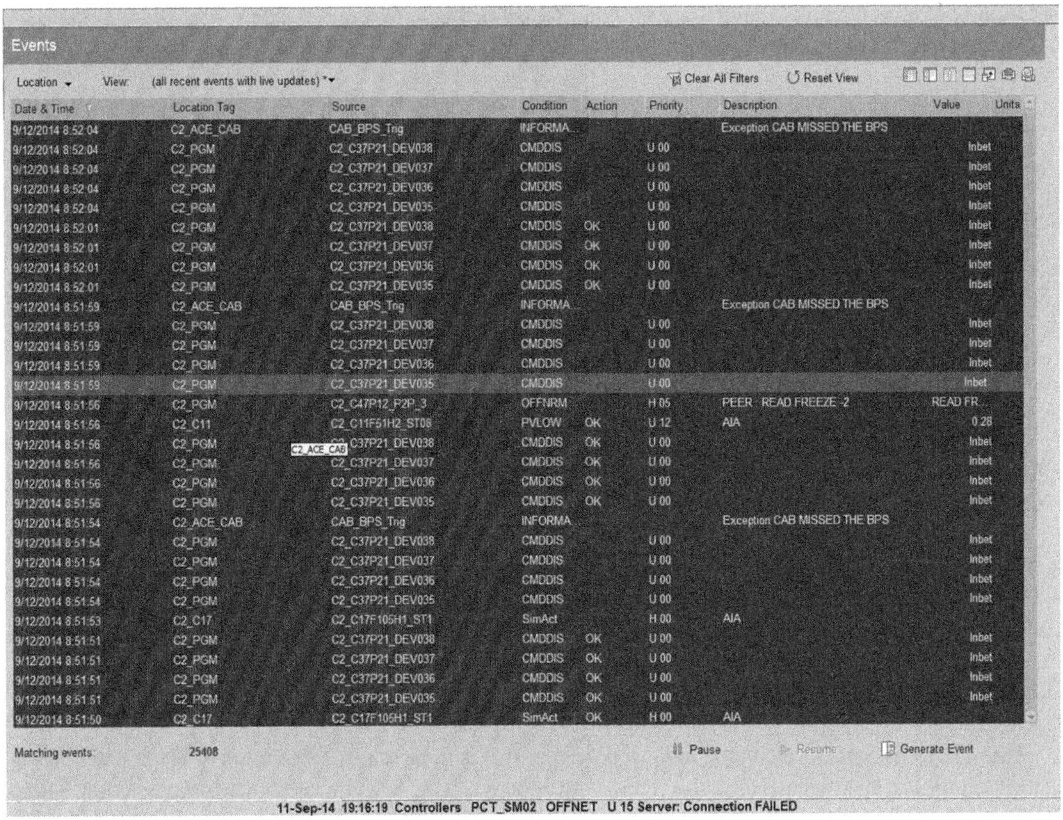

Events

Location View: (all recent events with live updates)

Clear All Filters Reset View

Date & Time	Location Tag	Source	Condition	Action	Priority	Description	Value	Units
9/12/2014 8:52:04	C2_ACE_CAB	CAB_BPS_Trig	INFORMA...			Exception CAB MISSED THE BPS		
9/12/2014 8:52:04	C2_PGM	C2_C37P21_DEV038	CMDDIS		U 00		Inbet	
9/12/2014 8:52:04	C2_PGM	C2_C37P21_DEV037	CMDDIS		U 00		Inbet	
9/12/2014 8:52:04	C2_PGM	C2_C37P21_DEV036	CMDDIS		U 00		Inbet	
9/12/2014 8:52:04	C2_PGM	C2_C37P21_DEV035	CMDDIS		U 00		Inbet	
9/12/2014 8:52:01	C2_PGM	C2_C37P21_DEV038	CMDDIS	OK	U 00		Inbet	
9/12/2014 8:52:01	C2_PGM	C2_C37P21_DEV037	CMDDIS	OK	U 00		Inbet	
9/12/2014 8:52:01	C2_PGM	C2_C37P21_DEV036	CMDDIS	OK	U 00		Inbet	
9/12/2014 8:52:01	C2_PGM	C2_C37P21_DEV035	CMDDIS	OK	U 00		Inbet	
9/12/2014 8:51:59	C2_ACE_CAB	CAB_BPS_Trig	INFORMA...			Exception CAB MISSED THE BPS		
9/12/2014 8:51:59	C2_PGM	C2_C37P21_DEV038	CMDDIS		U 00		Inbet	
9/12/2014 8:51:59	C2_PGM	C2_C37P21_DEV037	CMDDIS		U 00		Inbet	
9/12/2014 8:51:59	C2_PGM	C2_C37P21_DEV036	CMDDIS		U 00		Inbet	
9/12/2014 8:51:59	C2_PGM	C2_C37P21_DEV035	CMDDIS		U 00		Inbet	
9/12/2014 8:51:56	C2_PGM	C2_C47P12_P2P_3	OFFNRM		H 05	PEER : READ FREEZE -2	READ FR...	
9/12/2014 8:51:56	C2_C11	C2_C11F51H2_ST08	PVLOW	OK	U 12	AIA	0.28	
9/12/2014 8:51:56	C2_PGM	C2_ACE_CAB C37P21_DEV038	CMDDIS	OK	U 00		Inbet	
9/12/2014 8:51:56	C2_PGM	C2_C37P21_DEV037	CMDDIS	OK	U 00		Inbet	
9/12/2014 8:51:56	C2_PGM	C2_C37P21_DEV036	CMDDIS	OK	U 00		Inbet	
9/12/2014 8:51:56	C2_PGM	C2_C37P21_DEV035	CMDDIS	OK	U 00		Inbet	
9/12/2014 8:51:54	C2_ACE_CAB	CAB_BPS_Trig	INFORMA			Exception CAB MISSED THE BPS		
9/12/2014 8:51:54	C2_PGM	C2_C37P21_DEV038	CMDDIS		U 00		Inbet	
9/12/2014 8:51:54	C2_PGM	C2_C37P21_DEV037	CMDDIS		U 00		Inbet	
9/12/2014 8:51:54	C2_PGM	C2_C37P21_DEV036	CMDDIS		U 00		Inbet	
9/12/2014 8:51:54	C2_PGM	C2_C37P21_DEV035	CMDDIS		U 00		Inbet	
9/12/2014 8:51:53	C2_C17	C2_C17F105H1_ST1	SimAct		H 00	AIA		
9/12/2014 8:51:51	C2_PGM	C2_C37P21_DEV038	CMDDIS	OK	U 00		Inbet	
9/12/2014 8:51:51	C2_PGM	C2_C37P21_DEV037	CMDDIS	OK	U 00		Inbet	
9/12/2014 8:51:51	C2_PGM	C2_C37P21_DEV036	CMDDIS	OK	U 00		Inbet	
9/12/2014 8:51:51	C2_PGM	C2_C37P21_DEV035	CMDDIS	OK	U 00		Inbet	
9/12/2014 8:51:50	C2_C17	C2_C17F105H1_ST1	SimAct	OK	H 00	AIA		

Matching events: 25408 Pause Resume Generate Event

11-Sep-14 19:16:19 Controllers PCT_SM02 OFFNET U 15 Server: Connection FAILED

FIGURE 3.16

Events display

In addition to the core design-related recommendations, following are the human engineering recommendations for better perception of displays:

- Use a minimum of color codes consistently across display hierarchy levels. Consistent, distinguishable color codes allow operators to learn the codes and their meanings. For example, an unacknowledged, high alarm could be a brighter, more saturated yellow, which is distinct from an acknowledged high alarm that would be a paler, less saturated yellow. These two "states" of high alarms are distinct from red color–coded emergency alarms, which are either more or less saturated depending on their state of acknowledgment.
- Ensure that color combinations provide acceptable and sufficient contrast. Good contrast reduces the likelihood of eye fatigue and increases readability. Color combinations are composed of individual colors that provide sufficient contrast when objects with those colors are adjacent or layered onto one another, so distinctions can be made between the objects. One color affects another. Adjacent or background colors affect the perceived brightness or shade of a particular foreground color. A neutral color (e.g., medium gray) is often the best background color. Opposite colors paired as background/foreground, such as red and green, can make it difficult for the eye to focus.

- Avoid color combinations that are confusing for colorblind perception. Red/green, green/yellow, and white/cyan color combinations that have the same saturation and brightness levels are avoided.

3.8.5 REPORTS AND LOGS

Reports provide information about the occurrence of multiple events in a specified time frame; this can be autogenerated or on demand. Reports generally depend upon the events database that is generated. The events in database are preserved in generic database such a SQL, when a report is pulled it is presented in a user-readable format.

Logs, on the other hand, are the ones that provide insights into the operations happening internal to the system and their visible consequences in a well-arranged format for getting better insight into the system; they are slightly more sophisticated and often needs skill to interpret them.

Reports normally have lot of data in them and user can customize the way he can generate a report. Customization includes but not limited to

- Generate report on the basis of a location tag.
- Generate reports on the basis of specific condition, descriptions, and so on.
- Generate reports that are periodic in nature.
- Generate reports as and when user requires.
- Generate report in the format defined by the user; this includes lengths of certain fields in report, and kind of events for which report is generated.

3.8.5.1 Supported formats for reports

Users can generate a report in the following areas.

Alarms and Events: As the name indicates the report is generated purely on basis of alarms and events in the system. The significance of such a report is to understand the SOE in case of any process upset. There is a feature called as periodic reporting wherein the user can customize a report to generate report of certain values at user-specified periods. These alarms and events are recorded in the system for a certain period of time. As and when the user wants to have a view on happenings in the system they can generate the reports; although they are not directly available for view in the station, in one way it is analogous to search operation performed in a desktop. You will not know the location of the file/folder; however, its existence can be found in the system by search. Similar to this the alarms might have gone by after acknowledging them or the events might have gone out of station window after the live buffer getting filled, but they do exist on the systems and can be retrieved on demand. At times there are SOEs configured in the system that gives the event information with the resolution of milliseconds in case of critical processes such as turbines. A turbine operates at high speeds and any control mechanism associated thereby has to be very responsive in a very short time. It is difficult to notice the events and analyze them on the fly, especially when there are faulty conditions such as trip. In such cases exclusive reports for the events are dedicated and made available for users to see the sequence that has caused the trip, thereby known as SOE. These alarms- and events-based reports can also be generated by means of scripts by which the user can define own version of detail capture and deploy it in the system. Scripting is capable of extending the available functionality above the traditional capabilities offered to user by the system. Alarm and event handling along with periodic or on-demand request can be termed as traditional reporting because any basic system should have the capability to interact with the system at that level. A sample alarm's event report is shown in Figure 3.17.

However, there are advanced features available in report generation, which as discussed below.

31-Aug-14 08:54:32 RPT003: ESVT : Demanded : Page 1

FTE

Category: All Activity
Event location: On-Line
Asset:
Source: *C2ES*
Description: *FTE*
Operator: Period: 27-Aug-14 08:54:32 to 31-Aug-14 08:54:32

Event file 1: 17-Aug-14 00:00:18 to 31-Aug-14 08:54:20

Time	Source	Condition	Action	Level	Description	Value	Units	Operator
8/27/2014 16:43:56.722	C2EST01.A	NETREDERR~		L 00	FTE device C2EST01 interface ~			
8/27/2014 16:43:56.743	C2EST01.B	NETREDERR~		L 00	FTE device C2EST01 interface ~			
8/27/2014 16:43:56.744	C2EST01	NETREDERR~		H 00	FTE device C2EST01 disjoined ~			
8/27/2014 16:44:44.603	C2EST01	NETREDERR~	OK	H 00	FTE device C2EST01 disjoined ~			
8/27/2014 16:44:44.622	C2EST01.B	NETREDERR~	OK	L 00	FTE device C2EST01 interface ~			
8/27/2014 16:44:44.640	C2EST01.A	NETREDERR~	OK	L 00	FTE device C2EST01 interface ~			
8/27/2014 16:44:44.640	C2EST01.A	NETREDERR~	OK	L 00	FTE device C2EST01 interface ~			
8/27/2014 16:44:44.640	C2EST01.A	NETREDERR~	OK	L 00	FTE device C2EST01 interface ~			
8/27/2014 16:44:44.640	C2EST01.A	NETREDERR~	OK	L 00	FTE device C2EST01 interface ~			
8/27/2014 16:44:44.664	C2EST01.B	NETREDERR~	OK	L 00	FTE device C2EST01 interface ~			
8/27/2014 16:44:44.691	C2EST01.A	NETREDERR~	OK	L 00	FTE device C2EST01 interface ~			
8/29/2014 18:46:29.900	C2EST03.A	NETREDERR~	DISCARD	L 00	FTE device C2EST03 interface ~			
8/29/2014 18:46:29.900	C2EST03.B	NETREDERR~	DISCARD	L 00	FTE device C2EST03 interface ~			
8/29/2014 18:46:29.900	C2ESVTA.B	NETREDERR~	DISCARD	L 00	FTE device C2ESVTA interface ~			
8/29/2014 18:46:29.900	C2ESVTA.A	NETREDERR~	DISCARD	L 00	FTE device C2ESVTA interface ~			
8/29/2014 21:20:41.800	C2EST01.B	NETREDERR~	DISCARD	L 00	FTE device C2EST01 interface ~			
8/29/2014 21:20:41.800	C2EST01.A	NETREDERR~	DISCARD	L 00	FTE device C2EST01 interface ~			
8/30/2014 16:54:09.147	C2EST01.A	NETREDERR~		L 00	FTE device C2EST01 interface ~			
8/30/2014 16:54:09.148	C2EST01	NETREDERR~		H 00	FTE device C2EST01 disjoined ~			
8/30/2014 16:54:09.148	C2EST01.B	NETREDERR~		L 00	FTE device C2EST01 interface ~			
8/30/2014 16:54:10.672	C2EST01	NETREDERR~		H 00	FTE device C2EST01 disjoined ~			Auto Dele~
8/30/2014 16:54:14.170	C2EST02.A	NETREDERR~		L 00	FTE device C2EST02 interface ~			
8/30/2014 16:54:14.171	C2EST02.B	NETREDERR~		L 00	FTE device C2EST02 interface ~			
8/30/2014 16:54:14.172	C2EST02	NETREDERR~		H 00	FTE device C2EST02 disjoined ~			
8/30/2014 16:54:55.061	C2EST02	NETREDERR~	OK	H 00	FTE device C2EST02 disjoined ~			
8/30/2014 16:54:55.078	C2EST02.A	NETREDERR~	OK	L 00	FTE device C2EST02 interface ~			
8/30/2014 16:54:55.078	C2EST02.A	NETREDERR~	OK	L 00	FTE device C2EST02 interface ~			
8/30/2014 16:54:55.095	C2EST02.B	NETREDERR~	OK	L 00	FTE device C2EST02 interface ~			
8/30/2014 16:54:55.095	C2EST02.B	NETREDERR~	OK	L 00	FTE device C2EST02 interface ~			
8/30/2014 16:54:56.438	C2EST01	NETREDERR~	OK	H 00	FTE device C2EST01 disjoined ~			
8/30/2014 16:54:56.456	C2EST01.B	NETREDERR~	OK	L 00	FTE device C2EST01 interface ~			
8/30/2014 16:54:56.459	C2EST01.B	NETREDERR~	OK	L 00	FTE device C2EST01 interface ~			
8/30/2014 16:54:56.480	C2EST01.A	NETREDERR~	OK	L 00	FTE device C2EST01 interface ~			
8/30/2014 16:57:43.000	C2EST03.A	NETREDERR~	DISCARD	L 00	FTE device C2EST03 interface ~			
8/30/2014 16:57:43.000	C2EST03.B	NETREDERR~	DISCARD	L 00	FTE device C2EST03 interface ~			
8/30/2014 16:57:43.000	C2ESVTA.A	NETREDERR~	DISCARD	L 00	FTE device C2ESVTA interface ~			

FIGURE 3.17

Sample Report

Reference Reports: In a system a point might be used in multiple places, and a spurious alarm might be generated in the station; directly looking in the associated control might not yield the reason. There is a probability that the point used in an algorithm is causing the alarm, or the point used in a user-defined custom block led to this alarm.

Reference reports come handy in such situations; a report is generated with specific intention of identifying its reference in multiple areas on the system. This would help the user keep track on the flow and the dependencies on a particular point. This ensures availability of proper information to user when needed and that helps in tracking the process upsets due to certain points with ease.

Data Exchange Reports: Data exchange reporting is a method in which the system interacts with third-party components and generates or provides information to third-party components for better visualization of information. This interaction happens at two levels: one is interaction of an application with another application, and another one is interaction of application with a database.

An example of application interacting with application is the data from DCS being written to spreadsheet and vice versa. Basic reporting structure as stated above does not have the capability to derive a data from the available data, whereas in this scheme if we consider that the DCS application interacts with Excel, then the report is capable of the following tasks:

- Perform calculations on the points; all calculations supported by Excel are supported.
- History can be recovered and used for calculating multiple statistical parameters such as mean, median, standard deviation, and so on.
- Generating graphs on the basis of available data.

This helps the user to make accurate calculations all the time and process it thereafter. This can be customized to provide the information needed for business decisions directly from the process box.

ODBC Reports: ODBC stands for open database connectivity. This is a way of generating reports by querying databases supported by a specific DCS vendor. This is an example of application interacting with database, and following is an illustration of an ODBC reporting structure (Figure 3.18).

The illustration depicts the application pulling data directly from databases; these databases are the commonly used ones, and they can be SQL/Oracle/Access, and so on. The only criterion is proper driver connectivity for the application to interact with the database. The report generation also needs a certain level of user privileges. This privilege restriction is given while defining the report. Apart from generating the reports in the format visible to users, ODBC reports also have a provision to forward these reports in the form of a data to databases at other levels in DCS. These databases have special applications that exclusively process the information obtained, and if there is any abnormality or deviation, it is reported in the form of an SMS/Text message/Email to the configured users.

In addition to above functions, ODBC reports are capable of performing basic functions such as generating process values and history reports; however, configuration is slightly different from Excel data exchange or basic report generation and needs a skilled person to define the reports for extracting the information without overloading the system.

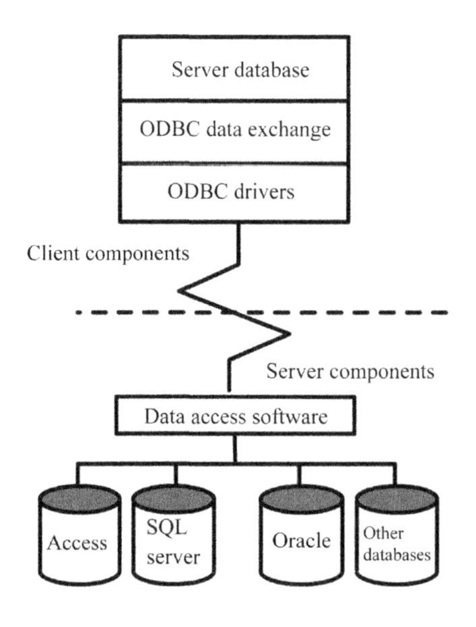

FIGURE 3.18

Sample Data Hierarchy

3.8.6 DIAGNOSTICS

Integrated diagnostics is an important feature of the DCS. The diagnostics cover hardware, redundancy, communications, control, and, to some extent, the software that makes up the DCS. Usually a system alarm is reported on the failure or malfunction of any of these components and the necessary log messages are recorded.

The tests built into the control room equipment are designed to analyze a high proportion of all failures, diagnose the problem, and pinpoint the logical replaceable unit (LRU) or optimum replaceable unit without intervention by the operator or a maintenance technician while the system is online and controlling the process.

3.8.6.1 Redundancy

Redundancy is an important requirement for any critical process control application using DCS. Several DCS redundancies are built at the level of communication media, controllers, I/O cards and I/O communications/connections, and workstations. In very critical systems, where safety is a big concern, it is also possible to take redundant or preferably two out of three voting measurements and discard the defective or inaccurate one during control execution. Redundancy at various levels helps the user to upgrade components online in the control system. However, in critical processes this has to be very carefully planned so that neither safety is compromised nor does the enterprise suffer a production loss.

3.8.7 HISTORICAL DATA

The DCS usually includes the ability to collect batch, continuous, and event data. A centrally defined history database is available for the storage of historical data. The value of any attribute, alarm or any control strategy, alert, or process condition can be recorded in the history database along with its status. In modern control systems, the data values are collected as an integrated feature of the system. Events are collected and time-stamped at their source – in some cases down to a resolution of few milliseconds. Users and layered applications can retrieve the batch, continuous, and event data in a time-ordered fashion. For security reasons, values cannot be edited without leaving behind an audit trail. The engineering tools and operator tools enable selection of points for history storage.

3.8.8 HISTORY AND TREND

History has a special place in process control; a data is historized for the usage of equipment at higher levels and also to have the record at lower levels and see how things are working at the equipment end.

Why equipment at a higher level does need history? In process plants, the production happens as a result of multiple processes and parameters governing these processes. Historians are used to provide the management a view of the history of certain production details. Historians, combined with other applications, can draw data in a presentable format, to understand the performance of specific process or plant and the factors that are boosting or degrading the production. So it is extremely important for making business decisions such as planning finances, production, and so on.

For people working on the shop floor, history is important to understand the process behavior under certain circumstances. If changing an operating procedure or performing an operation is resulting in a process glitch that was unnoticed earlier, how must the user track it? The best way is through history. Normally operating procedures in plants are repetitive in nature. If an abnormality is detected during one such phase, the user can track the earlier instance of such operation and see what happened to the process around that time and confirm if that is a problem and can get to the root of it.

However, the question that is difficult to address is what needs to be historized and how long. There are thousands of tags and parameters, but not everything needs to be historized. Historizing takes data from the parameters and stores it in a specific format in the machines. Historizing all parameters consumes excessive disk space of the machine; therefore, proper discretion has to be exercised against the parameters that need to be captured as history. Historizing can be done at various intervals; it can vary from 1 s to 1 h.

Figure 3.19 is a sample of history configuration.

Along with the number of parameters captured, the interval at which they are captured also determines the disk space usage. Hence, this should also be chosen judiciously. The number of parameters that can be captured increases with increased interval of history; for example, at an interval of 1 s the user may be provisioned to have 1,000 parameters, at 5 s the number may be 5,000 parameters, and at the interval of 1 min the number may be 20,000 parameters thereby reducing the load on the machine capturing the data. There are additional provisions to reduce load on the server and controller at the point of data capture. For example, if there are 20,000 parameters requested to multiple controllers at every 1 min, the server can be overloaded for every 1 min. Hence, an offset can be provided with a certain interval so that the load on server is distributed, For example, consider 20,000 parameters data

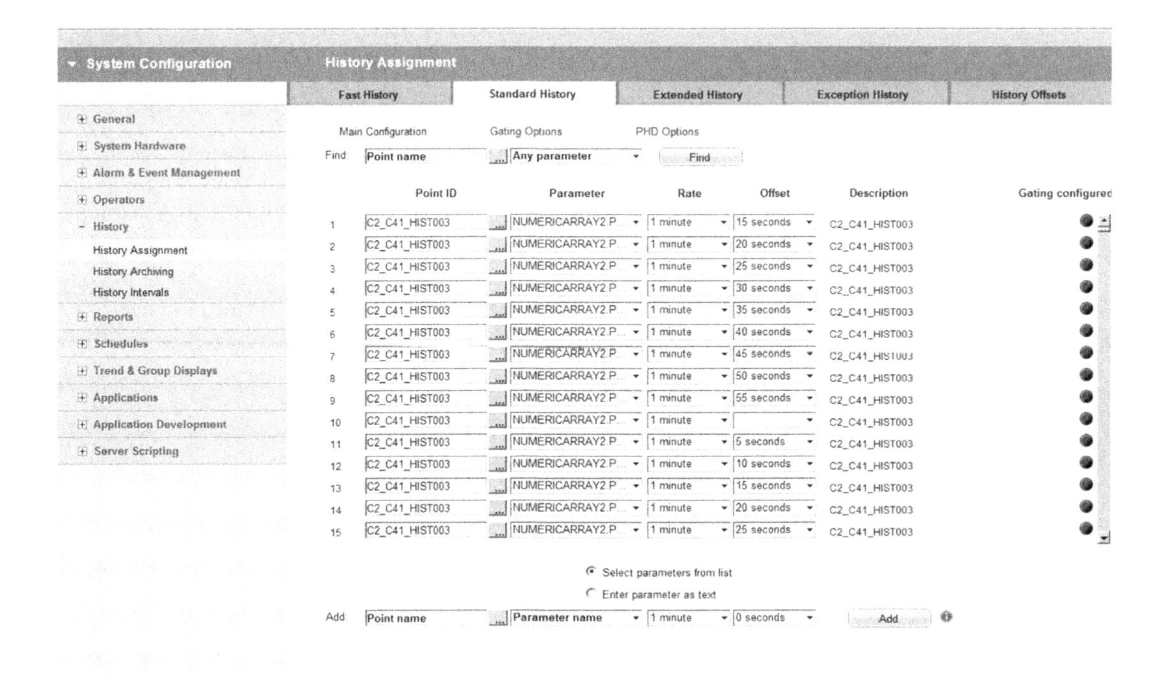

FIGURE 3.19

Sample History Configuration

collection at every 1 min and an offset of 5, 10, 15, 20, 25, 30, 35, 40 s; then each group data is collected at the interval of 1 min wherein each group has information of 2500 items to collect. Therefore, it serves the data collection within 1 min and, in one shot, picks data of only 2500 items, which reduces the load on the machine significantly.

Trends are the face of the process. Whatever is happening in the field can be verified in the form of a trend. Although a mimic diagram is good, it comes with the disadvantage of not being able to track the change throughout. While trend gives the value continuously over a period of time, if a parameter is plotted in trend the live values are shown in record for a certain period of time. Once the period elapses the trend just goes off without being recorded anywhere. That is where history becomes important; if the parameter is captured as history then a trend can be plotted with the help of live values or older values can also be retrieved. Hence, capturing history helps in maintaining the record of elapsed values.

Trend has the following key attributes that help the user.

- It has a number of descriptions by means of which it can be uniquely identified.
- In one trend, the trend of multiple parameters can be configured; this number is limited by vendor and often the parameters specific to a certain area can be joined in one or more trends.
- It has scaling capability wherein the trend can be scaled on the basis of the data being captured. For example, if the raw value of PV is captured and it is >100 it can be scaled accordingly to fit in the trend. All parameters in the trend can be scaled together under common scale or individual scale of each parameter can be applied.
- There is period and interval, wherein period is the tenure for which a trend plot can be viewed. Interval is the frequency at which trend is plotted, that is, value is captured for the plot at that interval. Interval also varied with plot. Period is proportional to interval.

Figure 3.43 diagram illustrates a sample trend with the attributes.

ASM guidelines for trends are as follows:

- Use trend displays when operators need to make decisions about the performance of a variable or variables over time. As discussed earlier, this facilitates the user to understand the upsets in a plant happening over a period of time and act against rectifying them.
- Ensure that trend displays provide the operator with the flexibility to change the display features. The user should be able to customize the trends from the value being read and the interval that is monitored. The color for plotting the trend and the background should also fall in line of better human engineering aspects such as contrast and color.

3.8.9 SECURITY

Security is essential in process control. The DCS system must be able to limit access to the various parts of the control system to authorized people only. This is done by user, plant area, and workstation. Layered applications have to establish a session before they are allowed access into the system. There are several aspects to security as summarized below, in addition to the normal physical security measures:

- Authentication: Access to the DCS for human users and layered applications users is controlled by password-protected user accounts.
- User: A human user of the DCS must have a user account on the system to gain access. All user accounts are named. User accounts have unique names within the scope of a site. All user

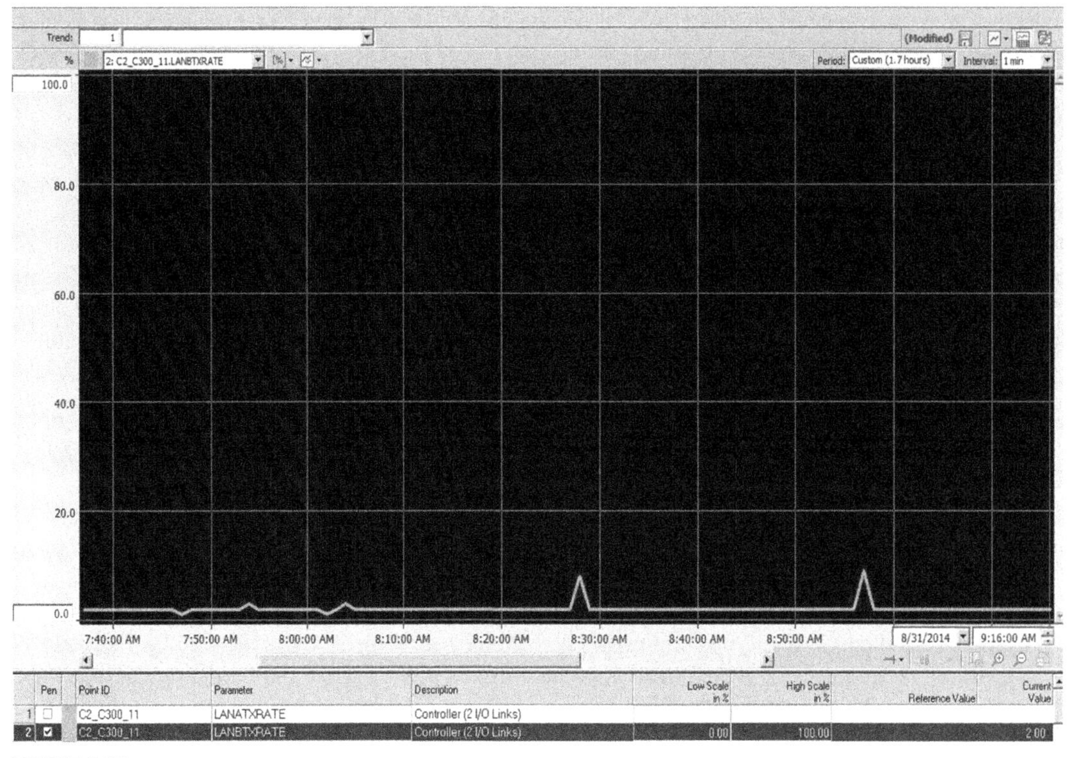

FIGURE 3.20

Sample Trend

accounts have a password, which must be provided in conjunction with the account name in order to start a DCS session.

- Plant area security: A user account can be permitted or denied access to make changes within one or more plant areas within a site. The user account can also be denied access to any of the plant areas. For each plant area where access is permitted, access can be restricted at runtime according to the classification of the runtime attribute data. For each plant area where access is permitted, the ability to make configuration changes can be restricted. A user account can be permitted or denied access to view or modify user account and privilege information. In some systems, it is also possible to enable authorization as an additional security mechanism. In these cases a user, or in some cases several users, need to confirm by password the changing of certain parameters, starting/stopping a batch, and so on. This, in addition to the password, is used for logging into the system.

In addition, access to specific areas of the plant or specific field equipment is governed by permissions from operations personnel and interlocks are created in the control room to ensure safety of plant personnel.

Security for operator interfaces is one of the most important aspects of any control system especially in the current world where we are more prone to cyber threats than ever.

Security on an operator interface can be classified in two ways:

- Platform-based security
- Interface-based security

Platform-based security is based on the privileges provided for the entire platform where the operator interface application resides. In other words, if an application resides on machines running on a Windows 2003/XP/Win7 platform, and if security is provided by users available on the platform then by default they are restricted to the privileges available on the system. These restrictions are local to the machine and can also apply to a group of machines. Local settings related to machine are applied when the machines are in a workgroup; administrator can configure the settings for privileges a user can have. These privileges are applicable mostly toward accessing the applications. When a user is restricted in launching the application the probability of breaching is less.

In case of groups of machines they are controlled on group-based policy restrictions; this is achieved by usage of domain controller. Domain controller can enforce the policies to a group of computers intended to perform certain functions.

Domain controller maintenance has an advantage of concentrating on a single node for maintaining the security. However, on the flipside, adequate care should be taken to protect this domain controller from breach because access to all systems can be controlled from this system and the location domain controller sits is generally above the process control level, which makes it more vulnerable.

Interface-based security is another way of securing the system. Interface security is purely a built-in feature for the interface software. Interface security can be operator based or interface based. Operator-based security works as follows. Similar to Windows users, one can create operator, supervisor, engineer, and manager with a certain level of control. Actually the control is built-in by the application, so if an engineer creates a user with name "x" and assigns a privilege of operator then the interface itself allocates the pages that the operator can act upon and what pages he cannot, along with the actions he is supposed to do. For example, an operator is allowed to view the alarms but he cannot acknowledge them. Similarly, he may not be able to do anything on report generation and configuration as the page is totally restricted.

As the level increases to manager the system becomes more open to operations. Therefore, one has to be specific and cautious while creating the users and providing access. Apart from that there is interface-based security. In interface security the roles are predefined, wherein the access levels are readily defined and the restrictions are provided accordingly. So the fundamental difference between operator-based and interface-based security is the restriction in first case has to be defined by the user. In the second case the security role and user are built in and the user does not have an option to change that. Therefore, based on the convenience one can opt for the type of security.

Security levels in interface are for the whole interface; however, there is a provision to restrict the access to certain components only. For example, if multiple operator interfaces are connected to a single server, then all interfaces can access the entire server. This is not needed in most of the cases; therefore, the user can be restricted by areas. An area or an asset is a name given to describe an area in a plant. When a control is designed, the tags that control the process are classified by an attribute called area or asset. This is an attribute that is loaded along with the tag into the database. Now when an access has to be given for a particular station one can configure the station to have access only to certain areas. This can be explained with a better example. Consider a plant that host turbine, boiler, and conveyor. The tags associated with these areas will be given an area/asset name called namely turbine, boiler, and conveyor.

During the design of interfaces, access to the interface should also be taken care. In case of interface-based security it can be defined on what areas can this interface have access. In such cases no matter who launches the interface, certain areas cannot be accessed by them. This avoids one area's control going to the hands of other area operators. On the other hand, the restriction can be given on per operator basis. An operator/his profile can have access to limited area/assets, thereby allowing only privileged users to have access to certain areas hence restricting unwarranted access. Among the protections above there is no defined rule that one can follow for a better protection. The recommendations are provided by DCS suppliers. However, it is better to have a security restriction that follows the hierarchy of platform-based security -> interface-based security -> area restriction.

Methods of securing system can be done in two ways:

- Authentication
- Encryption

Authentication is used for gaining access to platform/interface. This involves keying in a password and unlocking the needed privilege levels. Common guidelines that are followed for passwords are that they should be alphanumeric with a mix of special characters, with no commonly used words such as Password or Password1.

It is always recommended to have unique passwords for each login for better protection. Often authentication can be the most vulnerable protection because certain people may share the passwords across without realizing the importance of protecting it.

Encryption is purely based on data being made to unread by anyone but the destined device with the help of digital trust certificates. Encryptions in industrial grade are generally proprietary and can be decrypted only by the end devices. A mix of authentication with encryption is a fairly secure system.

And when it comes to security it is always a journey but not a destination, and as a saying in the cyber world goes, the most secured machine in the world is the one that is shutdown; therefore, everyone should pursue security on a continual basis.

3.8.10 INTEGRATION

When a new plant area is added or expanded, the operators of the new area may need some information about the existing plant to provide a coordinated operation. Similarly, the operators of the existing plant may need feedback from the new process area in making decisions on how best to run the balance of the plant. In most cases, only a small fraction of the information in either system must be communicated to support such coordination between these areas. Several techniques are used to integrate systems. The OPC foundation has defined an industry standard for accessing information within a control system. Thus, many control systems provide OPC server capability in workstations designed for interfacing to the plant LAN. Several DCS also communicate using several other hardware connections (RS232, RS485, USB, etc.) and software protocols (Modbus, LON, etc.) including hardwired serial and parallel communication lines.

The most popular techniques for integration use the following technologies, through gateways or other interface techniques given by any vendor.

- OPC
- Profibus and Foundation FieldBus
- Serial

A typical interface diagram is shown in Figure 3.21.

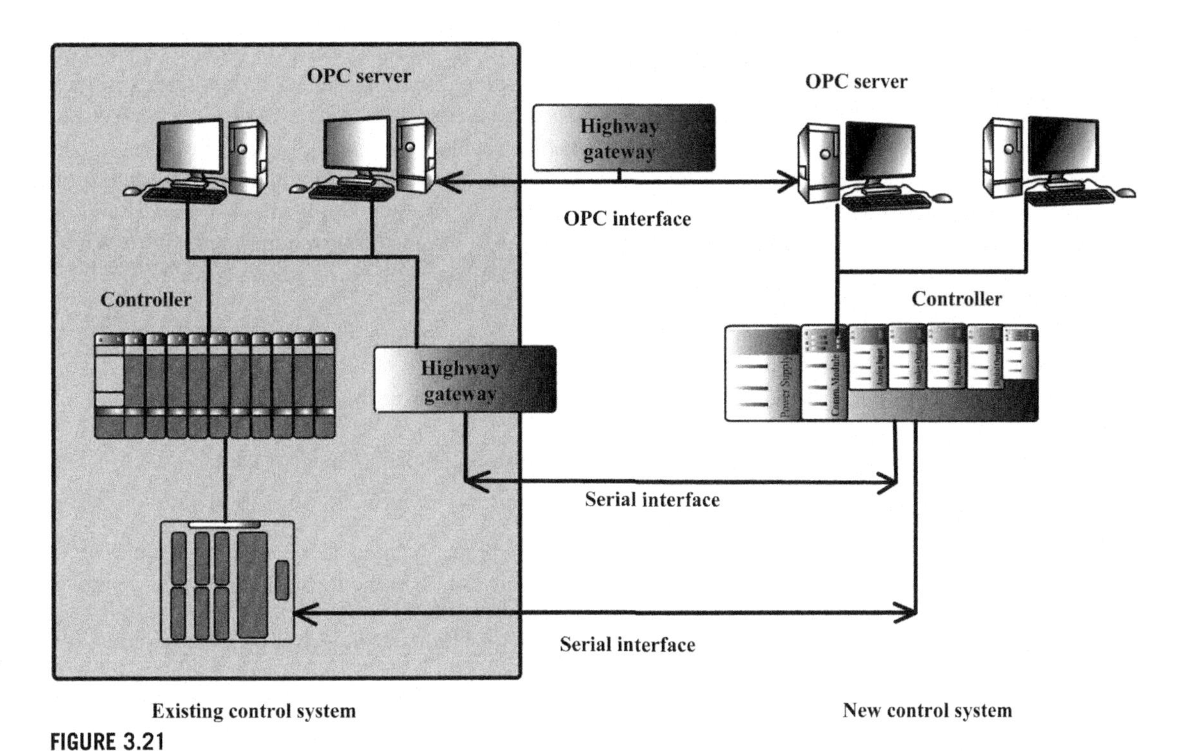

FIGURE 3.21

Integration in automation

3.8.10.1 OPC (OLE for process control)

- OLE means object linking and embedding is an open industrial standard for process control and information system interconnectivity designed on the basis of Microsoft's ActiveX (OLE), Component Object Model (COM), and Distributed COM (DCOM) technology. Data access and exchange by using MS COM/DCOM defines common set of interfaces, methods, and properties for use in industrial automation applications and follows client/server architecture. It is often interfaced through gateway across multiple systems or through customized techniques that a DCS vendor provides.
- OPC follows server/client architecture. Server is a slave device – it responds to requests from a master OPC server and understands how to communicate with the vendor device. Client is a master device – it requests data from a slave. OPC client understands how to communicate to the vendor host server and client. DCOM is the vital design for OPC communication, and offers communication across two subsystems and management consoles. It is used for monitoring applications and I/O specific software drivers and low-level sensors typically use a dedicated hardware interface and protocol.
- OPC is capable of providing real-time sensor data – temperature, pressure, flow, and control parameters such as open, close, run, stop, and status information along with
  - Status of the hardware connection
  - Status of the local software and subsystem

Each OPC server such as data access is a separate object. The data access server provides a window into existing data. Data is accessed by name (a string), which is generally vendor or hardware specific. Data for lists of items can be read explicitly (polled) or subscriptions can be created. It is perfect for merging DCS alarms and events, logging network events, replacing sequence of event recorders, loggers, or any other type device that sends text-based messages. Applications are interested in a subset of the data items (tags) available within the underlying control subsystem. Applications are interested in many different subsets of data items at different times and may have variable requirements for response and resolution. Applications want to be independent of the data structures (or objects) used by the subsystems (i.e., they want symbolic access to the data).

3.8.10.2 Limitations of OPC COM/DCOM DA

- DCOM cannot pass firewalls, thereby making communication over internet impossible.
- Not supporting internet-based HMI.
- Not supporting web-enabled field devices and enterprise-wide management systems.

3.8.10.3 OPC XML DA dominates above limitations

- This facilitates OPC data exchange protocol for communicating data and information seamlessly and transparently between devices independent of OS and networking technology that the specific device employs.
- To develop flexible, consistent rules and formats for exposing plant floor data using XML.
- XML is text-based and can pass through firewalls.
- OPC XML DA (Protocol) OPC- OLE for Process Control XML-extensible markup language DA- data access for horizontal movement of data between control platform but not only for transferring data from DCS/PLC to HMI/SCADA

However, there are other supported standards of OPC used in industrial control.

- HDA control is purely meant for historizing the data that can be retrieved later.
- Alarms and events are purely for the generated alarms and events of the OPC system and only used for better alarm handling capabilities.

However, of late the DCOM dependencies were considered seriously by the OPC foundation and a standard emerged called as OPC UA (unified architecture) following a service-based model architecture that reduces the platform dependency and firewall issues. So, the starting point for the OPC UA discussion relates to the scope of where it will be used. This time, the OPC foundation expanded its horizons enterprise wide.

As a typical example, often HMI application becomes the alarm server and operates on the data from the factory floor; it transposes the data via rule processing, to determine if the data from the factory floor represents some sort of alarm condition that needs to be passed out to the works between different applications. When alarm information needs to be passed between applications that are distributed and connected via the internet, the best technology to do that is by using XML and web services.

A service-oriented architecture no longer uses an OPC client call method (functions) directly on the OPC server, which starts to place limitations on what languages/operating systems can be used.

Instead, just as a web client and web server communicates by exchanging messages, OPC UA does the same thing. This means that an OPC client no longer cares how the OPC server is written, what operating system it runs on, or what parameters its internal functions need to pass. Instead, it sends a well-defined service request that the OPC server listens to and then the server acts on the message. This creates complete independence between OPC clients and OPC servers.

3.8.10.4 Foundation FieldBus

Fieldbus subsystem is the gateway. Fieldbus networks are designed to provide highly reliable bidirectional communication between "smart" sensors and actuators and a control system in time-critical applications. Fieldbus messages contain the readings of several floating-point process variables all sampled at the same time and their respective status (Figure 3.22).

One of the main objectives for fieldbus is to become the digital replacement for analog 4- to 20-mA transmission of process variables in the process industries. Examples of implemented low-speed fieldbus networks include foundation FieldBus H1 and Profibus PA. Due to the differing requirements for process versus discrete installations, fieldbus networks typically have slower transmission rates than device-bus or sensor-bus networks. Another key differentiator is the addition of a user layer, or layer 8, on top of the typical three-layer communications stack. The user layer portion of Foundation FieldBus includes standard and open FBs that can be used to implement distributed field control systems.

3.8.10.5 Serial communication

One of the most popular applications for serial communication is through RS232/RS485 through Modbus communication. Modbus protocol is a messaging structure developed by Modicon in 1979. It is used to establish master–slave (serial)/client–server (TCP) communication between intelligent devices. It is a truly open and the most widely used network protocol. It has been implemented by hundreds of vendors on thousands of different devices to transfer discrete/analog I/O and register data between con-

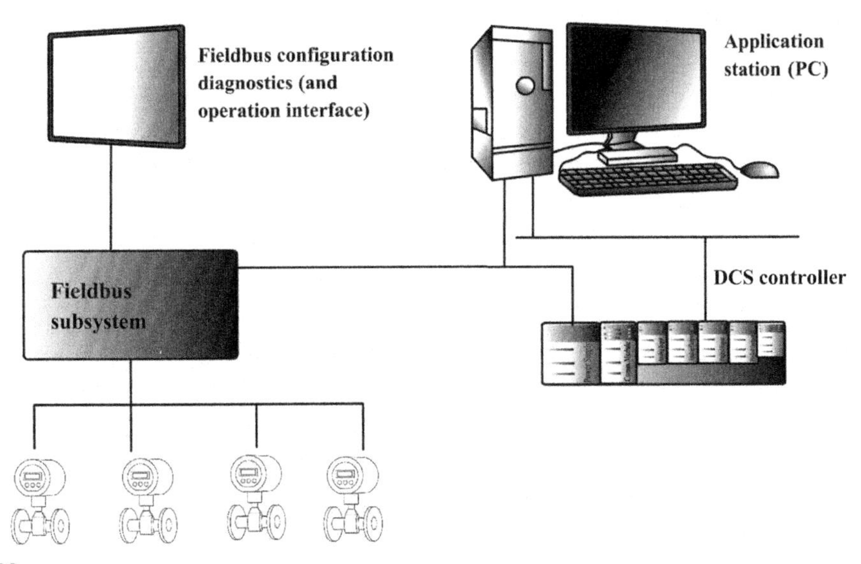

FIGURE 3.22

Foundation Fieldbus System

trol devices. It is used to monitor field devices using PCs and HMIs. Modbus is also an ideal protocol for RTU applications where wireless communication is required.

The physical medium used for Modbus is:

- Modbus serial: RS232 (or) RS485
- Modbus TCP: Ethernet and it uses transmission modes of
  - Modbus RTU – Messages are coded in hexa decimal
  - Modbus ASCII – Messages are coded in ASCII characters (only 0–9 and A–F characters are used)

As per the OSI model Modbus fits in the following layers (Figure 3.23).

As per the Modbus protocols the limitations of stations and other specification are as follows.

Normally as per Modbus protocol each RTU/device is treated as a station; for Modbus serial following is the specification.

Modbus RTU works on master–slave architecture. SCADA is always the master; subsystem is always the slave. SCADA initiates the communication. SCADA sends the request to the slave in Modbus messaging format. Subsystem-Modbus slave responds to the request in Modbus messaging format. Modbus-TCP means that the Modbus protocol is used on top of Ethernet-TCP/IP, the specifications for which are as follows.

SCADA sends the request to the server in Modbus messaging format. Subsystem-Modbus TCP server responds to the request in Modbus messaging format. It uses TCP Port Number-502 for the communication (Figure 3.24).

This client/server model is based on four types of messages:

- MODBUS Request – A MODBUS Request is the message sent on the network by the client to initiate a transaction.
- MODBUS Confirmation – A MODBUS Confirmation is the Response Message received on the client side.
- MODBUS Indication – A MODBUS Indication is the Request message received on the server side.
- MODBUS Response – A MODBUS Response is the Response message sent by the server.

OSI	MODBUS Serial	MODBUS TCP/IP
Application layer	MODBUS application protocol client/server	MODBUS TCP
Presentation layer	NULL	NULL
Session layer	NULL	NULL
Transport layer	NULL	TCP
Network layer	NULL	IP
Data link layer	MODBUS serial line protocol	IEEE 802.3/IEEE 802.2 Ethernet
Physical layer	EIA/TIS-485(232)	10/100 Base T

FIGURE 3.23

OSI Layers in MOdbus

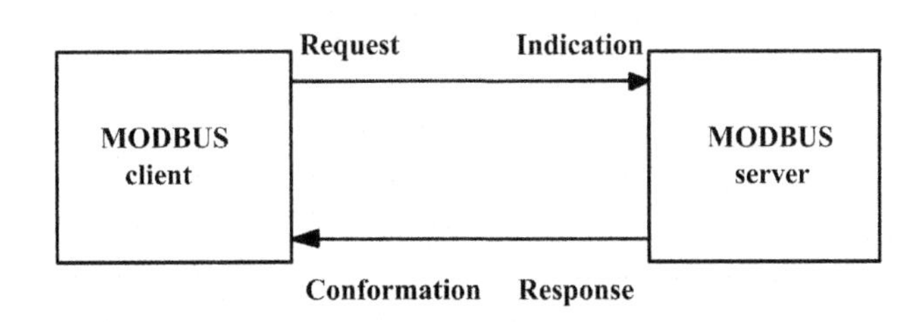

FIGURE 3.24

Serial Modbus Communication

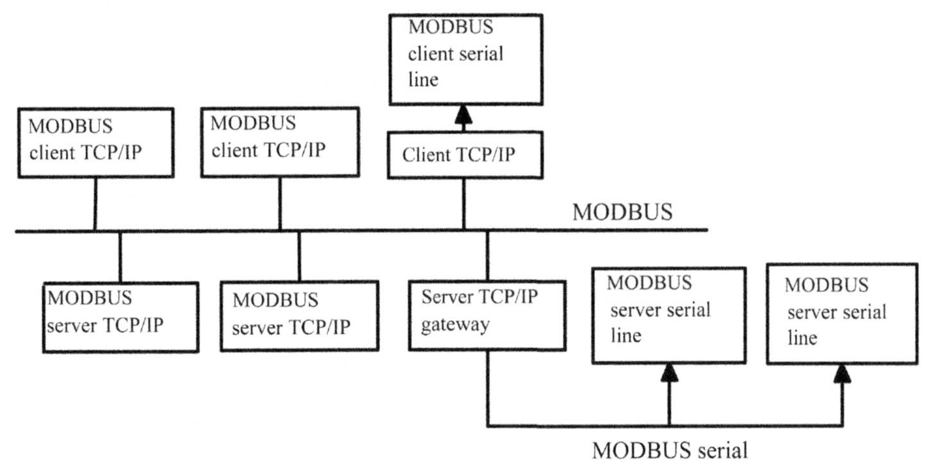

FIGURE 3.25

Modbus on Ethernet

The MODBUS messaging services (client/server model) are used for real-time information exchange:

- between two device applications,
- between device application and other devices,
- between HMI/SCADA applications and devices,
- between a PC and a device program providing online services.

Figure 3.25 shows the architecture of the Modbus client/server TCP communication.

FURTHER READINGS

Barnes, G.F., 1977. Single loop microprocessor controllers. Instrument. Technol. December.
Bibbero, R.J., 1977. Microprocessors in Instruments and Control. John Wiley & Sons, New York.
Lukas, M.P., 1986. Distributed Control Systems. Van Nostrand Reinhold Co, New York.
Studebaker, P.S., 1993. The state of the architecture. Control December.

CHAPTER

BATCH AUTOMATION SYSTEMS

4

4.1 INTRODUCTION

The discipline of process control is an amalgamation of statistics and engineering that deals with architectures, mechanisms, and algorithms for maintaining the output of a specific process within a desired range.

Process control is the mass production of continuous processes such as oil refining, paper manufacturing, chemicals, petrochemical, pharmaceutical, biopharmaceutical, power plants, and many other industries. Extensively used in process-based industries, process control enables automation, with the help of which a small staff of operating personnel can operate a complex process from a central control room.

For example, raising the temperature in a room is a process that has the specific outcome to reach and maintain a defined temperature (e.g., 20°C), kept constant over time. Here, the temperature is the controlled variable. At the same time, it is the input variable since it is measured by a thermometer and used to decide whether to heat or not to heat. The desired temperature (20°C) is the set point. The state of the heater (e.g., the setting of the valve allowing hot water to flow through it) is called the manipulated variable since it is subject to control actions.

Programmable logic controller, or a PLC, is a commonly used control device used to read a set of digital and analog inputs, apply a set of logic statements, and generate a set of analog and digital outputs. Using the example in the previous paragraph, the room temperature would be an input to the PLC. The logical statements compare the set point to the input temperature and determine whether more or less heating is necessary to keep the temperature constant. A PLC output then either opens or closes the hot water valve, depending on whether more or less hot water is needed. Larger more complex systems can be controlled by a distributed control system (DCS) or SCADA system.

Process control systems can be characterized as one or more of the following forms:

- Discrete – Discrete process control is mostly used in manufacturing, motion, and packaging applications. Robotic assembly, such as that found in automotive production, can be characterized as discrete process control. Most discrete manufacturing involves the production of discrete pieces of product, such as metal stamping.
- Continuous – Often, a physical system is represented through variables that are smooth and uninterrupted in time. The control of the water temperature in a heating jacket, for example, is an example of continuous process control. Production of fuels, chemicals, and plastics are good examples of a continuous process. Continuous processes, in manufacturing, are used to produce very large quantities of a product per year (millions to billions of pounds).
- Batch – Some applications require that specific quantities of raw materials be combined in specific ways for particular durations to produce an intermediate or end result. One example is

the production of adhesives and glues, which normally requires the mixing of raw materials in a heated vessel for a period of time to form a quantity of end product. Other important examples are the production of food, beverages, and medicine. Batch processes are generally used to produce a relatively low to intermediate quantity of product per year (a few pounds to millions of pounds).

- Hybrid – Applications having elements of discrete, batch, and continuous process control are often called hybrid applications.

4.1.1 BATCH PROCESS

Batch process is a process that leads to the production of finite quantities of material by subjecting quantities of input materials to an ordered set of processing activities over a finite period of time using one or more pieces of equipment.

4.1.1.1 Batch control

Batch processes are a mixture of continuous and discrete processes. The differences between batch control and continuous control are suitably summarized as

"Batch is very different from a continuous control process. Continuous control is like a tanker; difficult to get running and steer, but once running not a problem to hold on course. Batch is like a fighter airplane; inherently unstable and unpredictable, very responsive, very agile and powerful."

4.1.1.2 Importance and characteristics of batch

A batch process is very useful, as it uses finite quantities, thereby maximizing profits.

Characteristics of a batch process are as follows:

- Liquid-phase reactions.
- Most of the batch chemical reactors are operated at constant temperature or at a constant feed rate.
- A recipe, which consists of a series of time-related steps, is followed in open-loop fashion.
- Temperature, pressure, pH, and viscosity are measured on-line; periodic samples are taken to measure concentration offline.
- Highly flexible – suits the manufacture of a variety of products.
- Exhibits nonlinear phenomena.
- Absence of well-defined operating point like continuous process.

4.1.1.3 Batch industry classification

Based on the consideration of the production paths required for the products, the batch industry is classified into the following:

- Single-product batch plant: A plant which produces the same product in each batch is called single-product batch plant. For example, sugar industry.
- Multiproduct batch plant: In a multiproduct batch plant, multiple products are produced following the same production path.
- Multipurpose batch plant: In a multipurpose batch plant, the same equipment is used for multiple purposes.

- Depending on the complexity of processing sequences employed to produce products we have two subdivisions.
  - Sequential processing plant: Different products follow the same processing sequence.
  - Network-represented processing plant: When production recipes become more complex and/or different products have low recipe similarities, processing networks are used to represent the production sequences in which equipment units are flexibly connected.

4.1.2 SPECIAL OPERATIONS IN BATCH

4.1.2.1 Intermediate storage facility

- Unlimited intermediate storage (UIS): In this case, there is no need to model inventory levels.
- No intermediate storage (NIS): There are no storage tanks available for intermediate materials. However, the materials can be held in the processing unit after the task is finished before they are transferred into the next unit.
- Zero wait (ZW): Relevant intermediate materials are required to be consumed immediately after being produced. Special timing constraints are required to be incorporated into scheduling formulations to address the same.
- Finite intermediate storage (FIS): This corresponds to the most general case.

4.1.2.2 Cleaning operation

- Sequence-dependent cleaning: The type and duration of cleaning required between two tasks depend on the identity and relative order of two tasks. This is more general case of cleaning requirement.
- Frequency-dependent cleaning: In some batch plants, the need for cleaning a particular unit of equipment is expressed in terms of the frequency of their utilization. For example, cleaning after using the equipment for three times, etc.

4.1.3 ARCHITECTURE OF BATCH CONTROL SYSTEMS

Architecture of a typical batch system is given in Figure 4.1. There is no hard rules on the batch control systems architecture and they vary significantly from one vendor to another vendor. But the most common among them is, they utilize the control systems components like sensors and controls and a specialized batch control software is used to configure and operate, report out the batch operations in the central processor.

Detailed overview is given in Figure 4.2. The detailed operations and state of the components and their traversing paths are represented.

4.1.4 STANDARDS RELATED TO BATCH INDUSTRIES

In 1995, the Instrument Society for Automation (ISA) launched the S88.01 and S88.02 standard to standardize the terminology, models, data structures, and functionality in batch control. Hierarchical structure of ISA S88 extends from control of units at the lowest level to management functions at the highest level.

a. ANSI/ISA-S88.01-1995 Batch Control, Part 1: Models and Terminology
b. ANSI/ISA-S88.00.02-2001 Batch Control Part 2: Data Structures and Guidelines for Languages

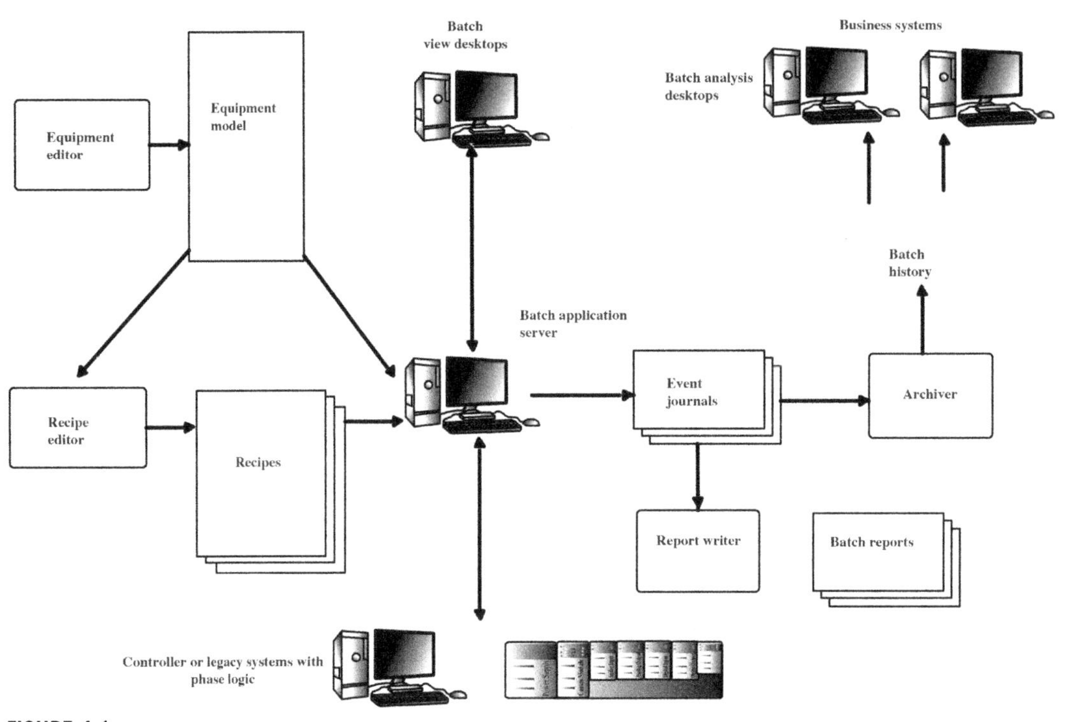

FIGURE 4.1

Architecture of a batch system.

c. ANSI/ISA-S88.00.03-2003 Batch Control Part 3: General and Site Recipe Models and Representation
In 2000 and 2001 ISA launched S95.01 and S95.02, which are advanced versions of the S88 standard and deal with enterprise-control system integration.

d. ANSI/ISA-95.00.01-2000 Enterprise-Control System Integration Part 1: Models and Terminology

e. ANSI/ISA-95.00.02-2001 Enterprise-Control System Integration Part 2: Object Model Attributes

f. PAT – Process Analytical Technologies
In September 2004, the Food and Drug Administration (FDA) issued a "Guidance for Industry" called PAT – A Framework for Innovative Pharmaceutical Manufacturing and Quality Assurance for the Pharmaceutical Industry.
- IEC 61512-1-1998 Batch Control, Part 1: Models and Terminology
- IEC 61131-3 SFC Programming
- FDA GAMP Practices V x.x and cGMP Practices

g. Compliance to the 21CFR Part 11: Electronic Signatures and Data Security
21 Code of Federal Regulations (CFR) Part 11 applies to electronic records and electronic signatures created, modified, archived, retrieved, or transmitted under any records or signature requirement set forth in the Federal Food, Drug, and Cosmetics act, the Public Health Service act, or any FDA regulation.

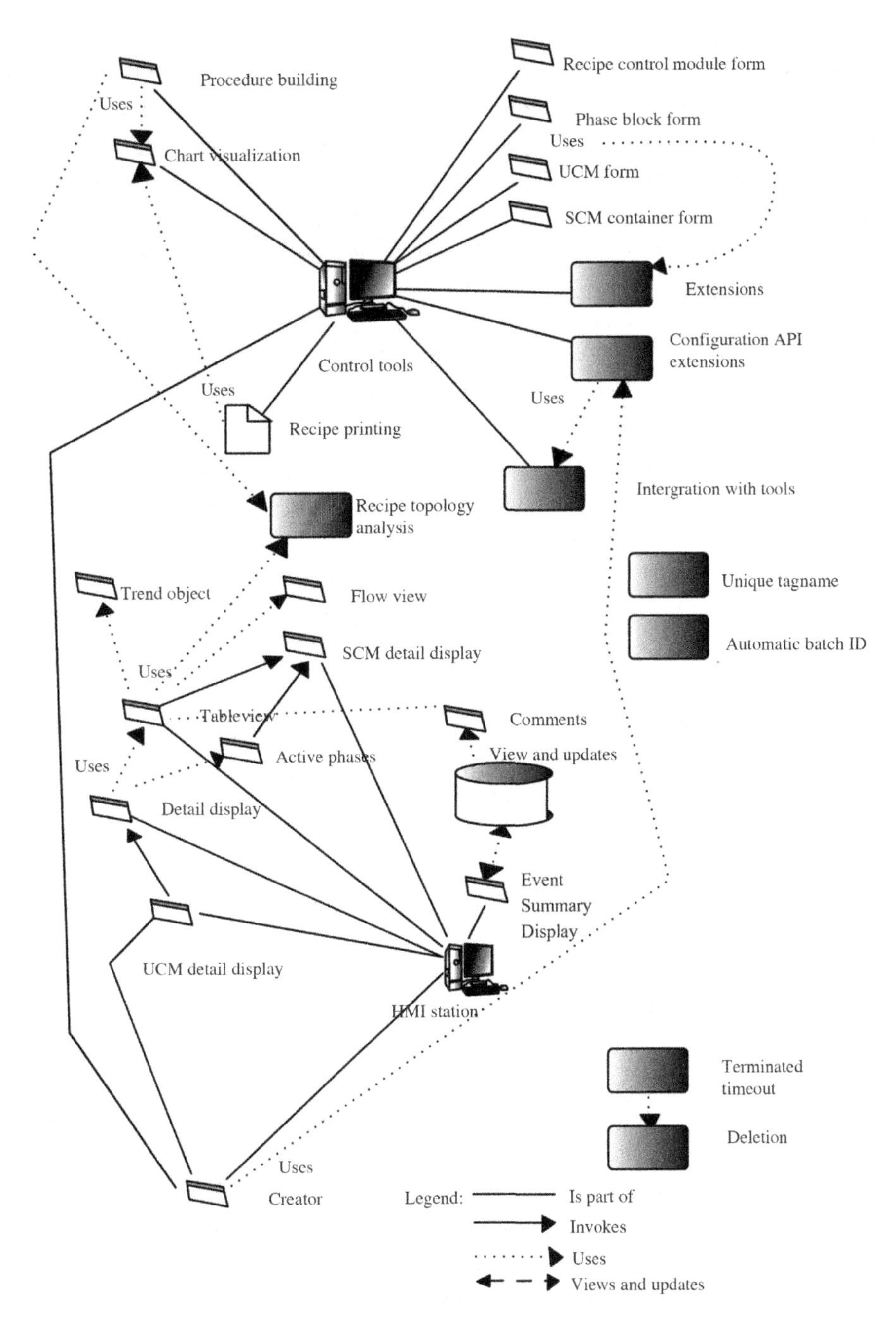

FIGURE 4.2

Example of functional relationships in batch engineering

4.1.4.1 S 88 Batch Control Part 1: Models and Terminology

Introduction to S88

S88 is a standard for batch processing industries, with a focus on product and production. It is applied in discrete and continuous processes that require a certain amount of flexibility. The S88 models and terminology are developed for the control of production processes. A standard does not tell you what you have to do, but rather what you should do in order to be successful. A standard contains common and good knowledge developed with lots of thought and with input from many people with different knowledge and experience. S88 is the standard. SP88 is the committee that wrote it (standards and practices).

The S88 standard is intended for people:

1. Involved in the design, construction, or execution of batch processes.
2. Responsible for the development of batch control and the associated software.
3. Involved in the design and marketing of products in the area of batch control.
   - SP88 mission statement

     "To make the Standard(s) comprehensive enough to be of value and to accomplish its goals, yet flexible enough in the right areas to allow continuous system management to be innovative. They (Standards) must be sophisticated enough to apply to the most complex Batch process, and be able to collapse to fit to the Simplest of them. They must give guidance to achieve understanding and uniformity without being restrictive and cumbersome. They should apply equally as well to the plant that is run manually as to one that is fully automated."

Batch control is a very important aspect of batch manufacturing. For years, companies faced four challenges with batch control:

1. Lack of a universal model existed for batch control
2. Difficulty in communicating the batch processing requirements
3. Engineers found it hard to integrate solutions from different vendors.
4. Engineers and users had a difficult time configuring batch control solutions.

All of these problems led to expensive batch control systems that often failed to meet all the needs of the users and were difficult to maintain. And so, in 1988, ISA formed the S88 committee to address batch control.

Benefits of the S88.01 Standard

The S88.01 standard offers the following benefits:

- Reduces cost of automation
- Reduces life cycle engineering effort
  - Reduces time to market
  - Improves flexibility
  - Improves process quality.
- Enables rapid batch recipe development.

These benefits are made possible by the following:

- Isolating equipments from recipes
- Providing guidelines for abnormal events
- Tracking the historical data

- Making requirements identification easier
- Enabling modularity, thereby improving the ROI
- Making the validations and startups easier and faster.

S88.01: Models and terminology

The S88 standard consists of models and terminology for structuring the production process. The main idea of S88 is to provide a methodology to separate equipment control from the procedure that describes how to make the end product. This makes it possible to use the same equipment to execute different procedures to manufacture different end products.

Physical model

S88.01 provides a physical model, which is a hierarchical model for structuring the physical equipment. The enterprise is the highest level in this physical model. An enterprise can have one or more sites and each site can have one or more area's that can consist of one or more process cells. For batch control and S88, the process cell level and successive levels: the unit, the equipment module, and the control module, are more important.

- Enterprise: The enterprise is responsible for determining the products that must be manufactured, the sites where they must be manufactured, and how they must be manufactured. The enterprise works like an organization and coordinates the operation of one or more sites. These sites may contain areas, process cells, units, equipment modules, and control modules.
- Site: A site is a physical, geographical, or logical grouping that is a component of a batch manufacturing enterprise. The boundaries of a site are usually defined, based on organizational or business criteria rather than on technical criteria. A site may contain several areas.
- Area: An area is a component of a batch-manufacturing site that is identified by physical, geographical, or logical grouping within a site. An area may contain several process cells.
- Process cell: A process cell is a component of an area and is logical grouping of equipment required for production of one or more batches. It defines the span of logical control of one set of process equipment within an area. The existence of a process cell enables scheduling production on a process cell basis, and also enables the design of process cell-wide control strategies, which is particularly useful in emergency situations.
- Unit: A unit is a grouping of equipment modules and control modules. The modules in the unit may be configured as part of the unit or may be acquired temporarily to carry out specific tasks. The unit carries out one or more major processing activities – such as react, crystallize, and make a solution. A unit is capable of performing activities as an independent equipment grouping by combining all necessary physical processing and control equipment required. It is usually focused on a major piece of processing equipment, such as a mixing tank or reactor. To be able to complete the major processing task(s) required of a unit, it physically includes or can acquire the services of all the necessary logically related equipment. Units operate relatively independently of each other.
- Equipment module: The equipment module may be made up of control modules and subsidiary equipment modules. An equipment module may be part of a unit or a stand-alone equipment grouping within a process cell. An equipment module is a functional group of devices capable of carrying out a finite number of specific minor processing activities such as dosing and weighing. It combines all necessary physical processing and control equipment to perform and execute those activities. It is usually centered on a piece of processing equipment such

as a filter. Functionally, the scope of the equipment module is defined by the finite tasks it is designed to carry out.

- Control module: A control module is typically a collection of sensors, actuators, other control modules, and associated processing equipment. The control module is operated as a single entity from the point of view of control. A control module can also contain other control modules. A control module is the lowest level of equipment in the physical model.

In summary, an enterprise can have one or more sites, which can have one or more areas that can consist of one or more process cells that must contain one or more units that contains an equipment module and control module.

Procedural control model

The equipment capabilities are independent of the product. The product information is maintained in recipes. To make a batch of a product, recipes are run in the process cell using the defined units and the unit's equipment phases. The procedure complies with the procedural model, that is, a procedure can be broken down into unit procedures, a unit-procedure into operations, and an operation into phases. A recipe also contains the formula, that is, parameter values useful in the procedure.

- A procedure is the highest level in the procedural control hierarchy, consisting of a sequential set of unit procedures. It defines the overall strategy for making a batch of a product.
- A unit procedure is a sequential set of operations carried to completion on a single unit. A unit procedure is contiguous production sequence acting on one unit only at a time. Only one unit procedure is allowed to be active on a unit at any time. Multiple unit procedures can run concurrently as part of the same procedure, if they are active on different units.
- An operation is a sequential set of phases carried to completion within a single unit. Operations entail taking the material being processed through some type of physical, chemical, or biological change. Only one operation is allowed to be active on a particular unit at a time.
- A phase, the smallest element of procedural control, is responsible for accomplishing process-oriented tasks. Phases perform unique and generally independent, basic process-oriented functions, such as charging an ingredient or agitating a tank. Phases may be deemed, as the workhorses of recipes. All other elements (procedures, unit procedures, and operations) simply group, organize, and direct phases.

In summary, a procedure consists of a sequential set of unit procedure of a sequential set of operation of a sequential set of phases.

Recipe model. There are four types of recipes defined in the S88.01 model. Each recipe type has a different purpose in a company, and is generally created and maintained by different people. As per S88 there are four types of recipes:

- General recipe: The general recipe is used at the company level and is the basis for lower-level recipes. It defines raw materials and their quantities, and the processing required for making the product. General recipes do not include specifics about geography or the equipment required for processing.
- Site recipe: A site recipe is manufacturing site-specific. The site recipe is derived from the general recipe but takes into account the geography of the site (there may be different grades of raw materials in different countries or continents) and the local language.

- Master recipe: Master recipes are derived from site recipes and are targeted at the process cell. A master recipe is of great importance as it is not possible to create control recipes or produce batches without a master recipe. A master recipe takes into account its equipment requirements within a given process cell. It includes the following information categories:
  - Header: A typical header contains the recipe name, product identification number, version number, author, approval for production, and other administrative information.
  - Formula: The formula of a master recipe contains raw materials with their respective target amounts, and process parameters such as temperature and pressure.
  - Procedure: Recipe procedures are built and implemented on equipment (units or classes of units).
  - Equipment requirements: Equipment requirements provide the information necessary to constrain the choice of equipment when implementing procedures.
- Control recipe: A control recipe is used to create a single, specific batch. It starts as a copy of a master recipe and is modified as necessary to create a batch. The modifications may be to account for batch size, characteristics of raw materials on-site or actual equipment to be used. A master recipe may have one or more control recipes on the batch list or in running status.

General recipe may be transformed into a site recipe; site recipes may be transformed into a master recipe; and master recipes may be transformed into control recipes.

Process model. A process model consists of the following:

- Process
- Process stage
- Process operation
- Process action

Linking the physical, procedural control, and process models:

"To accomplish process functionality, you need equipment and a procedure to control that equipment." The process model defined the process required to make a batch of a product. The physical model defines the equipment needed to carry out the batch process. The procedural control model defined the procedure to control the equipment used to carry out the batch process.

Procedural control/equipment relationships necessary to achieve process functionality:

- The procedural control model, in combination with the physical model provides process functionality to carry out a process model.
- Units run unit procedures, operations, and phases to accomplish process stages, process operations, or process actions.
- Phases run against a unit or an equipment module to accomplish a process action.

Linking recipes and equipment control:

The procedural control model links a control recipe procedure with equipment control:

- A recipe procedure must exist in a control recipe procedure.
- An equipment phase must exist in an equipment control.
- Control recipe and equipment control can be linked at phase level, operation level, unit procedure level, or procedure level.

4.1.4.2 S88.02: Batch Control Part 2: Data Structures and Guidelines for Languages

ISA-S88.02 is focused on data that must move into, out of, and within a batch process cell. The on-going work of the SP88 committee as documented in S88.02 Batch Control, Part 2: Data Structures and Guidelines for Languages, draft 14, dated May 1999 defines a method for exchanging batch control information between computer programs or systems.

- Data structure was addressed by: Relational tables for information exchange
- Languages were addressed by: Recipe depiction methodology
- The S88.02 draft standard does not propose to define the internals of batch control, or other related systems.
- It states that only an interface specification is being defined, not the internal requirements for a system using the interface.

Therefore, the local data stores may have different structures and contents in different tools. The exchange tables represent a common format that can be used to exchange data.

The S88.02 exchange tables support four types of batch data:

1. Master and control recipe information
2. Process cell equipment information
3. Schedule information
4. Production information.

S88.02 defines a DATA MODEL as describing batch control as applied in the process industries. DATA STRUCTURES facilitate communication within and between batch control implementation. LANGUAGE guidelines are for recipe implementation.

4.1.4.3 S88.03 Batch Control Part 3: General and Site Recipe Models and Representation

This standard on batch control defines a model for general and site recipes; the activities that describe the use of general and site recipes within a company and across companies; a representation of general and site recipes; and a data model of general and site recipes. It basically talks about the following:

- Recipe types,
- Equipment-independent recipe contents
- Equipment-independent recipe object model
- Equipment-independent recipe representation
- Transformation of equipment-independent recipes to master recipes.

S95

S95.01 and S95.02 are advanced versions of the S88 standard and deal with enterprise-control system integration. S95.01-Enterprise-Control System Integration Part 1: Models and Terminology. S95.02-Enterprise-Control System Integration Part 2: Object Model Attributes.

It is a standard that defines the information flow and business processes between the enterprise systems, ERP, and the control systems, DCS, which run a manufacturing facility.

- There are two published standards:
  - ANSI/ISA-95.00.01-2000 – Models and Terminology.
  - ANSI/ISA-95.00.02-2001 – Object Model.

- There are three other pending draft standards:
  - ANSI/ISA-95.00.03 – Draft 18 – Activity Models of Manufacturing Operations
  - ANSI/ISA-95.00.04 – Draft 7 – Object Model of Activity Models from Part 3.
  - ANSI/ISA-95.00.05 – Draft 1a – Business to Manufacturing Transactions.

4.1.4.4 FDA GAMP practices V x.x and cGMP practices

Good automated manufacturing practice (GAMP) is a trademark of the International Society for Pharmaceutical Engineering (ISPE). The ISPE's guide "The Good Automated Manufacturing Practice (GAMP) Guide for Validation of Automated Systems in Pharmaceutical Manufacture" describes a set of principles and procedures to ensure required quality of pharmaceutical products. One of the core principles of GAMP is to build quality into each stage of the manufacturing process rather than testing quality of a batch of products. As a result, GAMP covers all aspects of production, including raw material, facility and equipment to the training, and hygiene of staff. Standard operating procedures (SOPs) are essential for processes that can affect the quality of the finished product.

A group of pharmaceutical professionals have created the GAMP forum, which is now a technical subcommittee, known as the GAMP community of practice (COP) of the ISPE. The community aims to promote the understanding of the regulation and use of automated systems in the pharmaceutical industry. The GAMP COP organizes discussion forums for its members. ISPE organizes GAMP-related training courses and educational seminars. Several local GAMP COPs, such as GAMP Americas, GAMP Nordic, GAMP DACH (Germany, Austria, Switzerland), GAMP Francophone, GAMP Italiano, GAMP Benelux (Belgium, Netherlands, Luxembourg), and GAMP Japan bring the GAMP community closer to its members in collaboration with ISPE's local affiliates in these regions.

"Current good manufacturing practice" or "cGMP" refer to the practices and systems required to be adapted in pharmaceutical manufacturing, quality control, quality system covering the manufacture and testing of pharmaceuticals or drugs including active pharmaceutical ingredients, diagnostics, foods, pharmaceutical products, and medical devices. GMPs outline the aspects of production and testing that can impact the quality of a product. Many countries have legislated pharmaceutical and medical device companies to follow GMP procedures, and have also created their own modified GMP guidelines corresponding with their legislation. The basic concept of all of the guidelines remain mostly similar to the ultimate goals of safeguarding the health of the patient and producing good-quality medicine, medical devices, or active pharmaceutical products. In the United States a drug may be deemed adulterated if it passes all of the specifications tests but is found to be manufactured in a condition that violates current good manufacturing guidelines. Therefore, complying with GMP is a mandatory aspect in pharmaceutical manufacturing.

All guidelines follow a few basic principles:

- Manufacturing processes must be clearly defined and controlled. All critical processes must be validated to ensure consistency and compliance with specifications.
- Manufacturing processes must be controlled, and any change to the processes must be evaluated. Changes that have an impact on the quality of the drug must be validated as necessary.
- Instructions and procedures must be written in clear and unambiguous language (good documentation practices).
- Operators must be trained to carry out and document procedures.

- Records must be made, manually or by instruments, during manufacture that demonstrate that all the steps required by the defined procedures and instructions were taken and that the quantity and quality of the drug is as expected. Deviations must be investigated and documented.
- Records of manufacture (including distribution) that enable the complete history of a batch to be traced must be maintained in a comprehensible and accessible form.
- The distribution of the drugs must minimize any risk to their quality.
- A system must be available for recalling any batch of drug from sale or supply.
- Complaints about marketed drugs must be examined, the causes of quality defects investigated, and appropriate measures taken with respect to defective drugs and to prevent recurrence.

GMP guidelines are not prescriptive instructions on how to manufacture products. They are a series of general principles that must be observed during manufacturing. When a company is setting up its quality program and manufacturing process, it is the company's responsibility to determine the most effective and efficient quality process to fulfill GMP requirements.

Multiunit recipe control

The regulatory nature of GMP manufacturing facilities requires a high degree of focus on product quality, system validation, and manufacturing records. While these objectives can be met with manual operations or unit-based automation using PLCs, a multiunit batch recipe control approach based on the S88 model using an integrated system including data management provides highly significant benefits over the unit-based PLC approach. The DCSs these days can result in dramatic improvements in the way batch automation is implemented, validated, and executed.

When planning automation for a life sciences project, a decision needs to be made if multiunit S88 batch automation will be implemented. It is often viewed as a simpler proposition to implement unit-based automation using only PLCs that are delivered with the equipment skids. The advantages of multiunit, batch automation needs to be considered when making this decision. The following benefits should be evaluated when considering whether a process should include full recipe automation:

Product quality: Operations that depend on people for executing manual recipes are subject to human variability. How precisely are the operators following the recipe? Processes that are sensitive to variations in processing will result in quality variation. Full recipe automation that controls most of the critical processing operations provides very accurate, repeatable material processing. This leads to very highly consistent product quality.

Improved production: Many biotech processes have extremely long cycle times (some up to 6 months), and are very sensitive to processing conditions. It is not uncommon for batches to be lost for unexplained reasons after completing a large portion of the batch cycle time. The longer the batch cycle time and the more sensitive production is to processing conditions, the more batch automation is justified. Imagine losing a batch of very valuable product because the recipe was not precisely followed!

Process optimization: Increasing the product yield can be done by making small changes in processing conditions to improve the chemical conversions or biological growth conditions. Manual control offers a limited ability to finely implement small changes to processing conditions due to the inherent lack of precision in human control. Conversely, computers are very good at controlling conditions precisely. In addition, advanced control capabilities such as model predictive control can greatly improve process optimization. This results in higher product yield and lower production cost. This consideration is highly relevant to pilot plant facilities where part of the goal is to learn how to make the product.

Record keeping: A multiunit recipe control system is capable of collecting detailed records as to how a batch was made and relates all data to a single batch ID. Data of this nature can be very valuable for QA reporting, QA deviation investigations, and process analysis.

Safety: Operators spend less time exposed to chemicals when the process is fully automated as compared to manual control. Less exposure to the process generally results in a safer process.

GMP pilot plant automation strategies: Most major GMP manufacturing plants today will include DCS-class automation because the benefits stated previously can clearly justify the costs. Pilot plants and small manufacturing facilities should also consider value propositions for full recipe control and data management.

The economics of automation investment have changed: One perception of DCS is that it is expensive and hard to justify in a smaller facility. Times have changed. The starting price of a typical "DCS" hardware and software package became less. With the introduction of scalable technology and systems, smaller applications can be much more economically addressed. Digital systems are now available starting at tens of thousand dollars and can be economical for small manufacturing, pilot plants, and even research facilities.

Reducing capital requirements: Many of the new biotech companies are now bringing their first products through regulatory approval and into manufacturing. Initially, manufacturing is required to support clinical trials, and pilot plants are often used to meet these production requirements. One advantage in adopting batch automation and data management capabilities in the pilot plant is that the facility will be better able to meet manufacturing needs after the new drug is approved and introduced to the market. A pilot plant's production capacity may not be an important consideration during clinical trials. But if the pilot plant is used to satisfy production demands after the drug is approved, maximizing the equipment's capacity will be strategic and can delay the requirement to invest more capital for a new major manufacturing plant.

Improving time to market: Once the production requirements of a new drug outgrow the capacity of the pilot plant, the product will have to move from the pilot plant to the manufacturing plant. Adopting the same automation strategy in the pilot plant and the manufacturing plant can reduce the time and cost of the transition from the pilot plant to the manufacturing plant. The batch recipes and control strategies will have been developed in the pilot plant, and the time and cost for developing automation for the manufacturing plant will be greatly reduced.

Production record keeping: GMP manufacturing requires that records be kept to document significant processing events and quality indications. Regulators place a high value on accuracy and completeness of production records, and failure to maintain adequate records can result in an enforcement action. Quality record keeping can be a challenge in a manual record-keeping environment, and automated batch record keeping is justified for the following reasons:

- Failure to collect required information might result in scrapping a perfectly good batch of product.
- Good record keeping can provide information needed to explain processing deviations and result in saving a batch that otherwise may need to be scrapped.

- Electronic records provide quick access to information needed for QA release of products resulting in faster release of batches and less QA resources required to review production reports and quality data.
- Electronic record keeping is more accurate and complete, reducing the risks of regulatory enforcement action.

A good batch historian should be able to collect records for a production run to include the following information:

- Product and recipe identification
- User-defined report parameters
- Formulation data and relevant changes
- Procedural element state changes (operations, unit procedures, procedures)
- Phase state changes
- Operator changes
- Operator prompts and responses
- Operator comments
- Equipment acquisitions and releases
- Equipment relationships
- Campaign creation data (recipe, formula values, equipment, etc.)
- Campaign modifications
- Campaign execution activity
- Controller I/O subsystem events from the continuous historian
- Process alarms
- Process events
- Device state changes.

4.1.4.5 21CFR Part 11: Compliance to the 21CFR Part 11 Electronic Signatures and Data Security

21CFR Part 11 applies to electronic records and electronic signatures that are created, modified, archived, retrieved, or transmitted under any records or signature requirement set forth in the Federal Food, Drug, and Cosmetics act, the Public Health Service act, or any FDA regulation.

According to cGMP regulations in parts 210 and 211 of the CFR, the U.S. FDA calls for maintaining strict records during the manufacture and inspection of products manufactured under its control. These records were originally created and submitted only on paper, but with computerization of inspection systems, electronic record keeping and submittal became more common.

In 1997 the FDA issued the final rule for 21CFR Part 11, taking into account comments from human and veterinary pharmaceutical companies, and biological products, medical device, and food interest groups. The rule provides criteria for "acceptance by FDA, under certain circumstances, of electronic records, electronic signatures, and handwritten signatures executed to electronic records as equivalent to paper records and handwritten signatures executed on paper."

Part 11 is divided into three subparts, dealing with the scope of the ruling: general (including definitions and implementation), electronic records, and electronic signatures. Within subpart B, electronic records, sections 11.10 and 11.30 define the requirements for closed and open systems. This document addresses how these requirements apply to PPT VISION's IMPACT machine vision system. As per Part 11, it is not the equipment manufacturer's sole responsibility to meet the requirements. Those "who use closed systems to create, modify, maintain, or transmit electronic records are also required to employ procedures and controls designed to ensure the authenticity, integrity, and, when appropriate, the confidentiality of electronic records …"

Subpart A – General Provisions
- 11.1 – Scope.
- 11.2 – Implementation.
- 11.3 – Definitions.

Subpart B – Electronic Records
- 11.10 – Controls for closed systems.
- 11.30 – Controls for open systems.
- 11.50 – Signature manifestations.
- 11.70 – Signature/record linking.

Subpart C – Electronic Signatures
- 11.100 – General requirements.
- 11.200 – Electronic signature components and controls.
- 11.300 – Controls for identification codes/passwords.

Class-based engineering (CBE)

In almost all processes, there are many identical pieces of equipment, instruments, logical sequences, and recipes. Additional equipment items may be very similar or have only small differences. An implementation of CBE allows one-time configuration of identical or similar equipment within a unit, control and sequential logic, as well as recipes; then it allows for "replication" of the subsequent instances. The advantage is the need to design, implement, test, and document only once. This results in a greatly reduced cost of validation. This CBE method is significantly different from merely copying configuration code. Copying configuration code does not eliminate the need to completely test and validate the subsequent instances of the code, since it still is different code that potentially could be changed. In addition, a change in the original must be made manually to all instances of the copy. CBE uses the concept that there is only one instance of the code that is referenced by subsequent equipment, control logic, sequence logic, and recipes. Since only one set of code exists for the identical or similar elements, it only needs to be validated once. The following provide details regarding the CBE used in DCS.

Module classes: Equipment modules and control modules are the lowest two levels of the S88 physical model. Control modules consist of sensors and other control modules that together perform a specific task. Control modules perform regulatory or state control over their constituent parts. Equipment modules consist of equipment and control modules that together perform a minor processing task. Equipment and control module classes allow generic regulatory controls to be created and validated. Then instances are easily created from the module class. An example of a module class might be called On–Off valve. The class creates the necessary interface with the valve, as well as any logic that would tell the valve when to open and close. When instances of On–Off valve are created, they are given specific I/O channel references and individual tag names.

Phase classes: A phase is the detailed logical sequence defined by the S88 procedural model such as fill tank. It is very common for identical sequential logic to be used on many pieces of equipment in a process. A phase class allows a logical sequence to be written and validated once and reused on all equipment that requires the use of that logical sequence. For example, tanks within a process will likely require a fill tank phase. The phase class allows a fill tank phase to be created once and used with any tank that requires a fill tank phase. From tank to tank, only the specific tag name of the instrumentation will change. The phase class is written with generic tag names (aliases) and when the phase is run on any specific tank, an alias table is defined as part of the tank configuration. The same configuration code is used to fill all tanks. This configuration code is validated once with only the alias tables having to be validated on each tank. This is different from *copying* code. If code were copied, all copies would have to be tested and validated, since there is no way to be sure changes were not made.

Unit classes: The unit class allows the configuration and validation of a generic unit equipment type with multiple instances of the unit equipment type created for all the specific equipment of that type. For example, a unit class could be media prep tank. This unit class is defined including unit parameters, phase, and module classes that may operate on the type of unit and the instrumentation alias names. When this is completed, specific instances of media prep tanks are created simply by defining the tag names for the units and modules, referencing the alias names to actual tags, and enabling the phases that will run on the instances of the equipment class.

Recipe formulas: Many times the recipes for different products are logically identical but differ in ingredient amounts and values of processing parameters (temperature, agitation time, etc.). The recipe formula allows the development and validation of a single recipe and allows variations of the recipe by defining recipe formulas. This allows a single recipe configuration to easily make different products. The basic recipe is developed and validated once, with the different formulas requiring only the recipe parameters to be validated. This decreases the total number of recipes that have to be configured and validated.

Challenges of a batch process: Some of the major challenges faced in a batch process are:

- Control:
  - Dynamic and asynchronous operations.
  - A combination of discrete and continuous control.
  - Need for constant tracking of process variable changes.
  - Managing the change in process variables over a wide range, exposing the nonlinearities inherent in the process dynamics.
  - Time-invariant linear models often fail to describe process dynamics adequately, rendering traditional linear control techniques ineffective.
- Scheduling and optimization:
  - Flexibility in equipment assignment to tasks.
  - Diversity:
    - Products (single, multiple)
    - Grades (single, multiple)
    - Path (single/parallel streams, multipath)

- Product-dependent processing path.
- Hybrid topology of batch and semicontinuous units.
- Difficulties in solution of the scheduling problems in a multipurpose plant.
- Multiplicity of processing tasks in equipment.
- Quality:
  - Batch-to-batch variation in product yield and quality.
  - Need to align the batch operation to the requirements of total process network.

4.1.4.6 IEC 61131: Programming Languages of Batch Control Systems

Different programming languages as per IEC 61131-3

- Ladder diagram
- Function block diagram
- Structured text
- Instruction list
- Sequential function chart (SFC)

Ladder diagram

Ladder logic is a programming language that represents a program by a graphical diagram based on the circuit diagrams of relay logic hardware. It is primarily used to develop software for PLCs used in industrial control applications. Programs in this language resemble ladders, with two vertical rails and a series of horizontal rungs between them, and hence the name ladder.

Ladder logic is widely used to program PLCs, which require sequential control of a process or manufacturing operation. Ladder logic is useful for simple but critical control systems or for reworking old hardwired relay circuits. As PLCs became more sophisticated, it has also been used in very complex automation systems. Often the ladder logic program is used along with a human machine interface (HMI) program operating on a computer workstation.

Function block diagram

A function block diagram describes a function between input variables and output variables. It is a block diagram, where a function is described as a set of elementary blocks. Input and output variables are connected to blocks by connection lines. An output of a block may also be connected to an input of another block.

Inputs and outputs of the blocks are wired together with connection lines, or links. Single lines may be used to connect two logical points of the diagram:

- An input variable and an input of a block
- An output of a block and an input of another block
- An output of a block and an output variable.

The connection is oriented, meaning that the line carries associated data from the left end to the right end. The left and right ends of the connection line must be of the same type.

Multiple right connections also called divergence can be used to transfer information from its left end to each of its right ends. All ends of the connection must be of the same data type.

Structured text

Structured text mostly used to program PLCs. It is a high-level language that is block structured and syntactically resembles Pascal. The variables and function calls are defined by the common elements so that different languages can be used in the same program. It also supports complex statements and nested instructions:

- Iteration loops (REPEAT-UNTIL; WHILE-DO)
- Conditional execution (IF-THEN-ELSE; CASE)
- Functions (SQRT(), SIN())

Instruction list

Instruction lists are also designed for PLCs. It is a low-level language and resembles assembly. The variables and function call are defined by the common elements so different languages can be used in the same program.

Program control (control flow) is achieved by jump instructions and function calls (subroutines with optional parameters).

Sequential function charts

SFC is a graphical languagethat provides a diagrammatic representation of sequences and flowcharts. It is based on the French Grafcet (IEC 848). The main components of SFC are as follows:

- Steps with associated actions
- Transitions with associated logic conditions
- Directed links between steps and transitions

Steps in an SFC diagram can be active or inactive, where actions are only executed for active steps. A step can be active for one of two motives: (1) it is an initial step as specified by the programmer and (2) it was activated during a scan cycle and not deactivated since.

Steps are activated when all prior steps are active and the connecting transition is superable (i.e., its associated condition is true). When a transition is passed, all steps above are deactivated at once and after all steps below are activated at once.

4.1.5 SEQUENTIAL AND PROCEDURAL CONTROL

In most of the control systems, with its new controller level batch execution software, in combination with batch software, provides the batch processing industries with increased functionality and ease of use. The batch software consisting of the Sequential Control Module (SCM), the Recipe Control Module (RCM), the phase function block, and the Unit Control Module (UCM), forms the heart of the system. Working together, they provide full execution of unit procedures for multiple recipes, in the manual, semiautomatic, or fully automatic mode within a controller. This includes the ability to collect recipe data at the controller level. In addition, the server batch execution software works with the RCMs to control unit-level operations and perform traditional phase-level control.

Controls in System can be designed to execute sequential and procedural process activities such as start up, shut down, react, and crystallize in addition to other complex procedures. In all instances, controls interact with one or more control modules and SCMs. DCS systems typically have containers called control module (CM) in which discrete and continuous devices are configured for specific control functionality. Control modules can be configured to control a PID loop, discrete valves, accumulated flows, and so on. It is important to understand the functions of a control module for creating effective sequential and procedural control strategies.

4.1.6 STRUCTURE OF AN SCM/RCM

The SCM is a container composed of one or more handlers. Each SCM handler can contain one or more transitions, steps, and/or synchronization blocks. Each RCM handler can contain one or more phases, transitions, steps, and/or synchronization blocks.

- A transition block waits for specific process conditions to be true. When the logical statement formed by a transition's conditions and logic gates is true, the SCM and RCM execute the next step, phase or synchronization block.
- A step block executes output expressions that change process parameters and/or provide instructions to the operator.
- A synchronization block enables the SCM and RCM to execute parallel paths contained within the handler.
- A PHASE block is used in a RCM procedure to monitor and control the execution of a linked equipment module that can be implemented as an SCM.

A handler is an empty space to which transitions, steps, and synchronization blocks can be added and configured as a sequential set of actions designed to carry out the activity required by the process operation. While only one handler can actively execute, exception handlers can be waiting for specified process conditions. When the specified process condition occurs, the main handler's process actions cease execution and the exception handler (Abort, Stop, Hold, and Interrupt) begins its sequential set of process actions.

A handler must have exactly one invoke transition. The invoke transition is not a mandatory entry or gating condition for commanded handler invocation. It enables the definition of additional criteria for handler invocation, but it can never prevent handler invocation.

Configurable handlers fall into the following categories:

- Check for normal exception handling and execution of normal process actions prior to Main.
- Main for primary process actions.
- Interrupt for normal exception handling and processing of Interrupt process actions.
- Hold for abnormal exception handling and processing of Hold process actions.
- Restart for normal exception handling and execution on return from SCM/RCM Hold.
- Stop for abnormal exception handling and processing of Stop process actions.
- Abort for abnormal exception handling and processing of Abort process actions.

4.1.6.1 States of an SCM/RCM

The state of an SCM/RCM is directly related to current handler activity as demonstrated in Table 4.1.

4.1.6.2 SCM/RCM commands

- **Start:** This command orders the procedural element to begin executing the normal.
- **Running logic:** This command is only valid when the procedural element is in IDLE state.

Table 4.1 Different states of SCM/RCM

SCM/RCM State	Handler Activity
Checking	Check Handler executing
Idle	Check Handler complete
Running	Main Handler executing
Complete	Main Handler complete
Holding	Hold Handler executing
Held	Hold Handler complete
Interrupting	Interrupt Handler executing
Interrupted	Interrupt Handle complete
Restarting	Restart Handler executing
Restarted	Restart Handler complete
Stopping	Stop Handler executing
Stopped	Stop Handler complete
Aborting	Abort Handler executing
Aborted	Abort Handler complete
Inactive	Edit Handler executing
Validated	Edit Handler complete with errors identified

- **Stop:** This command orders the procedural element to execute the Stopping logic. This command is valid when the procedural element is in Running, Holding, Held, or Restarting state.
- **Hold:** This command orders the procedural element to execute the Holding logic. This command is valid when the procedural element is in the Running or Restarting state.
- **Restart:** This command orders the procedural element to execute the Restarting logic to safely return to the Running state. This command is only valid when the procedural element is in the Held state.
- **Abort:** This command orders the procedural element to execute the Aborting logic. The command is valid in every state except for Idle, Completed, Aborting, and Aborted.
- **Reset:** This command causes a transition to the Idle state. It is valid from the Complete, Aborted, and Stopped states.
- **Resume:** This command orders the procedural element to transition to Validated state or a Single Step mode to resume execution.

4.1.6.3 SCM/RCM state/command interaction

It is important to understand the SCM/RCM states and conventions for transitioning from one state to another for proper planning of sequential control. Figure 4.3 illustrates the conventions for transitioning between states and the interaction between SCM/RCM STATE and the SCM/RCM COMMAND parameter. COMMAND changes the SCM/RCM from one STATE to another.

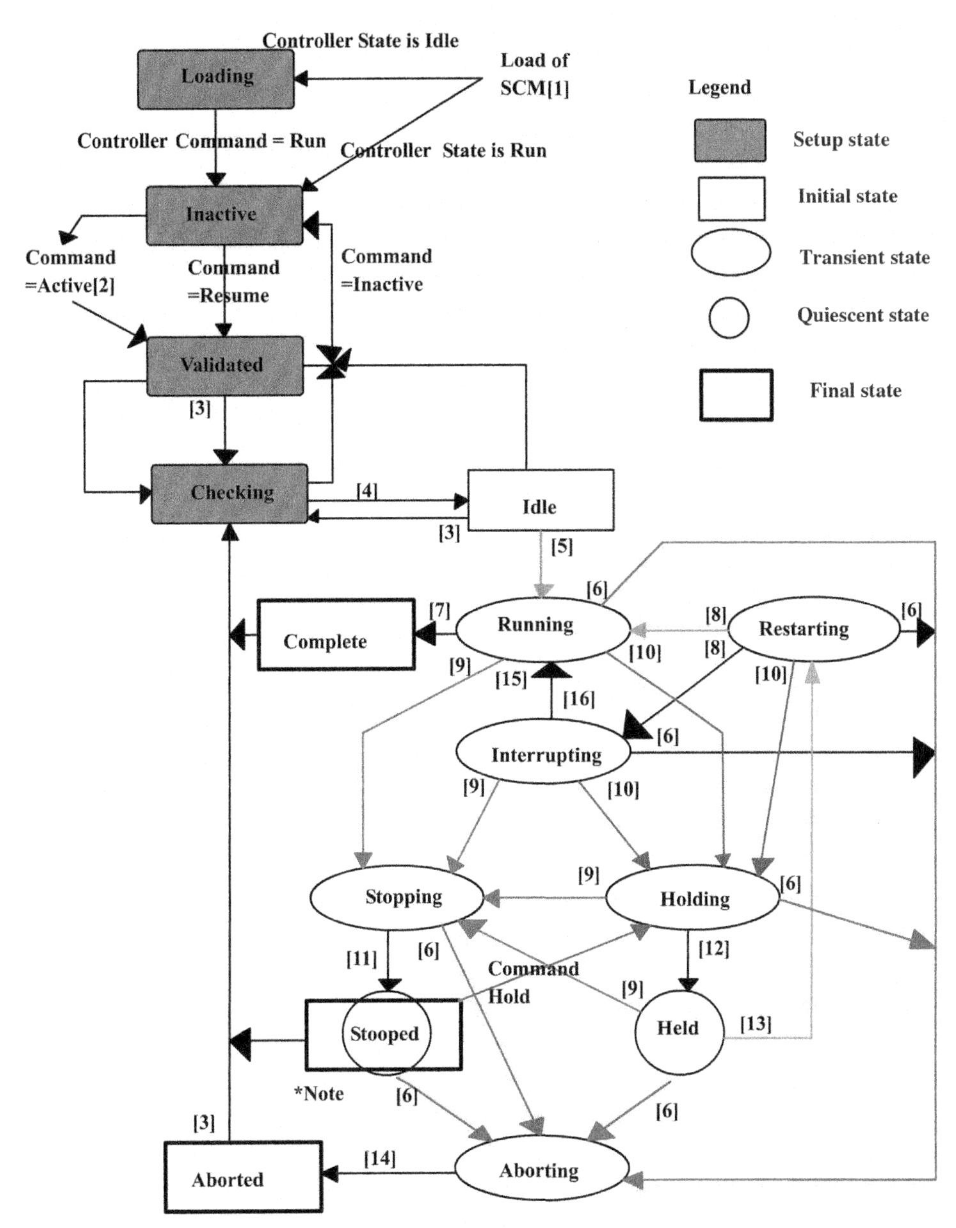

FIGURE 4.3

Example state diagram of a batch process

4.1.6.4 Typical state diagram for procedural elements

Where a "COMMAND =" is shown in the diagram, the transition must be commanded.

The explanations for the notes [ ] in the overview diagram of SFC state and command interaction are as follows.

Explanation

1. SFC STATE after load depends on the Controller STATE.
2. Active command is a macro command that internally issues Resume and Reset commands.
3. Transition to Checking is by Reset Command or Invoke Condition.
4. Transition to Idle is on completion of Check Handler.
5. Transition to Running is by Start Command or Invoke Condition.
6. Transition to Aborting is by Abort Command or Invoke Condition.
7. Transition to Complete is on completion of Main Handler.
8. Transition to Running/Interrupting is on completion of Restart Handler.
9. Transition to Stopping is by Stop Command or Invoke Condition.
10. Transition to Holding is by Hold Command or Invoke Condition.
11. Transition to Stopped is on completion of Stop Handler.
12. Transition to Held is on completion of Hold Handler.
13. Transition to Restarting is by Restart Command or Invoke Condition.
14. Transition to Aborted is on completion of Abort Handler.
15. Transition to Interrupting is by Interrupt Command or Invoke Condition.
16. Transition to Running is on completion of Interrupt Handler.

FURTHER READINGS

ANSI/ISA-88.01-1995, Batch Control Part 1 Models and Terminology.

ANSI/ISA-88 series of batch control standards.

WBF.org – "The Forum for Manufacturing and Automation Professionals" (formally World Batch Forum). A wealth of papers and tutorials on batch related topics.

www.batchcontrol.com – Technical information with a sense of humor.

http://www.batchcentre.tudelft.nl/ – A batch knowledge center maintained by Delft University in the Netherlands.

Parshall, J.H., Lamb, L.B., 2000. Applying S88: Batch Control from a User's Perspective. ISA.

Fleming, D.W., Pillia, V., 1999. S88 Implementation Guide: Strategic Automation for the Process Industries. McGraw-Hill.

Fisher, T.F., Hawkins, W., Forthcoming. Batch Control Systems, 2nd ed. ISA.

CHAPTER

FUNCTIONAL SAFETY AND SAFETY INSTRUMENTED SYSTEMS

5

5.1 FUNCTIONAL SAFETY: AN INTRODUCTION

The purpose of this chapter is to introduce the concept of functional safety and give an overview of the safety instrumented system and the international standard International Electrotechnical Commission (IEC) 61508/IEC61511.

This chapter gives an informal definition of functional safety; describes the relationship between safety functions, safety integrity, and safety-related systems; gives an example of how functional safety requirements are derived; and lists some of the challenges in achieving functional safety in electronic or programmable systems. Section 3 gives details of IEC 61508, which provides an approach for achieving functional safety. The section also describes the standard's objectives, technical approach, and parts framework. It explains that IEC 61508/IEC61511 can be applied as is to a large range of industrial applications and yet also provides a basis for many other standards.

5.2 WHAT IS FUNCTIONAL SAFETY?

We begin with a definition of safety. The functional safety implies freedom from unacceptable risk of physical injury or of damage to the health of people, either directly or indirectly, as a result of damage to property or to the environment. Functional safety is part of the overall safety that depends on a system or equipment operating correctly in response to its inputs.

For example, an overtemperature protection device, using a thermal sensor in the windings of an electric motor to de-energize the motor before they can overheat, is an instance of functional safety. However, providing specialized insulation to withstand high temperatures is not an instance of functional safety (although it is still an instance of safety and could protect against exactly the same hazard).

5.3 SAFETY FUNCTIONS AND SAFETY-RELATED SYSTEMS

Generally, the significant hazards for equipment and any associated control system should be identified by the specifier or developer via a hazard analysis. The analysis determines whether functional safety is necessary to ensure adequate protection against each significant hazard. If so, then it has to be taken into account in an appropriate manner in the design. Functional safety is just one method of dealing with hazards, and other means for their elimination or reduction, such as inherent safety through design, are of primary importance.

The term safety-related is used to describe systems that are required to perform a specific function or functions to ensure that the risks are kept at an accepted level. Such functions are, by definition, called safety functions. Two types of requirements are necessary to achieve functional safety:

- Safety function requirements (what the function does)
- Safety integrity requirements (the likelihood of a safety function being performed satisfactorily).

The safety function requirements are derived from the hazard analysis, and the safety integrity requirements are derived from a risk assessment. The higher the level of safety integrity, the lower the likelihood of dangerous failure.

Any system, implemented in any technology, which carries out safety functions, is a safety-related system. The safety-related system may be separate from any equipment control system or may be included within it. Higher levels of safety integrity necessitate greater rigor in the engineering of the safety-related system.

5.4 EXAMPLE OF FUNCTIONAL SAFETY

Consider a machine with a rotating blade that is protected by a hinged solid cover. The blade is accessed for routine cleaning by lifting the cover. The cover is interlocked so that whenever it is lifted an electrical circuit de-energizes the motor and applies a brake. In this way the blade is stopped before it could injure the operator.

In order to ensure that safety is achieved, both hazard analysis and risk assessments are necessary.

1. The hazard analysis identifies the hazards associated with cleaning the blade. For this machine it might show that lifting of the hinged cover more than 5 mm should not be possible without the brake activating and stopping the blade. Further analysis could reveal that the time for the blade to stop must be 1 second or less. Together, these describe the safety function.
2. The risk assessment determines the performance requirements of the safety function. The aim is to ensure that the safety integrity of the safety function is sufficient to ensure that no one is exposed to an unacceptable risk associated with this hazardous event.

The safety integrity of the safety function will depend on all the equipment that is necessary for the safety function to be carried out correctly, that is, the interlock, the associated electrical circuit, and the motor and braking system.

The harm resulting from a failure of the safety function could be large injuries to the operator's hand or just a bruise. The risk also depends on how frequently the cover needs to be lifted, which may be many times during daily operation or may be less than once a month. The level of safety integrity required increases with the severity of injury and the frequency of exposure to the hazard.

To summarize, the hazard analysis identifies what has to be done to avoid the hazardous event, or events, associated with the blade. The risk assessment gives the safety integrity required of the interlocking system for the risk to be acceptable. These two elements, "what safety function has to be performed" – the safety function requirements – and "what degree of certainty is necessary that the safety function will be carried out" – the safety integrity requirements – are the foundations of functional safety.

5.5 LEGISLATION AND STANDARDS

Different types of documentation serve different purposes. As the following list explains, some documentation is internally driven and some is externally driven. One must understand the differences between legislations, regulations, standards, and recommended practices.

The American Model of Legislation and Standards:

- ***Legislations:*** Laws enacted by elected officials, federal, state, or local. Clean air act amendments and local building codes are examples of such legislations.
- ***Regulations:*** Rules, which have the weight of law, through the delegation of the authority. Occupational safety and health administration (OSHA), environmental protection agency (EPA), petroleum and natural gas regulation board (PNGRB), and oil industry safety directorate (OISD) are few examples for such regulations.
- ***Standards:*** Consensus of an industry group on lowest level of engineering acceptable. IEC61508 and IEC61511 are few examples.
- ***Recommended practices:*** Recommendations of an industrial group such as API RP14C "Recommended Practice for Analysis, Design, Installation and Testing of Basic Surface Safety Systems for Offshore Production Platforms" and API RP14G "Fire Prevention and Control on Open Type Offshore Production Platforms."

Standards and recommended practices can have the weight of the law but only when "Incorporated by Reference."

The ANSI/ISA-84.00.01-1996 was the first standard developed by the ISA SP84 committee to address the need to increase safety by systematically reducing the process risk. The standards are updated approximately every 5 years. The new version adopts the IEC 61511 standard and adds the original 1996 version of ANSI/ISA 84 as a "grandfather clause." IEC 61508 is an umbrella standard released by the International Electrotechnical Committee in 1998 that is based on ANSI/ISA 84 and applies to all industries. The IEC 61508 standard is viewed as the document that vendors follow to receive certification for safety integrity level (SIL) suitability ratings for products and system components. IEC 61511 was released in 2003 and developed specifically to address functional safety in the process industry. This document is used by end-users to successfully implement system safety throughout the entire life cycle of a system. Please note that if a system was implemented prior to the update of the standards in 2004 and has been operating in a safe manner without an increase in risks or safety incidents, then the system can be "grandfathered" and does not need to comply with the ANSI/ISA 84 2004 update. However, once significant changes are made to the system or if a safety issue arises, the system will have to comply with the most recent version of the standards. It is important to note that none of the standards are prescriptive, but are performance based. The standards present guidelines for best practices, but do not identify procedures for specific implementation (Figure 5.1).

5.5.1 THE EUROPEAN MODEL OF LEGISLATION AND STANDARDS

Directives: A directive is a legislative act of the European Union, which requires member states to achieve a particular result without dictating the means of achieving that result. It can be distinguished from regulations that are self-executing and do not require any implementing measures. Directives normally leave member states with a certain amount of leeway as to the exact rules to be adopted.

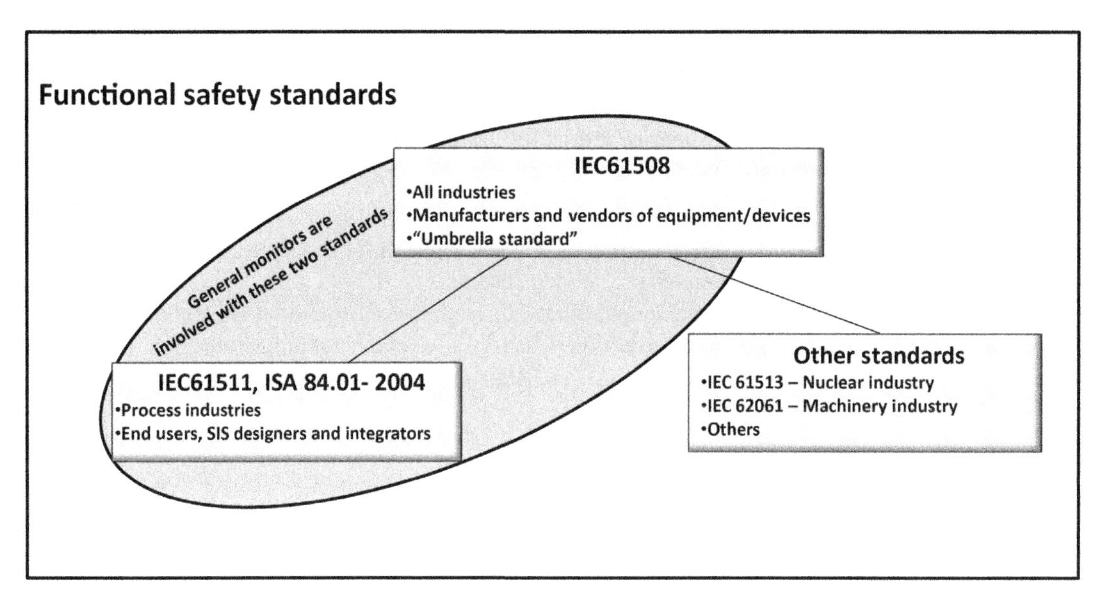

FIGURE 5.1

Functional safety standards.

Directives can be adopted by means of a variety of legislative procedures depending on their subject matter.

Standards: These are normative (informative) and referenced by the directives.

Seveso II directives:

Major accidents in chemical industries have occurred worldwide. In Europe, the Seveso accident in 1976 prompted the adoption of legislation aimed at the prevention and control of such accidents. In 1982, the first EU Directive 82/501/EEC – so-called Seveso Directive – was adopted. On December 9, 1996, the Seveso Directive was replaced by Council Directive 96/82/EC, so-called Seveso II Directive.

This directive was extended by the Directive 2003/105/EC. The Seveso II Directive applies to some thousands of industrial establishments where dangerous substances are present in quantities exceeding the thresholds in the directive.

5.6 IEC 61508/IEC 61511: AN INTRODUCTION

5.6.1 GENERAL STANDARDS ON SAFETY INSTRUMENTED SYSTEMS

In a processing plant there is no such thing as risk-free operation or 100% reliability. Therefore, one of the first tasks of the SIS system designer is to perform a risk–tolerance analysis to determine what level of safety is needed. IEC Standard 61508 (functional safety of electric, electronic and programmable electronic systems) is a general standard that covers functional safety related to all kinds of processing and manufacturing plants. IEC Standard 61511 and ISA S84.01 (replaced by ISA 84.00.01-2004) are standards specific to the process industries. These standards specify precise levels of safety and quantifiable proof of compliance.

By the 1980s, software had become the first choice of most designers of control systems. Its apparent speed of production, the cost of its reproduction, and the ease with which it facilitates the introduction of new facilities made it more attractive than purely hardware solutions. Its increased use included more and more safety-related applications, and there were uncertainties about the wisdom of this. It was recognized that it was almost impossible to prove software correct, and even if it were correct with respect to its specification, the difficulty of getting the specification correct was well known.

At that time, software engineering was still more of art than engineering, and safety engineering was unknown in the software development community. Furthermore, because safety was not in most cases studied in its own right, there was an implicit assumption that if a product or plant functioned reliably it would be safe. But safety and reliability are not synonymous. Moreover, with systems becoming larger and more complex, questions started to be asked about how safety might be "proved" and how the use of software in safety-related applications might be justified.

The questions and uncertainties were not limited to software. At the same time, hardware, in the form of microelectronics, was also becoming extremely complex and difficult to prove correct. The awareness of these issues led to two studies being set up by the IEC, one on "systems" (hardware) and the other on software, both within the context of the functional safety of modern programmable electronic systems. The purpose behind each was the development of a standard to guide system designers and developers in what they needed to do in order to claim that their systems were acceptably safe for their intended uses.

IEC 61508 is a "generic" standard, intended to satisfy the needs of all industry sectors. It is a large document, consisting of seven parts. Ideally, it should be used as the basis for writing more specific (e.g., sector-specific and application-specific) standards, but it is also intended to be used directly wherein these do not exist. It has become a requirement of many customers, and its principles are perceived as defining much of what is considered to be good safety-management practice.

The physical form of the IEC61508 standard:

The standard consists of seven parts. The first four are "normative" – that is, they are mandatory – and the fifth, sixth, and seventh are informative – that is, they provide added information and guidance on the use of the first four.

- **Part 1:** (general requirements) defines the activities to be carried out at each stage of the overall safety life cycle, as well as the requirements for documentation, conformance to the standard, management, and safety assessment.
- **Part 2:** (requirements for electrical/electronic/programmable electronic (E/E/PE) safety-related systems).
- **Part 3:** (software requirements) interpret the general requirements of Part 1 in the context of hardware and software, respectively. They are specific to Phase 9 of the overall safety life cycle, illustrated in Figure 5.4.
- **Part 4:** (definitions and abbreviations) gives definitions of the terms used in the standard.
- **Part 5:** (Examples of Methods for the Determination of SILs) gives risk-analysis examples and demonstrates the allocation of SILs.
- **Part 6:** (guidelines on the application of Parts 2 and 3) offers guidance as per its title.
- **Part 7:** (overview of techniques and measures) provides brief descriptions of techniques used in safety and software engineering, as well as references to sources of more detailed information about them. In any given application, it is unlikely that the entire standard would be relevant. Thus, an important initial aspect of use is to define the appropriate part(s) and clauses.

5.7 SCOPE OF THE STANDARD

IEC 61508 is not merely a technical guideline; indeed, its primary subject is the management of safety. Within this context, IEC 61508 addresses the technical issues involved in the design and development of systems. The standard seeks to introduce safety management and safety engineering, not only into software and system engineering but also into the management of all aspects of systems. The standard embraces the entire life cycle of a system, from concept to decommissioning (Figure 5.2).

Although the standard formally limits itself to the aspects of safety that depend on the hardware and software of electrical/electronic/programmable electronic (E/E/PE) systems, its principles are general and form a framework for addressing all aspects of the safety of all systems.

The wording of the standard is based on the model of Figure 5.2. In this, there is "equipment under control" (EUC) that, with its control system, provides a utility (e.g., electricity generation, railway signaling), but which, in order to do this, poses one or more risks to the outside world.

The standard requires that each risk posed by the EUC and its control system should be identified, analyzed, and tested against tolerability criteria. All risks found to be intolerable must be reduced, as shown in Figure 5.3.

A risk-reduction measure may be to change the design of the EUC or its control system, but there comes a point when it is not effective to make further improvements, or when, even if they have been made, the required level of safety cannot be demonstrated. If any of the residual risks is still intolerable (or cannot be shown to be tolerable), then "safety functions" must be incorporated either within the control system or in one or more added "protection systems" (see Figure 5.2). In principle, their separation from the control system is preferred.

The model of Figure 5.2 is based on the process industry, and it may not be perceived as representing many modern systems – for example, information systems whose handling of data is safety-related, such as medical databases.

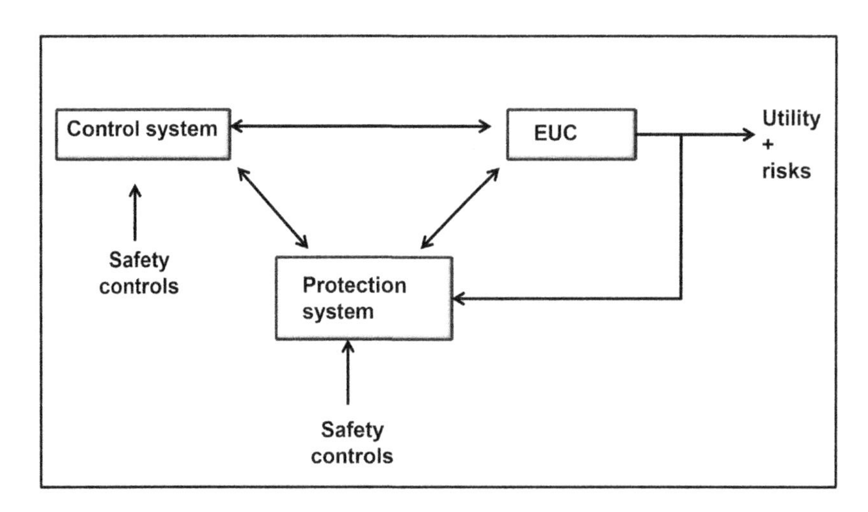

FIGURE 5.2

Risk and safety functions to protect against it.

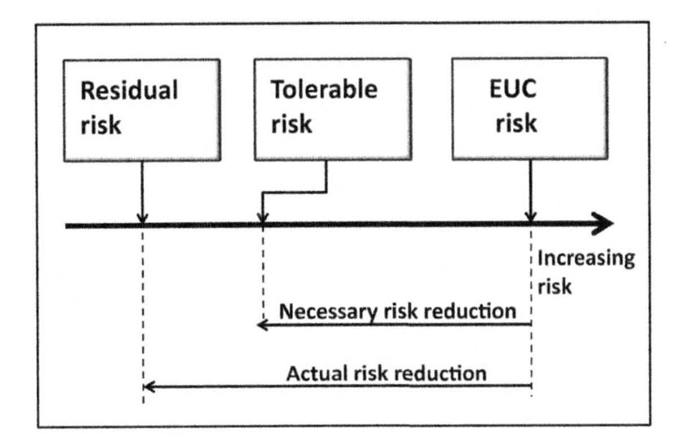

FIGURE 5.3

The determination of the necessary risk reduction.

Even though the wording of the standard does not obviously refer to such systems, the standard's principles do. Figure 5.3 shows that the risk reduction that must necessarily be achieved is the difference between the risk posed by the EUC (and its control system) and the level of risk that is deemed, in the given circumstances, to be tolerable.

The risk reduction is achieved by "**safety functions**," and these must be based on an understanding of the risks. However, risk values are always approximate, and the actual reduction achieved by risk-reduction measures can never be determined exactly, so it is assumed in Figure 5.3 that the achieved risk reduction will be different from (and greater than) the reduction deemed to be necessary. The figure thus shows that the residual risk is not exactly equal to the tolerable risk – and nor is it zero.

The development of safety functions, which embody the main principles of the standard, requires the following steps (Figure 5.4):

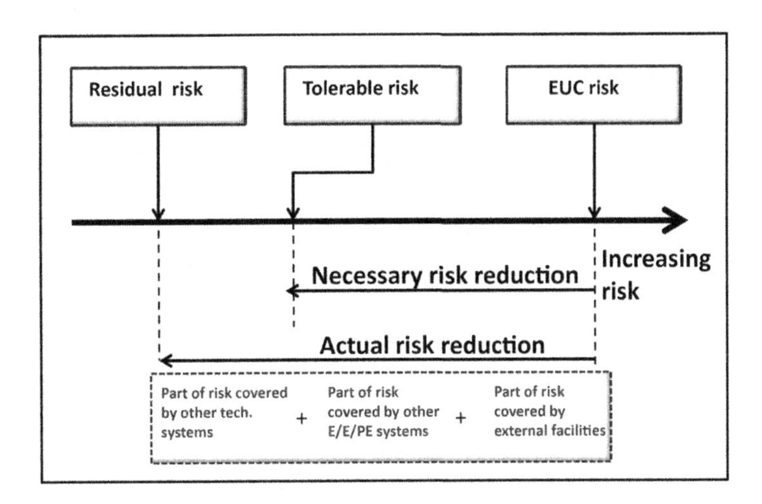

FIGURE 5.4

Understanding of risks and the means of reduction.

- Identify and analyze the risks.
- Determine the tolerability of each risk.
- Determine the risk reduction necessary for each intolerable risk.
- Specify the safety requirements for each risk reduction,
- Include their SILs.
- Design safety functions to meet the safety requirements.
- Implement the safety functions.
- Validate the safety functions.

Although the standard formally addresses only safety-related E/E/PE systems, it points out (see Figure 5.4) that safety functions may also be provided in other technologies (such as hydraulic systems) or external facilities (e.g., management procedures). The principles of the standard should be applied to all cases.

5.8 THE OVERALL SAFETY LIFE CYCLE (SLS)

The overall safety lifecycle (Figure 5.5) is crucial to IEC 61508. Not only does it offer a model of the stages of safety management in the life of a system but it also forms the structure on which the standard itself is based. Thus, the standard's technical requirements are stated in the order defined by the stages of the overall safety life cycle.

The purpose of the overall safety life cycle is to force safety to be addressed independently of functional issues, thereby overcoming the assumption that functional reliability will automatically produce safety. Then, specifying separate safety requirements allows them to be validated independent of functionality, thus giving higher confidence of safety under all operating and failure conditions. The paradox, however, is that safety activities should not be carried out, or thought of, as totally disconnected from other project or operational activities. They need to be integrated into a total perspective of the system at all life cycle phases.

In the overall safety life cycle, Phases 1 and 2 indicate the need to consider the safety implications of the EUC and its control system, at the system level, when first they are conceived of. In Phase 3, their risks are identified, analyzed, and assessed against tolerability criteria. In Phase 4, safety requirements for risk-reduction measures are specified, and in Phase 5, these are translated into the design of safety functions, which are implemented in safety-related systems, depending on the selected manner of implementation, in Phases 9, 10, and 11. However the safety functions are realized, no claim for safety can be made unless its planning considers the overall safety context, and this is reflected in Phases 6, 7, and 8. Then, again, carrying out the functions of installation and commissioning, safety validation, and operation and maintenance, is shown in Phases 12, 13, and 14 to be on the overall systems, regardless of the technologies of the safety-related systems. Phases 15 and 16 cover later modification and retrofit of the system and decommissioning, respectively.

The overall safety life cycle covers not merely the development of a system, but its entire life cycle, and this is illustrated by the inclusion of Phases 12 to 16. At the same time, like all models, this one is an approximation. Its phases are shown sequentially, so the iteration between them is not portrayed. For example, if modification and retrofit (Phase 15) is carried out to an operational system, all activities, from risk analysis, through specification, to revalidation, would need to be carried out, but this is not explicit in the model. A lesson from this is that a model cannot be a substitute for

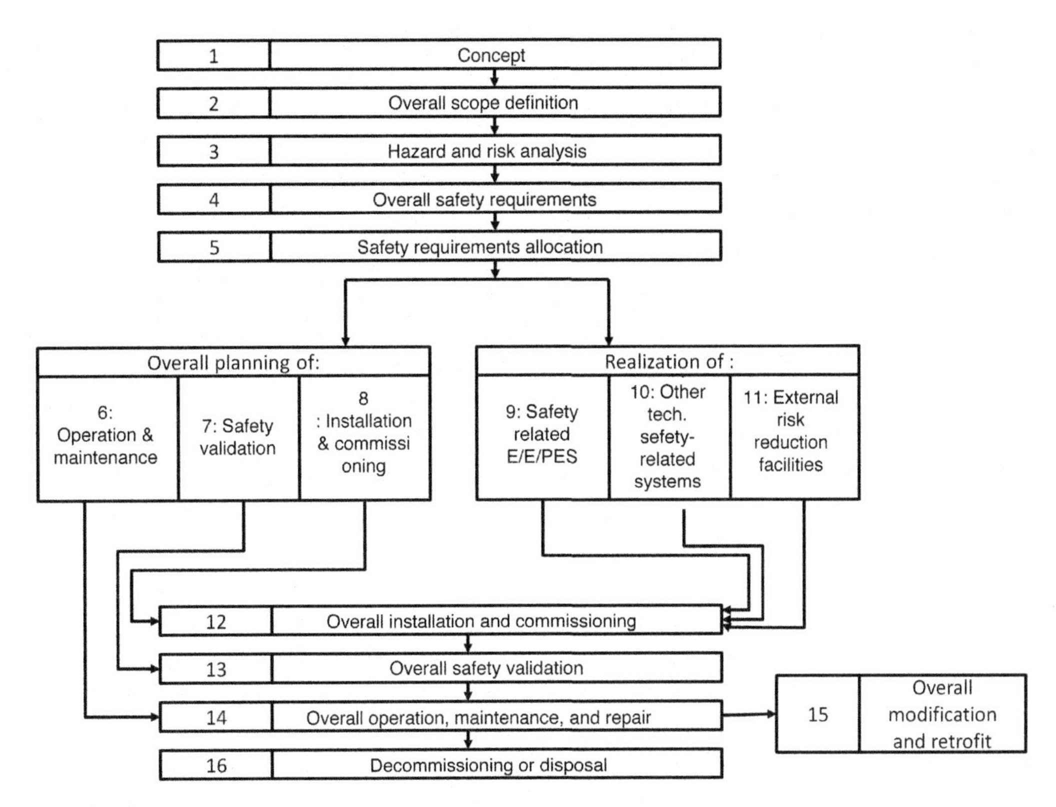

FIGURE 5.5

The overall safety life cycle (from IEC 61508-1).

good engineering or good management, should never be relied on entirely as a guide to what to do, and should only be used in support of well-understood good practice. Other omissions from the life cycle are activities, such as management, documentation, verification, quality assurance, and safety assessment, that are essential to all phases – but these are set out as requirements in the clauses of the standard.

5.9 RISK AND ITS ANALYSIS AND REDUCTION

A fundamental principle of the standard is that the measures taken to ensure safety should be commensurate with the risks posed by the EUC and its control system. Thus, a thorough risk analysis must be carried out, as required by Phase 3 of the overall safety life cycle and by Clause 7.4 of Parts 1, 2, and 3 of the standard. Risk analysis is normally defined as consisting of three stages – hazard identification, hazard analysis, and risk assessment – and some examples of how it may be carried out are offered in Part 5 of the standard.

Hazard identification consists of an attempt to identify the potential sources of harm. For simple systems that have already been in operation for some time, methods such as brainstorming and the use of a checklist may be adequate. But for systems that are novel or complex, a team effort is required. An EUC and its control system may pose many hazards, and as many as possible must be identified, for the risks associated with unidentified hazards will not be analyzed or reduced. The importance of hazard identification cannot be emphasized too strongly, and the standard points out that identifying hazards concerned only with normal operation is not sufficient. Those arising from failures and "reasonably foreseeable" misuse must also be identified. For this, professionals in the domain, functioning in a carefully chosen and well managed team, are required.

Hazard analysis is the study of the chains of cause and effect between the identified hazards and the hazardous events (accidents) to which they might lead. The analysis is intended to determine causes and consequences, so that the risk attached to each hazard can be derived. It may be quantitative or qualitative. In a quantitative analysis, the probabilities of events are estimated, as are numeric values of their consequences. Then, the risks are calculated by multiplying the two. But qualitative analysis is also admissible, and the standard's definition of risk, as the combination of likelihood and consequence, facilitates this. Various qualitative methods of analysis, using the risk matrix and the risk graph, are illustrated in Part 5 of the standard. In the case of simple hardware with a history of use in conditions that are the same as those of the safety-related application, the probabilities of certain events, such as equipment failures, may be estimable from data on past frequencies. Similarly, consequences may also be expressed numerically, for example, as the number of lives lost, or some financial value of the total resulting losses. However, the standard recognizes that because software failure is systematic and not random, qualitative methods must be used in the case of software. In the risk-assessment stage of risk analysis, the risk values determined in the previous stage are compared against tolerability criteria to determine whether they are tolerable as they are, and, if not, by how much they need to be reduced. There is necessarily a great deal of subjectivity in this process, not least in the decision of what level of risk is tolerable. It should be noted that tolerability may be different for each risk posed by the EUC and its control system, for it depends not only on the level of risk but also on the benefits to be gained by taking the risk and the cost of reducing it.

5.10 SAFETY REQUIREMENTS AND SAFETY FUNCTIONS

The safety requirements are those requirements that are defined for the purpose of risk reduction. Like any other requirements, they may at first be specified at a high level, for example, simply as the need for the reduction of a given risk. Then they must be refined so that their full details are provided to designers. The totality of the safety requirements for all risks forms the safety requirements specification. At the design stage, the safety requirements are provided by means of safety functions. These are implemented in "safety-related systems" which, as seen in Figure 5.3, need not be restricted to any given technology.

For example, a braking system may be hydraulic. A safety requirement may be met by a combination of safety functions, and these may be implemented in systems of different technologies – for example, a software-based system along with management procedures, checklists, and validation procedures for using it. When a safety function is implemented via software, there also needs to be a hardware platform, in which case a computer system is necessary. Then, the same demands are made of the entire system as of the software. Furthermore, the standard allows for more than one software safety function to be implemented on the same hardware platform, and it imposes rules for this.

5.11 SAFETY INTEGRITY LEVELS (SIL)

If there is an important job to be done, the means of doing it must be reliable, and the more important the job, the more reliable they should be. In the case of a safety-related system, the job is to achieve safety, and the greater the system's importance to safety, the lower should be the rate of unsafe failures.

A measure of the rate of unsafe failures is the safety integrity of the system, which is defined in Part 4 of IEC 61508 as "the likelihood of a safety-related system satisfactorily performing the required safety functions under all the stated conditions, within a stated period of time." If the rate of unsafe failures could always be measured numerically, there would be no need for SILs because SILs are categories of safety integrity – and categories would be unnecessary if exact values were available. In the standard a SIL is defined as "a discrete level (one of 4) for specifying the safety integrity requirements of safety functions." Thus, a SIL is a target probability of dangerous failure of a defined safety function, and was originally intended for use when qualitative hazard analysis has been carried out and numerical risk values are not available, as in the case of software.

The standard demands that whenever a safety requirement is defined it should have two components: functional component and safety integrity component. The safety requirement arises out of the need for risk reduction. Thus, at the highest level, the functional component is to "reduce the risk." The safety integrity component consists of a SIL (between 1 and 4), and this is related to the amount of risk reduction that is required. As said earlier, the more important the job, the more reliable the system must be. Here, the greater the risk reduction needed, the greater the extent to which safety depends on the system that provides the risk reduction, so the higher the SIL.

The standard equates SILs with numeric probabilities of unsafe failures in two tables (Tables 5.1 and 5.2), one for systems whose operation is continuous and one for on-demand (or low-demand) systems. The standard's definition of low-demand is "no greater than one demand per year," hence, the difference of 104 in the values in the two tables (where 1 year is approximated to 104 h).

Assuming a failure rate of once per year, the SIL 4 requirement for the low-demand mode of operation is no more than one failure in 10,000 years. As stated in their definition, SILs are intended to provide targets for developers. In the case of simple electromechanical hardware, it may be possible to claim achievement of the SIL, using historic random-failure rates. But for complex systems, and for software, whose failures are systematic and not random, such a claim is unsupportable by evidence. Thus, SILs are used to define the rigor to be used in the development processes. In other words, because evidence of the rate of dangerous failures of the product cannot be determined with confidence, attention is turned to the development process. Here, SIL 1 demands basic sound engineering practices and

Table 5.1 Safety integrity levels for continuous operation for continuous and high demand mode

Safety integrity level	Continuous/high-demand mode of operation (prob. of a dangerous failure per hour)
4	$\geq 10^{-9}$ to $< 10^{-8}$
3	$\geq 10^{-8}$ to $< 10^{-7}$
2	$\geq 10^{-7}$ to $< 10^{-6}$
1	$\geq 10^{-6}$ to $< 10^{-5}$

Table 5.2 Safety integrity levels for continuous operation for low demand mode

Safety integrity level	Low demand mode of operation (average prob. of failure on demand)
4	$\geq 10^{-5}$ to $<10^{-4}$
3	$\geq 10^{-4}$ to $<10^{-3}$
2	$\geq 10^{-3}$ to $<10^{-2}$
1	$\geq 10^{-2}$ to $<10^{-1}$

adherence to a quality management standard, such as ISO 9000. Higher SILs, in turn, demand this foundation plus further rigor, and guidance on what is required is found for hardware and software in Parts 2 and 3 of the standard, respectively.

5.11.1 IEC 61511

This international standard, entitled, "Functional Safety: Safety Instrumented Systems for the Process Industry Sector," is currently under development by the same IEC committee that produced IEC 61508. It is defined as being "process industry specific within the framework of the International Electro technical Committee (IEC) Publication 61508." It is therefore intended to perform the same function internationally that S84 performs in the United States. It defines SISs as including sensors, logic solvers and final elements, and states that it covers all these components of an SIS as well as all technologies by which they may be constructed.

This standard is broader in scope than S84, for it covers the early hazard and risk analysis and the specification of all risk-reduction measures, which S84 assumes to have been done. It also contains sections on such issues as how to show conformity with the standard, so it is of much greater length than S84. IEC 61511 follows the IEC 61508 overall safety life cycle and uses the system of SILs described in that standard. In short, it is a sector-specific interpretation of the generic standard.

IEC 61511 covers the design and management requirements for SISs from cradle to grave. Its scope includes initial concept, design, implementation, operation, and maintenance through to decommissioning. It starts in the earliest phase of a project and continues through startup. It contains sections that cover modifications that come along later, along with maintenance activities and the eventual decommissioning activities. The standard consists of three parts:

1. Framework, definitions, system, hardware, and software requirements
2. Guidelines in the application of IEC 61511-1
3. Guidance for the determination of the required SILs.

ISA 84.01/IEC 61511 requires a management system for identified SIS. An SIS is composed of a separate and independent combination of sensors, logic solvers, final elements, and support systems that are designed and managed to achieve a specified SIL. An SIS may implement one or more safety instrumented functions (SIFs), which are designed and implemented to address a specific process hazard or hazardous event. The SIS management system should define how an owner/operator intends to assess, design, engineer, verify, install, commission, validate, operate, maintain, and continuously improve their SIS. The essential roles of the various personnel assigned responsibility for the SIS should be defined and procedures developed, as necessary, to support the consistent execution of their responsibilities.

ISA 84.01/IEC 61511 uses an order of magnitude metric, the SIL, to establish the necessary performance. A hazard and risk analysis is used to identify the required safety functions and risk reduction for specified hazardous events. Safety functions allocated to the SIS are SIFs; the allocated risk reduction is related to the SIL. The design and operating basis is developed to ensure that the SIS meets the required SIL. Field data are collected through operational and mechanical integrity program activities to assess actual SIS performance. When the required performance is not met, action should be taken to close the gap, ensuring safe and reliable operation.

5.12 FUNCTIONAL SAFETY MANAGEMENT

The functional safety life cycle management audit is a mechanism used to help reduce systematic problems from appearing in the design of a product. In the case of the manufacturing process, the quality control measures that go into the process dictate the quality of the product coming out. Many of these measures may be procedural in nature, and tied to documentation related to the product specification or functional safety standard to which the product aspires. The functional safety life cycle management audit looks at those elements of the manufacturer's process that may impact the quality of the safety of the product being produced.

5.12.1 HAZARD AND RISK ANALYSIS

A hazard analysis is used as the first step in a process used to assess risk. The result of a hazard analysis is the identification of risks. Preliminary risk levels can be provided in the hazard analysis. The validation, more precise prediction, and acceptance of risk are determined in the risk assessment (analysis). The main goal of both is to provide the best selection of means of controlling or eliminating the risk. The term is used in several engineering specialties, including avionics, chemical process safety, safety engineering, reliability engineering and, food safety. Alternative definitions include the following:

- Identification, studies, and monitoring of any hazard to determine its potential, origin, characteristics, and behavior.
- The process of collecting and evaluating information on hazards associated with the food under consideration to decide which are significant and must be addressed.
- An analysis or identification of the hazards that could occur at each step in the process, and a description and implementation of the measures to be taken for their control.

Hazard is defined in FAA Order 8040.4 as a "Condition, event, or circumstance that could lead to or contribute to an unplanned or undesirable event." Seldom does a single hazard cause an accident. More often, an accident occurs as the result of a sequence of causes. A hazard analysis will consider system state, for example, operating environment, as well as failures or malfunctions.

While in some cases safety risk can be eliminated, in most cases a certain degree of safety risk must be accepted. In order to quantify expected accident costs before the fact, the potential consequences of an accident, and the probability of occurrence must be considered. Assessment of risk is made by combining the severity of consequence with the likelihood of occurrence in a matrix. Risks that fall

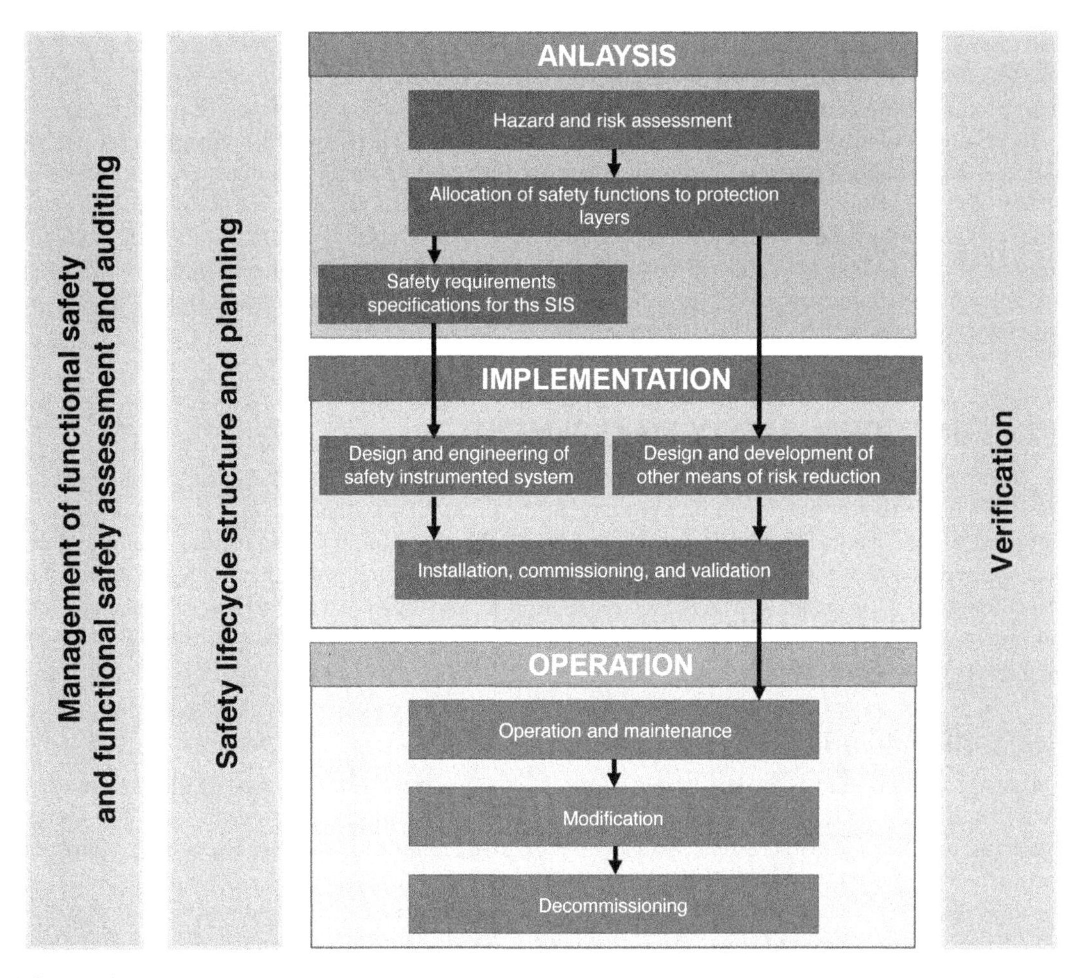

FIGURE 5.6

IEC61511 safety life cycle.

into the "unacceptable" category (e.g., high severity and high probability) must be mitigated by some means to reduce the level of safety risk.

IEC 61511 covers the whole life cycle as shown in Figure 5.6, but this chapter is concerned only with Phases 1 through 3, leading to the "Safety Requirements Specification for the Safety Instrumented System."

5.13 LAYERS OF PROTECTION

The introduction of the layers of protection concept shown in Figure 5.7 originates from the American approach to SIS in ANSI ISA-SP 84.01-19963. This American standard has been the major influence in the differences between IEC 615084 and IEC 61511 and the importance of independence between layers and the implications of common cause issues between layers is emphasized. The allocation of

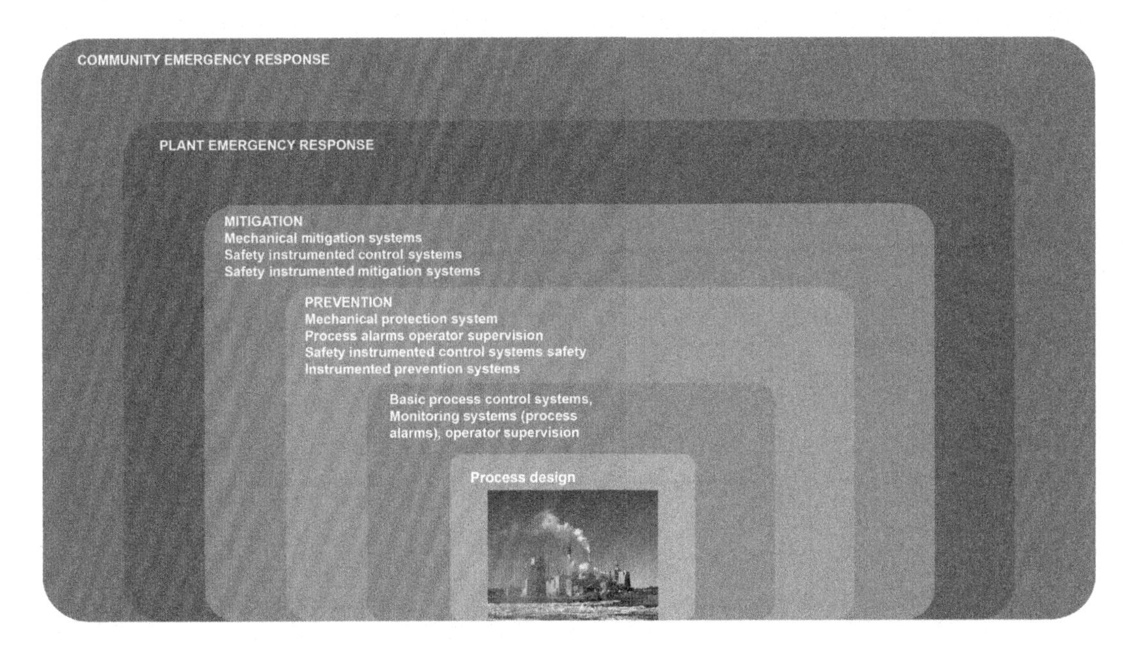

FIGURE 5.7

Layers of protection.

safety functions to specific layers or systems (e.g., a hazard may be protected by a combination of relief valves, physical barriers and bunds, and an SIS), and the contribution required of each element to the overall risk reduction should be specified as part of the transfer of information from the risk analysis to those responsible for the design and engineering (Figure 5.7).

5.14 RISK ANALYSIS TECHNIQUES

5.14.1 CONCEPTS OF RESIDUAL RISK, RISK REDUCTION, AND REQUIRED SIL

Both IEC 61508 and 61511 imply that the only action of an SIS is to reduce the frequency or likelihood of a hazard. Thus, the model of risk (reproduced in Figure 5.3) is one-dimensional. All the methods of determining SIL are based on similar principles (Figure 5.8):

- Step 1 Identify the "process risk" from the process and the basic process control system (BPCS).
- Step 2 Identify the "tolerable risk" for the particular process.
- Step 3 If the process risk exceeds the tolerable risk, then calculate the necessary risk reduction and whether the protection layers will operate in continuous or demand mode.
- Step 4 Identify the risk reduction factors (RRFs) achieved by other protection layers.
- Step 5 Calculate the remaining RRF or the failure rate that should be achieved by the SIS and thus from Table 5.1 or 5.2 the required SIL.

The residual risk is the process risk reduced by all the RRFs and will normally be less than the tolerable risk. Identifying the tolerable risk is a major issue that is discussed in Reducing Risks, Protecting People and is beyond the scope of this paper. Identifying the frequencies of all initiating causes (or the

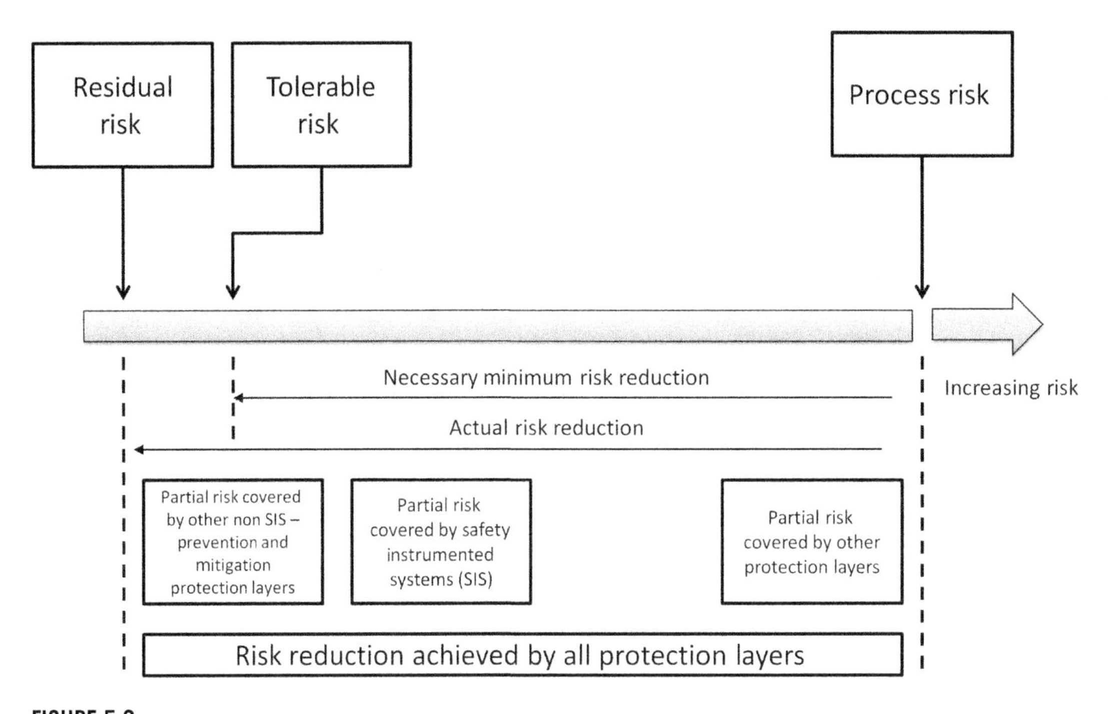

FIGURE 5.8

IEC61511 risk reduction model.

demand rates used in Steps 1 and 3 above) is also difficult unless excellent records of all incidents are available.

Some methods of determining SIL requirements

IEC 61508 offers three methods of determining SIL requirements:

- Quantitative method.
- Risk graph, described in the standard as a qualitative method.
- Hazardous event severity matrix, also described as a qualitative method.

IEC 61511 offers:

- Semiquantitative method (incorporating the use of fault and event trees).
- Safety layer matrix method, described as a semiqualitative method.
- Calibrated risk graph, described in the standard as a semiqualitative method, but by some practitioners as a semiquantitative method.
- Risk graph, described as a qualitative method.
- Layer of protection analysis (LOPA). (Although the standard does not assign this method a position on the qualitative/quantitative scale, it is weighted toward the quantitative end.)

These are developments and extensions of the methods originally outlined in IEC 61508-5. They have all been used by various organizations in the determination of SILs, but with varying degrees

Table 5.3 Typical results of SIL assessment

SIL	Number of functions	% of total
4	0	0
3	0	0
2	1	0.3
1	18	6.0
None	281	93.7
Total	300	100

of success and acceptability, and do not provide an exhaustive list of all the possible methods of risk assessment. All of these methods require some degree of tailoring to meet the requirements of an individual company, together with training of the personnel who will apply them, before they can be used successfully. Quantitative risk assessment (QRA), risk graphs, and LOPA are established methods for determining SIL requirements, particularly in the process industry sector, but LOPA is less well known in the UK and is the focus of this chapter.

5.14.2 TYPICAL RESULTS

As one would expect, there is wide variation from installation to installation in the numbers of functions that are assessed as requiring SIL ratings, but the numbers in Table 5.3 were assessed for a reasonably typical offshore gas platform. Typically in the process control area there might be a single SIL 3 requirement in an application of this size, while identification of SIL 4 requirements is very rare. If a SIL 3 or SIL 4 requirement is identified, it is reasonable to investigate the use of the basic process design and other protection layers in risk reduction and whether undue reliance is being placed on the SIS, and indicates a serious need for redesign detection and emergency shutdown are the principal examples of such functions. An assessment of the required SILs of such functions presents specific problems.

5.14.3 QUANTITATIVE RISK ANALYSIS (QRA)

Quantitative risk analysis (QRA) is usually done with Fault Trees and Event Trees or reliability block diagrams. Some people refer to a combination of Fault and Event Tree as a cause–consequence diagram. Figure 5.4 shows an example of an Event Tree, and Figures 5.5 and 5.6 show a Fault Tree and a reliability block diagram (Figure 5.9).

Normally the "Top Event" will be a particular hazard and provided that

- appropriate failure models are chosen for each basic event or block;
- accurate data are available for the particular environment for each of the failure modes, repairs, and tests;
- all the relationships are correctly modeled; then the frequency at which the hazard occurs, and hence the risk can be calculated.

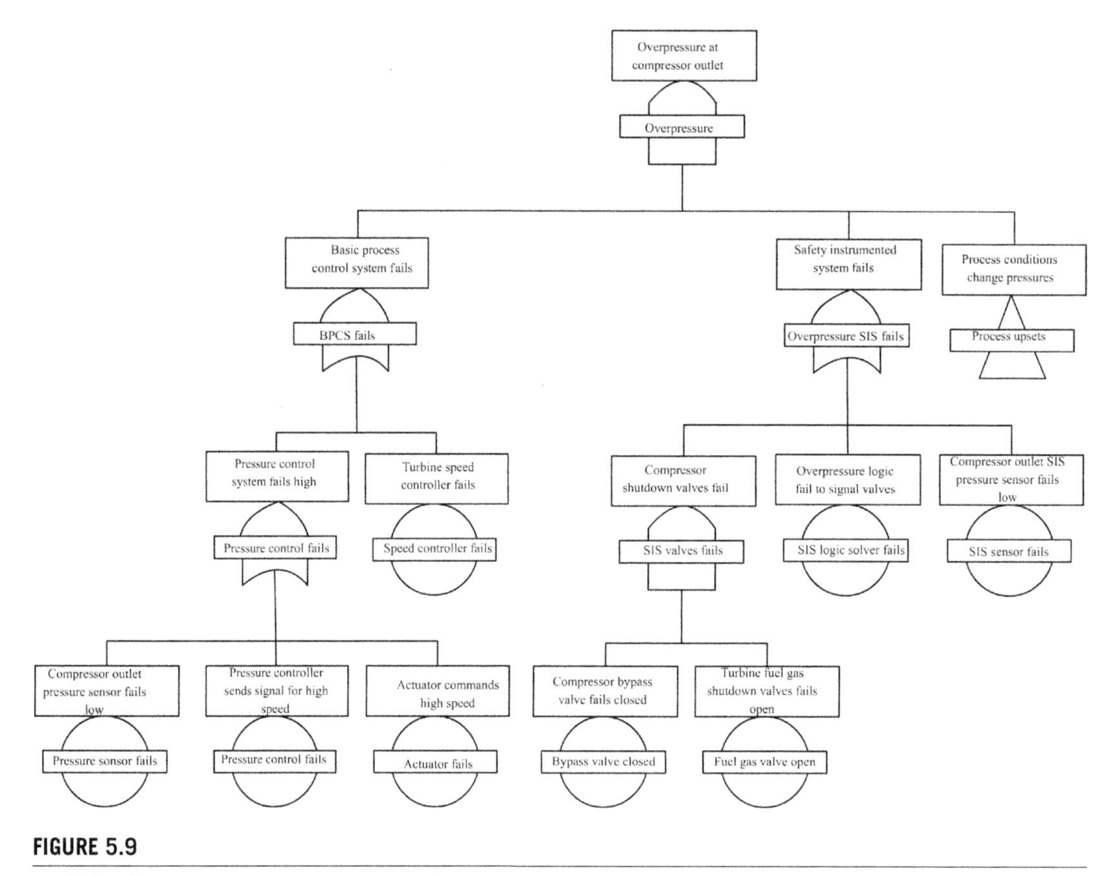

FIGURE 5.9

Fault Tree for overpressure at compressor outlet.

The successful outcome of a QRA is highly dependent on the assumptions that are made, the detail of the model developed to represent the hazardous event, and the data that are used. However well a QRA has been done it does not provide an absolute indication of the residual risk. A sensitivity analysis of the data and assumptions is a fundamental element of any QRA.

5.14.4 RISK GRAPH METHODS

Figure 5.10 shows a typical risk graph. The risk graph method is described in both IEC 61508 and 61511 and is an excellent means of quickly assessing and screening a large number of safety functions so as to allow effort to be focused on the small percentage of critical functions.

A serious limitation of the risk graph method is that it does not lend itself at all well to assessing "after the event" outcomes:

- Demand rates would be expected to be very low, for example, 1 in 1000 to 10,000 years. This is off the scale of most of the risk graphs used.

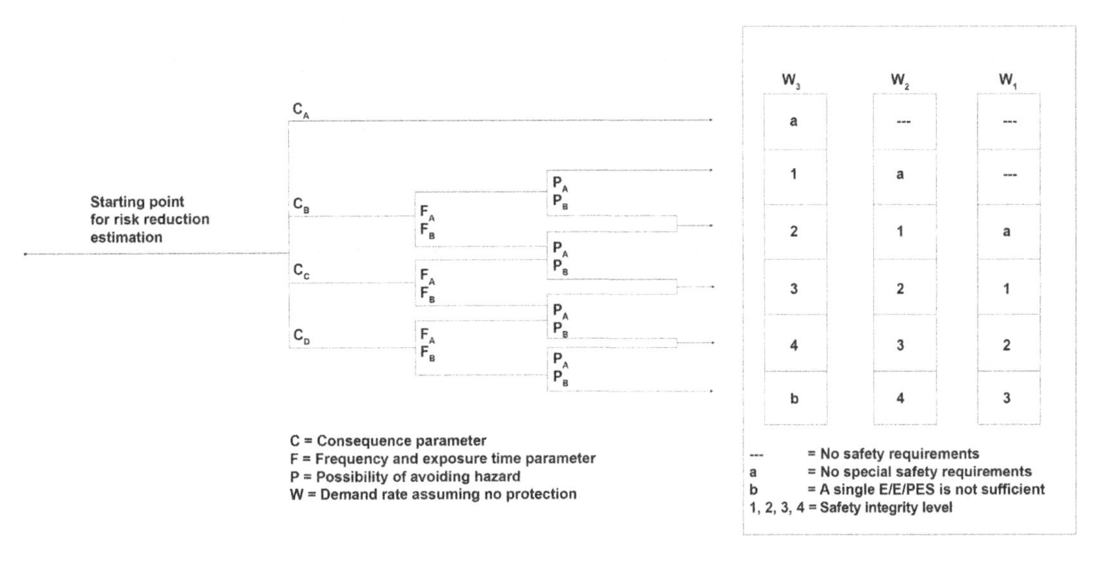

FIGURE 5.10

Typical risk graph.

- The range of outcomes from function to function may be very large, from a single injured person to major loss of life. The outcomes are also potentially random depending on a wide range of circumstances. Where large-scale consequences are possible, use of such a coarse tool such as the risk graph method can hardly be considered "suitable" and "sufficient."

5.14.5 SIL EVALUATION

IEC 61508 contains guidance on using both qualitative and quantitative methods to determine the SIL for a system based on risk frequency and consequence tables and graphs. This book will focus on a simple quantitative method as an illustrative example, and reference should be made to the actual standard for further details on alternative methods. Assuming the hazards analysis and risk assessment phase reveals that overall risk reduction is required it may be determined that an SIS is necessary. It follows that the amount of risk reduction to be provided by the SIS must be determined and this will in turn determine the SIL level for the intended SIS. The following steps illustrate application of the general guidelines contained in IEC 61508:

1. Set the target tolerable risk level (Ft), where Ft is the risk frequency, often determined as hazardous event frequency × consequence of hazardous event expressed numerically.
2. Calculate the present risk level (Fnp) for the EUC, which is the risk frequency with no protective functions present (or unprotected risk).
3. The ratio Fnp/Ft gives the RRF required to achieve the target tolerable risk.
4. Determine the amount of RRF to be assigned to the SIS. The reciprocal the same gives the target average probability of failure on demand (PFDavg) the SIS must achieve.
5. Translate the PFDavg value into a SIL value (using guidance tables).

Table 5.4 Risk classification of accidents (Table B1 of IEC 61508-5)

Frequency	Catastrophic > 1 death	Critical 1 death or injuries	Marginal Minor injury	Negligible Production loss
1 per year	I	I	I	II
1 per 5 years	I	I	II	III
1 per 50 years	I	II	III	III
1 per 500 years	II	III	III	IV
1 per 5000 years	III	III	IV	IV
1 per 50000 years	IV	IV	IV	IV

Consider a system with EUC that has an unprotected risk frequency (Fnp) of one hazardous event per 5 years (Fnp = 0.2/year) with a consequence classified as "Critical." Tables 5.4 and 5.5 show examples of guidance tables used for risk classification and class interpretation of accidents from IEC 61508-5.

Using Tables 5.4 and 5.5, the unprotected risk is determined as class I. The target is to reduce this risk to a tolerable risk of class III, that is, 1 hazardous event per 500–5000 years. If we consider the safest target, Ft = 1 hazardous event in 5000 years, this represents a frequency of 0.0002 events/year.

This gives a target RRF of Fnp/Ft = 0.2/0.0002 = 1000

If there are no non-SIS protective layers assigned to the system, the SIS must fulfill the total RRF of 1000. So, in this case the total RRF = RRFSIS.

Now PFDavg = 1/RRFSIS = 1/1000 = 0.001 = 1×10^{-3}

5.15 SAFETY REQUIREMENT SPECIFICATIONS

5.15.1 INTRODUCTION

In order to fulfill the requirements of the standard IEC 61511, a safety requirement specification (SRS) is needed. Most companies must develop their own format of SRS; this document provides a

Table 5.5 Risk classification of accidents (Table B2 of IEC 61508-5)

Risk class	Interpretation
I	Intolerable risk
II	Undesirable risk, tolerable only if risk reduction is impracticable or if the cost are grossly disproportionate to the improvement gained
III	Tolerable risk if the costs of risk reduction would exceed the improvement gained
IV	Negligible risk

guideline on how to write an SRS. In order to fulfill the requirements, the standard IEC 61511 has to be used.

The main purpose with the SRS is to identify and present the safety requirements for the SIFs. The development of the SRS is one of the important activities during the design of SISs. If important information regarding safety issues is missing, the design of the SIS may not be performed. The SRS is an important document for personnel dealing with the validation process (or validation activities). Validation personnel have often no detailed knowledge about the design of the SIS, and therefore the SRS must cover all safety aspects for the actual SIS. During the validation phase the SRS is used as a reference to check that the safety requirements are implemented in the SIS.

The SRS shall specify all requirements of SIS needed for detailed engineering and process safety information purposes.

"Specification that contains all the requirements of the safety instrumented functions that have to be performed by the safety instrumented systems."

The requirements need to be documented during the safety planning. The SRS is created after the hazard and risk analysis and the allocation of safety functions to protective layers in the safety life cycle according to IEC61511-1 (Figure 5.11).

The safety requirements shall be derived from the allocation of SIFs.

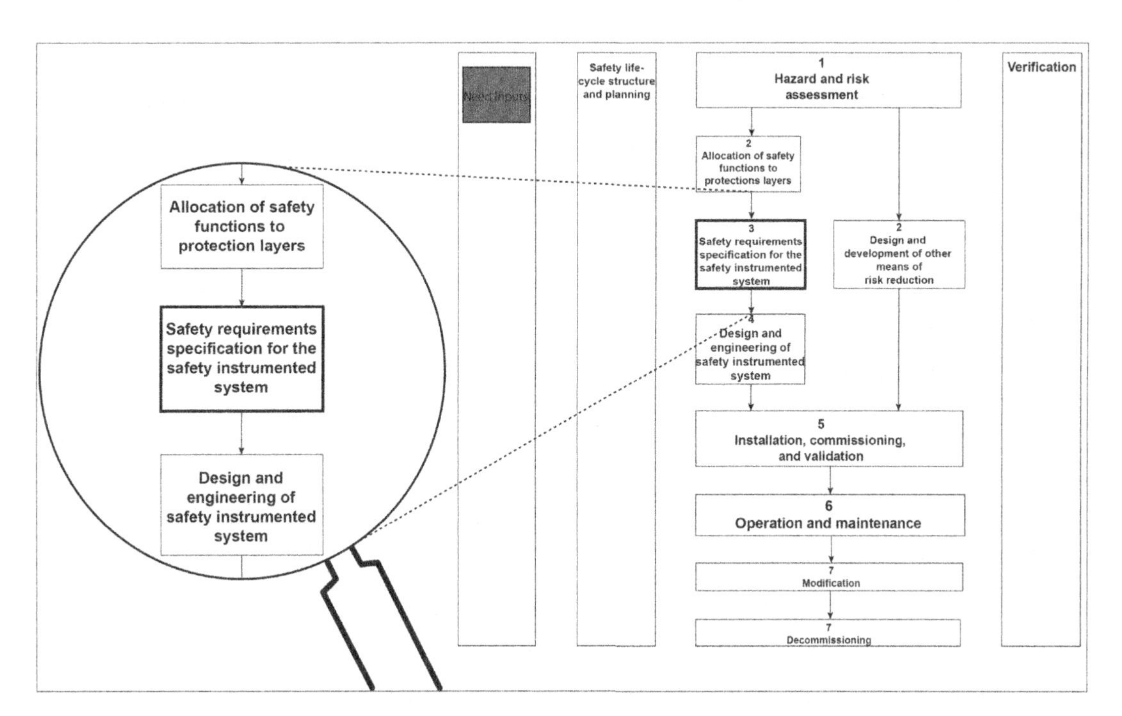

FIGURE 5.11

SIS safety life cycle phases (IEC 61511-1:2003).

5.16 GENERAL REQUIREMENTS

5.16.1 SAFE STATE

The safe state is defined as "state of the process when safety is achieved" [IEC 61511-1:2003]. In order to set a process to a safe state the knowledge of the process is very important. In some cases the safe state exists only if the process is continuously running; in other cases the process may have to go through a number of states before the process enters the final safe state.

Actions necessary to achieve or maintain a safe state in the event of detected fault(s) shall be described [10.3.1 IEC 61511-1:2003]. The relevant human factors that can affect the safe state shall be taken in account.

The description shall address safe state details regarding process actions needed, for example,

- Sequential shutdown
- Which process valve(s) is needed to perform a specific action during the safe state? Shall the valve open or close?
- Which flows should be started or stopped
- Stop, start, or continue operation of rotating elements (motors, pumps, etc.)

5.16.2 PROOF-TEST INTERVALS

The proof-test interval shall be defined [10.3.1 IEC 61511-1:2003]. It is important that the proof-test interval is taken into account during the design of the process application because the proof-test interval affects the design of the application. The proof-test idea is to test the function as far as possible. It is more advisable to perform a proof test is when the process (factory) is stopped.

Important activities:

- Describe the proof-test procedures
- Investigate whether additional safety measures (monitoring, redundancy, etc.) have to be adapted during the proof-test interval
- Investigate whether human aspects (forgotten bypass, etc.) could affect the safety during the proof test especially if the consequences could be catastrophic if the proof-test goes wrong
- Specify the required proof tests during the life cycle
- The proof-test activity shall be documented (the final result of the proof test)

5.17 RESPONSE TIME

The response time requirements for the SIS, to bring the process to a safe state, shall be stated. [10.3.1 IEC 61511-1:2003]. Parameters that affect the response time are as follows.

- Process Related - Time constants in the process, dead time in the process response.
- Control System (Electrical) - Time delay in control system, the sampling time of the controller.
- Other (Mechanical) - Inertia, friction, wear and tear.
- Reset - The reset after shutdown shall be defined [10.3.1 IEC 61511-1:2003].
- Spurious Trips -Define the maximum allowable spurious trip rate [10.3.1 IEC 61511-1:2003].

5.17.1 SIS PROCESS MEASURES AND TRIP POINTS

Describe SIS process measurements and their trip points [10.3.1 IEC 61511-1:2003].

Information regarding the inputs to the SIS: a description of the following:

- Every measurement circuit
- The architecture
- Number of inputs
- Type of input
- Range of measurement
- Accuracy of measurement
- Trip levels

5.17.2 SIS PROCESS OUTPUT ACTIONS

Describe SIS process output actions and the criteria for successful operation, for example, requirements for tight shut-off valves [10.3.1 IEC 61511-1:2003].

Information regarding the outputs from the SIS; a description of the following:

- Every measurement circuit
- The architecture, for example, block- and-bleed
- The number of outputs
- Type of input
- Range of measurement
- Trip levels
- Feedback

5.17.3 MANUAL SHUTDOWN

The requirements for manual shutdown shall be described [10.3.1 IEC 61511-1:2003].

Operator actions shall be defined. For example, if there is a requirement that the operator is able to manually shut down the process, this action shall be defined.

Specify requirements for independence of manual shutdown devices. For example, the manual shutdown concerns only some parts of the plant. Detailed description of the involved parts should be given.

Specify the location of manual shutdown devices (e.g., control room, field location).

The manual shutdown command from the operator shall not create any other hazards.

5.17.4 INTERFACES

All interfaces between SIS and any other system (including BPCS and operators) shall be described.

5.18 SIF SPECIFICATION

The requirement is quite clear in the IEC 61511-1 standard, Chapter 10, part 1. The SIS safety requirements specification shall include the following:

A description of all the SIFs necessary to achieve the required functional safety [10.3.1 IEC 61511-1:2003];

The IEC 61511-1 standard does not give precise instructions for the design of the SRS other than the SRS shall be expressed in a clear, precise, verifiable and maintainable way.

5.18.1 FUNCTIONAL REQUIREMENTS

The functional requirements for the SIF shall be described. The SRS input requirement documentation is used to give detailed information regarding functional requirements. The functional requirement describes, "How it should work":

- Definition of safe state
- Process inputs and their trip points
- Process parameters normal operating range
- Process outputs and their actions
- Relationship between inputs and outputs
- Selection of energize-to-trip or de-energize-to-trip
- Consideration for manual shutdown
- Consideration for bypasses
- Action on loss of power
- Response time requirements for the SIS to bring the process to a safe state
- Response actions for overt fault
- Operator interface requirements
- Operator actions
- Reset functions
- Response time requirements

5.18.2 INTEGRITY REQUIREMENTS

The SRS includes the integrity requirements as described below:

- The requested SIL for each SIF
- Requirements for diagnostics coverage to achieve the required SIL
- Requirements for maintenance and testing to achieve the required SIL
- Reliability requirements if spurious trips may be hazardous
- High or low demand mode
- Requirements for proof testing
- Environmental stress

5.18.3 FUNCTIONAL DESCRIPTION

Each identified SIF shall be expressed in a general way since the description shall be easy to understand by other persons involved during the safety life cycle. Below is a list of important issues to be taken into account during the creation of the SRS.

Functional description:

A functional description shall describe why the SIF is needed [10.3.1 IEC 61511-1:2003]. The functional description of the SIF includes words such as "prevent," "protect," or "mitigate."

Example: "The SIF protects the tank from overpressure by opening the release valve on high pressure."

Defined safe process state:

Define the safe state for each SIF [10.3.1 IEC 61511-1:2003]. The description of the safe state for the SIF describes the process action needed to prevent an accident and additional actions needed to maintain the safe state.

The description shall explain safe state details regarding needed process actions, for example:

- If sequential shutdown is needed
- The process valve(s) needed to perform a specific action (open or close)
- Which flows should be started or stopped
- Stop, start, or continue operation of rotating elements (motors, pumps etc.)

Example: When an abnormal situation occurs (hazardous event), which measure should be taken, for example, "On low level alarm the valve V10 shall be closed."

Which measures must be taken when:

- Power supply is missing
- Air supply is missing
- Fault/faults occur (hardware or software)

Example: If the air supply is missing, the valve has a mechanical return spring that closes the valve and prevents overflow.

5.18.4 PRIMARY ACTIONS/SEQUENCE (FOR BRINGING THE PROCESS TO THE DEFINED SAFE STATE)

The actions that are needed to prevent the hazardous event shall be described [10.3.1 IEC 61511-1:2003]. The primary actions describe the measures that are necessary to bring the process to the defined safe state. A primary action could be "open the relief valve in order to reduce too high pressure."

Some more examples:

- "reduce the pressure within a specific time"
- "reduce the flow by 10%"
- "open or close the valve"
- "measures to prevent additional hazardous conditions"

The threshold value of the parameter at which an action should be taken is also needed. This value will need to be outside the normal operating range and less than the value that will result in a hazardous condition. The response time of the system has to be taken into account; allowance will need to be made for the response of the system and the accuracy of measurement.

5.18.5 SECONDARY ACTIONS (FOR OPERATIONAL REASONS)

In some cases actions for operational reasons are needed. These actions may involve other measures, for example, "stop the inlet flow to the tank in order to eliminate overflow and give an operator alarm."

Some more examples:

- Actions that enables faster startup
- Shutdown of upstream or downstream units to reduce demands on other protection systems
- Operator alarm

5.18.6 DEMAND RATE AND SAFETY INTEGRITY

Specify the estimated demand rate and the target safety integrity level, SIL, for the SIF. The assumed sources of demand and demand rate on the SIF shall be specified [10.3.1 IEC 61511-1:2003].

Estimated demand sources:

Sources of demand are the source of events leading to the hazardous event.

Some examples:

- "malfunction of the inlet valve V6, valve jammed in open position, leading to over pressure"
- "malfunction of the temperature transmitter T2 e.g. too low indication, leading to over temperature"

Estimated SIF demand rate:

- Specify the SIF demand rate.
- Low demand, high demand, or continuous mode of operation.
- Specify if the SIF uses demand or continuous mode of operation.

Demand mode:

The "need" of safety appears when a certain level is reached. In demand mode, the safety is not always dependent on the SIF.

Example: A SIF is used to protect a tank for overpressure "when the pressure is above 5 bar the relief valve is opened."

Continuous mode:

In continuous mode of operation, the safety entirely depends on the SIF. If the SIF gets some kind of failure, it will result in a hazardous event.

In the process industry the SIFs are generally of demand mode type.

5.18.7 TRIGGING/TRIPPING

The trigging modes (automatic, manual) for the SIF and trigging detection need to be explained. The goal of this activity is to describe the conditions that affect the trigging of the SIF.

Automatic mode of trigging and trigging detection: explain briefly, what shall be detected? Describe the level for detection and accuracy.

Example: "High pressure in the extractor tank shall automatically open the relief valve V14. The set point for maximum pressure is 5 bar and the accuracy must be within +/- 0.2 bar."

Manual mode of trigging:

The manual trig mode of the SIF needs to be described. Are there any restrictions to activate the manual trigging? Describe the use of manual trigging during different modes.

Example: "Manual trigging, pushbuttons mounted near the control panel, shall open the relief valve. The relief valve shall automatically close when the pressure is below 1 bar. During process shutdown the manual trigging shall be disabled. In all other modes the manual trigging is needed."

Trigging response and delay time requirements

Specify the requirements regarding response and time delay.

5.18.8 RESET/RESTART

Describe the reset functions (automatic mode, manual mode) [10.3.1 IEC 61511-1:2003]. Explain the conditions that affect the reset.

Automatic reset:

Example: "The relief valve shall automatically close when the pressure is below 1 bar." "The emergency draining shall stop when low level switch L2 is affected."

Manual reset:

- Explain the conditions that affect the manual reset.
- Reset response and delay time requirements.
- Specify response time requirements. The response time shall not affect the reset or restart.

Example: The relief valve V23 shall close within 5 s when the pressure in tank T22 is below 1 bar.

5.18.9 OVERRIDING, INHIBITING, AND BYPASSING

In some process applications the need of overrides/inhibits and bypass may appear. Describe the requirements regarding overrides, inhibits, and bypasses including how they will be cleared [10.3.1 IEC 61511-1:2003].

Important issues regarding overrides, inhibits, and bypass functions:

- How should the SIF be tested during normal operation
- Are there any requirements regarding key lock or password
- The need of instructions

5.18.10 SPURIOUS TRIPS AND RESET FAILURES

The SRS shall include the maximum allowable spurious trip rate [10.3.1 IEC 61511-1:2003].

Estimated conscience of nuisance trips

The maximum allowable spurious trip is an economical issue. Describe the losses. Specify the estimated consequence and the effort to restore the process to normal conditions.

Maximum allowable reset failure rate

Specify maximum allowable reset failure rate if the SIF uses automatic reset function. The reset function is important in case of preventing hazards during trip conditions, for example, avoid complete draining of the vessel.

5.18.11 FINAL ELEMENTS DESCRIPTION

The typical final element attributes needs to be documented.

Description of output actions

Give a brief explanation of the output action.

Defined fail-safe position of final elements.

Describe the final element and its fail-safe position (open or close).

Justification of the defined fail-safe positions.

Explain why the final element has to be in the defined fail-safe position.

Final elements specification

Specify the final element:

- TAG name
- Type

- Required number
- Actuator action

Requirements for successful operation of final elements

Specify whether there are any specific requirements regarding environmental quality (e.g., temperature, humidity) of the final element.

5.18.12 FAIL-SAFE PROCESS OUTPUT DESCRIPTION

The typical fail safe process output attributes needs to be documented.

- Number of outputs
- I/O name, the name of output
- Device, the connected device to the output
- Trip action (energize, de-energize)

Output circuit requirements

Specify requirements regarding the output circuit safety measures:

- Periodic tests
- Alarm actions
- Feedback

5.18.13 FAIL-SAFE PROCESS INPUT AND TRIP LIMIT DESCRIPTION

The typical fail safe process input and trip limit attributes needs to be documented:

- Type (digital, analog)
- Number of inputs
- Name
- Voting
- Open or closed work circuit (digital input)
- Trip limit (analog input)

Input circuit requirements

Specify requirements regarding the input circuit safety features:

- The need of wire breaks detection
- The need of failure detection

5.18.14 REQUIREMENTS FOR PROOF-TEST INTERVALS

The following details on desired proof-test interval (months) needs to be documented.

Is it possible to execute a fully proof test during operation (yes/no)? If no, is it possible to execute a partial proof test (yes/no).

- Special proof test design requirements.
- Specify the requirements for the proof test.
- Describe the test sequence.

5.18.14.1 Relationship between process inputs and outputs

Give a logical description of the SIF. The description shall be easy to understand.

Trigging and reset:

- Describe the architecture (1oo1, 1oo2, 2oo3, etc.)
- Describe the conditions that trig the SIF (inputs or other communication signals that trig the SIF)
- Describe the conditions that reset the SIF (inputs or other communication signals that reset the SIF)
- Provide time and delay requirements

Actuating:

- describe the actuating of the output
- provide time ad delay requirements
- forced energized or de-energized
- describe bypass modes

5.19 OPERATOR INTERFACES (HMI)

Panels/ buttons:

Describe the use of pushbuttons, key switches, indicators, and others included in the SIF.

Graphics:

Provide a description of the graphics representation (picture) of the SIF. The graphic representation shall indicate the following:

- Included components (switches, transmitters, etc.)
- The position of the included components
- Abnormal modes
- Alarms/warnings

Generation of alarms

Describe the different failure modes that activate alarms (high temperature, low level, high pressure, abnormal conditions, detected errors, valve in an abnormal position, hardware or software errors, etc.)

Generation of events

Important events shall be displayed for the operator, for example, automatic or manual trig of the SIF, affected switches, bypasses, valve position.

Alarm and event logging

Provide a description of alarm and event logging.

5.19.1 REQUIREMENTS FOR PROTECTING THE SIF FROM SPECIAL ENVIRONMENTAL CONDITIONS

Describe the requirements regarding environmental aspects that affect the SIF (temperature, humidity, etc.).

5.19.2 REQUIREMENTS FOR PROTECTING THE SIF FROM MAJOR ACCIDENTS

Specify the requirements that protect the SIF in case of major accidents (fire, explosion, etc.):

- Resisting fire in XX minutes
- The need of instrumentation air
- The need of redundancy (air supply, power supply, etc.)
- Safety devices (relief valves, etc.) and manual safety devices

5.19.3 CONSEQUENTIAL HAZARDS (DUE TO IMPLEMENTATION OF THE SIF)

- Discovered consequential hazards
  Describe consequential hazards that could occur, for example:
  - Mechanical faults (e.g., valve jam)
  - Human behavior (e.g., operation by accident, lack of knowledge)
- Hazards due to concurrently occurring events:
  Describe possible hazards due to concurrently occurring events, example:
  - Fire (pool fire, flash fire, jet fire)
  - Explosion (fireball, physical explosion, vapor cloud explosion)
- Possible risk-reducing measures:
  Describe possible risk-reducing measures, for example,
  - Indicators (level, pressure, temperature, etc.)
  - Monitoring of manual bypasses

5.20 SAFETY INSTRUMENTED SYSTEMS

An SIS is a form of process control usually implemented in industrial processes, such as those of a factory or an oil refinery. The SIS performs specified functions to achieve or maintain a safe state of the process when unacceptable or dangerous process conditions are detected. SISs are separate and independent from regular control systems but are composed of similar elements, including sensors, logic solvers, actuators, and support systems.

The specified functions, or SIFs, are implemented as part of an overall risk-reduction strategy, which is intended to reduce the likelihood of identified hazardous events involving a catastrophic release. The safe state is a state of the process operation where the hazardous event cannot occur. The safe state should be achieved within the process safety time. SIFs are focused on preventing hazardous events with a health and safety or environmental consequence.

The correct operation of an SIS requires a series of equipment to function properly. It must have sensors capable of detecting abnormal operating conditions, such as high flow, low level, or incorrect valve positioning. A logic solver is required to receive the sensor input signal(s), make appropriate decisions based on the nature of the signal(s), and change its outputs according to user-defined logic. The logic solver may use electrical, electronic, or programmable electronic equipment, such as relays, trip amplifiers, or programmable logic controllers. Next, the change of the logic solver output(s) results in the final element(s) taking action on the process (e.g., closing a valve) to bring it to a safe state. Support systems, such as power, instrument air, and communications, are generally required for SIS operation. The support systems should be designed to provide the required integrity and reliability.

5.20.1 PROCESS CONTROL SYSTEMS AND SIS

As illustrated in the below figure, it is generally preferable that any protection system (including an SIS) be kept functionally separate from the BPCS in terms of its ability to operate independent of the state of the BPCS. The operating equipment is also known as the EUC. In essence, protection systems should be capable of functioning to protect the EUC when the process control system is in fault. Where separation is not possible because the safety functions are integral with the process control system (increasingly common in modern complex systems), all parts of the system that have safety-related functions should be regarded as an SIS for the purposes of safety integrity assessment (Figure 5.12).

Following figure shows the basic layout of a typical SIS (in this case controlling a shutdown valve as the final control element) (Figure 5.13).

The basic SIS layout comprises:

- Sensor(s) for signal input and power
- Input signal interfacing and processing
- Logic solver with associated communications and power
- Output signal processing, interfacing, and power
- Actuators and valve(s) or switching devices to provide the final control element function.

The scope of an SIS encompasses all instrumentation and controls that are responsible for bringing a process to a safe state in the event of an unacceptable deviation or failure.

Following design parameters must be considered for an SIS to realize the required SIL.

- The application conditions
- Environment circumstances
- Number and voting of applied components (HFT – hardware fault tolerance)
- Are the components identical? (redundant vs. divers)
- Failure rate of components (failures/h) (λ)
- Safe and dangerous failure fraction %
- Diagnostics coverage % (DC) for safe and dangerous failures

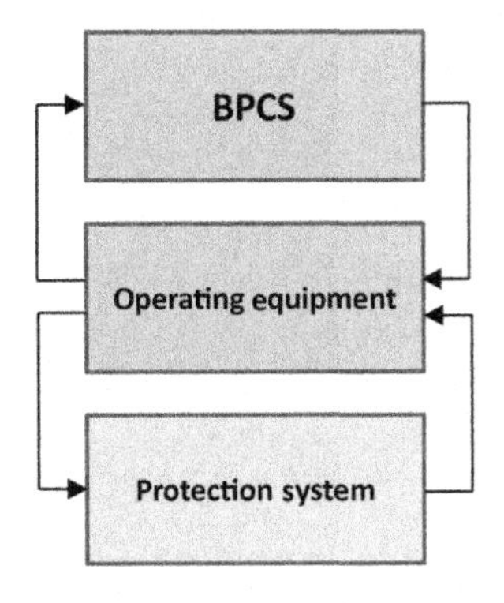

Separation of BPCS and protection system

FIGURE 5.12

Separation of BPCS and protection system.

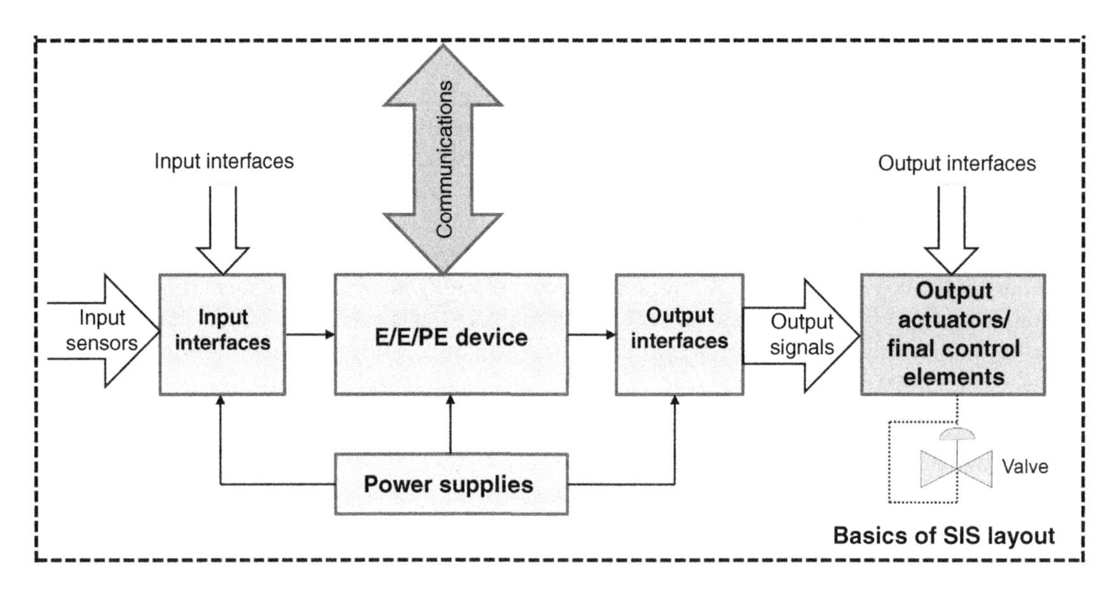

FIGURE 5.13

Basic SIS layout.

- Safe failure fraction % (SFF)
- Mean time to repair and mean time between failures (MTBF)
- Off-line proof test interval
- Proof test coverage
- Common cause factor% (β)
- PFD – probability of failure on demand
- Low-demand/high-demand mode

Safe failure fraction:

Percentage of failures without the potential to put the safety-related system into a dangerous or fail-to-function state.

Probability of failure on demand:

- PFDav average PFD
- λ_{SD} failure rate for all safe detected failures
- λ_{SU} failure rate for all safe undetected failures
- λ_{DD} failure rate for all dangerous detected failures
- λ_{DU} failure rate for all dangerous undetected failures

Hardware fault tolerance:

The ability of a functional unit (hardware) to continue to perform a required safety functions in the presence of faults or errors. A HFT of N means that N+1 fault could cause a loss of the safety function.

Proof test interval:

Periodic test performed to detect failures in a safety-related system.

Mean time between failure

Mean time between failures

Mean time to repair

Mean time to restoration

Voting (MooN):

M out of N channel architecture: Classification and description of safety-related systems regarding redundancy and applied selection process. "N" denotes how often the safety function is performed (redundancy). "M" denotes how many channels have to work properly. Pressure measurement example: 1oo2 architecture – A safety-related system decides that a predefined pressure limit is exceeded when one of two pressure sensor reaches this limit. If a 1oo1 architecture is used, there is only one pressure sensor available.

MooND:

M out of N channel architecture with diagnostics

Low-demand mode:

Mode of operation where the frequency of demands for operation made on a safety-related system is no greater than one per year and no greater than twice the proof-test frequency.

High-demand mode:

Mode of operation where the frequency of demands for operation made on a safety-related system is greater than one per year or greater than twice the proof-test frequency.

Classification of SIS devices:

SIS devices can be classified as Type A or Type B based on following factor.

Type A: The behavior of "simple" (type A) devices under fault conditions can be completely determined. The failure modes of all constituent components are well defined. Such components are metal film resistors, transistors, relays, and others.

Type B: The behavior of "complex" (type B) devices under fault conditions cannot be completely determined. The failure mode of at least one component is not well defined. Such components are, for example, microprocessors and ASICs.

5.21 RELIABILITY AND DIAGNOSTICS

5.21.1 SIS RELIABILITY

What an SIS shall do (the functional requirements) and how well it must perform (the safety integrity requirements) may be determined from hazard and operability studies, LOPA, risk graphs, and so on. All techniques are mentioned in IEC 61511 and IEC 61508. During SIS design, construction, installation, and operation, it is necessary to verify that these requirements are met. The functional requirements may be verified by design reviews, such as failure modes, effects, and criticality analysis, and various types of testing, for example, factory acceptance testing, site acceptance testing, and regular functional testing.

The safety integrity requirements may be verified by reliability analysis. For SIS that operates on demand, it is often the PFD that is calculated. In the design phase, the PFD may be calculated using generic reliability data. Later on, the initial PFD estimates may be updated with field experience from the specific plant in question.

It is not possible to address all factors that affect SIS reliability through reliability calculations. It is therefore also necessary to have adequate measures in place (e.g., procedures and competence) to avoid, reveal, and correct SIS-related failures.

5.21.2 SAFETY FAILURES AND THEIR CAUSES

Failures in a functional safety system can be broadly classified into two categories: systematic and random failures.

Systematic failures

- Result from a failure in design or manufacturing
- Often a result of failure to follow best practices
- Rate of systematic failures can be reduced through continual and rigorous process improvement

Random failures

- Result from random defects inherent to process or usage condition
- Rate of random failures cannot generally be reduced; focus must be on the detection and handling of random failures in the application.

The term SIS has been introduced in the international standard IEC 61511 and covers the equipment from sensors, logic solver, and final elements that is needed to realize the SIF, another IEC term. Reliability with respect to these systems is defined by its ability to command an output to a safe state on a process demand and to function within a required time span without causing a spurious action (e.g., nuisance process trip). The first term has to do with safety integrity as meant by IEC 61508; the second is often presented as process availability, in short availability. The latter is not formally defined in international standards.

IEC 61508 part 2, § 7.4.3.2.1 prescribes: "The probability of failure of each safety function due to random hardware failures, estimated according to 7.4.3.2.2 and 7.4.3.2.3, shall be equal to or less than the target failure measure as specified in the safety requirements specification (see 7.2.3.2)." And adds in note 3: "In order to demonstrate that this has been achieved it is necessary to carry out a reliability prediction for the relevant safety function using an appropriate technique (see 7.4.3.2.2) and compare the result to the target failure measure of the safety integrity requirement for the relevant safety function (see IEC 61508-1, Tables 2 and 3)."

Systematic failures and the human factor are also mentioned in this standard; however, they will not be considered in this context for the sake of clearness.

Failures of SIS can be categorized into two types:

- ***Dangerous failure*** – has the potential to put the SIS in a hazardous or fail-to-function state.
- ***Safe failure*** – does not have the potential to put the SIS in a hazardous or fail-to-function state.

A detected failure is a failure that is detected by the diagnostic tests or through normal operation. Failure rate is represented with the Greek character lambda, λ, and can be broken into many categories.

- λ_S: rate of "safe" failures that do not affect safety function
- λ_{SD}: safe, detected failure rate
- λ_{SU}: safe, undetected failure rate
- λ_D: rate of "dangerous" failures that compromise the safety function
- λ_{DD}: dangerous, detected failure rate
- λ_{DU}: dangerous, undetected failure rate

Failure rate is often expressed in "FITs." One FIT (failure-in-time) = 1 failure per billion hours of operation (1×10^{-9} failures/h).

Both can be demonstrated by calculation; both calculations need the failure rates of components involved as input parameter. Be aware that literature mentions a lot of databases with huge deviation in output. A so-called failure mode effect analysis and/or reliable field experience failure figures are needed to split up the total failure rate of components (λ) into safe (λ_S) and dangerous (λ_D) fractions (Figure 5.14).

The main goal for designing inherent fail-safe systems is the reduction of λ_D, without using any additional test circuitry. However, in general additional diagnostics will be required to make failures manifest. Therefore, it is necessary to make a further division of the failure rates. This is shown in Figure 5.14.

λ_D in above figure is split up in a detected (revealed) part (λ_{DD}) and an undetected (unrevealed) part (λ_{DU}). The relation between these gives the so-called DC factor:

$$\lambda_{DU} = (1 - DC) \times \lambda_D$$

$$SFF = \frac{(\lambda_{SD} + \lambda_{SU} + \lambda_{SD})}{\lambda}$$

Assuming that detected (revealed) failures can be considered as safe, only λ_{DU} can lead to unsafe action. IEC introduced – with respect to this subject – the factor SFF to define the required HFT [ref. IEC 61508-2 Tables 2 and 3].

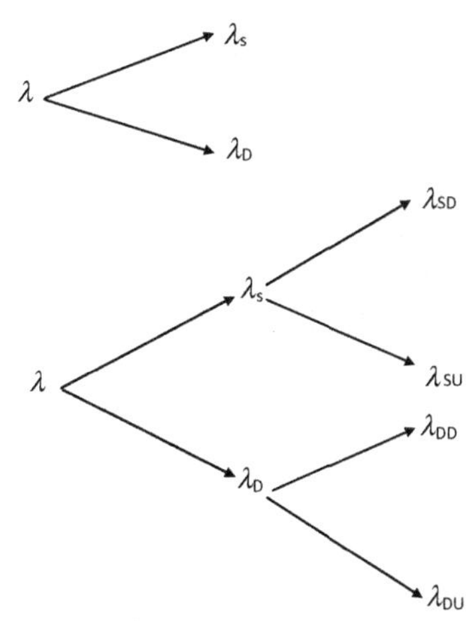

λ = overall failure rate

λ_s = rate of safe failure

λ_D = rate of dangerous failures

λ_{SD} = rate of safe failures, detected

λ_{SU} = rate of safe failures, undetected

λ_{DD} = rate of dangerous failures, detected

λ_{DU} = rate of dangerous failures, undetected

FIGURE 5.14

Division in safe and dangerous failure rates, diagnostics included.

λ_{DU} is used to calculate the PFD, that is, the chance that the safety system will miss the ability to command the output to a safe state in case there is a demand from the process." λ_{SD}, λ_{SU}, and λ_{DD} can cause a before-mentioned spurious action and therefore the sum of these fractions can be used to calculate the process availability.

5.21.3 PROBABILITY OF FAILURE ON DEMAND: THE UNRELIABILITY APPROACH

An unreliability function is calculated as a function of time interval for a specified mission time usually equal to a "proof test" interval for industrial equipment. Then the function is "averaged" over the entire mission time.

This model is used for safety-related systems with the assumption that the system is periodically inspected and tested. It is often assumed that the periodic test will detect all failed components and the system will be renewed to perfect condition (Figures 5.15).

Therefore, the unreliability function is perfect for the problem. It is further reasoned that the system may fail right after the inspection, right before the inspection, or at any time in between. Therefore, PFDavg is the average value of the unreliability function plotted over the inspection period.

It is a well-known equation for a single-channel system with a constant failure rate that: unreliability for a specified mission time, *t*: $F(t) = 1 - e^{-\lambda D} \times t$. This is sometimes called probability of failure, PF. $PFD(t) = 1 - e^{-\lambda d}$.

For one failure mode, fail-danger, l = ld, and the probability of failure in the dangerous mode:

This is approximated by $PFD(t) = \lambda_d t$

The approximation works acceptably when the result is small with a result >0.1, having an error of <3%. Since all SILs require a PFDavg value >0.1, the approximation is acceptable.

PFDavg is obtained by arithmetic average during the time interval *T* $PFD_{avg=1/T \int_0^T pfd(t)dt}$.

Using the approximation:

This *T* is sometimes the proof test interval T_i of the devices. $PFD_{avg} = \lambda_d T/2$

Type A: Simple devices where all faults known and describable

Safe failure fraction (%)	Hardware fault tolerance		
	0	1	2
<60	SIL 1	SIL 2	SIL 3
60–90	SIL 2	SIL 3	SIL 4
90–99	SIL 3	SIL 4	SIL 4
>99	SIL 3	SIL 4	SIL 4

Type B: Complex devices where not all faults known and describable

Safe failure fraction (%)	Hardware fault tolerance		
	0	1	2
<60	Not allowed	SIL 1	SIL 3
60–90	SIL 1	SIL 2	SIL 3
90–99	SIL 2	SIL 3	SIL 4
>99	SIL 3	SIL 4	SIL 4

FIGURE 5.15

HFT on Safe failure fraction

5.21.4 COMPONENT RELIABILITY AND UNRELIABILITY

The reliability and unreliability of SIS components can be described via the following equations:

Reliability $R(t) = e^{-\lambda t}$

Unreliability $F(t) = 1 - e^{-\lambda t}$ (Figure 5.16).

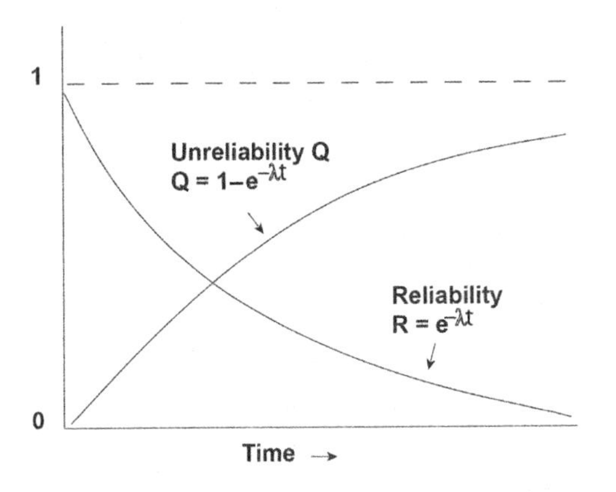

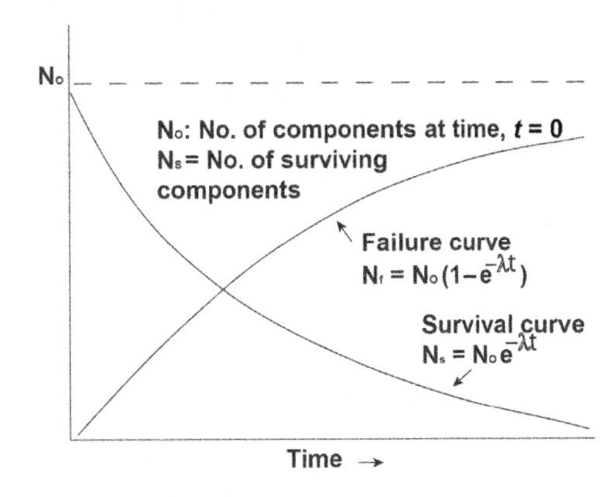

FIGURE 5.16

Reliability curves with time

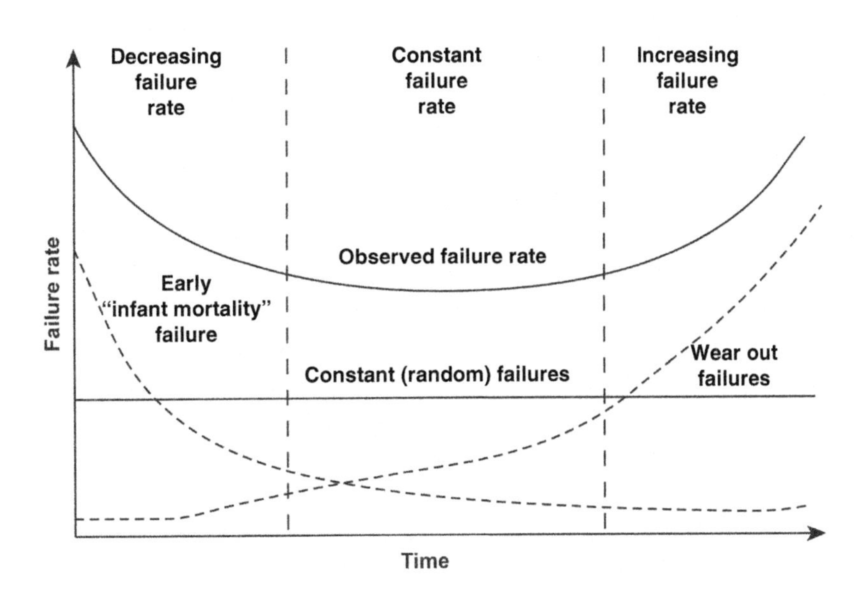

FIGURE 5.17

Failure approximation over time

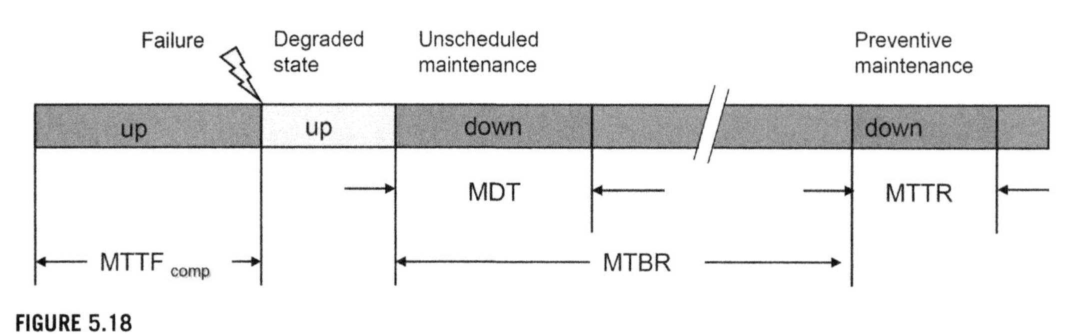

FIGURE 5.18

Repair and maintenance

5.21.5 HARDWARE RELIABILITY

The fundamental principles of reliability engineering are common to both hardware and software. However, software and hardware have basic differences that make them different in failure mechanisms. Hardware faults are mostly physical faults, whereas software faults are design faults, which are harder to visualize, classify, detect, and correct. Design faults are closely related to human factors and the design process. In hardware, design faults may also exist, but physical faults usually dominate. The quality of software will not change once it is uploaded into the storage and start running. Trying to achieve higher reliability by simply duplicating the same software modules will not work.

The following figures depict the profiles of software and hardware reliability against time. In the below diagram, it is seen that hardware passes through three phases of reliability, namely, burn-in phase, useful life phase, and wear-out (Figure 5.18).

5.21.6 FAILURE RATE, REPAIR RATE, AND THE STABILITY

The mean time between repairs (MTBR) is the average time between human interventions. The MTBF is the average time between failures. If the systems can autorepair itself, the MTBF is smaller than the MTBR.

5.22 SIS VOTING PRINCIPLES AND METHODS

Probably one of the most important design parameters to realize a highly reliable PLC is the concept of fault tolerance. Systems become fault-tolerant if multiple channels are used and not all of them are needed to fulfill the safety function. Fault tolerance is normally expressed as a specific voting scheme. For example, 1oo2 voting (one-out-of-two) implies that only one channel of a two-channel system is required to fulfill the intended safety functions.

In case of a “safe” failure of one the channels (e.g., switching from logic 1 to logic 0), the system will bring the process to a safe state. Similarly, a 2oo2 voting system requires healthy operation of both channels to fulfill the safety function. However, one safe failure does not lead to a spurious process trip.

FIGURE 5.19

FIGURE 5.20

As defined in ANSI/ISA S84.01-96, an MooN voting system requires at least M of the N channels to be in agreement before the SIS can take action (M out of N). The following voting principles are most commonly used:

- 1oo1 (one out of one)
- 1oo2 (one out of two)
- 2oo2 (two out of two)
- 2oo3 (two out of three) – TMR (triple modular redundant)
- 1oo2D (one out of two with diagnostics)
- 2oo4D (two out of four with diagnostics) – QMR (quad modular redundant)

1oo1: The 1oo1 voting principle involves a single-channel system, and is normally designed for low-level safety applications. If such a system is characterized by a high level of DC, this voting is expressed as 1oo1D. The "D" denotes the applied concept of automatic failure diagnostics. Nevertheless, this voting system is zero-fault tolerant, which means that a system failure will always and immediately result in the loss of the safety function or shutdown of the process (Figure 5.19).

1oo2: The 1oo2 voting principle was developed to improve the safety integrity performance of 1oo1-based safety systems. If one channel fails in a dangerous mode, the other one is still able to fulfill the safety function. Unfortunately, this concept does not improve the spurious trip rate. Even worse, the probability of a spurious trip is almost doubled (Figure 5.20).

2oo2: The major disadvantage of a single (i.e., nonredundant) safety system is that a single failure in a safety mode immediately results in a process trip. Duplication of the channels and application of the 2oo2 voting concept significantly reduces the probability of a spurious trip, because both channels must fail in a safe mode before the system shuts down the safeguarded process. On the contrary, the system does have the disadvantage that the PFD is two times higher than that of a single system.

2oo3 (TMR): In 2oo3 voting (also called TMR, or triple modular redundant), there are three channels, two of which need to operate healthy in order to fulfill the safety functions. This voting concept is therefore also one-fault-tolerant for safety. The 2oo3 voting principle is best applied if there is a clear and thorough physical separation of the microprocessors. However, this does require them to be located on three different modules, which results in a "heavily equipped" hardware system (Figures 5.21 and 5.22).

And the conceptual diagram is (Figure 5.23)

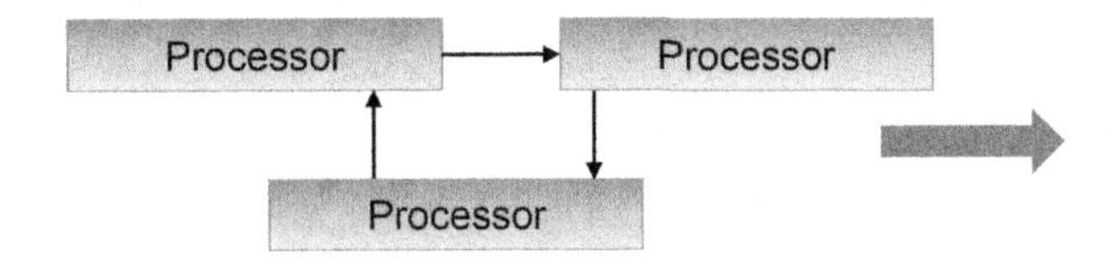

FIGURE 5.21

2003 Processor voting logics

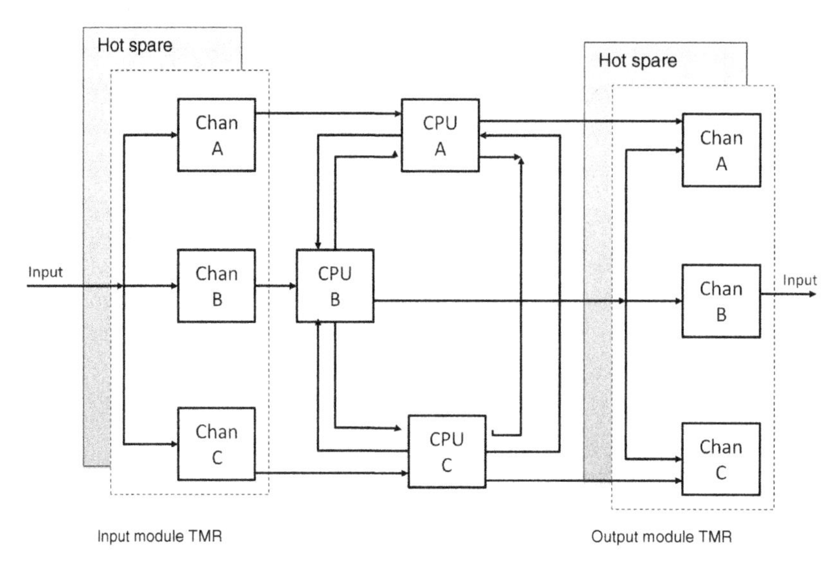

FIGURE 5.22

Functional model of a TMR

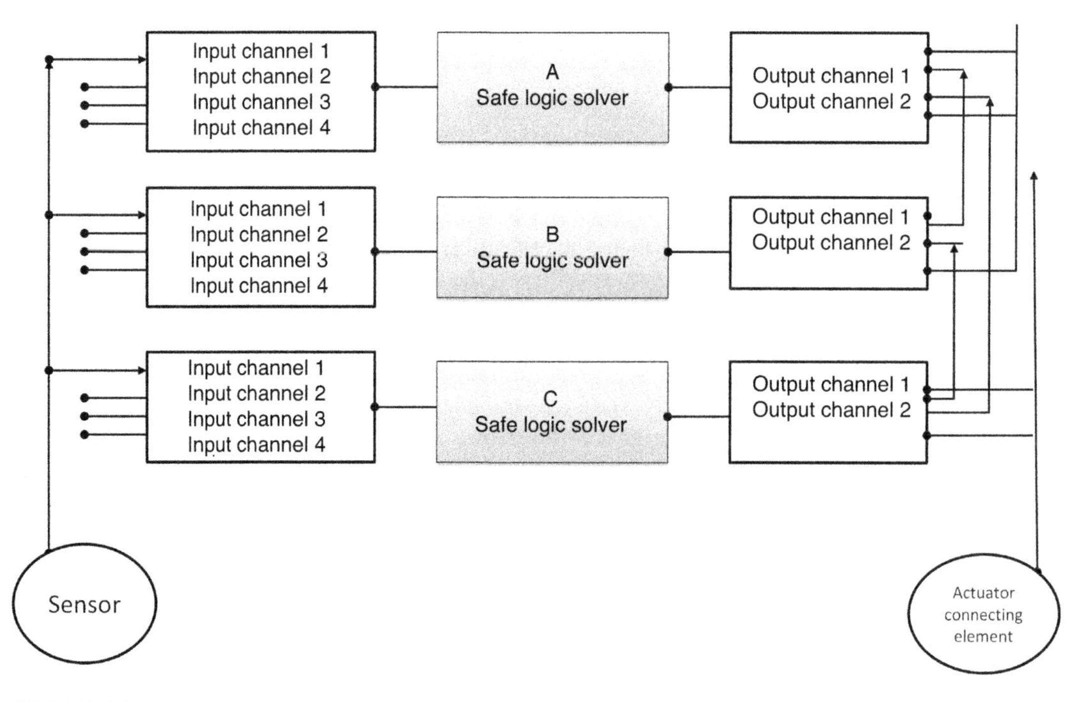

FIGURE 5.23

Conceptual model of a triple modular system

Although the latest systems have been enhanced by an increasing level of diagnostics, 2oo3-based (TMR) safety systems still retain the disadvantage of having a PFD, which is about three times higher than that of 1oo2-based systems. Furthermore, some manufacturers of TMR-based safety logic solvers have unfortunately designed their system in such a way that all three CPU microprocessors are located on a single printed circuit board. It is not hard to imagine the impact this has on the probability of a common cause failure.

1oo2D: During the development of the second-generation safety-dedicated PLCs, the advantages of the 1oo2 and 2oo2 voting principles were combined, without the disadvantages of the less reliable and "heavy" 2oo3 voting system. The 1oo2 concept has an excellent performance with regard to safety, but its availability performance is not fault-tolerant. Therefore, a new voting concept was designed called 1oo2D. As mentioned before, "D" stands for "Diagnostics," because such systems are characterized by a high level of automatic system self-testing. The impact on voting is such that a single, automatically detected failure will not immediately lead to loss of the safety function or a process trip, but the affected channel will be isolated, and operation will continue through the healthy channel. As soon as the DC approaches really high levels (as is the case with 1oo2D), the negative impact of common-cause failure for the 2oo3 concept will exceed the probability of a spurious trip owing to a safe undetected channel failure. This is why extensive calculations in the past have shown that the 1oo2D concept performs better with regard to safety integrity as well as system availability using less hardware. 1oo2D systems are therefore often said to achieve the safety levels of a 1oo2 system and the availability levels of a 2oo2 system (Figure 5.24).

2oo4D (QMR): During the development of the second-generation safety-dedicated PLCs, the advantages of the 1oo2 and 2oo2 voting principles were combined. A new, third-generation breed of safety PLCs is currently emerging, which is characterized by a two-fault-tolerant, two-level system. Redundant central parts each contain two main processors, and because only two modules are used to achieve quadruple redundancy, the probability of common-cause failure is even further reduced compared with previous voting principles. This architecture is called 2oo4D (with "D" again signifying "Diagnostics" to indicate the high level of DC) (Figure 5.25).

Section 5 on page 8 explains the concept and advantages of the new generation 2oo4D PLCs in more detail (Figure 5.26).

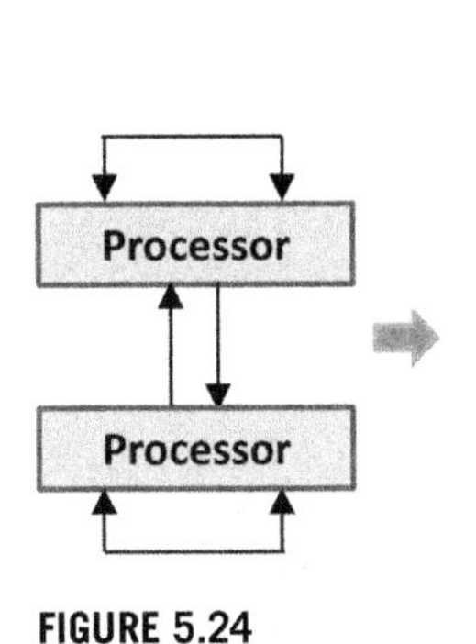

FIGURE 5.24

Conceptual 1oo2D

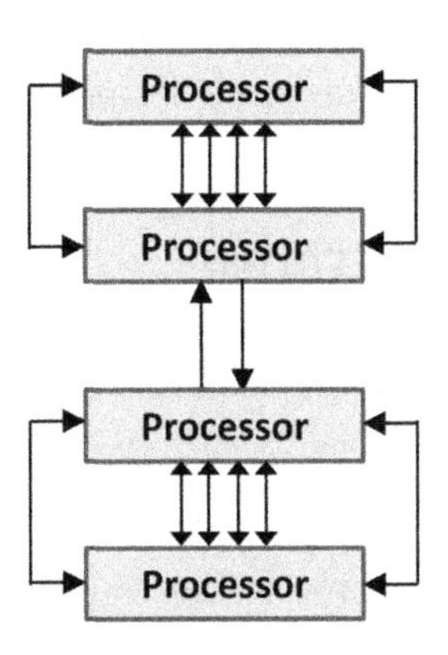

FIGURE 5.25

Conceptual 2oo2 System

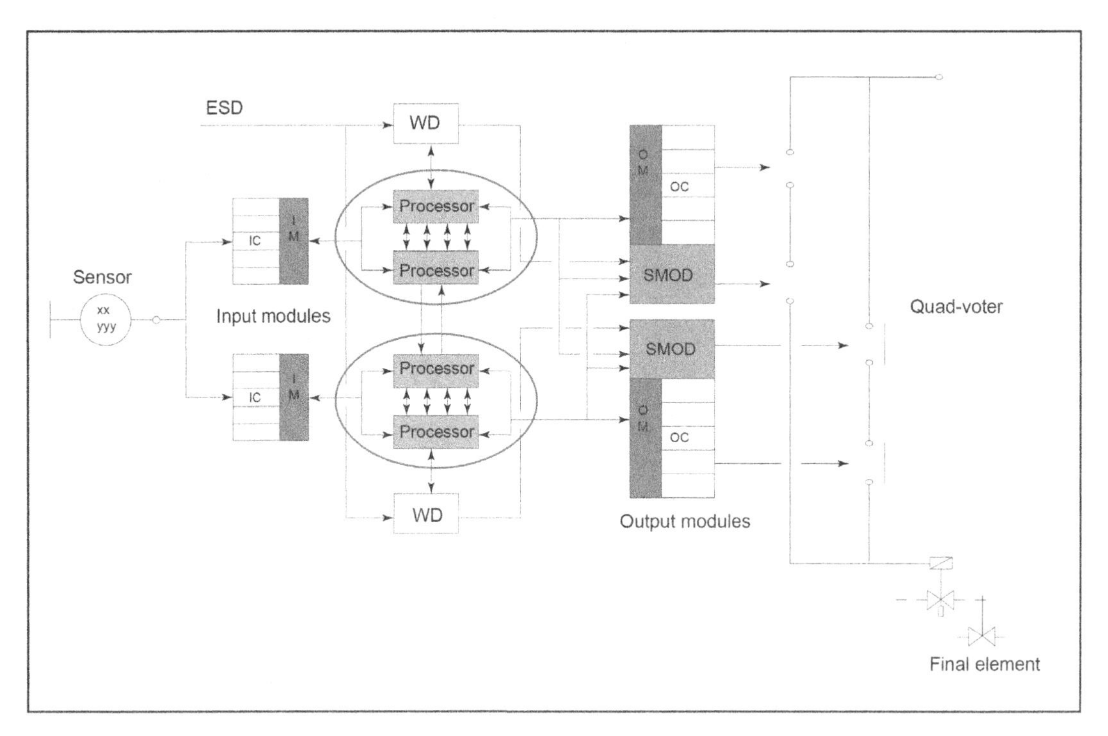

FIGURE 5.26

2oo4D system architecture.

5.22.1 QMR SYSTEM

In QMR systems, the safety-related PLC design has been enhanced to support two CPU modules, each equipped with two main processors. The 2oo4D voting is realized by combining 1oo2 voting for both main processors on one module and 1oo2D voting between the two modules (i.e., between central parts) (see Figure 5.4). Voting is therefore applied on two levels: on a module level and between the central part modules (Figure 5.27).

5.22.2 TMR SIS ARCHITECTURE

Fault tolerance TMR is achieved by means of triple-modular redundant architecture. The controller provides error-free, uninterrupted control in the presence of either hard failures of components or transient faults from internal or external sources (Figure 5.28). The controller is designed with a fully triplicated architecture throughout, from the input modules through the main processors to the output modules. Every I/O module houses the circuitry for three independent legs. Each leg on the input modules reads the process data and passes that information to its respective main processor. The three main processors communicate with each other using a proprietary high-speed bus system. Once per scan, the three main processors synchronize and communicate with their two

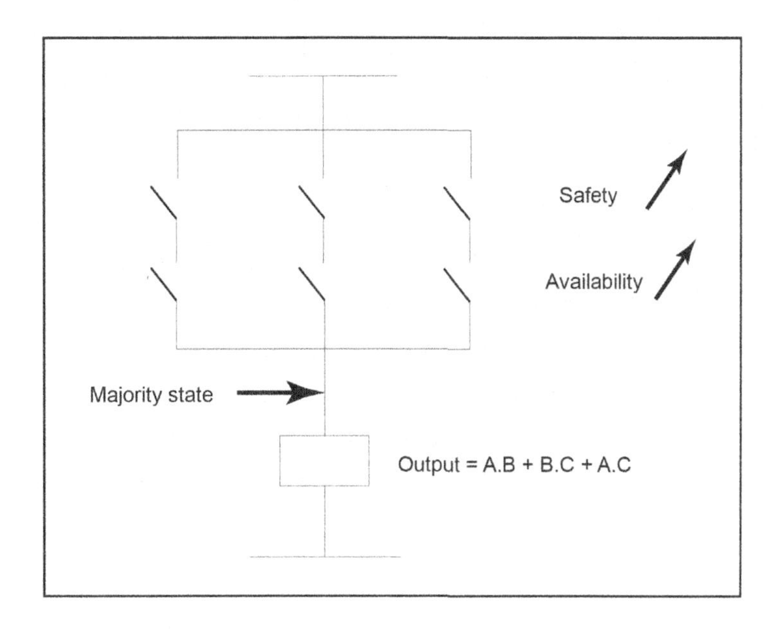

FIGURE 5.27

Logic in 2oo3 System

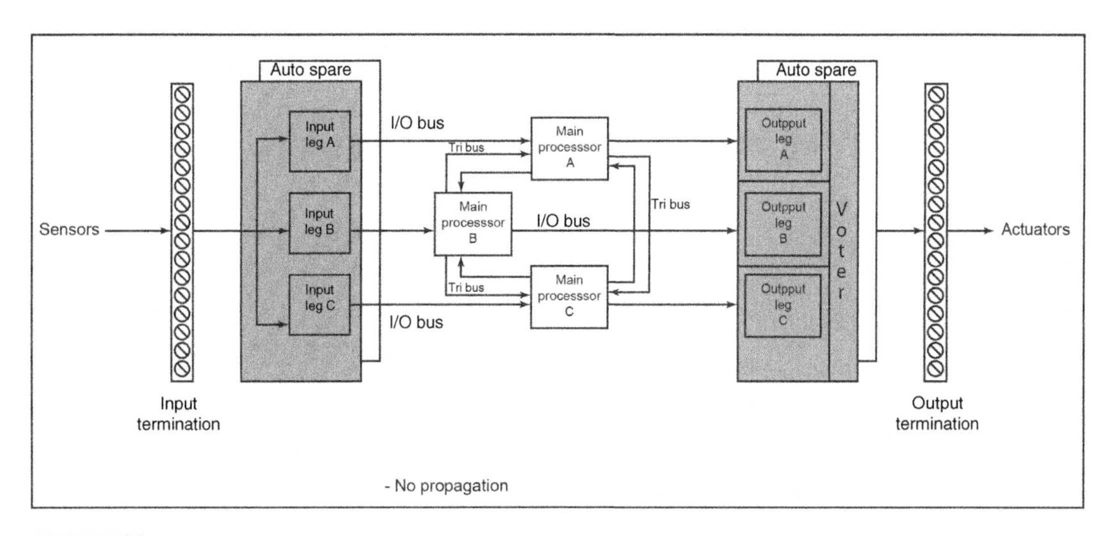

FIGURE 5.28

TMR Architecture

neighbors. The controller votes digital input data, compares output data, and sends copies of analog input data to each main processor. The main processors execute the user written application and send outputs generated by the application to the output modules. In addition to voting the input data, the controller votes the output data. This is done on the output modules as close to the field as possible, in order to detect and compensate for any errors that could occur between the controller voting and the final output driven to the field.

Similarly, if the system is returned to state $P(n)$ from state $P(n + 1)$ at an average rate of μ_n (by repair or replacement), the rate at which the system returns to the unfailed state is given by the product $P(n + 1)^* \mu_n$.

5.23 SIS SIL LEVEL CALCULATION TOOLS

There are a number of software available from different SIS manufacturers for calculating the SIL level of the SIFs. Most of these software tools assist in calculating functional safety factors other than SIL, such as

- Target PFD calculation(s)
- Testing interval calculation(s)
- Mean time between failures (MTTR)
- Mean time to repair (MTBR)
- LOPA

5.24 SIS COMMUNICATION PROTOCOLS AND FIELD-BUSES

The usage of communication protocols in safety instrumented systems and SIFs is a very recent development in functional safety domain. There are protocols that provide very high data reliability while transferring data between field devices and the SIS or between two different SIS. Following are few field-level and SIS–SIS communication protocols that are available in the market.

Safety-related field buses:

1. Profisafe
2. FF-SIS
3. CIP Safety (DeviceNet Safety)
4. MTL SafetyNet

Safety-related communication protocols:

1. SafeNet
2. CIP safety
3. Sercos-I, II, III
4. SafetyBUSp

The type of communication protocol that is suitable for a SIL 2 or SIL 3 system is really dependent on the type of platform that is being used. Options include, but are not limited to, 4–20 mA output signal, ControlNet (Allen Bradley), DeviceNet Safety (Allen Bradley), SafetyNet (MTL), and PROFIsafe. Currently, the ISA SP84 committee is working on developing guidelines for a safety bus, to make sure that the foundations comply with IEC 61508, and IEC 61511 standards. The first devices with a safety bus were made available by 2008. The Fieldbus Foundation is actively involved in the committee and working on establishing Foundation Fieldbus Safety Instrumented Systems (FF-SIS) project to work with vendors and end-users to develop safety bus specifications.

The following protocols will be discussed in detail in coming sessions:

- FF-SIF
- ProfiSafe

5.25 FF-SIS: FOUNDATION FIELDBUS FOR SAFETY INSTRUMENTED SYSTEMS

In January 2006, the Fieldbus Foundation announced that TÜV Rheinland Industrie Service GmbH, Automation, Software and Information Technology, a global, independent, and accredited testing agency, had granted Protocol Type Approval for its safety specifications. The foundation technical specifications – SIFs are in compliance with IEC 61508 standard (functional safety of electrical/electronic/programmable electronic safety-related systems) requirements up to, and including, SIL 3.

With the TÜV Protocol Type Approval, FOUNDATION technology has been extended to provide a comprehensive solution for safety-instrumented functions in a wide range of industrial plant applications. The specifications enable manufacturers to build FOUNDATION-registered devices in compliance with IEC 61508. Third-party test agencies such as TÜV will certify whether these devices are suitable for use in safety-instrumented functions. End-users will be able to choose devices meeting the requirements of IEC 61511 (functional safety: SIFs for the process industry sector) from multiple suppliers, instead of being restricted to devices designed specifically for a proprietary safety system platform. IEC 61511 is also available as an ANSI/ISA standard: ANSI/ISA-84.00.01-2004.

The SIF project was initiated by end-users and approved by the Fieldbus Foundation's board of directors in October 2002. Companies participating in the project included ABB, BIFFI, BP, Chevron, Dresser-Masoneilan, Emerson Process Management, Endress + Hauser, Fieldbus Diagnostics, HIMA, Honeywell, Invensys, Magnetrol International, Metso Automation, Moore Industries, MTL, Pepperl + Fuchs, Risknowlogy B.V., Saudi Aramco, Shell Global Solutions, Siemens, Smar, Softing, TopWorx, TÜV Rheinland, TUV SUD, Westlock Controls/Tyco, and Yokogawa. The development team achieved its first major milestone at the end of 2003 with TÜV approval of the overall system concept. The development team met with external experts at a meeting hosted by Shell Global Solutions in Amsterdam, the Netherlands, in March 2004 to review the initial specifications. Comments from this review were resolved and the management team developed the top-level project plan for laboratory validation testing. During the laboratory-test phase, conducted at the BIS Prozesstechnik GmbH facility in Frankfurt, Germany, each prototype supplier independently implemented the foundation's safety instrumented systems specifications. In parallel, the test team separately developed test cases and prepared expected test results.

Extensive laboratory testing and application analysis has verified that the foundation's safety instrumented systems technology meets the needs of industrial end-users, who regard these systems as critical to their overall plant operating strategy. TÜV Type Approval will help meet the growing worldwide demand for commercial, standard-based, safety instrumented system products incorporating FOUNDATION technology. End-users can now adopt the powerful FOUNDATION diagnostics, and at the same time, maintain the protection in up to a SIL 3 environment. No changes were required to the

	OSI Model	Fieldbus model	FF_SIS model
		User layer	User layer
Layer 7	Application	Communications stack FMS FAS Data link layer	FF-SIS diagnostics Communications stack FMS FAS Data link layer
Layer 6	Presentation		
Layer 5	Session		
Layer 4	Transport		
Layer 3	Network		
Layer 2	Data Link		
Layer 1	Physical	Physical	Physical

FIGURE 5.29

OSI Communication Model of SIS

existing H1 protocol to add the safety instrumented systems protocol extensions, clearly indicating the value of the comprehensive, forward-thinking design of FOUNDATION technology.

5.25.1 FF-SIS BLACK CHANNEL APPROACH

The following picture depicts how the different layers in OSI network model, the Foundation Fieldbus and FF-SIS can be arranged (Figure 5.29).

As you could see, the FF-SIS diagnostic is one major distinguishing feature between an FF segment and an FF-SIS segment (Figure 5.30)

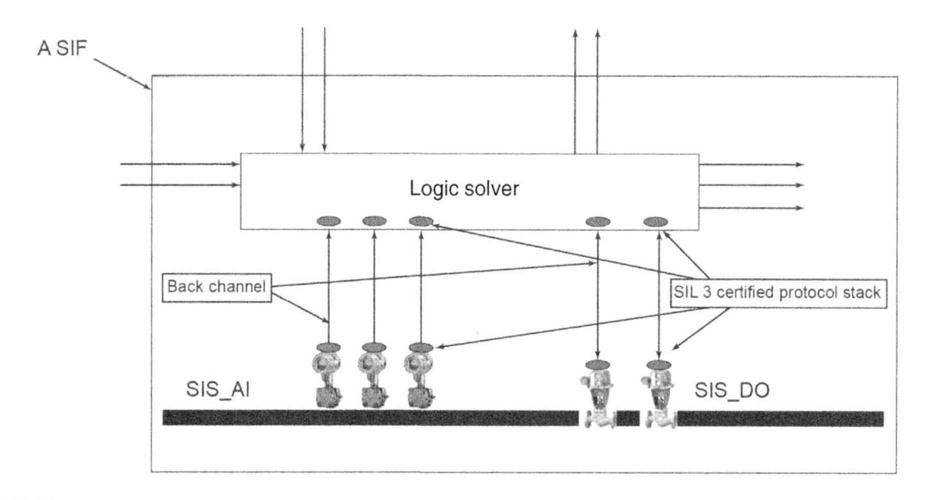

FIGURE 5.30

Architecture of SIF

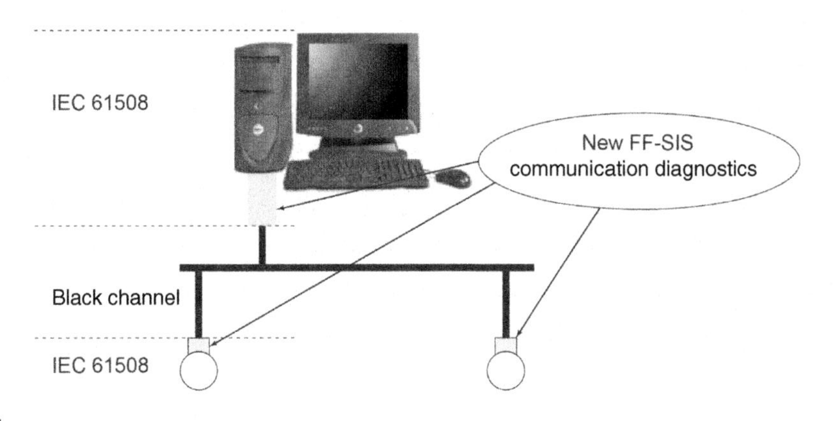

FIGURE 5.31

Black Channel Approach

The two primary approaches for communication networks in safety systems are the white channel and the Black Channel. The white channel is essentially an entire communication scheme that is designed from the bottom up for safety applications. Of course, all of this specialized equipment requires certification and implies added cost, including hubs switches, and so on. The Black Channel comes from the "black box" concept and uses a regular transmission system and regular network hardware. The safety layer is essentially built into the network and exists between the communication stack and the application according to the IEC 62280-1 standard. The safety layer of the Black Channel approach handles all the safety-related applications and requirements. With the Black Channel concept, end-users can even connect non-safety devices to the safety network because the protection layers are built into the network. Everything can share a common network and costs are greatly reduced.

FF-SIF uses the existing H1 protocol as the "Black Channel" network between the logic solver and the SIS devices. It is important to note that no changes were made to the fundamental H1 protocol for implementation in SIS, but additional device diagnostic functions and fault detection capabilities were required. In addition to the device diagnostics, the primary advantage to using FF-SIF as the Black Channel for safety applications is the network diagnostics that are possible. Traditional analog-based networks lack the ability to detect noise, corruption, and faults in the network (Figure 5.31).

5.25.2 IN BLACK CHANNEL APPROACH

- H1 communication system (Black Channel) is unchanged.
- A new FF-SIS protocol above the Black Channel detects network faults and appropriate action is taken without human intervention.

5.25.3 FF-SIS NEW FUNCTION BLOCKS

FF-SIS has come up with New Function Blocks for FF-SIS Applications – FBAP Part 6. New FBAP diagnostics detect application faults and appropriate action is taken without human intervention.

Following are the new Function Blocks added:

- Function Block Application Process – Part 6
    - SIS Write Lock
    - SIS Discrete Input
    - SIS Analog Input
    - SIS Discrete Output
    - SIS Analog Voter
    - SIS Discrete Voter
    - SIS AND/OR/XOR

Diagnostics and statistics added to device resource block.

New FF EDDL development and researches are going on currently, and in near future, FF-SIS has a good potential to become widely used safety-related Fieldbus with its capabilities.

5.26 PROFISAFE

Industrial automation has profited tremendously from technology advances in recent years. Industrial Ethernet, distributed intelligence, "smart" safety, and wireless networking are just a few of the enabling technologies that have brought tangible value to end-users. However, manufacturers are traditionally arch-conservative decision-makers and do not accept new a technology until it is well proven, or until investments in these technologies can be corroborated with solid business arguments. Faced with a landscape of both fact and hype, manufacturers are learning how to match technology benefits to specific business metrics.

Two topics in particular – industrial Ethernet and safety – have caught the attention of manufacturers in recent years thanks to their potential to cut costs and improve business performance. Industrial Ethernet has unified network architectures and created a high level of data transparency from which applications such as Asset Management or Manufacturing Execution Systems profit tremendously. Safety has evolved from being a cost burden and "necessary evil" to a strategy for improving productivity and reducing downtime.

Profibus International, an industrial consortium of automation suppliers, has mated safety with industrial networking to create solutions for "networked safety" based on all available media, including Profibus DP and PA, Profinet (industrial Ethernet), and wireless. End-users are now taking advantage of these solutions to unify their network architectures and eliminate the need for a second, parallel bus while maintaining conformity with safety standards up to SIL 3. Benefits of networked safety include shortened start-up times, lower wiring costs and, in the long term, faster and more efficient maintenance (Figure 5.32).

On the industrial networking side, a parallel development to this consolidation has taken place as well. Thanks to developments in safety technology, running two separate busses for safety and nonsafety data is simply no longer commensurate with modern automation philosophy. A two-bus architecture requires double the amount of training, twice the network access hardware, and makes startup and troubling shooting tasks unnecessarily complex.

With the introduction of PROFIsafe, the safety protocol that is part of the communication protocols of both Profibus and Profinet, end-users can now eliminate the need for a separate safety network and

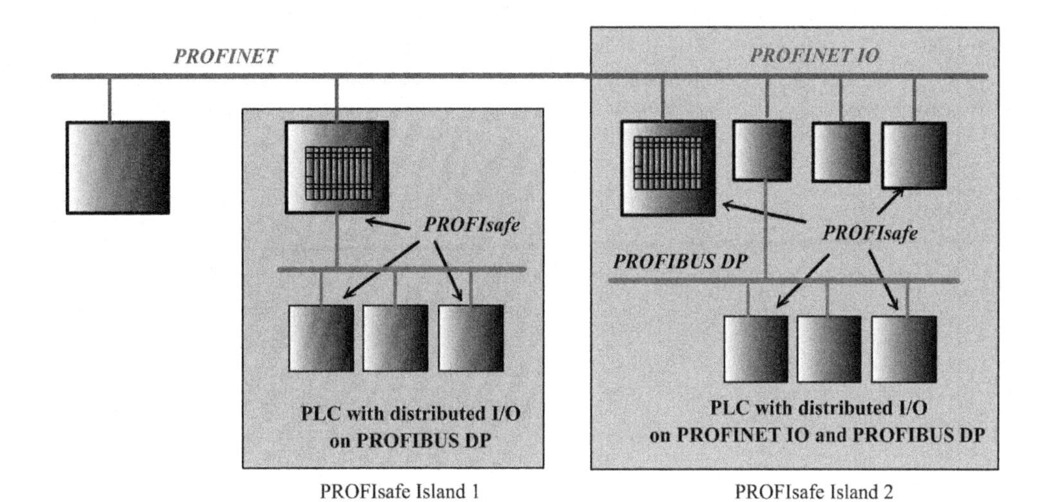

FIGURE 5.32

Architecture of Profibus based Safety System

reduce their industrial network architectures to a single bus. PROFIsafe extends the standard Profibus communications protocol to address special requirements for safety-related information necessary to conform to strict safety standards. For example, PROFIsafe adds elements such as message numbering and data consistency checks to rule out typical network messaging faults, enabling networked safety devices to meet the reliability requirements of SILs (to SIL 3) prescribed by international safety standards. Because PROFIsafe is built into the communications protocol, it can be used by devices connected to any PI network: Profibus DP and PA or Profinet, as well as Profinet over wireless. PROFIsafe works with safety devices on Profinet in the same way as with Profibus. Because it is part of the communications protocol, transmission of safety data occurs automatically and seamlessly, even between different network media (Figure 5.33).

A large portion of the classic process industries actually contain many applications typically associated with discrete or factory automation. Chemical plants or wastewater treatment plants, for example, often employ motor control centers or discrete I/O modules together with process instrumentation. The higher the discrete application content, the more the industry is classified as "hybrid" rather than pure process or discrete.

Profibus PA offers communication to field instruments in process applications that require a different network medium, for example, for use in explosive environments. Because Profibus PA uses the same communications protocol as Profibus DP, devices on both networks can communicate safety data via Profisafe seamlessly without having to worry about bridges or gateways.

In the process industries, PROFIsafe plays an important role in supporting highly available, redundant systems using flexible modular redundancy (FMR). Using a Fieldbus allows a system to tolerate multiple faults without interruption and provides for I/O redundancy independent of CPU redundancy. FMR architectures with PROFIsafe are certifiable up to SIL 3 according to IEC 61508 (Figure 5.34).

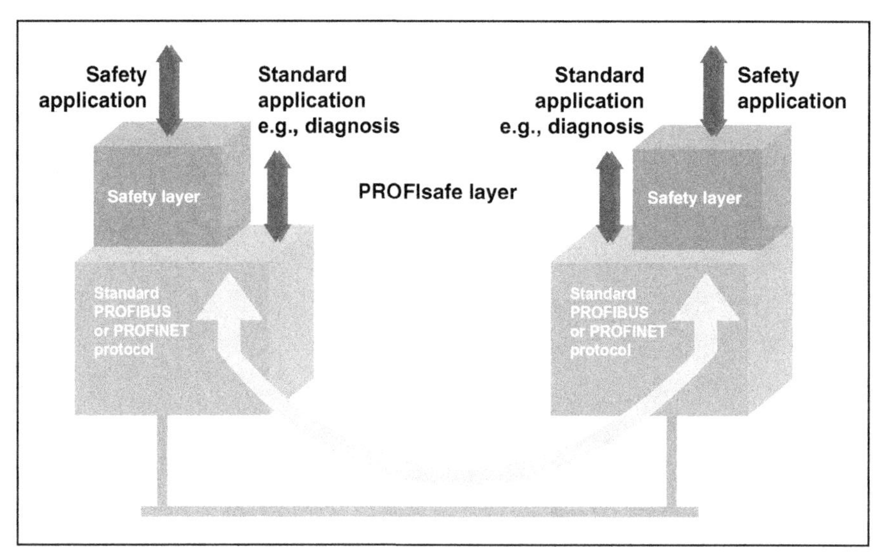

FIGURE 5.33

Profinet based safety system

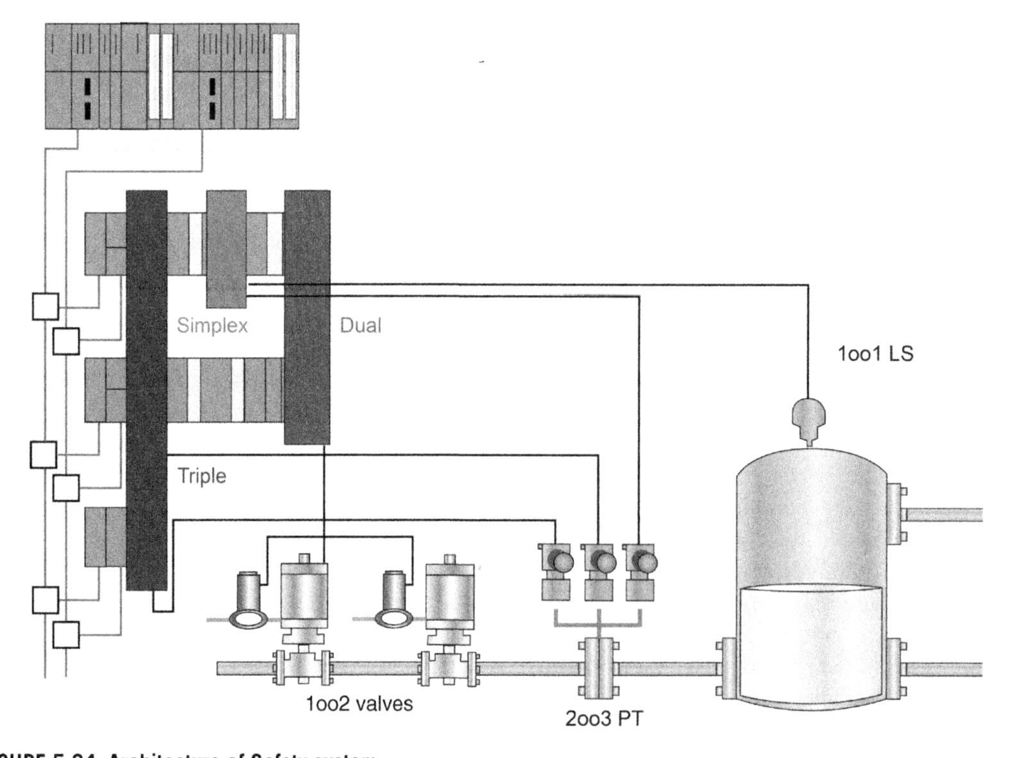

FIGURE 5.34 Architecture of Safety system

FMR architectures tolerate multiple faults without interruption and allowed independent I/O redundancy.

5.27 PROFIsafe PROTOCOL

The PROFIsafe system is an extension of the Profibus and PROFINET system. Freely programmable safety functions can be executed with the system and the required safe input and output data can be transmitted from and to the safe I/O devices (Figure 5.35).

5.27.1 PROFIsafe PROTOCOL

The safe controller and the safe bus devices communicate with one another via the PROFIsafe protocol, which superposes the standard Profibus or PROFINET protocol and contains the safe input and output data as well as data security information.

5.28 BLACK CHANNEL PRINCIPLE

PROFIsafe uses the Black Channel principle exactly the same way as INTERBUS-Safety for the transmission of safe data via a standard network.

The safe data, consisting of the purely safety-related user data and the protocol overhead required for protection, are transmitted via Profibus or PROFINET together with data that are not safety-related.

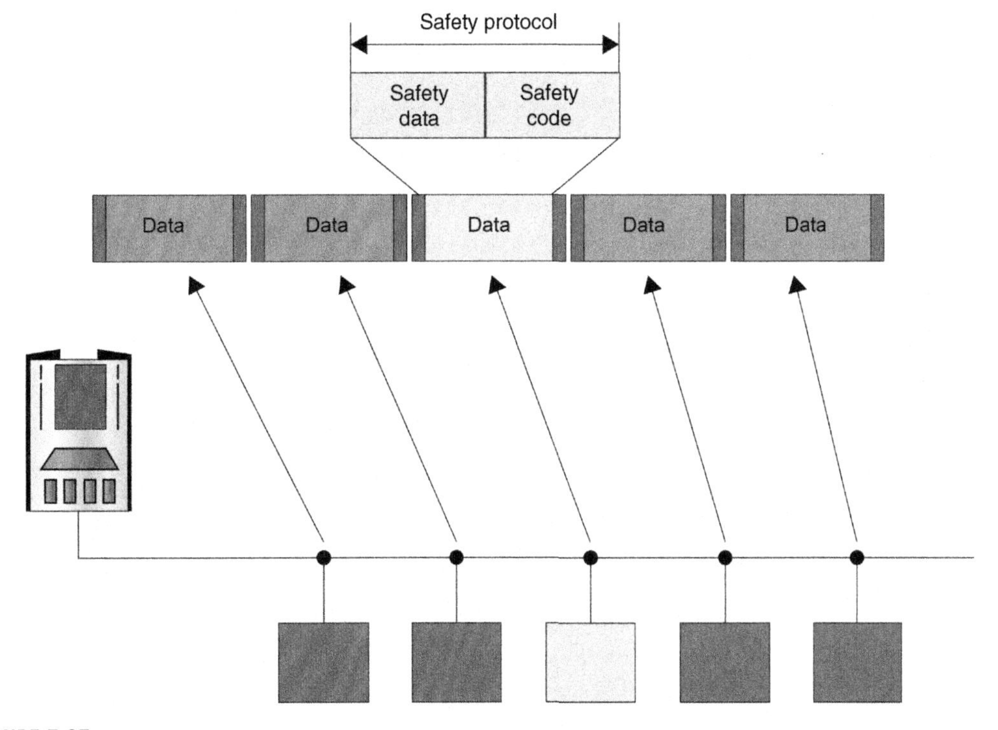

FIGURE 5.35

Profisafe system

5.29 INTEGRATED SAFETY DATA COMMUNICATIONS

The integrated safety mechanisms protect against the following possible errors (Figure 5.36):

- Repetition of messages
- Loss of messages
- Insertion of messages
- Incorrect order of messages
- Data corruption
- Delay of messages
- Recurring memory error in switches
- Reversal of devices

Response time

The minimum distance of the safety equipment to danger depends on the following factors (Figure 5.37):

- Delay time of the sensor
- Processing time of the safety program in the safety controller including the network transmissions

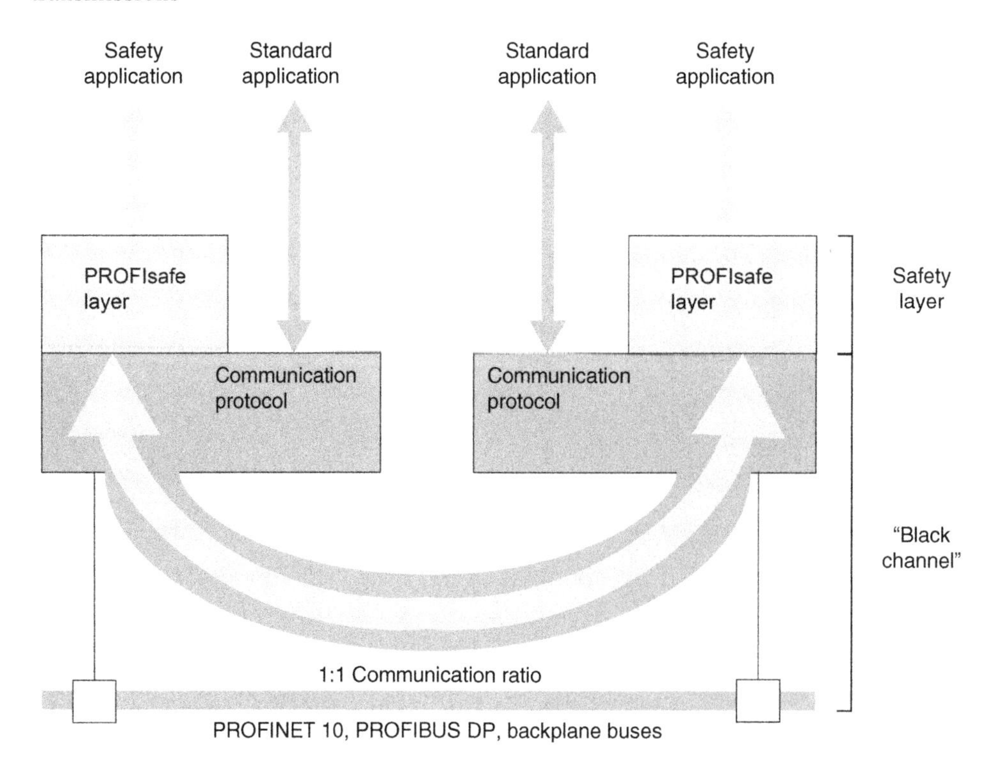

FIGURE 5.36

Safety Comunication

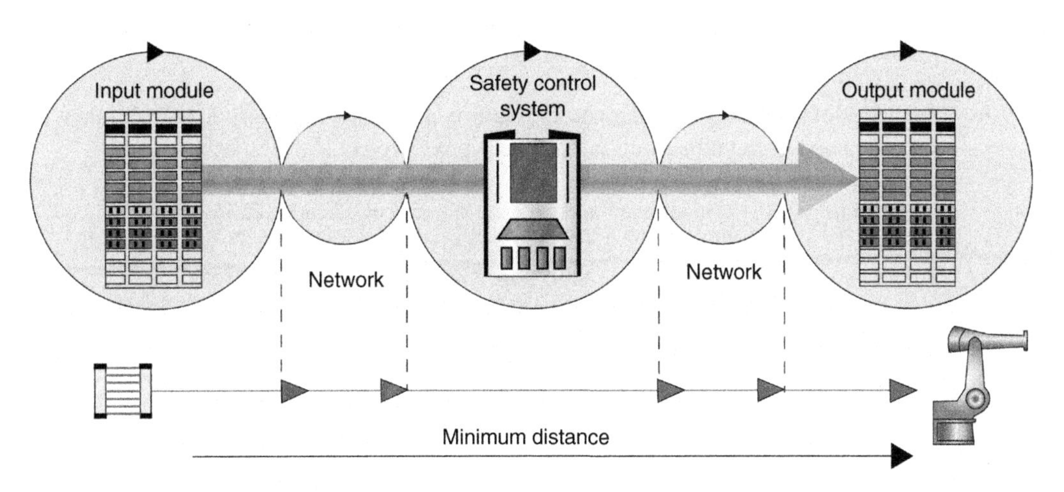

FIGURE 5.37

Safety Communication response times

- Processing and filter times in the input and output modules
- Delay or over travel time of the actuator

Response times with the PROFIsafe system

The delay times of the sensors and actuators are independent of the safety protocol used and must be viewed separately. Owing to unsolicited, unsynchronized individual processes, the response time is determined as the sum of the following values:

- Watchdog time of the input module
- Set watchdog time of the network from the input module to the controller
- Set watchdog time of the safe controller
- Set watchdog time of the network from the controller to the output module
- Watchdog time of the output module

5.30 SELECTION OF SAFETY INSTRUMENTED SYSTEM

The scope of the criteria covers the needs of SIS systems including communications with other systems, instrumentation, and networks. The guide allows those involved in an SIS selection process to make quicker and improved decisions.

The need for an SIS has never been more important than it is today. In addition, a modern SIS needs to be able to easily integrate with other systems and technologies from a distributed control system to HMI and SCADA. It is necessary to ensure not only the uninterrupted operation but also the reliability and safety of processes under all conditions.

An SIS selection has become mission critical. The total solution has become complex and functionality involves the combination of an expanded range of capabilities and specific technology requirements. Although product plays a major role, suppliers have specific domain expertise, geographical presence, and knowledge of certain industry dynamics. These must all be evaluated in a supplier selection process.

A. Company issue

1. Knowledgeable staff
   - Knowledge of marketing and engineering staff on available/proven product or solution
   - Experience and strength in executing similar type of project with respect to quantity, quality, and schedule
   - Understanding of the project need and develop the execution methodology
   - Capability to work with licensor, DEC, PMC, and customer
   - Number and type of system supplied till now and feedback from the user
   - Adherence with the latest codes/standard recommendations and regulatory requirements
   - Certified trainers for providing the exhaustive training on the system
   - Knowledge service engineers for after installation services
   - Knowledge of implementation and validation of SIS system and SIL loops
2. Relationship with the vendor
   - Vendor–customer relationship – past experience
   - Past experience with the vendor in terms of response, updating the user regarding the upgradation of the product firmware
   - Number of failures and rate of failures occurred since the product installation
   - Line of actions suggested by the supplier and help extended to troubleshoot the issues
   - Knowledge sharing
3. Support
   - Vendor's knowledge and support experience
   - Support capabilities – local (near plant site) and global base
   - Online service request support services
   - Remote access support availability
   - Off-the-shelf product availability
   - Spare part support period for the series of the system selected
   - Previously supplied spare parts interchangeability, if possible
   - Free upgradation of the firmware
4. Integration by third parties
   - Openness of the solution provided
   - Seamless integration possibilities for any third-party interfaces
   - Majority of interface gateways are a part of product ranges
   - Inhouse availability of the programming experts

B. Hardware Issue

1. Fault tolerance
   - Single or dual – level of fault tolerance
   - Hardware or software fault tolerance techniques used
   - Other diagnostics related to system fault
   - Watchdog timer continuously monitoring system
   - System shall go to fail-safe mode in case of total system failure
   - Check of all the possibility of failure conditions
   - System availability and reliability calculation to decide the level of fault tolerance
   - System shall meet the minimum requirement of fault tolerance as per project specification

2. Configuration flexibility
   - Ease of configuration and flexibility (Macroprograms are not allowed)
   - Addition of the new I/Os online
   - Controller memory segregation
   - Flexibility of configuration through portable device (such as laptop)
   - Logic shall be configured in functional logic diagram, or ladder logic program
   - On line configuration as per standard
   - Different configuration techniques/option available
3. Size
   - Controller of different capacity (I/O handling) availability
   - Capacity in terms of I/O handling, system database, system nodes, and so on
   - Capability of each module (number of I/P or O/P)
   - System loading capacity
   - Proven track record on biggest system delivered as on date
   - System expandability
   - Memory capacity and auto backup facility
   - Architecture and sizing of the system without diluting the safety requirements
4. Distributiveness
   - I/O bus design, distance limitation, node limitation, redundancy availability
   - Remote I/O availability, and any limitation
   - Possibility of placement of any I/O card in any slot based on requirement
   - Redundancy availability (power, I/O, communication)
   - Communication subsystem design in terms of data highway and information network
   - Manufacturing and integration facility and distribution network capability
   - Network loading requirements
5. Environmental ruggedness
   - Suitability of every component/card for extreme condition
   - Hardware temperature rating
   - Environmental class protection (IP classification) as per the standard
6. Scan time
   - Scanning time and execution time detail
   - Tentative overall execution time (including I/O scan time, processor execution time, and others)
   - System scan time shall be less than process safety time
7. SOE resolution
   - SOE storing capacity
   - SOE data retrieval
   - SOE resolution and scan time
   - Special requirements (e.g., 100 ms for the specific I/Os), if any
   - Type of model numbers of these special I/O modules
8. Range of I/O modules
   - I/O density (universal or different type)
   - Measurement capability
   - Variety of I/O cards and its function

- Different range of I/O modules based on the I/O type and quantity requirement
- Open- and short-circuit protection
- Protection against reverse polarity
- Fault diagnostic feature availability
- Individually isolated inputs and outputs
- Field termination block design, overall wiring system
- Inbuilt or external isolators detail
- Digital cards detail (relay contact or solid state I/O)
- Analog cards detail (4–20 mA or 1–5 VDC)
- Analog cards A/D conversation resolution
- HART compliance analog cards
- Galvanic isolations between each channel of I/O card

9. Configuration restrictions
   - Any restriction in configuration of module for any specific application
   - Restriction in logic or function usage
   - Memory restriction (for configuration, execution or storing), if any
   - Any restriction on customer-specific requirement (e.g., logic, reports, data storage, archiving)
10. Module DIP switches or settings
    - Modules free of any DIP switches
    - Identification of such cards and DIP switch setting detail on such cards
11. Ease of changing a module
    - Possibility of hot swapping (RIUP – removal and insertion under power) facility shall be available for easy replacement of the modules
    - Capability to diagnose whether a wrong card is placed in the predefined slots
    - Auto I/O scheduling irrespective of the change in slot positions of the newly inserted module
    - Minimum steps involved for auto upload
    - Consideration of static charge
12. Time synchronization
    - Time synchronization capability
    - Different time synchronization technique offered by the supplier
    - Dependent or independent of external synchronization
    - As per the project philosophy, capability to accept global positioning system or pulse over the serial link once in a day at fixed time
    - Additional hardware required, if any
13. Same modules for remote I/O
    - Line monitoring – option availability to monitor line health
    - Watchdog timer functionality
    - Module size/form factors
    - Identical hardware across the project and complex
    - Minimum types of hardware and modules
14. Certified safety communications
    - Communication protocols used as per safety standards

15. Connectivity.
    - Connectivity options (different scheme for establishing communication with the host system)
    - Proven interface module with the similar host system
    - Different protocols, speed, redundancy
    - Redundant I/O bus, HMI interface
    - Various types protocol communication
    - Past proven record of the system interface with various other types of systems
    - Any uncommon method of connectivity requirement for any specific application
    - OPC connectivity option
    - Redundant connectivity options for third-party/other system interface
    - Gateway requirement
16. Remote I/O capabilities.
    - Remote I/O types and capabilities
    - Distance limitation, if any
    - Number of remote I/O connection capability
    - Remote configuration functionality
17. Automatic built-in valve testing
    - Availability of automatic built-in valve testing
    - Possibility of generating it
18. Capability of adding modules online
    - Modules addition and its configuration online, without stopping or rebooting the system
    - Minimum steps involved in module addition (user-friendly system)
    - Output monitoring option in case of online addition
    - Self detection of I/O modules
19. Conformal coating
    - Type of conformal coating available (e.g., G2, G3)
    - Reports of conformal coating
20. Busses.
    - Number and types of busses
    - Redundancy requirement
21. Ability to detect field faults.
    - Field fault detection and diagnostics availability
    - Earth leakage monitoring
22. Size of installed base
    - System vendor's local infrastructure and capability to handle the size and complexity of projects
    - Client list and detail of installed system
    - Details of installation in last 5 years
23. Years the product has been in service
    - Life cycle period of product
    - Assurance from vendor for its availability for next 15 years
    - Product launched details, proven track record
    - Future plans for the upgradation
    - Past record of product support after phase-out

24. Documentation
    - Documentation procedure – from engineering to as-built
    - Online availability of all relevant documentation, for example, installation, maintenance manuals, and certification
    - Any project binder used to develop the deliverables, database handling capacity
    - Integration with client automation software
25. Power consumption
    - Power consumption of each individual modules in proposed system
    - Power consumption of the overall system solution offered
    - UPS and non-UPS requirement and distribution
    - Other power required for system functionality (such as +12VDC, +5V DC and its redundancy)
    - Separation of levels of powers as per standard
    - Heat dissipation calculation
    - Earthing requirement

C. Software issue
1. Programming languages
    - Number of programming language used. Its features and its capability
    - Ease of upgradation of programming language in future in case of version change
    - Programming languages to be used to implement the application code
    - Supplier's flexibility to adopt the change in usual techniques to satisfy client's requirements
2. Ease of programming/configuration.
    - Configuration steps and ease of programming
    - Capability to view status of logics online
    - Capability to stimulate the functional logic in the system
    - User-friendly software development techniques
    - Server administration and redundancy
    - Standard predefined logic block
3. Version control and comparison
    - Programming language versions
    - Version control procedure
    - Version upgradation procedure and requirement
    - Availability of proper diagnostics in case of two different versions in two different processors in the same system
    - Notification system for version change control
    - Old version retention features
4. Online changes (programs and firmware)
    - Online changes features
    - Possibility of online changes of programs and firmware
    - Knowledge and steps requirement for online changes
    - Troubleshooting
    - Online firmware upgradation
5. Simulation capabilities
    - Inbuilt simulation capability (process simulation, logic simulation capabilities)

- Any extra hardware/software requirement
- Capability to view logic on line, simulation of logic

6. Ability to see field and logic values of forced tags
   - In-built capacity to see field and logic values of forced tags simultaneously
   - Facility to force and remove logics with security
7. Security/access control
   - Security levels available
   - Access control level defining procedure
   - Definition of all security levels, for example, VIEW mode only for monitoring, ENGINEERING mode for configuration
8. Diagnostic reporting and asset management
   - Extent of diagnostic capability available and ease of use
   - Asset management capability if data available from the field devices
   - Functionality of reset function of system alarms
   - Availability of system diagnostic alarms (before and after fault rectification)
   - Availability of diagnostic on system level, card level, power, and others as per safety standards
   - Self-diagnostic and reporting capabilities
   - Any special software requirement

FURTHER READINGS

IEC61508 Parts 1–7: 1998, Functional Safety of Electrical/Electronic/Programmable Electronic Safety-Related Systems, International Electrotechnical Commission, Geneva, Switzerland.

ANSI/ISA Standard S84.01-1996, Application of Safety Instrumented Systems to the Process Industries, International Society for Measurement & Control, Research Triangle Park, NC; 1996.

Safety Shutdown Systems – ISA, Gruhn and Cheddie; 1998.

Out of Control – UK Health & Safety Executive Publication; 1995.

Programmable electronic systems in safety related applications: an introductory guide – Health & safety executive publication; 1990.

Functional safety: a straightforward guide to IEC61508 and related standards. Butterworth-Heinemann Publications; D.J. Smith & K.G.L. Simpson; 1985.

http://www.sp.se/safeprod – Management of functional safety guideline.

CHAPTER

FIRE AND GAS DETECTION SYSTEM

6

6.1 INTRODUCTION TO THE FIRE AND GAS (F&G) DETECTION SYSTEM

6.1.1 NEED FOR AN F&G SYSTEM

In today's competitive marketplace, maintaining a high level of process and plant safety is a critical concern. Manufacturers can reduce costs by minimizing the damage to equipment and eliminating incidents that impact people and the environment. At the same time, they can maintain a positive image as a company that is aware of its corporate responsibility and acts accordingly. An industrial area is always prone to hazards such as fire and toxic gas leaks, which can cause damage to assets, humans, or environment. There is a historical record of many such incidents, which led to the loss of many lives. Therefore, it becomes imperative to ensure early detection and mitigation of a fire or gas leakage in industrial areas to avoid disasters, minor or major injuries to persons, or damage of equipment leading to production loss.

Fire and gas detection systems are key to maintaining the overall safety and operation of industrial facilities. F&G systems include offshore petroleum exploration and production, onshore oil and gas facilities, refineries and chemical plants, marine operations, tank farms and terminals, pipelines, power plants, mining, and paper mills. An F&G safety system continuously monitors for abnormal situations such as a fire, or combustible or toxic gas release within the plant, and provides early warning and mitigation actions to prevent escalation of the incident and protect the process or environment. By implementing an integrated F&G strategy based on the latest automation technology, plants can meet their safety and meet critical infrastructure protection requirements while ensuring operational and business readiness at project start-up.

Throughout the process industries, plant operators are faced with risks. For example, a chemical facility normally has potential hazards ranging from raw material and intermediate toxicity and reactivity, to energy release from chemical reactions, high temperatures, and high pressures.

According to international standards, safety implementation is organized under a series of protection layers, which include, at the base levels, plant design, process control systems, work procedures, alarm systems, and mechanical protection systems. The safety shutdown system is a prevention safety layer, which takes automatic and independent action to prevent a hazardous incident from occurring, and to protect personnel and plant equipment against potentially serious harm. Conversely, the F&G system is a mitigation safety layer tasked with taking action to reduce the consequences of a hazardous event after it has occurred. The F&G system is used for automating emergency actions with a high-integrity safety and control solution to mitigate further escalation. It is also important for recovering from abnormal situations quickly to resume full production.

6.1.2 CHALLENGES IN PROCESS PLANTS

Now, more than ever, industrial companies are concerned with fire and gas in their production operations. Manufacturing plants must cope with business challenges ranging from increased accident, incident, and insurance costs, to compliance with strict standards and codes such as NFPA, API, and OSHA in USA and BS EN and SEVESO II in Europe. Also, issues related to corporate image and environmental stewardship have growing implications in the global market. Industrial plants need effective solutions for improving a wide range of process-safeguarding practices. This requires a control system architecture allowing engineers to design and build standalone safety applications, as well as distributed plant-wide safety topologies. Plants must find ways to improve F&G system effectiveness through optimization of F&G detector coverage, system safety availability, and mitigation effectiveness, and at the same time, reduce the cost of ownership for safety equipment. Many facilities are also dealing with the cost of upgrading and refurbishing existing, nonintegrated F&G systems. As more plant owners move toward "smart" plants, appropriate integration with other systems will play an important role in increasing safety as well as efficiency. The F&G systems have communications integration with the DCS in order to have F&G graphics and alarms for display to the operator. Meanwhile, there shall be independent displays such as independent HMIs to draw attention to F&G excursions when the DCS HMIs are not available. The plant F&G system, with fire system for occupied buildings, shall be integrated with plant evacuation and system/site security center for plant evacuation. As part of an overall plant safety strategy, end users are using a unified platform for emergency shutdown, F&G detection to have a single window for operators, and common tool for engineering and maintenance to drive down operational risk and costs.

6.1.3 CAUSES OF FIRE

Fire can be caused due to many reasons such as electrical short circuits, rise in the temperature of a combustive material, and leakage of combustible gases or liquid. Typically, electrical or instrumentation cables in control room or plant interface building are prone to catching fire, or fire could be caused by an ignition in a generator room of a nonplant building. Leakage of combustible fluids outside process line, which are already in autoignition temperature, can also be a possible hazard.

One of the major and most common causes of fire accidents in process industries such as refinery or petrochemical plants is leakage of combustible gases or liquids. Just the leakage of combustible fluids cannot cause fire, as it needs the mixture of three components that cause fire, as explained in the following illustration (Figure 6.1).

For a fire to start, in addition to fuel, there must be enough percentage of oxygen and an ignition source in the particular area. The right combination of the three components is important to cause fire. The ratio of air–fuel mixture required to cause fire (if ignition source is present) is decided by the upper and lower exposure limits of that particular fuel (Figure 6.2).

Lower Explosive Limit (LEL) and Upper Explosive Limit (UEL) of a material are mentioned in %v/v (volume by volume) with respect to air. If the amount of gas present in the atmosphere is below LEL then the mixture is too lean to cause a fire. If the amount of gas present in the atmosphere is above UEL then the mixture is too rich to cause a fire. Fire can occur only if the gas in the atmosphere is in between LEL and UEL, and then it has enough potential to cause a fire in the presence of an ignition source.

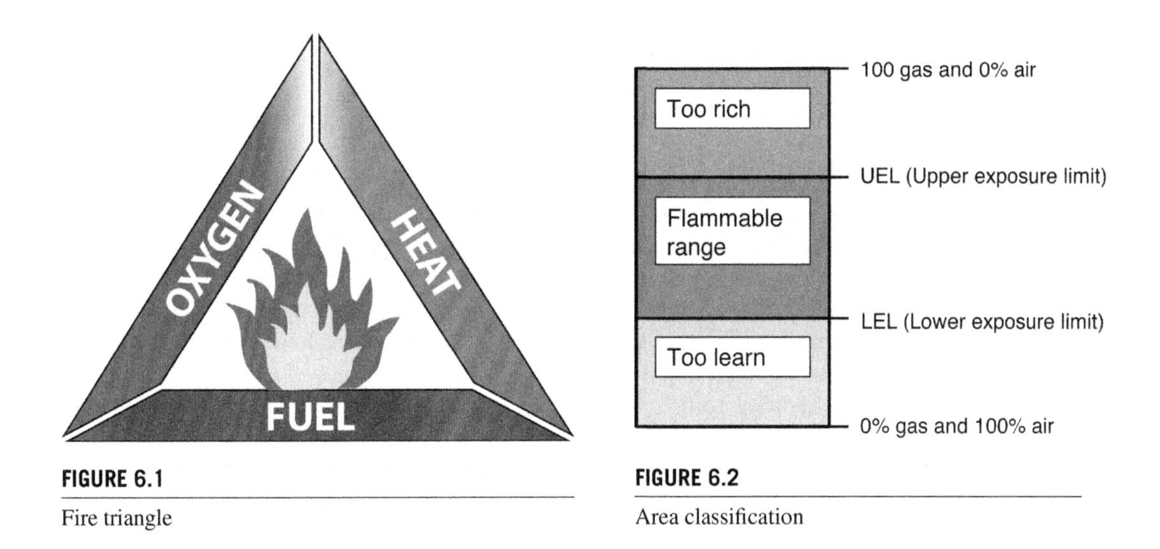

FIGURE 6.1

Fire triangle

FIGURE 6.2

Area classification

6.2 UNDERSTANDING INDUSTRY SAFETY PERFORMANCE STANDARDS

Industrial standards are playing a large role in developing, implementing, and installing F&G systems. The IEC 61511 standard (ANSI/ISA S84.01 in USA) is a major step toward protecting industrial plants. The overall safety lifecycle model described in the IEC standard lists all the necessary project activities, from the concept (definition) phase to the decommissioning phase, necessary to ensure the functional safety of equipment under control (EUC). These activities can be divided over a wide range of categories such as procedures, documentation, testing and validation, planning, hardware and software development, and risk assessment.

More recently, there have been discussions over whether F&G detection systems should contribute to risk reduction or be considered as a protection for the installation only. The implementation of the IEC 61511 and S84.01 standards is becoming increasingly prevalent for F&G detection systems. An ISA technical report TR84.00.07 to provide guidance on the evaluation of F&G system effectiveness is currently in draft review.

The IEC 61511 standard concerns the determination and development of Risk Reduction Measures (RRMs) required as the outcome of the EUC risk assessment. The basic principle of risk assessment is that all potential risks to the EUC are identified and analyzed. This includes calculating the probability of each potential EUC hazard and determining the risk reduction measures required to achieve an acceptable safety integrity level (SIL).

The potential risk of EUC hazards can be considered as the outcome of the probability that the hazard occurs, and the consequences of the hazard:

$$\text{Risk} = \text{Probability} \times \text{Consequences}$$

The prescribed reduction measures either decrease the risk probability (e.g., emergency shutdown (ESD) systems) or mitigate their consequences (e.g., F&G systems). The risk of EUC hazards can be reduced by a combination of several RRMs, where each measure takes care of a part of the total required risk reduction factor (RRF).

6.3 CRITICAL COMPONENTS

A typical F&G safety system comprises detection, logic control, and alarm and mitigation functions. Logic solver is the central control unit of the overall F&G detection and control system. The controller receives alarm and status or analog signals from field monitoring devices required for F&G detection. The controller handles the required actions to initiate alarms and mitigate the hazard. F&G detection devices have developed greatly over recent years. Using new techniques and adding intelligence to these instruments to reduce the number of spurious alarms has greatly improved detection rates. Correct and proven connection of F&G detectors to plant safety systems is an important factor in reliable performance of the F&G system and for establishing the desired SIL. In the past, proprietary F&G systems were standalone equipment or a hardwired mimic overview panel via relays. Mitigation of the risk would take place via manual activation of fire control measures. These methods are not considered best practice. Today, F&G detection systems are generally programmable electronic systems (PES) type with high safety availability and mitigation effectiveness. As modern F&G systems are tightly integrated with the overall process safety strategy, mitigation takes place either via the emergency shutdown system or directly from the F&G system itself (Figure 6.3).

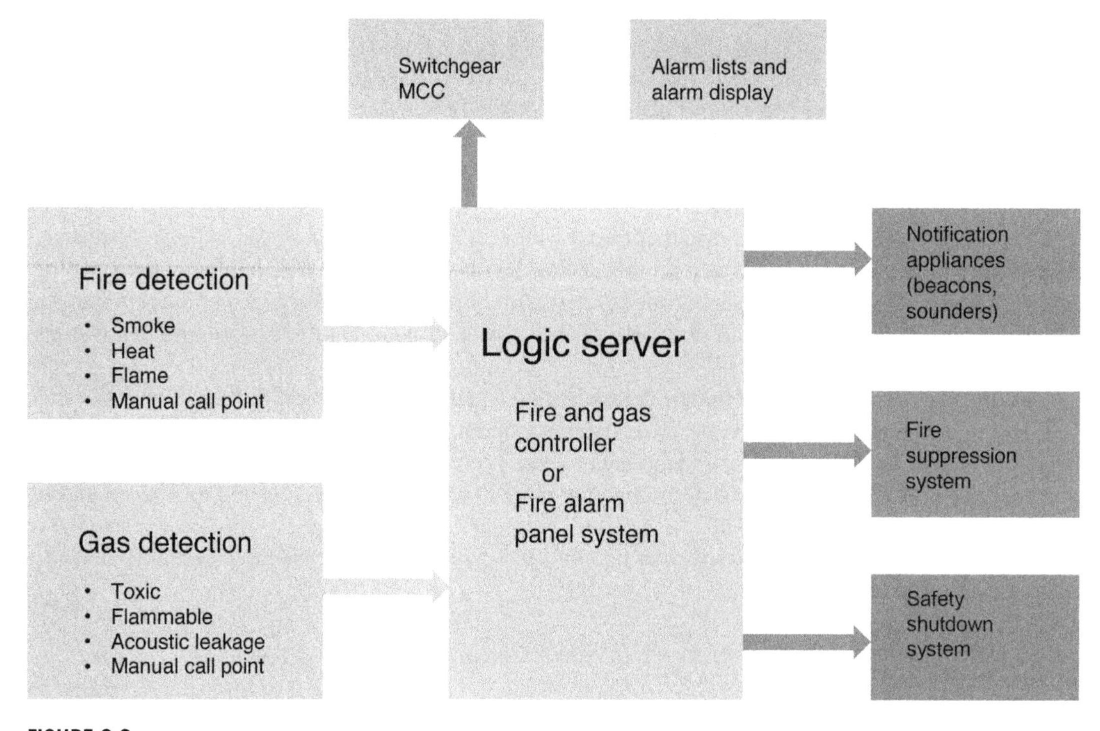

FIGURE 6.3

Sub systems of F&G System

6.4 F&G DETECTORS

6.4.1 MANUAL CALL POINT

It requires manual action to initiate an alarm. In case of fire or any abnormality in the area, breaking the glass results in micro switch operation.

6.4.2 OPTICAL SMOKE DETECTOR

Optical Smoke Detector, an infrared light-emitting diode (IR LED) within its collimator is arranged at an obtuse angle to the photodiode. The photodiode has an integral daylight-blocking filter. The IR LED emits a burst of collimated light every second. In clear air, the photodiode receives no light directly from the IR LED because of the angular arrangement and the dual mask. When smoke enters the chamber, it scatters photons from the emitter IR LED onto the photodiode in an amount related to the smoke characteristics and density. Optical application specific integrated circuit (ASIC) application specific integrated circuit processes the photodiode signal and passes it to the A/D converter on the communications ASIC ready for transmission when the device is interrogated. Optical Smoke Detector are generally used on the ceiling of buildings (Figure 6.4).

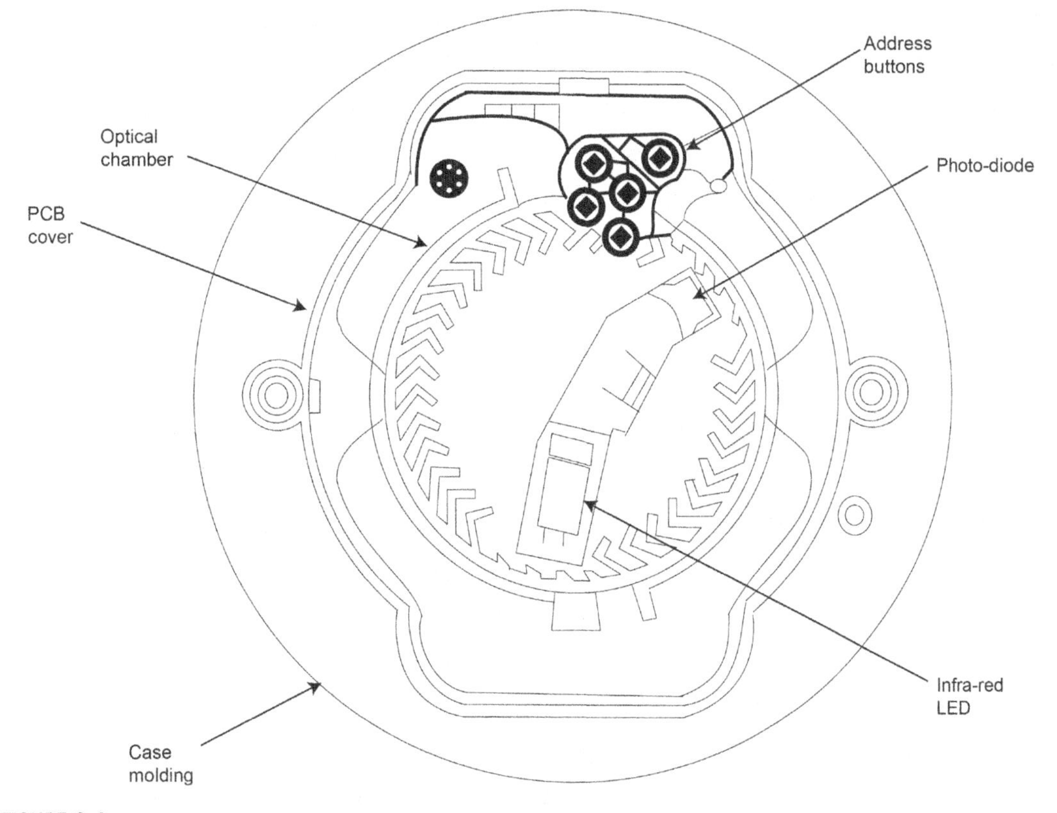

FIGURE 6.4

Optical Smoke Detector

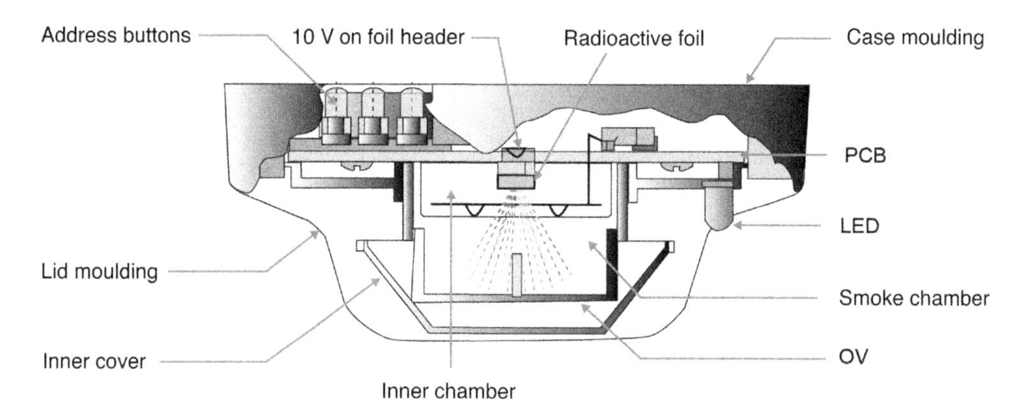

FIGURE 6.5

Ionization Smoke Detector

6.4.3 IONIZATION SMOKE DETECTOR

The ionization chamber system is an inner reference chamber contained inside an outer smoke chamber. The outer smoke chamber has smoke inlet apertures fitted with an insect-resistant mesh. The radioactive source holder and the outer smoke chamber are the positive and negative electrodes, respectively. An Americium 241 radioactive source mounted within the inner reference chamber irradiates the air in both chambers to produce positive and negative ions. On applying a voltage across these electrodes, an electric field is formed as shown in Figure 6.5. The ions are attracted to the electrode of the opposite sign. Some ions collide and recombine, but the net result is that a small electric current flows between the electrodes. At the junction between the reference and smoke chambers is the sensing electrode that is used to convert variations in the chamber currents into a voltage.

When smoke particles enter the ionization chamber, ions attach to them resulting in a decrease in the current flowing through the ionization chamber. This effect is greater in the smoke chamber than in the reference chamber and the imbalance causes the sensing electrode to go more positive (Figure 6.6). The voltage on the sensing electrode is monitored by the sensor electronics and is processed to produce a signal that is translated by the A/D converter in the communications ASIC ready for transmission when the device is interrogated. This device is generally used in the false floor of the buildings where some instrument and electrical cables are laid.

6.4.4 HEAT DETECTOR

The heat detectors have a common profile with ionization and optical smoke detectors but have a low airflow resistance case made of self-extinguishing white polycarbonate. The devices monitor heat by using a single thermistor network that provides a voltage output proportional to the external air temperature (Figure 6.7).

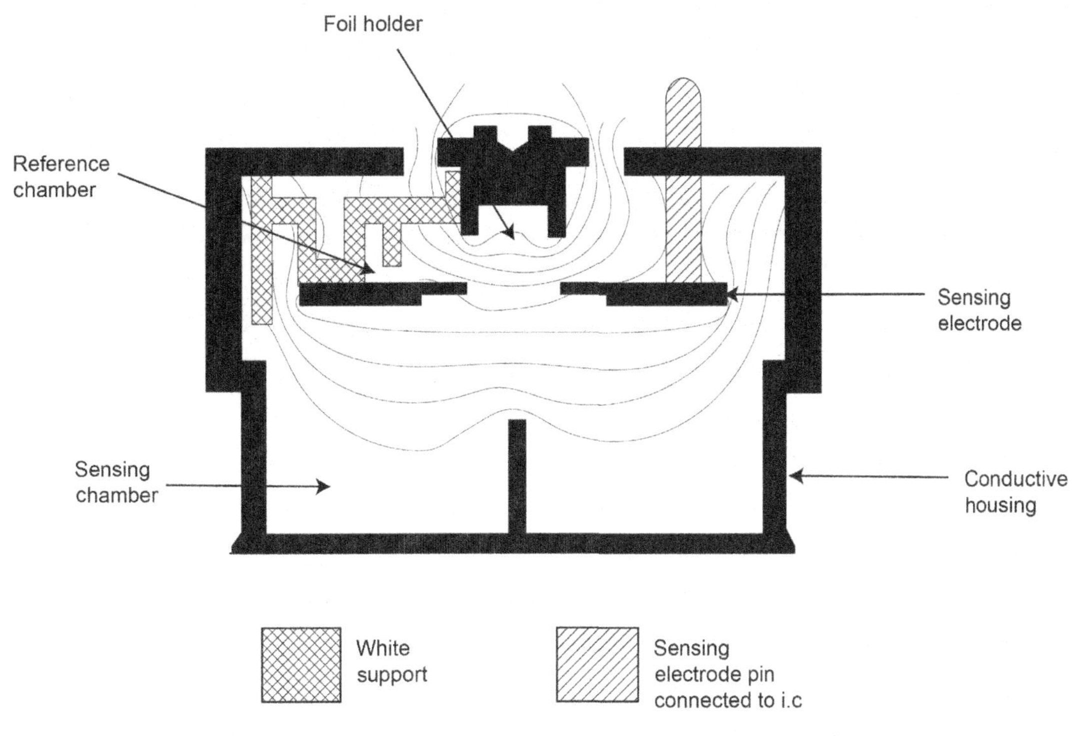

FIGURE 6.6

Heat Detector

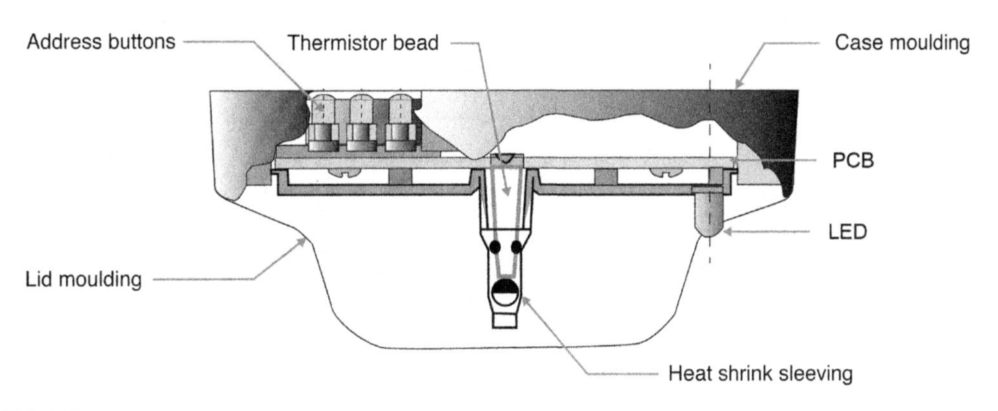

FIGURE 6.7

Multi Sensor with Light and Heat

6.4.5 MULTISENSOR DETECTOR

The multisensor detector contains an optical smoke sensor and a thermistor temperature sensor whose outputs are combined to give the final analog value (Figure 6.8). The multisensor construction is similar to that of the optical detector but uses a different lid and optical moldings to accommodate the thermistor temperature sensor. Figure 6.9 shows the arrangement of the optical chamber and thermistor.

The signals from the optical smoke sensing element and the temperature sensor are independent and represent the smoke level and the air temperature, respectively, in the vicinity of the detector. The detector's microcontroller processes the two signals. The temperature signal processing extracts only rate of rise information for combination with the optical signal. The detector does not respond to a slow temperature increase – even if the temperature reaches a high level. A large sudden change in temperature can, however, cause an alarm even in the absence of smoke, if sustained for 20 s. The processing algorithms in the multisensor incorporate drift compensation if the control panel does not have a drift compensation algorithm enabled. The sensitivity of the detector is considered the optimum for most general applications, since it offers good response to both smoldering and flaming fires. This device is generally used in the false floor of the buildings where instrument and electrical cables are laid.

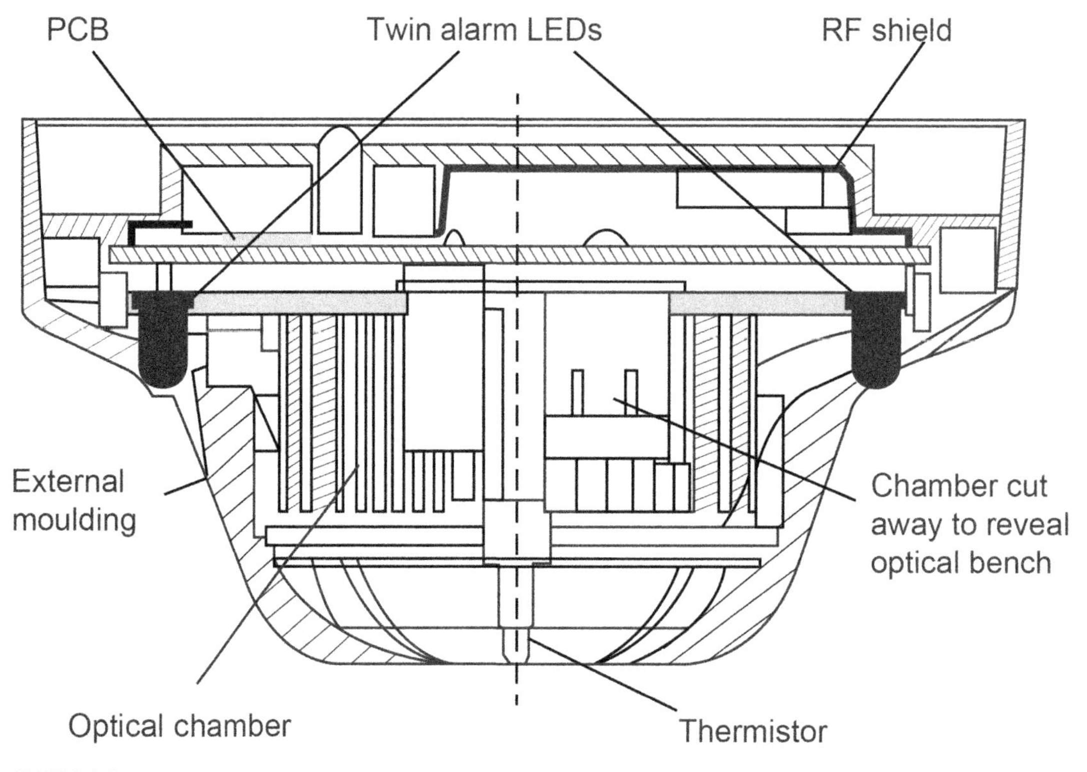

FIGURE 6.8

Multi Sensor Arrangement

6.4.6 FLAME DETECTORS (IR/UV)

The detectors work on the flame-flicker principle, that is, the detector responds to the flickering of most hydrocarbon fires. In the event of a fire, it emits IR/UV radiation. This light is passed through IR/UV filters and reaches the photodetector.

6.4.7 TOXIC GAS DETECTOR (H_2S, CHLORINE, ETC.)

- Threshold limit value (TLV) – time-weighted average (TWA) represents the time-weighted average concentration of a toxic substance over a normal 8-h workday and 40-h workweek, to which nearly all workers may be repeatedly exposed, every day, without adverse health effects.
- Short-term exposure limit (STEL) is a 15-min time-weighted average exposure that should not be exceeded during the workday even if the 8-h TWA is within the TLV.
- Exposures at the STEL should not be repeated more than four times per day (Figure 6.9).
  Zone A – irritation
  Zone B – coughing
  Zone C – breathing difficulty
  Zone D – collapse
  Zone E – death

6.4.7.1 Electrochemical sensors

Electrochemical sensors are used to detect the presence of toxic gases such as H_2S, Cl_2, and SO_2, and variation of oxygen in the air.

- It consists of two electrodes immersed in common electrolyte medium in the form of gel.
- The electrolyte is isolated using a membrane.
- A voltage is applied between the two electrodes. When gas enters a chamber through the membrane, oxidation or reduction takes place due to reaction of gas with gel, which causes a small current to flow, which is linear to gas concentration.

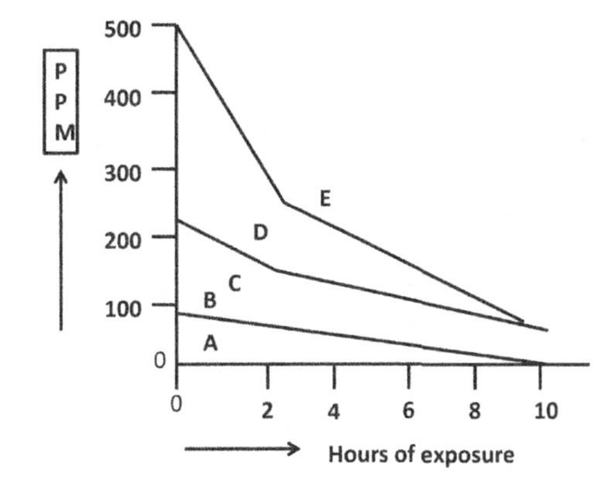

FIGURE 6.9

6.4.7.2 Combustible gas detector

Infrared absorption principle is used to detect the presence of combustible gas (HC). Light passes through the sample mixture at two wavelengths, one of which is set at the absorption peak of the gas to be detected. Two light sources are pulsed alternatively and guided along a common optical path to emerge via flameproof window and sample gas. These beams are reflected back by a retro-reflector. These beams pass through sample gas once again and then into the unit. Detector compares the signal lengths of sample and reference beams. Difference of sample and reference gives the measure of gas concentration. This can be used to generate an alarm after converting in analog form.

6.4.7.3 Hydrogen gas detector

Used to detect hydrogen concentration. Electrocatalytic type sensor is used, which consists of a small sensing element called pellistor made of electrically heated platinum wire coil, covered with the coating of ceramic bases (alumina), and finally outer coating of palladium or rhodium catalyst. When combustible gas/air mixture passes over the hot catalyst surface, combustion occurs, which in turn increases the temperature of the pellistor. This causes the change in resistance of platinum coil. This resistance change can be measured through bridge configuration, which is directly proportional to gas concentration. It displays on the detector window also.

6.4.7.4 Open path gas detector

The open path gas detector eliminates the disadvantages with point detectors. This covers the entire area between transmitter and receiver. It is generally used between a bay of pumps or compressors, where it can detect leakages missed by point gas detectors.

In open path IR measurement, dual-wavelength beams are used (Figure 6.10). These coincide first with absorption band peak of the sample gas and then with reference beam that lies close by in an unabsorbed band area. When gas cloud is formed in the beam path, the strength of the beam is reduced as the gas molecules absorb the IR light. Any changes in the relative signal strength between the two beams can be attributed to the presence of an interfering target gas and indicates the total number of gas molecules within the beam. The measuring unit for open ath gas detector is LELm (Figure 6.11).

6.5 F&G NETWORK ARCHITECTURE

A good F&G system combines innovative F&G detectors, conventional and analog addressable fire panels, clean agent and inert gas fire suppression systems, and a SIL 3-certified F&G logic solver into a consistently designed and executed solution. An integrated system provides common tools, operating interface, and networking, resulting in a common platform with independent systems (Figure 6.12). Thanks to advancements in F&G detectors, F&G systems can detect early warnings of explosive and health hazards, including combustible and toxic gas releases, thermal radiation from fires, and minute traces of smoke in sensitive equipment enclosures. They also provide audible and visual alarm indications that help in ensuring that operators and personnel are informed of potentially hazardous situations. The F&G system automatically initiates executive actions; minimizes escalation of safety incidents; and protects personnel, property, and the environment. Integration at the controller level provides plant-wide safety instrumented system (SIS) point data, diagnostics and system information, as well as alarms and events, operator displays, and sequence of event information to any station. This

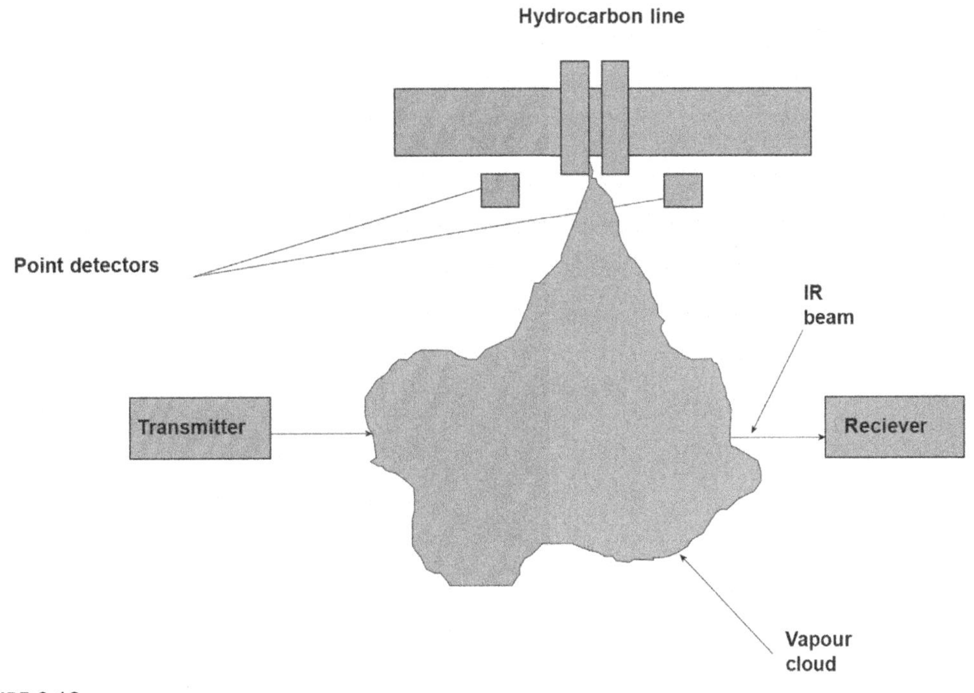

FIGURE 6.10

Open path Gas detection.

minimizes intervention and shutdowns, reduces hardware costs, and allows plants to recover more easily from process upsets. The new generation of F&G solutions provides alerts of abnormal situations in a fast, accurate, and structured way, giving personnel time to decide upon the correct course of action. These solutions include new integration capabilities with process simulation tools, F&G detectors, and control communication protocols, enabling safety engineers to design and build large integrated and distributed plant-wide safety strategies.

Overall, SIS technology integrates safety measures dispersed throughout a plant to reduce risk to employees and assets, increase process availability, and improve regulatory compliance. SIS solutions can be integrated with F&G detectors for increased protection and unified with third-party applications and systems to reduce validation and acceptance testing costs. Modern safety systems, seamlessly integrated with the plant automation system through a secure communication network, transfer alarm signals, fault signals, and system diagnostics. Information from all related systems can be transferred, gathered, and handled at the same location, and an additional layer can be achieved to monitor the status and operability of the total F&G detection and control system. Safety system platforms based on various diagnostic technologies execute automated safety functions and provide the interfaces and input functions for standard connection of a wide range of F&G detection devices. Applications include emergency shutdown, process shutdown, F&G systems, burner management, compressor control, pipeline management, or any critical safeguarding in the process industry.

With innovative simulation solutions, safety engineers can easily test the impact of safety strategies on the overall plant design and operations before implementation. This reduces overall risk and

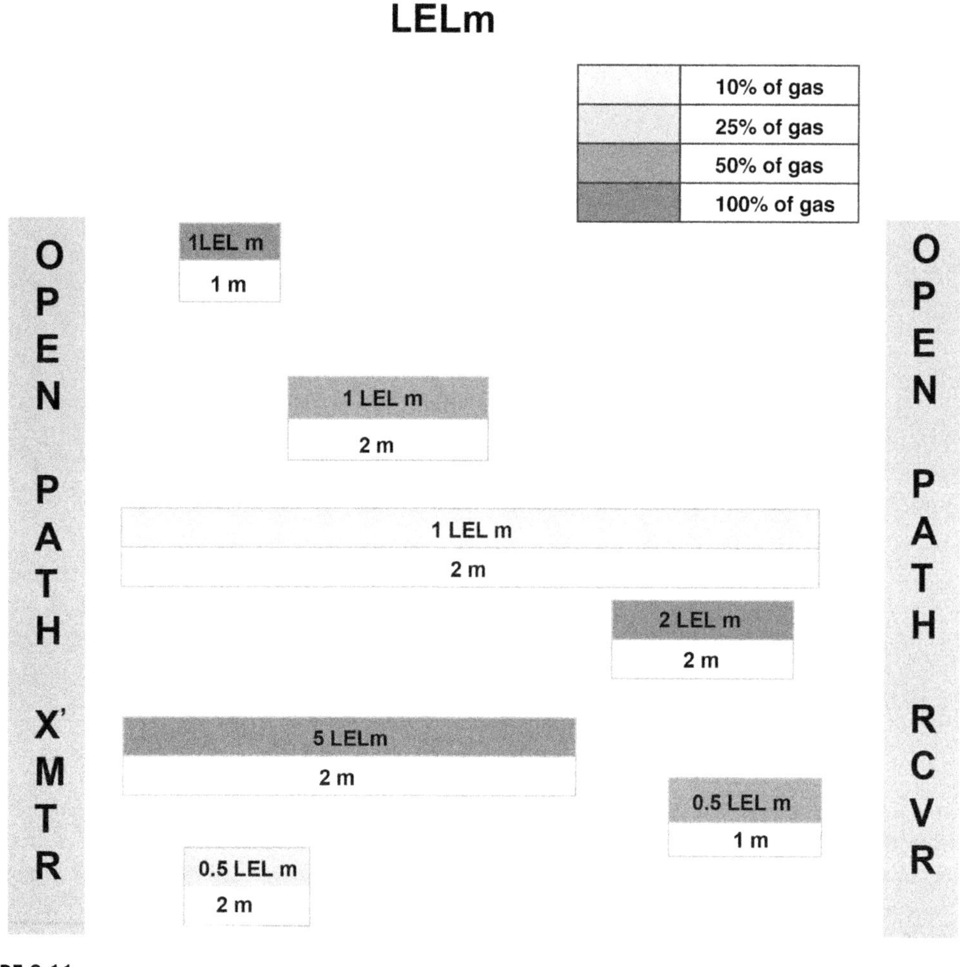

FIGURE 6.11

Measuring Units of Open path Gas detection

the impact of system modifications and, ultimately, increases profitability by bringing new plants into full production much faster. In addition, new field device configuration tools allow plant personnel to automatically configure intelligent safety devices and integrate them into the control system database. Facilities subsequently save money by using a single tool to manage all equipment assets.

Figure 6.13 shows a typical F&G network. All F&G detectors are connected to marshalling cabinet terminal blocks from field. Cables from the detectors directly come from device or from a junction box in between.

F&G detectors are basically classified into two types:

- Addressable devices
- Nonaddressable devices

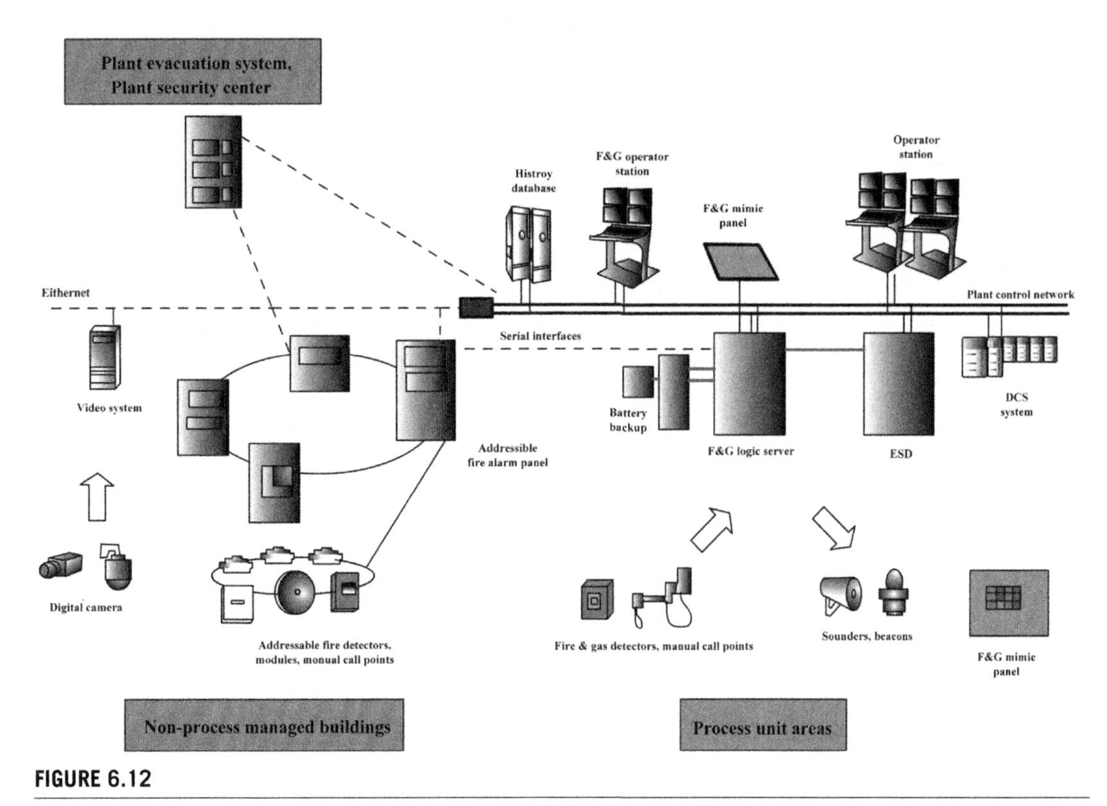

FIGURE 6.12

A sample Integrated System Architecture for Safety

6.5.1 ADDRESSABLE DEVICES

Addressable devices are connected in a loop, and they are connected to a fire alarm panel (FAP). Two wires go from OUT of the detection cards (Loop OUT) of FAP, which is connected to the IN of Device 1, and OUT of Device 1 is connected to IN of Device 2, and so on. The OUT of last device in loop is connected to the IN of Detection IP card (Loop IN). These devices are identified by the binary addressing given to them such as DIP switch for manual call point or address strip for smoke detectors. Addressable devices are mainly used in building areas such as PIBs, substations, office buildings, and control rooms where maximum detectors are covered in minimum area. Connecting all these devices in loop saves a lot of cabling and space. It also has its limitations, for example, any loop short can take all the devices in that loop to fault state (this can be avoided by placing some loop isolators between some devices in case of big loops; this will restrict the loop fault in case of short to those devices between the consecutive devices). Any loose connections at ends of the loop may bring all devices down.

These addressable loops are again classified into two types:

1. IS loop (intrinsically safe)
2. NIS loop (nonintrinsically safe)

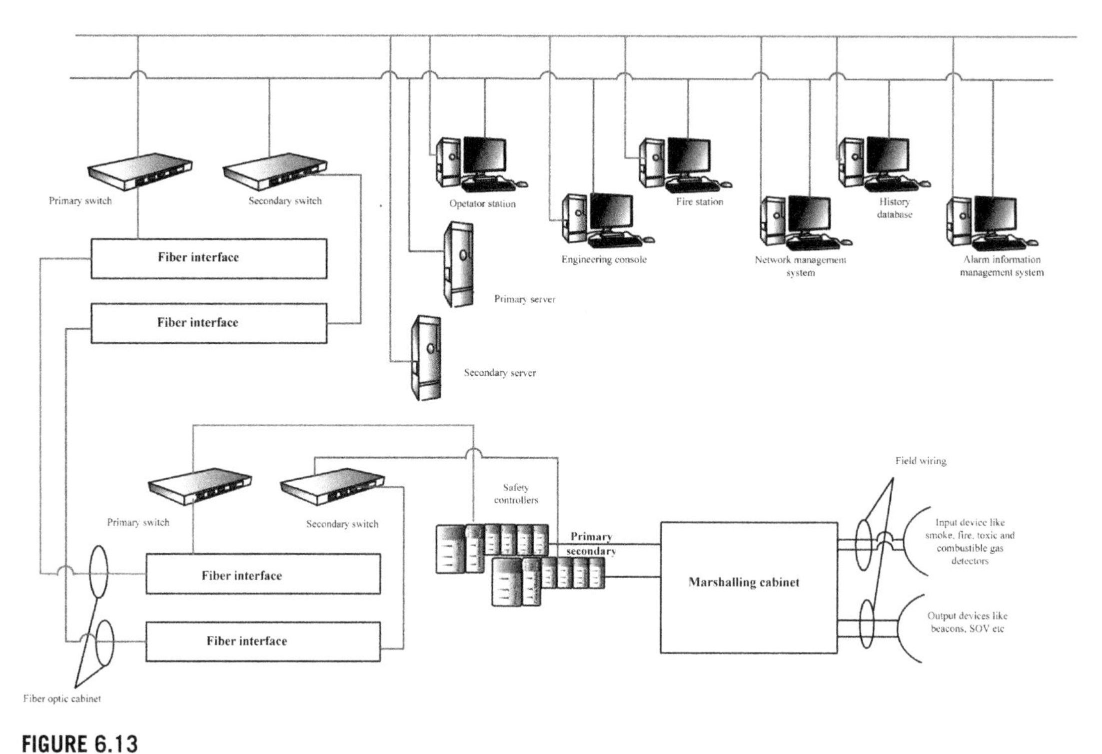

FIGURE 6.13

A typical F&G architecture

The loops going to nonhazardous areas such as office areas and instrument panel rooms are directly connected to FAP, but the loops going to hazardous areas such as operational area (field) or battery storage area are connected to FAP via barriers. In a nonIS loop, 120 devices can be connected with maximum loop distance of 2500 m. In case of IS loops, 80 devices can be connected in one loop with maximum distance of 2000 m.

The FAP panel can itself act as a master to control some outputs in cases of small buildings and also can be integrated to Safety PLC via Modbus.

6.5.2 NONADDRESSABLE DEVICES

Nonaddressable devices are one-to-one detectors. These detectors are used when distance between each detector is more and in critical areas where availability on demand is very much required. Common nonaddressable devices are gas detectors, flame detectors, and output devices (though some addressable gas and flame detectors are available in market). Generally, all these F&G detectors are three wired. These detectors are connected to the marshalling cabinet in PIBs by direct run of three core cables to terminal blocks of marshalling cabinet, or a group of detectors from one area are connected to junction box (JB) in the field area, from JB, a multicore cable, to marshalling cabinet. Barrier or isolator will be added before connecting to system cabinet in case of devices placed in a hazardous area. From marshalling cabinet, these devices will be connected to terminal assembly and from there these will be connected to IO chassis in system cabinet via interconnection cable, which are compatible with IO cards. Each input is taken by redundant input cards, and the input data are sent to redundant controllers via IO bus. The controller scans the inputs, processes the logic, and updates the output in every cycle. The updated outputs are sent back to output cards via the same IO bus. Output cards are connected to interconnection cables that are connected to terminal assemblies. Output devices are connected to terminal assemblies.

6.5.3 F&G NETWORK

The plant operator and fire officer monitor the F&G data, so all PLCs in the plant are connected to the control room via a fiber optic network. Both redundant PLCs in system cabinet are connected to the redundant network switches in the network cabinet (primary and secondary) via Ethernet. Each switch is connected to its light interface unit (LIU) via a fiber optic patch cord (single mode). On the other end of the LIU, a multicore fiber optic cable is spliced to pig tails (that are connected to the switch). These multicore cables from primary and secondary switches are laid to the control room via different set routes (S1 & S2). Different set routes ensure that at least one link is healthy in case of some damages to the route due to a fire or damage to the cables during digging. On the control room end again, there is a redundant pair of switches. The other end of the multicore fiber optic cable is connected to LIU in the control room network cabinet and further connected to the network switch. Each PLC can be connected to the control room in a star or ring network. Ring network is the preferred network. F&G servers in the server room and operating stations in the panel area are connected to these switches via Ethernet. The fire panel in the fire stations is connected to the same switch via Ethernet or a fiber optic network depending on the distance from the control room. In case of big plants, there are multiple redundant switches in the control room that are connected to different PLCs across the plant. All these switches are connected in a daisy chain and further connected to a backbone switch.

From the backbone, all nonF&G servers and stations such as Historians, alarm management systems, and network management systems are connected. All these machines are placed in the engineering room where system engineers sit. Any modification to the database or graphic development of server is done from here. Also, it has a separate station for PLC programming. Other systems such as Historians are mainly for gas detectors. Historian database is used to view the trend of each detector, and helps to analyze the presence of gas impact areas in cases of leakage and also to determine the trueness of the alarm. A spike or very nonlinear trend is mainly due to short circuits or cartridge failure of gas detector. Alarm information management system is mainly for maintaining the history of alarms (comparable to a network printer). Maintenance relies on this for generating daily reports and monthly reports for fire or gas leakage incidents. Network management server monitors the health of entire F&G network and other systems interfaced with F&G network.

6.5.4 FIRE ALARM PANEL

Many systems are integrated with F&G system such as the fire alarm panel (FAP) is connected to the PLC via Modbus. As both PLC and FAP act as slaves, a Modbus master device is placed in between. The master device has a pack of registers and decides from where FAP and PLC can read/write device data. PLC unpacks the Modbus data to get status of each nonaddressable device. Each register has data of four nonaddressable devices, which means that one nibble contains data of one device, where each bit has a specific meaning – one bit for health status, one bit for alarm status, and one for fault status.

6.5.5 PUBLIC ANNOUNCEMENT SYSTEM

PA (public announcement) system, in any F&G system, is used to communicate to the affected area immediately when true alarm occurs. COM DO point is generated every time an alarm is generated. When this DO is generated in the server, the point data are transferred to PA system server via Modbus through a terminal server. Once it is received by the PA system server, it asks for operator permission before announcing in the field. The operator can confirm the authenticity of the alarm before pressing the confirm button in the PA system board. Then a voice alarm is announced in the respective fire zone. In case of toxic gas alarm, it gives field announcement without operator intervention.

6.5.6 LARGE SCREEN MONITORS

Large screen monitors are generally used for critical plant data, F&G monitoring, and process CCTV. Also, when some data goes beyond its limits, some features such as abnormal situation management (ASM) graphics automatically display the corresponding graphics on the large screen. F&G data are sent to large screens via an OPC server through firewall (router will come into picture in case of different IP network).

6.5.7 THIRD-PARTY INTERFACE SYSTEMS

Many packaged systems such as deluge systems, turbine systems, analyzer systems, fire hydrant systems and HVAC, emergency shutdown, and CCTV systems are integrated into the FGS systems for better safety and control.

6.6 INTEGRATED APPROACH FOR F&G

Industrial operations benefit from a holistic approach to safety that supports a secure process control network within the perimeter of the plant to protect people, assets, and profitability. A layered safety strategy encompasses process and system technology – and the people who interact with that technology – to help plants achieve their safety objectives. A layered safety strategy unifies all plant protection layers (i.e., basic control, prevention, and mitigation as outlined in IEC 61511 standard) required for achieving optimum functional safety. In addition, it provides the required functional safety with a high SIL. This includes superior visualization and logging facilities enabling optimal operator response and accurate evaluations. By integrating basic control, prevention and mitigation components, overall project costs, and ongoing maintenance expenses can be vastly reduced.

A truly integrated safety system delivers the following:

- Integrated operational interface
- Integrated peer control
- Integrated diagnostics
- Integrated postmortem analysis
- Integrated F&G system
- Integrated power supplies
- Integrated modifications
- Integrated simulation and optimization

Operational integration allows plant personnel to have a seamless interface to the process under control, and at the same time, maintain safe separation. From an operational perspective, it makes no difference where the application is running. All required information is available to the operator. This allows applications ranging from rotating equipment and compressor protective systems to emergency shutdown systems and large plant-wide F&G applications to be monitored from any operator console.

Integrated control and safety systems (ICSS) provide multiple benefits to process plants. For instance, they help operators to minimize intervention and shutdowns and recover more easily from process upsets. They also allow facilities to reduce hardware and installation cost, and ensure easier system configuration with preconfigured Function block selections.

Plants implementing an ICSS platform for F&G, ESD, and DCS systems can significantly lower their operation and maintenance costs, and in many cases, reduce overall wall-to-wall project costs. Seamless integration with the ESD and DCS via a common network protocol also provides a safe landing in case of emergencies and eliminates the need for additional equipment or engineering. Integration of fire detection and security systems for off-sites and utilities with the plant automation infrastructure further improves operator efficiency through single-window access for alarm visualization, diagnostics, and events/historians.

At the core of a layered safety strategy is process design – the embodiment of the business, safety, and production considerations necessary for effective operations. At the next layer, this approach implements tools and procedures for managing abnormal situations and reducing incidents. When an abnormal situation occurs, alarm management, early event detection, and ASM-designed displays ensure operators have the information available in the context they need it. This enables faster reaction to hazardous situations, thereby avoiding safety incidents.

Next, properly designed emergency shutdown systems and automated procedures can move a plant to a safe state in the event an incident escalates beyond the inner sphere of protection. Should an incident occur, F&G detection solutions coupled with rapid location of individuals and a carefully designed emergency response procedure will help contain the impact. Finally, a layered approach to safety protects the perimeter of the plant using physical security that safeguards access to structures and monitors traffic approaching the facility.

6.6.1 SUPPLIER SUPPORT INCREASES LIFECYCLE SUSTAINABILITY

Industrial facilities can leverage the benefits of their F&G strategy by employing an single automation contractors to meet critical asset protection needs – and ensure operational and business readiness at project start-up. With the single automation contractor approach, plant management has a single point of contact throughout the entire system lifecycle. This results in optimized risk reduction and operational performance, better compliance with safety standards, and increased lifecycle sustainability. End users should select a single automation contractor who brings together all necessary expertise in F&G detection for a complete, integrated solution. Certified device connections to the F&G safety system improve reliable performance of the overall mitigation function and establish the desired SIL. The chosen supplier should have global capabilities with consistent local support and implementation, as well as consistent engineering tools and processes for each project phase. An effective project strategy starts with an assessment of future or existing fire and gas performance according to functional safety standards. Based on this assessment, users have a detailed roadmap for installing new equipment or updating obsolete infrastructure to an optimal level of safety. The main automation contractor can help to identify F&G hazard points and possible risks, and develop basic design packages and related acceptance test criterions to meet safety requirements (Figure 6.14).

By partnering with a knowledgeable, experienced automation contractor, industrial plants can develop an IEC 61511/ISA S84-compliant F&G detection and suppression capability, as well as solutions meeting desired international standards such as NFPA or EN. Supplier assistance can extend to implementing SIS solutions; live hot cutover, implementation, and execution of revamps; and installation, commissioning, and safety validation. To sustain the end user's F&G system performance, leading automation contractors also provide lifecycle support services that include periodic proof testing; system maintenance; training programs on safety, code, and standard compliance; and spare parts management.

6.6.2 TYPICAL INDUSTRY APPLICATION

Like other process industry operations, oil and gas terminals present difficult challenges for automation and safety technology. Tank farms, storage areas, and loading/unloading operations all require fire & gas and safety systems (FGS) to protect personnel, assets, and the environment. The consequences of incidents at oil and gas terminals can be enormous.

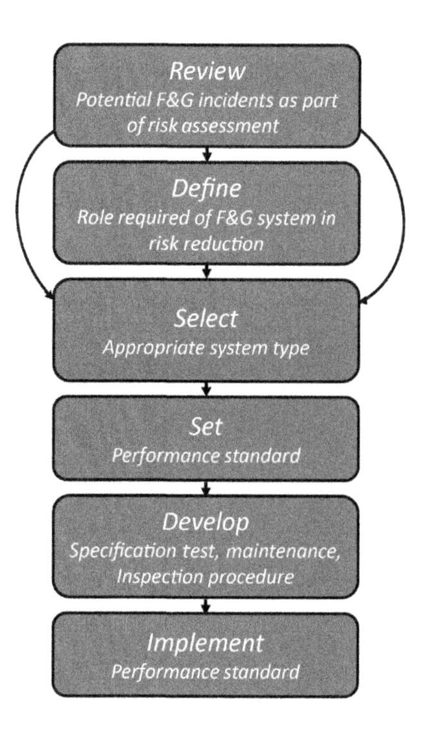

FIGURE 6.14

Layered Safety Development Strategy

In oil and gas terminal applications, operators need an integrated fire and life safety system allowing proactive response to alarms and events and a single real-time view to any potential threat. Industrial plants have procedures and safety systems designed to bring operations to a safe state in the event of equipment malfunctions and other operational problems. In the case of a significant security incident, an integrated system can activate these same procedures and systems. An integrated system also leads to less expensive implementation and maintenance because all the pieces work together.

An integrated fire and life safety solution for terminals typically includes high tank alarms and loading system interlocks (compliant with IEC 61511) that prevent overflow through shutdown of pumps. Likewise, it incorporates the latest technology gas/vapor detectors, integrated with the overfill shutdown system, to detect incidents before they escalate. The system will mitigate safety incidents by ensuring all personnel are informed of hazardous situations in a clear and unambiguous manner, and providing fast and efficient response to associated risks. An effective terminal automation and safety/security solution will also integrate security access control, personnel mustering systems, and video monitoring to reduce the possibility of unauthorized access or intrusion. Integration of technologies such as security biometrics and wireless mesh networks enhance the operation and the lower cost of implementation of these systems.

6.7 CONCLUSION

For today's manufacturers, the safety of their facilities, personnel, production processes, and the environment is crucial to achieving on-time delivery and minimizing any potential losses. Plants must meet their safety needs while ensuring operational and business readiness at project start-up. Faced with this reality, they are seeking the lowest risk, and highest value protection, from their safety system and F&G technology.

FURTHER READINGS

Integrated Fire and Gas Solutions – Improves Plant Safety and Business Performance, ISA.

American Petroleum Institute (API): RP 2001, 2005. Design, Construction, Operation, Maintenance and Inspection of Terminal and Tank Facilities, 2nd ed. API, Washington, DC.

National Fire Protection Association (NFPA). 2008a. NFPA 30, Flammable and Combustible Liquids Code. NFPA, Quincy, MA.

National Fire Protection Association (NFPA). 2008b. NFPA 58, Liquefied Petroleum Code. NFPA, Quincy, MA.

National Fire Protection Association (NFPA). NFPA 329, 2010. Handling Releases of Flammable and Combustible Liquids and Gases. NFPA, Quincy, MA.

CHAPTER

SCADA SYSTEMS

7

7.1 OVERVIEW OF SCADA SYSTEMS

Supervisory control and data acquisition (SCADA) systems are widely used to monitor and remotely control the processes in utility industries such as electric power transmission and distribution, water distribution, and liquid and gas cross-country pipelines. SCADA is being increasingly used in modern-day industries such as environmental monitoring, cell phone tower monitoring, well head monitoring, etc.

A SCADA system enables an operator in a location central to a widely distributed process, such as an oil- or gas-filled, pipeline system, irrigation system, or hydroelectric generating complex, to make set point changes on distant process controllers, to open or close valves or switches, to monitor alarms, and to gather measurement information. The benefits of SCADA systems are best appreciated when the process is very large – hundreds or even thousands of kilometers from one end to the other in terms of reducing the cost of routine visits to monitor facility operation. The value of these benefits grows even more if the facilities are very remote and require extreme effort to visit.

It is interesting to study and understand the evolution of the SCADA systems, how they were earlier, how they are designed now according to different needs from different users, and industry segments that pushed the architectures to evolve. The needs of the industry are augmented by the support from technologies such as personal computers (PCs), networking and embedded microcontrollers, standards and protocols, and others, leading to the development of mature products in the market. During the 1960s, mainframe computers were used as central supervisory control systems and the term SCADA gained popularity in the 1980s. It was followed by the upgradation of older telecommunication technologies to new technologies and the addition of more logical and computational capabilities to the central supervisory system. Later, PCs were added as the central supervisory systems coupled with advanced telemetry for collecting, processing, manipulating, storing, and rendering of data. In general, the SCADA systems, in addition to providing near real-time information about the current state of the remote process, were also designed to manipulate/control the process remotely through a human operator.

The typical SCADA system is a combination of four different subsystems:

1. The central computer (host system).
2. A field-located remote measurement and control equipment called remote terminal unit (RTU).
3. A wide-area communication system (telephonic, Internet, or intranet) to connect all these equipment. Typically, a fiber-optic communication network is used to integrate the multiple pipelines and its related equipment as it is the most reliable medium for data transfer.
4. An operator interface that provides access to the system for the user.

Over the decades, the underlying technologies and operational use of the SCADA system have evolved with improved technology and topology, but the underlying architecture with the four critical subsystems has not changed much (Figure 7.1).

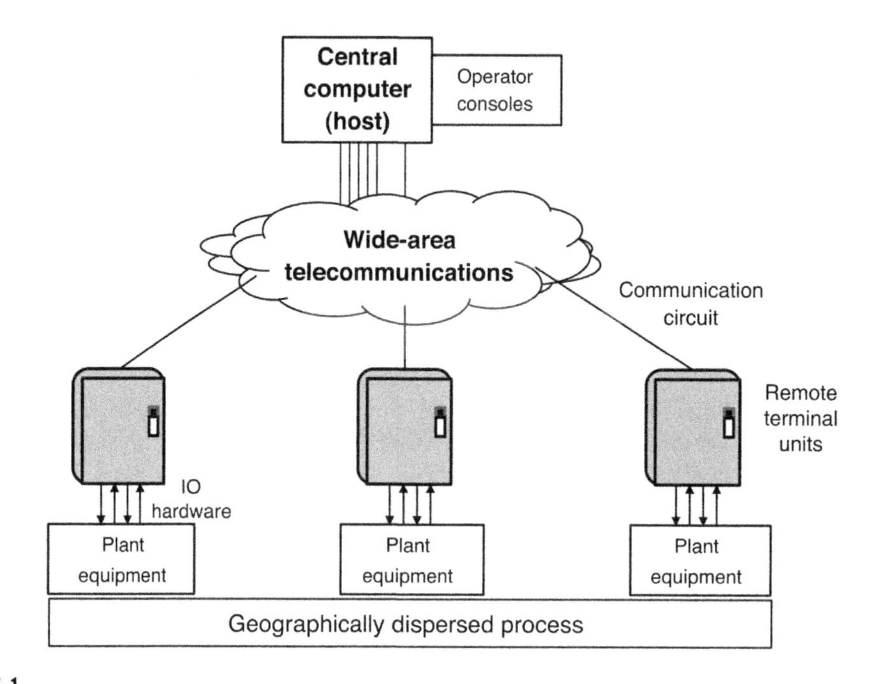

FIGURE 7.1

Simplified diagram of a SCADA

Core functionalities of a SCADA system are as follows:

1. Acquisition of data from field instrument devices via RTU.
2. Processing the field data to detect alarms and other significant process changes.
3. Providing a consistent database of process information about the pipeline and facility.
4. Presenting the data via casy-to-understand graphical user interface, alarms, trends, and reports.
5. Performing remote control of field devices.
6. Performing system monitoring and diagnosis and taking appropriate actions.
7. Historical archiving of data for recent and long-term historical storage and analysis.
8. Transferring real-time engineering data directly to and from the modeling system such as pipeline application system.
9. Providing system data to Management Information Systems (MIS) and supply chain management.
10. Providing integration with geographical information system (GIS) facilities.

To enable real-time data updates from the field and to command the operator instructions on field equipment, the central systems needs remote electronic input/output (I/O) devices with computation and communication capabilities. RTUs are used for the above purposes and are physically located near the process equipment where measurements are taken or control action is provided. The central supervisory system continuously sends request messages to these RTUs and collects the messages and infers the information. The message contains the configuration information, commands for the operator station, updated period values of the process measurements, and so on. The system sequentially repeats the operation for each RTU, until all have been processed. The sequence starts again for the next iteration and continues without an end. The RTUs are simple electronic devices with limited set of functionality. Normally, RTUs are designed with I/O hardware circuits for the measurement of electrical signals generated from devices that produce voltage or current in proportion to physical process parameters such as pressure, flow rate, level, temperature, frequency, pulses, voltage, and current. RTUs also generate control outputs, as either voltage or current signals or in the form of switch or contact singles. Output control signals are initiated on receipt of a command from the central computer. A communication system enables the RTU and the central host computer to exchange data and control commands. The communication system enables messages to be sent/received over large distances in an agreed-upon message format and predefined set of messages and responses (Figure 7.2). Telephones and radios were the only technology options available during the 1960s and therefore were the only design choice for the SCADA designers. In the cases of remotely located RTU equipment with no telephone service, the companies (e.g., a pipeline company or an electric utility) built their own private telephone system, using the same equipment used by telephone companies (microwave relay towers, signal multiplexers, etc.). However, this was an expensive option and designed only in the event of a strong business case for the value generated by the system. In general, telephone and radio technologies were designed for voice (sound) communications. Therefore, SCADA systems used modems to turn computer and RTU electrical signals into sound. In the late 1960s, modem technology was restricted to very low data transmission rates, typically 110–300 bits per second (bps). The RTUs in the 1960s were electronic, but not computer based, so they had to be hardwired to support a simple set of messages for exchanging data with the central host computer. Furthermore, since most communications would take place at 300 bps or less, keeping messages short was essential. The set of messages that could be sent to the RTUs and the set of messages that the RTUs could generate together define a communications protocol. The vendors of SCADA systems had to design and build their own RTUs and thus also defined their own (proprietary) communications protocol(s). In certain industries today (primarily, the electric utilities), there are still RTUs utilizing some of these old, obsolete protocols – often referred to as legacy protocols. In the 1960s, each vendor chose the format of their protocol messages (i.e., how

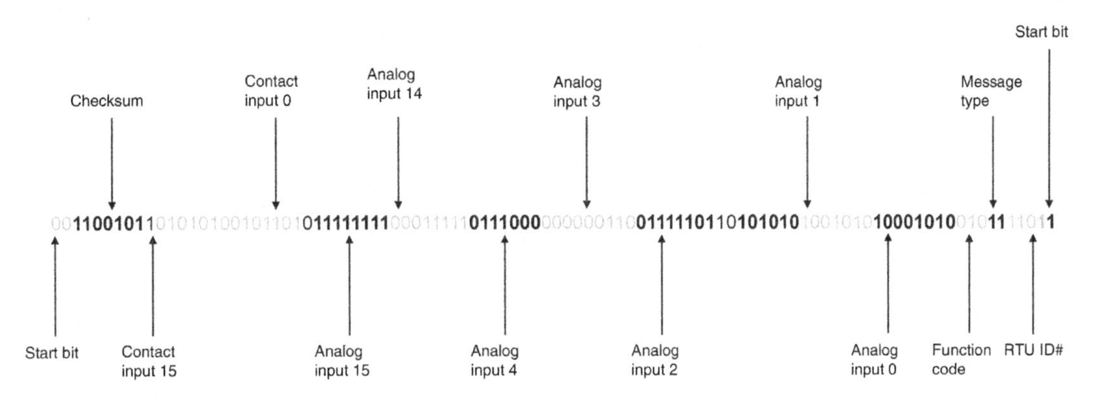

FIGURE 7.2

Example bit-oriented message format (starting and ending bits only, owing to actual number of bits required in a full length message

many bits in a message), in the absence of the universal asynchronous receiver transmitter (UART) chip. To simplify the electronic design of the RTUs, most vendors elected to send all available numeric (analog) or binary (status) data in a single message.

This would mean messages of 30, 48, 64, or some other extended number of bits. In essence, a response to a polling message from the SCADA host was the transmission of the current values of the inputs, sent as one long message. These bit-oriented protocols became unpopular with the invention of the UART chip (and microprocessors), but some of these legacy bit-oriented protocols remain in limited use even today. Most RTU protocols used today are based on the breaking of long messages into some integral number of 8-bit octets suitable for asynchronous serial transmission via UART circuits, which all modern computers employ in their serial ("COM:") ports. These are generally called character-oriented protocols.

A SCADA system is used to fetch and present current data values to a human operator. The time required to refresh the measurement data in a SCADA system that represent the current state of the remote process, through the polling of RTUs, depends on several factors:

- The bit rate (110, 300, or 1200 bps) of the communication circuit(s), also called throughput
- The number of RTUs sharing a given communications circuit
- The length (in bits) and number of messages exchanged in the polling process
- The number of communication circuits being used sequentially or concurrently
- The time delay characteristics of the communication circuits

The protocol used by an RTU can be designed to permit multiple RTUs to share a common communication circuit, similar to a party-line telephone. This means that the protocol incorporates some mechanism (usually an RTU identification number [ID] in the message) that enables a given RTU to identify messages intended for that RTU and to ignore messages addressed to other RTUs. Placing multiple RTUs on communication circuits reduces the required number of such circuits, but it can increase the time required to poll all RTUs for current data values. Most SCADA systems that support multiple polling circuits are designed to concurrently poll RTUs on all these circuits. Thus, if there are x circuits, the SCADA host can be polling x RTUs concurrently (i.e., one on each circuit). The process dynamics in some industries calls for a human operator (or supervisory application program) monitoring and controlling the process through a SCADA system to have fresh measurements more frequently than with less dynamic processes. In such applications (e.g., electric power transmission), it is common to see multiple communication circuits with little or no multi-dropping (sharing) of those communication circuits across multiple RTUs. In less dynamic processes (e.g., water/wastewater transportation), it is not uncommon for all RTUs to be polled on a single, shared radio channel, resulting in much longer times between field data updates. The presentation of information to a human operator is a fundamental feature of SCADA systems, and this could be accomplished in several ways. Initially, it was with a map board (mimic panel), a wall-sized drawing of the process with indicator lights and numeric meter-like read-outs mounted on a wall display at appropriate positions representing the physical process areas. The high-resolution video, projection, and flat-panel display technologies of today were not available to the system designers of the 1960s to 1980s.These lights and numeric displays were electrically controlled by I/O signals from the host computer and updated based on data received from the RTUs. In many SCADA system control rooms today, you will still find the equivalent of a mimic panel/ map board display, but implemented with modern video projection technology. Information presentation in the 1960s also included printed reports, logs, and alarm messages. It also included very basic

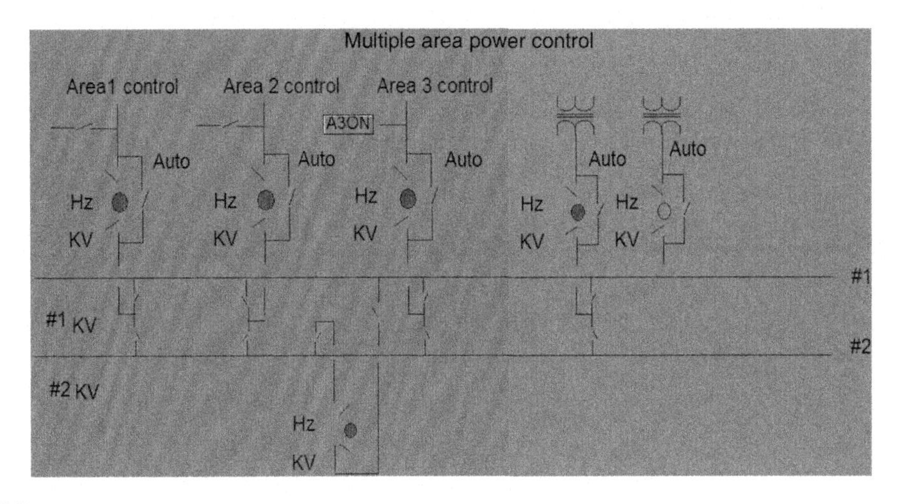

FIGURE 7.3

Example graphical operator displays

cathode-ray tube (CRT) displays with tabular, alphanumeric information or very primitive and simplistic, monochromatic, semigraphic displays (similar in nature to Figures 7.3 and 7.4, respectively). Such CRT-based displays used hardware and software developed by the SCADA system vendor. These early SCADA systems were programmed in assembly language (with an occasional application written in Fortran), and essentially 100% of the software (and a good bit of the hardware) was proprietary to, and developed by, the vendor, including the operating system software.

7.2 MINICOMPUTERS AND MICROPROCESSORS

The introduction of mainframe computers into telemetry systems, to create the SCADA system, was a significant technological advance. It enabled the development of custom application programs to make use of the real-time process data – for trending, performing reasonability/alarm checking, and generating additional information by making calculations, and for presenting the data to the operator in a variety of formats. High cost and reliability were the biggest drawbacks of mainframe computers. Therefore, SCADA systems were restricted to applications that could justify the high implementation costs involved. The introduction of the (relatively) low-cost, 16-bit "minicomputer" and then of 8-bit microprocessors in the 1970s were two major advances in SCADA systems technology. The minicomputer was a much more cost-effective computing platform as compared to the mainframe computers of the time. They were also much simpler devices and easier to connect and configure together into a redundant configuration. The use of minicomputers as the host computers drastically reduced the cost of SCADA systems. This made possible the financial justification of more applications. And then, there were hybrid systems – in which redundant minicomputers performed the real-time polling of the RTUs and provided the operator display and remote control functions, and a nonredundant mainframe computer was used to run nonessential advanced applications and models. The software

sysTnd03.htm | sysTnd03.htm | sysTnd03.htm | sysTnd03.htm | sysTnd03.htm | sysTnd03.htm | enchdiag_disp.htm | Channel Status

CPU Free	CPU FreeMin	RAMTotal
95.34	33.77	126348
RAM Free	**RAM FreeMin**	**ConfigDataMemTotal**
43.45	42.09	1604
ConfigDataMemFree	**LogDataMemTotal**	**LogDataMemFree**
932	10720.00	7264
SDCardMemTotal	**SDCardMemFree**	**EnclosureTemperature**
31154688	2448640	34.75

CPUusedfor eCLR	RAM used for eCLR
3.75	26412.00

	Reset	FDM req	CMD3 Req	CMD3 Resp Succ	CMD3 Resp Fail	CMD48 Req	CMD48 Res Succ	CMD48 Resp Fail
HART	-------	2089.00	??????	-1.00	??????	??????	??????	??????

	Req	Res	Valid Req	Succ Res	ReadcoilReq	ReadcoilRes	WritecoilReq	WritecoilRes
MB SIAVE	118859.00	118859.00	118859.00	118859.00	12445.00	12445.00	7612.00	7612.00
	ReadDIReq	**ReadDIRes**	**ReadHRReq**	**ReadHRRes**	**WriteHRReq**	**WriteHRRes**	**ReadIRReq**	**ReadIRRes**
	7185.00	7185.00	51391.00	51391.00	19228.00	19228.00	20998.00	20998.00

	Reset	ASDUSent	ASDUSentPerMin	ASDURecv	ASDURecvPerMin	EventSent	EventSentPerMin
DNP SIAVE	-------	5780.00	3.52	6136.00	??????	14131.00	3.83

	Reset	Req	Succ Res	Timeout Res	Fail Res	Req PerMin	MaxReq PerMin	MaxSuccResPerMin
MB MSTR	-------	273.00	191.00	0.00	82.00	0.10	3.52	3.37

17-Sep-14 11:10:37 PCT_ASSET MB_DI997 ALARM U 15 AO_OutofRange2 SW_ON

FIGURE 7.4

Example tabular operator displays

that ran the minicomputer-based SCADA systems was generally written in assembly language and was usually 100% proprietary to the system vendor. Eventually, minicomputers evolved into "super" (32-bit) minicomputers, and computer vendors started to provide general-purpose operating systems that were suitable for real-time applications and operating system interoperability standards (e.g., portable operating system interface and XWindows) were developed. In parallel, the PC came into being and ultimately evolved into the powerful devices we have today, with commercially available operating systems, networks, relational databases, peripherals, and much more. The 8-bit microprocessor revolutionized the design of the RTUs in the 1980s because RTUs based on microprocessor technology could be (re)programmed to add or expand their functions, perform much more sophisticated functions, and even provide local, closed-loop control. Sometimes, pre-microprocessor RTUs were referred to as dumb remotes and those with microprocessors as smart remotes. That delineation has become blurred, as the computing power of RTUs has kept pace with overall microprocessor technology. A modern RTU could be considered brilliant in comparison to the earliest 8-bit microprocessor models. One major difference between smart and dumb RTUs was in the sophistication of communications protocols they could support. To perform remote control functions with a dumb RTU, SCADA systems often employed a select-check-operate message sequence. Messages sent between the host computer and the RTUs were subject to interference and distortion from any number of environmental or communications system sources. Protocol messages are, in actual form, just a string of binary numbers represented by 1s and 0s. If a message was electrically distorted (i.e., if one or more bits were changed), this could mean that a command that was supposed to operate device X would appear to the RTU as a command to operate device Y. The error checking and detection capabilities of premicroprocessor RTUs were very limited and not always reliable. Early RTU protocols called for a strict sequence of multiple messages (called a check-before-operate or select-check-operate sequence) to effect a control output. This ensured the prevention of a communications error from causing an improper or dangerous control action. The human operator was an integral part of the message error-checking process. In a select-check-operate protocol, several messages and responses are exchanged in a defined sequence.

1. The operator sends a "select control output n" message to the specific RTU.
2. The RTU returns a "control output n is selected" message and starts a countdown by a dead-man timer in the RTU.
3. The operator sends a "prepare to perform operation x on selected output" message to the RTU.
4. The RTU returns an "operation x is to be performed" message.
5. The operator sends a "perform the operation" message to the RTU.
6. The RTU performs operation x on control output n and returns an "operation successful" message to the operator (and also cancels the dead-man timer).

At any point in the sequence, if the operator is in disagreement with the RTU recommendation (e.g., if it responds that it has selected output o rather than n as requested, or if a different RTU responds), the operator can send a "cancel" message – or just let the dead-man timer expire, causing the RTU to automatically cancel the activity. The human operator provides message verification and error checking through this process. The microprocessor-based RTUs enabled the implementation of protocols that incorporated comprehensive message error checking and correction schemes, thus eliminating the need for operator intervention to validate messages. (However, in the electric utility industry, the operator still verifies control actions in this manner.) Having reliable protocols is even more important when supervisory control commands are to be issued automatically by application programs running in the host computer; operator interaction would not be practical in those instances. Microprocessor-based RTUs introduced a whole range of additional capabilities including a very significant one of having the ability to download parameters, program logic, and calculations into the RTU, via the communications (polling) channel. This capability enabled remote modifications of the RTU's functionality, although it required implementing a more sophisticated RTU protocol that included message types for performing such downloading or parameter modification functions.

7.2.1 CENTRAL ARCHITECTURES

From the 1960s into the early 1980s, most SCADA systems used a central architecture in which a single powerful central computer managed and performed all the functions, including RTU polling, data processing, display generation, report generation, data archiving, and running application

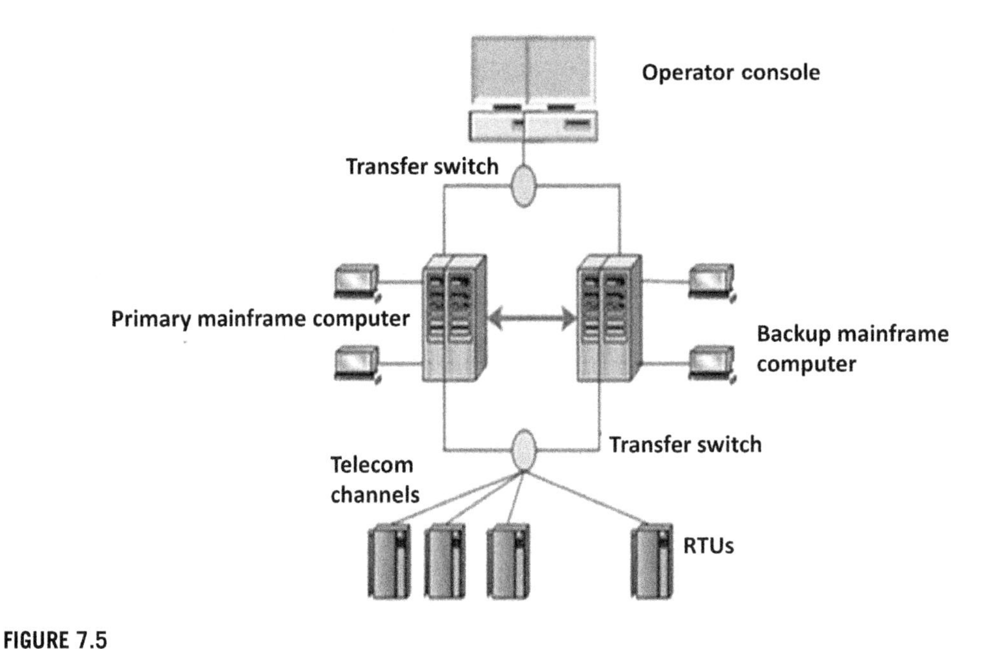

FIGURE 7.5

Centralized redundant SCADA system architecture

programs. To improve the reliability of the system, these centralized architecture systems invariably employed a second, identical (redundant) computer, and some form of dialog was maintained to keep the redundant computer synchronized and updated (Figure 7.5). In a redundant computer configuration, one computer is designated the primary unit and the other the backup unit. The primary unit performs all the SCADA functions and then transfers updated information to the backup computer, so that on any failure of the primary unit, the backup unit could take over and continue operations. The goal of a redundant design is to render a computer failure as transparent as possible to system users (with no loss of data, no loss of control capacity, no missed alarms, no loss of application programs, etc.)

Making redundancy work successfully was the biggest challenge for most SCADA system vendors. Early schemes that involved high-speed copying of memory areas from the primary to the backup computer (via direct memory access) would instantly corrupt the backup if the primary became corrupted. Redundancy schemes were often complicated because of the need to transfer communication circuits, key peripherals, and operator console hardware from the primary to the backup computer. These early systems often used electromechanical transfer switches to implement such peripheral transfers. In other instances, the design of shared equipment, such as operator video displays, would incorporate a dual-ported capability, so that either computer could interact with those devices. For the triggering of an automatic transfer to the backup computer, some means was needed to automatically determine that the primary was inoperative. Most systems included a special hardware device/circuit called a watchdog timer – a circuit that would generate a trigger signal, when a countdown by its hardware timer reached zero, unless that timer was constantly reset by commands from the computer. Both the primary and backup computers would incorporate a watchdog circuit, and each would have an application that was run periodically to reset their respective timers. If either computer stalled or had a hardware failure, or if its programming went into an infinite loop, the watchdog would count down to zero and generate an electrical signal that would cause the transfer of peripherals and notify the alternate computer to take over operations. At least, that is what was supposed to happen. In reality, such redundancy schemes often failed to operate or mis-operated (e.g., triggered a transfer when the primary computer had not actually failed). Many redundancy designs were implemented, but few worked as advertised. One measure of the quality of a redundancy scheme was the time lag between the data in the primary system and the backup. Some vendors claimed that the backup would lag the primary by seconds; others updated the backup only every few minutes. Another differentiation in redundancy designs was their classification as fully automatic, bump-less, or hot-standby schemes, as compared to offering a warm-backup or cold-backup scheme. The last two of these schemes are less than perfect and possibly even require some level of manual intervention.

In a centralized SCADA system architecture, in addition to doing all the work of polling RTUs, processing and alarming the incoming data, generating and updating operational displays and reports, etc., it still has to allocate central processing unit (CPU) resources and time to update the backup computer with the newest data. Over time, the number of RTUs, I/O points, calculations, and other functions grew beyond the initially implemented design capacity, and the redundancy scheme was taxed and degraded. With a centralized design, the only way to add capacity to the system was to replace the computers with newer, more powerful models. This is one of the primary reasons why a popular replacement design eventually supplanted centralized architecture.

7.2.2 DISTRIBUTED ARCHITECTURE

From the 1980s to now, the majority of SCADA systems followed the trend of computer technology and erected a distributed architecture in which multiple computers were networked together and specific functions were assigned to particular computers. The introduction of local area networking (LAN) technologies (particularly Ethernet) made this possible and practical.

One of the most common architectural strategies was to separate out the RTU polling process, placing this function in dedicated front-end computers (Figure 7.6). This strategy also simplified the problem of computer redundancy because front-end computers could be stripped down to their basics – RAM, serial ports, and LAN controllers – making primary backup updating much less complicated. The front-end computers received data from RTU polling and sent data-update messages to all other computers on the LAN, thus keeping them synchronized. Another distributed-architecture strategy was to use a separate, dedicated computer to run large, complex, processor-intensive applications.

This computer was not necessarily redundant because a temporary loss of those applications does not affect the operator's basic ability to control and monitor the process.

Another distributed strategy arose with the growing importance of historical archiving of data, in which a specific computer was assigned the responsibility for long-term data-archiving/trending functions. This computer would need much more online storage than the rest, as well as removable media storage devices for long-term off-line archiving of historical data. A distributed architecture was the next logical step from a centralized design because it increased overall computing power by spreading the various SCADA functions into additional computers dedicated to performing

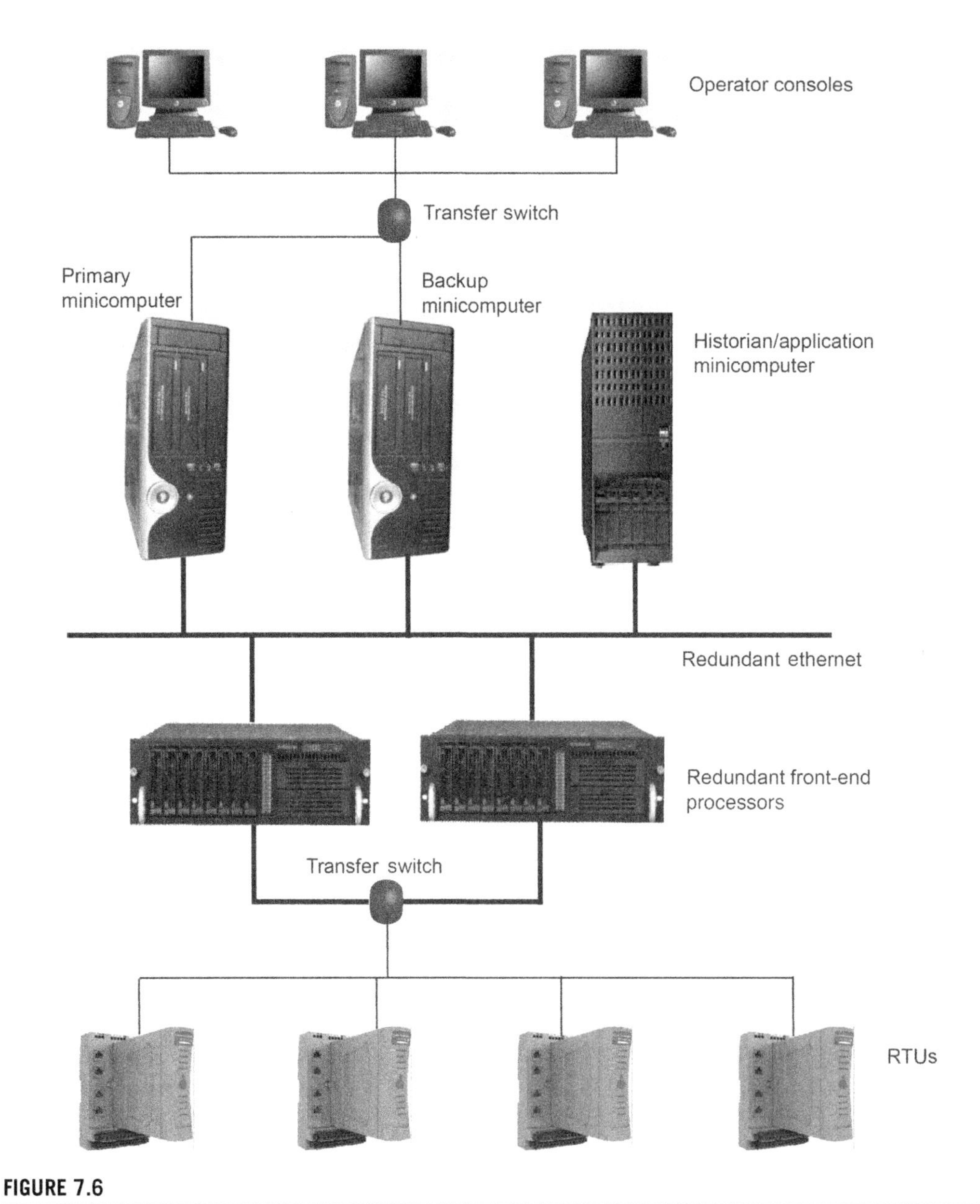

FIGURE 7.6

Distributed SCADA system architecture

those specific functions. It also improved overall system reliability, since only certain SCADA functions were lost in the event of a fault or malfunction in one of the computers, rather than losing the total system. This ability to suffer a partial functional loss is usually described as graceful degradation.

7.2.3 CLIENT/SERVER ARCHITECTURE

Because of the advent of powerful 64-bit microprocessors and PC in the 1990s – and transmission control protocol/Internet protocol (TCP/IP) networking technology, high-speed Ethernet, and commercial real-time operating systems – the design of SCADA systems took another evolutionary step. In effect, SCADA systems began to look very much like the corporate information technology (IT) systems of the time. These new systems were designed to be scalable and fault tolerant, and the software that performed the SCADA-specific functions was redesigned and rewritten for distribution across multiple computer servers. Software that had been monolithic in prior generations of SCADA (e.g., operator console display software) was broken into interacting components that could be run in the same computer or in different ones, with cross-LAN communications. Peripheral transfer switching and complex redundancy schemes were replaced with smarter software. Clients (e.g., any program that generates operational display updates) could seek out their respective servers (e.g., an RTU polling program) across the LAN to get the required data. If for some reason the server became unavailable, these clients could automatically switch to an alternative server if one was available. If more computing capability was required, additional servers could be connected to the high-speed network and more copies of the client/server software could be loaded and run (Figure 7.7). SCADA became a special set of applications that could be run on any conventional, commercial computing platform. In fact, there are several commercially available SCADA software packages that can be purchased as packaged software and loaded onto conventional PCs and servers, as a do-it-yourself SCADA system.

7.2.4 GENERAL SOFTWARE ARCHITECTURE

The technology underlying SCADA systems has evolved corresponding with advances in digital electronics, communications, and computing. The architectural designs of SCADA systems have been changed to take advantage of these new technologies. Regardless of the hardware architecture and technology employed, SCADA systems still perform a basic set of functions:

- Communicate with field-located RTUs for retrieval of up-to-date process information.
- Maintain a constantly updated set of this information in a database.
- Provide an interface that enables human operators to examine and display information.
- Provide mechanism for a human operator sending a control command to RTU.

In addition, the SCADA systems also provide at least a few additional functions, such as automatic comparison of field measurements against user-defined alarm limits and against their prior values, automatic annunciation and logging on detection of alarms and events, and long-term recording of the value/status history of selected field inputs. The generation of user-defined reports and logs that incorporate available measurements, calculations, and historical/recorded data, and the calculation of new values based on user-defined equations that incorporate available historical and real-time measurements.

In addition to the discussed features and functions, all modern SCADA systems support a whole lot more, including industry-specific applications that customize the SCADA system for the unique requirements of individual market segments. Figure 7.8 provides a simplified information flow diagram for the basic functions and features of a generic SCADA system. The actual implementations depend on the specific architectural and technology choices made by the individual SCADA system vendors.

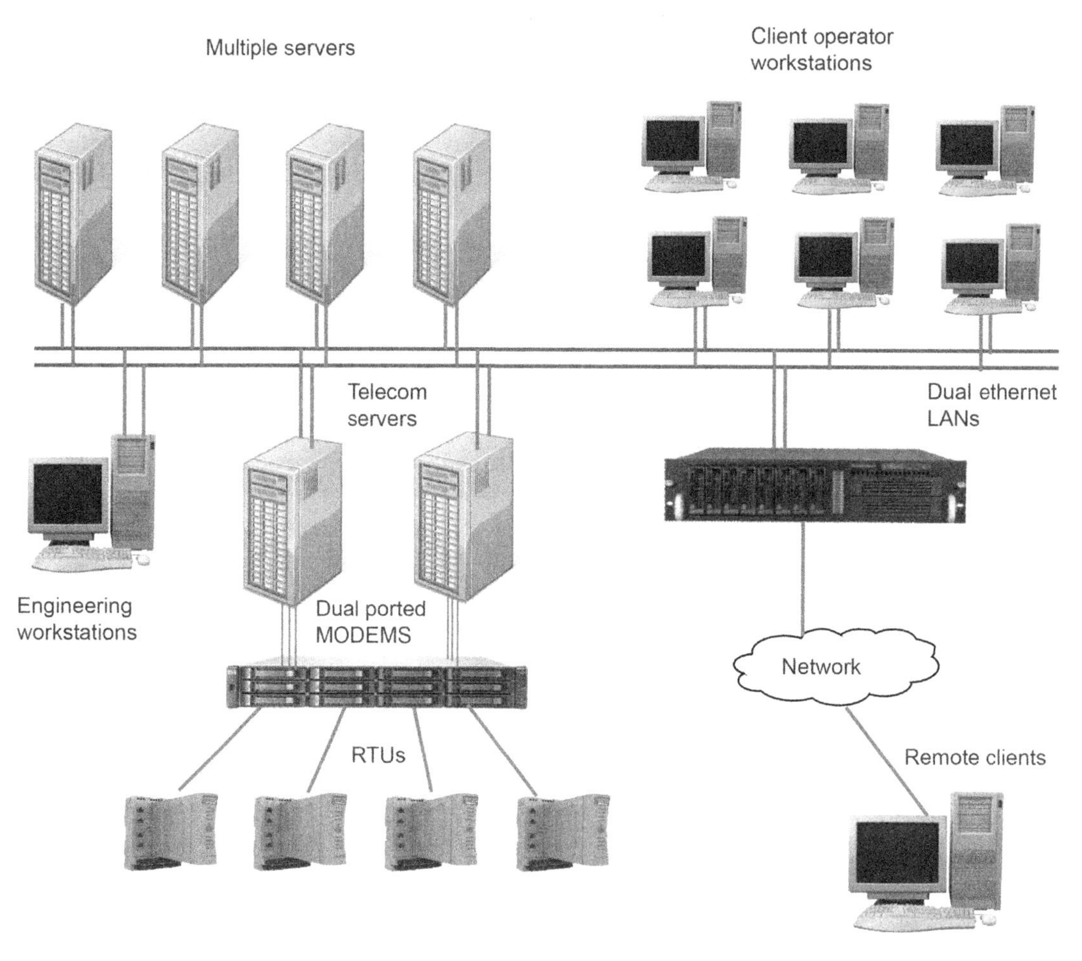

FIGURE 7.7

Client server SCADA system architecture

7.3 REMOTE TERMINAL UNITS

7.3.1 BASIC FEATURES AND FUNCTIONS

The primary purpose of a SCADA system is the remote measurement and control of geographically distributed processes/equipment. A SCADA system incorporates multiple RTUs, physically placed in the field at key measurement and control locations. The RTU began as a hardwired, fixed-function electronic device that could respond to simple commands received as serially transmitted numeric codes. Early RTUs were used for remote telemetry purposes, even before SCADA systems were created, through the integration of a central computer. RTUs were connected, through serial communication channels, to master terminal units (MTUs) – electronic devices that provided a human interface,

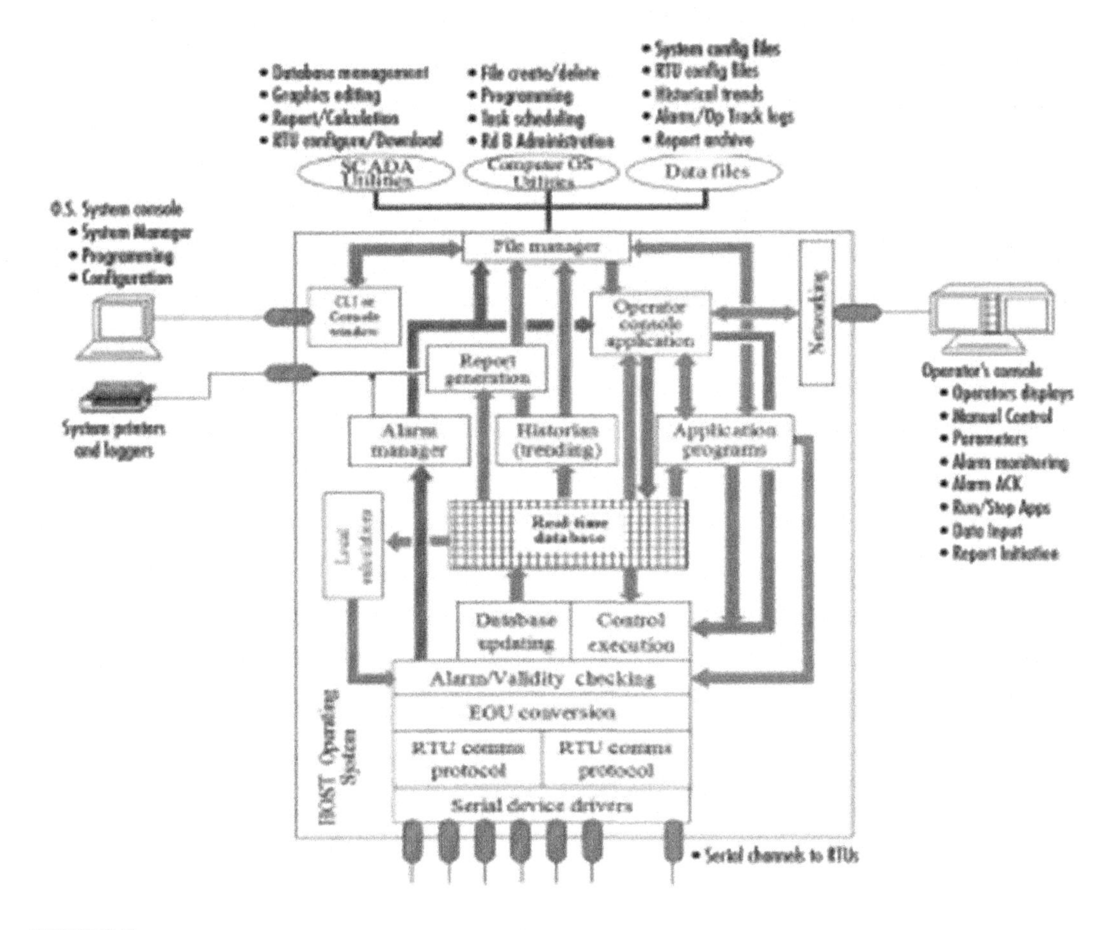

FIGURE 7.8

Generalized information flow in a generic SCADA system

to enable display of data from the RTUs and manipulation of control outputs of the RTUs. An MTU generally consisted of an alphanumeric readout or display and a keypad for entering parameters and commands. An operator could enter an RTU ID and an input number and press a poll key. The MTU would send a serial numeric code (message) to the RTU(s), receive the response from the RTU (in the form of a coded numeric message), and display the value of the requested input point on its readout. RTU control outputs could be manipulated in a similar fashion. More sophisticated MTUs supported multiple numeric readouts and continuous, automatic polling. Similar to their corresponding RTUs, the MTUs were electronic devices with hardwired functions and features, rather than programmable computer-based devices. When computers were introduced to replace the electronic MTUs (and programmed to replicate the communication protocol messages), the term MTU was often thereafter applied to the computer. SCADA central computers are sometimes still called the "host," the "central unit," or occasionally even the "MTU".

When the first SCADA systems were developed, the obvious strategy was to replace the MTU with a computer, programming that computer to send and receive the same simple numeric messages that had previously been generated and processed by the MTU. The set of messages to and responses from the RTUs constitute a communications protocol. Because the early RTUs and MTUs were electronic, but not based on any form of programmable computer technology, they had to be designed with electronic circuits that could generate and interpret these simple numeric message codes. As might be expected, this kept the number and complexity of such messages to a minimum. The only functions supported by early RTUs were reading and transmitting the current value of analog and status inputs and generating analog and contact outputs in response to a properly formatted control message. RTUs are supported with a defined set of I/O options and vendors provide various options and combinations to choose based on the need and application. The I/O types such as analog input, analog output, status input, contact output, pulse input, special inputs such as RTD and TC, and others, are dealt in other chapters, such as DCS (Chapter 6), in much more detail. In an RTU, although the types of I/O and method of scanning may be different, they are conceptually same with different scan times based on the type of analog-to-digital (A/D) converters used.

7.3.2 RTU TECHNOLOGY

RTUs underwent a fundamental change with the introduction of microprocessors. Earlier, adding any features and functions to an RTU involved a massive hardware redesign and changes to the protocol messages. With respect to RTUs, most of the energy was channeled into just expanding the maximum quantity of I/O and the types of I/O supported by those RTUs. However, with a microprocessor running the RTU, changes were mainly a function of software modifications, including protocol enhancements. The early smart RTUs used 8-bit microprocessors such as the Intel 8080 and 8085. These RTUs were limited in memory capacity (4–16 kilobytes was typical) and processing power; still, they were capable of supporting functions that were not possible with dumb RTU technology, such as simple calculations (Figure 7.9). As microprocessor technology advanced and CPUs became faster and were expanded to include math coprocessors and more memory, it became possible to add more features and functions to the RTUs.

RTU feature evolution was nonuniform across industry sectors using SCADA technology. SCADA system vendors maintained a primary presence in one key segment (pipelines, electric power, transportation, etc.) and focused product developments, including RTU developments, on the needs of that market segment. Many SCADA system vendors that attempted to cross over into other market segments to expand their business base discovered that their systems and RTUs lacked features and capabilities that had become standard in these other market segments. Probably, the most advanced RTUs, from a feature and function perspective, are those used in the oil and gas pipeline market. RTUs in the pipeline industry were used to perform local regulatory and sequential control functions as early as the mid-1980s. These RTUs could download user-defined control schemes and logic over the polling communication channel and perform advanced calculations, such as the American Gas Association (AGA) volumetric computations (AGA-3, -5, and -7 and NX-19). They also had to support serial communications with other smart instrumentation to get the data to perform some of those calculations. These RTUs, once downloaded with the appropriate logic, functioned as independent, autonomous local controllers, capable of performing their assigned functions without operator intervention, even if communications to the host were lost. As memory capacity expanded, some of these RTUs incorporated the ability to

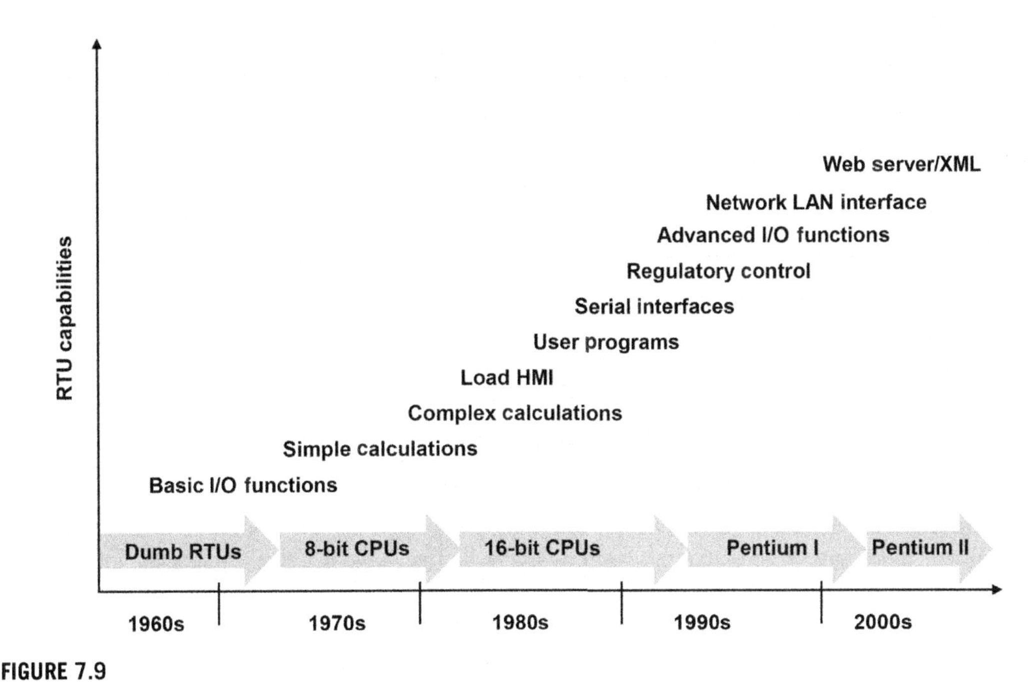

FIGURE 7.9

Evolution of smart RTU technologies and capabilities

buffer data in the event of extended communication outages and then to upload this information to the host when communication was restored. All these capabilities required extensive modifications to the RTU communications protocols. Even as vendors moved rapidly to match the capabilities of their competitors, protocol versions were released at an even faster rate. If the revisions merely added functions, without changing any preexisting functions, then it was possible to mix RTUs with different protocol versions on the same communication channel. However, a protocol enhancement would sometimes modify the protocol sufficiently rendering mixing of RTUs nonviable, and SCADA system owners would have to send technicians to the field, to make firmware changes to all of the existing RTUs or at least to those on a shared communication circuit. Keeping track of protocol versions and compatibilities was a major overhead for SCADA system owners and vendors. Over time, it became typical for SCADA vendors to have a wide range of RTUs with a combination of features and functions (and prices), and a set of different protocols that would be used, based on the particular RTU being supplied. Although there have been successful cases of multiple RTU serial protocols coexisting on the same communication channel, this is not recommended. It is all too possible for messages and data sent using one protocol to be accidentally interpreted as a valid message by the other protocol. Serial ports RTUs normally come with at least one serial port, through which the SCADA host communicates with the RTU. This serial port is typically dedicated to supporting the protocol messages, either by hardwired logic or by program logic. With the introduction of microprocessor-based RTUs, it also became typical to have a second serial port, dedicated to local interrogation of the RTU for diagnostic and configuration purposes (often referred to as the console port). The second serial port was usually managed by a special program in the RTU called a command line interpreter (CLI), as in Microsoft DOS. The CLI understood simple text commands and enabled a service technician to examine and modify internal settings and parameters and to run available diagnostic routines. A typical use of this console port would be to tell the RTU how to configure the host polling port (in terms of baud rate and number of stop bits, etc.) and what its ID was; the console port also allowed for manual reading and controlling of the RTU's I/O. In some industrial applications, sometimes two separate organizations, with their own SCADA systems, needed to use the same field measurements. Initially, this meant placing two separate RTUs at the same location and wiring the same inputs to both units. This worked well for inputs, but if both organizations needed to control the same field equipment, double wiring presented a problem. The solution eventually developed was to use multiported RTUs. Additional serial ports were added to RTUs, and separate protocol drivers were assigned to each port. Each SCADA system could interrogate the RTU (possibly using different protocols) on their port, and both could issue control commands to the RTU. A mechanism for coordination for systems with multiple ports was devised in which a parameter or flag in the RTU was used to identify the SCADA system that currently had control rights. The protocol software in the RTU had to be modified to examine these flags when receiving a control command message, to determine whether to accept and execute the command. RTUs with multiple host connections are most common in the electric utility market. Dual (two) connections are the most common, but there have been situations with as many as five separate hosts connected to a single, shared RTU. Commonly, the RTU and host computer are said to be in a master and slave protocol arrangement. The master in this style of protocol is the computer that initiates all communications, and the slave responds to messages from the master. This type of communication scheme is also known as a poll-response architecture. For polling and communications to work, one device must follow the master sequence of the protocol definition, and the other must follow the slave sequence. Normally, the polling ports on the RTU are configured to support the slave portion of the specific protocol. However, in many applications, it is impractical to provide direct communication channels to every RTU, and in those cases, selected RTUs are used as a sub-master or concentrator. It is not unusual to have a large, higher-capacity RTU act as both slave and master, on separate communication ports. This concentrator RTU might have one port connected to a radio transceiver and be running a master protocol on that port, in order to poll several smaller RTUs in the general vicinity (Figure 7.10); its other port would run the slave protocol, to respond to host polling. In this sort of architecture, the concentrator RTU typically combines all the inputs received from its slave RTUs into its own input table and presents all this aggregated information to the host as a single, virtual RTU. For control outputs, the concentrator RTU identifies commands that are actually intended for control points in its slave RTUs and initiates the equivalent output control message sequence to the respective slave RTU. With a concentrator RTU arrangement, different protocols may be used on each of the ports, so that the concentrator also acts as a protocol translator or converter. Effectively, it is really functioning like a small SCADA master because polling of its own slave RTUs is normally done asynchronously to the polling for the data from those other RTUs by the central SCADA system. (The concentrator RTU polls its slaves and retains a local copy of their most recent data, which it uses to answer the central SCADA system's requests for such data.) There is some form of coordination only with outgoing control commands that must pass through the concentrator RTU.

Lately, it has become commonplace for RTUs to support a large number (16–32) of serial ports especially in the electric utility market (while the amount of physical I/O has decreased). Many of the measurements and status points that would normally be wired to RTU analog and status inputs are being wired into other microprocessor-based devices, and the value of these inputs can be obtained by serial

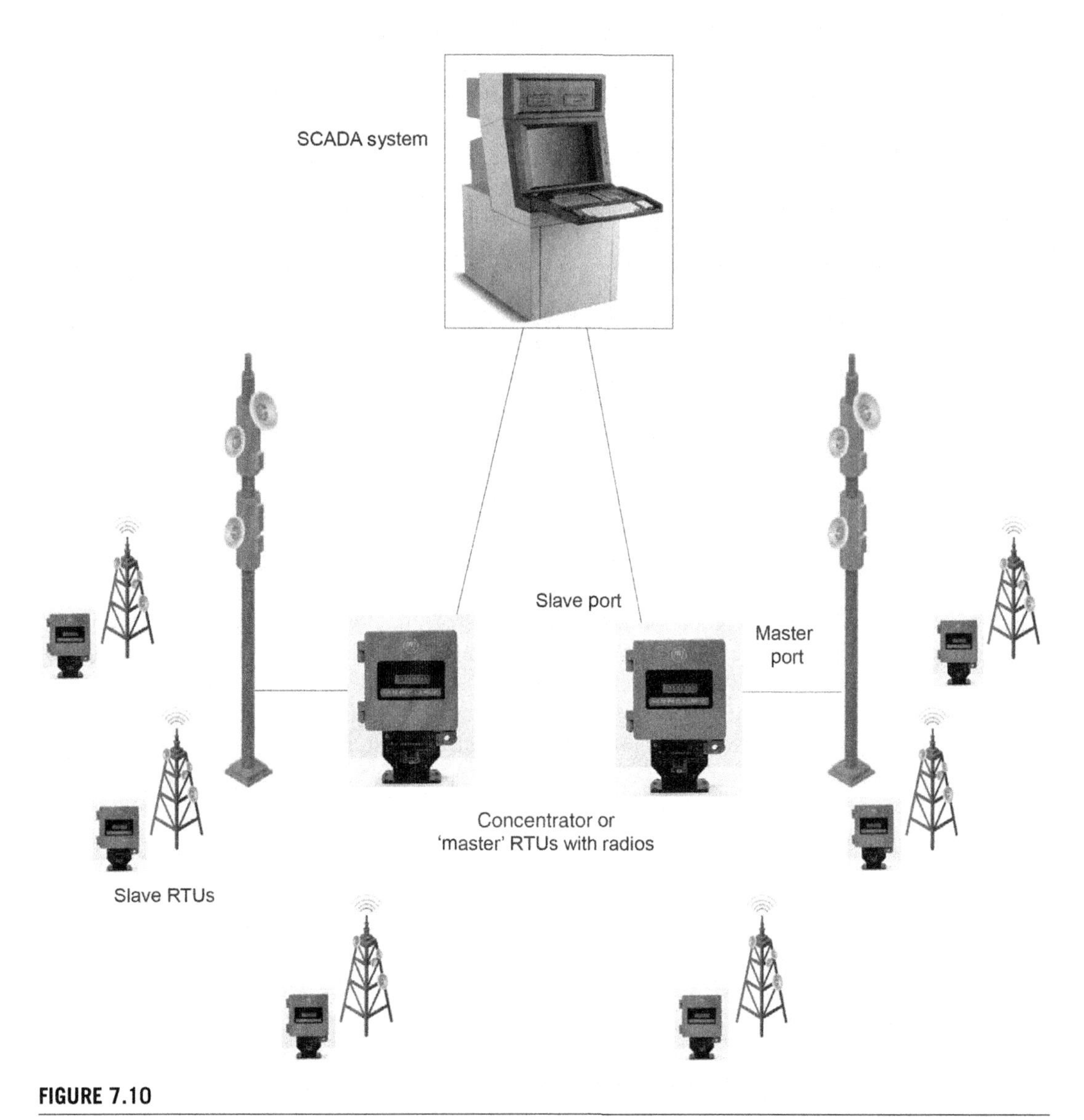

FIGURE 7.10

RTU hierarchy using master and slaves protocols combinations

communications with those other devices. Such microprocessor-based equipment are called intelligent electronic device (IED). Substation automation or substation data concentration is the process of using an RTU with a large number of serial ports (and possibly no physical I/O at all) to collect all the data from the IEDs in a substation, to be sent to a SCADA system over a single communication circuit. Most RTUs manufactured today support multiple serial ports, and can be configured as either a master or a slave port, using any number of available serial protocols. Intercommunication is sometimes a problem because of the complex types of data in the IEDs. RTUs were initially designed to support numeric and binary data type values. Initially, the numeric values were integer only, but today many RTUs can support floating-point or long integer values. Similarly, binary values were originally single-bit only; today multi-bit versions are supported (to represent devices with multiple states: e.g., a valve could be open, closed, or traveling). Modern RTUs also support date/time information and quality tags that indicate the validity of the data to which they are attached. However, there are many other types of data, and data structures that combine multiple values and different data types. Most RTU protocols do not support advanced or sophisticated data types and therefore are useless for extracting all but simple data from IEDs. A common IED in the electric utility industry is a protective digital relay, the device that tells circuit breakers to open when bad things are happening to the power line. Inside a digital relay, there can be dozens of measured and calculated numeric values and status bits. These relays also do Sequence of Events (SOE) recording, power totalization over time intervals, and high speed recording and capture of the AC voltage and current waveforms on all three phases. It is extremely difficult to get this complex data using a conventional serial RTU protocol.

7.3.3 LOCAL DISPLAY

Since RTUs contain a lot of real-time process information, it becomes important to display this information to a service technician or process engineer in the field. This helps in calibrating the instruments. Smart RTUs enables providing a range of electronic displays that could be driven and updated under software control. Today, RTUs have an Ethernet port (or Bluetooth wireless capability) and provide dynamic Web pages on the laptop PC. In the early days of smart RTUs, display capability had to be built into the RTU hardware itself. An RTU might have a simple number/function keypad on its front panel and a multiline liquid crystal display (LCD) screen. The user could punch in an input number and see it displayed on the panel, or if no keypad was provided, the LCD might be set to cycle through all inputs on a continuous basis. A keypad of some sort would be necessary to enable output control. Importantly, the analog inputs and status inputs could be displayed only in an unconverted form (e.g., as raw counts or percentage of range), unless there was sufficient configuration data to allow a conversion calculation. When an RTU reads an analog input, the A/D converter generates a binary number proportional to the range of the input value. Thus, if the input signal is 4–20 milliamps and the A/D converter is an 8-bit version, then at 12 milliamps (50% of the input range), the binary number from the converter should read "064" (i.e., half of the 000 to 127 conversion range). In this case, the RTU could present either the raw counts of "064" or the percentage of range this represents: namely, 50%. No additional data is needed to provide/compute these two display values.

However, if the input is actually a temperature, and the 4–20 milliamp signal range has been set for 150–350°F, the host computer displays a value of 250°F for this input reading, which is the engineering unit (EGU) value for the input. The host has additional information about the EGU range to compute this from the raw counts; to supply EGU displays at the RTU, the RTU also needs a copy of that additional information. A similar situation occurs for status inputs. In the RTU, these will be seen as a 0 or 1 raw value, but in the host computer (because of additional data), a 0 means the pump is stopped and a 1 means that the pump is running (and that is how the input would be displayed to the operator at the host computer). The early smart RTUs had insufficient memory to hold tables of EGU conversion factors, and the protocols were inherited from dumb RTUs, which sent only raw counts and binary values. It is still common for RTUs to send all their analog and status inputs as raw counts in the electric utility market and all EGU conversions are done at the host level. Many of the protocols still in use in that segment support only raw counts. In the pipeline and water/wastewater industries, performance of calculations – and even sequential and regulatory control – is more common at the RTU level. Many of the calculations used are based on thermodynamics or physics and require that the equation elements be in defined EGUs: pounds per square inch, degrees Fahrenheit, cubic feet per minute, and so forth. This makes it necessary to perform at least some EGU conversions within the RTU itself. In some RTUs, this is accomplished by user-written calculations or program logic in which the conversion factors are built into the logic or calculations. In other RTUs, the vendors have allowed the EGU conversion tables to be downloaded into the RTUs and have modified their communications protocols to support floating-point numbers. An additional extension was the downloading of the alarm limit value tables, so that the EGU values could be checked against these limits. This required even more memory in the RTU and appropriate protocol extensions. RTUs that supported this full range of input processing were capable of providing local information displays that were much like those provided at the host (if equipped with the necessary display hardware). Such RTUs sometimes have printers attached to them, to print out alarm logs and reports.

7.3.4 DOWNLOADED LOGIC AND PARAMETERS

The first smart RTUs were preprogrammed at the factory, and all their functions were set into the firmware forever. The evolution of smart RTUs has seen three levels of remote configuration, each being successively more complex to implement but more flexible in application:

1. Preprogrammed functions with remotely adjustable parameters and flags
2. Downloadable library of user-selected and connected function blocks
3. User-written program download and execution

Most RTUs were provided with one or more of these capabilities. The more advanced capabilities often incorporated the lesser functions. SCADA vendors soon started adding simple function blocks and calculations that could be enabled and (re)parameterized via the communications port. The logic and programming to perform the functions was still factory installed, but the user could adjust supervisory points (binary and analog values stored in the software of the RTU) via the RTU protocol (Figure 7.11). The RTUs supported real analog and binary I/O points, while the smart RTUs added a new type of point, which enabled the host/operator to manipulate RTU functions by sending down supervisory point control commands. The idea of supervisory (software) points was eventually expanded to include both incoming points (to receive data and parameters from the host) and outgoing points (so that RTU logic could send values to the host computer). Some obvious uses for supervisory points include parameter setting in simple logic and equations and changing details such as meter factors in simple totalization calculations, the timing or duration of momentary contact outputs, and calibration scaling values. Using supervisory points eliminated the need to go to the field and change RTU firmware in order to make simple adjustments. Support for supervisory points required very minor enhancements to an RTU protocol. These points looked very much like conventional I/O except that they had no physical hardware associations.

In addition to supervisory points, many SCADA vendors provided host-level configuration tools, for defining simple calculations and logic that could then be downloaded to the RTU for execution. SCADA systems generally have a utility that permits the user to define multiple computations (using

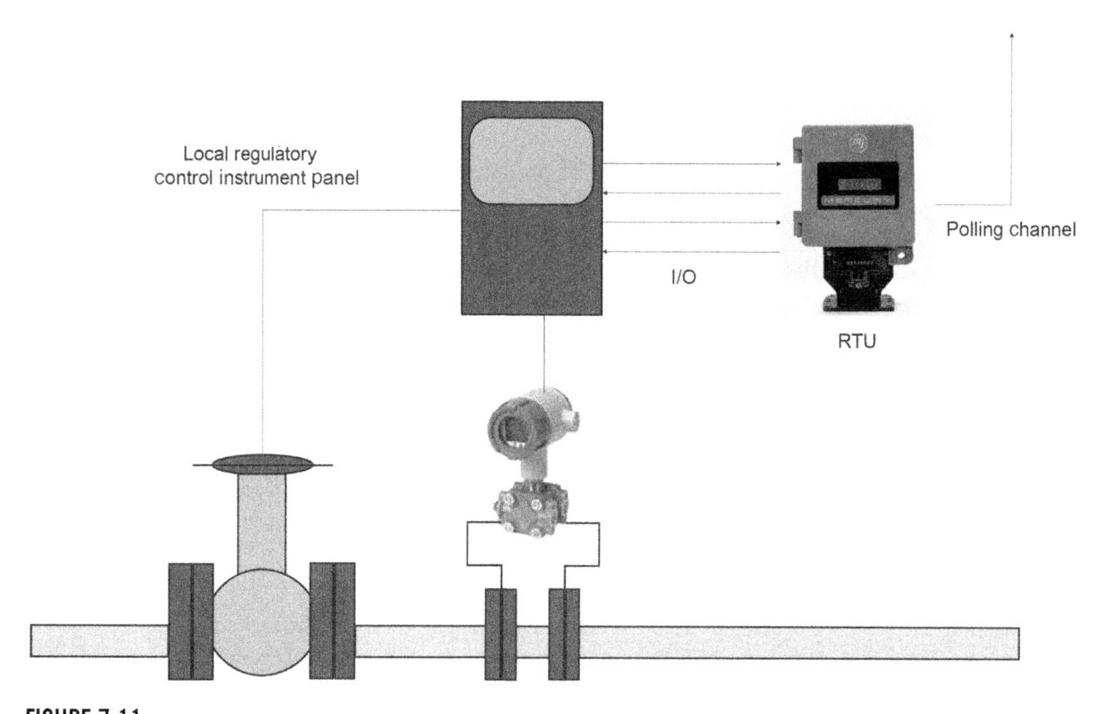

FIGURE 7.11

RTU protocols with adjustable supervisory points

real inputs and the results of other calculations) at the host level. However, in the pipeline and water/wastewater industries, there are many instances where some level of calculation capability is needed at the RTU level. The addition of downloaded calculations and logic required further enhancements to protocols, so that there were message types for sending (and receiving) these data blocks and confirming successful delivery. Until recently, this kind of capabilities involved vendor-developed proprietary host-based utility programs that would input and process such logic and mathematical calculations, and convert them into data (or actual program code) that could then be downloaded to the specified RTU, either locally or over the polling circuit using protocol extensions devised for such purposes. RTUs would be equipped with software to deal with the data or install and execute the programming. Today, there are some widely adopted standards that have evolved for accomplishing the same results.

7.3.5 REGULATORY AND SEQUENCE CONTROL

In the pipeline and water/wastewater industries, the RTUs are installed in the physical field locations around the process calling for some level of local regulatory control and sequence logic. Basic SCADA systems are supervisory control systems, where decisions to take a control action are made at the host level and then dispatched to the RTU to execute. This is fine as long as the communications bandwidth is sufficient and reliable. If gas pressure needs to be controlled at a delivery point and the target pressure needs to be adjustable based on conditions, then control adjustments to the pressure regulating valve

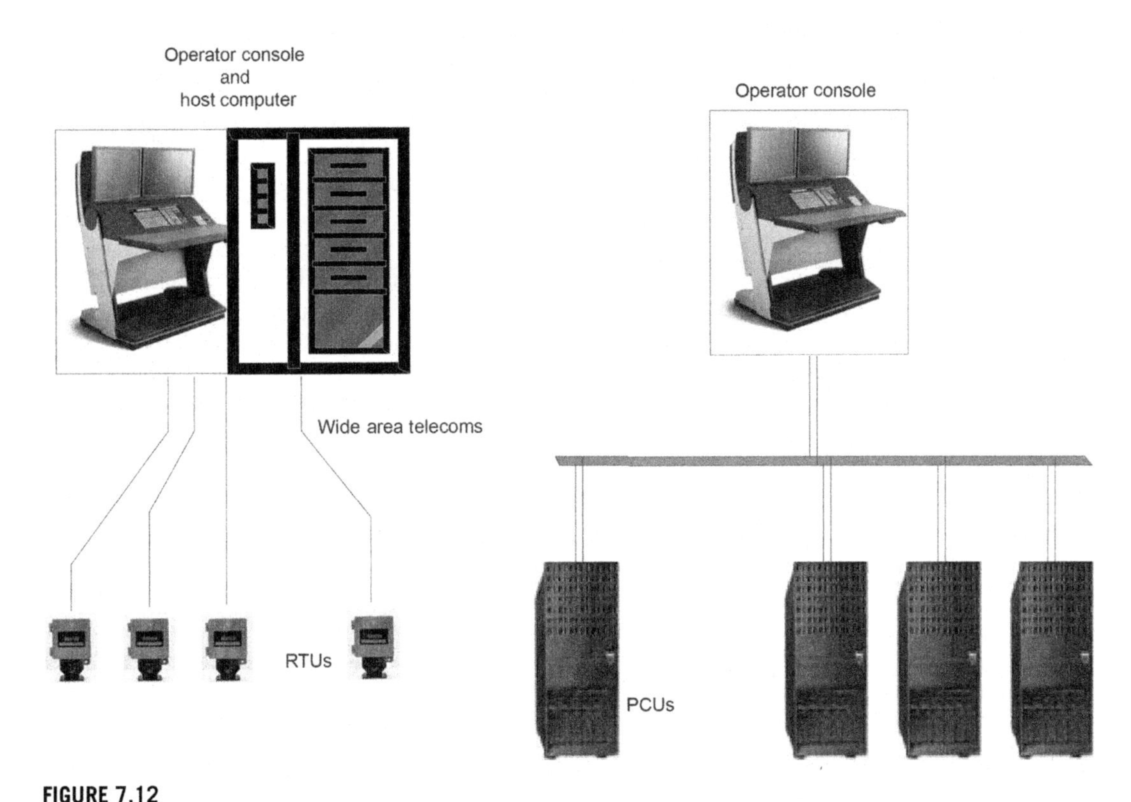

FIGURE 7.12

Supervisory control of local regulatory control

may need to be made every second or faster. With the low-baud-rate serial communication schemes used by most SCADA systems (not to mention channel sharing across multiple RTUs), it is typically not possible to read the gas pressure, send it to the host, make a control adjustment calculation in the host, and send a control output command to the RTU all within a second. In the past, in order to provide local regulation for such applications, it was typical to have some form of instrumentation and control panel at the site, to perform the regulatory control, and the RTU would merely interface this panel via analog, pulse, and contact I/O points (Figure 7.12).

The 1980s saw a revolution in the process control industry, while microprocessor-based RTUs also became reasonably powerful. Traditional analog instrumentation and control panels were being replaced with computer-based technology. The distributed control system (DCS) had been introduced and, was attaining wide acceptance as a replacement for conventional instruments and controls for financial and technical benefits. DCS consists of distributed process controllers, all tied to a high-speed (for the 1980s) LAN, or data highway, and operator consoles. The distributed process controllers were microprocessor based and had a full complement of I/O. Architecturally, a DCS is quite similar to a SCADA system, other than the use of LAN communications instead of WAN communications.

One fundamental difference between these two types of systems is that distributed process controllers were intended to perform local regulatory and sequential control from the outset. The microprocessor technology and I/O technology of a process controller is very similar to an RTU, although process controllers usually came in fully redundant or one-for-N fault tolerant designs. It did not take SCADA vendors long to realize that they could add value to their RTUs by incorporating some of the control functions used in the process control industry. Computer-based process control systems were in use since the 1970s, although they were used in centralized architectures. Control system vendors developed a set of algorithms to mimic the actions of the analog controllers replaced by computer technology. The basis for most regulatory control functions was provided by algorithms such as the proportional-integral-derivative (PID) controller, the lead–lag algorithm, the ratio–bias controller, and the on–off ("bang–bang") controller. These were all numerical algorithms that could be used individually or combined mathematically to process analog inputs and generate analog outputs, performing the analogous functions of the older analog panel instruments. SCADA system vendors, at least those serving the pipeline and water/wastewater industries, started offering these algorithms in their RTUs. That eliminated the need for the local panel instruments, where customers were willing to accept an RTU as a process controller.

RTUs are reliable and have proved to operate flawlessly 24/7 for 10 years or more after installation.

In contrast, many DCS vendors found making a process controller fully redundant just as difficult as making the host computers of a SCADA system redundant. Today, many pipeline and water/wastewater SCADA systems still rely on the local regulatory control capabilities of their RTUs. In addition to regulatory control (which generally involves analog inputs and outputs), many local control problems needed the capability of sequence and state-driven control, which is based on Boolean logic. Actually, a hybrid of regulatory control, altered and manipulated by Boolean logic, tends to serve best for sophisticated control. Many SCADA system vendors expanded their RTU features and functions to support both capabilities. Not all vendors could support top–down configuration (developed in the host and downloaded to the RTU), and not all vendors took a block approach. Several vendors developed their own special process control programming language (often a BASIC-like language), which included PID as a library function. Users developed sophisticated controls by writing programs in these languages and then loading these programs into the RTUs, through the console port, prior to field installation.

7.3.6 ACCUMULATOR FREEZE

A common use of pulse inputs is to count them over time to determine material or energy quantities that have flowed past a particular measuring point. In pipelines (and electric power applications), this is an important task because what goes in should come back out elsewhere (adjusted for parameters such as pressure and temperature). Pipelines generally count pulses as material (gas or liquid) moves past a measurement point. All inflows and outflows are metered in this manner. Therefore, if information about what is currently passing in, out, and through the various measurement points in the pipeline can be captured, it will enable in the detection of a problem. If such measurements are not taken at the same time, then by the time all accumulators have been collected, something more will have moved past the measurement point(s), rendering it difficult to balance the inputs and the outputs. The vast majority of RTU protocol messages are destined for a particular RTU, as designated by the RTU ID contained in the message. However, a few messages need to be directed to all RTUs simultaneously. Such messages are called broadcast messages in protocol terminology. One common function of RTUs and their protocols is the ability to send out an accumulator freeze broadcast message. The purpose of such a message is to get all RTUs to take snapshots of their accumulators, at the same instance in time, and hold them for collection by the host computer. This becomes more complicated when there are multiple polling channels, as a mechanism is needed to stop all polling and, then, once all polling channels are idle, to send out this freeze message on all of them at the same time. With smart RTUs, each RTU was equipped with a local clock. A clock is just a counter that, if properly designed, counts at an exact rate, so that time changes can be related to the count change. Early clock circuits were bulky and temperature sensitive and could drift off by a significant amount if not reset frequently. One broadcast message type supported by most RTUs was a time broadcast that could be used to (re)set the local clock of all the RTUs. Most SCADA systems broadcast the time from the host on a periodic basis, to keep RTU clocks in reasonable synchronization with the host clock. This was an effective scheme when a time skew of many milliseconds was acceptable, either between the host and a given RTU or between numerous RTUs themselves. Schemes that timed the message transit time from host to RTU were employed for improved accuracy.

7.3.7 GLOBAL POSITIONING SYSTEM TIME RECEIVERS

Sometimes, the status inputs of an RTU are used to time events or to show the SOE. The local time tagging of a contact input state change merely indicates saving a record of what bits changed, to which state, at a given millisecond in time. An RTU examines all its status inputs every millisecond and determines whether any have changed from their value at the prior millisecond for the tagging. When changes are found, a record is made, showing the time (to the millisecond) and the inputs that have changed. Although RTUs for years had sufficient computing power to provide a 1-millisecond time scan of all status inputs, a problem remained in that the local time of every RTU could not be easily synchronized to the exact same millisecond. The only solution was to use a highly accurate time source that could be made available to all RTUs. GPS technology has enabled global placement of RTUs that are time synchronized to 1-millisecond (or better) accuracy. Furthermore, the host can use this technology to enable clocks in a SCADA system keep the exact time. This brings up another idiosyncrasy of SCADA systems. Because they cover large geographic areas, it is quite probable that the local time where one RTU is situated may be in a different time zone than another RTU and even the host. Rather than keeping local time, most SCADA systems elect either to keep time based on where the operators (and host) are located or to use Greenwich mean time (GMT) for the system time. This second choice may be taken to deal with another time-related problem: the daylight savings time adjustments made each autumn and spring.

7.3.8 DAYLIGHT SAVINGS TIME

Computer systems often store data chronologically (by date/time). The date/time information is stored in data tables. This causes a problem when, in the autumn and the spring, the time suddenly changes by an hour. This time-shift issue has plagued SCADA systems since they started collecting and archiving data. To avoid such problems, many SCADA vendors use GMT in their systems (at least for time tagging of data) and never make those spring and autumn time adjustments.

7.3.9 TRANSDUCER-LESS AC INPUTS

The electric utility industry has a few RTU features and functions that are unique to itself. High-resolution (1 millisecond) SOE recording is one. While AC 50/60-Hertz electric power is a source of noise to be filtered out for all other industries, it is what the electrical industry measures. Until the mid-1990s, RTUs used in the electric utility industry had to employ external devices called transducers to convert AC voltage, current, power, and other attributes into a DC voltage or current value that could be read by a conventional A/D converter. In the 1990s, the digital signal processor was invented, making it economically possible to take high-speed samples of the actual AC waveform and then apply calculations, such as the fast Fourier transform (FFT) or the discrete Fourier transform, to extract frequency-domain information. This also enabled to compute real and reactive values – and determine spectral component energy levels – for the electrical signals being measured. Today, transducer-less RTUs are common in the electric utility industry. Few electric utilities have the requirement of RTUs performing calculations or local control. RTUs in that industry have had few feature or functional enhancements (other than being forced to upgrade to newer microprocessor technology when older parts became unavailable). Nevertheless, over the past few years, the electric utilities have suddenly been looking at multiserial-port RTUs as data concentrators for collecting (simple) IED data in a single device.

7.3.10 LEGACY PROTOCOLS

The SCADA industry has produced a huge number of bit- and character-oriented serial protocols. In effect, all the protocols perform the same basic operations: send input values to a host computer and generate control outputs when commanded to do so by a host computer (Figure 7.13). The different protocols are divided into classes or categories, based on the more advanced features and functions supported.

The modem technology offered a mere 110- or 300-baud rate in the early days of RTUs. Shortly thereafter, the rates went up to 1200 baud that continues to be used by a lot of SCADA systems today, even though modem technology has jumped well past those speeds. Many early RTUs had modem

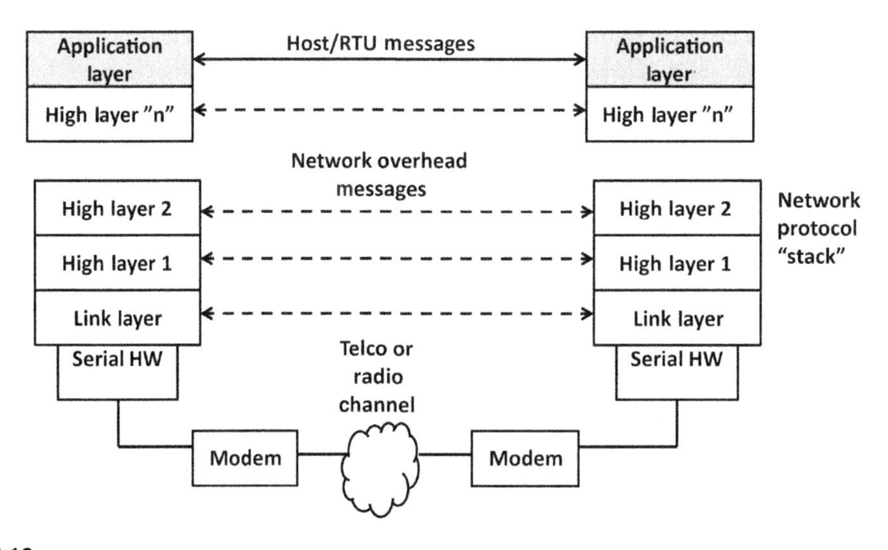

FIGURE 7.13

Categories of typical RTU protocol message types

circuitry built into the RTU circuitry, so upgrading to a faster modem is not possible unless the RTU is replaced. In case of multiple such RTUs on a polling circuit, all need to be replaced to upgrade the channel bit rate. Many of the older RTUs have very limited computing power and cannot actually support high data rates, and some old SCADA systems have front-end processors that do not have the computing power or the serial port clock circuitry to handle higher data rates. Many improvements have been implemented to make RTU protocols as efficient as possible, so as to reduce polling-response times and to enable more RTUs to share a common channel without extending the overall polling time cycle. Early RTUs generally did a full report of the current readings for all inputs (or possibly all analog or all status inputs) every time they were polled. In many cases, these inputs had not changed since the previous polling request (especially the status inputs). With smart RTUs, it was possible to use the logic in the RTU to keep track of prior and current values for all inputs and to identify those that had changed. More advanced protocols support report-by-exception polling, wherein the host does not ask for a report of values, but rather asks for a report of changed values since the prior poll request. A host could still ask for a full poll every so often (often called an integrity poll) to ensure that nothing has been missed. The problem with polled exception reporting is that when all data is not reported, additional information must be collected to identify the changed values. There is a threshold above which this added information makes it better to do a full report rather an exception report. (For example, if all but one point has changed, the added data to identify all of the other values would be more bytes of data than reporting the one value that did not change.) In a more recent development of unsolicited report by exception, the host computer does no polling (except an occasional integrity poll) and leaves it to the RTUs to send messages when values change. This scheme is best for processes where values change slowly and where you never see a lot of the measured parameters changing in lots of places at once. Another problem with unsolicited exception reporting is collision recovery, which is when multiple RTUs decide to report their changes at the same time. This type of reporting is popular in applications in which a lot of RTUs have to share a single, low-speed channel (e.g., a radio channel). If multiple RTUs transmit at the same time, the host will just hear garbled noise. A commonly used mechanism is having smart radios that use out-of-band signaling to coordinate with each other to ensure that only one RTU initiates transmission at any given instance in time. The primary mission of a communications protocol is to successfully and accurately deliver a message from one computer/device to another. All RTU protocols include some scheme that enables the message receiver to verify that the message arrived uncorrupted (by including something extra [a check code] with the message). The robustness of these schemes was greatly improved with the introduction of the cyclic redundancy check code calculation in the 1980s.

7.3.11 STANDARD PROTOCOL

The SCADA system customer base has been soliciting a standard protocol that provides an all-encompassing set of features and functions, to replace all of the older, sometimes incompatible proprietary protocols. The hope is to provide for compatibility and interoperability between and among vendors and products. Some popular proprietary serial protocols, such as Modbus and Bristol standard asynchronous/synchronous protocol, have become *de-facto* standards because of their publication, ease of implementation, and widespread adoption. Others, such as DNP3.0, were brought into the public domain and turned over to committee for oversight. With the rapid evolution of communications technology over the past two decades – particularly, the development of LANs and WANs – other protocols suitable for use across networks have emerged. Protocols such as IEC870-5-101 and IEC870-5-103, IEC61850, inter-control center communications protocol (ICCP), also called TASE.2 or UCA1.0, and UCA2.0 were defined by a committee, to run over LANs and/or WANs, and were then published and promoted by standards organizations.

7.3.12 NETWORK VERSUS SERIAL PROTOCOLS

Currently, existing protocols are classified as those that are standalone and only require a voice-grade communication circuit, and others that expect a functioning digital network (LAN or WAN) of some type, composed of a communications circuit and an underlying basic networking protocol to provide transport services. Modern, digital communications protocols generally refer to a class of protocols that have been broken down into functional layers, each of which provides services for the layers above and employ services from the layers below, all the way down to the drivers that deal with the actual communications hardware. Such protocols often expect an underlying network to deal with intermediate transport of messages. For example, if there are IEDs in a substation, remotely accessible via the IP version of DNP protocol, these IEDs probably use Ethernet (with its SDLC protocol) as the connection mechanism to the communications interface device in the substation (e.g., a router) and the interface device delivers messages to the wide-area digital network (e.g., frame relay or a private IP network) that delivers them to another Ethernet LAN where they get routed to an engineer's desktop PC. Some digital communications protocols can be used across the voice-grade communication circuits used by traditional serial RTU protocols. For example, in a TCP/IP network, the lowest functional layer of the protocol, the one that deals with the hardware connection out of the given computer/RTU, can be the point-to-point protocol (PPP). In case of a dial-up Internet service provider, your PC is probably utilizing PPP to move bits over the telephone line (using a modem as the electrical interface to the phone system). However, there are layers above PPP that want to send their own messages to the other computer to manage the functions they are responsible (and consume bandwidth and time in this process). In addition, PPP carries a lot of overhead in its base message structure, to deal with more sophisticated communication issues not relevant to RTU polling. Therefore, a simple analog input value report message requires many more total bits than with a conventional serial RTU protocol (Figure 7.14).

7.3.13 ENCAPSULATED PROTOCOLS

The introduction of LAN and WAN technology called for an upgrade of serial protocols (DNP3.0 and Modbus) to make them available for use over such digital networks. Both protocols were rereleased, in the early 2000s, in an IP version suitable for use over any IP-based network. The most popular usage of these two protocol variations has been on high-speed Ethernet LANs. In fact, they are often referred to as Ethernet Modbus or Modbus TCP or Ethernet DNP. TCP (part of the IP suite) is commonly used to transport and deliver messages (data) over LANs and WANs. TCP is just responsible for delivering the data without errors, irrespective of the type of data. In this case, the data being delivered are the actual bytes that would have been sent over a serial communication channel, when the normal versions of these protocols are employed. Rather than sending these bytes over a wire, they are delivered as the data portion of TCP/IP messages. The process of sending one type of message as data within another type of message is called encapsulation.

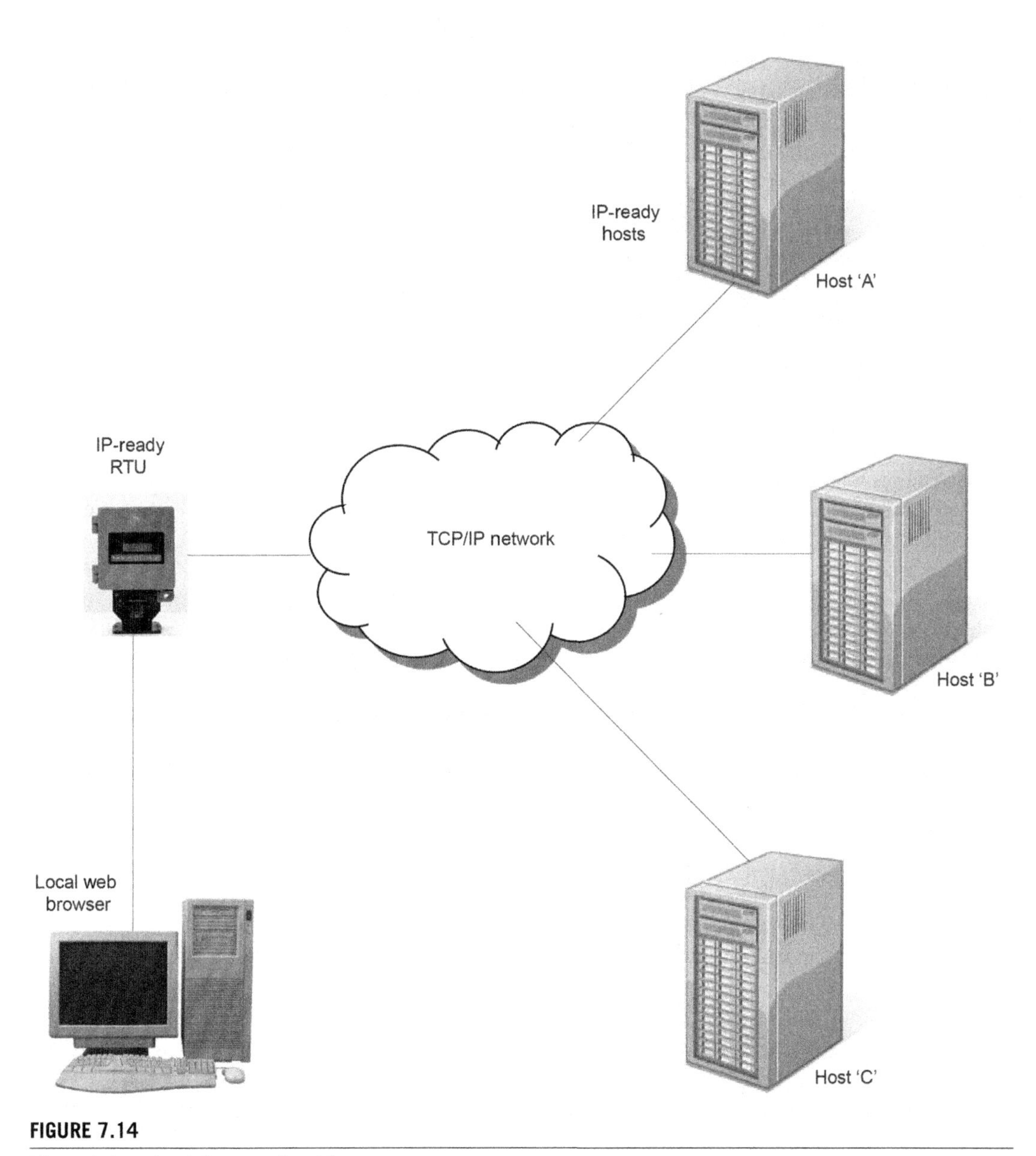

FIGURE 7.14

Network-based serial protocol architecture

7.3.14 IP-READY RTUs AND PROTOCOLS

The 2000s witnessed a great leap in the computing power of RTUs, with the availability of lower-power-consumption (higher temperature range) versions of the popular Intel Pentium processors (and compatibles). Many full-featured commercial operating systems, with real-time extensions, were also introduced to run on these computing platforms. This resulted in the introduction of high-performance IP-ready RTUs that can support both legacy and standard serial protocols, and having the LAN/WAN interface and computing capability to deal with the more sophisticated, network-based protocols. These RTUs come with integral Ethernet and TCP/IP support and offer a platform on which a wide range of applications can be deployed. These new RTUs offer promise to industries that are converting from traditional, analog serial communications and replacing them with some form of digital networking. The new generation of RTUs extend IP networking to the field, thereby offering the possibility of employing the same network security technologies on these communication links as are used to secure other corporate networks. Another interesting capability of these RTUs (if supported by the vendor and an appropriate IP-based protocol) is the ability to have peer-to-peer communications directly between and among RTUs, independent of the host computer(s). With traditional serial protocols, the RTUs could communicate only with the host computer. These RTUs also support multisession communications (which comes for free with a good TCP/IP implementation). When a SCADA operator has replicated control centers, or just a fully redundant system design, most IP-ready RTUs can simultaneously process and respond to polling messages from multiple sources. This eliminates the need for one host (in a redundant pair) to poll all RTUs and send updates to the backup unit, or for messy communication switching between a primary and an alternate operating site.

Currently, the list of IP-compatible RTU protocols is fairly short: DNP3.0, ICCP, UCA2.0, and Modbus. There are many well-proven computer protocols that can be used for data transport, eliminating the need to restrict oneself to SCADA protocols. Some RTUs use extensible markup language (XML)-based Web pages for data exchanges, transported by hypertext transfer protocols (HTTP and HTTPS). There are also implementations of the IP version of DNP3.0 that have been deployed with virtual private network-based security (which adds encryption, session keys, and authentication functions).

If an RTU (and the host with which it is communicating) can support an IP-based protocol, the SCADA system owner has a wide range of commercial communication service providers and technologies available. The Internet could also be used as a communications system. Another interesting aspect of TCP/IP networking is the ability to share a single communication channel across multiple applications. There are SCADA applications in which multiple host computers need to access a common RTU. With conventional serial communications, this requires individual communication circuits from each host to the RTU site and an RTU with a sufficient number of serial ports. If digital (TCP/IP) networking is used, the RTU can concurrently handle communications transactions from all host computers, via its single connection into the digital network (similar to a Web site concurrently servicing multiple browsers). Some of the more advanced IP-ready RTUs today are able to support a local Web server function so that the RTU can offer a range of Web pages to anyone with a laptop PC, a Web browser, and an Ethernet cable (Figure 7.15). Such Web page services may be used for data transfer purposes (e.g., using XML pages) or for human interaction with the RTU.

7.4 COMMUNICATION TECHNOLOGIES

When SCADA systems emerged, there were only two choices for long-distance communications:

- Telecommunications technologies – the telephone system (or telephone system technology).
- Voice-grade (analog) telephony – private, licensed radio.

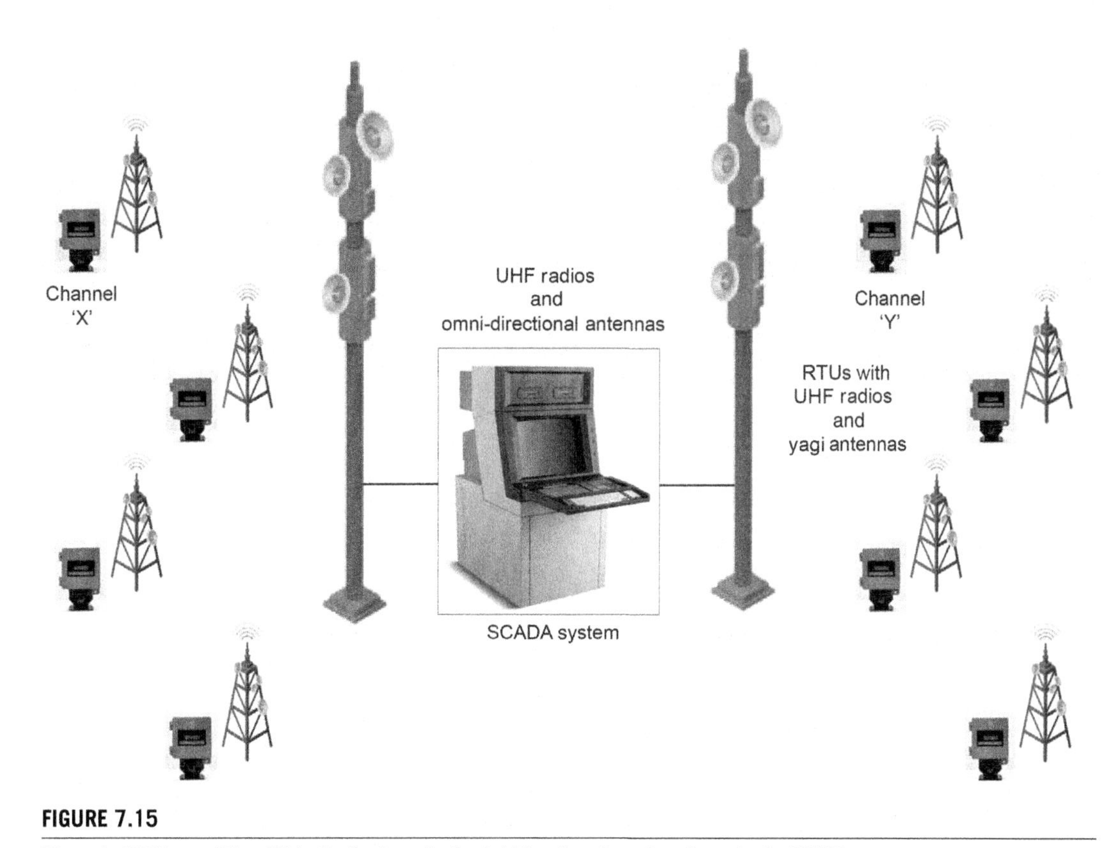

FIGURE 7.15

IP-ready RTU providing Web displaying via the LAN and exchanging data via the WAN

Both these technologies (radio and telephone) were designed for long-distance, bidirectional transmission of voice/sound. Most of the communications technologies we take for granted today – satellites, cell phones, and the Internet, among others – did not exist or were inappropriate for SCADA usage (although the first satellite-based telephone links, between Europe and the United States, were established in the 1960s).

7.4.1 TELEPHONE TECHNOLOGY

The (analog) telephone system of the 1960s provided no value to SCADA systems. (Today, we have call forwarding, voice mail, caller ID, etc., as part of the service.) The telephone company provided a reasonably reliable, somewhat noisy, low-bandwidth (voice-grade) channel over which electronic signals in the frequency range of the average human voice could be transmitted. Those signals could be generated with the telephone handset they provided (and by speaking) or with some other electronic device, such as a modem. Establishing a connection (dialing), maintaining a connection, selecting a signaling mechanism and symbol set, detecting and correcting errors, managing and pacing the data flow, and terminating the connection were all considered to be the responsibility of those who wanted to use the telephone system. For most consumers, the establishment of a call began with dialing a telephone number and waiting for someone to answer. The process of dialing takes a surprising amount of time, and this was even more time-consuming when pulse dialing was used. Dialing a telephone number to reach an RTU (and then do it again and again) was therefore inefficient for a SCADA system.

A SCADA system needs a dedicated telephone connection to its various RTUs. The telephone company accommodated SCADA system owners by providing private leased lines. These were telephone circuit connections that had the same appearance and functioned as if a number had been dialed and answered, but the circuit was locked such that it never disconnected or broke any of the intervening connections.With older electromechanical telephone switching equipment, this was implemented by going to the various switches and manually placing actual clips on the connections so that they were mechanically latched. The telephone company was providing a metallic circuit between the SCADA host and the RTU; the leased line was a voice-grade circuit with amplifiers and coupling transformers along the way to boost the signal, not just a set of wires.

These leased circuits – and the telephone system in general – were analog in the 1960s (and well into the 1970s). This meant that a telephone modem (or even a handset) could be connected to the telephone line at any point along the leased circuit and leaves drop on the SCADA–RTU messages going back and forth (or even inject some false messages). Even today, many SCADA systems still depend on leased lines provided by the local telephone company, although the underlying technology of the telephone is no longer analog. Starting in the 1970s and accelerating in the 1980 and 1990s, the telephone system was converted from analog to digital. Only the local loop, from the nearest telephone switching center to your home, is analog, with everything else being digital. With readily available equipment, it was possible to record, replay, or simulate valid-appearing message traffic at either of the two end points. A typical office had have its own private branch exchange (PBX) that deals with analog telephones but makes a digital connection (via one or more T1/T3s) into the public telephone system. For SCADA systems that still use analog leased telephone lines, the telephone system is digital up to the panel in the telecommunications room of the facility where the SCADA host is located and analog from there up to the room where the SCADA system front end and modems are physically placed. In the field, the circuit is digital to the nearest switching center and analog to the field RTU site. For electric utility substations, many telephone companies have a policy of running an analog telephone line to the demarcation point outside the substation facility (usually a well-marked and obvious box) and making it the responsibility of the utility to bring the line into the substation and to the RTU. This means that the analog (and easily tapped) leased telephone circuit exists only at each end of the line (but often in unprotected areas). The telephone system provides a full-duplex communication channel: sound can be sent in both directions at the same time – unlike half-duplex channels, as provided by a walkie-talkie or an intercom, and simplex channels such as television, which only sends information in one direction.

In most simple, serial SCADA protocols, that capability is not utilized. The vast majority of serial protocols work totally on a half-duplex basis: a message goes out from the SCADA host to the RTUs, and the appropriate RTU formulates a response and sends it back. This is not an efficient use of the full-duplex nature of the telephone system, and modern computer networking protocols usually support data flowing in both directions at once. SCADA serial protocols do not utilize this full-duplex capability because a protocol needs to be far more sophisticated to manage two independent, concurrent data streams. Another reason is that most such protocols, when they were developed, needed to work over the other major available communications technology, radio, which tends to be a half-duplex channel. Although analog telephone lines restrict data transmission rates to low speeds (generally under 34 kilobits per second [kbps]), they have been sufficient for basic SCADA applications because traditional SCADA systems and RTU protocols were designed for even lower transmission rates (1.2–9.6 kbps).

7.4.2 LICENSED RADIO

Radio communications technology from the 1960s fall into two categories:

- point-to-point,
- high-bandwidth microwave transmission, and point-to-multipoint ultrahigh frequency (UHF), and very-high frequency (VHF) radio transmission.

The telephone company pioneered the use of microwave transmissions to carry many multiplexed communication channels over long distances. For some vital applications that use SCADA technology, the owners/operators had to make a decision about relying on third parties (e.g., the telephone company) for critical SCADA communications. At many instances, electric utilities and water companies, whose service areas are covered by a local telephone company, construct and maintain their own communications infrastructure rather than being dependent on a third party. Water utilities often implement a UHF or VHF radio-based SCADA communications system, with one or more master radios used to send/receive messages to RTUs operating on the same frequency. To cover more area or to speed up polling times, multiple master radios, operating on different frequency channels, are used, each with its own set of RTUs with radios set to their particular master radio's frequency (Figure 7.16).

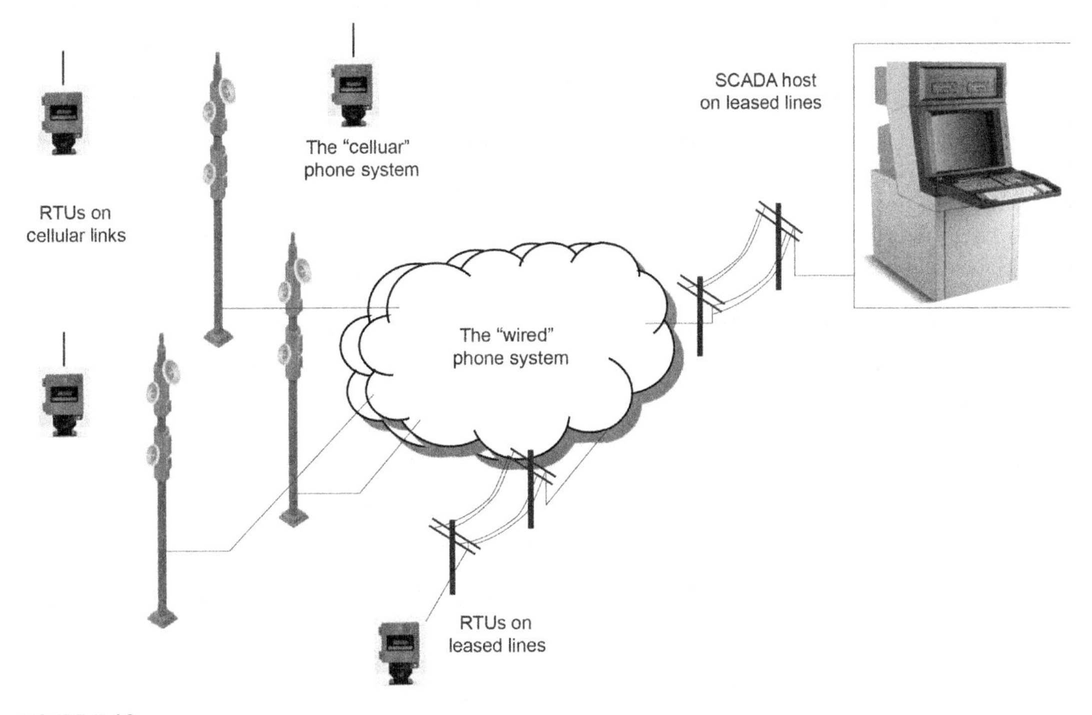

FIGURE 7.16

SCADA host with multiple master radios on separate frequencies

A general problem faced by people trying to use licensed radio is the congestion on the available frequency slots, particularly in urban areas. It is not always possible to find an available frequency slot. The Federal Communications Commission maintains a geographic directory of frequency assignments, to avoid overlap and interference. Just as RTUs became intelligent with the infusion of microprocessor technology into their designs, radio equipment also became smarter. Originally, radios were just a transmitter and receiver that were controlled by the SCADA host or RTU, just like a person controls a walkie-talkie.

A modem circuit was used to convert the binary digits into sounds, and the RTUs and SCADA hosts had program logic to deal with keying the radio (switching from receive to transmit mode) and providing time delays between the end of a receiving a message and starting the transmission of a reply. Every RTU was also required to listen to (and interpret) every host message, so that it could check the RTU ID in each message and find out if it was the recipient of that message.

With the integration of microprocessors into radios, the radios could use a small portion of their bandwidth to exchange messages (a process called out-of-band signaling), and they (the radios) could make decisions and identify messages that the RTU actually needed to hear. Radio systems became a bit more flexible with the introduction of the multiple address system and trunked radio systems. The connection between the radio and the computer became a serial (RS-232/C) connection (i.e., computer to computer), and the radios could identify that messages were being directed to them (the radios had a unique ID); thus, RTUs received and had to process only those messages sent specifically to them by the host. A conventional radio is still used when a SCADA system owner wants a privately operated and maintained telecommunications system.

7.4.3 COMMUNICATIONS BACKUP

Although utilities might elect to construct, operate, and maintain their own private radio communications systems, they could also use the telephone system as a backup in case of radio failures. It was not uncommon, once multiported RTUs became available, to have critical RTUs configured with an auto-answer modem and dial-up telephone line, as a backup to the primary radio polling channel. If the SCADA host could not reach a critical RTU via the radio system, it could dial a telephone number and reconnect to the RTU via the public telephone system, to poll the RTU or to send a control command. If there were multiple critical RTUs and none could be reached via radio, the SCADA host could cycle through the set of RTUs by dialing them consecutively, polling and then hanging up and dialing the next in sequence. This was not very fast, but was much better than having no communications at all. The opposite strategy of using leased telephone lines as the primary polling and communication mechanism is also used, and falls back to radio to reach the critical RTUs if the telephone system goes down.

7.4.4 WiFi AND WiMAX

One particular category of spread-spectrum radio that has been standardized and commercialized is IEEE 802.11 wireless Ethernet, or WiFi. Although initially designed for LANs or personal area networks, a version with much higher power and greater range called 802.16, or WiMAX, is now becoming commercially available. WiMAX is essentially wireless Ethernet over a large area (creating a municipal area network [MAN]) and is intended to provide high-bandwidth data rates (244 million bits per second [Mbps]) over municipal areas. WiMAX is Ethernet-ish and thus is intended for computers that have digital networking capabilities. It includes advanced encryption capabilities and supports point-to-multipoint communications. A modern IP-ready SCADA system, with IP-ready RTUs, would be able to use this type of communications infrastructure for real-time polling and supervisory control and still have lots of bandwidth left for deployment of other applications and systems. Currently, the SCADA system owner is responsible for building and maintaining a WiMAX system, just like any other radio communications system. However, in the future, municipalities may construct WiMAX systems to offer Internet connectivity to inhabitants and visitors and for their own municipal networking needs. SCADA system owners operating within the WiMAX service area can probably utilize these systems for their communications (although security and reliability may dictate having a backup strategy). The downside of this technology, however, is that anyone in the service area with appropriate commercial hardware and software can use this wireless MAN and could attempt to attack the SCADA host or RTUs.

7.4.5 CELLULAR

The various digital cellular telephone systems installed in major areas can be used for data and voice communications. A modern cell phone actually sends and receives data that represent digitized speech (using spread-spectrum technology). The cellular system is a wireless mechanism for putting data onto the wired telephone system, where it can be transported in a conventional manner (possibly to another service area where it leaves the wired telephone system and gets onto another cellular system). The cellular telephone system offers a flexible means for establishing field communications to SCADA systems installed in a municipal area, such as water/wastewater, electric power, or gas distribution. RTUs can be equipped with commercially available cellular modems that can be dialed just like any conventional telephone number by a SCADA host. It is also possible to have the equivalent of leased connectivity (non-dialed, continuously connected) via the cellular system, in which case a wireless connection is maintained continuously; the cost for this type of service is based on traffic (messages sent/received), and the telephone company routes this traffic back to the SCADA system via conventional wired (or wireless) means. Most cellular service providers charge by the message, therefore making this technology most suitable for RTUs that employ a report-by-exception protocol design. These would be used in a modern RTU only if it was intended to utilize the Internet as the basic means for communications. Using cellular communications comes with the challenge of dealing with incompatible cellular technologies used by the competing cellular service providers.

7.4.6 DIGITAL NETWORKING TECHNOLOGIES

Over the past few years, most SCADA system vendors have enabled digital networking to allow for communications between the host and the RTUs. This is partially because of the migration, by conventional telecommunications service suppliers, from analog communications to digital communication. The proliferation of available digital networking technologies and the widespread adoption of IP networking and the Internet have also boosted the adoption of digital technology. One of the qualities of most digital networking technologies is the ability to support multiple, concurrent communication sessions (conversations) over a single physical communication link and to provide a flat communication architecture. In other words, any computer can directly communicate with any other computer, if this is needed (as when smart RTUs exchange data directly with each other or with multiple hosts). These are both useful capabilities for SCADA systems and are not generally available with conventional analog technologies. A SCADA system owner can use any of the digital networking technologies depending on the desire to use commercial suppliers or to construct a proprietary network. The network-ready SCADA systems and RTUs refer to the systems that support IP networking with one of the currently available IP SCADA protocols: IP-Modbus, IP-DNP3.0, ICCP, or UCA2.0.

Oil pipelines are spread across the ranges from few kilometers to hundreds of kilometers. Maintaining the pipeline with continuous health monitoring is the biggest challenge. A fiber-optic communication backbone is built to collect control data from programmable logic controller PLC/RTU and monitor fire and gas systems, heating, ventilation and air conditioning (HVAC) system, and leakage monitoring. To achieve the above transmission to the central control room, proper bandwidth and fiber-optic availability is the key. Consolidating the system requirements for above applications is as follows:

- Remote monitoring, control, and management of oil pipelines from the control center
- Uses fiber optics for long-distance transmission of data across the country
- High-speed redundant ring for high data availability
- Rugged design to operate in harsh environments

For successful data transmission, a network consisting of multiple Ethernet switches is needed that can provide multiple ports of gigabit capability for making more bandwidth available. Based on the switches capability, they are able to transfer data over fiber optic in a range of 25–80 km. SDH multiplexing one of the popular methods is used for data transfer for such a long distances. Basic communication normally happens this way; data from RTU/PLC is collected by the switch using twisted pair coaxial cable. After that, the output of the switch is converted to data that can be transmitted over fiber optic and sent across to the control room. Apart from transmitting the data from PLC and RTUs, the fiber-optic technology can be used to monitor the pipeline general health and leakages. This is useful where there are extreme weather conditions such as heavy rains and landslides that may impact position of the pipeline and to monitor human intervention accompanied by malicious attempts of tampering the pipeline with intent of theft. A single fiber-optic cable serves as both the sensor and the path for data flow. Sensor cable, lead cable, and optical junction boxes are all optical to reduce maintenance costs.

General fiber-optic cable design is intended to withstand all external influences such as external humidity, side pressures, and multidimensional strains. These are effectively used for sensing extreme temperatures between subzero to above 50°C. Multidimensional strain is monitored using Brillouin scattering that enables a reliable transfer of strain to the optical fiber. Finally, when sensing distributed strain, it is necessary to simultaneously measure temperature to separate the two components.

While sensing the strain, temperature has to be measured by installing strain and temperature sensing cables in parallel. It would, therefore, be desirable to combine the two functions into a single packaging.

7.4.7 EXTREME TEMPERATURE SENSING CABLE

The extreme temperature sensing cables are designed for temperature monitoring over long distances that is distributed in nature. They consist of single-mode or multi-mode optical fibers contained in a stainless steel tube, protected with stainless steel armoring wires and optionally a polymer sheath.

The use of appropriate optical fiber coating allows the operation over large temperature ranges; the stainless steel protection provides high mechanical and additional chemical resistance while the polymer sheath guarantees corrosion protection. The length of the optical fibers is selected in such a way that the fiber is never pulled or compressed, despite the difference in thermal expansion coefficients between glass and steel. This is extremely helpful in leak detection because when the liquid is out of the pipe, the temperature difference can be sensed and therefore can be detected that there is a leak.

Pipeline application software modules comprises:

1. Real-time transient pipeline modeling
2. Batch scheduling
3. Batch tracking
4. Scraper tracking
5. Leak detection and location
6. Shut in leak detection
7. Pipeline inventory analysis
8. Look-ahead model
9. Training simulator

7.4.8 THE INTERNET

Internet has successfully been used as a communications system for real-time supervisory and control applications, although it is relatively rare. The Internet can be considered as a communication system, even if for financial reasons, as SCADA host computers and RTUs with IP support are increasingly available. Thus, some knowledge of the Internet (as opposed to IP networking) is essential, when trying to understand SCADA.

7.4.9 SUPERVISORY CONTROL APPLICATIONS

Host system has the software that runs the SCADA, plus the data and configuration information it needs to perform its functions. The software in a SCADA system can be classified into layers. The first layer is the software that owns the computer and its resources and makes them available to application programs. That is the operating system layer. The next layer is the basic SCADA system functions – including RTU polling and communications, basic display generation, alarming and reporting, and other fundamental SCADA capabilities. The top layer consists of the advanced supervisory application programs that make use of the collected information to perform more advanced calculations and potentially send control commands back down to the RTUs in the field.

7.4.10 OPERATING SYSTEM UTILITIES

In the early days of SCADA system development, there were many vendors of computer hardware, each with their own proprietary designs. Most did not supply an operating system, with their computer equipment. A SCADA manufacturer would receive a set of program development and debugging tools (text editor, assembler, compiler, and simple debugger) and possibly a simple set of utility programs (file utility, diagnostic routines, copy program, etc.). The SCADA supplier was expected to develop the software needed to perform SCADA functions and to control and manage the computer hardware from scratch. In the 1980s, this changed as computer manufacturers began to provide at least rudimentary operating systems – and then more advanced, multiuser operating systems – along with their hardware. From the 1970s until today, the software supplied by a SCADA vendor has probably increased, based on the growing list of advanced applications most vendors offer.

At the same time, the percentage of basic software, including the functions performed by a modern operating system, has dropped to a minimal level. Also, the migration toward a standardized platform, has led to the development of more third-party applications. Today, nearly every SCADA system sold runs on an Intel Pentium series-based (or compatible) computer platform, with either a Microsoft Windows or a Unix/Linux operating system. There are a few VAX/VMS and Sun/Solaris-based systems still being produced, but they are rapidly falling to the wayside. This consolidation of technologies and platforms has a beneficial side: very good application portability; availability of third-party vendor; "best of breed" layered applications; readily available technological resources for software support; lower cost of ownership; and a good migration path for long-term system hardware support. All of these provide good, financially sound business reasons to explain why this consolidation has occurred.

7.4.11 SCADA SYSTEM UTILITIES

A basic SCADA system comes with specialized utilities necessary for SCADA system configuration, maintenance, modification, and expansion (not for normal operational use). These are utilities specific to the SCADA system vendor and particular SCADA software package(s). The SCADA functions of a system normally provide for a wide range of operational configuration flexibility, since a given vendor's SCADA product needs to be able to accommodate a wide range of application and industry variations. Many of the SCADA utilities exist for the purpose of setting up customer-specific and application-specific configuration tables that direct the actions of the generic SCADA software modules (Figure 7.17).

For example, the customer is expected to define configuration parameters such as number of polling channels, the number of RTUs per channel, baud rate per channel, and protocol to be used on each channel to enable the SCADA system to perform its RTU polling duties. Configuration activities go well beyond merely setting up the polling channels. It is necessary to describe each and every I/O that the SCADA system is to process for each RTU. The description for each I/O signal (point) includes assigning a tag name to the signal, defining what type of signal it is, providing all of the processing and alarm-checking information necessary to manipulate the point, defining the frequency at which the point is to be processed, the actions to take if the point's value exceeds acceptable limits, and many other items. The process of filling in all of the necessary tables with detailed configuration data used to direct the collection and processing of RTU I/O is called building the tag/point database. In addition to the actual physical inputs from the field, SCADA systems also generally compute a large number of other values using the inputs from the field-based RTUs, manually entered parameters and even archived data. These may be simple or complex numeric calculations or even logical (Boolean) calculations that produce a true/false result. In many SCADA systems, the point definition database creation utility may also be used to define these pseudo-points whose values are derived by user-defined calculations (Figure 7.18).

If not the calculations themselves, the point definition creation utility may at least be used to assign alarm checks and display parameters to these pseudo-points. The point (or tag) database contains

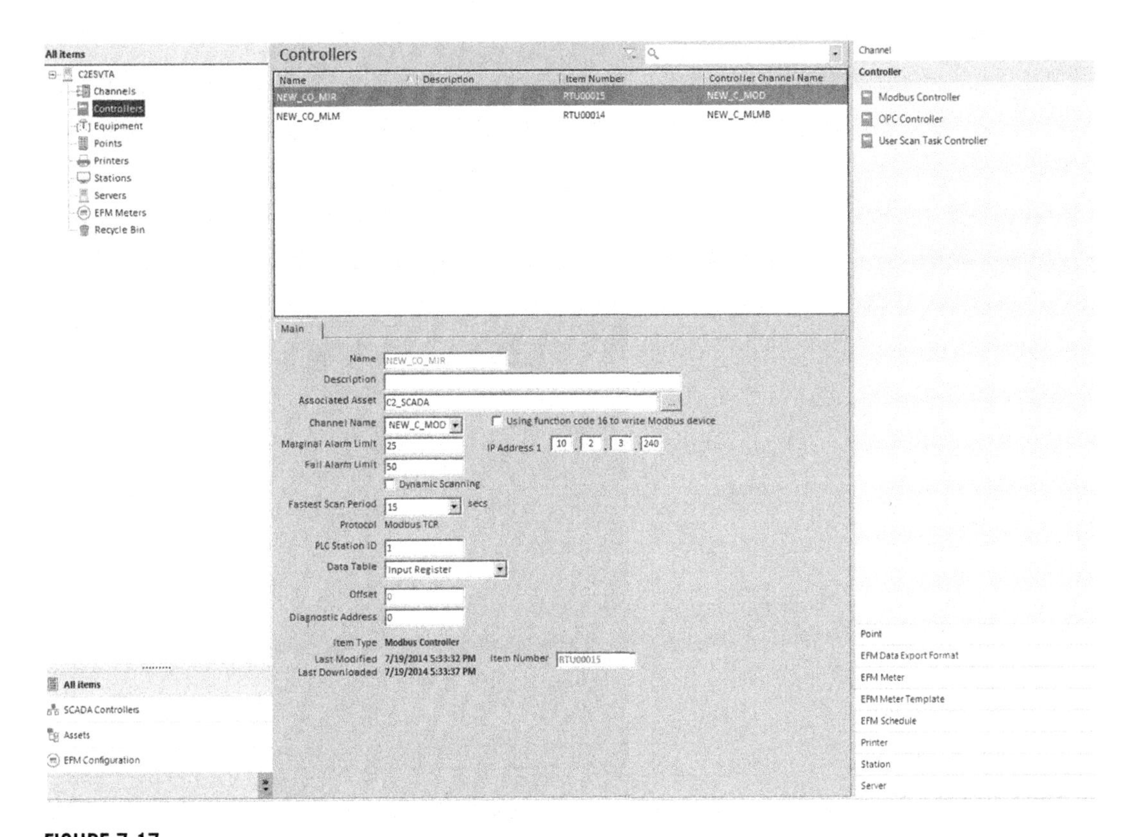

FIGURE 7.17

SCADA configuration utilities

information that is used to direct the host computer's collection and processing of the RTU inputs, and in the case of smart RTUs, some of this information may be downloaded into the RTUs themselves. If any of the logic or calculations needs to be performed in the RTU, then these definitions would need to be sent to the respective RTUs. If an RTU requires EGU values for local calculations, control, or display purposes, then the RTUs will need to download EGU conversion factors, and possibly even alarm limits, would need to be downloaded into the respective RTUs. Control and sequence logic must be defined and loaded into the respective RTUs for them to perform control. All this information definition and data entry is time and energy consuming and a large portion of the overall manpower is invested in system implementation. Intentional or accidental corruption of the SCADA system's configuration tables can cause a partial or total loss of a SCADA system's operability. Undetected corruption of or modification to configuration tables can cause bad data to be presented to operational personnel and supervisory application programs. Most SCADA systems have some ability to check their databases for corruption and errors. In some cases, this can be done online (while the system is operating), possibly even on a continuous basis. In other cases, an off-line diagnostic check may be required. Incorrect (accidental) configuration information has been known to cause mis-operation of equipment and even

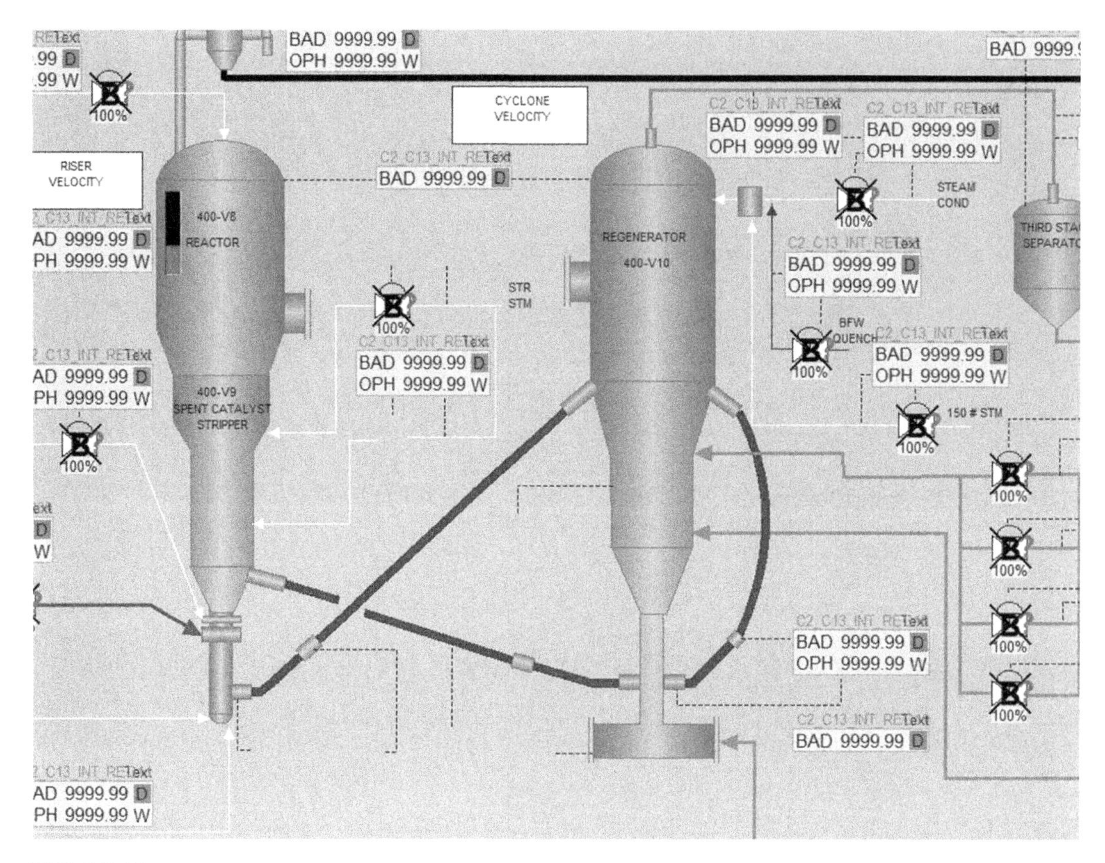

FIGURE 7.18

Custom graphic

equipment damage. One of the most important tests performed on a SCADA system, prior to its being put into production, is a point-by-point verification of each I/O definition, spanning from the actual RTU I/O all the way to the Human machine Interface (HMI) displays. Many SCADA systems maintain a tracking/audit log of modifications made to the various databases (if modifications were made using vendor supplied SCADA system utilities). Some SCADA systems use commercial relational database packages to hold their configuration information. In such cases, the table entries can be modified without using the SCADA system utilities, thus circumventing the modification-tracking process. These modification-tracking processes were established primarily to aid in correcting human errors. Full configuration of a SCADA system involves much more than just building the point database. SCADA systems have to present information to human operators, and although most systems support some level of automatically created displays, most SCADA system operational personnel work from customized, graphical displays that have to be created using an editor utility. SCADA systems have included graphical editors for creating custom displays, although the graphical quality and sophistication have obviously improved over years.

7.4.12 GRAPHICAL DISPLAY EDITOR

The graphical display editor enables the development of customer-specific displays without any programming. Custom graphical displays can incorporate actual images, schematic representations, a variety of drawing elements, and even audio and video clips. Such displays generally present a set of user-selected I/Os, with color coding and animation to indicate potential problems, in addition to the use of graphical elements to convey the process/plant operating conditions. These same displays usually provide control access to physical outputs via a select-and-command sequence built into the graphical display. In addition to direct controls, the displays may offer access to user adjustable parameters and even to alarming functions.

7.5 PROGRAM DEVELOPMENT TOOLS

System and application programming tools are included with most SCADA systems, apart from operating system utility programs and SCADA system utility programs. These are generally a combination of programming tools provided by the operating system vendor (compilers, editors, debuggers, etc.) plus additional extensions added by the SCADA system vendor. A system may provide general-purpose programming tools and specialized application-oriented programming languages that are unique to a given SCADA system product. For the most part, developing and running user-defined application programs is a tricky business that requires knowledge of both the operating system and the SCADA system. It is important to ensure that customer-developed applications do not interfere with the essential functions of the SCADA system.

7.5.1 STANDARDIZED APIS

The 1990s witnessed the consolidation of the SCADA products onto standard operating system and hardware platforms. This led to third-party vendors stepping in to develop application software – in the same way that anyone can develop software for a Microsoft Windows/Intel (Wintel) platform. All SCADA systems have to deal with the collection and movement of both real-time and historical data, including moving these data (bidirectionally) between applications and the places where the SCADA system maintains this data. Most SCADA systems include some form of application process interface (API), offering a well-defined means for application programmers to request and exchange real-time and historical data, leading to a uniform program design. Originally, vendors devised their own proprietary APIs, since no one else was going to supply software to run on their systems! In the 1990s, standards for APIs were defined on the commercial/business side of computing, and the standards that resulted eventually found their way into the SCADA systems because of their adoption by the operating system vendors. Although there was often a goal of nonpartisanship when defining standards, they all eventually fell into either the Unix/Linux camp or the Microsoft camp. Most of today's SCADA systems incorporate one or more of the following popular APIs.

7.5.2 OPC

OLE for Process Control (OPC) is a data-exchange standard that came out of the process control world, resulting from the attempts to interface PC-based graphical display packages with PLCs and other smart

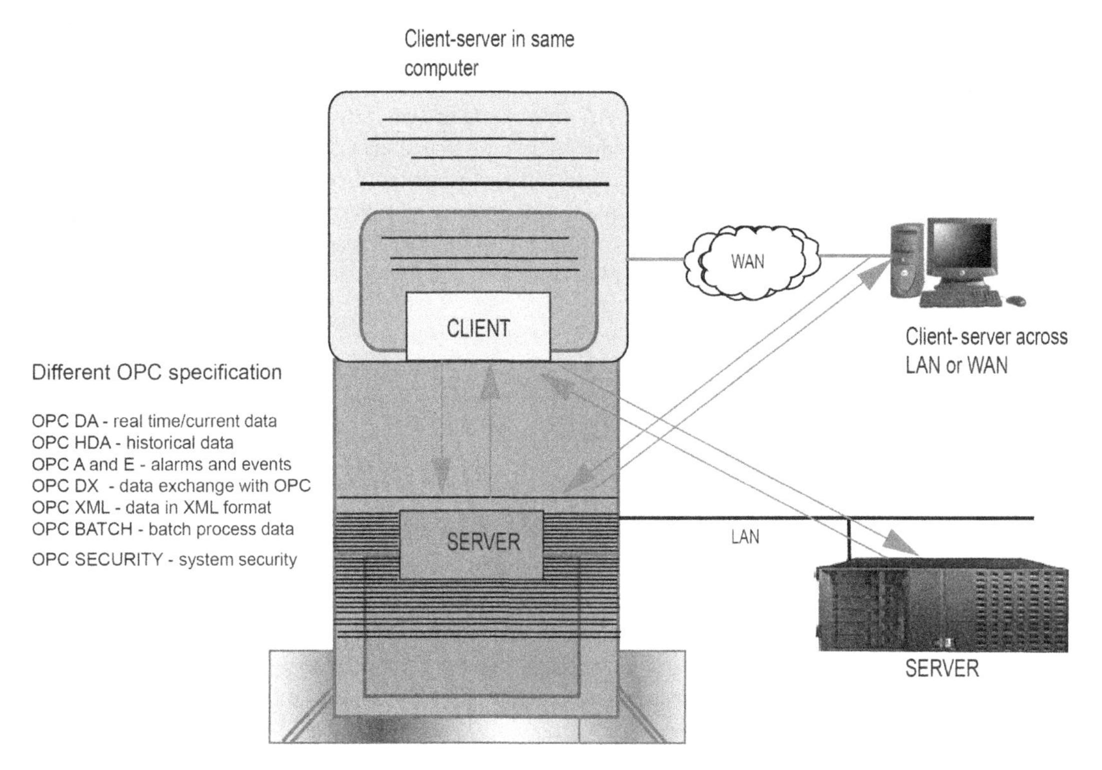

FIGURE 7.19

OPC client server types and configuration alternatives

devices. In the beginning, these devices mainly used slow-speed, serial communications to exchange data, often with the well-known Modbus protocol. This protocol was ill designed for handling the more complex types of data that were commonly available in such devices and originally did not support an Ethernet LAN version. OPC defined a more sophisticated set of client/server interface mechanisms so that complex data (data with time and quality tags, sequence of events, etc.) could be exchanged among applications running on the same (or different) computer and so that multiclient/multiserver equipment configurations could be built on top of a high-speed Ethernet LAN (Figure 7.19).

Several commercially available SCADA systems today are built from hardware and software components from multiple manufacturers, with inter-component data exchanges (often over LANs and even WANs) accomplished via OPC.

7.5.3 STANDARD QUERY LANGUAGE

The standard query language (SQL) is not an API *per se*, but in many SCADA systems, the mechanism for exchanging data with other systems, or between sophisticated application programs, is to pass it through tables in a relational database server (Figure 7.20).

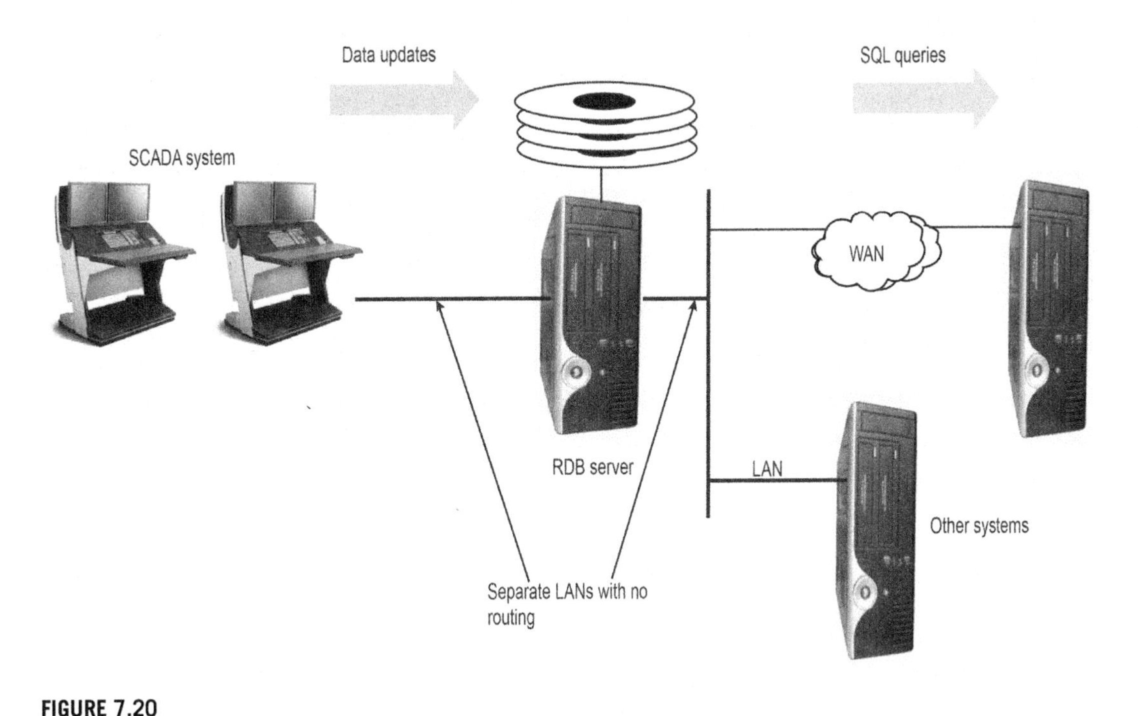

FIGURE 7.20

Using SQL compliant relational database server to exchange SCADA exchange

(Some types of applications, such as pipeline models, involve so much data and configuration information that the application uses relational tables for its own storage. GIS applications normally use relational database tables to hold their data as well.) This scheme is facilitated through the availability of a platform-independent database query language (i.e., SQL) and the migration of popular SQL-compliant relational database packages onto Wintel and Linux/Unix platforms.

7.5.4 COMMON OBJECT REQUEST BROKER ARCHITECTURE

The Unix world was a much more diverse place in the late 1980s and 1990s, as many different computer hardware manufacturers felt that that a Unix implementation might be a good marketing move. Digital, IBM, and Hewlett-Packard (among others) came out with their own versions of Unix. Although these systems did share a Unix heritage, they were nevertheless incompatible hardware platforms. The computing world has always aimed to allow cross-platform, distributed computing, irrespective of the make or model of the computers involved. We are approaching that point today, but only because all of the computers are actually the same (Intel or equivalent). One effort that came out of work toward that end was the common object request broker architecture (CORBA). This was a set of cross-platform services and communication mechanisms, and an API, that enabled programs running on one computer to request services (look up a value in a database, perform a statistical calculation, etc.) from another computer, without having to deal with the incompatibility issues. Larger, multicomputer SCADA systems, based on Unix variants, have employed CORBA for their data exchange mechanisms.

7.5.5 DCOM

As CORBA was being promoted in the Unix world, Microsoft fought back with its own version: the common object model, which was for a single computer on which applications needed to communicate; this was followed by DCOM, which worked across networks. The OPC standard was built upon the DCOM services offered by Microsoft

7.5.6 ICCP

Ever since the early days of SCADA, the electric power industry has needed to exchange information between local SCADA systems and regional/district control centers. In an effort to standardize these data exchanges, the industry devised ICCP (also called TASE.2 and UCA1.0) to provide a mechanism for automatic data exchanges. Initially, ICCP was designed to run as an application layer on top of an Open System Interconnect OSI/ISO seven-layer protocol stack, but in the last 10 years, with the success of TCP/IP, almost all implementations now use that protocol (IP) as the underlying network stack.

7.5.7 UCA2.0

In the last decade or so, the electric and water utility industries have progressed in defining an all-encompassing data and communications architecture that can connect any data source to any user of that data. Although progress in fully defining and promoting this standard has not been without problems, it has started making inroads with some utilities and product manufacturers. Various manufacturers of protective relays and substation equipment now offer UCA2.0 communications, and there have been numerous demonstrations of the various components of this rather broad standard. Within a facility, UCA2.0 utilizes high-speed (100 Mbps, switched) Ethernet LAN technology. Between facilities and to upper-level supervisory systems, the standard provides for transport over IP networks and even low-bandwidth serial connections.

7.6 OPERATOR INTERFACE

7.6.1 ACCESS-CONTROL MECHANISMS

Once a SCADA system is installed, commissioned, and placed into continuous operation, it is primarily the system operators who interact with the system and use the system to monitor and control the target process and field equipment. Other personnel still have to maintain and administer the SCADA system and perform routine housekeeping and support functions, but it is the operational staff who predominately access the system on a constant basis. When you enter the control room of a SCADA system, depending on the industry and the age of the system, you are frequently faced with rows of equipment consoles filled with color CRT displays. You may also see wall-sized informational displays based on projection video technology or mosaic-tile panel board technology. You may even find instrument panels with chart/pen recorders, indicator lamps, and manual, push-button controls.

Figure 7.21 shows a simple control room design that includes two separate consoles filled with CRT displays, a multi-window projection video wall, onto which live data from any of the CRT screens can be directed, and a master clock. The two consoles allow two operational groups (or two operators) to deal with different problems simultaneously (and provide equipment redundancy). For logging and reporting purposes (and so that hard copy printouts of screen images can be made), the operating

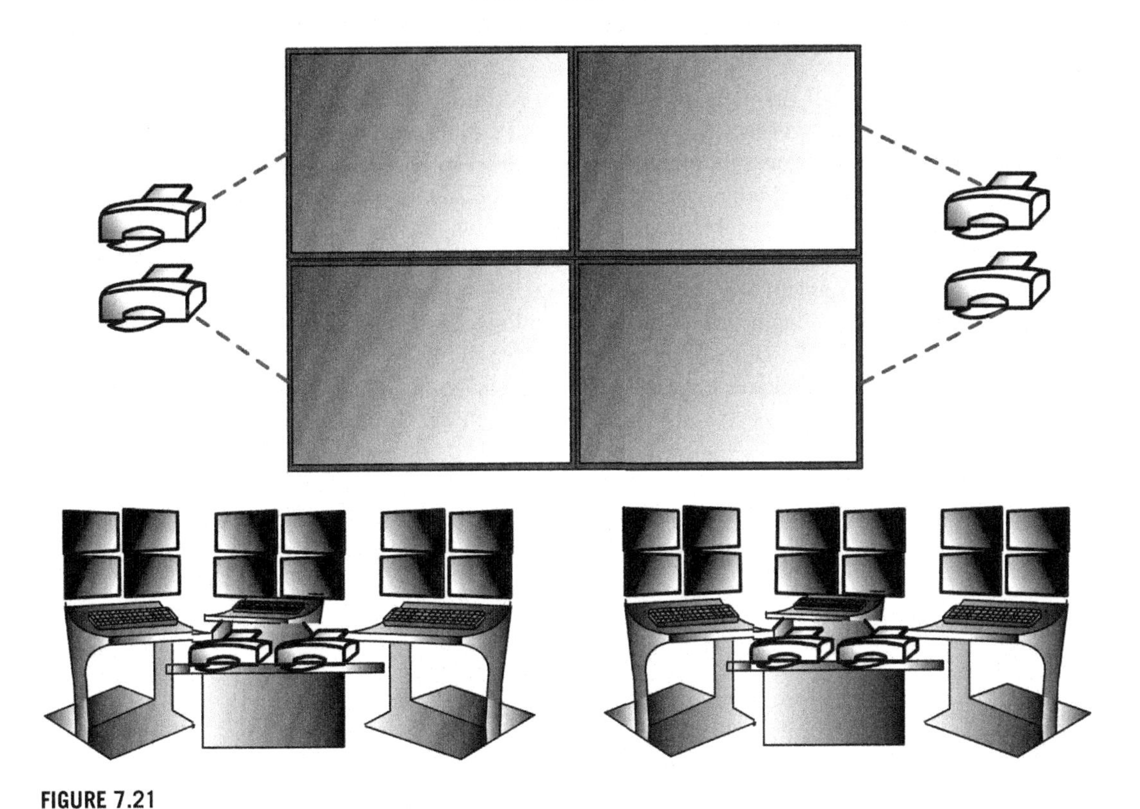

FIGURE 7.21

Example SCADA system control room console design

consoles each include a set of printer/loggers. There also are small SCADA systems consisting of a desktop PC, a printer, and a master radio, all sitting on a desk. Pipeline and electric transmission utilities tend to have the big, fancy control rooms. Small water utilities and rural electric cooperatives tend to have the desktop SCADA systems. The purpose of all this equipment is to give operational personnel a real-time view of the state of the process they are monitoring and to provide them with a means for initiating control actions, as necessary.

7.6.2 STANDARD SYSTEM DISPLAYS

All SCADA systems collect real-time and historical information and then provide operational personnel with a wide range of modes in which this information can be displayed and accessed. Most SCADA systems offer process-related (operational) displays and system-related (diagnostic) displays. There are a wide range of operating system and SCADA system utility programs that are used to initially configure the SCADA system to perform the necessary tasks. The displays and presentation modes discussed in this section have been created using those utility programs or automatically created using the data provided during the configuration process.

7.6.3 DIAGNOSTIC DISPLAYS

It is necessary to periodically examine the operational performance of the SCADA system and verify that it is functioning properly or to make adjustments or modifications as needed.

All SCADA systems provide a set of system diagnostic displays that enable operational and maintenance personnel to check and adjust various aspects of SCADA system operation. A very common SCADA diagnostic display is an RTU polling channel status display – showing the current polling status of the individual RTUs and the communications robustness of the various polling channels, as well as diagnostic error counters and indicators. Such a display also often provides a means for altering polling assignments and polling priorities, placing a polling channel in and out of service, and placing RTUs on and off polling.

Figure 7.22 gives an example of such an RTU polling and communications diagnostic and configuration display. Most SCADA systems also provide a system operational status display, a diagnostic

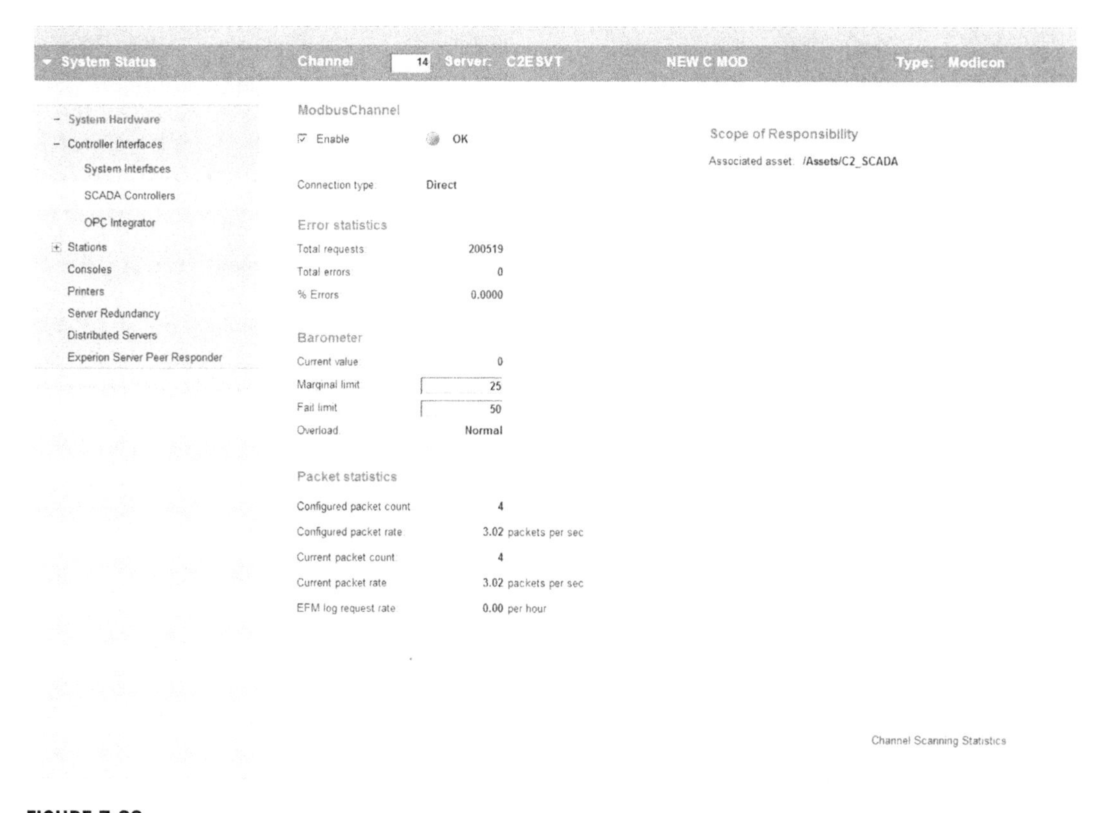

FIGURE 7.22

Typical RTU polling and communications diagnostics display

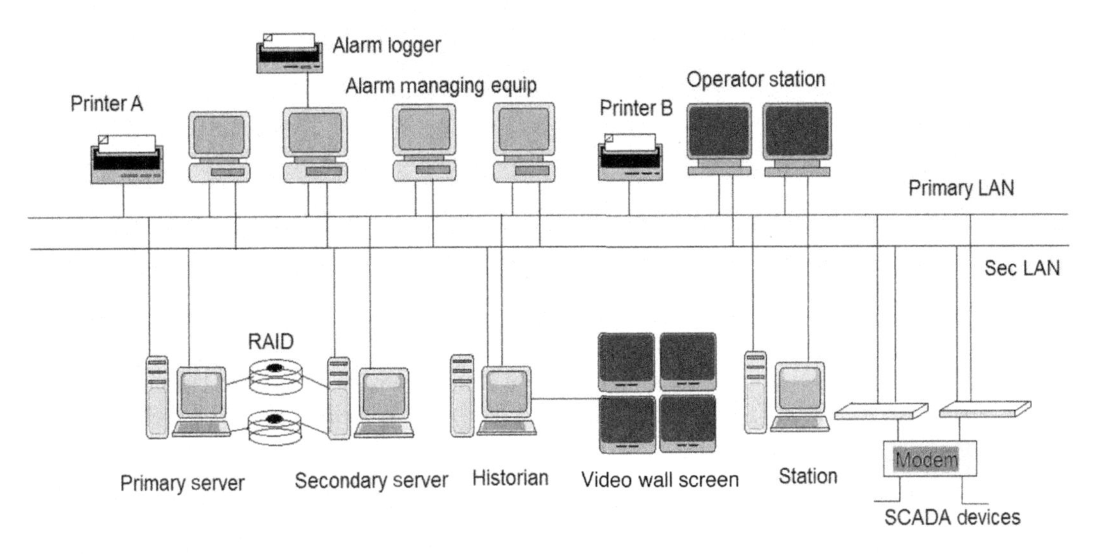

FIGURE 7.23

SCADA system operational status display

display that shows the gross operational status of the equipment, and peripherals that form the system's hardware basis. With some systems, this is an automatically generated tabular display, and with others a (manually created) graphical pictorial representation is provided, with color encoding and equipment status legends (see Figure 7.23).

Just like the RTU polling and communication channel diagnostic display, this diagnostic display enables operational personnel to identify and isolate system problems. Since the vast majority of SCADA systems are built with some level of redundancy, these displays also provide operational personnel with assurances that the backup equipment is actually functioning – and even with a means for manually initiating an operational transfer to the backup equipment. It is common practice to perform software updates to redundant SCADA systems by first loading the new software (after sufficient testing on a nonproduction system) into the backup equipment and then initiating a transfer to this equipment, which then enables the loading of the new software into the former primary equipment. This strategy also allows a return to the old software and equipment, in case the transfer to the new software and equipment is unsuccessful for any reason.

As with all such system-level displays, access to, and manipulation of, the controls and settings in such diagnostic displays needs to be controlled through suitable authorization mechanisms. All SCADA systems incorporate other types of standard (automatically created) displays, and displays that require a limited amount of configuration effort. A good example of another standard display that is usually automatically created would be an RTU current value display. Such a display provides a tabular list of the current I/O values for any selected RTU and possibly additional RTU diagnostic information. An operator can usually select an RTU (from an automatically generated list or via an on-screen poke point in some other display) and get an immediate RTU value status display. In some systems, the operator can demand an immediate integrity scan of the selected RTU, thus guaranteeing that the data values are current. Figure 7.22 shows what a typical RTU summary display might look like. Other types of automatically generated standard displays include alarm summaries and equipment tagging lists, as well as displays that list available historic trend pages, custom graphical displays, and log/report pages. Semiautomatically generated displays (requiring a minimal amount of configuration) include point group displays (tabular, bar graph, etc.), where a user must merely define the points/tags to be in each group and possibly a group descriptive heading. These automatically generated and semiautomatically generated displays often compose the largest percentage of the overall display pages available to system users. Since SCADA systems supplanted older telemetry technology in which numeric values were often presented as a voltage signal into an actual panel meter, it was (and still is) common to simulate the same type of informational display on CRT screens.

Figure 7.24 shows a meter-like point group display. SCADA vendors have explored a multitude of presentation formats for offering information to operational personnel; meters, bar graphs, slider controls, line plots, and textual presentation formats are normally available with any SCADA system.

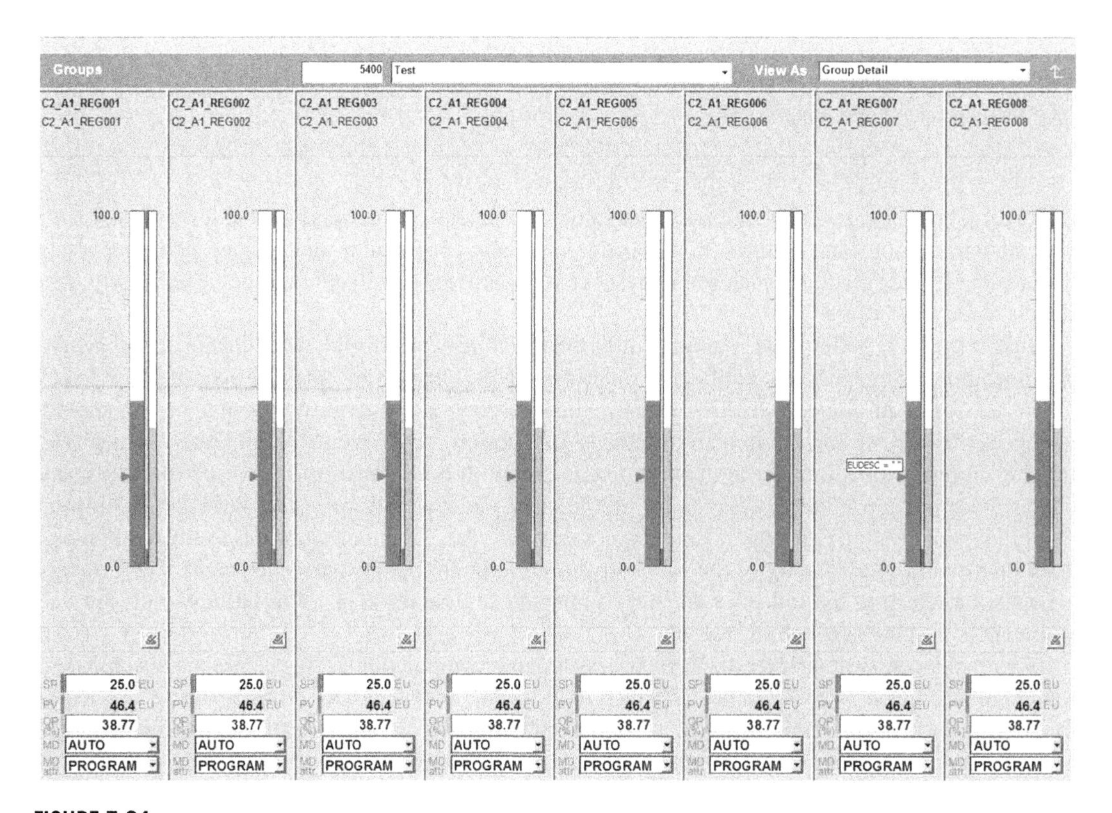

FIGURE 7.24

Point group display

7.6.4 SITE/INDUSTRY-SPECIFIC DISPLAYS

Most modern SCADA systems (since the 1990s) have used semigraphical and now fully graphical custom-developed displays as the primary way of presenting information to the system operators. These displays typically require a good deal of time and effort to design, create, edit, and test. These displays are created using a special (vendor-specific) graphical editor utility that offers the ability to assemble graphical elements, dynamic data elements, control elements, and other components into a user-defined display. Depending on the industry and application, these user-defined display pages may be in the form of process-flow diagrams, map displays, or plant layout displays. The ease of creation and editing vary greatly between products from different vendors, but all graphical editors since about 1990 have supported drag and drop of graphical elements from a menu or table. Most current SCADA system graphics packages come with drawing capabilities similar to those in a PC package such as Microsoft PowerPoint™ or Corel Draw™. Most customized graphical editor packages for SCADA systems support importing of standard graphic images in a standard file format such as JPEG or TIFF. These images may be used as wallpaper on which other dynamic and static drawing elements can be placed. The graphical editor enables the selection of data points for presentation, in some form, on the graphical display page. Numeric data may be presented as a textual numeric value, a bar graph, a meter display, a pie chart, or a trend plot or even used to alter the physical placement and/or color of other graphical elements. Status information is presented as textual descriptions, multistate objects, or color changes or even used to make other graphical elements appear or disappear from the display. The most popular Microsoft Windows-based SCADA packages include custom graphic editors that support Visual Basic (VB) scripts being attached to specified graphical elements and controls. VB scripts can be used to add animation, user interactivity, input validity checking, and display flexibility to custom graphic displays. VB scripts are treated like a supervisory control program and validated prior to being placed into a production system since they can potentially interact with system settings, controls, and resources. Since graphic displays are expected to enable operator supervisory control actions, the graphical editor must offer mechanisms for selecting control points and issuing valid control commands. The graphical editor is also linked with the operator access control mechanisms to ensure authorized control actions. In most if not all SCADA systems today, there are really two parts to user-defined graphical displays. The first part is the graphic editor, which provides for graphic design, editing, and modification. This editor actually builds a data file that tells the second part what to draw, much like an HTML file tells a Web browser how to draw a Web page. The second part is the graphical display execution package, which actually draws and updates the displays and interacts with the user. This display execution package may perform the access-control checks when a user requests a display or attempts to initiate a supervisory control action or alter a protected system or operational parameter.

7.6.5 GRAPHICAL DISPLAYS

Most modern SCADA systems make extensive use of graphical data presentation technologies. Graphical data presentation – particularly when there is a physical, geographic, or process-flow relationship – is the clearest and least ambiguous way to deliver information to operational personnel. User-defined graphic displays begin as a blank page onto which a user decides what to place and where and how the display will interact (if at all) with the user. There are many ways to graphically present individual data elements. A custom graphical display helps in organizing and presenting a lot of data in a way that an operator can immediately comprehend. Since most applications that utilize SCADA technology are geographically dispersed, a common graphical display strategy is to offer data arranged in a geographically aligned manner.

The pipeline, transportation, and water/wastewater industries tend to make extensive use of geographic displays and the electric utility industry as well, to a lesser degree. Such displays are often part of a hierarchy of displays that permit a broad, all-encompassing informational view with drilldown capability that takes operational personnel to whatever level of detail is required. For example, an entire pipeline could be displayed on such a display, and using poke points, an operator could rapidly zoom in to a specific pump station and a specific control point or loop in that pump station. A process-flow diagram is another commonly employed graphical presentation design that provides a simplified physical interconnection drawing. Initially, these diagrams were stick figure drawings, but today, with libraries of graphical images and the ability to import actual photographs of equipment, such process-flow diagrams are far more realistic and interesting. Process-flow diagrams tend to be used to display specific equipment sets or facilities within the overall process being monitored by the SCADA system. An operator might watch a high-level overview display that is geographic in nature and then zoom in to a more detailed process-flow graphic that shows the specifics of the selected substation, pipeline pump station, or water-pumping/storage facility. Figure 7.25 shows a simple example of a multi-window process

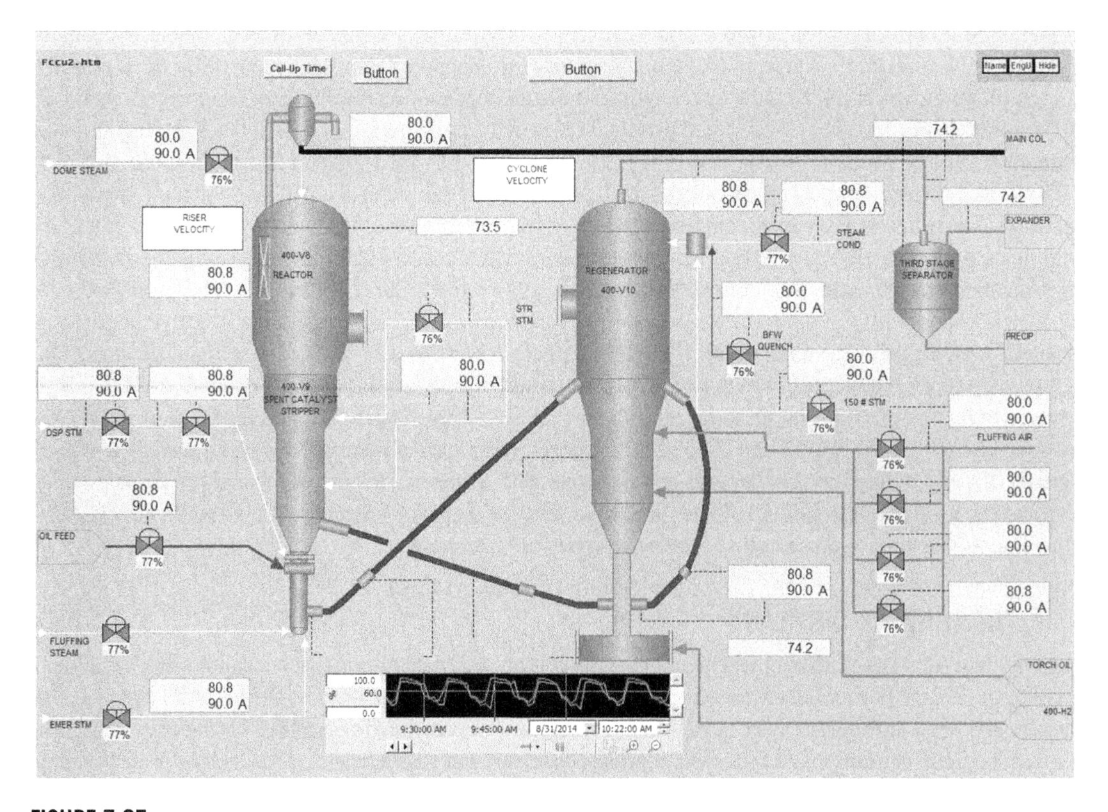

FIGURE 7.25

Process flow operational graphical display

graphical display that is designed to represent the physical process and equipment for superior operator clarity and comprehension.

This figure could represent the lowest level of display drill-down for an operator watching over a large oil/gas offshore production field. Going from a field-wide overview display, to the platform specific display shown in Figure 7.25, might only require one or two mouse clicks. Thus, an operator can rapidly go to a display level that affords the desired level of information detail. This is essential with SCADA systems that have lots of display pages.

7.6.5.1 Display hierarchy

In most SCADA systems, there are hundreds of display pages available to the operational personnel. Even the best operator cannot remember and mentally organize that many pages, let alone the naming scheme used to uniquely identify each. All SCADA systems provide a method for quickly and efficiently locating and navigating between these pages. One commonly used approach is to build displays into a hierarchy, or tree structure, so that any page can be found by starting at the top of the hierarchy (the bottom of the tree) and following linkages down or up to the desired page. Such a hierarchy often begins at a home page that is brought up as a result of the user's login. This home page incorporates links to logically adjacent or logically dependent display pages. A SCADA system has a set of inter-display linkages, similar to navigating on the World Wide Web using hyperlinks. Some SCADA systems now offer their displays in the form of Web pages, thus allowing Web browser and hyperlink technology to be employed. Other systems with non–Web-based operational interfaces employ similar navigational schemes. Navigation may be accomplished by using poke points on displays that transfer to other displays, pull-down menus of available pages, or for geographically associated pages, a north/south/east/west navigation control (see Figure 7.26) that calls up the display page that corresponds to the geographically adjacent area. One use for this type of navigational control would be to follow a pipeline or

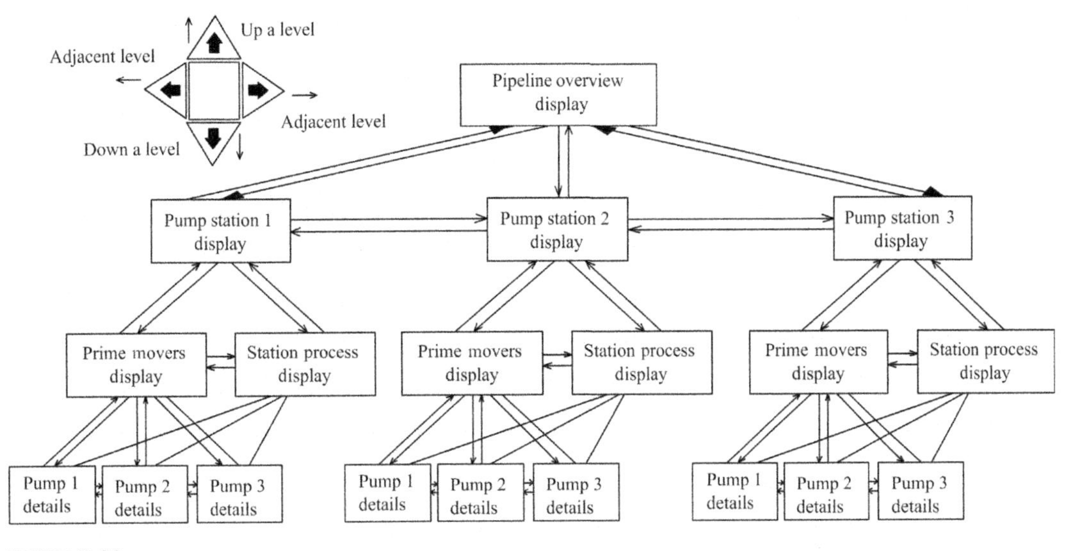

FIGURE 7.26

Display hierarchy and Inter display navigation

a power line across a series of map pages to the desired location. Display pages in SCADA system are set up by establishing the navigational links between and among the various pages. With automatically generated displays (such as diagnostic displays and status displays), the SCADA system usually automatically creates a directory of all such pages. But graphical pages, with associations to specific (semi) automatically generated pages, usually require some user input to define these navigational linkages.

7.6.6 DISPLAY NAVIGATION

GIS-based graphical displays are based on geographic considerations. Thus, navigation around the display is also in terms of geographic positioning. Most GIS displays offer some form of compass point scheme with north/south/east/west selection icons (panning). There are also usually control icons for zoom manipulation and additional controls for layering and decluttering. Also, you can click on objects and locations within the display and call up additional detailed information about those items or to cause a transfer to a different display that is specific to the selected item or location. The GPS technology enabled storing GPS coordinates in their geographic/asset databases, in many water, gas, and electric utilities, so that an alarm received from a given location provides the display system with GPS coordinates and the display can automatically pan to the source of the alarm.

7.6.7 ALARMS AND INDICATORS

SCADA systems deal with large volumes of constantly changing data (2,000–70,000 tags, depending on the application), making it infeasible, and far too time-consuming, for a human operator to constantly cycle through all of the data looking for problems. Therefore, it is important that the SCADA system provide at least basic validity and alarm checking on all of the data. Detecting alarm and abnormal conditions – and bringing these to the attention of the operators – is a primary function of SCADA systems. Configuration to commission a SCADA system involves defining the alarm thresholds to be used for each point. Some systems allow for a default set of alarm limits, and most also have built-in checks for invalid inputs. (For example, if an analog input is to be a 4–20-milliamp signal, then readings much outside of that range would indicate a potentially invalid and probably useless input signal.)

Figure 7.27 shows how a signal value traverses its expected (or possible) value range, and how it goes in and out of differing levels of alarm severity and indication. In most SCADA systems, value transition across a predefined alarm or validity limit causes an alarm indication and recording unless some form of alarm inhibiting or disabling has been placed on an input. This usually results in some form of visual indication on all displays containing that input, and a corresponding entry being made in a chronological log of all alarm events. The inputs value returning to a valid (normal) value is also usually accompanied by changes in the visual presentation and by another log entry. On powering up and initializing of SCADA systems, with no prior data, an integrity poll of all of their RTUs is initiated, to develop a complete set of real-time data. This process is usually time-consuming and generates a lot of spurious alarms unless the system has a means for suppressing them. On completing this process, most systems are generally stable and just continue the polling and alarm-checking process. During the initial startup process, many SCADA systems are essentially unusable and the operators blinded. In the event of a communication failure with an RTU, many systems mark all the associated points in the real-time database

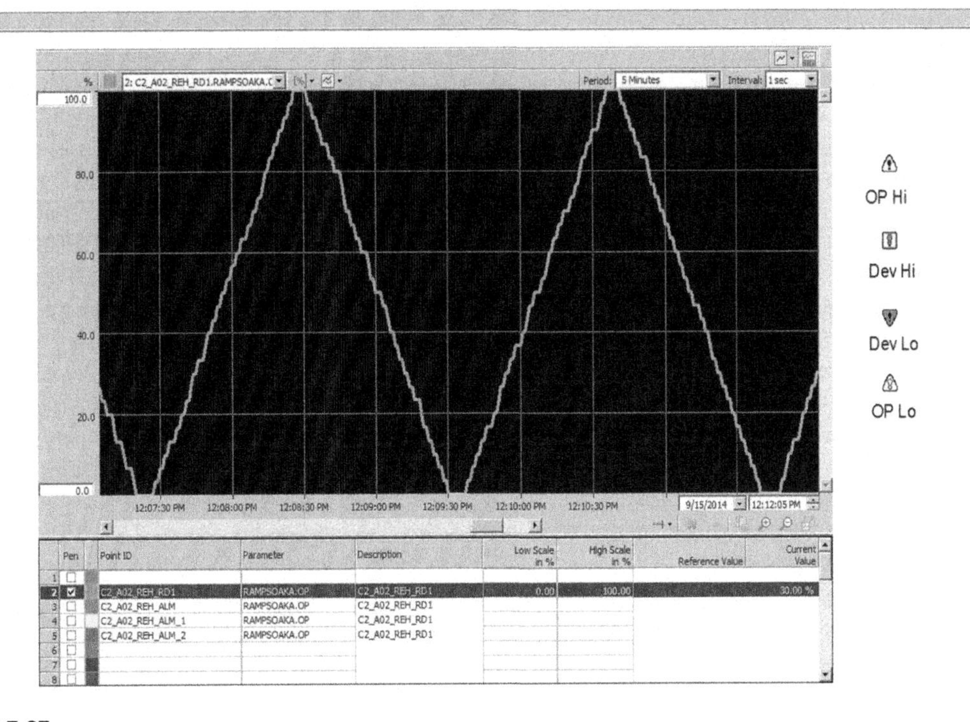

FIGURE 7.27

Alarm limit checking on a typical analog input point

in some manner that indicates that the data contained in that database, for those points, may not be trustworthy. (Some systems have an off-line flag, some use a "questionable" flag, and some mark the data as "old.") In a less sophisticated system, this may also cause a flurry of alarms on all the RTU's points. When the communication with the RTU is restored, the flags have to be removed, the RTU is commanded to perform an integrity poll, and again, there can be a flurry of alarms, depending on how things changed during the communication outage. Managing the way a SCADA system deals with alarms, and its impact on the operators, is an important consideration (and potential vulnerability).

Analog inputs are generally given a validity check (to determine if the input can be trusted), followed by one or more sets of upper and lower allowable value range checks (often called high- and low-level checks) and possibly a rate-of-change check and an expected value (set point) deviation check. Most inputs to a SCADA system are assigned a set of fixed alarm limits, but some alarm limit checking may be based on the operational state of the process being monitored and use some form of arithmetical model to dynamically set and adjust the alarm limit values. The violation of alarm and validity limits is of differing levels of importance, based on the measurement in question. Alarming functions must operate properly and limits must be properly set since operators usually rely on alarm checking to alert them of potential problems. If alarming functions were disabled, or limits changed to extreme values, dangerous conditions could go unnoticed by the operational personnel, resulting in damage, outages, and even threats to health and safety. Another aspect of system configuration is the assignment of these alarm limits and of a severity level to each of the potential alarm conditions for each input. A SCADA system can generate a lot of nuisance alarms or fail to alarm critical conditions if it is not properly configured. When a system generates a lot of alarms, the operational staff tend to become insensitive to these alarms and either ignore them or disable the alarming functions, neither being a safe operational situation. Contact and status inputs have similar alarm-checking provisions. Contact inputs can be checked for their current state (on/off) and for unexpected state changes. Contact input status changes may be considered as either alarms or events, depending on the circumstances. If the operator sends a command to start a pump, and receives a confirmation of the pump's status changing to RUN, then that status (contact input) change is not an alarm, but rather an event. On the contrary, if no such command was sent and the pump status were to change to RUN, that would be an alarm condition (uncommanded change of state). Operational personnel typically ignore events (which may not even be annunciated in any manner by the system). It is important to ensure that the alarm checking is properly performed on contact inputs. For some applications, in which precise (1-millisecond accuracy or more) time tagging is used on contact inputs, the determination, from a group of contacts that may be in different physical locations, of the first contact in the group that changed determines the alarm severity and type (called a first-out alarm). In most SCADA systems, user-defined computed values (numeric and Boolean) are checked for alarm conditions each time they are recomputed just like physical inputs. When an input (or computed value) goes out of its normal operating range leading to an alarm condition, the operational personnel are alerted, with the method used dependent on the severity of the alarm condition detected. Proper alarm checking on computed values can be as critical as that performed on physical inputs. Most systems have positive acknowledgment of alarms to ensure that alarms are actually noticed by operational personnel. When a point enters an alarm state, it is, by default, unacknowledged until an operator acknowledges the alarm. This unacknowledged condition may be specifically indicated with a color or blinking condition on any display where the point is presented. Most SCADA systems differentiate between acknowledged (definitely seen by the operational personnel) and unacknowledged (ambiguous as to having been seen by the operational personnel) alarms. They may even be placed into different display pages to differentiate them even further.

Figure 7.28 shows a typical alarm summary window with a few active, serious alarms, as well as the highlighted alarm, which has returned to normal without being acknowledged by the operator. (The active alarms would be in red, and the unacknowledged one would be blinking in green.) Such displays usually present alarms in either chronological order or sorted by severity, and the display window alarm list will grow and shrink as alarms enter and exit. Many systems provide scrolling lists to accommodate large numbers of alarms. Many systems use blinking to indicate unacknowledged alarms, and this persists even if the point returns to a normal value, until acknowledged by an operator.

7.6.7.1 Alarm filtering

In the event of certain types of operational or process-upset conditions, there may be a flood of alarms from the field, and this can distract operational personnel. This could be caused by an actual problem or by such activities as shutting down a process area or turning off an RTU. In a large SCADA system that supervises an extensive geographically distributed process, different operational personnel might be responsible for a subset of the overall process. In those cases, it is important that a given operator not be distracted by alarms and notifications related to a subset for which he or she has no responsibility.

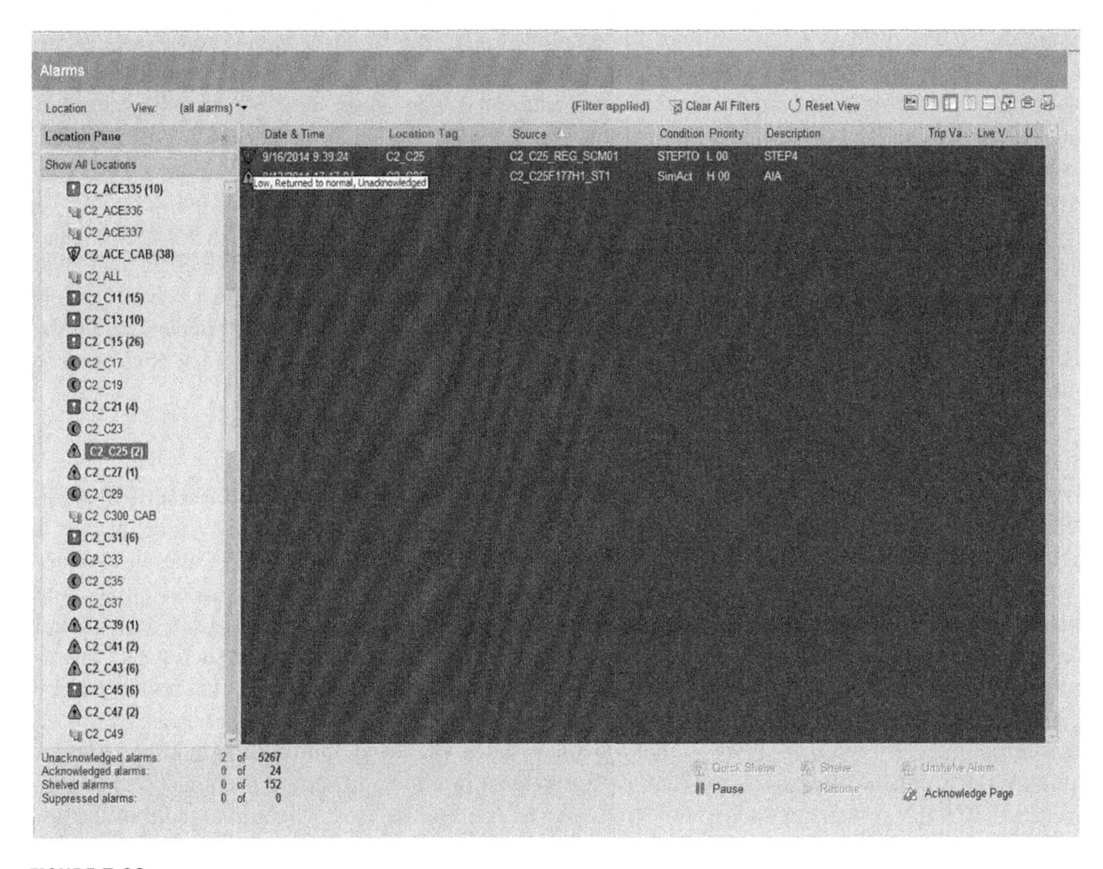

FIGURE 7.28

Typical current alarm summary display alarms

SCADA systems aim at reducing and focusing the information presented to operational personnel. Alarm filtering is one such mechanism. Most systems enable the operational personnel to designate the logical or physical process areas from which they do, or do not, wish to receive alarms. Categories or severity levels of alarms to be excluded from their displays can also be defined. When a field site is disrupted by onsite work, it is important to prevent the generation of spurious alarms by the SCADA system every time an on-site worker turns the power to a piece of equipment or an RTU on or off. A SCADA operator can select predefined filter options from a list or define customized filter options. It is important that operational personnel be aware that alarms are being filtered and that filters require some periodic reapplication so they are not forgotten and left in effect by accident.

7.6.7.2 Alarm annunciation

The SCADA system is expected alert operational personnel, in the event of detecting new alarm conditions. This is accomplished by placing the new alarm information into the active/current alarm summary list and display. Another commonly employed mechanism is to generate some form of audible signal that will attract the attention of operational personnel. This may be a set of unique sounds played through the speakers of the operational workstation(s) or an alarm signal generated by an external Klaxon or bell. The operational personnel would acknowledge the new alarms and possibly operate a separate control to silence the annunciation. For particularly serious alarm conditions, some SCADA systems also extend the range of personnel to be notified (outside the immediate control room), by using mechanisms such as personal pagers, cell phone text messaging, and e-mail to send automated alarm messages to relevant senior technical, management, and supervisory personnel. E-mail (and some forms of cell phone) notification calls for IP connectivity – from the SCADA system to a corporate intranet and mail server, and maybe through that server to the Internet itself, depending on the recipients. IP networking, having other IP-based systems between you and the Internet, could lead to vulnerability to attacks coming from the Internet.

7.6.8 ALARM HISTORY FILE

There are typically a subset of the entire set of inputs and calculated values that are currently in alarm (outside their normal, acceptable operating range), at any point. This list varies over time, as inputs return to a normal (non-alarm) condition and others enter an alarm condition. An alarm summary display (Figure 7.28) usually lists the signals currently in an alarm state. (Most systems purge from this list points that return to normal, unless they have not been acknowledged by the operational staff.) It is also useful to keep track of all the alarm comings and goings, and the actions of operational personnel. Most SCADA systems have some form of historical logging function, whereby alarm transitions, events, alarm acknowledgments, operator actions, and system messages are placed into chronologically ordered tables so that this data can be reviewed and (data) mined, to look for situations that indicate a process problem, an operational problem, a training problem, or a system problem. SCADA systems maintain an operator action (or tracking) log wherein a record of the parameter adjustments, alarm acknowledgements, control commands, and other actions taken by each operator is maintained. This log is especially helpful when keeping track of different operational shifts, evaluating the performance of operator trainees and keeping an audit trail of remote personnel with operator control authority.

An intrusion detection system makes use of this log in order to learn what is normal for each operator and to identify unusual and abnormal operator actions. SCADA systems usually provide a means for displaying, sorting, filtering, and printing these tables. Deletion or alteration of such tables requires a higher level of access and different (administrator level) tools.

The term log is usually used to describe the printed results of such operations. Most SCADA systems provide a range of predefined logs and even some level of configurability in formatting the resulting printed documents. The older SCADA systems came configured with several printers, each dedicated to a particular class of log – with boxes of form-feed, fanfold paper providing a continuous, printed, chronological audit trail of alarms, events, operational actions, and system messages. Storing this information in relational database tables, rather than printing, has undoubtedly saved a great number of trees and made the information much easier to sort, organize, search, and utilize. Nevertheless, some SCADA system owners still prefer the use of printed logs, because they can be harder to manipulate and falsify (or delete – but not destroy) than data stored in a computer file or relational database table.

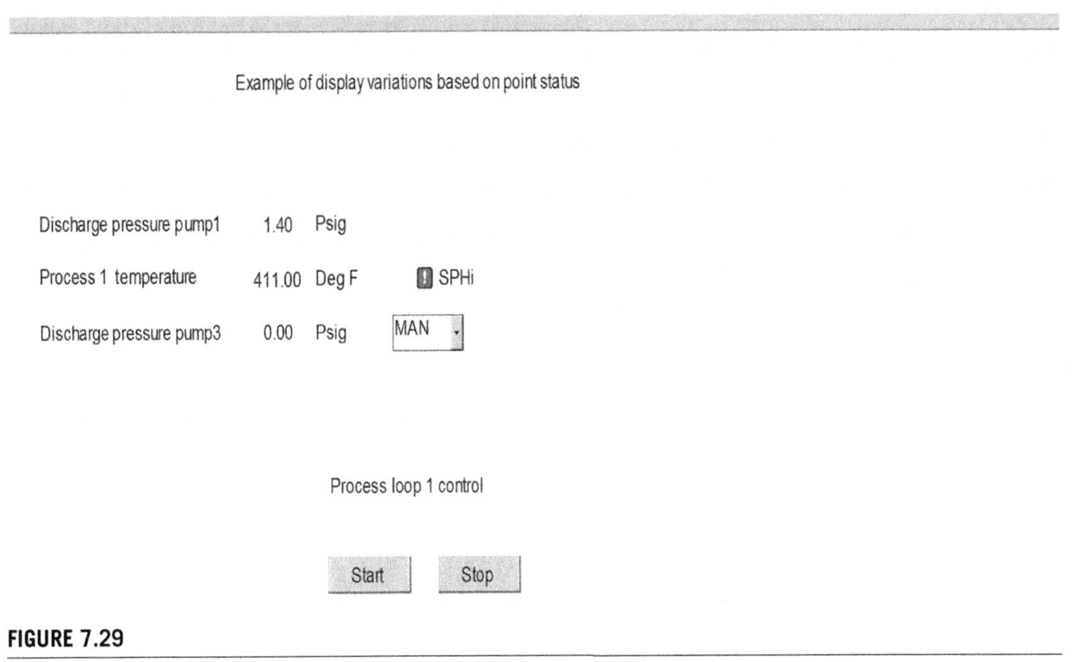

FIGURE 7.29

SCADA system use symbols or code letters to indicate measurement conditions

7.6.8.1 Alarm-state visual indication

In every SCADA system, color and blinking are generally used in the operational displays to indicate current alarm status. If a measurement is present on an operational display and is within its normal operating range, the system may be configured to display the value of that measurement in green (although this varies a bit by industry). Red, yellow, and other colors are often used to indicate parameters that have entered an alarm, or questionable, condition, or that are disabled. In the early SCADA systems (1970s), alarm conditions were indicated by using reverse video and by adding code letters adjacent to the displayed value. The current SCADA systems append special icons or symbols to a displayed value, in addition to color and blinking to indicate specific conditions. As an example, if an input point has been taken off scan, or is having its value manually set by the operator, then those conditions are also indicated. Every display in which a given parameter or point is presented should incorporate the same set of color/symbol/letter codes for that parameter or point. Since many conditions can be present concurrently, combinations of color, icons, and code letters are often employed.

Figure 7.29 shows examples of point-presentation variations. For control output points, a tagged indicator next to the control signal is used to indicate on all displays where that control is presented and available for control action initiation. Tagging is the process of disabling a control-output-form supervisory control for the purpose of equipment and/or personnel safety. The term derives from the practice of placing actual physical, paper tags onto control handles on equipment, so that people would see the tags and avoid operating or activating the equipment. In theory, tagged control outputs cannot be operated through the SCADA system until all user-assigned tags are removed. Tagging is common in the electric utility industry and has become more common in other industries over the past decade, as SCADA vendors introduced their technologies into other market segments.

Enforcement of tagging can be implemented at various levels on a SCADA system's software. The most robust implementations use information written directly into the configuration database of the RTU that contains a tagged control point, so that checking for a tagged condition happens right at the level on which actual manipulation of those physical outputs takes place. SCADA systems usually enable multiple tags to be assigned to any control point, by different personnel, and all such tags must be removed to restore supervisory control of any given control output. SCADA systems that implement tag enforcement at the host level (which is the most commonly used design) or the operator permission-checking level are potentially vulnerable to tag bypassing. An attacker is not prevented from sending control commands to the RTU and having them acted upon because the RTU is unaware of tagging. Likewise, an application program could send such commands because tags are only checked on manually initiated (through the HMI) control actions.

Figure 7.30 shows tagging activity being initiated, via the operator's display, on a circuit breaker in a substation. Violation of tagging opens up the distinct possibility of causing damage, serious injury, or even death to onsite personnel, in extreme cases. Most experienced organizations do not fully trust the SCADA system and require manual disabling of controls in the field, possibly by switching off control loop power at the RTU.

7.6.8.2 Historical trending

SCADA systems collect a constant stream of updating real-time data from the RTUs in the field and update (overwrite) their real-time database with these new values. However, just as important as knowing the current value of a measurement is knowing the past value changes (history) of many of those same measurements. Knowing that a tank level is currently at 5.56 feet does not provide the same understanding of what is happening as being able to see that this level has dropped (or risen) to that point over the past few minutes. All SCADA systems incorporate some level of historical data recording, whereby the operational personnel have the capability of reviewing the past history of a selected set of key measurements (or calculated values) for some predefined time span extending into the past. Prior to SCADA systems, one of the primary mechanisms for monitoring a remote measurement was to use that measurement value to move a pen up and down across a strip or disk of moving paper.

The historical trending functions and displays of most SCADA systems essentially mimic the actions of those obsolete (but still used) electromechanical devices. Strip-chart and circular-chart recorders were invented for pre-computer data recording purposes and have remained a staple of the process-monitoring and control industry up to the present day. The historical trending and display functions of a SCADA system are an electronic reproduction of that older technology. Today, historical trending packages are available from third-party vendors, particularly for systems based on the Microsoft Windows operating system and the OPC data-interchange standard. In prior decades, SCADA system vendors generally had to develop their own proprietary versions. A major aid in doing this came in the mid-1990s with the evolution of commercial relational database packages that had adequate performance capabilities and massive increase in the storage capacity of disk drives. Historical trending involves four aspects: initial configuration of the data collection and display

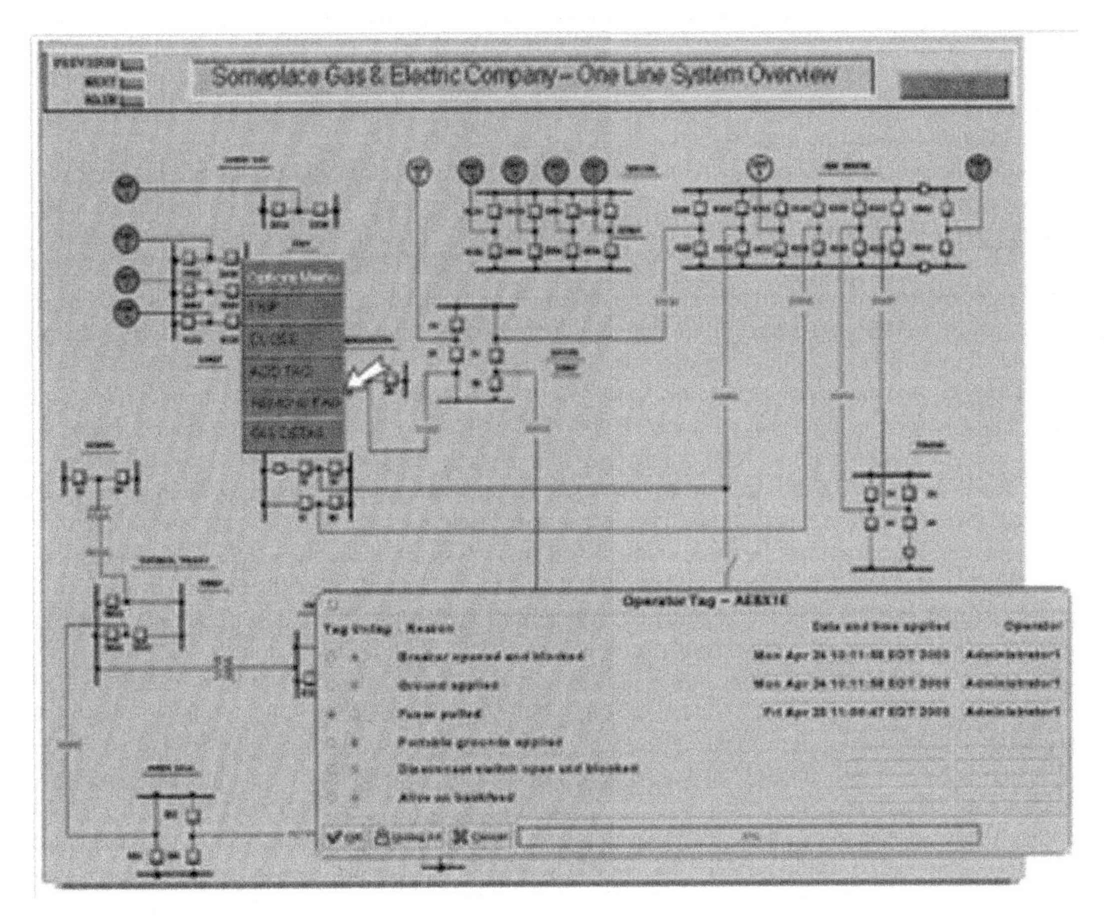

FIGURE 7.30

Control point tagging display

functions; actual collection of the specified data values; storage and management of the collected data; and finally, the presentation of the recorded data to operational personnel. Computers have dramatically increased in computing power and storage capacity over the past few decades. In spite of those advances, computers still have practical and physical limitations. It is often not physically possible or practical to attempt to collect and continuously store every measurement, status input, and computed value within the SCADA system. No matter how powerful your computer, you can bring it to a halt by placing upon it too great a processing and data manipulation burden. Similarly, it is not possible to equip a computer with an infinite amount of mass storage. For those reasons, all SCADA system historical trending packages require the user to pick a subset of the available database points. These selected points will be the only measurements assigned to historical trending. Similarly, SCADA systems limit the user in the amount of such data that can be physically accommodated by available online storage. All field measurements are not equal in their importance or in the rate at which they can change. A bearing temperature could be very important and might change rapidly. The water level in a lake might also be important, but is unlikely to change very rapidly under any reasonable conditions. Therefore, a user may wish to store samples of these different measurements at a different periodicity.

Historical trending packages often manage the storage and collection process by offering prespecified trend groups. For example, a trending package might allow data to be written to historical storage, from the real-time database, at a 1-, 5-, 15-, or 30-minute collection rate (i.e., one sample stored to the trend files every 1, 5, 15, or 30 minutes). Obviously, over any fixed time period, the data collected once per minute will be of a much greater quantity (60 times as much) than that collected once per hour, if the number of trended parameters is the same for each collection rate. Such a trending package might place maximum limits on the number of parameters that could be assigned to each collection rate, thus predefining the maximum amount of disk space that could be used by the historical trending function. (For group 1 [data collection every 1 minute], up to 200 points may be assigned; for group 2 [data collection every 5 minutes], up to 1000 points may be assigned; and for group 3 [data collection every 15 minutes], up to 3000 points can be assigned.) Once the predefined maximum amount of data has been collected, there is, in theory, no further storage space available, so either historical trending has to cease or additional room must be made available. In many systems, the storage that is pre-allocated is treated as if it were logically structured into a circular buffer: when the end of the buffer is reached, the subsequent data are stored at the beginning of the buffer, overwriting the oldest previously recorded data values. Using storage in this manner prevents the filling of all available storage space (and shutting down of the system), while ensuring that there is a specified amount of prior history available for any given measurement assigned to historical data collection. Another way in which trending packages save space is by reducing data quantities through statistical data manipulation (e.g., taking all of the data samples for a 15-minute interval and reducing them to just the average, maximum, and minimum values for that time span).

Scada systems perform variety of data archiving, storage, and reduction methods, for long-term historical recording purposes. The purpose of collecting and archiving data is to provide a historical perspective on the actions of critical measurements and values. Historical trending applications provide a limited look back in time for the values being archived. It may be important to have 1-minute data samples on a given point x for the most recent 24 hours. Beyond that time span, it may not be as critical to collect data that often, so sampling and collection of x values could be reduced to once per hour. One can achieve this by placing point x into both a 1-minute collection group and a 60-minute collection group. This approach leads to the problem of storing more data than is necessary (although when storage is plentiful, this may not matter). When sampling is extended over a long time interval (an hour, a day, or a week), there is always the potential problem of missing something important during the intervening time between sample captures. Some SCADA systems address this problem by computing, for example, the hourly value (rather than fetching the current value at the hour demarcation), using the minute data collected in the prior hour. This approach facilitates deriving a more representative value, rather than merely saving the instantaneous reading at each hourly point. The average/mean, maximum value, or minimum value (or all of them) can be computed from the prior hour's data and used as the hourly recorded value(s). Many SCADA systems offer an alternative to merely overwriting (and losing) the oldest historical data as additional data is stored. In these systems, prior to overwriting (and losing) the oldest data values, those values are copied onto some form of removable storage media, such as a magnetic tape cartridge, a compact disc (CD), or these days, a digital versatile disc (DVD) (Figure 7.31).

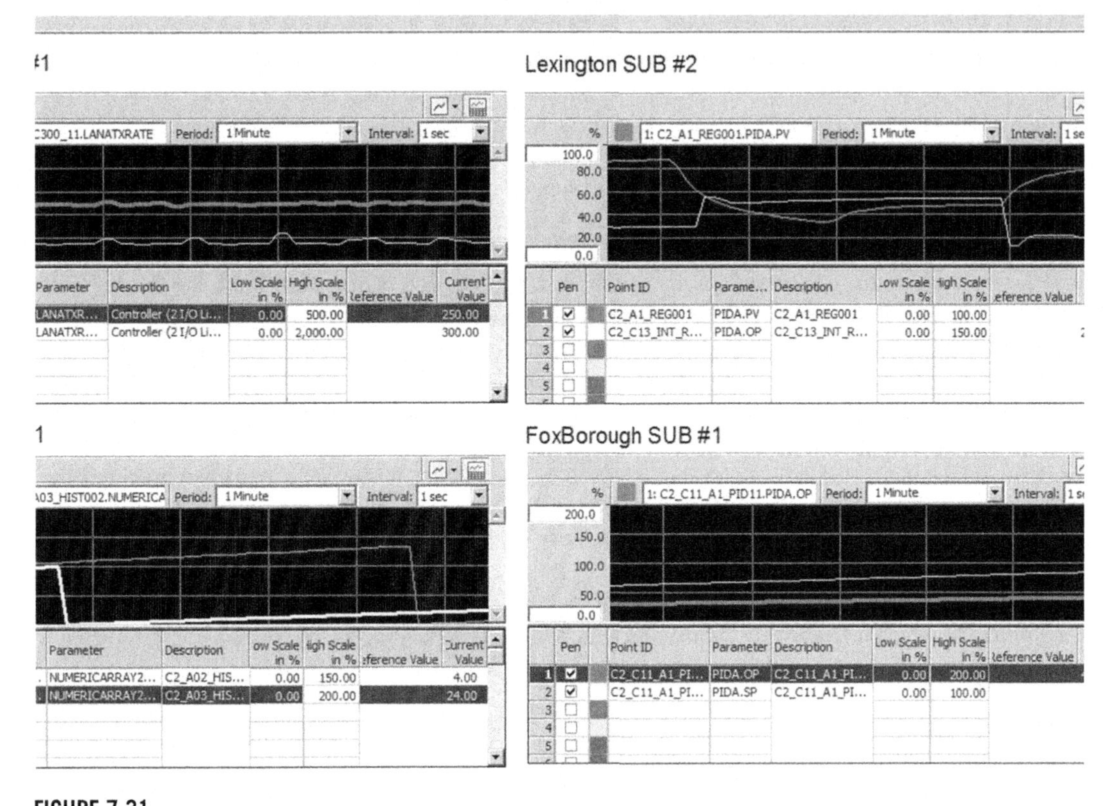

FIGURE 7.31

Typical historical trending display

The data preserved in this manner are no longer immediately available for display on the SCADA system, but if needed at some point in time, the data can be copied (temporarily) onto the system disk, or the removable media 'mounted' in the appropriate removable media drive, and made available for display. Data that are not immediately available but must be brought back into a system via some manual intervention (and possibly a data transfer process) are usually called off-line storage. Data that reside on the hard drive(s) and can be immediately accessed by application programs are usually called online storage. Removable, directly readable media, such as CD and DVD, have blurred this distinction a bit. The data contained on them is not immediately available, but there is no requirement to copy the data onto a hard drive in order to access this data. In some industries, it is of regulatory and legal value to maintain historical data records with no limit on the time span being preserved. Using off-line storage as a permanent means for data archiving addresses this need, without placing undue demands on the production system. In some cases, the historical trend data may be of financial importance (e.g., used for documenting that contract pressures or flow rates were maintained or for deliver totalizing computations) and, therefore, its loss or alteration could have a potential financial impact.

7.6.8.3 Historical trending displays

Once historical data are collected by a SCADA system, they are available to the operational personnel, usually in the form of trend (time plot) displays. A trend display usually provides some description of the data being presented, as well as some form of scale or grid to determine the values being plotted. Data are presented in the form of an *X–Y* plot of values (on the *Y* axis) against time (on the *X* axis). Most historical trending packages allow multiple signals to be co-plotted on the same display grid for comparison purposes. Most historical trending packages allow the user to rescale the *X* and *Y* axes to obtain the desired resolution and data clarity. Most also permit the scrolling of data forward and backward in time, within the constraints of the available data. (No, you cannot scroll into the future to see what is going to happen.) Figure 7.31 shows a multi-pane, multi-trace historical trending display of various measurements associated with transformer performance. There are a wide range of more advanced functions that can be found in various historical trending packages, and it is not the goal of this book to enumerate all of these features. A historical trending package deals with fixed data, recorded at some time in the past, possibly up to the present moment. For that reason, such trend displays generally are static. (Once the requested data has been presented, the display has no reason to change or update.) Most packages also allow for some form of quality code to be attached to the stored data. This is used to indicate any issues concerning the validity or integrity of the stored data values. For example, if stored values were from physical inputs that were being manually overridden, a code might be used to indicate this state. If no data was recorded over a time interval, the display may leave a corresponding gap in the trend plot. The specific features, formats, and functions will depend upon the SCADA system vendor, unless they use a third-party commercial historian package.

7.6.8.4 Real-time trending

There are also occasions when it is useful to continuously monitor a measurement (or several measurements) over a period of time – watching what the values have been, as well as the current value(s). Such trending can be supplied either as an extension of the historical trending package or as a separate capability. The term real time has to be interpreted in light of the processes being monitored and the communication capabilities of the SCADA system. Because of the generally slow-speed communications between the SCADA host and the field-based RTUs in some industries, fetching new data values is usually what limits the ability to provide real-time trending. In the water/wastewater industry, a real-time trend may update only every few minutes, whereas in the electric utility industry, trend updates may be possible every few seconds – simply because of the differences in the communication architectures and technology. An important consideration when configuring a real-time trending package is to avoid attempting to collect and store data at a rate that is faster than the actual values are being updated by polling of the RTUs. The process of updating the real-time database is typically asynchronous of (disconnected from) the process of writing database values to historical files. An RTU could scan and update a particular measurement every 10 seconds and be polled for that value every 30 seconds (so the data could be up to 30 seconds old in the real-time database of the SCADA system), and yet the operator has it on a trending display that updates at a 10-second rate. An operator might be fooled into thinking that a process or particular measurement was stable because a real-time trend plot of values was holding steady. However, in reality, communications to the particular RTU might have been lost. In most SCADA systems, the historical and real-time trending packages normally fetch their data from the real-time database that is refreshed by polling RTUs. Therefore, it is important that the status of data in the real-time database be included and updated, and not just the value. As with any operational display, it is important to somehow alert a system user that the data being viewed could possibly be invalid or questionable or is just not available. In SCADA control rooms, real-time trending displays, just like alarm summary displays, are a critical component of the overall operational data presentation. Many SCADA system operational personnel were originally trained with older panel instrumentation (a wall filled with chart recorders and indicator lights) and telemetry and prefer trending displays, based on their training. In many Microsoft Windows-based systems, OPC is used to link many trending packages to their data sources. Thus, a third-party commercial historian/trending package might receive data updates from one or more sources, over a LAN and/or a WAN connection. Other third-party historian packages employ Modbus-over-IP or other IP-based communications to receive their data updates.

7.6.9 LOGS AND REPORTS

SCADA systems provide real-time monitoring and supervisory control, in addition to utilizing the collected data to automatically generate (printed) reports and/or logs – for corporate purposes, regulators, and government agencies and other organizations that require regular reporting. In the water/wastewater market, it is not uncommon for daily, weekly, quarterly, and annual reports to be required by the appropriate regulatory agencies. All businesses need regular reporting on operating costs, inventory, actual production, asset productivity, and asset utilization. Since a SCADA system collects and processes data from many sources, it is a logical place to generate the reports that require these data. It is a good idea to differentiate between logs and reports. Logs are a chronological accumulation of associated information, sorted by category – generated by the occurrence of conditions that are defined as abnormal or worthy of recording. They are, in effect, an audit trail of monitored conditions. If, in a given time interval (e.g., a day or a week), no occurrences of monitored conditions are detected, then there is nothing in the log. Reports, by contrast, are a predefined set of data and computations based on those data that are to be put into a predefined format (and probably printed) either on a scheduled basis or on an event-triggered basis. When a report is generated, it should not be blank unless problems prevented the collection of the specified data. An empty log may be a good thing; an empty report (one with no data) is not. Figure 7.32 shows a demand log for a given operating area (the user-selected filter criteria) listing control events recorded over a user-specified time interval.

7.6.9.1 Calculated values

Although most reports involve collecting and presenting a predefined set of data, most useful reports call for some level of calculations on the data. An hourly list of water quantities delivered to custody points typically includes a total for each custody point and a total across all delivery points for the day. Addition is a simple mathematical function, but some mechanism is needed to perform the calculation. In some older SCADA systems, all such calculations needed to be created using a separate user-defined calculated point facility. Any computed values needed for report generation had to be defined in the form of analog or binary database points, and these could then be referenced in any subsequent need.

7.6.9.2 Statistical calculations

Frequently, the mathematical processing of data involves statistical calculations, such as computing the mean, the median, or the standard deviation for a set of values. This set of values may actually be the historical set of values for a single measurement, over some selected time interval (e.g., the prior day or week). In these cases, the reporting package needs to be able to interact with the historical trending package to fetch the necessary data points. Again, in some systems, this can be handled in the reporting package, but in others, such computations need to be performed with separate capabilities and made available to the reporting package.

A1 OP	A2 OP	SP	DO	TEMP
23.33	25.00	40.00	58.33	101.00
25.00	21.67	40.00	58.33	100.00
21.67	23.33	40.00	58.33	100.00
21.67	25.00	13.00	58.33	100.00
23.33	21.67	79.00	58.33	41.00
25.00	21.67	12.00	10.00	40.00
21.67	23.33	101.00	10.00	40.00
21.67	25.00	12.00	10.00	40.00
23.33	-5.00	40.00	8.33	13.00
25.00	-5.00	13.00	8.33	12.00
-5.00	-5.00	12.00	8.33	12.00
-5.00	-5.00	12.00	100.00	58.00
-5.00	0.00	100.00	100.00	58.00
-5.00	20.00	41.00	100.00	79.00
0.00	-5.00	101.00	65.00	79.00
-5.00	20.00	100.00	65.00	79.00
20.00	20.00	100.00	65.00	79.00
20.00	21.67	100.00	30.00	79.00
20.00	21.67	41.00	30.00	79.00
21.67	21.67	40.00	53.33	79.00
21.67	1.67	40.00	96.67	79.00
21.67	21.67	12.00	96.67	100.00
1.67	3.33	79.00	96.67	100.00
3.33	21.67	79.00	58.33	41.00
21.67	21.67	79.00	58.33	100.00

FIGURE 7.32

Example of a spreadsheet-based reports

7.6.9.3 Spreadsheet report generators

Commercial spreadsheet software packages are being used for report generation, with the migration of SCADA systems to the Microsoft Windows and Linux operating system platforms. These packages all have several forms of data import mechanism and allow vendor-written add-ins that can be used to connect them to SCADA system data sources (or by collecting and placing such data into a file where the spreadsheet can use its import capability). The computational and output-formatting capabilities of such packages are prodigious and very flexible. These packages can be used to define any reasonable report. Most current SCADA systems make use of (integrate) these commercial spreadsheet packages. Figure 7.32 shows a daily water production report (taken partway through the day) giving hourly breakdowns of volumes and water-quality information. Most commercial relational database packages support an Open Database Connectivity interface as do most commercial spreadsheet packages. SCADA

```
:RDB Release EXP430.1-74.0</DbDesc>
115D622E-ECFB-412A-9298-ECF6D490F06B</DbGUID>
>ENU</DbLangID>


lame>C2_A02_DEH_RD1</BlockName>
Name>C2_A02_DEH_RD1</EntityName>
d>20265526</BlockId>
UID>DFCC5D7E-E050-4C55-A71C-BE47594DDF8A</BlockGUID>
Desc>Exported on (MM-DD-YY HH:MM) 10-23-2013 15:50</BlockDesc>
ateName>SYSTEM:CONTROLMODULE</TemplateName>
lame>CONTROLMODULE</ClassName>
emplateName>SYSTEM:CONTROLMODULE</BaseTemplateName>
Type> </CreateType>
ite>1610613248</Attribute>
>
ft>0</Left>
p>0</Top>
ght>192</Right>
ttom>-120</Bottom>
Order>0</ZOrder>
ordVersion>0</CoordVersion>
l>
cleState>Assigned</LifeCycleState>
edTo>C2_CEETACE_332</AssignedTo>
ner>C63_EHPM_DEH</Container>
>
s>
eter>
ramName>ALIASOPT</ParamName>
ramValue>OFF</ParamValue>
neter>
eter>
ramName>DISCOVORDER</ParamName>
ramValue>"TPN"</ParamValue>
neter>
eter>
ramName>METHODSCOPE</ParamName>
ramValue>"ALL"</ParamValue>
neter>
eter>
ramName>EENAME</ParamName>
ramValue>"C2_CEETACE_332"</ParamValue>
neter>
```


FIGURE 7.33

Example of an XML data file

systems that maintain all of their data, even the real-time data, in relational database tables make it easy to select and fetch the desired data into a report spreadsheet. Most SCADA systems trigger reports at prespecified times and dates, even in nonspecific forms such as last day of the month, second Tuesday of each month, last day of the year, and so on.

7.6.9.4 Reports as data-exchange mechanism

Reports are primarily designed for presenting information to human beings in a format that makes the data understandable. However, reporting packages, particularly spreadsheet-based reporting packages, have also been used to produce data that are intended as input for other programs and applications. It is very easy to create a report with the textual output in the form of a series of numeric and status values, represented as ASCII strings and separated in the output stream by either a comma or a space. This is, of course, the definition of a comma-separated value (CSV) or file, which is commonly used as a mechanism for moving data between spreadsheet packages. A major challenge, for many years, was establishing reliable, automatic, and flexible (reconfigurable) data exchanges between SCADA systems and other systems that needed access to some subset of the SCADA system's data. Complex and messy data-exchange protocols, such as ICCP (also called TASE.2 and UCA1.0), were developed to address this requirement. But these protocols were often expensive and cumbersome to implement. Some SCADA system vendors and customers found that they could use their spreadsheet reporting packages to collect the required data and produce a CSV file, which could automatically be sent to another computer using such readily available IP networking applications as FTP. Developing applications that could parse and utilize such files as their data inputs was normally considered a simple programming task. The problem with such a scheme was that any change required reprogramming.

Lately, the XML data file, another type of ASCII text file, is being used for intersystem data exchange. Most spreadsheet reporting packages are capable of creating XML data files.

Figure 7.33 shows an example of an XML Web page that might be generated either by an IP-ready RTU or through a properly designed and cleverly configured spreadsheet reporting package. A recent extension of the OPC standard is also based on using XML files for data exchange. The XML file can be displayed in a standard Web browser (as seen in Figure 7.33), although that is not the actual goal. The XML document contains both information and descriptive tags (delineated by matching pairs of angle brackets) that help clarify the information; in this example, the RTU is providing its own identification and the time and date corresponding to the data. It then provides two analog and two status values, along with their descriptions, EGUs, and alarm status. If just the data were provided, the file might look like a string of numbers. The descriptive tags, although somewhat wasteful of space and transmission bandwidth, make it clear to a viewer (or an application program) what the data mean and to what each value corresponds. XML files are sometimes called self-descriptive data files since they contain actual data values, plus lots of descriptive information that aids in identifying and interpreting that data.

FURTHER READINGS

Stuart A. Boyer SCADA Supervisory Control and Data Acquisition 3rd edition, ISA.
Shaw, William T., 2006. Cyber Security for SCADA Systems. Pennwell Publishing.
Gordon Clarke, Deon Reynders, Practical Modern Scada Protocols.

CHAPTER

PROGRAMMABLE AUTOMATION CONTROLLER

8

8.1 MODERN INDUSTRIAL APPLICATION

It is challenging to implement a modern industrial application as it sometimes comes with a daunting mix of requirements. For example, a typical control system must be able to interface with signals from simple sensors and actuators, yet for many modern applications this is merely the starting point. Modern industrial applications often need capabilities such as advanced control features, network connectivity, device interoperability, and enterprise data integration. These modern requirements extend far beyond the traditional discrete-logic-based control of input/output (I/O) signals handled by a programmable logic controller (PLC). Most PLCs are programmed using ladder logic, which has its origins in the wiring diagrams used to describe the layout and connections of discrete physical relays and timers in a control system. Applications that diverge from or expand beyond this model become increasingly hard to program in ladder logic. For example, mathematically complex applications such as proportional–integral–derivative (PID) loops used for temperature control involve floating-point arithmetic. PLCs must often be enhanced with separate – and separately programmed – hardware cards to perform these operations.

8.1.1 MAKING A PLC MORE LIKE A PC

It is a challenge to use a PLC to meet modern application requirements for network connectivity, device interoperability, and enterprise data integration. These types of tasks are usually more suited to the capabilities of a computer (PC). To provide these capabilities in a PLC-based application, additional processors, network gateways or converters, "middleware" software running on a separate PC, and special software for enterprise systems must often be integrated into the system.

A programmable automation controller (PAC) is a compact controller that combines the features and capabilities of a PC-based control system with that of a typical PLC.

- PLC feel
  - Modular footprint, industrial reliability
  - Wide array of I/O modules and system configurations
- PC power
  - Large memory and high-speed processing
  - High-level data handling and enterprise connectivity
  - Extensive communications capability, multiple protocols, and field networks.

PACs are most often used in industrial settings for process control, data acquisition, remote equipment monitoring, machine vision, and motion control. Additionally, because PACs function and communicate over network interface protocols such as TCP/IP, OLE for process control (OPC) and SMTP,

PACs are able to transfer data from the machines they control to other machines and components in a networked control system or to application software and databases.

PACs are most often used for complex machine control, advanced process control, data acquisition, and equipment monitoring and motion control. The term PAC was given by ARC (Automation Research Corporation). A PAC offers the following features:

1. Helps users of automated hardware to define the applications they need.
2. Gives the vendors a term to effectively communicate the characteristics and abilities of their product.

ARC also made and explained a few rules or guidelines for a device to be considered as a programmable automation controller.

- Single platform: The device must operate using a single platform. It should be true for single or multiple domains and in drives, motions, and process controls.
- Single development platform: The device must employ a single development platform, using a single database for different tasks in all the disciplines.
- The device must tightly integrate controller hardware and software.
- The device must be programmable by using software tools that can design control programs to support a process that flows across several machines or units.
- The device must operate on open, modular architectures that mirror industry applications, from machine layouts in factories to unit operation in process plants.
- The device must employ *de facto* standards for network interfaces, languages, and protocols, allowing data exchange as part of networked multivendor systems.
- The device must provide efficient processing and I/O scanning.

8.1.1.1 Functional benefits

The characteristics used to define a PAC also explain the benefits that can be obtained from its industrial installation and application. A PAC can meet complex requirements without the need for additional components such as a PLC. Improved control system performance is experienced through high integration of hardware and software. Integrated development environment (IDE), which is used in the manufacturing of a PAC, uses a tag name database that is used and shared by all the development tools.

A PAC only needs one software package to cover all the existing automation needs and the ones that may arise in the future and does not need utilities from different vendors. The control systems can be upgraded easily. Owing to its compact size, a PAC uses lesser space compared to other options.

8.1.1.2 Financial benefits

PACs offer multiple financial advantages. The overall cost of the control system is lowered because hardware is less expensive and less development and integration time is needed. Deploying a PAC is often more affordable than augmenting a PLC to have similar capabilities. There is also an increased return on assets, reduced lifecycle costs, and lower total cost of ownership (TCO) due to extending an automation system's range of applications (also known as its domain expertise). The ability to add I/O as separate modules means that a minimum number of modules needed for initial development can be used during design, and the remaining modules added toward the end of the project.

8.1.2 PAC VERSUS PLC

Generally, PACs and PLCs serve the same purpose. Both are primarily used to perform automation, process control, and data acquisition functions such as digital and analog control, serial string handling, PID, motion control, and machine vision. The parameters within which PACs operate to achieve this, however, sometimes run counter to how a PLC functions.

Unlike PLCs, PACs offer open, modular architectures, the rationale being that because most industrial applications are customized, the control hardware used for them needs to allow engineers to pick and choose other components in the control system architecture without having to worry about compatibility with the controller.

PACs and PLCs are also programmed differently. PLCs are often programmed in ladder logic, a graphical programming language resembling the rails and rungs of ladders that is designed to emulate old electrical relay wiring diagrams. PAC control programs are usually developed with more generic software tools that permit the designed program to be shared across several different machines, processors, HMI terminals, or other components in the control system architecture.

8.1.3 APPLICATION OF PAC

PACs meet the complex demands of modern industrial automation applications because they combine features of more traditional automation technologies such as PLCs, distributed control systems (DCSs), remote terminal units (RTUs), and personal computers (PCs).

8.1.4 APPLYING THE PAC TO A MODERN INDUSTRIAL APPLICATION

Let us look more closely at how a PAC is applied to a modern industrial application.

8.1.4.1 Single platform operating in multiple domains

The single PAC can be used for operating in multiple domains to monitor and manage a production line, a chemical process, a test bench, and shipping activities. To do so, the PAC must simultaneously manage analog values such as temperatures and pressures; digital on/off states for valves, switches, and indicators; and serial data from inventory tracking and test equipment. At the same time, the PAC is exchanging data with an OLE for process control (OPC) server, an operator interface, and a SQL (structured query language) database. Only PAC is capable of handling these tasks at the same time without need for additional processors, gateways, or middleware.

8.1.4.2 Support for standard communication protocols

The PAC, operator, and office workstations; testing equipment, production line, and process sensors and actuators; and barcode reader are connected to a standard 10/100 Mbps Ethernet network installed throughout the facility. In some instances, devices without built-in Ethernet connectivity, such as temperature sensors, are connected to I/O modules on an intermediate Ethernet-enabled I/O unit, which in turn communicates with the PAC.

Using this Ethernet network, the PAC communicates with remote racks of I/O modules to read/write analog, digital, and serial signals. The network also links the PAC with an OPC server, an operator interface, and a SQL database. A wireless segment is part of the network, so the PAC can also communicate with mobile assets such as the forklift and temporary operator workstation.

The PAC can control, monitor, and exchange data with this wide variety of devices and systems because they use a common standard network technology and protocol. This example includes wired and wireless Ethernet networks, Internet protocol (IP) network transport, OPC, and SQL. In another control situation, common application-level protocols such as Modbus®, SNMP (simple network management protocol), and PPP (point-to-point protocol) over a modem could be required. The PAC has the ability to meet these diverse communication requirements.

8.1.4.3 Data exchange with enterprise systems

The PAC exchanges manufacturing, production, and inventory data with an enterprise SQL database. This database in turn shares data with several key business systems, including an enterprise resource planning (ERP) system, operational equipment effectiveness (OEE) system, and supply chain management (SCM) system. The PAC constantly and automatically updates data from the factory floor, ensuring availability of timely and valuable information for all business systems.

8.1.5 SOFTWARE FOR PACs

Because a key defining characteristic of PACs is that the same hardware can be used in multiple domains, including logic, motion, drives, and process control, it follows that the software must be capable of programming all control and monitoring tasks that must be done in multiple domains.

The PAC software must handle discrete control, process control, motion control, remote monitoring, and data acquisition. And the software must let the developer mix and incorporate these as needed into control programs, so these programs can "flow" as the requirements of the application dictate.

For example, the following are typical requirements for a small production facility with a combination of process and discrete control (e.g., micro brewery) for producing the end product (Figure 8.1):

- Water is piped in from a spring a couple of miles away, so you need to monitor the pressure and flow of that water and security at the spring (remote monitoring using analog and digital devices).
- You measure water quality as it enters your facility, track these data over time, and store it in your company database (data acquisition, database connectivity).
- There may be more than one microbrew, so recipes, temperatures, and processing must vary (batch process control, PID loop control, and distributed control).
- Operator interfaces mimic the process, providing secure interactive controls for technicians and operators.
- Quality control is ensured by testing all products at several stages. Quality data are kept as required by government health authorities (monitoring, more data acquisition, and database connectivity).
- In another building, the bottling line requires discrete control. As bottles come off the line, they are boxed and identified with radio-frequency identification (RFID) tags, then sent to shipping.
- In the separate shipping area, boxed stock automatically moves via conveyors (discrete control) based on RFID tags (serial device connectivity).
- Temperatures in the storage area are controlled and monitored. Energy usage is monitored and building systems are controlled throughout all buildings (remote I/O, distributed intelligence).
- Production and inventory data go directly from machines and barcode readers to company computers; customer and shipping data flow in the opposite direction (database connectivity).

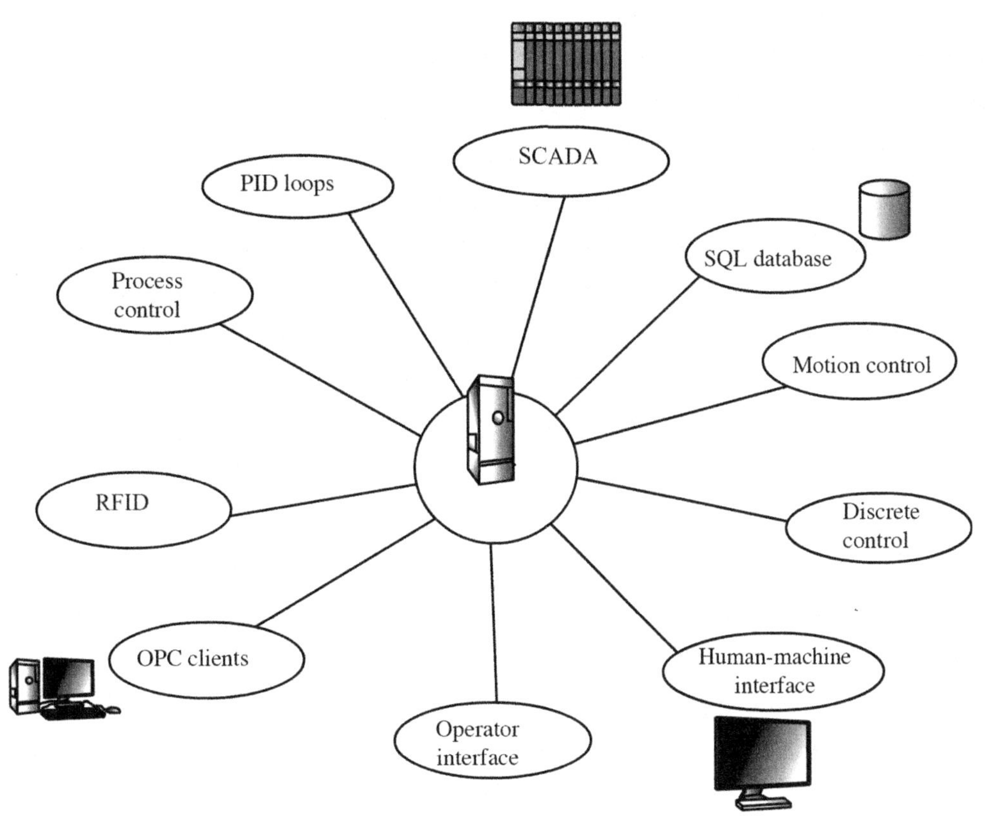

FIGURE 8.1

PAC Integration with multiple systems

This microbrewery is just an example of how several different types of control in different domains are required by a modern industrial automation application. Most industrial applications today are similarly varied. While the number of PACs needed depends on application requirements, each PAC can be used in any domain or in multiple domains. Because the application requires processes that flow into each other over space and time, the PAC software accommodates that flow and integrates these multiple domains into one system.

I/O points and variables defined while building a control program (called a control strategy) are stored in a single tag name database. When you open the HMI development, OPC server, or database connectivity software, those defined items are immediately available for use.

In summary, a PAC provides advanced control features, network connectivity, device interoperability, and enterprise data integration capabilities found in PLC- or PC-based automation controllers, all in a single compact controller. These features make the PAC essential for meeting the new and diverse requirements demanded in a modern industrial application.

FURTHER READINGS

The Future of Control by IEE Manufacturing Engineer August/September 2005.
"PACs for Industrial Control, the Future of Control" Tutorial, National Instruments (Feb. 28, 2012).
"Programmable Automation Controller" Information Presented in Wikipedia.
New Generation Programmable Automation Controller. ETHERNET DIRECT.
Columba Sara Evelyn (Ed.). Programmable Automation Controller.

CHAPTER

SERIAL COMMUNICATIONS

9

9.1 RS232 OVERVIEW

Serial communications is used extensively within the electronics industry due to its relative simplicity and low hardware overhead (as compared to parallel interfacing). EIA/TIA-232-E specification is the most popular serial communications standard. This standard, developed by the Electronic Industry Association and the Telecommunications Industry Association (EIA/TIA), is more popularly referred to simply as "RS232" where "RS" stands for "recommended standard." In recent years, this suffix has been replaced with "EIA/TIA" to help identify the source of the standard. This book uses the common notation of "RS232."

The EIA-232 interface standard was developed for interfacing data terminal equipment (DTE) and data circuit terminating equipment (DCE) (earlier called as Data Communication Equipment) employing serial binary data interchange. EIA-232 was particularly developed for interfacing data terminals to modems. The engineering department of the EIA issued the EIA-232 interface standard in the United States in 1969.

The first layers of the OSI model include functions of the physical, data link, and network layers. It is important to understand these layers in order to comprehend the various other protocols. Let us begin by examining the first level of interaction with the transmission medium itself – the physical layer and its specifications.

9.2 RS232 SIGNAL INFORMATION

Most equipments using RS232 serial ports use a DB25 type connector even if the original documents didn't specify a specific connector; many PCs today use DB9 connectors since all you need in asynchronous mode is nine signals. The document does specify the amount of pins and their assignment; 20 assigned to different signals, 3 are reserved, and 2 are not affected. Normally the male connector is on the DTE side and the female connector is on the DCE side even if this is not always the case (Figure 9.1).

9.2.1 SIGNAL STATE VOLTAGE ASSIGNMENTS

Voltages of −3 V to −25 V with respect to signal ground (pin 7) are considered logic "1" (the marking condition), whereas voltages of +3 V to +25 V are considered logic "0" (the spacing condition). The range of voltages between −3 V and +3 V is considered a transition region for which a signal state is not assigned (Figure 9.2).

In serial communications the terminal end (PC) is called the DTE and the modem end is called the data communications equipment (DCE) as illustrated in Figure 9.3.

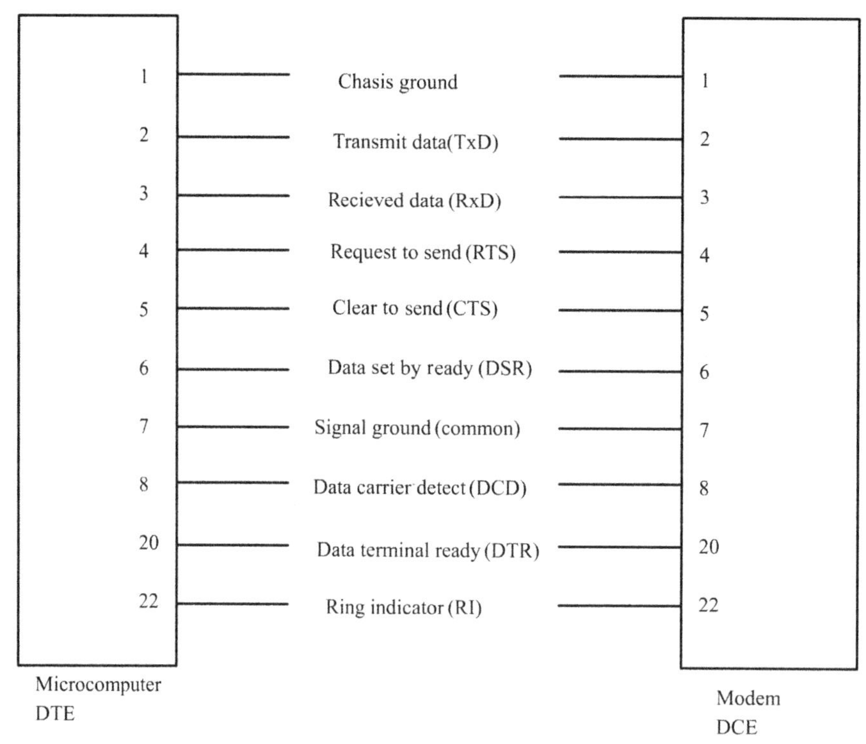

FIGURE 9.1

Example of a serial communication connections

RS232 signals have a direction (in or out) depending on whether they are with respect to a DTE or a DCE. In all the following pinout diagrams the signal direction is with respect to the DTE (PC) end.

9.2.2 NULL MODEM CONNECTION

When PCs are connected back-to-back, each end acts as a DTE (there is no DCE in this case) and consequently certain signals may have to be looped in the connection to satisfy any input signal requirement. This is called a NULL (no) modem configuration. For example, when the DTE raises Request to Send (RTS), it typically expects Clear to Send (CTS) from the DCE. Since there is no DCE to raise CTS, the outgoing RTS signal is looped in the NULL modem cable to the incoming CTS to satisfy the DTE's need for this signal, as illustrated in Figure 9.4.

9.2.3 DB9 AND DB25 MALE AND FEMALE PIN NUMBERING

These following diagrams show the male (gray background) and female (black background) pin numbering for DB9 and DB25 subminiature connectors. Generally Pin 1 is marked on the front of the connector right next to the pin, although you may need a magnifying glass to read it. Some manufacturers mark each pin number on the plastic housing at the rear of the connector. The male connector has the pins sticking out! (Figures 9.5–9.8).

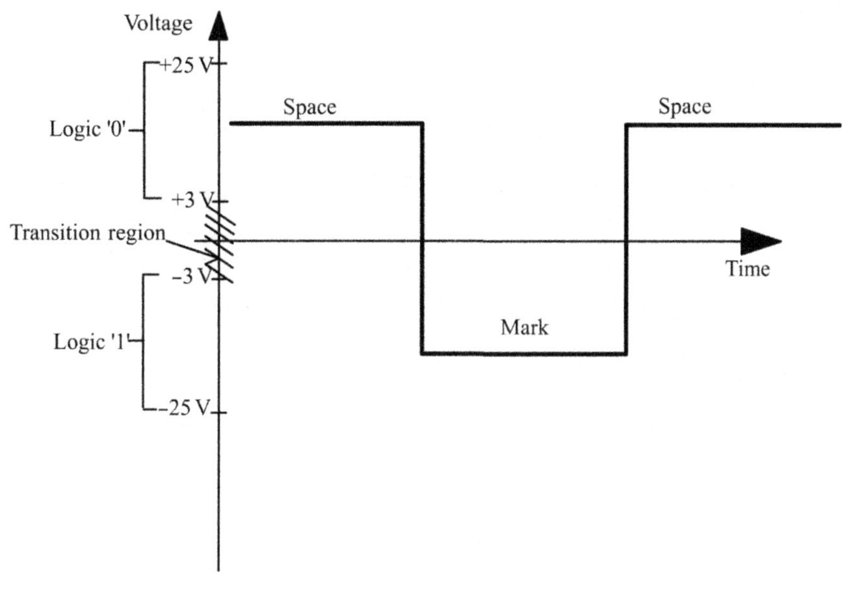

FIGURE 9.2

Signal Voltage State

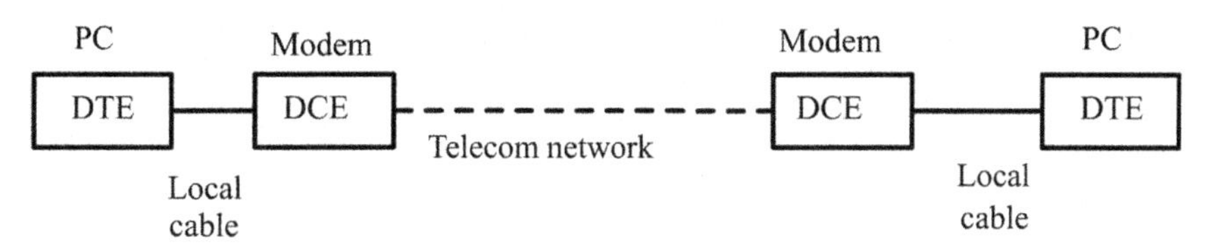

FIGURE 9.3

Serial communications with a modem

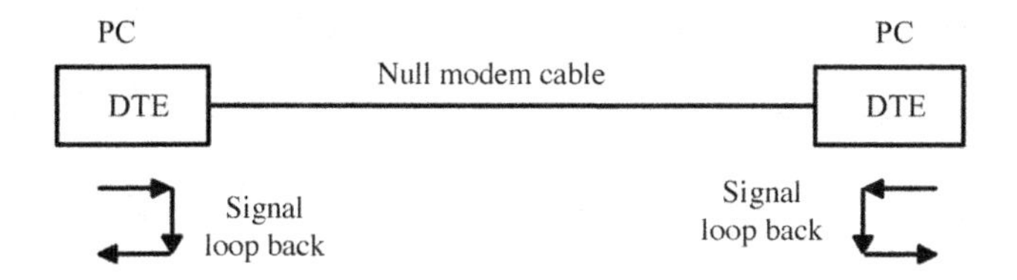

FIGURE 9.4

Serial communications with a NULL modem configuration

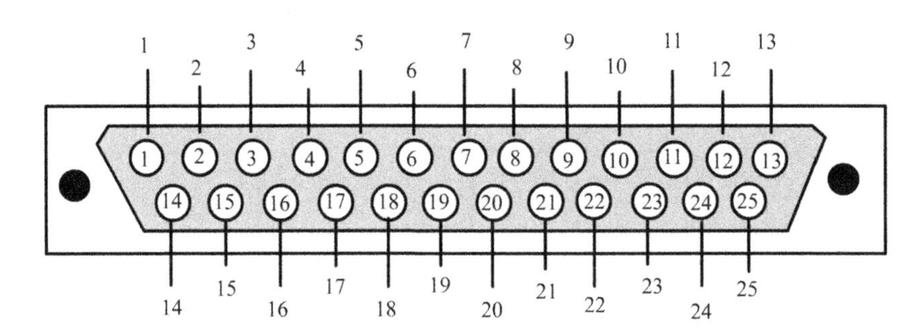

FIGURE 9.5

DB25: View looking into male connector

9.2.4 RS232 ON DB25 (RS232C)

The RS232 DB25 connector is capable of supporting two separate connections – each with its own optional clock when used in synchronous mode or bit-synchronous mode. If the interface is purely used for asynchronous communications then you only need those marked with (ASYNC) in Table 9.1 or you

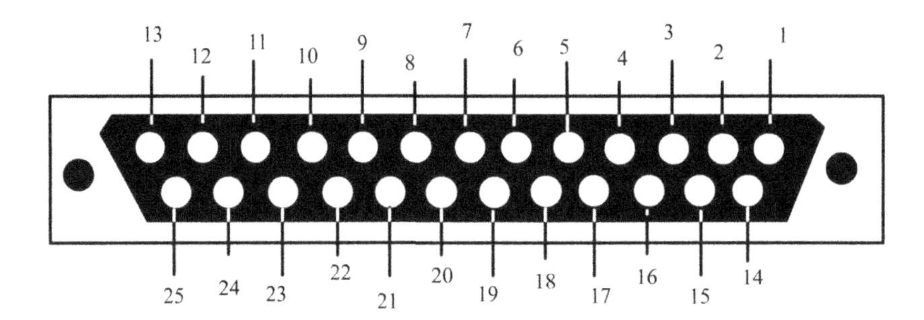

FIGURE 9.6

DB25: View looking into female connector

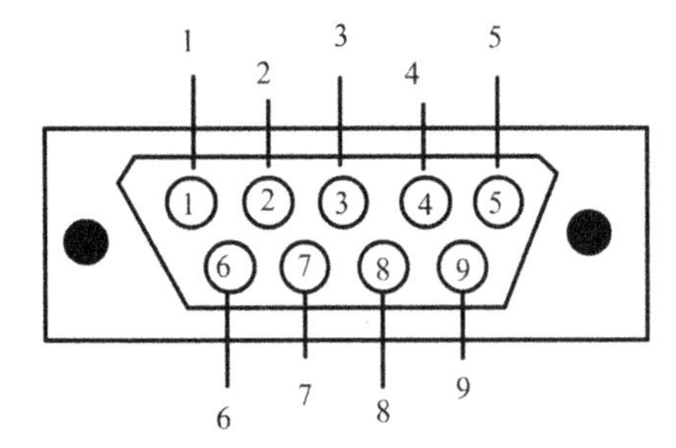

FIGURE 9.7

DB9: View looking into male connector

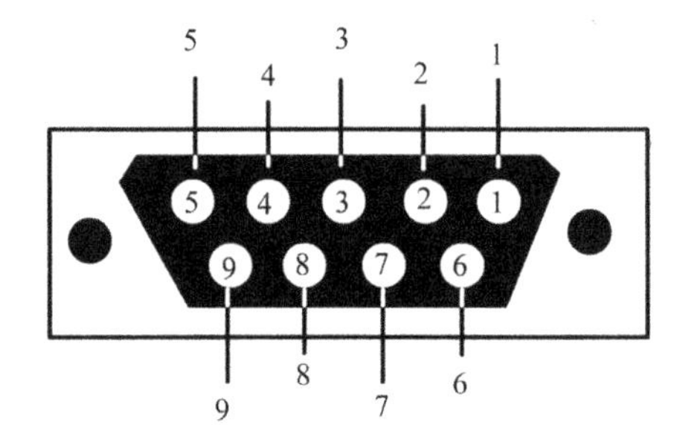

FIGURE 9.8

DB9: View looking into female connector

Table 9.1 Signal/Pin information on DB25

Pin No.	Name	Dir.	Notes/Description
1	–	–	Protective/shielded ground
2	TD	OUT	Transmit data (a.k.a. TxD, Tx) (ASYNC)
3	RD	IN	Receive data (a.k.a. RxD, Rx) (ASYNC)
4	RTS	OUT	Request to Send (ASYNC)
5	CTS	IN	Clear to Send (ASYNC)
6	DSR	IN	Data Set Ready (ASYNC)
7	SGND	–	Signal Ground
8	CD	IN	Carrier Detect (a.k.a. DCD)
9	–	–	Reserved for data set testing.
10	–	–	Reserved for data set testing
11	–	–	Unassigned
12	SDCD	IN	Secondary Carrier Detect. Only needed if second channel being used.
13	SCTS	IN	Secondary Clear to Send. Only needed if second channel being used.
14	STD	OUT	Secondary Transmit Data. Only needed if second channel being used.
15	DB	OUT	Transmit Clock (a.k.a. TCLK, TxCLK). Synchronous use only.
16	SRD	IN	Secondary Receive Data. Only needed if second channel being used.
17	DD	IN	Receive Clock (a.k.a. RCLK). Synchronous use only.
18	LL	–	Local Loopback
19	SRTS	OUT	Secondary Request to Send. Only needed if second channel being used.
20	DTR	OUT	Data Terminal Ready (ASYNC)
21	RL/SQ	–	Signal Quality Detector/Remote Loopback
22	RI	IN	Ring Indicator. DCE (modem) raises when incoming call detected used for auto answer applications.
23	CH/CI	OUT	Signal Rate Selector.
24	DA	–	Auxiliary Clock (a.k.a. ACLK). Secondary channel only.
25	–	–	Unassigned

can use even fewer (if you understand what is happening). The column marked Dir shows the signal direction with respect to the DTE (Figure 9.9).

9.2.5 RS232 ON DB9 (EIA/TIA 574)

The column marked Dir in Table 9.2 shows the signal direction with respect to the DTE (Figure 9.10).

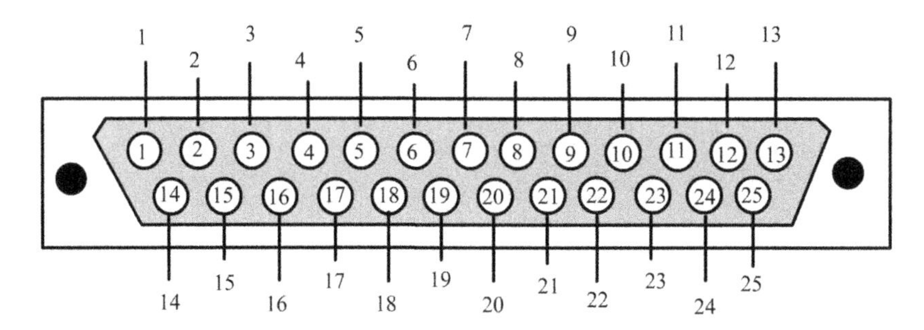

FIGURE 9.9

View – looking into male connector

9.2.6 THE MAJOR ELEMENTS OF EIA-232

The EIA-232 standard consists of three major parts, which define:

- Electrical signal characteristics
- Mechanical characteristics of the interface
- Functional description of the interchange circuits

9.2.6.1 Electrical signal characteristics

EIA-232 defines electrical signal characteristics such as the voltage levels and grounding characteristics of the interchange signals and associated circuitry for an unbalanced system.

The EIA-232 transmitter is required to produce voltages in the range ±15 to ±25 V as follows:

- Logic 1: –5 V to –25 V
- Logic 0: +5 V to +25 V
- Undefined logic level: +5 V to –5 V

Table 9.2 Signal/Pin information in DB9

Pin No.	Name	Dir.	Notes/Description
1	DCD	IN	Data Carrier Detect. Raised by DCE when modem synchronized.
2	RD	IN	Receive Data (a.k.a. RxD, Rx). Arriving data from DCE.
3	TD	OUT	Transmit Data (a.k.a. TxD, Tx). Sending data from DTE.
4	DTR	OUT	Data Terminal Ready. Raised by DTE when powered on. In autoanswer mode raised only when RI arrives from DCE.
5	SGND	–	Ground
6	DSR	IN	Data Set Ready. Raised by DCE to indicate ready.
7	RTS	OUT	Request to Send. Raised by DTE when it wishes to send. Expects CTS from DCE.
8	CTS	IN	Clear to Send. Raised by DCE in response to RTS from DTE.
9	RI	IN	Ring Indicator. Set when incoming ring detected – used for autoanswer application. DTE raised DTR to answer.

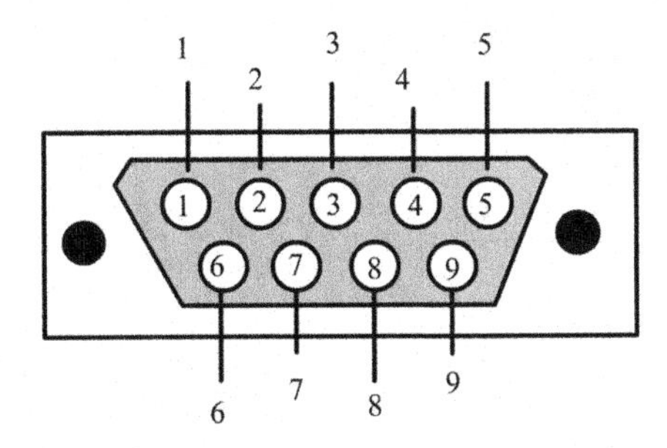

FIGURE 9.10

DB9 (EIA/TIA 574): View – looking into male connector

At the EIA-232 receiver, the following voltage levels are defined:

- Logic 1: –3 V to –25 V
- Logic 0: +3 V to +25 V
- Undefined logic level: –3 V to +3 V

Note: The EIA-232 transmitter requires a slightly higher voltage to overcome voltage drop along the line.

The voltage levels associated with a microprocessor are typically 0 V to +5 V for transistor–transistor logic (TTL). A line driver is required at the transmitting end to adjust the voltage to the correct level for the communications link. Similarly, a line receiver is required at the receiving end to translate the voltage on the communications link to the correct TTL voltages for interfacing to a microprocessor. Despite the bipolar input voltage, TTL compatible EIA-232 receivers operate on single +5 V supply.

Modern PC power supplies usually have a standard +12 V output that could be used for the line driver.

The control or "handshaking" lines have the same range of voltages as transmission of logic 0 and logic 1, except that they are of opposite polarity. This means that:

- A control line asserted or made active by the transmitting device has a voltage range of +5 V to +25 V. The receiving device connected to this control line allows a voltage range of +3 V to +25 V.
- A control line inhibited or made inactive by the transmitting device has a voltage range of −5 V to +25 V. The receiving device of this control line allows a voltage range of −3 V to −25 V (Figure 9.11).

At the receiving end, a line receiver is necessary in each data and control line to reduce the voltage level to the 0 V and +5 V logic levels required by the internal electronics (Figure 9.12).

The EIA-232 standard defines 25 electrical connections. The electrical connections are divided into four groups, namely:

- Data lines
- Control lines
- Timing lines
- Special secondary functions

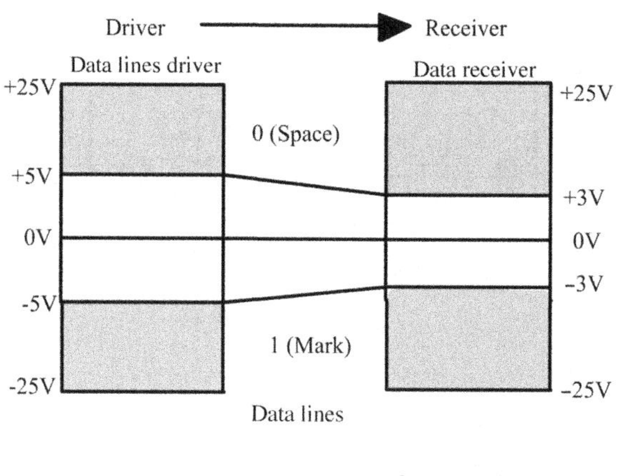

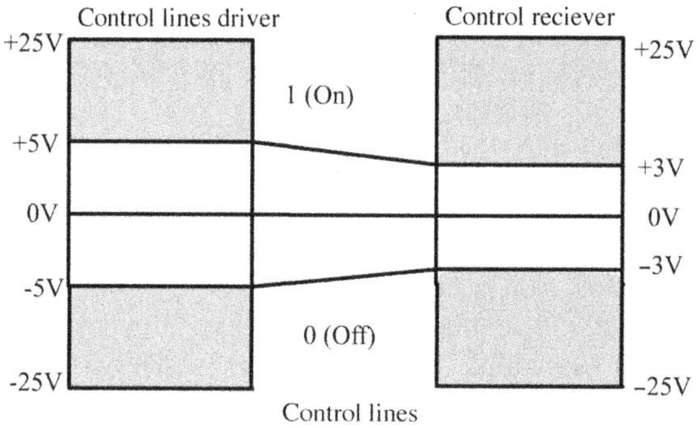

FIGURE 9.11

Voltage levels for EIA-232

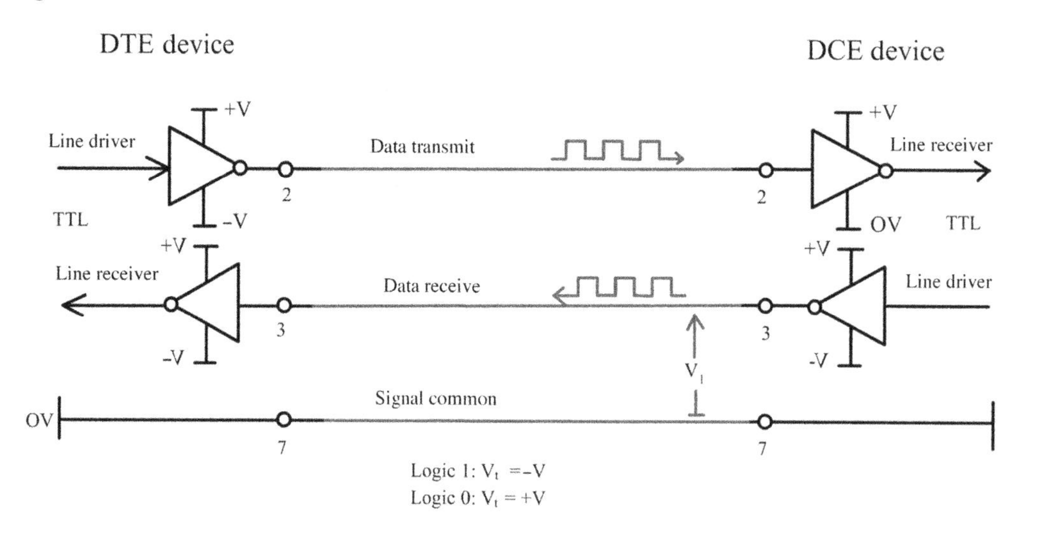

FIGURE 9.12

EIA-232 transmitter and receiver

Data lines are used for the transfer of data. Data flow is designated from the perspective of the DTE interface. The transmit line, on which the DTE transmits and the DCE receives, is associated with pin 2 at the DTE end and pin 2 at the DCE end for a DB25 connector. These allocations are reversed for DB9 connectors. The receive line, on which the DTE receives and the DCE transmits, is associated with pin 3 at the DTE end and pin 3 at the DCE end. PIN 7 is the common return line for the transmit and receive data lines.

Control lines are used for interactive device control, which is commonly known as hardware handshaking. They regulate the way in which data flows across the interface. The four most commonly used control lines are as follows:

- RTS: Request to Send
- CTS: Clear to Send
- DSR: Data Set Ready (or DCE ready in EIA-232D/E)
- DTR: Data Terminal Ready (or DTE ready in EIA-232D/E)

It is important to remember that with the handshaking lines, the enabled state means a positive voltage and the disabled state means a negative voltage.

Hardware handshaking is the cause of most interfacing problems. Manufacturers sometimes omit control lines from their EIA-232 equipment or assign unusual applications to them. Consequently, many applications do not use hardware handshaking but, instead, use only the three data lines (transmit, receive, and signal common ground) with some form of software handshaking. The control of data flow is then part of the application program. Most of the systems encountered in data communications for instrumentation and control use some sort of software-based protocol in preference to hardware handshaking.

There is a relationship between the allowable speed of data transmission and the length of the cable connecting the two devices on the EIS-232 interface. As the speed of data transmission increases, the quality of the signal transition from one voltage level to another, for example, from −25 V to +25 V, becomes increasingly dependent on the capacitance and inductance of the cable.

The rate at which voltage can "slew from one logic level to another depends mainly on the cable capacitance and the capacitance increases with cable length. The length of the cable is limited by the number of data errors acceptable during transmission. The EIA-232 D& E standard specifies the limit of total cable capacitance to be 2500 pF. With typical cable capacitance having improved from around 160 pF/m to only 50 pF/m in recent years, the maximum cable length has extended from 15 m (50 ft) to about 50 m (166 ft).

The common data transmission rates used with EIA-232 are 110, 300, 600, 1200, 2400, 4800, 9600, and 19,200 bps. For short distances, however, transmission rates of 38,400; 57,600; and 115,200 can also be used. Based on field tests, Table 9.3 shows the practical relationship between selected baud rates and maximum allowable cable length, indicating that much longer cable lengths are possible at lower baud rates. The achievable speed depends on the transmitter voltages cable capacitance as well as the noise environment.

In the context of the NRZ-type of coding used for asynchronous transmission on EIA-232 links, 1 baud = 1 bit per second.

9.2.7 ELECTRICAL CHARACTERISTICS

The electrical characteristics section of the RS232 standard specifies voltage levels, rate of change for signal levels, and line impedance. As the original RS232 standard was defined in 1962 and before the days of TTL logic, it is no surprise that the standard does not use 5 V and ground logic levels. Instead,

Table 9.3 Demonstrated maximum cable lengths with EIA-232 interface

Baud Rate	Cable Length (m)
110	850
300	800
600	700
1,200	500
2,400	200
4,800	100
9,600	70
19,600	50
115 k	20

a high level for the driver output is defined as between +5 V and +15 V, and a low level for the driver output is defined as between −5 V and −15 V. The receiver logic levels were defined to provide a 2 V noise margin. As such, a high level for the receiver is defined as between +3 V and +15 V, and a low level is between −3 V and −15 V.

Figure 9.13 illustrates the logic levels defined by the RS232 standard. It is necessary to note that, for RS232 communication, a low level (−3 V to −15 V) is defined as a logic 1 and is historically referred to as "marking." Similarly, a high level (+3 V to +15 V) is defined as a logic 0 and is referred to as "spacing.")

The RS232 standard also limits the maximum slew rate at the driver output. This limitation was included to help reduce the likelihood of crosstalk between adjacent signals. The slower the rise and fall time, the less the chance of crosstalk. With this in mind, the maximum slew rate allowed is 30 V/ms. Additionally, the standard defines a maximum data rate of 20 kbps, again to reduce the chance

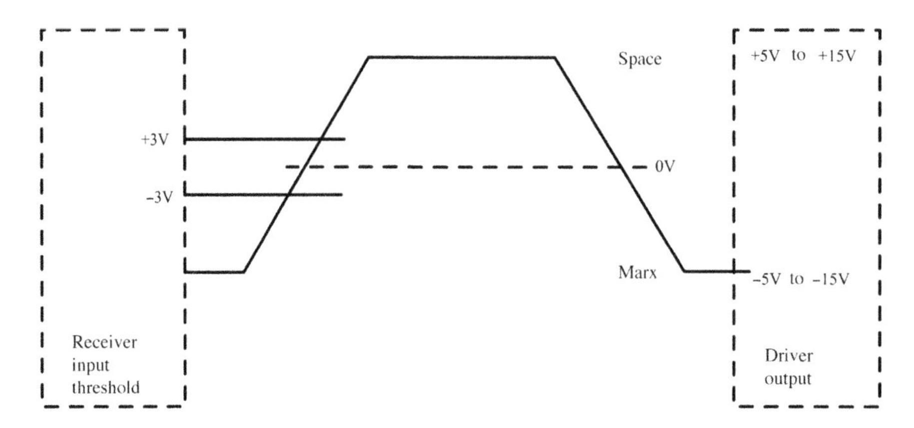

FIGURE 9.13

RS232 logic-level specifications

Table 9.4 RS232 specifications

Cabling Single	Ended
Number of devices	1 transmit, 1 receive
Communication mode	Full duplex
Distance (max)	50 ft at 19.2 kbps
Data rate (max)	1 Mbps
Signaling	Unbalanced
Mark (data 1)	−5 V (min), −15 V (max)
Space (data 0)	5 V (min), 15 V (max)
Input level (min)	±3 V
Output current	500 mA (note that the driver ICs normally used in PCs are limited to 10 mA)
Impedance	5 k (Internal)
Bus architecture	Point-to-point

of crosstalk. The impedance of the interface between the driver and receiver has also been defined. The load seen by the driver is specified at 3–7 k. In the original RS232 standard, the cable length between the driver and the receiver was specified to be 15 m maximum. Revision "D" (EIA/TIA-232-D) changed this part of the standard. Instead of specifying the maximum length of cable, the standard specified a maximum capacitive load of 2500 pF, clearly a more adequate specification. The maximum cable length is determined by the capacitance per unit length of the cable, which is provided in the cable specifications. Table 9.4 summarizes the electrical specifications in the current standard.

9.2.8 FUNCTIONAL CHARACTERISTICS

RS232 is a complete standard, and includes more than just specifications on electrical characteristics. The standard also addresses the functional characteristics of the interface, #2 on the previous list. This essentially means that RS232 defines the function of the different signals used in the interface. These signals are divided into four different categories: common, data, control, and timing. The standard provides abundant control signals and supports a primary and secondary communications channel. All these defined signals are rarely required at a time by a few applications. For example, only eight signals are used for a typical modem. The complete list of defined signals is included in Figure 9.14 as a reference.

9.2.8.1 RS232 in modem applications

Modem applications are one of the most popular uses for the RS232 standard. Figure 9.15 illustrates a typical modem application. In the diagram, the PC is the DTE and the modem is the DCE. Communication between each PC and its associated modem is accomplished using the RS232 standard. Communication between the two modems is accomplished through telecommunication. It should be noted that, although a microcontroller is usually the DTE in RS232 applications, this is not mandated by a strict interpretation of the standard.

Circuit pneumonic	Circuit name	Circuit direction	Circuit type
AB	Signal common		Common
BA	Transmitted data (TD)	To DCE	Data
BB	Received data (RD)	From DCE	
CA	Request to send (RTS)	To DCE	Control
CB	Clear to send (CTS)	From DCE	
CC	DCE ready (DSR)	From DCE	
CD	DTE ready (DTR)	To DCE	
CE	Ring indicator(RI)	From DCE	
CF	Received line signal detector (DCD)	From DCE	
CG	Signal quality detector	From DCE	
CH	Data signal rate detector from DTE	To DCE	
CI	Data signal rate detector from DCE	From DCE	
CJ	Ready for receiving	To DCE	
RL	Remote loop back	To DCE	
LL	Local loop back	To DCE	
TM	Test mode	From DCE	
DA	Transmitter signal element timing from DTE	To DCE	Timing
DB	Transmitter signal element timing from DCE	From DCE	
DD	Receiver signal element timing from DCE	From DCE	
SBA	Secondary transmitted data	To DCE	Data
SBB	Secondary received data	From DCE	
SCA	Secondary request to send	To DCE	Control
SCB	Secondary clear to send	From DCE	
SCF	Secondary received line signal detector	From DCE	

FIGURE 9.14

RS232 connector pin assignments. RS232 defined signals

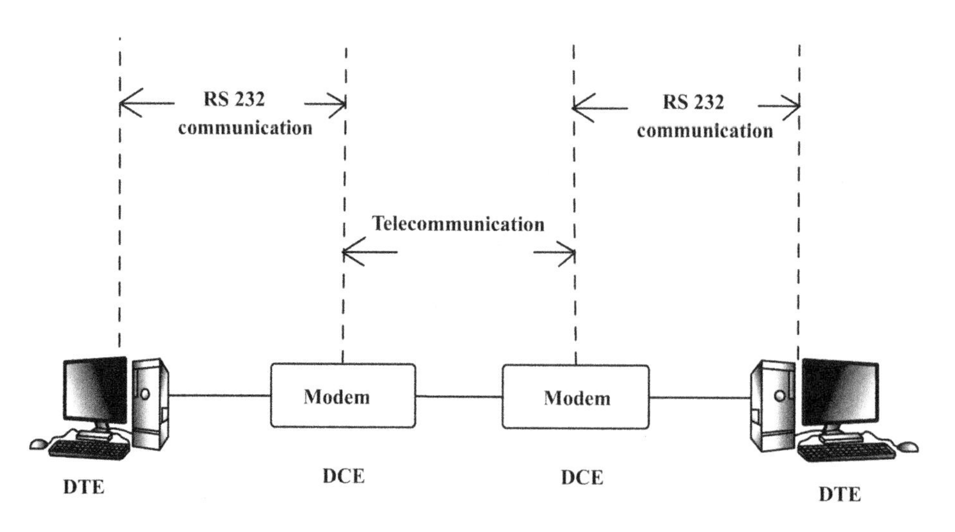

FIGURE 9.15

Modem communication between two PCs

Although some designers choose to use a 25-pin connector for this application, it is not necessary as there are only nine interface signals (including ground) between the DTE and DCE. With this in mind, many designers use 9- or 15-pin connectors. The "basic nine" signals used in modem communication are illustrated in the previous figure; three RS232 drivers and five receivers are necessary for the DTE. The functionality of these signals is described as follows. Note that for the following signal descriptions, ON refers to a high RS232 voltage level (+5 V to +15 V), and OFF refers to a low RS232 voltage level (−5 V to −15 V). Keep in mind that a high RS232 voltage level actually represents a logic 0, and that a low RS232 voltage level refers to a logic 1.

- Transmitted Data (TD): One of two separate data signals, this signal is generated by the DTE and received by the DCE.
- Received Data (RD): The second of two separate data signals, this signal is generated by the DCE and received by the DTE.
- RTS: When the host system (DTE) is ready to transmit data to the peripheral system (DCE), RTS is turned ON. In simplex and duplex systems, this condition maintains the DCE in receive mode. In half-duplex systems, this condition maintains the DCE in receive mode and disables transmit mode. The OFF condition maintains the DCE in transmit mode. After RTS is asserted, the DCE must assert CTS before communication can commence.
- CTS: CTS is used along with RTS to provide handshaking between the DTE and the DCE. After the DCE sees an asserted RTS, it turns CTS ON when it is ready to begin communication.
- DSR: This signal is turned on by the DCE to indicate that it is connected to the telecommunications line.
- Data Carrier Detect (DCD): This signal is turned ON when the DCE is receiving a signal from a remote DCE, which meets its suitable signal criteria. This signal remains ON as long as a suitable carrier signal can be detected.
- DTR: DTR indicates the readiness of the DTE. This signal is turned ON by the DTE when it is ready to transmit or receive data from the DCE. DTR must be ON before the DCE can assert DSR.
- Ring Indicator (RI): RI, when asserted, indicates that a ringing signal is being received on the communications channel.

The signals described form the basis for modem communication. Perhaps the best way to understand how these signals interact is to examine a step-by-step example of a modem interfacing with a PC. The following steps describe a transaction in which a remote modem calls a local modem.

1. The local PC uses software to monitor the RI signal.
2. When the remote modem wants to communicate with the local modem, it generates an RI signal. This signal is transferred by the local modem to the local PC.
3. The local PC responds to the RI signal by asserting the DTR signal when it is ready to communicate.
4. After recognizing the asserted DTR signal, the modem responds by asserting DSR after it is connected to the communications line. DSR indicates to the PC that the modem is ready to exchange further control signals with the DTE to commence communication. When DSR is asserted, the PC begins monitoring DCD for an indication that data is being sent over the communication line.
5. The modem asserts DCD after it has received a carrier signal from the remote modem that meets the suitable signal criteria.

6. At this point data transfer can begin. If the local modem has full-duplex capability, the CTS and RTS signals are held in the asserted state. If the modem has only half duplex capability, CTS and RTS provide the handshaking necessary for controlling the direction of the data flow. Data is transferred over the RD and TD signals.
7. On completing the transfer of data, the PC disables the DTR signal. The modem follows by inhibiting the DSR and DCD signals. At this point, the PC and modem are in the original state described in step 1.

9.3 LIMITATIONS OF RS232 APPLICATIONS

The electronics industry has evolved immensely in the over four decades since the RS232 standard was introduced. There are, therefore, some limitations in the RS232 standard. One limitation – the fact that over 20 signals have been defined by the standard has already been addressed. Designers simply do not use all the signals or the 25-pin connector. Other limitations in the standard are not necessarily as easy to correct.

9.3.1 GENERATION OF RS232 VOLTAGE LEVELS

RS232 does not use the conventional 0 and 5 V levels implemented in TTL and CMOS designs. Drivers have to supply +5 V to +15 V for a logic 0 and −5 V to −15 V for a logic 1. This implies the need for extra power supply to drive the RS232 voltage levels. Typically, a +12 V and a −12 V power supply are used to drive the RS232 outputs. This is a great inconvenience for systems that have no other requirements for these power supplies. With this in mind, RS232 products manufactured by Dallas Semiconductor have on-chip charge-pump circuits that generate the necessary voltage levels for RS232 communication. The first charge pump essentially doubles the standard +5 V power supply to provide the voltage level necessary for driving a logic 0. A second charge pump inverts this voltage and provides the voltage level necessary for driving a logic 1. These two charge pumps allow the RS232 interface products to operate from a single +5 V supply.

9.3.2 MAXIMUM DATA RATE

Another limitation of the RS232 standard is the maximum data rate. The standard defines a maximum data rate of 20 kbps, which is slow for many of today's applications. RS232 products manufactured by Dallas Semiconductor guarantee up to 250 kbps and can communicate up to 350 kbps. While providing a communication rate at this frequency, the devices still maintain a maximum 30 V/ms maximum slew rate to reduce the likelihood of crosstalk between adjacent signals.

9.3.3 MAXIMUM CABLE LENGTH

The cable-length specification once included in the RS232 standard has been replaced by a maximum load-capacitance specification of 2500 pF. To determine the total length of cable allowed, one must determine the total line capacitance. Figure 9.16 shows a simple approximation for the total line capacitance of a conductor. The total capacitance is approximated by the sum of the mutual capacitance between the signal conductors and the conductor to shield capacitance (or stray capacitance in the case of unshielded cable).

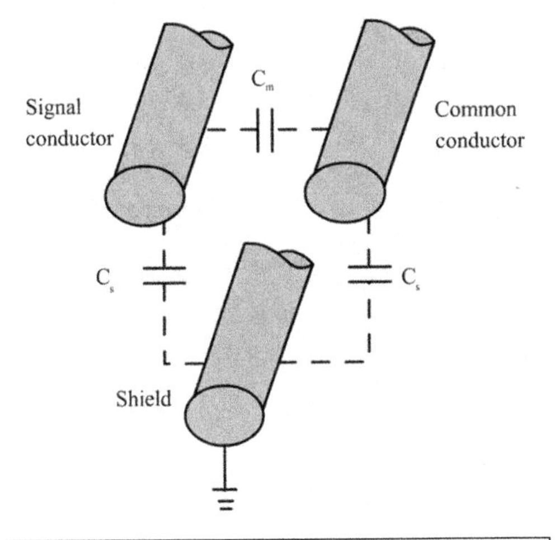

Data rate calculation:

$$\text{Unit interval} = \frac{1}{3}\frac{0.04}{C_C \ln\left(\frac{I_O+1}{I_O-1}\right)}$$

Io= Short circuit current of driver

Line length calculation:	
Maximum capacitance:	= 2500 pF
Receiver I/P capacitance:	< 20 pF
Maximum line capacitance:	2480 pF
Total line capacitance /mts	C_C=C_M+C_S
Mutual capacitance /mts	C_M= 100 pF
Stray capacitance /mts	C_S = 200 pF
Maximum line length	2480 C_C
Standard cable C/M	24 Pf/m
Maximum line length shielded:	100 mts

FIGURE 9.16

Interface cable-capacitive model, per unit length

For example, if the user decided to use nonshielded cable when interconnecting the equipment. The mutual capacitance (C_m) of the cable is found in the cable's specifications to be 20 pF per foot. Assuming that the receiver's input capacitance is 20 pF, this leaves the user with 2480 pF for the interconnecting cable. The total capacitance per foot is 30 pF as per the equation given in Figure 9.16. Dividing 2480 pF by 30 pF indicates the maximum cable length is approximately 80 ft. If a longer cable length is required, the user must find a cable with a smaller mutual capacitance.

9.4 OVERVIEW OF EIA-485

Data transmission between computer system components and peripherals over long distances and under high-noise conditions are usually very difficult, if not impossible, with single-ended drivers and receivers. EIS standards for balanced digital voltage interfacing enable the design engineers to come up with a universal solution for long-line system requirements.

The EIA-485 [TIA-485] balanced (differential) digital transmission line interface was developed to incorporate and improve upon the advantages of the current loop configuration and improvements to 232 limitations.

EIA-485 is also called the RS485 standard, but the term RS485 is out-dated. EIA-485 specifies bidirectional, half-duplex data transmission. It enables the interconnection of up to 32 transmitters and 32 receivers in any combination, including 1 driver and multiple receivers (multidrop), or 1 receiver and multiple drivers. A maximum of 32 devices is defined based on the unit load [UL] of each device [12K Ω]. The maximum devices on the net may be increased if the devices represent less than the UL [fractional unit load]. A number of devices are being produced which represent 1/4 or 1/8 the unit load. A maximum of 256 devices can exist on the bus when each is at 1/8 the UL [96K Ω].

EIA-485 requires a 120 Ω line impedance (normally shielded twisted pair "STP"), 120 Ω (10%, 1/2 watt) terminations at both ends of the line (at the receivers). Pull-up/pull-down resistors (idle-line failsafe) at one end of the 485 bus may be used to bring the line voltage to a steady state (200 mV) value at the end of a transmission; when all drivers are in the passive state.

There is no maximum bus length recommendation for RS485, but ends up around 1200 m at 200 kbps or 50 m at 10 Mbps. The speed of the system and the distance between devices are determined in large amount by the interconnecting cable. Single board solutions may be connected, but chassis to chassis systems require a common ground connection to run between the driver and the receiver.

EIA/TIA-422 define a balanced (differential) interface; specifying a single, unidirectional driver with multiple receivers (up to 32). RS422 supports point-to-point, multidrop circuits, but not multipoint [EIA-485]. EIA-485 devices may be used in 422 circuits, but EIA-422 may not be used in 485 circuits (because of the lack of an enable line). EIA-422 is the differential "pair" to EIA-423. One application note indicated that the combination of cable length (in meters) and data signaling rate (in bps) for RS422 must not exceed 108. The differences between 485 and 422 lie primarily in the driver features that allow reliable multipoint communications.

Owing to its versatility, an increasing number of standards committees are embracing the 485 standard as the physical layer specification of their communications standard. Examples include the American National Standards Institute (ANSI) small computer systems interface (SCSI) that is featured in the Interface Circuits for SCSI Applications Report (Literature Number SLLA035), the Profibus standard, and the DIN Measurement Bus (Figure 9.17).

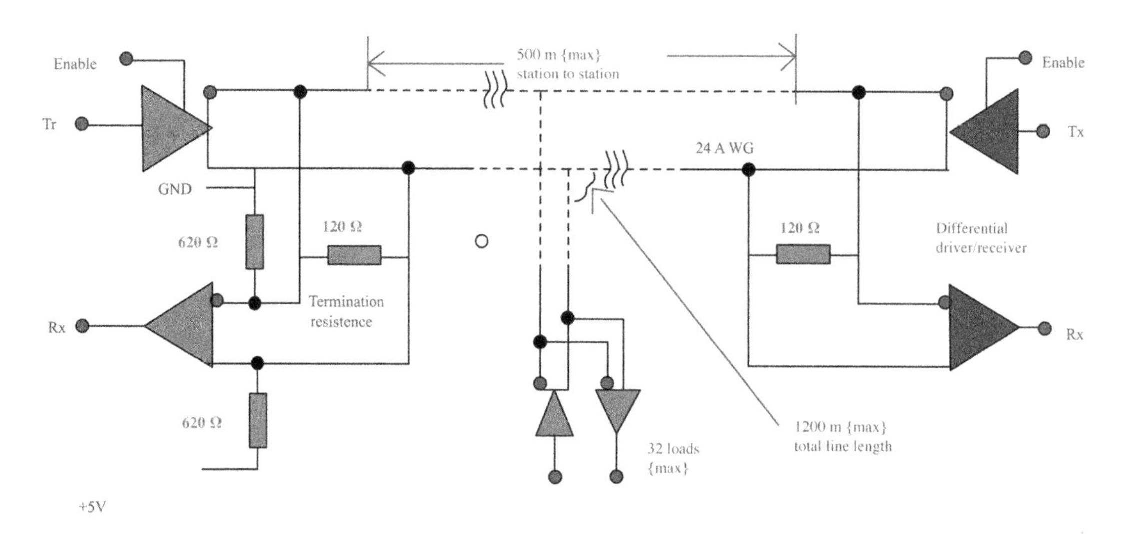

FIGURE 9.17

EIA/TIA-485 interface circuit (RS485)

9.4.1 EIA STANDARD RS485 DATA TRANSMISSION

The RS485 standard permits a balanced transmission line to be shared in a party line or multidrop mode. A maximum of 32 driver/receiver pairs can share a multidrop network. The RS485 and RS422 standards share many common characteristics of the drivers and receivers. The range of the common mode voltage *V*cm that the driver and receiver can tolerate is expanded to +12 to −7 V. Since the driver can be disconnected or tristated from the line, it must withstand this common mode voltage range while in the tristate condition. Some RS422 drivers, even with tristate capability, do not withstand the full *V*cm voltage range of +12 to −7 V.

The major enhancement of EIA-485 is that a line driver can operate in three states called tristate operation:

- Logic 1
- Logic 0
- High-impedance

Figure 9.18 shows a typical two-wire multidrop network. Note that the transmission line is terminated on both ends of the line but not at drop points in the middle of the line. Termination should only be used with high data rates and long wiring runs. RS422 systems require a dedicated pair of wires for

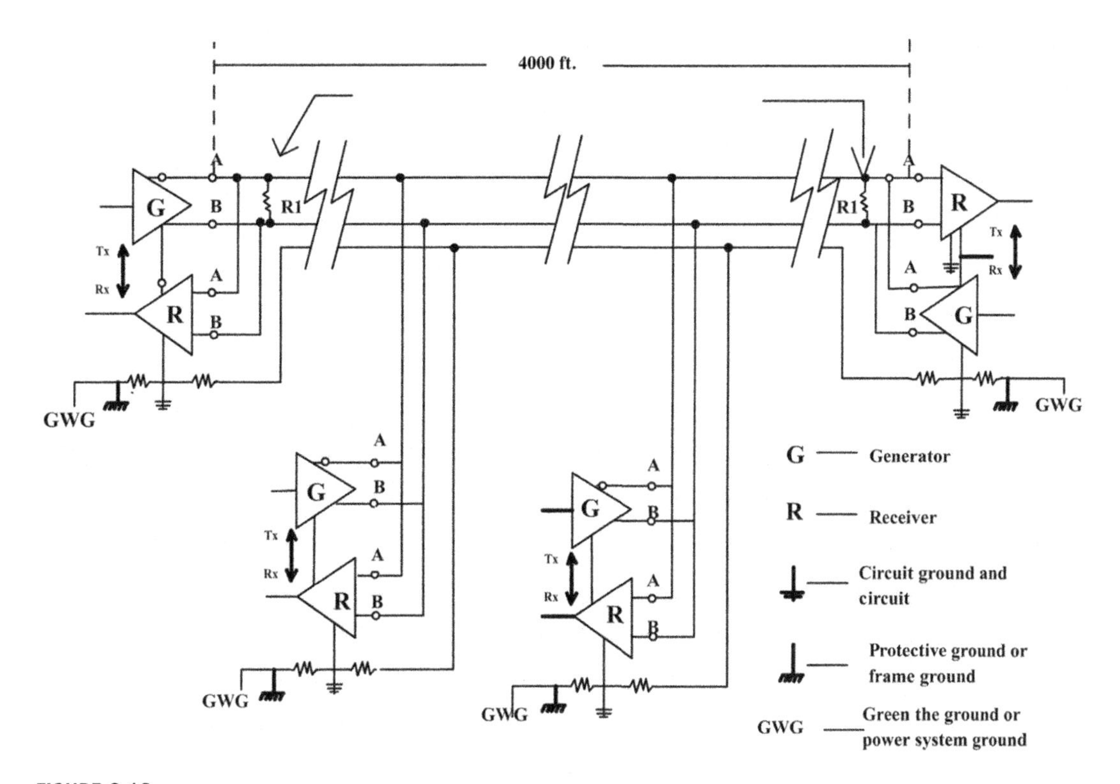

FIGURE 9.18

Two-wire multidrop network

each signal, a transmit pair, a receive pair, and an additional pair for each handshake/control signal used (if required). The tristate capabilities of RS485 allow a single pair of wires to share transmit and receive signals for half-duplex communications. This "two wire" configuration (note that an additional ground conductor should be used) reduces cabling cost. RS485 devices may be internally or externally configured for two-wire systems. Internally configured RS485 devices simply provide A and B connections sometimes labeled "−" and "+"). The signal ground line is also recommended in an RS485 system to keep the common mode voltage that the receiver must accept within the −7 to +12 V range. When an RS485 network is connected in a two-wire multidrop party line mode, the receiver at each node is connected to the line. The receiver can often be configured to receive an echo of its own data transmission. This is desirable in some systems, and troublesome in others. It is a good practice to check the data sheet for converter to determine how the receiver "enable" function is connected.

An RS485 network can also be connected in a four-wire mode as shown in Figure 9.19. Note that four data wires and an additional signal ground wire are used in a "four-wire" connection. In a four-wire network it is necessary to designate one node as master node and all others as slaves. The network is connected so that the master node communicates to all slave nodes. All slave nodes communicate only with the master node. This network has some advantages with equipment with

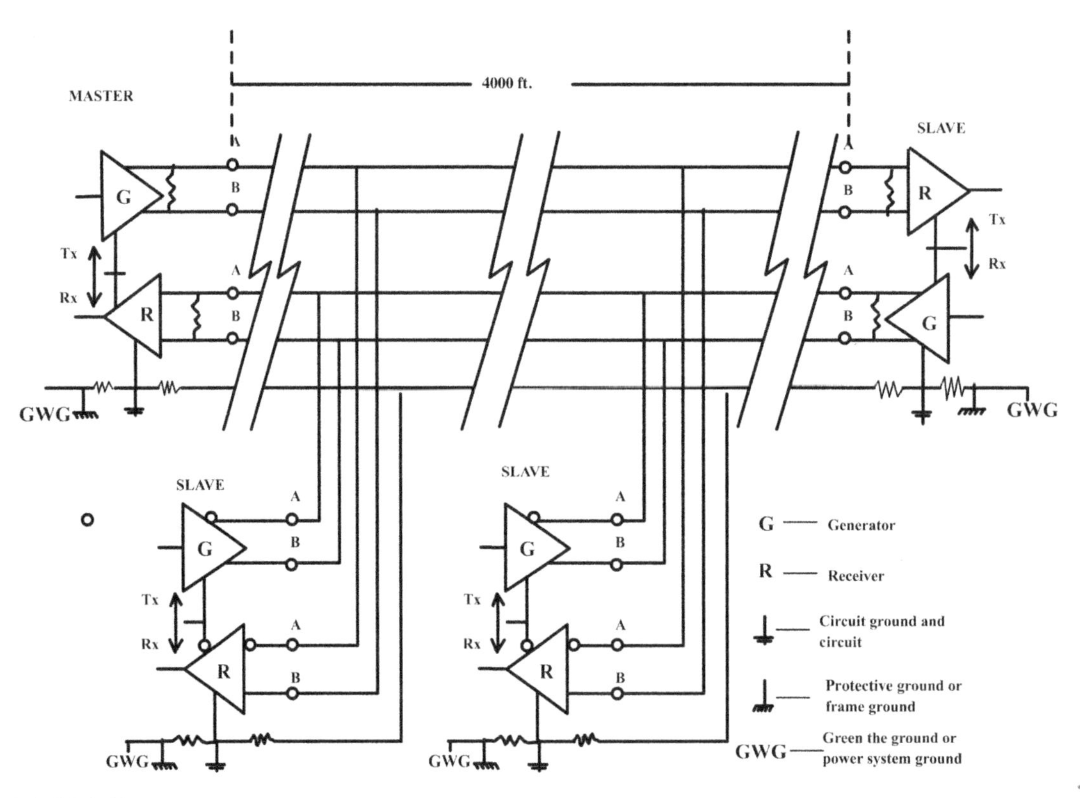

FIGURE 9.19

Four-wire multidrop network

mixed protocol communications. Since the slave nodes never listen to another slave response to the master, a slave node cannot reply incorrectly to another slave node. Devices configured for four-wire communication bring out A and B connections for both the transmit and the receive pairs. The user can connect the transmit lines to the receive lines to create a two-wire configuration. The latter type device provides the system designer with the most configuration flexibility. Note that the signal ground line should also be connected in the system. This connection is necessary to keep the *Vcm* common mode voltage at the receiver within a safe range. The interface circuit may operate without the signal ground connection, but may sacrifice reliability and noise immunity.

9.4.2 TERMINATION

Termination is used to match the impedance of a node to that of the transmission line being used. A mismatched impedance leads to the transmitted signal not being completely absorbed by the load resulting in a portion to reflect back into the transmission line. These reflections can be eliminated by ensuring equal impedance of the source, transmission line, and load. Terminations come with the disadvantages of increasing load on the drivers, increasing installation complexity, changing biasing requirements, and rendering system modification more difficult.

The decision whether or not to use termination should be based on the cable length and data rate used by the system. A good rule of thumb is if the propagation delays of the data line are much less than one-bit width, termination is not needed. This rule makes the assumption that reflections damp out in several trips up and down the data line. Since the receiving UART samples the data in the middle of the bit, it is important that the signal level be solid at that point. For example, in a system with 2000 feet of data line the propagation delay can be calculated by multiplying the cable length by the propagation velocity of the cable. This value, typically 66–75% of the speed of light (c), is specified by the cable manufacture. For our example, a round trip covers 4000 feet of cable. Using a propagation velocity of $0.66 \times c$, one round trip is completed in approximately 6.16 μs. If we assume the reflections to damp out in three "round trips" up and down the cable length, the signal stabilizes 18.5 μs after the leading edge of a bit. At 9600 baud one bit is 104 μs wide. Since the reflections are damped out much before the center of the bit, termination is not required. There are several methods of terminating data lines. A resistor is added in parallel with the receiver's "A" and "B" lines in order to match the data line characteristic impedance specified by the cable manufacture (120 Ω is a common value). This value describes the intrinsic impedance of the transmission line and is not a function of the line length. It is recommended to use a terminating resistor of more than 90 Ω. Termination resistors should be placed only at the extreme ends of the data line, and no more than two terminations should be placed in any system that does not use repeaters. This type of termination clearly adds heavy DC loading to a system and may overload port powered RS232 to RS485 converters. Another type of termination, AC-coupled termination, adds a small capacitor in series with the termination resistor to eliminate the DC loading effect. Although this method eliminates DC loading, capacitor selection is highly dependent on the system properties.

9.5 THE DIFFERENCE BETWEEN RS232/RS485/RS422

9.5.1 SIMPLEX AND DUPLEX

One of the most fundamental concepts of communications technology is the difference between simplex and duplex.

Simplex can be viewed as a communications "one-way." Data only flows in one direction. That is to say, a device can be a receiver or a transmitter exclusively. A simplex device is not a transceiver. A good example of simplex communication is an FM radio station and your car radio. Information flows only in one direction where the radio station is the transmitter and the receiver is your car radio. Simplex is not often used in computer communications because there is no way to verify when or if data is received. However, simplex communication is a very efficient way to distribute vast amounts of information to a large number of receivers.

Duplex communications overcome the limits of simplex communications by allowing the devices to act as transceivers. Duplex communication data, flow in both directions thereby allowing verification and control of data reception/transmission. Exactly when data flows bidirectionally further defines duplex communications.

Full duplex devices can transmit and receive data at the same time. RS232 is a fine example of full duplex communications. There are separate "transmit" and "receive" signal lines that allow data to flow in both directions simultaneously. RS422 devices also operate full duplex.

Half-duplex devices allow both transmission and receiving, but not at the same time. Essentially only one device can transmit at a time while all other half duplex devices receive. Devices operate as transceivers, but not simultaneous transmitter and receiver. RS485 operates in a half-duplex manner (Table 9.5).

One of the major differences between RS232 and RS422/RS485 is the signaling mode. RS232 is unbalanced while RS422/RS485 is balanced. An unbalanced signal is represented by a single signal wire where a voltage level on that one wire is used to transmit/receive binary 1 and 0: it can be considered a push signal driver. A balanced signal is represented by a pair of wires where a voltage difference is used to transmit/receive binary information: sort of a push–pull signal driver. In short, unbalanced voltage level signal travels slower and shorter than a balanced voltage difference signal.

Table 9.5 Difference among serial communications

	RS232	RS422	RS485
Cabling	Single ended	Single-ended multidrop	Multidrop
Number of devices	1 transmitter 1 receiver	5 transmitters 10 receivers	32 transmitters 32 receivers
Communication mode	Full duplex	Full duplex half duplex	Half duplex
Max. distance	50 feet at 19.2 kbps	4000 feet at 100 kbps	4000 feet at 100 kbps
Max. data rate	19.2 kbps for 50 feet	10 Mpbs for 50 feet	10 Mpbs for 50 feet
Signaling	Unbalanced	Balanced	Balanced
Mark (data 1)	−5 V min. −15 V max.	2 V min. ($B > A$) 6 V max. ($B > A$)	1.5 V min. ($B > A$) 5 V max. ($B > A$)
Space (data 0)	5 V min. 15 V max.	2 V min. ($A > B$) 6 V max. ($A > B$)	1.5 V min. ($A > B$) 5 V max. ($A > B$)
Input level min.	±3 V	0.2 V difference	0.2 V difference
Output current	500 mA	150 mA	250 mA

Table 9.6 25 Pin Male and Female Pinout

DTE	DCE
25-pin male pinout	25-pin female pinout
Pin 1 – Shield Ground	Pin 1 – Shield Ground
Pin 2 – Transmitted Data (TD) output	Pin 2 – Transmitted Data (TD) input
Pin 3 – Receive Data (RD) input	Pin 3 – Receive Data (RD) output
Pin 4 – Request to Send (RTS) output	Pin 4 – Request to Send (RTS) input
Pin 5 – Clear to Send (CTS) input	Pin 5 – Clear to Send (CTS) output
Pin 6 – Data Set Ready (DSR) input	Pin 6 – Data Set Ready (DSR) output
Pin 7 – Signal Ground	Pin 7 – Signal Ground
Pin 8 – Carrier Detect (CD) input	Pin 8 – Carrier Detect (CD) output
Pin 20 – Data Terminal Ready (DTR) output	Pin 20 – Data Terminal Ready (DTR) input
Pin 22 – Ring Indicator (RI) input	Pin 22 – Ring Indicator (RI) output

9.5.2 DTE AND DCE

The difference between DCE and DTE devices is largely in the plug and the direction of each pin (input or output). A desktop PC is termed as a DTE device.

DCE devices use a 25-pin female connector while a DTE device uses a 25-pin male connector. Also, complimentary signal lines like transmit and receive are "swapped" between the two types. Therefore, a straight-through cable can be used to connect a DCE device to a DTE device (Table 9.6).

DCE/DTE devices can be effectively converted using a NULL modem cable. The null modem cable swaps the complimentary signals and allows a DCE device to act like a DTE and vice versa.

Table 9.7 depicts the cabling of the DB9 connector found on an IBM-PC type computer.

9.6 MODBUS SERIAL COMMUNICATIONS

9.6.1 INTRODUCTION TO MODBUS

Modbus is a serial communication protocol developed and published by Modicon® in 1979 for use with its programmable logic controllers (PLCs).

In simple terms, it is a method used for transmitting information over serial lines between electronic devices. The device requesting the information is called the Modbus master and the devices supplying information are called Modbus slaves. In a standard Modbus network, there is one master and up to 247 slaves, each with a unique slave address from 1 to 247. The master can also write information to the slaves.

Table 9.7 9 Pin male Pinout

DTE
9-pin male pinout
Pin 1 – Carrier Detect (CD) input
Pin 2 – Receive Data (RD) input
Pin 3 – Transmitted Data (TD) output
Pin 4 – Data Terminal Ready (DTR) output
Pin 5 – Signal Ground
Pin 6 – Data Set Ready (DSR) input
Pin 7 – Request to Send (RTS) output
Pin 8 – Clear to Send (CTS) input
Pin 9 – Ring Indicator (RI) input

Transactions are either a Query or Response type (where one device is accessed at a time) or a Broadcast or No Response type (where all slaves are accessed at the same time).

9.6.2 FUNDAMENTALS OF MODBUS

Modbus is an open protocol, free for manufacturers to build into their equipment without having to pay royalties. It has become a standard communications protocol in the industry, and is now the most commonly available means of connecting industrial electronic devices.

It is widely used by many manufacturers across many industries. Modbus is typically used to transmit signals from instrumentation and control devices back to a main controller or data gathering system, for example, a system that measures temperature and humidity and communicates the results to a computer. Modbus is often used to connect a supervisory computer with a remote terminal unit (RTU) in supervisory control and data acquisition (SCADA) systems. Versions of the Modbus protocol exist for serial lines (Modbus RTU and Modbus ASCII) and for Ethernet (Modbus TCP).

9.6.3 MODBUS AND OSI MODEL

9.6.3.1 OSI model

The Open Systems Interconnection (OSI) model is a product of the OSI effort at the International Organization for Standardization (ISO).

It entails characterizing and standardizing the functions of a communications system in terms of abstraction layers. Similar communication functions are grouped into logical layers. An instance of a layer provides services to its upper layer instances while receiving services from the layer below.

The MODBUS standard defines an application layer messaging protocol, positioned at level 7 of the OSI model that provides "client/server" communications between devices connected on different types

Layer	ISO/OSI model	Modbus layers
7	Application layer	Modbus application protocol
6	Presentation layer	Empty
5	Session layer	Empty
4	Transport layer	Empty
3	Network layer	Empty
2	Data link layer	MODBUS serial line protocol
1	Physical layer	EIA/TIA-485 (EIA/TIA-232)

FIGURE 9.20

MOdbus OSI Layers

of buses or networks. It standardizes a specific protocol on serial line to exchange MODBUS requests between a master and one or several slaves.

At the physical level, MODBUS over serial line systems may use different physical interfaces (RS485, RS232). TIA/EIA-485 (RS485) two-wire interface is the most common. As an add-on option, RS485 four-wire interface may also be implemented. A TIA/EIA-232-E (RS232) serial interface may be used as an interface, when only short point-to-point communication is required.

Figure 9.20 gives a general representation of MODBUS serial communication stack compared to the seven layers of the OSI model.

9.6.4 MODBUS OPERATION

Modbus is transmitted over serial lines between devices. The simplest setup would be a single serial cable connecting the serial ports on two devices, a master and a slave (Figure 9.21).

The data is sent as series of ones and zeroes called bits. Each bit is sent as a voltage. Zeroes are sent as positive voltages and ones as negative. The bits are sent very quickly. A typical transmission speed is 9600 baud (bits per second).

9.6.5 HEXADECIMAL

When troubleshooting problems, it can be helpful to see the actual raw data being transmitted. Long strings of ones and zeroes are difficult to read, so the bits are combined and shown in hexadecimal. Each block of 4 bits is represented by one of the 16 characters from 0 to F.

1

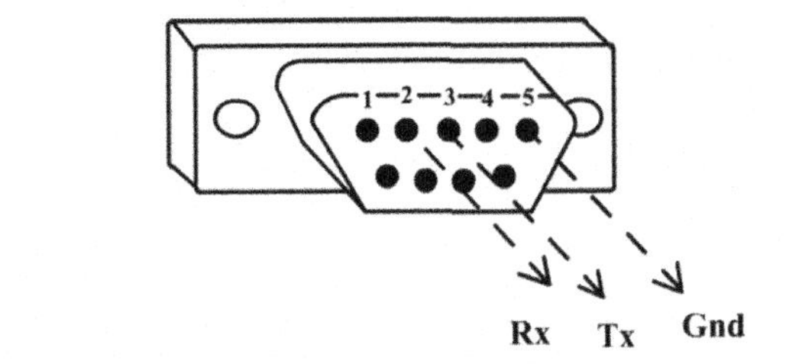

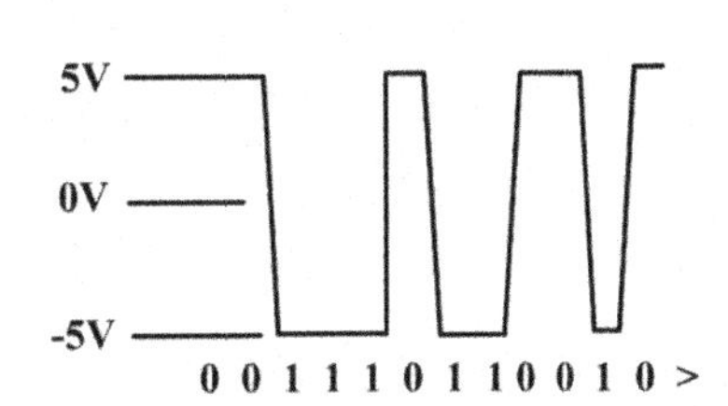

FIGURE 9.21

Physical view of Modbus

Table 9.8 Modbus address

Coil/Register Numbers	Data Addresses	Type	Table Name
1–9999	0000 to 270E	Read-write	Discrete output coils
10001–19999	0000 to 270E	Read-only	Discrete input contacts
30001–39999	0000 to 270E	Read-only	Analog input registers
40001–49999	0000 to 270E	Read-write	Analog output holding registers

0000 = 0	0100 = 4	1000 = 8	1100 = C
0001 = 1	0101 = 5	1001 = 9	1101 = D
0010 = 2	0110 = 6	1010 = A	1110 = E
0011 = 3	0111 = 7	1011 = B	1111 = F

Each block of 8 bits (called a byte) is represented by one of the 256 character pairs from 00 to FF.

9.6.6 DATA STORAGE IN STANDARD MODBUS

Information is stored in the slave device in four different tables. Two tables store on/off discrete values (coils) and two store numerical values (registers). The coils and registers each have a read-only table and read-write table (Table 9.8).

- Each table has 9999 values.
- Each coil or contact is 1 bit and assigned a data address between 0000 and 270E.
- Each register is 1 word = 16 bits = 2 bytes and also has data address between 0000 and 270E.

Coil/register numbers can be considered as location names since they do not appear in the actual messages. The data addresses are used in the messages. For example, the first holding register, number 40001, has the data address 0000.

The difference between these two values is the offset. Each table has a different offset: 1, 10001, 30001, and 40001.

9.6.7 MASTER–SLAVE PROTOCOL

The MODBUS serial line protocol is a master–slave protocol. At a particular time, only one master is connected to the bus, and one or several (247 maximum number) slave nodes are connected to the same serial bus. The master always initiates a MODBUS communication. The slave nodes never transmit data without receiving a request from the master node. The slave nodes never communicate with each other. The master node initiates only one MODBUS transaction at a time.

The master node issues a MODBUS request to the slave nodes in two modes:

- In unicast mode, the master addresses an individual slave. After receiving and processing the request, the slave returns a message (a "reply") to the master. In that mode, a MODBUS transaction consists of two messages: a request from the master and a reply from the slave. Each slave must have a unique address (from 1 to 247) so that it can be addressed independently from other nodes.

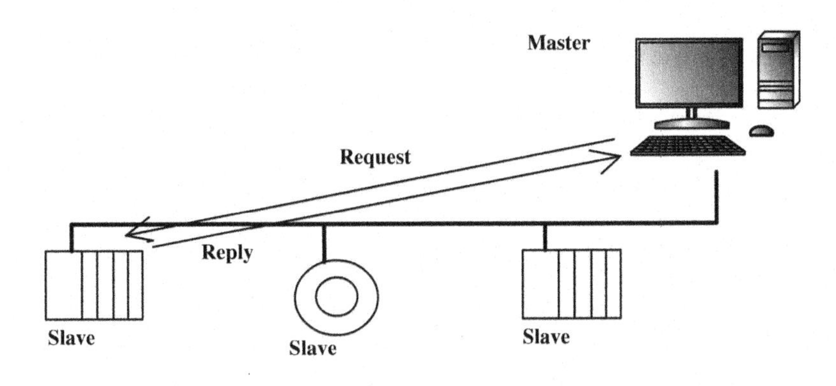

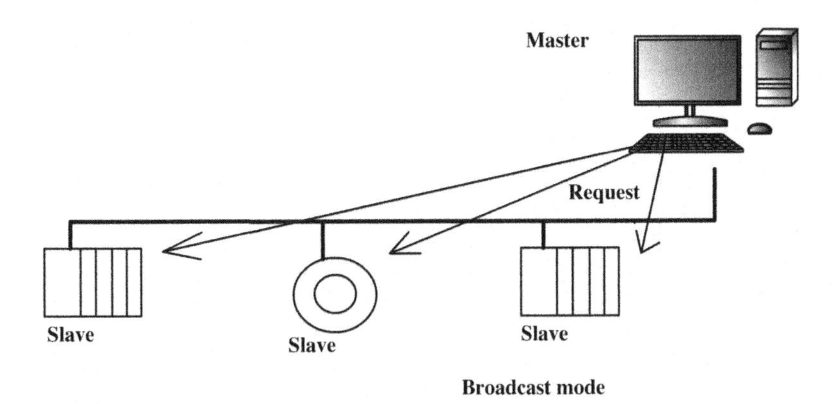

FIGURE 9.22

Communication methods in Modbus

- In broadcast mode, the master can send a request to all slaves. No response is returned to broadcast requests sent by the master. The broadcast requests are necessarily writing commands. All devices must accept the broadcast for writing function. The address 0 is reserved to identify a broadcast exchange (Figure 9.22).

9.6.7.1 MODBUS addressing rules

0	From 1 to 247	From 248 to 255
Broadcast address	Slave individual addresses	Reserved

- The address 0 is reserved as the broadcast address.
- All slave nodes must recognize the broadcast address.

Table 9.9 Typical Function Codes in Modbus

Function Code	Action	Table Name
01 (01 hex)	Read	Discrete output coils
05 (05 hex)	Write single	Discrete output coil
15 (0F hex)	Write multiple	Discrete output coils
02 (02 hex)	Read	Discrete input contacts
04 (04 hex)	Read	Analog input registers
03 (03 hex)	Read	Analog output holding registers
06 (06 hex)	Write single	Analog output holding register
16 (10 hex)	Write multiple	Analog output holding registers

- The MODBUS master node has no specific address; only the slave nodes must have an address. This address must be unique on a MODBUS serial bus.

9.6.8 FUNCTION CODE

The second byte sent by the master is the function code. This number tells the slave which table to access and whether to read from or write to the table (Table 9.9).

9.6.9 CYCLIC REDUNDANCY CHECK (CRC)

CRC stands for cyclic redundancy check. It is 2 bytes added to the end of every modbus message for error detection. Every byte in the message is used to calculate the CRC. The receiving device also calculates the CRC and compares it to the CRC from the sending device. If even 1 bit in the message is received incorrectly, the CRCs are different and result in an error.

9.6.10 SERIAL TRANSMISSION MODES

Two different serial transmission modes are defined:

- RTU mode
- American Standard Code for Information Interchange (ASCII) mode

It defines the bit contents of message fields transmitted serially on the line. It determines how information is packed into the message fields and decoded.

The transmission mode (and serial port parameters) must be the same for all devices on a MODBUS serial line. Although the ASCII mode is required in some specific applications, interoperability between MODBUS devices can be reached only if each device has the same transmission mode: All devices must implement the RTU mode. The ASCII transmission mode is an option. The users must set up devices to the desired transmission mode, RTU or ASCII.

With parity checking

START	1	2	3	4	5	6	7	8	PARITY	STOP

FIGURE 9.23

Bit pattern with parity

9.6.11 RTU TRANSMISSION MODE

When devices communicate on a MODBUS using the RTU mode, each 8-bit byte in a message contains two 4-bit hexadecimal characters. The main advantage of this mode is that its greater character density allows better data throughput than ASCII mode for the same baud rate. Each message must be transmitted in a continuous stream of characters.

The format for each byte (11 bits) in RTU mode is:

Coding system:	8-bit binary
Bits per byte:	1 start bit 8 data bits, least significant bit (LSB) sent first 1 bit for parity completion 1 stop bit

Even parity is required; other modes (odd parity, no parity) may also be used. In order to ensure a maximum compatibility with other products, it is recommended to support "no parity" mode also. The default parity mode must be even parity.

9.6.11.1 How characters are transmitted serially

Each character or byte is sent in this order (left to right): from the least significant bit (LSB) to Most Significant Bit (MSB) (Figure 9.23).

Devices may accept by configuration either even, odd, or no parity checking. If no parity is implemented, an additional stop bit is transmitted to fill out the character frame to a full 11-bit asynchronous character (Figure 9.24).

9.6.11.2 CRC checking

The RTU mode includes an error-checking field that is based on a CRC checking method performed on the message contents.

The CRC field checks the contents of the entire message. It is applied regardless of any parity checking method used for the individual characters of the message. The CRC field contains a 16-bit value implemented as two 8-bit bytes.

With out parity checking

START	1	2	3	4	5	6	7	8	STOP	STOP

FIGURE 9.24

Modbus RTU bit pattern

The CRC field is appended to the message as the last field in the message. When this is done, the low-order byte of the field is appended first, followed by the high-order byte. The CRC high-order byte is the last byte to be sent in the message.

The CRC value is calculated by the sending device, which appends the CRC to the message. The receiving device recalculates a CRC during receipt of the message, and compares the calculated value to the actual value it received in the CRC field. If the two values are not equal, it results in an error.

The CRC calculation is started by first preloading a 16-bit register to all 1's. Then the process of applying successive 8-bit bytes of the message to the current contents of the register begins. Only the 8 bits of data in each character are used for generating the CRC. Start and stop bits and the parity bit, do not apply to the CRC.

9.6.12 THE ASCII TRANSMISSION MODE

When devices are set up to communicate on a MODBUS serial line using ASCII mode, each 8-bit byte in a message is sent as two ASCII characters. This mode is used when the physical communication link or the capabilities of the device does not allow the conformance with RTU mode requirements regarding timers management.

The format for each byte (10 bits) in ASCII mode is:

Coding system:	Hexadecimal, ASCII characters 0–9, A–F One hexadecimal character contains 4-bits of data within each ASCII character of the message
Bits per byte:	1 start bit 7 data bits, LSB sent first 1 bit for parity completion 1 stop bit

Even parity is required, while other modes (odd parity, no parity) may also be used. In order to ensure maximum compatibility with other products, it is recommended to support "no parity" mode also. The default parity mode must be even parity.

Note: The use of no parity requires 2 stop bits.

9.6.12.1 How characters are transmitted serially

Each character or byte is sent in this order (left to right): from Least Significant Bit (LSB) to Most Significant Bit (MSB) (Figure 9.25).

With parity checking

START	1	2	3	4	5	6	7	8	PARITY	STOP

FIGURE 9.25

Modbus ASCII bit pattern with parity

With out parity checking

START	1	2	3	4	5	6	7	8	STOP	STOP

FIGURE 9.26

Modbus ASCII bit pattern with out parity

Devices may accept by configuration either even, odd, or no parity checking. If no parity is implemented, an additional stop bit is transmitted to fill out the character frame (Figure 9.26).

9.7 MODBUS MAP

A Modbus map is simply a list for an individual slave device that defines the following:

- What the data is (e.g., pressure or temperature readings)
- Where the data is stored (which tables and data addresses)
- How the data is stored (data types, byte, and word ordering)

Some devices are built with a fixed map that is defined by the manufacturer, while other devices allow the operator to configure or program a custom map to fit their needs.

9.7.1 EXTENDED REGISTER ADDRESSES

The range of the analog output holding registers is 40001 to 49999, implying that there can be a maximum of 9999 registers. Although this is usually enough for most applications, there are cases where more registers would be beneficial.

Registers 40001 to 49999 correspond to data addresses 0000 to 270E. If we utilize the remaining data addresses 270F to FFFF, over six times as many registers can be available, 65536 in total. This would correspond to register numbers from 40001 to 105536.

Many Modbus software drivers (for master PCs) were written with the 40001 to 49999 limits and cannot access extended registers in slave devices. And many slave devices do not support maps using the extended registers. But, on the other hand, some slave devices do support these registers and some master software can access it, especially if custom software is written.

9.7.2 2-BYTE SLAVE ADDRESSING WORK

Since a single byte is normally used to define the slave address and each slave on a network requires a unique address, the number of slaves on a network is limited to 256. The limit defined in the Modbus specification is even lower at 247.

To get beyond this limit, the protocol can be modified to use 2 bytes for the address. The master and the slaves would all be required to support this modification. Two bytes addressing extends the limit on the number of slaves in a network to 65535.

By default, the simply Modbus software uses 1 byte addressing. When an address greater than 255 is entered, the software automatically switches to 2 byte addressing and stays in this mode for all addresses until the 2 byte addressing is manually turned off.

9.8 ERROR CHECKING METHODS

The security of standard MODBUS serial line is based on two kinds of error checking:

- Parity checking (even or odd) should be applied to each character.
- Frame checking (longitudinal redundancy checking, LRC or CRC) must be applied to the entire message.

Both the character checking and message frame checking are generated in the device (master or slave) that emits and applied to the message contents before transmission. The device (slave or master) checks each character and the entire message frame during receipt.

The master is configured by the user to wait for a predetermined timeout interval (response timeout) before aborting the transaction. This interval is set to be long enough for any slave to respond normally (unicast request). If the slave detects a transmission error, the message is not to be acted upon. The slave does not construct a response to the master. Thus the timeout expires and allows the master's program to handle the error. Note that a message addressed to a nonexistent slave device also causes a timeout.

9.8.1 PARITY CHECKING

Users may configure devices for even (required) or odd parity checking, or for no parity checking (recommended). This determines how the parity bit is set in each character. If either even or odd parity is specified, the quantity of 1 bit is counted in the data portion of each character (seven data bits for ASCII mode or eight for RTU). The parity bit is then be set to a 0 or 1 to result in an even or odd total of 1 bit.

For example, these eight data bits are contained in an RTU character frame: 1100 0101.

The total quantity of 1 bit in the frame is four. If even parity is used, the frame's parity bit is a 0, making the total quantity of 1 bit still an even number (four). If odd parity is used, the parity bit is a 1, making an odd quantity (five).

When the message is transmitted, the parity bit is calculated and applied to the frame of each character. The device that receives counts the quantity of 1 bit and sets an error if they are not the same as configured for that device (all devices on the MODBUS serial line must be configured to use the same parity checking method).

Note that parity checking can only detect an error if an odd number of bits are picked up or dropped in a character frame during transmission. For example, if odd parity checking is employed, and two 1 bit are dropped from a character containing three 1 bit, the result is still an odd count of 1 bit.

If no parity checking is specified, parity bit is not transmitted and parity checking is not made. An additional stop bit is transmitted to fill out the character frame.

9.8.2 FRAME CHECKING

Two kinds of frame checking are used depending on the transmission mode: RTU or ASCII.

In RTU mode, messages include an error-checking field, based on a CRC method. The CRC field checks the contents of the entire message. It is applied regardless of any parity checking method used for the individual characters of the message.

In ASCII mode, messages include an error-checking field that is based on a LRC method. The LRC field checks the contents of the message, exclusive of the beginning "colon" and ending carriage return, line feed Carriage Return Line Feed (CRLF) pair. It is applied regardless of any parity checking method used for the individual characters of the message.

9.9 MODBUS EXCEPTION CODES

9.9.1 OVERVIEW

When a client device sends a request to a server device it expects a normal response. One of the four possible events can occur from the master's query:

- If the server device receives the request without a communication error, and can handle the query normally, it returns a normal response.
- If the server does not receive the request due to a communication error, no response is returned. The client program will eventually process a timeout condition for the request.
- If the server receives the request, but detects a communication error (parity, LRC, CRC, ...), no response is returned. The client program will eventually process a timeout condition for the request.
- If the server receives the request without a communication error, but cannot handle it (e.g., if the request is to read a nonexistent output or register), the server will return an exception response informing the client of the nature of the error.

The exception response message has two fields that differentiate it from a normal response.

Function Code Field: In a normal response, the server echoes the function code of the original request in the function code field of the response. All function codes have a MSB of 0 (their values are all below 80 hexadecimal). In an exception response, the server sets the MSB of the function code to 1. This makes the function code value in an exception response exactly 80 hexadecimal higher than the value would be for a normal response. With the function code's MSB set, the client's application program can recognize the exception response and can examine the data field for the exception code.
Data Field: In a normal response, the server may return data or statistics in the data field (any information that was requested in the request). In an exception response, the server returns an exception code in the data field. This defines the server condition that caused the exception.

Example of a client request and server exception response (Figure 9.27).

Request		Response	
Field name	**Hex**	**Field name**	**Hex**
Function	01	Function	81
Starting address Hi	04	Exception code	02
Starting address Lo	A1		
Quantity of outputs Hi	00		
Quantity of outputs Lo	01		

FIGURE 9.27

Exception code

Table 9.10 List of Modbus exception Codes

Code Dec/Hex	Name	Meaning
01/0x01	Illegal function	The function code received in the query is not an allowable action for the server (or slave). This may be because the function code is only applicable to newer devices, and was not implemented in the unit selected. It could also indicate that the server (or slave) is in the wrong state to process a request of this type, for example, because it is not configured and is being asked to return register values.
02/0x02	Illegal data address	The data address received in the query is not an allowable address for the server (or slave). More specifically, the combination of reference number and transfer length is invalid. For a controller with 100 registers a request of offset 96 and a length of 5 will generate exception 02.
03/0x03	Illegal data value	The value contained in the query data field is not an allowable value for the server (or slave). This indicates a fault in the structure of the remainder of a complex request, such as that the implied length is incorrect. It specifically does NOT mean that a data item submitted for storage in a register has a value outside the expectation of the application program, since the MODBUS protocol is unaware of the significance of any particular value of any particular register.
04/0x04	Failure in associated device	An unrecoverable error occurred while the server (or slave) was attempting to perform the requested action (see Note 1).
05/0x05	Acknowledge	Specialized in conjunction with programming commands. The server (or slave) has accepted the request and is processing it, but long duration of time will be required to do so. This response is returned to prevent a timeout error from occurring in the client (or master). The client (or master) can next issue a poll program complete message to determine if processing is completed.
06/0x06	Busy, rejected message	Specialized use in conjunction with programming commands. The server (of slave) is engaged in processing a long-duration program command. The client (or master) should retransmit the message later when the server (or slave) is free.
07/0x07	NAK – negative acknowledgment	The program function just requested cannot be performed. Issue poll to obtain detailed device dependent error information. Valid for program/poll 13 and 14 only.
08/0x08	Memory parity error	Specialized use in conjunction with function codes 20 and 21 and reference type 6, to indicate that the extended file area failed to pass a consistency check. The server (or slave) attempted to read record file, but detected a parity error in the memory. The client (or master) can retry the request, but service may be required on the server (or slave) device.
10/0x0A	Gateway path unavailable	Specialized use in conjunction with gateways. Indicates that the gateway was unable to allocate an internal communication path from the input port to the output port for processing the request.
11/0x0B	Gateway target device failed to respond	Specialized use in conjunction with gateways. Indicates that no response was obtained from the target device. Usually means that the device is not present on the network.

In this example, the client addresses a request to server device. The function code (01) is for a read output status operation. It requests the status of the output at address 1185 (04A1 hex). Note that only one output is to be read, as specified by the number of outputs field (0001).

If the output address is nonexistent in the server device, the server will return the exception response with the exception code shown (02). This specifies an illegal data address for the slave. Table 9.10 is a listing of exception codes that are supported in current Modbus specifications.

FURTHER READINGS

MODBUS Application Protocol Specification V1.1b – modbus IDA.
MODBUS Messaging on TCP/IP Implementation Guide V1.0b – modbus IDA.
Modbus_over_serial_line_V1_02 – Modbus IDA.
http://www.modbus.org.
Process Software And Digital Networks – bela g liptak 3 edition.

CHAPTER

INDUSTRIAL NETWORKS

10

10.1 INTRODUCTION TO INDUSTRIAL NETWORKS

The two most important layers of industrial network are process/plant control network (PCN) and plant information network (PIN). Before getting into the details of industrial network and its layers (PCN and PIN), it is important to understand the basics of networking. A network is a collection of computer hardware and software along with cabling, networking devices such as switches, routers, and the like that enables computers to communicate with each other. A network provides a means of sharing information between computers. A network is formed when two or more computers are connected over a shared medium to share files or peripherals such as modems, printers, and others.

Connecting two computers using a cable is the simplest example of a simple network. The two computers communicate over the cable connecting them and share files or resources and thereby form a simple network (Figure 10.1).

The "topology" of a network is the first point that comes up when one considers how the devices are interconnected.

10.1.1 NETWORK TOPOLOGY

"Topology" refers to the pattern of interconnection of different elements of computers. There are two types of network topologies, namely "physical" and "logical."

The physical topology refers to the layout of network cables, its interconnection with computers, and/or other networking devices. Logical topology defines how data is transferred from one device to another irrespective of network hardware connectivity. In general, logical topology is defined by network protocol whereas physical topology is mainly defined by physical layout of networking cables and devices.

The following are some of the popular physical network topologies:

- Point to point topology
- Bus topology
- Ring topology
- Star topology
- Extended start topology
- Mesh topology

The physical topologies are illustrated in Figure 10.2.

In the early days, computer vendors developed their own standards and protocols to communicate between their own products, forming a proprietary network. But with growing usage of computers and increasing demand for networking to share resources, eventually the computer world shifted its

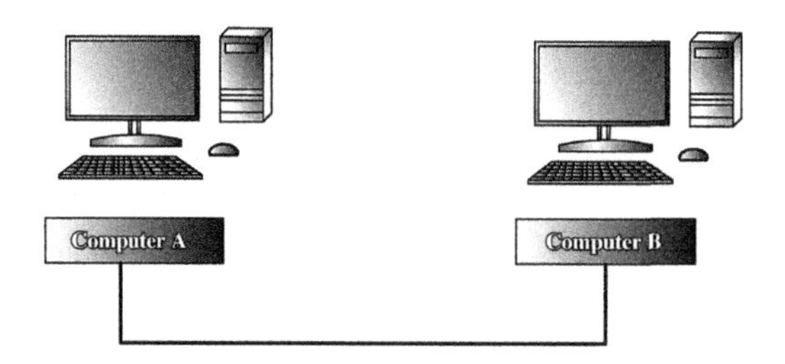

FIGURE 10.1

Simple network

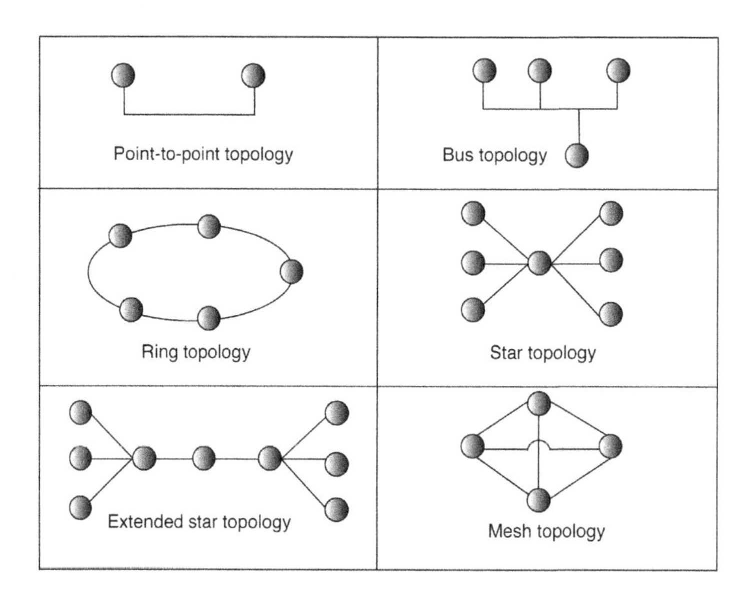

FIGURE 10.2

Networking topologies

paradigm from proprietary technology to open standards and protocols. Open standards are not owned by any one company or vendor but are developed by standard bodies such as International Organization for Standardization, Institute of Electrical and Electronic Engineers (IEEE), and others.

The network models define standards and protocols of networking.

10.2 THE OSI NETWORK MODEL

With the evolution of networking and increasing demand for networks to share resources, the need for standard and protocol for intercommunication and data transfer between different hosts emerged as one of the top priority in the computer world. The open system interconnection (OSI) network model was aimed at guiding host vendors to create interoperable network devices and software to achieve the goal of networking between disparate hosts.

OSI model consists of seven different layers:

- Layer 7: Application
- Layer 6: Presentation
- Layer 5: Session
- Layer 4: Transport
- Layer 3: Network
- Layer 2: Data link
- Layer 1: Physical

Layers 7, 6, and 5 can be grouped together and called as "Upper Layer." These three layers are related to application and operating systems. Layers 4, 3, 2, and 1 can be grouped together and called as "Lower Layer." This is the place where actual networking takes place. The data coming from "upper layer" gets logical and physical addresses in this layer and sends the data out as electrical signals.

Upper Layer	7: Application 6: Presentation 5: Session
Lower Layer	4: Transport 3: Network 2: Data link 1: Physical

Layer 7 – Application Layer
The application layer represents the entry point for different applications on the computer or host, which want to send data out over a network to some other host.
This layer acts as an interface between the actual application program and the layer below this (session layer). Telnet and hypertext transfer protocol (HTTP) are examples of layer 7 applications.

Layer 6 – Presentation Layer
The presentation layer presents data to the application layer (at the receiving computer) after translating and formatting the data received from the application layer of the host that wants to transmit data. Data encryption and decryption are some of the important tasks of this layer.

Layer 5 – Session Layer
The communication connection between two hosts is initiated in the session layer. This layer is responsible for establishing, managing, and then closing the sessions between two hosts.

Layer 4 – Transport Layer
The transport layer receives data from the session layer, segments that data into smaller chunks as per network restrictions, and hands over the segmented data to the next layer. This layer is

responsible for error detection and recovery and also controls the flow of the packets. Transmission control protocol (TCP) and user datagram protocol (UDP) are examples of layer 4 protocols.

Layer 3 – Network Layer
The network layer is a complex layer, and the source and destination addresses are assigned in this layer. The segmented data received from the transport layer are again segmented in this layer based on the capacity of layer 2 technologies, and source and destination addresses are attached with this segmented data. This layer 3 chunked data is called "packets." Additionally, the network layer utilizes the address resolution protocol (ARP) protocol to determine the source and destination media access control (MAC) addresses and pass this information to layer 2 data link layer. Internet protocol (IP) is a widely used example of layer 3 protocol.

Layer 2 – Data Link Layer
The data link layer enables the physical transmission of data. The data protocol unit name at data link layer is Frame. This layer mainly works with physical addressing (MAC address) and error notification. Ethernet, PPP, and Frame Relay, and so on are examples of layer 2 protocols.

Layer 1 – Physical Layer
The physical layer sends and receives bits and converts the bits into electrical signal for transmission over the physical network cables. The physical layer defines different attributes such as maximum transmission distance, voltage levels, and others.
When two machines communicate over a network, the seven layers of the sending PC's process, format, encrypt, and chunk the data to be sent based on the role defined for that layer. The layer also adds its own header and trailer and sends out the packets over the network. Layer 1 at the receiving PC receives the data and moves upward, allowing each layer to reverse whatever the corresponding layer at sending machine did and finally hand over the data to receiving machine's application (Figure 10.3).

10.3 TCP/IP

TCP/IP was first developed by U.S. Department of Defense (DoD) and therefore called TCP/IP model or DoD model. This is again an open standard. Committees such as Internet Architecture Board and the Internet Engineering Task Force define the TCP/IP standards. If we keep the OSI layer and TCP/IP layer side by side and try to correlate the layers, it would look as shown in Figure 10.4.

Let us start with the bottom most layers.

10.3.1 NETWORK ACCESS LAYER

The network access layer is also known as the network interface layer. It monitors data exchanges between the host and the network, and defines how computers access the physical networking medium similar to OSI layer 1 and 2 (physical and data link layer). The TCP/IP model does not directly define anything new for the network access layer but uses the other defined networking standards such as Ethernet. Internet Engineering Task Force, instead of defining layer 1 and layer 2 standard,

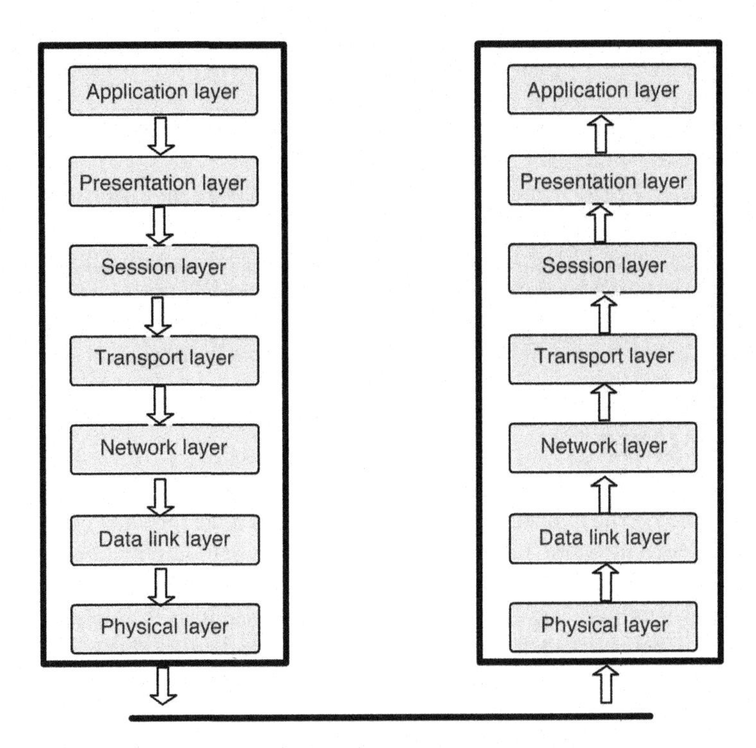

FIGURE 10.3

Layered view of communication

OSI	TCP/IP
Application layer	Application
Presentation layer	
Session layer	
Transport layer	Transport
Network layer	Internet
Data link layer	Network access
Physical layer	

FIGURE 10.4

Comparision of OSI layers with TCP/IP

to ensure all new standards defined by other layer 2 and layer 1 committees (such as IEEE) can be supported by TCP/IP as a network access layer protocol.

Ethernet, Fast Ethernet, PPP, FDDI, and the like are some examples of network access layer protocols that are supported by TCP/IP.

Before discussing about other TCP/IP layers, it is important to understand a few basic networking terms.

PDU: protocol data unit is a set of data bytes that can be sent from one host to other over a network.
Frame: This is a layer 2 PDU that contains the data that could be sent over a network, which includes end-user data + all headers and/or trailers from higher layers + layer 2 header and trailer.
Packet: A packet is a layer 3 PDU that contains the data that to be sent over a network, which includes end-user data + all headers and/or trailers from higher layers + layer 3 header.
Segment: A segment is a layer 4 PDU that contains the data that can be sent over a network, which includes end-user data + all headers and/or trailers from higher layers + layer 4 header.

10.3.2 INTERNET LAYER

The network access layer works with physical addressing and therefore determines how data transfer is to take place between one host to another when both the hosts are connected over the same physical network such as local area network (LAN). The Internet layer comes into picture when hosts are connected over two different physical networks, that is, data from one host to another has to be transferred over multiple physical networks. This layer separates the logical address from physical one (as defined by layer 2 standards) to define a way to deliver data from one host to the other.

Some examples of Internet layer protocols are IP, ARP, Internet control message protocol, and others. IP is the most prominent among all Internet layer protocols.

"Routing" is the process of transferring data between two hosts connected over different physical media. A device, called the "router," is used to connect two different networks. Let us consider a simple network connection where two different networks are connected via a single router (Figure 10.5).

To understand IP routing, it is essential to first understand what is an IP address, a subnet mask, and use of default gateway. In Figure 10.5, these three are defined for all four computers (Computer1, Computer2, Computer3, and Computer4). In the diagram, the following acronyms have been used.

- IP: Internet protocol address
- SM: Subnet mask
- DG: Default gateway

The IP address is the logical address of any host on the network. IP addresses are 32-bit binary numbers. For better readability (and easier for us), IP addresses are written in dotted decimal form, that is, 32-bit address is represented by four decimal numbers, each number representing eight bits.

Example:

Computer1 IP address is 10.1.1.5

10,1,1,5 each decimal represents 8 bits.

So, in binary form, the address is $\underbrace{00001010}_{10}.\underbrace{00000001}_{1}.\underbrace{00000001}_{1}.\underbrace{00000101}_{5}$

Subnet mask defines the network or subnet address and the host part from the IP address. Subnet mask is also a 32-bit number similar to IP address.

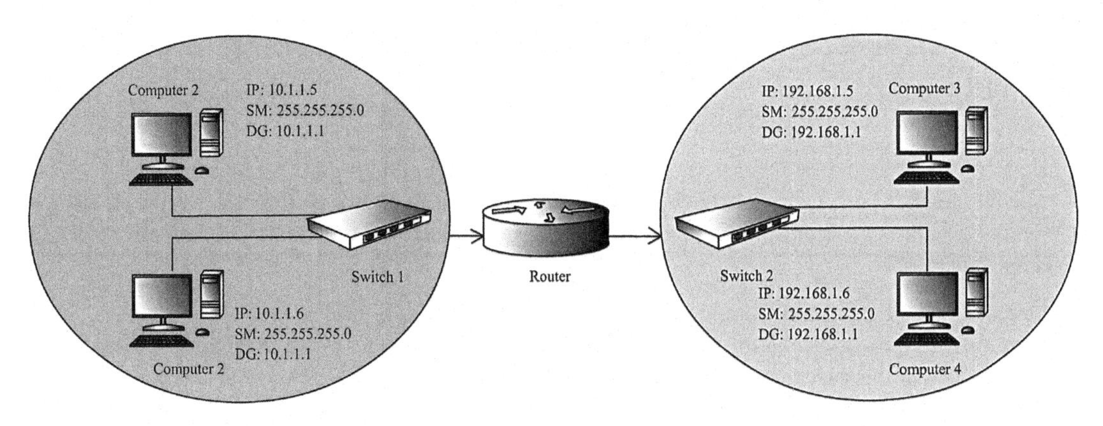

FIGURE 10.5

Communication between two networks

Subnet mask defines the network/subnet and host address just by doing bit-wise ANDing. The result of bit-wise AND between IP address and subnet mask represents the network or subnet part.

In above diagram Computer1's subnet mask is 255.255.255.0.

In binary format this is 11111111.11111111.11111111.00000000

255 . 255 . 255 . 0

By doing bit-wise AND we get

```
00001010.00000001.00000001.00000101
11111111.11111111.11111111.00000000
-----------------------------------
00001010.00000001.00000001.00000000
   10   .    1   .    1   .    0
```

Network/subnet address of Computer1 is 10.1.1.0, and the host address is "5." As any bit doing AND with binary 1 always results in the same bit, the part of IP address with subnet mask binary 1's represents the network/subnet and with binary 0's represents the host part.

All computers connected to same physical network must have the same network/subnet address. In Figure 10.5 Computer1 and Computer2 form one physical network and have the same network address 10.1.1.0, and Computer3 and Computer4 form another physical network with a shared network address 192.168.1.0. The left side circle in Figure 10.5 represents one network (10.1.1.0) and the right side circle represents the other (192.168.1.0). Two different networks can communicate only if they are connected via router(s). In a simple network as shown in Figure 10.5, two networks have been connected using only one router but in a complex network there could be multiple numbers of routers in between. The third part is default gateway, which is nothing but the IP address of the router port where the physical network is connected.

IP header	TCP/IP transport and application layer headers	Data

FIGURE 10.6

The data transfer mechanism between two hosts, which reside in two different networks, is shown in Figure 10.5. To send data from Computer1 to Computer3, use the following procedure (from the IP layer's perspective).

- Decides to send data from Computer1 application to Computer3
- Identifies the network ID of the host by using IP address and subnet mask of the host machine (Computer1 in this case and network ID is 10.1.1.0)
- Compares the same with destination machines IP address (as host now knows its own network ID)
- Finds that destination machine is in different network
- Sends the packet to default gateway (router port IP 10.1.1.1)
- Router accepts the packets and identifies the destination network IP address
- Makes the routing decision (through which port to send out the packet?) by comparing the destination IP address with the IP routing table
- The packet gets routed to 192.168.1.0 network through the port at which 192.168.1.0 network is connected
- The packet reaches the destination Computer3

In Figure 10.6 IP, a packet refers to data received from one layer above it (transport) and the header IP layer adds to it. IP header usually consists of information such as IP version, header length, total length of the packet, time to live, cyclic redundancy check (CRC; only for header), source IP address, destination IP address, and so on.

10.3.2.1 Time to live

The "time to live" is set into a packet when the packet is first generated. Whenever any router forwards this packet, this value gets decremented by "1" by the router until this reaches "zero." Once this becomes zero, the router discards the packet. Time to live ensures that the packet does not indefinitely route through the network in a condition when network loop exists, which could result into infinite loop of the packet in the network and end up consuming network bandwidth.

IP addresses used in examples so far were classless. IP addresses were originally allocated as per "address classes." Smaller number of networks with a large number of hosts falls under "class A" category. "Class B" represents moderate number of networks with moderate number of hosts, whereas "class C" has large number of networks with smaller number of hosts. "Class D" was intended for multicasting and "class E" for research purpose.

Class A IP address:

0-------.--------.--------.--------

Network Host

Size of the network ID – 1st 8 bits (1st octet)
1st bit of network ID is fixed to "0"
Next 7 bits contribute to create network ID
2nd, 3rd, and 4th octet (3 × 8 = 24 bits) forms host ID
Subnet mask: 255.0.0.0
As 1st octet forms network ID and 1st bit is set to 0, range of class A IP address is 1–126.
127 (all 7 bits "1") is reserved for special purpose, for example, 127.0.0.1 is loopback IP address.
Total number of class A network ID: 27 – 2 = 126
Total number of hosts per network ID: 224 – 2 = 1,67,77,216

Class B IP address:

10 ------.--------.--------.--------

Network Host

Size of the network ID – 1st two octets (2 × 8 = 16 bits)
1st 2 bits of network ID is fixed to "1" and "0," respectively.
Next 14 bits contribute to create network ID
3rd and 4th octet (2 × 8 = 16 bits) forms host ID
Subnet mask: 255.255.0.0
1st octet network address range: 128–191
Total number of class A network ID: 214 = 16,384
Total number of hosts per network ID: 216–2 = 65,534

Class C IP address:

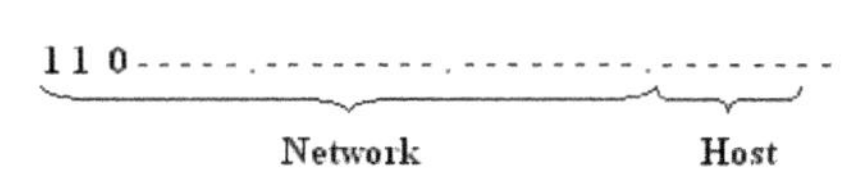

Size of the network ID – 1st three octets (3 × 8 = 24 bits)
1st three bits of network ID is fixed to "1", "1," and "0," respectively.
Next 21 bits contribute to create network ID
Only 4th octet (1 × 8 = 8 bits) forms host ID
Subnet mask: 255.255.255.0
1st octet network address range: 192–223
Total number of class A network ID: 221 = 20,97,152
Total number of hosts per network ID: 28 – 2 = 254

Note: Class D IP addresses (224–239) used for multitasking and Class E IP addresses (240–255) have been reserved for experimental use.

The following table summarizes the IP address class-related basics discussed so far:

Class	First octet	Initial fixed bits	Size of network (octet)	Size of host (octet)	Description
A	1–126	0	1st	2nd, 3rd and 4th	Unicast IP
B	128–191	10	1st and 2nd	3rd and 4th	Unicast IP
C	192–223	110	1st, 2nd and 3rd	4th	Unicast IP
D	224–239	1110	NA	NA	Multicast IP
E	240–255	1111	NA	NA	Experimental use

10.3.3 TRANSPORT LAYER (HOST-TO-HOST LAYER)

Transport layer is also known as host-to-host layer. This layer accepts data from the upper layer, that is, application layer, processes it, adds headers, and passes it on to next layer (Internet layer). Two most commonly used transport layer protocols are TCP and UDP.

10.3.3.1 Transmission control protocol (TCP)

TCP is a slightly complex layer 4 protocol. This accepts data from application, segments the data based on limitation in layers down below, sequences each segment so that the destination's TCP layer can put the segments back into sequence, waits for acknowledgment from the receiving computer's TCP, and finally, retransmits the data in case of loss during transmission.

TCP does a lot of processing before handing over the data to next layer. For example, if one computer application wants to send 3000 bytes data to another computer and underlying network layer protocols limit the maximum permissible data size to be sent over the network to 1000 bytes, the TCP segments the data into three parts, adds sequence numbers, headers, and then passes the data to network layer for further processing.

The TCP header contains critical information needed to successfully complete TCP protocol's task. The header size is 24 bytes (but used only 20 bytes today).

<table>
<tr><td colspan="2">Source port (16 bits)</td><td colspan="2">Destination ports (16 bits)</td></tr>
<tr><td colspan="4">Sequence number (32 bits)</td></tr>
<tr><td colspan="4">Acknowledgment number (32 bits)</td></tr>
<tr><td>Header length (4 bits)</td><td>Reserved (6 bits)</td><td>Flags (6 bits)</td><td>Window (16 bits)</td></tr>
<tr><td colspan="2">Checksum (16 bits)</td><td colspan="2">Urgent (16 bits)</td></tr>
<tr><td colspan="4">Options (0 or 32 bits)</td></tr>
<tr><td colspan="4">Data</td></tr>
</table>

1. ***Source and destination ports:*** To understand the importance of these two fields, let us consider an example where PC1's (computer) web-browser wants to send some data to PC2's web-browser. PC1 has multiple web-browsers open, and in PC2 too, user has opened multiple web-browsers. Once PC2 receives the IP packets from PC1, it must know to which web browser it must forward the data. To make this possible TCP (and UDP too) uses port numbers. Every time an application

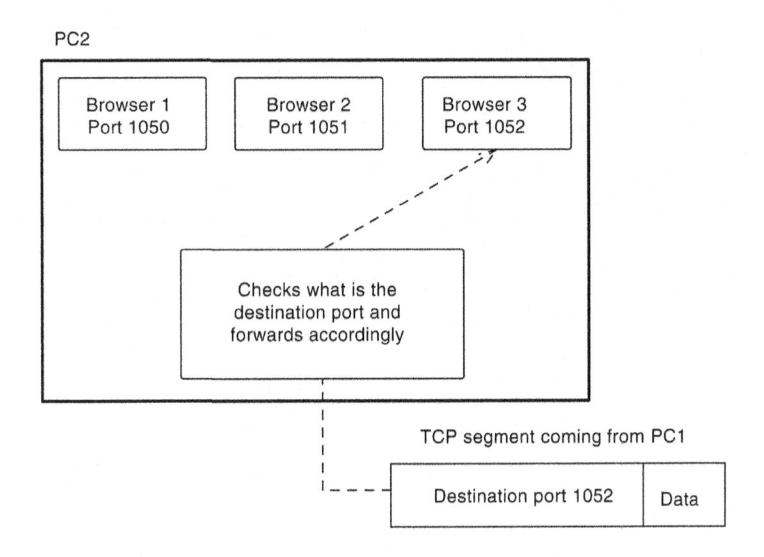

FIGURE 10.7

TCP Communications

is started on a computer, the host allocates a dynamic port number that the computer is not already using for another process. (Figure 10.7).

As shown in Figure 10.7, PC1 sends segment with destination port as "1052." PC2's TCP software helps in determining the application to which the received data needs to be forwarded by comparing the port number. Observe that the dynamic port number always starts from 1024 because most of the TCP/IP applications use client-server communication. When client wants to talk to server, client must in advance know for that particular application what port server uses. For this reason TCP/IP defines well-known ports each defined for specific applications. 0-1023 ports are reserved for that. Few examples of well-known ports are as follows:

Port 20 : FTP
Port 23 : Telnet
Port 80 : HTTP

Note: Internet Assigned Numbers Authority (IANA) assigns the values for well-known ports.

2. ***Sequence number:*** To understand the concept of sequence number, let us consider an example wherein PC1 is sending 3000 bytes data to PC2. The data is segmented into three 1000 bytes and being sent over the network owing to network limitations. If PC2 does not receive the segments in order, it becomes impossible for PC2 to merge those three segments and create 3000 bytes of meaningful data. That is why the sequence number is critical. Using this field PC2 can place the segments in order and retrieve the 3000 bytes data sent by PC1. There can be several reasons for the segments not reaching PC2 in order, even if they were sent out in order from PC1. The first or second segment could be lost in the network and PC2 requests for retransmission. Or, there could be two network paths available between PC1 and PC2 and for load balancing one segment might be sent through one path and other segment through another. There is no guarantee that both paths would have same network throughput, and hence the packets might not reach PC2 in order.

Features	TCP	UDP
Flow control and windowing	Yes	No
Connection oriented	Yes	No
Reliable (ACK and error recovery)	Yes	No
Segmentation and sequencing	Yes	No
Reassembly of data (in order)	Yes	No
Use port numbers to identify application	Yes	Yes

FIGURE 10.8

Feature comparison between TCP and UDP

3. ***Acknowledgment number:*** In Figure 10.8, observe the sequence number in each segment; it is nothing but the position of the first byte present inside the data. So, in the second segment SEQ: 1001 means the first byte of this segment is actually 1001th byte of the actual 3000 bytes of data sent from PC1. Similarly third segment's SEQ number is 2001. The acknowledgment is sent by PC2 upon receiving all 3000 bytes of data. Moreover, the acknowledgment number 3001 represents: PC2's expectation is to get 3001 onward data in next segment coming out from PC1. In case the second segment is lost during transmission then ACK should contain 1001 as the acknowledgment number so that the PC1 understands the loss of data and retransmits its second segment. Therefore, this also helps to achieve "error-recovery" feature of TCP, and hence makes this protocol reliable.
4. ***Header length:*** As the name suggests, this number indicates the length of the TCP header and from where the "Data" starts.
5. ***Reserved:*** This is reserved for future use.
6. ***Flags (6 bits):***
   - URG: Urgent pointer field significant
   - ACK: Acknowledgments field significant
   - PSH: Push function
   - RST: Reset the connection
   - SYN: Synchronize sequence numbers
   - FIN: No more data from sender

These are the six flags or binary bits. SYN and ACK bits are used to establish the communication between two computers sharing data over network using TCP. The process TCP follows to establish the communication is called "three-way handshake."

If PC1 wants to establish communication with PC2,

Step1: PC1 sends first segment with SYN flag ON
Step2: PC2 replies with another segment with both SYN and ACK bits ON
Step3: PC1 again sends one segment with only ACK bit ON

Moreover, the communication gets established between these two PCs. That is why TCP is called a "connection oriented" protocol.

- Window: This field tells the sending machine how many bytes it can send to the receiving machine without getting acknowledgment. This is mainly used for flow control. The receiver needs time to process the data it is receiving, and also it has finite memory to process the data. Flow control is introduced in order to ensure that the receiver can process the data based on its own limitation.

- If PC1 wants to send data to PC2, PC2 informs PC1 (using TCP segment) that window size is 3000 and then sends three segments 100 bytes each without waiting for acknowledgment. Once all three segments have been sent, PC1 stops and waits for acknowledgment from PC2 before sending next segment. PC2 sends the acknowledgment once it completes processing all three segments and might increase or decrease the window size depending on its capability to process new data in the acknowledgment message to inform PC1 how many bytes it can send now without waiting for acknowledgment. This method of increasing/decreasing window size is called as "dynamic sliding window."
- Checksum: The CRCs. TCP does not believe lower layers, and therefore has its own checksum to verify the correctness of header and data content.
- Urgent: This field points to the sequence number of sent data for which sending machine expects urgent acknowledgment from the receiver.
- Options: Not a frequently used field. This is additional header to expand the features of TCP in future.
- Data: Pure data that is being segmented by TCP based on limitations of lower-level protocols.

10.3.4 USER DATAGRAM PROTOCOL (UDP)

TCP supports lots of features for secured/reliable communication between two computers. But this does not come free. All these features add an overhead in terms of consumption of network resources, and also a complex protocol. Sometimes application developers build their own application to carry out similar tasks and using TCP as layer-4 protocol with that application does not make sense as most of the features become redundant and results into unnecessary overhead. In such cases, the use of UDP is recommended, which is a scaled-down version of TCP. Let us refer the UDP header first to understand what it supports.

Source port (16 bits)	Destination port (16 bits)
Length (16 bits)	Checksum (16 bits)
Data	

UDP is a thin protocol, with header field of just 8 bytes long. Source port, destination port, and checksum features are exactly the same as TCP. Length field indicates the size of UDP header and UDP data.

UDP does not support "three-way handshaking" to establish the connection and is therefore called as a "connectionless" protocol. It does not support windowing, does not wait for acknowledgment, cannot carry out error detection and error recovery (unreliable compared with TCP), and so on. UDP does not add much network overhead as TCP does. The table below summarizes the features supported by TCP and UDP:

After the upper layer protocols and Transport and Network layer protocols, let us understand the bottom most two layers (layer 2 and layer 1) of networking model. Let us understand the Ethernet standards, the most commonly used standard in networking world.

10.3.5 ETHERNET

Ethernet uses both layer 2 (data link layer) and layer 1 (physical layer) specifications. "Ether" means the air and "net" means network. Combining both, "Ethernet" means network over the air. Researchers (Robert Melancton Metcalfe) at University of Hawaii first set up an ALOHA network to connect sites on islands using radio broadcast in the early seventies. The idea was first documented in a memo that Metcalfe wrote on May 22, 1972. As the transmission media used was "air," it was named as "Ethernet." Later (1973–1975) Xerox Corporation took this forward. In 1975, Xerox filed a patent application listing Metcalfe, David Boggs, Chuck Thacker, and Butler Lampson as inventors. Later a team with Digital Equipment Corporation (DEC) and Intel was formed to establish Ethernet standards, and the standard used to be referred as DIX Ethernet. In early 1980s, IEEE (Institute of Electrical and Electronics Engineers) took over the development of Ethernet standards and formed 802 committee to work on LAN standards. Soon they realized the need for several LAN standards, and hence two subcommittees 802.2 and 802.3 had been formed. 802.3 subcommittee continued to work on Ethernet standards, whereas 802.2 subcommittee started working on several types of LANs such as Ethernet, token ring, and others.

10.3.5.1 CSMA/CD

When multiple computers are connected over a shared network, there is a probability of data collision because of two or more computers simultaneously sending data. Carrier sense multiple access/collision detect (CSMA/CD) was created to overcome this problem. Early Ethernet (10 mbps) uses CSMA/CD access method, which is an algorithm that defines rules each Ethernet NIC (network interface card) must follow to access an Ethernet LAN.

For a better understanding of CSMA/CD, let us consider 10BASE5 network that uses physical bus topology. Here 10 stands for 10 mbps (megabits per second) speed, BASE means baseband transmission, and 5 represents the maximum length of the network of 500 m. 10BASE5 is a physical bus topology as shown in Figure 10.9. Four computers are connected using physical bus topology.

Let us assume PC1 wants to send out data.

- First, PC1 checks for any activity (data transfer) over the network or not. This is called carrier sense.
- Only on confirmation of no activity PC1 starts sending data over network.

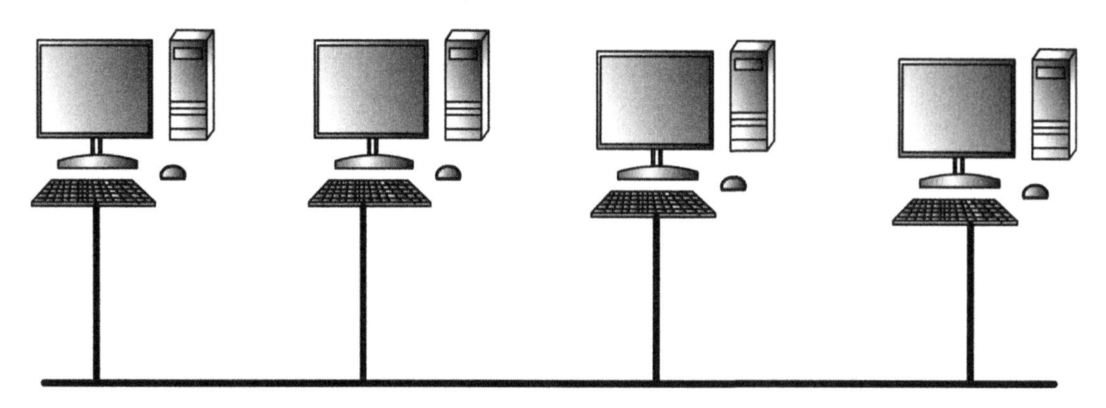

FIGURE 10.9

Simple network on TCP/IP

- There is a high probability of another computer detecting the LAN silence at the same time and start transmitting the data simultaneously (multiple access).
- Anticipating probability of multiple access at same point in time PC1 listens carefully to detect any collision in network after transmitting the data (collision detect).
- In the absence of collision, the data transmission is considered as completed.
- If sending computers detect collision, all sending devices send out "jamming signal" to inform all connected nodes about the collision.
- All sending computers whose data collided set a random back-off timer to delay the next retry.

10.3.5.2 Simplex, half duplex, and full duplex

In simplex communication, the network cable or channel transmits information in one direction only. Most of the fiber optic communications are simplex in nature.

In half-duplex communication, the node is capable of bidirectional data transmission (transmit or receive) but cannot do so simultaneously. At any point in time the node can either transmit or receive. Here "duplex" refers to the fact that communication can happen in both directions and "half" because it can either send or receive—cannot do both simultaneously. This is where CSMA/CD comes into picture. The main disadvantage of CSMA/CD with half-duplex communication is degradation of network throughput if more number of nodes are connected over the network and uses the network media for data transfer frequently (in other words, a busy LAN).

In full-duplex communication, connected nodes can simultaneously transmit and receive data. Most of the time full-duplex is achieved by using two simplex links together.

With full-duplex support and use of switches for interconnection, probability of collision in network has almost become zero, making CSMA/CD unnecessary. Switches prevent collision by buffering Ethernet frames, that is, if multiple nodes want to send data to a single node, switch makes sure that the data are transmitted from the switch port in series by buffering the data in switch memory (forms queue) and hence avoiding collision.

10.3.5.3 Layer 2 Ethernet specifications

At layer 2 (data link layer) Ethernet mainly talks about the MAC address and framing of data received from its top layer, that is, network layer.

MAC address is a 6-byte (48 bits) hexadecimal number that is used to identify NIC. Each NIC has a permanent unique MAC address (Figure 10.10).

First 24 bits used (3 bytes) are reserved for the vendor's number registered with IEEE. This field is called organizational unique identifier. Last 24 bits are used by the vendor to assign unique number to each and every physical NIC.

Example of MAC address in hexadecimal format:

C0-CB-38-97-D4-41

Organizational unique identifier vendor assigned number

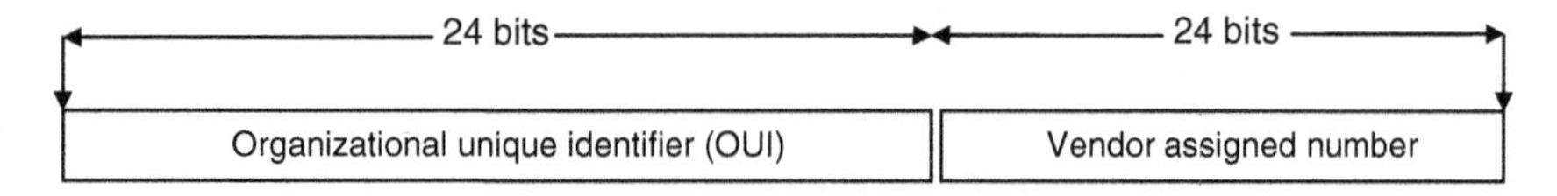

FIGURE 10.10

MAC address structure

Preamble	SFD	Destination	Source	Length/type	Data and pad	FCS
7 bytes	1 byte	6 bytes	6 bytes	2 bytes	46–1500 bytes	4 bytes

FIGURE 10.11

Ethernet Frame

The data link layer also forms Ethernet frames with the data (packet) received from network layer and adds its own header/trailer into that.

Layer 2 Ethernet also performs error checking for error detection only (not error correction).

Following diagram shows the contents of Ethernet frame (Figure 10.11):

- ***Preamble and SFD:*** Preamble and start frame delimiter (SFD) together use a "be aware" or "pay attention" kind of purpose. SFD signifies that the next byte would be the destination MAC address. When receiving nodes start getting preamble and SFD, they get ready for the data and start locking the incoming bit stream.
- ***Destination MAC address:*** This field contains the MAC address of the NIC for which the data is getting transmitted. All nodes listening to the Ethernet frame check this field. If it matches with its own MAC address, the node accepts the frame for further processing and rejects the frame otherwise. Remember MAC address can be any individual address (used for unicast), or could be multicast or broadcast address.
- ***Source MAC address:*** MAC address of the NIC that has transmitted the Ethernet frame. This field can never be multicast or broadcast address.
- ***Length/Type***: 802.3 uses this as a length field that defines the length of the frame's data field. Although as per Ethernet standard this field can also be used as type (either length or type, both cannot be used together) to define the network layer protocol embedded inside the frame.
- ***Data and pad***: This field holds data received from network layer. In case of TCP/IP, this is nothing but IP packet. Sometimes PAD is used if data size is small.
- ***FCS***: Frame check sequence (FCS) is the field at the end of Ethernet frame, which is used to store the CRC for error detection only.

10.3.6 LAYER 1 ETHERNET SPECIFICATIONS

When IEEE took over Ethernet standard development from DIX, Ethernet was a mere 10 Mbps network that used to run on coaxial cable. With time, Ethernet has also evolved and supports up to Gbps (Gigabit per second) and also supports different media such as twisted pair and optical fiber.

Different physical layer specifications (layer 1) of Ethernet are 10Base2, 10Base5, 10BaseT, 100BaseTX, 100BaseFX, and so on. EIA/TA (Electronics Industries Association/Telecommunication Industry Alliance) is the committee that creates and maintains layer 1 specifications of Ethernet.

The number before the word "Base" represents the speed of data transfer. The unit of speed is megabits per second (Mbps). The word "Base" signifies this is a baseband transmission, that is, a single frequency is used to encode the bits (the other technology is "broadband" where frequency band is used instead of using single frequency).

So, going by above explanation 10Base2 means:

- 10Base2: 10 Mbps Ethernet, baseband technology and 185 m maximum length of the cable supported (then why 10Base2? I interpret it this way, 2 represents 200 m length, nearest round figure of 185 m). This network is also known as "thinnet."
- 10Base5: 10 Mbps Ethernet, baseband technology and 500 m maximum length of the cable supported. This network is also known as "thicknet."
- 10BaseT: 10 Mbps Ethernet, baseband technology with copper made twisted pair cable as transmission media. "T" represents the cabling type. The maximum distance limitation is 100 m. Twisted pair cable used for this is "category 3" type.
- 100BaseTX: 100 Mbps Ethernet, baseband technology with copper made twisted pair cable as transmission media. Twisted pair cable used for this is "category 5/6/7" type.
- 100BaseFX: 100 Mbps Ethernet, baseband technology with optical fiber cable as transmission media. Maximum distance supported is 412 m.

10.3.7 INTRODUCTION TO HUB, SWITCH AND ROUTER

The study of TCP/IP and Ethernet is incomplete without the understanding of a few of the networking equipments such as hub, switch, and router. They are the backbone of networking.

10.3.7.1 Ethernet hub

An Ethernet hub is used to connect multiple nodes in a single LAN using twisted pair cables. Multiple computers can be connected to a single HUB, which forms a physical star topology as shown in Figure 10.12.

Hub is a layer 1 device. It receives signals from the transmitting device (say PC1 which is connected to port 1), regenerates the signal as the bit stream might get distorted because of transmission loss, and

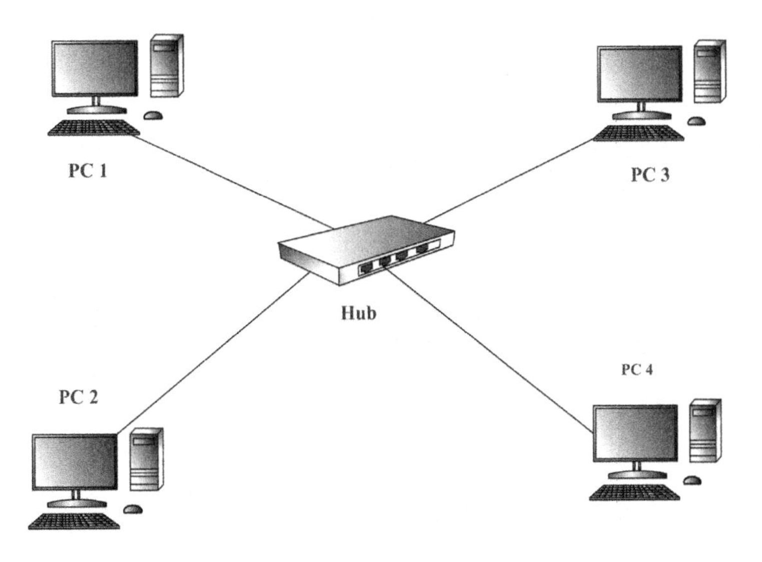

FIGURE 10.12

Network in star topology

sends the bit stream out of all other ports except the one through which hub received the data, that is, port 2, port 3, and port 4.

10.3.7.2 Ethernet switch

Ethernet switch is a layer 2 networking device. The physical topology of network using switch is exactly the same as hub (see Figure 10.12). However, switch is an intelligent device. It can read the Ethernet frame and identify the source and destination MAC addresses. Ethernet switch also maintains a table of MAC addresses Vs port number so that it can decide through which port the data needs to be sent out. The following table lists the MAC address of connected computers and its respective connecting port:

PC1 MAC Address	Port 1
PC2 MAC Address	Port 2
PC3 MAC Address	Port 3
PC4 MAC Address	Port 4

If PC1 wants to send data to PC4, the data comes to the switch through port 1. The switch reads layer 2 Ethernet frame and identifies the destination address as PC4 MAC address. From the addressable memory table, the switch gets the information that PC4 MAC address is connected to port 4. Therefore, the switch, unlike the hub that sends out the data through all the ports, only sends out the data through port 4. This implies that if PC2 wants to send data to PC3 simultaneously, it can be achieved since intelligent processing isolates the collision zone.

10.3.7.3 Router

Router is a layer 3 networking device, that is, it can read layer 3 (IP) packets and identify the source and destination IP addresses. It is used to connect two or more different subnets together. It is also used to form WAN (wide area network). Router maintains the routing table analogous addressable memory table that the switch maintains. Routing table contains subnet-related information (a particular subnet connections to a port of the router). This helps to decide how to route packets between different subnets.

Let us now understand the data transfer mechanism between two computers within the same network using Ethernet.

For example, Computer1 wants to send data to Computer2:

- The network ID of the host is determined using IP address and subnet mask of the host machine
- The network ID of the host and destination machines' IP address are compared to establish that the destination machine is in the same network
- Host refers the ARP table to find out the MAC address of the destination node (ARP table maintains IP address to MAC address mapping).
- After identifying destination MAC address, Ethernet forms the frame with source and destination MAC addresses.
- The frame reaches the switch and the switch reads the Ethernet Frame to identify destination MAC address.
- The Ethernet gets the port number for destination MAC address using its addressable memory table.
- The Ethernet regenerates the signal and sends out through the port where Computer2 is connected.

10.3.8 INDUSTRIAL NETWORK

The production place of a business activity is called as "industry." An industry takes in raw material as input and processes the input to produce some output for consumers. Automation and control plays a major role in any industry. It is almost impossible to manage a large industry and optimize its operations without automation and control solutions. Automation and control mainly includes sensing instruments for process variables (such as temperature, pressure etc.), control system to read the process variables and take the action based on the logic build into it, and output elements such as actuators, human machine interface for monitoring the process and take control actions, and the like. All these components are connected together by what is called an "industrial network."

ISA calls this as Manufacturing and Control Systems (M&CS). The formal definition of M&CS is "hardware and software system such as Distributed Control System (DCS), Programmable Logic Controller (PLC), Supervisory Control and Data Acquisition (SCADA) systems, networked electronic sensing, and monitoring and diagnostic system along with associated internal, human machine interface, network used to provide control, safety and manufacturing operations functionality to continuous, batch, discrete or other processes" (ISA-TR99.00.02-2004).

On a very high level, industry can be divided into three domains chemical processing, utilities, and discrete manufacturing. Chemical processing includes chemical plants, refineries, food and beverage, pulp and paper, and so on. Utilities include mainly distribution systems such as power, water, waste water, and others. Plants that manufacture discrete objects such as automobile parts fall under discrete manufacturing category.

10.3.8.1 Industrial Ethernet

Industrial network has undergone a major revolution since its inception. The change is happening very fast. The distributed control system was introduced in the market in the early 1970s. The network at that time was completely proprietary. The shift was mainly from proprietary network to open standards such as Ethernet. Commercial off-the-shelf products have replaced older proprietary components. Ethernet is being widely used as a communication medium and TCP/IP as communication protocol. Earlier, Ethernet used CSMA/CD impacting the data transfer throughput in a heavily loaded network. This was not a problem for office usage. The long delay (in seconds) was, however, not acceptable for critical process control and process safety applications. The rapid development of Ethernet over the past few decades has helped to gain widespread acceptance of Ethernet in industrial applications.

As soon as industrial network started using COTS products and open standards such as Ethernet, the next obvious requirement that came up was to connect industrial network to business network for ease of operation. This, however, made it vulnerable to cyber attacks, bringing into picture industrial network security. An attack on industrial network renders the process, equipments, and the people associated or, in single word, whole industry "unsafe."

10.3.8.2 Network separation (partitioning)

For security reasons as also for better performance at industrial network level, it is recommended to isolate business network from industrial network. The connectivity between business network and industrial network must be through a firewall to limit the access and provide security from outside cyber attacks.

Industrial network is also further separated into multiple subnetworks mainly to reduce unwanted network traffic in critical control network and enhance the speed of response by reducing the network latency. Reduction of network traffic in industrial network is very important, and to achieve this, most of the DCS and PLC vendors use exception-based reporting of process parameter. For example, the DCS/PLC writes to any digital point (ON of OFF) whenever there is any change in value such as ON to OFF transition or vice versa, instead of writing at regular frequency reducing the load on the network.

Switching technology is another important aspect widely used in industrial networks. Switches are intelligent layer 2 devices that can read the MAC address (layer-2) and take the decision through which port data needs to be sent out instead of sending out through all ports such as hub. This reduces the probability of collision and therefore increasing the overall speed of data transfer. Advanced switches also support VLAN (virtual LAN) that helps to create logical grouping of nodes, although connected to same switch physically. Nodes of one VLAN cannot communicate to nodes of another VLAN (creates separate broadcast domains), although all nodes are members of the same physical LAN.

Although industrial network best practices recommend isolating industrial network from business network, integration of business network with industrial network yields several benefits. ISA 95 recommends different level of industrial networks and their interconnectivity to achieve better integration of business to industrial network for optimum plant operation without compromising security. The following are the levels of network as recommended by ISA 95.

- Level 0: Defines the actual physical processes or the physical production process.
- Level 1: Defines the activities involved in sensing and manipulating the physical process and automatic control of process parameters. This includes sensors, input/output (I/O) modules, embedded controllers, and actuators.
- Level 2: Defines the activities of monitoring and controlling the physical processes. Human machine interface, SCADA, and others, are part of this level.
- Level 3: Defines the activities of the workflow to produce the desired end-products. In this level, the maintenance of production information is centralized to provide greater control and availability of the records.
- Level 4: Defines the business-related activities needed to manage a manufacturing organization. Enterprise resource planning is a critical component that resides in this level (Figure 10.13).

DCS/PLC forms the Level 1 and Level 2 networks. These two networks combined together is called " PCN." Level 3 and Level 4 combined known as " PIN."

10.3.8.3 Process control network (PCN)

Sensors/actuators are hooked with the control system (DCS/PLC) using a sensor network. The most conventional sensor network is 2-wire 4–20 mA connection. This is a point-to-point network. Each sensor/actuator gets connected to host's (PLC/DCS) I/O module using dedicated 1 pair of cables (2-wire). With time many other sensor network standards have evolved and slowly replaced conventional 4–20 mA connections.

AS-interface (AS-i) is the most popular sensor network for connecting discrete I/O to controllers. AS-i is a low-cost multidrop bus topology, using either a two-conductor round cable or unique two-conductor flat cables. Both cables deliver DC power to the node and the device using the same wires for both power and data. A single AS-i network can connect up to 124 inputs and 124 outputs and can be about 100 m long.

DeviceNet is another widely used sensor/actuator network. I/O modules similar in function to those of AS-i are available to terminate conventional sensors and actuators and communicate on the DeviceNet network. Although the higher-level application layer, called CIP, can be optionally used

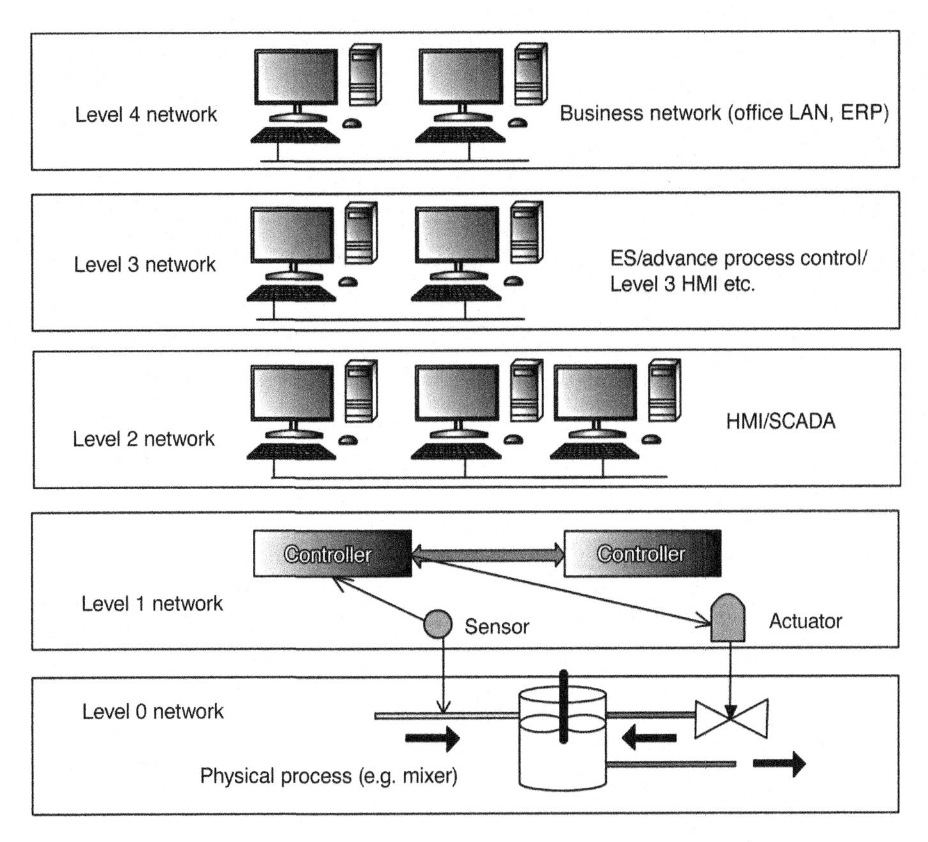

FIGURE 10.13

Network architecture of automation system

with DeviceNet as a field bus, making DeviceNet almost similar as AS-i network. Rockwell has developed DeviceNet, but afterward handed over the specifications to the open DeviceNet vendor association, which now controls any revisions/enhancement of this standard.

Ethernet industrial protocol (EtherNet/IP) has been created by open DeviceNet vendor association, ControlNet International, and Rockwell automation. EtherNet/IP has been designed to efficiently implement data transfer using CIP (Control Integration Protocol) at the application layer. It operates on commercial Ethernet, but uses alternative physical wiring and connectors better suited for industrial automation. One of the main objectives of EtherNet/IP is to provide complete industrial data communications using commercial off-the-shelf Ethernet hardware and cabling.

H1, H2, and HSE: H1 is a purely field-level network designed to connect field devices and final control elements. Speed of H1 is 31.25 kbps, intrinsically safe, and runs on existing field wiring. It delivers both power and field data over the same wires. H1 works on bus topology, that is, multiple devices can be connected to single bus (2-wire), which transfers the data from all the devices connected to the bus to the host or vice versa. H2 with 1 or 2.5 Mbps speed was designed to run at a higher speed and connect I/O subsystem with controllers. H2 is not intrinsically safe. The H2 specifications have been functionally replaced with HSE (high-speed Ethernet), which operates at 10 Mbps and has all the advantages Ethernet technology has. H1 at field level and Ethernet-based HSE at host level (controller-level network) is one of the most widely used network topology in industrial network today.

Communication between peer controllers called as peer-to-peer (P2P) communication is another important aspect of PCN. Controllers share critical to control data, and therefore technology used to achieve P2P communication is also important considering the high quality of service and speed of response requirement for such communications. In the 1970s and 1980s almost all DCS/PLC vendors used proprietary communication mechanism for P2P communications. With time semiproprietary network evolved and finally almost all vendors are coming up with open standard support such as Ethernet for process control.

The best practice is to put Level 1 nodes on a separate switch. This allows critical peer-to-peer traffic that cannot tolerate a communication delay. It also gives controllers a level of isolation from other nodes during catastrophic failure or network disturbance. The most critical elements of control must be connected to this switch pair. Because Level 1 nodes include controller nodes, the critical control traffic must have adequate bandwidth.

Level 2 network, which includes human machine interface/SCADA and gets connected to level 1 network to monitor plant data and supervisory control, has also evolved from proprietary to semi-proprietary to open standards. Ethernet is being widely used now in this part of PCN too. Level 2 switches are configured to have the same quality of service (QoS) approach as the Level 1 switches.

10.3.8.4 Plant information network (PIN)

Level 3 and Level 4 networks form a part of the PIN. From the security and quality of service perspective, Level 3 network must be isolated from Level 2 network by a router. The nodes that usually reside at Level 3 network are advanced process control, domain controller, applications, advanced alarming, and so on. The use of access list in the router that connects Level 2 to Level 3 is another recommended security best practice to limit access requests from Level 3 to Level 2, which in turn minimizes unwanted traffic in PCN.

Level 3 network is mainly used for manufacturing operation management such as dispatching, production scheduling, production tracking, and so on. Level 4 is for business logistics management (enterprise resource planning), such as inventory management, shipping and receiving, resource management, and so on. Level 4 or the business network must be isolated from Level 3 network using firewall. Access list should be written in such a way that only the required traffic can pass through the firewall and reach Level 3 network.

It is important to ensure connectivity between Level 4 and Level 3 and in turn between Level 3 and Level 2 for the following reasons.

- Business drivers: Key business drivers are the areas of performance that are most critical to an organization's success.
- Available to promise: Requires detailed knowledge of available capacity.
- Reduced cycle time: Major performance indicator with a direct impact on corporate profitability.
- Supply chain optimization: Optimizing the manufacturing link in the supply chain – agile and responsive.

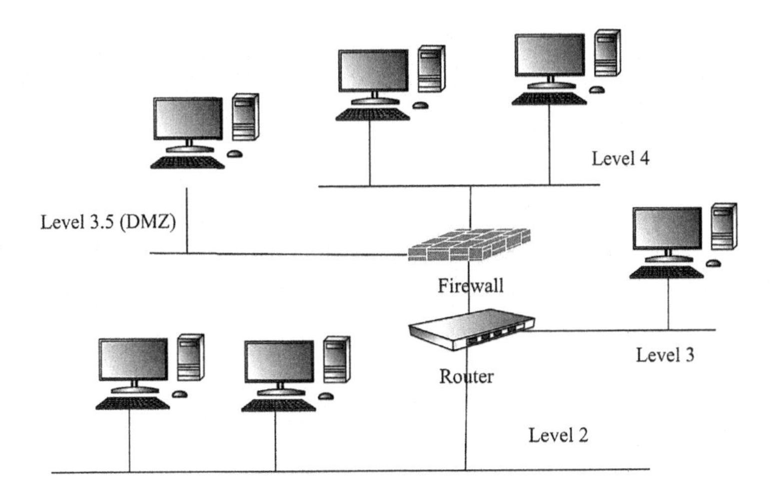

FIGURE 10.14

Architecture of secure automation system

- Asset efficiency: Requires detailed knowledge of actual use.
- Agile manufacturing: Requires ability to quickly synchronize planning and production.

In an industrial network, connectivity from Level 4 network directly to Level 2 network sometimes becomes a requirement for plant operation optimization. However, such connectivity is considered as violation of security norms. To avoid such violation of security standards, another level is introduced in between called as DMZ (demilitarized zone), as illustrated in Figure 10.14. In industrial network terminology this level is often also called as Level 3.5 network .

Users at Level 4 network get access to computers at Level 3.5 (DMZ). Through the DMZ nodes ONLY data from Level 2 can be pulled up and provided to Level 4. This ensures that the external attacker can only pass through till DMZ and not till Level 2/Level 1, which is the most critical level not only from plant operation perspective but also for safety of plant equipment and personnel associated with those equipment.

CHAPTER

HART COMMUNICATION

11

11.1 INTRODUCTION

Highway addressable remote transducer (HART) is the global standard protocol for sending and receiving digital information across analog wires between smart devices and control or monitoring system. It is a popular bidirectional digital communication protocol designed for industrial process measurement applications. It superimposes digital signal on top of the 4–20 mA analog loop. HART allows for a simultaneous analog signal with a continuous digital communication signal that has no effect on the analog signal.

HART provides data access between intelligent field instruments and host systems. A host can be any software application like a technician's hand-held device or laptop, a plant's process control, asset management, safety, or other system using any control platform.

HART was developed by Rosemount Inc. in the mid-1980s, but has been made completely open, and all rights now belong to the independent HART Communication Foundation (HCF). HCF is an international, not-for-profit, membership organization, and the technology owner and central authority on the HART protocol. The Foundation manages and controls the protocol standards including new technology developments and enhancements. HCF was founded in 1993 with the primary focus to advance the application of HART technology, strengthen its position in the global market, and help users maximize the value of their smart instrumentation investments.

A maximum of four measurements can be transmitted in a single message using digital HART communication. Multivariable instruments have been developed to take advantage of this. In addition, when using only digital communication, several instruments can be connected in parallel "multidrop" on a single pair of wires (with their analog currents each set to a minimum value, usually 4 mA). Each device has its own address, so a host can communicate with each one in turn.

11.2 TECHNOLOGY

The following section explains the basic principles behind the operation of HART instruments and networks.

11.2.1 FREQUENCY-SHIFT KEYING

The HART communication protocol is based on the Bell 202 telephone communication standard and operates using the frequency-shift keying (FSK) principle. The digital signal is made up of two frequencies – 1200 Hz and 2200 Hz representing bits 1 and 0, respectively. Sine waves of

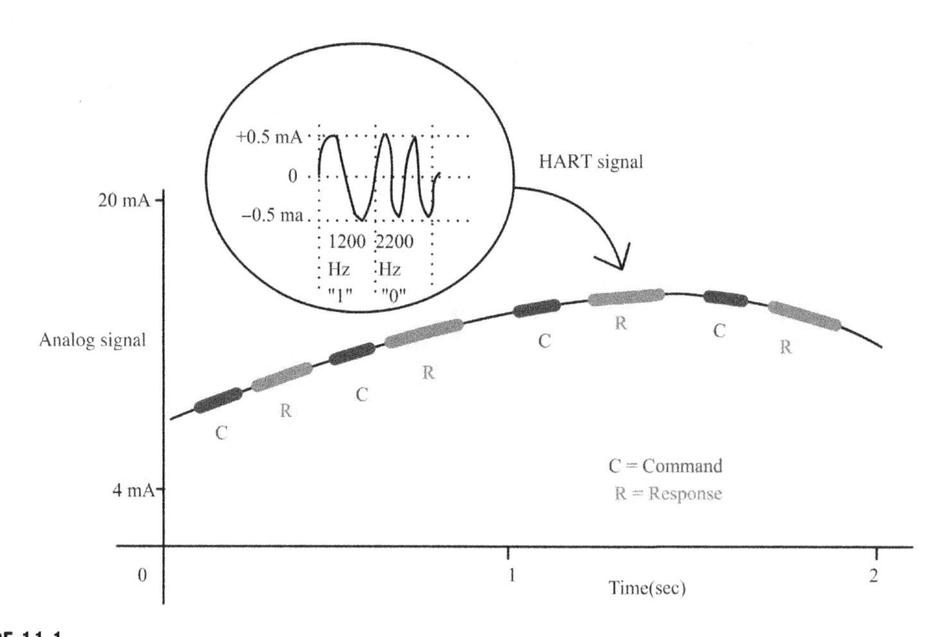

FIGURE 11.1

Digital over analog (frequency-shift keying).

these two frequencies are superimposed on the direct current (DC) analog signal cables to provide simultaneous analog and digital communications (Figure 11.1). Because the average value of the FSK signal is always zero, the 4–20 mA analog signal is not affected. The digital communication signal has a response time of approximately 2–3 data updates per second without interrupting the analog signal (Figure 11.2).

11.2.2 HART COMMUNICATION MODES

HART protocol supports two types of communication.

1. Master–slave mode: HART is a master–slave communication protocol. A master communication device initiates all slave (field device) communication during normal operation. Two masters can connect to each HART loop. The primary master is generally a distributed control system (DCS), programmable logic controller (PLC), or a personal computer (PC). The secondary master can be a handheld terminal or another PC. Slave devices include transmitters, actuators, and controllers that respond to commands from the primary or secondary master.
2. Burst mode: Burst mode enables faster communication (3–4 data updates per second). In burst mode, the master instructs the slave device to continuously broadcast a standard HART reply message (e.g., the value of the process variable). The master receives the message at the higher rate until it instructs the slave to stop bursting.

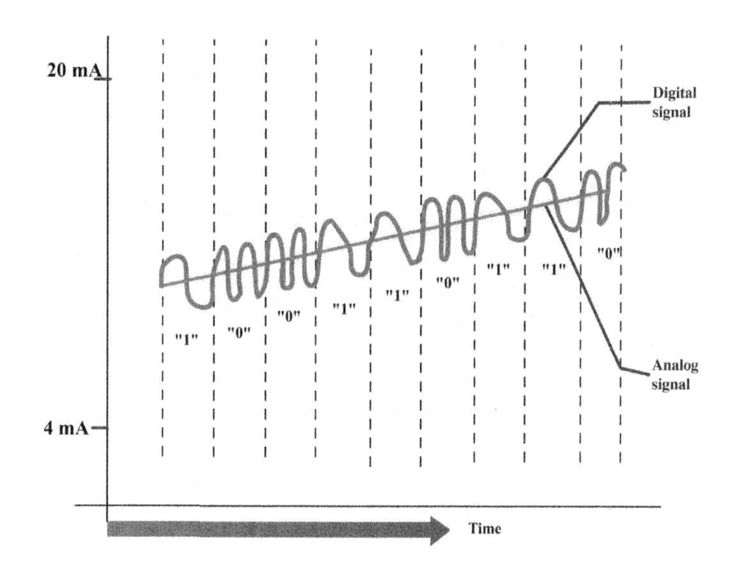

FIGURE 11.2

Example of phase-continuous frequency-shift keying.

11.2.3 HART NETWORKS

HART devices can operate in one of the two network configurations.

- Point-to-point: In point-to-point network mode, the traditional 4–20 mA signal is used to communicate one process variable, while additional process variables, configuration parameters, and other device data are transferred digitally using the HART protocol (Figure 11.3). The 4–20 mA analog signals are not affected by the HART signal and can be used for normal control. The HART communication digital signal gives access to secondary variables and other data that can be used for operations, commissioning, maintenance, and diagnostic purposes.
- Multidrop: The multidrop mode of operation requires only a single pair of wires and, if applicable, safety barriers and an auxiliary power supply for up to 15 field devices (Figure 11.4). All process values are transmitted digitally. In multidrop mode, all field device polling addresses are >0, and the current through each device is fixed to a minimum value (typically 4 mA).

11.2.4 HART COMMANDS

The HART command set provides uniform and consistent communication for all field devices. The command set includes three classes: universal, common practice, and device specific. Host applications may implement any of the necessary commands for a particular application.

- Universal commands (0–30): This command set must be supported by all HART devices and implemented exactly as specified by the HART foundation specification. Universal commands provide access to information useful in normal operations (e.g., read primary variable and units).

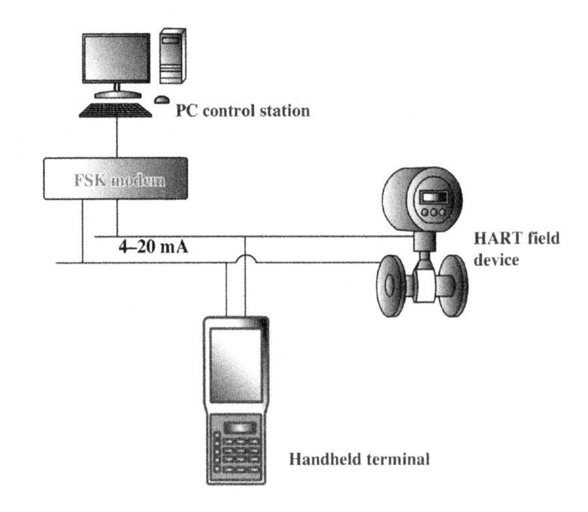

FIGURE 11.3

Point-to-point communication.

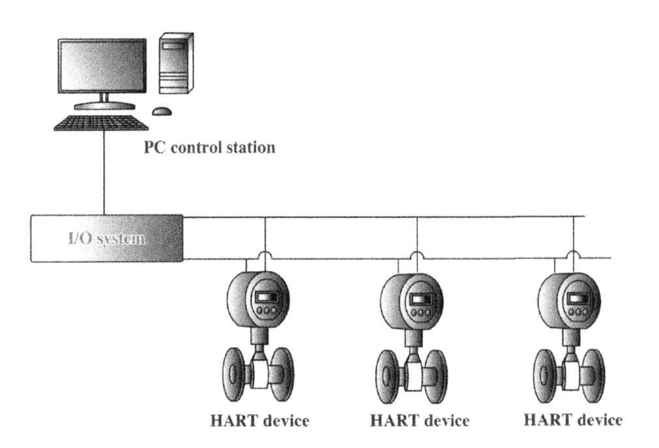

FIGURE 11.4

Multidrop mode.

- Common practice commands (32–121): This is a set of commands applicable to a wide range of devices. This set of commands must be supported by devices whenever possible. Although the function of each command is well defined by the HART foundation specification, the actual meaning of the response data may require the interpretation of vendor DD files.
- Device-specific commands (128–253): This command set is completely defined by the device vendor and each command performs a function specific to the particular device type and model. The use of these commands requires full interpretation of the vendor DD files.

11.3 HART TECHNOLOGY

HART communicating devices signal with either current or voltage, and all signaling appears as voltage when sensed across low impedance. For convenience, communicating devices are described in terms of data link and physical layers according to the OSI 7-layer communication model (Figure 11.5).

The HART protocol specifications directly address three layers in the OSI model: the physical, data link, and application layers.

11.3.1 PHYSICAL LAYER

The physical layer connects devices together and communicates a bitstream from one device to another. It is concerned with the mechanical and electrical properties of the connection and the medium (the copper wire cable) connecting the devices. Signal characteristics are defined to achieve a raw uncorrected reliability. The physical layer specifications specify the physical device types, network configuration rules, sample topologies, and common characteristics of all devices, which include analog signaling requirements, device impedance, test requirements, and cable characteristics.

The physical layer commonly uses twisted-pair copper cable as its medium and provides solely digital or simultaneous digital and analog communication. Maximum communication distances vary depending on network construction and environmental conditions. A representation of the relationship between the HART protocol communications layers (including firmware and hardware) is given in Figure 11.6.

OSI layer		Function	HART
7	Application layer	Provides the user with network capable applications	Provides the user with network capable applications
6	Presentation layer	Converts the application data between network and local machine formats	
5	Session layer	Connection management services for applications	
4	Transport layer	Provides network-independent transparent message transfer	
3	Network layer	End-to-end routing of the packets resolving network address	
2	Data link layer	Establishes data-packet structure, framing error detection, bus arbitration	A binary, byte-oriented token passing, master/slave protocol
1	Physical layer	Mechanical/electrical connection; transmits raw bit stream	Simultaneous analog and digital signaling-, normal 4–20 ma copper wiring

FIGURE 11.5

OSI 7-layer model.

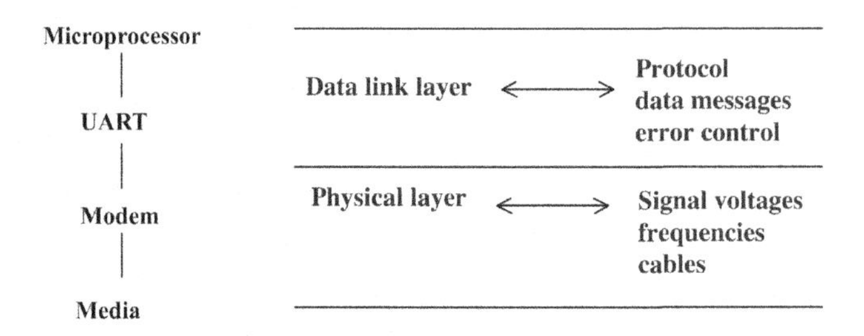

FIGURE 11.6

Data link layer and physical layer relationship.

11.3.2 DATA LINK LAYER

While the physical layer transmits the bitstream, the data link layer is responsible for reliably transferring that data across the channel. It organizes the raw bitstream into packets (framing), adds error detection codes to the data stream, and performs media access control (MAC) to ensure orderly access to the communication channel by both master and slave devices.

The bitstream is organized into 8-bit bytes that are further grouped into messages. A HART transaction consists of a master command and a slave response. Media access consists of token passing between the devices connected to the channel. The passing of the token is implied by the actual message transmitted. Timers are used to bound the period between transactions. Once the timer expires, the owner of the token relinquishes control of the channel.

11.4 APPLICATION ENVIRONMENT

11.4.1 OVERVIEW OF THE HART APPLICATION ENVIRONMENT

The protocol is intended to provide a reliable, transaction-oriented communication path to and from slave devices, such as field instruments, for digital data transfer. Communication is over twisted pair that may be simultaneously carrying 4–20 mA signaling. The protocol corrects errors due to noise on the communication links by using error-detecting information and an automatic repeat request (ARQ) protocol to request the retransmission of data blocks that may be corrupted by line noise or other disturbances.

If poll (short form) addresses are used, up to 15 slave devices may be multidropped on a single communication link. If unique (long form) addresses are used, the number of multidropped devices is essentially unlimited and is determined based on the application's required rate of scan of the devices on the communication link.

The protocol also arbitrates access to the field instrument between secondary master devices such as a handheld terminal and primary master devices such as a control system. The protocol gives equal access to the communication channel to both kinds of masters when they are being simultaneously used. The protocol does not arbitrate between two secondary or two primary masters trying to talk on the same link.

To support the regular transfer of information from slave device to master device, the protocol supports a mode of operation in which slave devices periodically broadcast information onto the communication link. A slave device is said to be in "burst mode" when it provides a synchronous cyclic broadcasting of data, without continuous polling by a master device. No matter how many slave devices are on a communication link, only one may be in burst mode.

Information transfer between devices on the communication link is through a defined message format. The entire message is protected by a single parity check product code (sometimes also known as vertical and longitudinal parity checking). Message framing is through a combination of a start of frame delimiter and a message length field.

The protocol provides services for device identification, segmented data transfer, and communication configuration.

11.4.2 DEVICE TYPES

The communications protocol recognizes three distinct device types among entities capable of using this protocol.

- Slave device: This is a device that accepts or provides a digital message carrying measurement or other data, but only when specifically requested, that is, this device always functions as a slave in a master/slave relationship.
- Burst mode device: A burst mode device is defined to be a slave mode device with burst mode capability (hence the use of the word "mode" in describing the device type). When such a mode is enabled, the device is said to "be in burst mode."
- Master device: A master device is responsible for initiating, controlling, and terminating transactions with a slave device or a burst mode device. Master devices are classified into primary master devices and secondary master devices to allow the simultaneous use of two master devices on a HART communication link. The same protocol rules are followed by a primary master device and a secondary master device except for customizing time-outs that differentiate between them.

11.4.3 PROTOCOL MODEL

Figure 11.7 is a conceptual model of a HART protocol implementation. It is intended as a frame of reference rather than as a description of an actual implementation. The purpose is to show the relationships between the various services, interfaces, and protocol components. As can be seen from the figure, the model distinguishes between five components. The user/protocol interface and the protocol/physical layer interface are descriptions of services that are provided to or required from adjacent protocol "layers." The other three components make up the protocol specification itself. This is broken up in terms of three corresponding state machines. Receive and transmit machines describe handling of message transmission and reception, while the main state machine uses invocations of these to implement the rest of the protocol.

The interface between the various components of the protocol is specified in terms of service primitives. The specification of a service consists of two parts, the first of which is a description of the primitive. This is a pseudo higher-level language module/procedure definition describing parameters used by the service. The second is a sequence diagram that shows the order of events at the interface.

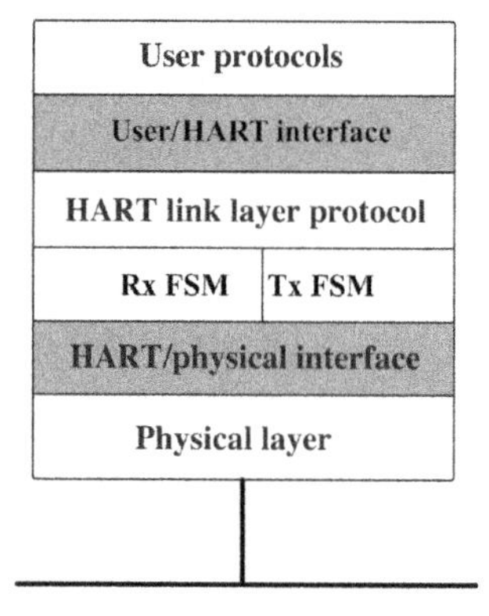

FIGURE 11.7

The HART protocol model.

11.4.4 FRAME FORMATS

All data transferred between entities involved in the protocol are transferred in the form of frames. A frame is an encapsulation of user data in control and addressing information. Protocol implementations shall not make any interpretation of user data (labeled as "data" in Figure 11.8). This means that frame recognition is disabled from the beginning of the user data field until the frame is complete as determined by a (correct) byte-count or by the physical layer signaling the end of a message (e.g., through absence of carrier detect).

A frame is delimited by the combination of characters called preambles, a unique "start of frame" character (delimiter) that identifies the frame's beginning, and a byte-count field that determines where the frame ends. Address fields determine the source and destination of data within a frame. The interface between the protocol and the user is through a command field, which identifies to the protocol whether this is a protocol command or a user command, and a response code field, which conveys the result of a protocol transaction back to the user. User data, if any, are transferred in the data field.

Master to slave frame

Preamble	Delimiter	Address	Command	Byte count	Data	Check byte

Slave to master frame

Preamble	Delimiter	Address	Command	Byte count	Response code	Data	Check byte

FIGURE 11.8

HART frame format.

All portions of the frame, including the delimiter, are protected by a combination of odd parity on each byte transmitted and a trailing check byte field. All fields in a frame are an integral number of bytes in length and all bytes of a frame must be transmitted in a single contiguous stream (i.e., there should be no more than a single-bit time gap between consecutive characters). Figure 11.8 shows the frame formats used by HART devices.

11.4.5 PREAMBLES

All frames transmitted by master, slave, or burst mode devices are preceded by a specified number of hexadecimal "FF" characters. These characters are called the preamble to the frame. They are required in some physical layer protocol implementations to train/synchronize the receiver.

Note: Many preambles may be received prior to the delimiter. However, two consecutive preambles followed by a leading delimiter define the start of a frame for the FSK physical layer.

While preambles are primarily a physical layer requirement, the data link layer provides management service support for specifying and determining the number of preambles required by an implementation. For the FSK physical layer, HART devices must transmit at least five preambles. This ensures sufficient time for asserting a carrier, for training the modem circuit, and for the listening device to begin receiving the message. HART devices are recommended to send a maximum of 20 preambles.

11.4.6 ADDRESSING

The protocol uses source and destination addresses in each frame. For both long and short frame addresses, the high-order bit of the address field indicates the master associated with this frame. A primary master uses the value "1" for this bit, and a secondary master uses the value "0." Slave devices must echo back this field unchanged. The next bit indicates whether the slave device is in burst mode. The slave devices must set this bit to "1" to indicate that the device is in burst mode or to "0" if the device is not in burst mode. Master devices must set this bit to "0" in requests to slaves.

The protocol supports long frame addresses and short frame addresses. The delimiter determines the use of short form addresses or long frame addresses in a message.

11.4.6.1 Long frame addresses

Except for command 0, all HART frames consist of a 5-byte address based on the slave device's unique identifier (see Figure 11.9). A unique identifier is associated with each slave device or field instrument manufactured. The protocol normally uses the lower 38 bits of the unique identifier to address data link layer requests to slave devices on the link. As a result, the long frame address must consist of the following:

- The master address and the burst mode bit as described.
- The least significant 6 bits of the manufacturer ID.
- A one-byte device type code: This code is allocated and controlled by the manufacturer. Since only the lower 6 bits of the manufacturer ID are used in the address, the device type code must be allocated as per specification.
- A three-byte device identifier: This is similar to a serial number in that each device manufactured with the same device type code must have a different device identifier.

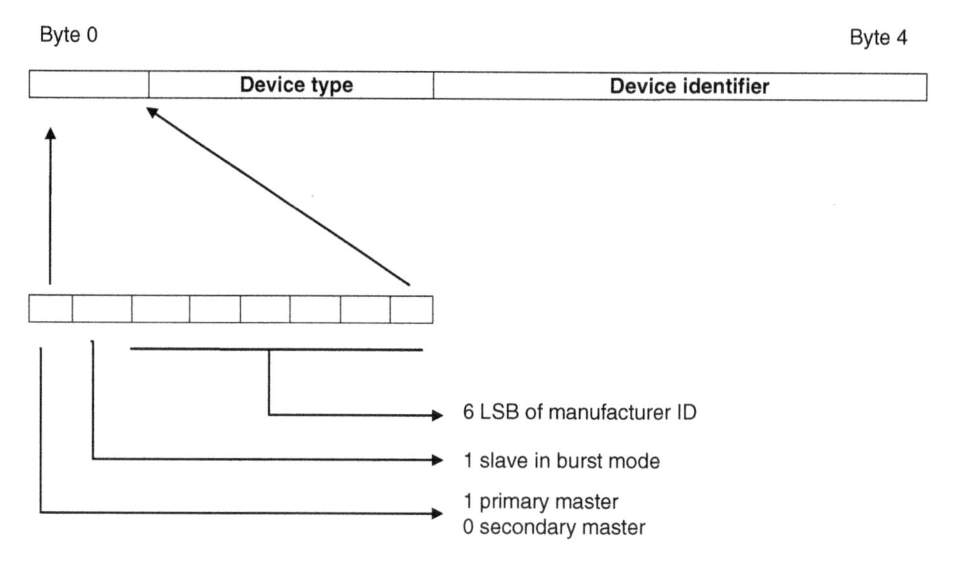

FIGURE 11.9

Long frame address.

In addition to long frame addresses based on the device's unique identifier, the protocol supports a broadcast address. A broadcast address is defined as 38 bits of zeros in place of the unique identifier in the long frame address. The devices treat a frame with this address as if the frame was addressed directly to them.

11.4.6.2 Short frame addresses

Previous, obsolete revisions of the protocol utilized short frame addresses. To maintain backward compatibility, only universal command 0 now supports short frame addressing. This enables the protocol to dynamically associate a short frame address with each slave device on the link. The short frame or polling addresses may be used during link initialization to rapidly scan the link address space. The protocol provides the capability necessary to manipulate these addresses on initialization and during normal operations. It also provides the capability to obtain the unique identifier associated with a particular short frame address when necessary.

Figure 11.10 shows the required format of the one-byte short frame address consisting of the following:

- The master address and the burst mode bit as described.
- The next two bits (4–5) must be set to zero.
- The least significant four bits specify the polling address.

11.4.7 ERROR DETECTION CODING

HART utilizes a single parity check product-coding scheme to detect errors. This allows HART to use parity checking in two dimensions: across the bits in a single byte and across the bit positions

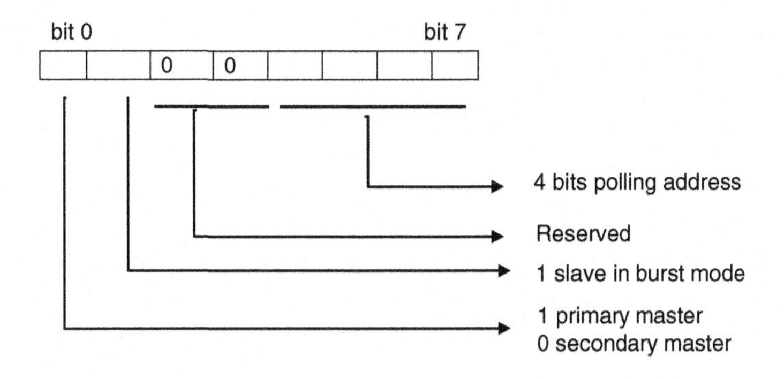

FIGURE 11.10

Short frame address.

in the transmitted message. The two dimensions of parity checking can be seen in Figure 11.11. Each byte consists of eight bits plus odd parity ("vertical parity"). The vertical parity can be automatically generated and checked by most UARTs. Figure 11.11 shows the organization of the individual bytes into a HART long address frame. The second dimension of error detection is generated by exclusive "OR"-ing all bytes from the delimiter up to and including the last data byte. The result is transmitted as the last byte ("check byte") of a message. There is odd parity on the check byte also.

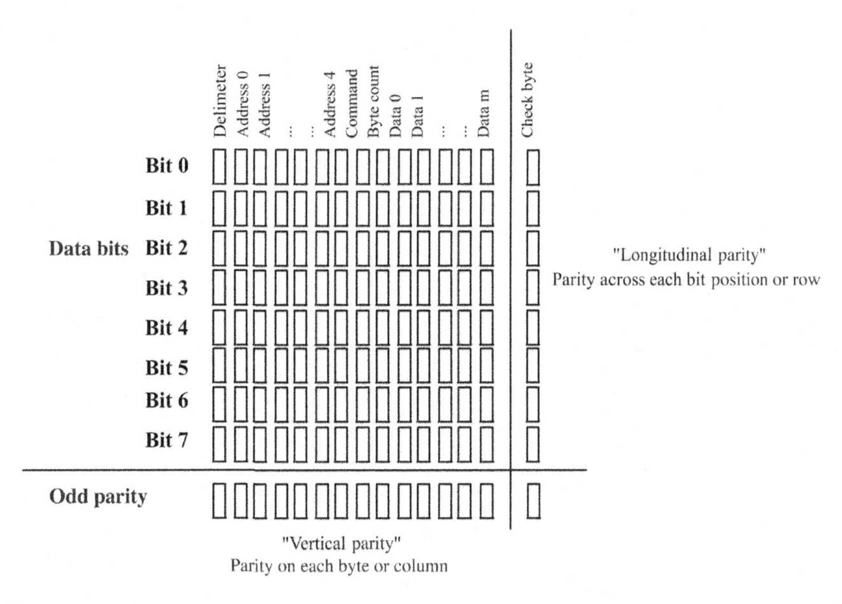

FIGURE 11.11

HART error detection scheme.

11.4.7.1 Master to slave or burst mode device frames

Figure 11.8 shows the format of a master to slave or burst mode device message. Reading from left to right, in the order in which the frame is transmitted, the fields are as given below:

1. Two preamble characters followed by the delimiter: The high-order bit of the delimiter determines whether the following frame is a short or long format frame (see Figure 11.12). The low-order three bits encode the type of frame. The reserved fields may be used as shown in Figure 11.12 and should be masked out in current implementations. The combination of these three characters is used to recognize the start of a frame.
2. Address: The high-order bit of the delimiter determines whether the following frame is a short or long addressed frame. The format of the address is described in Section 11.4.6.
3. Command: This field is one byte long. Command byte values are echoed back unchanged in responses from slave devices.
4. Byte count: This field is one byte in length and is the count of the number of bytes of user data that follow between the byte count and the parity check byte (both excluded from the count). All values between 0 and 255 (both inclusive) are legal in this field.
5. Data: The fifth field consists of an integral number of bytes of user data.
6. Check byte: This value is determined by a bitwise exclusive OR of all bytes of a transmitted frame beginning with the leading delimiter.

11.4.7.2 Slave to master frames

These frames are identical to master to slave frames except for the leading frame delimiter and the presence of a response code field. The fields are as given below:

1. Two preambles followed by the leading delimiter: The structure of the delimiter byte is identical to that of master to slave frames.
2. Address: The structure of the second field is identical to that of master to slave frames. The value of the burst mode bit is set or cleared depending on whether the device is in burst or nonburst mode. The rest of the address is echoed from the incoming master to slave frame.
3. Command: Echoed from the incoming master to slave frame.

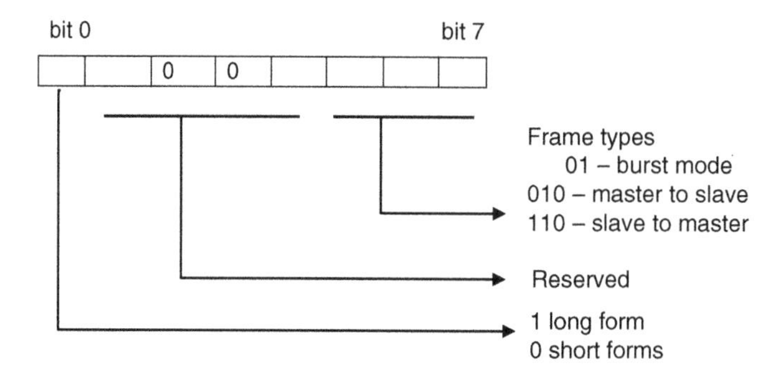

FIGURE 11.12

The delimiter character.

4. Byte count: This is set to the number of bytes between the byte count and the longitudinal parity or check byte at the end of the message. The minimum value of this field is 2, since all slave to master frames have at least two-byte response code.
5. Response code: A two-byte response code is unique to slave to master frames. This field holds information that describes the result of the attempted transaction.
   a. If bit 7 of the first response code byte is set, it indicates a communication error, and the remaining bits provide a summary of the communication error.
   b. If bit 7 of the first response code byte is clear, the remaining bits carry upper-level protocol information about the success or failure of the command.
   c. The second byte of the response code carries data link and device status. It may be set to zero if a communication error is reported in the first byte.
6. Data: The sixth field consists of an integral number of bytes of user data.
7. Check byte: This is calculated as described in Section 11.4.7

11.4.7.3 Burst mode device to master frames

Burst mode frames are generated without a corresponding master request to each burst "response" frame. Burst mode frames are distinguished by their unique delimiter. These frames are identical to the format of a slave to master frame (see Figure 11.12) except for the delimiter and the usage of the master address bit in the address field. The fields are as given below:

1. Two preambles followed by the delimiter indicating that this is a frame identical to a slave to master frame except that it was generated spontaneously by a burst mode device. Burst mode frames are generated in the long frame format only.
2. Address: The second field is identical to slave to master frames. The burst mode device alternates the value of the master address bit between "1" and "0." It also sets the burst mode device bit to "1," since it is in burst mode.
3. Command: The command byte is set by the burst mode device to the command that is being burst.
4. Byte count. This is set to the number of bytes between the byte count and the longitudinal parity byte at the end of the message. The minimum value of this field is 2, since all slave to master frames have at least two-byte response code.
5. Response code: The fifth field is identical to that used in slave to master frames. This field holds information that describes the result of the attempted burst transaction.
   a. Bit 7 of the first response code byte is always cleared, since there is no incoming message about which to report communication errors.
   b. The second byte of the response code carries data link and device status.
6. Data: The sixth field consists of an integral number of bytes of user data.
7. Check byte. It is calculated and checked as described previously.

11.4.7.4 HART protocol services

The services described in this section are used to obtain the following:

- A reliable, "at least once," transaction service between peer entities. The service is not designed to provide duplicate detection.
- An optional, reliable, transaction service between peer entities that provides end-to-end segmentation and duplicate detection.
- Management services for device identification and device configuration.

The primitives defined in this section fall into two classes. Those associated with the transfer of user data in the normal sequence of usage are termed user primitives. Those that are concerned with the initialization of the protocol, such as the setting up of addresses and establishing a unique relationship between addresses and peer protocol entities, are called management primitives. All primitives described here must be supported by a protocol implementation. The mapping of these primitives into an implementation is entirely a local matter and is in no way restricted by this specification.

11.4.8 MODEL FOR SERVICE SPECIFICATIONS

Figure 11.13 explains the structure of a typical sequence diagram. The two vertical lines represent the interfaces between users of a protocol service (which lie outside the lines) and the protocol service provider (which lies in between them). Arrows meeting these lines indicate events at each end of the communication link. Time increases down the page. Points on these axes mark when protocol transactions are initiated or received. Diagonal lines joining the time axes represent the propagation of messages or data between the two entities on the link and represent transactions that take place internal to the protocol. Diagonal lines outside the vertical axes represent signals or events that occur at the user interface to the protocol. The convention used in labeling these interface events is as follows: Transactions initiated by a user are labeled with a trailing ".request" to show that they are directed towards the protocol. The result of a request is communicated back to the user by the protocol in a transaction labeled with a ".confirm," to show that they are in response to a previous ".request." At the other end of the link, the protocol responds to an incoming transaction by providing an indication to the local user that data or control information has arrived. Such transactions are labeled with a trailing ".indication." The response of a peer protocol entity to an indication is labeled with a trailing ".response." The distinction between the use of a request/confirm pair and an indication/response pair to label a set of primitives is based upon whether the primitives are used to cause a protocol transaction or are caused by one.

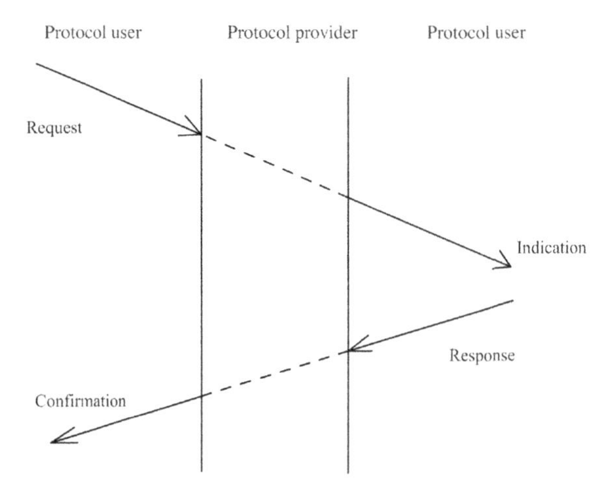

FIGURE 11.13

Model for protocol service specifications.

11.4.9 APPLICATION LAYER

The application layer defines the commands, responses, data types, and status reporting supported by the protocol. In addition, there are certain conventions in HART (e.g., how to trim the loop current) that are also considered part of the application layer. While the command summary, common tables, and command response code specifications all establish mandatory application layer practices (e.g., data types, common definitions of data items, and procedures), the universal commands specify the minimum application layer content of all HART-compatible devices.

11.4.9.1 Benefits of HART

HART offers numerous benefits as follows:

- Device calibration and diagnostics
  - Provides device status and diagnostic alerts
  - Provides process variables and units
  - Provides loop current and % range
  - Provides basic configuration parameters
  - Provides manufacturer and device tag
- Increases control system integrity
  - Gets early warning of device problems
  - Uses capability of multivariable devices
  - Automatically tracks and detects changes (mismatch) in range or engineering units
  - Validates PV and loop current values at control system against those from device
- HART is safe, secure, and available
  - Tested and accepted global standard
  - Supported by all major instrumentation manufacturers
- Saves time and money
  - Simple installation and commissioning of devices
  - Provides enhanced communications and diagnostics reducing maintenance and downtime
  - Low or no additional cost by many suppliers
- Improves plant operation and product quality
  - Additional process variables and performance indicators
  - Continuous device status for early detection of warnings and errors
  - Digital capability ensures easy integration with plant networks
- Protects your asset investments
  - Compatible with existing instrumentation systems, equipment, and people
  - Allows benefits to be achieved incrementally
  - No need to replace entire system

11.4.10 DEVICE DESCRIPTION LANGUAGE (DDL)

The device description language (DDL) is a language for specifying the external behavior of HART field devices. This language allows host devices to operate with field devices without prior knowledge of any particular field device.

Device description: A device description (DD) is an electronic data file prepared in accordance with DDL specifications that describes specific features and functions of a device including details of menus and graphic display features to be used by host applications (including handheld devices) to access all parameters and data in the corresponding device.

The HART protocol operates in a host/slave environment where an instrument is queried and responds to a host. HART has a command structure that allows device manufacturers in designing a device. All HART devices must have a minimum command set in order to be compliant. These commands are called universal commands. Devices may also have a number of optional commands called common practice commands. Many devices implement manufacturer specific commands called device-specific commands. The DD allows access to all commands. DD aware host systems can then use the data and display it as the user desires.

In summary, DDL is a language to describe devices. A DD is created as a text file and then translated into a standard binary file. Similar to the presentation of HTML that is independent of an operating system and browser, a DD is displayed with a DD application.

11.4.10.1 Benefits of DDL

- In engineering
  - Configuring of networks, gateways and, remote I/Os
  - Export/import of device data
  - Upload/download
  - Offline configuring
  - Comparison of parameters
- Commissioning
  - Address setting
  - Visualization of process data
  - Simulation
  - Calibration, adjustment
  - Online configuring
  - Identification of devices for a life list
- Operations and maintenance
  - Displaying device and diagnostic states
  - Support of device exchange
  - Identification check
  - Device state for maintenance, repair, and fault removal

FURTHER READINGS

Anon, HART Communication Foundation, HART Field Communications Protocol: An Introduction for Users and Manufacturers, July 1997.

Anon, Application Guide, HCF LIT 34, HART Communication Foundation, 1999.

Fieldbuses for Process Control: Engineering, Operation and Maintenance By Jonas Berge, ISA.

http://en.hartcomm.org/hcp/tech/aboutprotocol/aboutprotocol_what.html.

CHAPTER

PROFIBUS COMMUNICATION 12

12.1 OVERVIEW

PROFIBUS (process field bus) is a standard for field bus communication which is widely used for process automation. PROFIBUS was first promoted by BMBF (German Department of Education and Research) in 1989 in cooperation with several automation manufacturers.

Profibus uses 9-pin D-type connectors (impedance terminated) or 12-mm quick-disconnect connectors. The maximum number of nodes is limited to 127. The distance supported is up to several kilometers depending on various factors (described later) with speeds varying from 9600 bps to 12 Mbps. The message size can be up to 244 bytes of data per node per message with medium access control (MAC) mechanisms of polling and token passing. Profibus supports two types of devices namely – Masters and Slaves.

Master devices control the bus and when they have the right to access the bus, they may transfer messages without any remote request. These are referred to as active stations.

Slave devices are typically peripheral devices, that is, transmitters/sensors and actuators. They may only acknowledge received messages or, at the request of a master, transmit messages to that master. These are also referred to as passive stations.

12.1.1 ORIGIN OF PROFIBUS

In 1987, few companies and institutes started an association in Germany to implement and spread the use of a bit-serial field bus based on the basic requirements of the field device interfaces. First, the complex communication protocol – Profibus FMS (field bus message specification) used for nondeterministic data communication between Profibus Masters was specified. This was followed by establishing specifications for simpler, deterministic, and faster Profibus DP for communication between Profibus Masters and their remote IO slaves in 1993.

12.1.2 OSI MODEL

PROFIBUS networks make use of three separate layers of the OSI network model namely application layer, data link layer, and physical layer. Profibus does not define layers three to six (Table 12.1).

- Application Layer: There are multiple versions of PROFIBUS that handle different types of messaging at the application layer. PROFIBUS supports include cyclic and acyclic data exchange, diagnosis, alarm handling, and isochronous messaging.
- Data Link Layer: Communication efficiency is determined by level 2 functions, which specify the tasks of access to the bus, data frame structure, communication of basic services, and many other functions. The data link layer is completed through a Field Data Link (FDL). The FDL system combines two common schemes, master–slave methodology and token passing.

Table 12.1 OSI Layer Compared to Profibus

	OSI Layer	PROFIBUS
7	Application	FMS/DP-V0/DP-V1/DP-V2
6	Presentation	—
5	Session	—
4	Transport	—
3	Network	—
2	Data link	FDL
1	Physical	RS485/fiber optic/MBP

   - Master–Slave scheme: In a master–salve network, masters – usually controller send requests to slaves, sensors, and actuators. The slaves respond accordingly.
   - Token Passing scheme: In this scheme, a "token" signal is passed between nodes. Only the node with the token can communicate.
- Physical Layer: PROFIBUS defines a physical layer, although it leaves room for flexibility. One kind of media is a standard twisted pair wiring system (RS485). PROFIBUS system can also operate using fiber optic transmission if that is more appropriate. A safety-enhanced system called Manchester bus power (MBP) is also available in situations where chemical environment is prone to explosion.

12.1.3 THE APPLICATION LAYER

There are different variants of Profibus used in the application layer:

- PROFIBUS FMS (fieldbus message specification) was the initial version of Profibus and was designed to communicate between programmable controllers and PCs, exchanging complex information between them. This variant supports communication between automation systems besides the exchange of data between intelligent equipment, generally used in control level. The FMS technology was not as flexible to suit less complex messages or communication on a wider, more complicated network. Lately, it is being replaced by application in Ethernet since its primary function is peer-to-peer communication.
- PROFIBUS DP (decentralized peripherals) is the most commonly used low-cost, simple, and high-speed field level communication and generally preferred for critical time controls. It was developed specifically for communication between automation systems and decentralized equipment – to operate sensors and actuators via a centralized controller in factory automation applications. It uses the RS485 physical medium or fiber optics. For majority of the applications involving slaves, PROFIBUS utilizes PROFIBUS DP. There are three versions of PROFIBUS DP – DP-V0 (1993), DP-V1 (1997), and DP-V2 (2002). They provide newer, more complicated features, and have evolved alongside the technological advancement, in response to the long-time growing demand for the applications (Figure 12.1).
- PROFIBUS PA (process automation) is used to monitor measuring equipment via a process control system in process automation applications. This variant is designed for use in hazardous

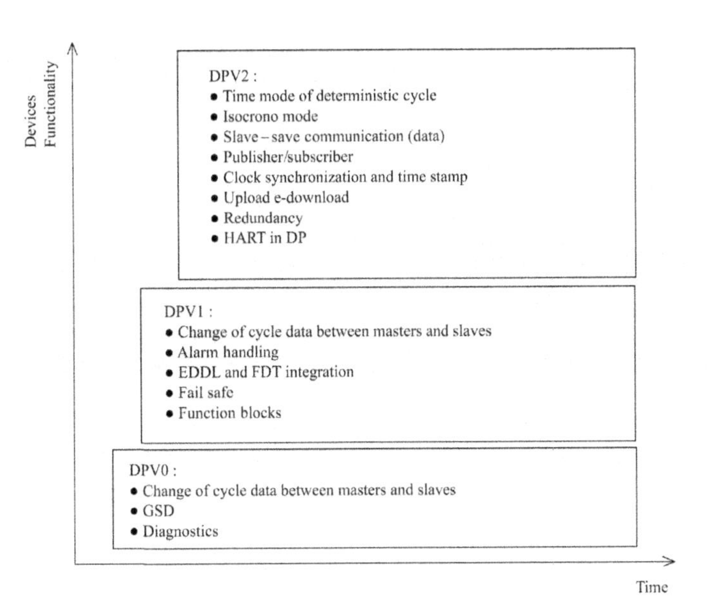

FIGURE 12.1 Profibus functions

areas. This was developed specifically for the process industry to replace 4–20 mA transmission and therefore, carries both power and data over the two wires. The number of instruments supported on a PA segment is limited due to the fact that the power delivered through the bus is limited to avoid explosive conditions during malfunction. PA is therefore used for process applications where speed is less important. The PROFIBUS PA communication protocol uses the same communication protocol as the PROFIBUS DP. This is because the communication and the messaging services are identical. In fact, PROFIBUS PA is equivalent to PROFIBUS DP communication protocol + extended acyclic services + IEC 61158 that is the physical layer, also known as H1.

Profibus PA can be linked to Profibus DP network using a coupler device. The much faster Profibus DP acts as a backbone network for transmitting process signals to the controller. Therefore, Profibus PA and Profibus DP can work tightly together in hybrid application where process and factory automation networks operate side by side.

PROFIBUS can be tailored to specific needs using application profiles. There are many profiles that combine standards for transmission media, communication protocol (FMS, DP-V0, etc.), and unique protocols. Each of these application profiles is tailored to specific use and new profiles appear regularly. Some application profiles are widespread. The following are some of the application profiles:

- PROFIsafe uses additional software to create high-integrity network. This network is useful in situations where high safety is a requirement. Suppliers and manufacturers maintains high standards in quality to be certified in PROFIsafe.
- PROFIdrive was created for motion control applications. Software added to the PROFIBUS DP stack allows the network to achieve precise control of servo motors and other equipment. Thus, PROFIdrive can achieve synchronization across network.

- HART on PROFIBUS is a solution for integrating HART devices installed in the field with the existing new PROFIBUS systems developed by collaborating with the HART foundation. The HART client application is integrated in a PROFIBUS master and the HART master in a PROFIBUS slave, where the latter servers as a multiplexer and takes over communication with the HART devices.

12.1.4 DEVICE INTEGRATION

A particular advantage of PROFIBUS is openness, which in turn brings with it compatibility with a large number of device and system manufacturers. Devices are usually integrated by means of mapping their functionality to operator software with identical data structures for all devices. All standards cited in the following can be used in conjunction with PROFIBUS.

General Station Description (GSD) – GSD is provided by the device manufacturer and is the electronic data sheet for the communication properties of each and every PROFIBUS device. It supplies all information necessary for cyclical communication with the PROFIBUS master and for the configuration of the PROFIBUS network in the form of a text-based description. It contains the key data of the device, information about its communication capabilities, and information about, for example, diagnostic values. The GSD alone is sufficient for the cyclic exchange of measured values and manipulated variables between field device and automation system.

Electronic Device Description (EDD) – The GSD alone is not sufficient to describe the application-specific functions and parameters of complex field devices. A powerful language is required for the parameterization, service, maintenance, and diagnostics of devices from the engineering system. The Electronic Device Description Language (EDDL) standardized in IEC 61804-2 is available for this purpose. An EDD is a text-based device description that is independent of the engineering system's OS. It provides a description of the device functions communicated acyclically, including graphics-based options, and also provides device information such as order data, materials, maintenance instructions, and so on.

Device-Type Manager (DTM) and Field Device Tool (FDT) interface – In comparison to the GSD and EDD technologies based on descriptions, the FDT/DTM technology uses a software-based method of device integration. The DTM is a software component and communicates with the engineering system via the FDT interface. A DTM is a device operator program by means of which device functionality (device DTM) or communication capabilities (communication DTM) are made operational; it features the standardized FDT interface with a frame application in the engineering system. The DTM is programmed on a device-specific basis by the manufacturer and contains a separate user interface for each device. DTM technology is very flexible in terms of how it is configured.

Tool Calling Interface (TCI) – TCI is an open interface between the engineering tool of the overall system and the device tools of complex devices, for example, with drives or laser scanners, which enable centralized parameterization and diagnostics from the engineering station during operation. TCI is nonmanufacturer-specific and allows dynamic parameters to be loaded to devices without having to exit the automation-system engineering tool.

12.1.5 THE DATA LINK LAYER

The data link layer tasks are executed by FDL and fieldbus management (FMA).

FDL performs the following tasks:

- Control of access to the bus (MAC)

- Telegrams structure
- Data safety
- Availability of data transmission services:
  - SDN (send data with no acknowledgment)
  - SRD (send and request data with reply)

The FMS provides several management functions:

- Configuration of operation parameters
- Events report
- Activation of services of access points (SAPs)

The PROFIBUS protocol architecture and philosophy ensures every station involved in exchange of cyclic data, the time sufficient to execute their communication task within a defined time period. For this, they use token passing from the bus master stations when communicating between themselves and the master–slave procedure to communicate between the master and the slave stations. The token circulates one time for each master within the maximum rotation time defined (configurable). On PROFIBUS, the token passing is used only for masters to communicate with each other (Figure 12.2).

The master–slave communication enables the active master (with the token) to access the slave through the reading and writing services (Figure 12.3).

PROFIBUS uses different subsets of level 2 services in its different variants as given in Table 12.2.

Addressing services with seven bits identify the network participant and on 0 to 127 ranges, the following addresses are reserved:

126 – Standard address attribute via master
127 – Used to send frames in broadcast

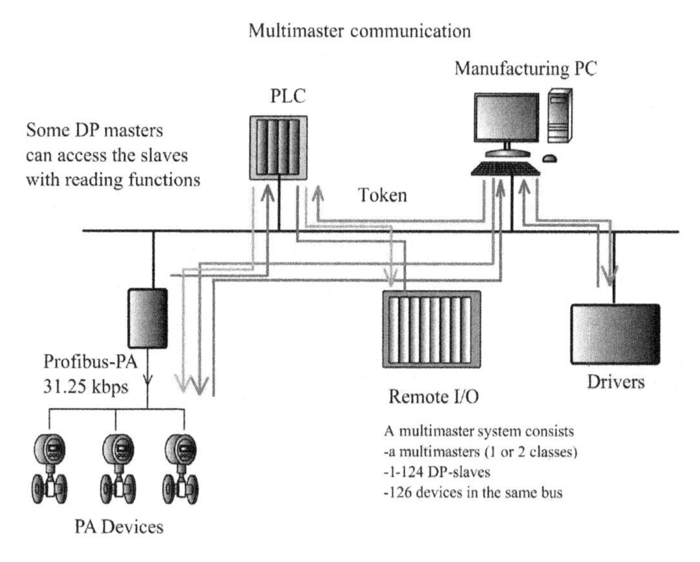

FIGURE 12.2 Multi master communications

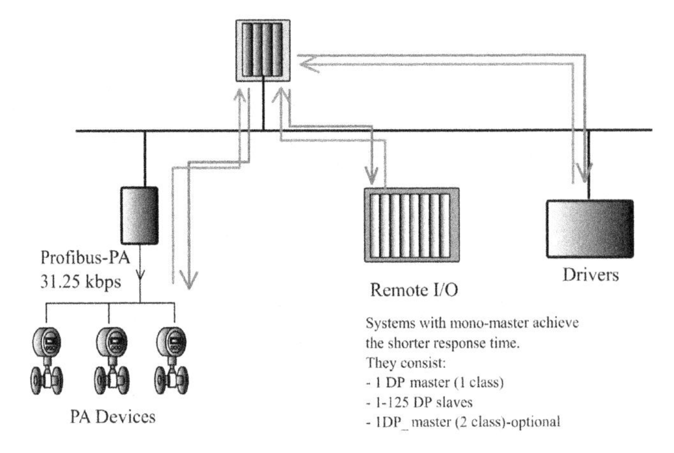

FIGURE 12.3 Profibus master and slave communication

12.2 SUPPORTED TOPOLOGY

Supported topologies are tree, bus, and point-to-point as illustrated in Figures 12.4–12.6.

- Tree topology (Figure 12.4)
- Bus topology (Figure 12.5)
- Point-to-point topology (Figure 12.6)

12.2.1 TYPES OF DEVICES

Each DP system has three different types of devices:

- CLASS 1 DP MASTER (DPM1): It is the main controller that cyclically exchanges information with the slaves. It is the central control unit of a system such as the programmable logic controllers,

Table 12.2 Different Profibus Level Services

Service	Function	DP	FMS	PA
SDA	Send data with acknowledgment	No	Yes	No
SRD	Send and request data with reply	Yes	Yes	Yes
SDN	Send data with no acknowledgment	Yes	Yes	Yes
CSRD	Cyclic send and request data with reply	No	Yes	No

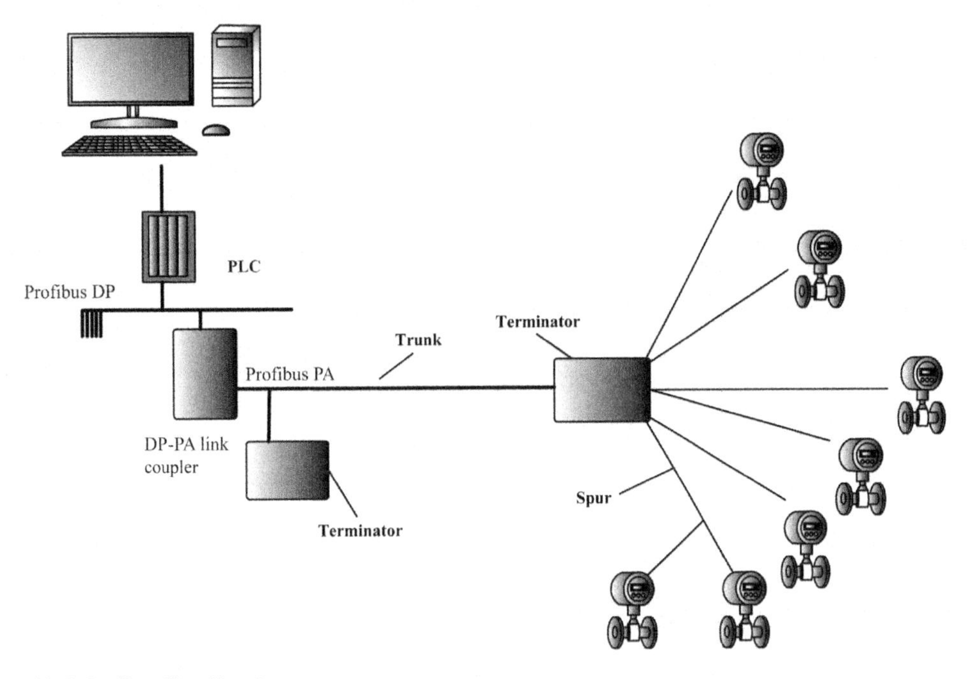

FIGURE 12.4 Profibus Tree Topology

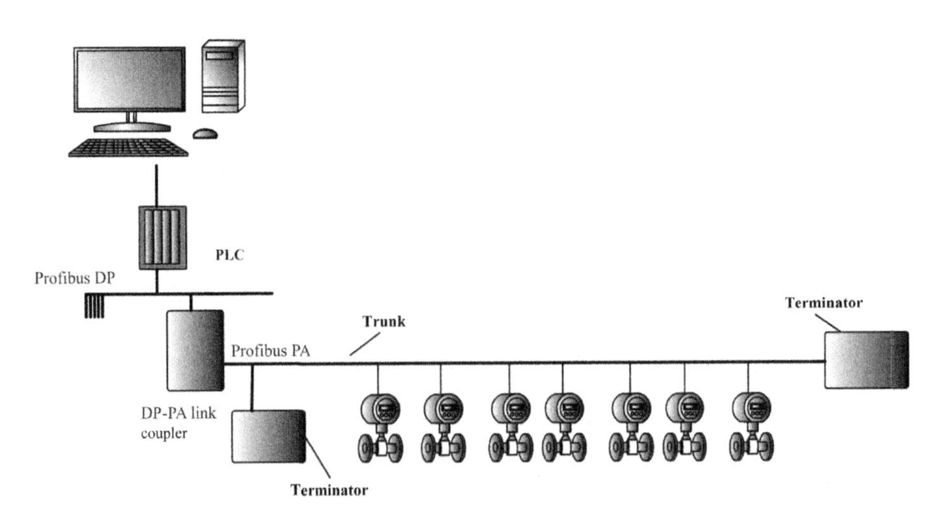

FIGURE 12.5 Profibus Bus Topology

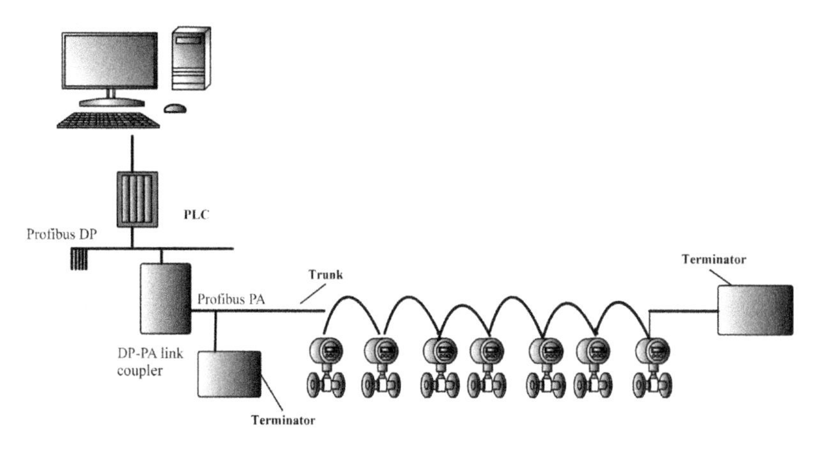

FIGURE 12.6 Profibus Point to Point Topology

which exchange data with the field devices within a specified message cycle. The DPM1 sets the baud rate and the slave devices auto-detect this rate. A DPM1 can passively (upon request) communicate actively with its assigned slave devices, with another class 2 master device.

- CLASS 2 DP MASTER (DPM2): They are used for operation and monitoring purpose and during start-up. Engineering stations used for configuration, monitoring, or supervisory systems such as Profibus View, Asset View, and so on are some examples of class 2 DP master. It acts like a supervisory master and actively communicates with class 1 masters and their slaves, in addition to its own slaves, but usually only for configuration, problem diagnosis, and data/parameter exchange. DPM2 may only briefly take over control of a slave. All exchanges between a class 2 master and class 1 master originate with the class 2 master.
- SLAVE: A DP slave is a passive station which can only respond as per a master request and acknowledge messages. Some examples of slave devices include peripheral device such as I/O device, valves, transducers, actuators, and so on.

12.3 DATA EXCHANGE

PROFIBUS DP normally operates using a cyclic transfer of data between master(s) and slave(s) on an RS485 network. That is, an assigned master periodically requests (polls) each node (slave) on the network. All exchange of data communication between a master and slave originate from the master device. Each slave device is assigned to one master and only that master can write output data to that slave. Other masters may read information from any slave, but can only write output data to their own assigned slaves.

Masters can address individual slaves, a defined group of slaves (multicast), or can broadcast a telegram to all connected slaves. Slaves return a response to all telegrams addressed to them individually, but do not respond to broadcast or multicast telegrams from a master device. ProfiBus sends broadcast and multicast messages as global control telegrams using address 127 and an optional group number for a targeted group of slaves.

PROFIBUS is deterministic as it uses a cyclic (periodic) polling mechanism between masters and slaves. Therefore, the behavior of a PROFIBUS system can be reliably predicted over time. In fact, PROFIBUS was designed to guarantee a deterministic response.

The length and timing of the I/O data to be transferred from a single slave to a master is predefined in the slave's device database or GSD file. The GSD files of each device connected via the network are compiled into a master parameter record, which contains parameterization and configuration data, an address allocation list and the bus parameters for all connected stations. A master uses this information to setup communication with each slave during startup.

After a master receives its master parameter record, it is ready to begin exchanging data with its slaves. During startup, after a system reset, or upon return to power, a master attempts to reestablish contact with all the slaves assigned to it before assuming the cyclic exchange of I/O data. Each slave must already have a unique valid address from 0 to 125 to be able to communicate with the master. Any slave that has a default address of 126 await the Set_Slave_Address command from a class 2 master before it can be parameterized. In attempting to establish communication, the master starts with the lowest address slave and ends with the highest address slave. A master sends parameterization and configuration telegrams to all of its assigned slaves (a slave may only be write-accessed by its assigned master, the master that parameterized and configured it during startup). The parameterization and configuration telegrams ensure that the functionality and configuration of a slave is known to the master. If an additional slave is added to the network bus and is not already accounted for in the master record, a new master record must be generated and a new configuration performed so that the master is informed of the status of the new device.

PROFIBUS DP most often uses a single class 1 master device (monomaster), cyclically polling many distributed slaves. However, PROFIBUS also allows for acyclic communication between class 2 masters and slaves, making more than one active station or master possible. A class 1 master automatically detects the presence of a new active station connected to the network bus (a class 2 master). When the class 1 master completes its polling cycle, it passes a "token" to the class 2 master granting it temporary access to the bus. Deterministic behavior is maintained because the class 2 master can only use the time allotted to it via the gap time specified. Although, monomaster operation is generally recommended, it is not mandatory. Therefore, a PROFIBUS system may have more than one class 1 master, but master-to-master communication is not permitted, except for granting bus access rights via token exchange.

To illustrate the idea of communication between masters in a PROFIBUS DP system, a class 1 master cyclically exchanges data with all the slaves assigned to it, one at a time, according to its list of assigned slaves taken from the master record. At the end of this data cycle, additional time (gap time) is allotted to provide for acyclic communication between a class 2 master and the same slaves. During this time, the class 1 master passes a token to the class 2 master granting it bus access rights. The class 2 master which currently holds the token has the opportunity to exchange data with all the slaves within a specific period of time called the token half-time or token hold-time (TH). The class 2 master may then proceed to read data or diagnostic information from any of the slaves, and then at the completion of its cycle, it passes the token back to the class 1 master.

Since there usually is not enough time during the gap to complete a full data exchange, this process of data retrieval by the class 2 master may continue over several cycles. At the end of record transfer, the class 2 master clears the connection. Note however, that the class 2 master may only establish communication with the slaves during the gap time.

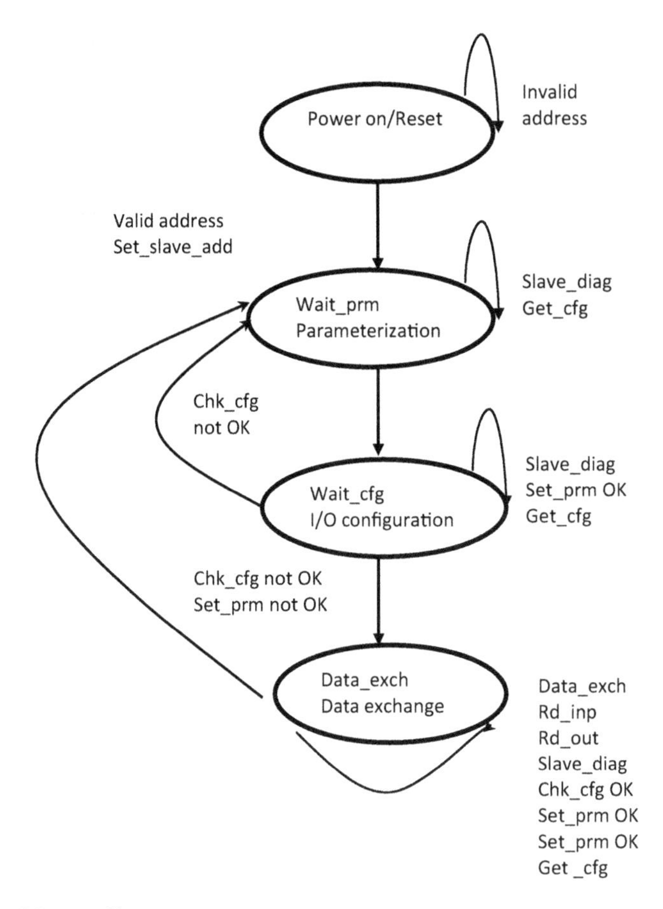

FIGURE 12.7 Profibus DP state diagram

It is possible for a class 2 master to temporarily take over control of a DP slave. During this time, the DP slave stops its normal data exchange with its class 1 master. The class 1 master recognizes this and cyclically requests diagnostics from the slave, checking the Master Address field for as long as another valid address is present. After the class 2 master finishes its communication with the slave, it sets the Master Address field of the slave to invalid (255). This leads to the class 1 master attempting to regain control of the slave and it reparameterizes and reconfigures the slave before resuming data exchange with it.

The following state machine helps illustrate how PROFIBUS DP operates with respect to the slaves (Figure 12.7):

- Power ON/RESET State: The power ON/RESET state is the initial state following power up of the DP slave. In this state, the slave may receive a telegram from a class 2 master to change its station address. A slave helps in this state if it does not have a valid address from 0 to 125. After completion of its power-ON initialization routine and if the slave has a valid station address, the slave proceeds to the wait for parameterization state.

- Parameterization State: The DP slave awaits the parameterization telegram from the master which identifies the slave's master and the mode the slave is to operate in. A slave in this state rejects all other telegrams except a request telegram for diagnostics or configuration. After its parameters have been set, the slave proceeds to the I/O configuration state.
- I/O Configuration State: In this state, the slave awaits a configuration telegram that specifies the number of input and output bytes to be exchanged in each data telegram cycle with the slave. The configuration telegram also causes the slave to check the configuration sent against the stored configuration. A slave in this state accepts a request telegram for diagnostics or configuration, or a set parameters telegram.
- Data Exchange State: After parameterization and configuration have been accomplished, the slave cyclically exchanges I/O data with the master. This is a cyclic transfer of I/O data and possible diagnostic information.

12.4 FAIL-SAFE OPERATION

A PROFIBUS master runs in two modes – Operate and Clear. A fail-safe situation for a class 1 master arises when it sends 0 length data or when the data is set to 0, when it is in Clear mode. For a DP slave, the term fail-safe refers to the way in which it processes the output telegrams with zero length data or if the master is in clear state. The combined master–slave system is considered fail safe when the failure of a master does not cause errors in any of its slaves and the slave outputs go to a predictable and predefined state.

A slave may go to a fail-safe state if its watchdog time expires without having received a message from its assigned master. Normally, this timer is reset every time the master talks to the slave. If this time expires, it is an indication that the master has not communicated with the slave recently and that the slave is not being controlled. The slave then leaves the data exchange mode and its output goes to the predefined fail-safe value.

The fail-safe state is usually set via user-defined parameters of the parameterization telegram and its GSD file or sometimes via hardware switches on the slave. Some slaves provide parameters or switches that also allow the slave output to retain their last state, but this is not considered fail-safe.

A slave may also assume a fail-safe state if its master switches from Operate Mode to Clear Mode. With normal operation in Data_Exchange mode, a class 1 master is in Operate Mode and cyclically exchanges I/O data with its assigned slaves. The class 1 master may use a global control telegram to inform the slaves that it is switching from Operate Mode to Clear Mode. A master may elect to switch to Clear Mode while it is bringing slaves online and not all slaves have been parameterized and configured yet. It may also switch to Clear Mode as a result of a run/stop switch on the master. In the Clear State, the master may attempt to parameterize and configure the remaining slaves assigned to it in an effort to reinitiate data exchange, while it continues data exchange with the other slaves (they receive output data of 0, or output data of zero length). Operate mode does not resume until all slaves are online and exchanging data, or until the master is told to resume operation via a run/stop switch or under program control. Further, some masters may go to Clear Mode if a slave is disabled, rather than continue to control a partial system (this response may be specified as a parameter in the master's GSD file and parameterization telegram, via a mechanical switch, or as part of its master program).

12.4.1 TRANSMISSION SERVICES

PROFIBUS DP uses two types of transmission services in sending message telegrams that are defined in the data link layer of the OSI model.

1. SRD (send and request data with acknowledge) – With SRD, data is sent and received in one telegram cycle. The master sends output data to the slave and receives input data from the slave within a specified period in response (if applicable) in a single telegram cycle. The SRD transmission service is the most often used service in PROFIBUS DP making the data exchange very efficient for mixed I/O devices.
2. SDN – This service is used when a message is to be sent simultaneously to a group of slaves (multicast) or all slaves (broadcast). Slaves do not respond or acknowledge multicast or broadcast messages.

A third type of transmission service used in PROFIBUS FMS is SDA (send data with acknowledge), with data sent to a master or slave and a short acknowledgment sent in response. This is not used in PROFIBUS DP.

12.4.2 PROFIBUS DP DATA CHARACTER FORMAT

All ProfiBus characters are comprised of 11 bits (1 start bit + 8 data bits + 1 even parity bit + 1 stop bit). ProfiBus DP exchanges data in NRZ code (nonreturn to zero). The signal form of binary "0" or "1" does not change during the duration of the bit. If nothing is being transmitted, the idle state potential on the line is "1." A start bit causes the line to go to "0" (Figure 12.8).

This character frame applies to all data/character bytes, including the telegram header bytes. Each character or data byte is sent in the sequence of least significant bit (lsb) to most significant bit (msb) during message transmission on PROFIBUS serial networks, as shown above. For word transfer (more than 1 byte), the high byte is transmitted first, followed by the low byte (Big-Endian/Motorola format).

PROFIBUS networks employ the even parity method of data error checking, which controls how the parity bit of a data frame is set. With even parity checking, the number of 1 bits in the data portion of each character frame is counted. Each character contains 8 bits. The parity bit is then set to a 0 or 1, as required in order to result in an even total number of 1 bits. For example, if a character frame contains the following 8 data bits: 1100 0011, then since the total number of 1 bits is 4 (already an even number), the frame's parity is set to 0 for even parity. When a message is transmitted, the parity bit is calculated and applied to the frame of each character transmitted. The receiving device counts the quantity of 1 bits in the data portion of the frame and sets an error flag if the count differs from that sent. As such, parity checking can only detect an error if an odd number of bits are picked up, or dropped off, from a character frame during transmission. For example, with even parity, if two 1 bits are dropped from a character, the result is still an even count of 1 bits and no parity error is detected.

Profibus NRZ coded character frame (even parity)

START	DO	D1	D2	D3	D4	D5	D6	D7	Parity	Stop
"0"	0	1	2	3	4	5	6	7	"even"	"1"
←	LSB	←	←	←	←	←	←	MSB	←	←

FIGURE 12.8 Profibus NRZ Coded Character Frame (Even Parity)

12.4.3 PROFIBUS: TELEGRAM STRUCTURE

The fieldbus data link defines the telegrams as follows:

- Telegrams without data field (6 bytes control)
- Telegrams with one fixed length data field (8 bytes data and 6 bytes control)
- Telegrams with variable field data (from 0 to 244 bytes data and 9 to 11 control bytes)
- Fast recognition (1 byte).
- Token telegram for bus access control (3 bytes)

The integrity and safety of the information are stored in all the transactions through the frame parity and block checking to reach the hamming distance of HD = 4, ensuring that a maximum of three errors are detected with certainty.

A PROFIBUS telegram may contain up to 256 bytes (up to 244 bytes of data per node per message and 11 bytes of overhead or Telegram Header). All telegram headers are 11 bytes except for data exchange telegrams, which have 9 bytes of header information (the destination service access point (DSAP) and source service access point (SSAP) bytes are dropped). Twelve bytes of overhead for a single message makes PROFIBIS less efficient for small amounts of data. However, since up to 244 bytes of data may occur per message and since the output data is sent and the input data is received in a single telegram cycle, this makes PROFIBUS more efficient when large amounts of data must be transferred. Also, an idle state of at least 33 Tbits (sync-time in bit time) must be present before every request telegram to be sent and all data is transferred without gaps between individual characters (Figures 12.9 and 12.10).

Figure 12.11 illustrates the telegram sequence between a class 1 master and a DP slave (Figure 12.11).

- Start Delimiter (SD): identifies the beginning of a telegram and its general format. PROFIBUS DP uses four types of SD for request and response telegrams, in addition to a fifth response for a short acknowledgment as shown. Note that the short acknowledge response does not use a SD. Also, a telegram response does not have to use the same SD as the request telegram (Table 12.3).
- Length of Telegram (LE & LEr): specifies the length of a telegram with variable data length (i.e., SD2 Telegrams) from the DA byte to the end DU byte (range is DU +5b to 249). Note that the length of the DU is generally limited to 32 bytes, but the standard allows for lengths up to 244 bytes. LE is repeated in the LEr field for redundant data protection.
- Destination Address (DA) and Source Address (SA): The master device addresses a specific slave device by placing the 8-bit slave address in the DA address field of the telegram. It includes its own address in the SA address field. Valid addresses are from 0 to 127 (00H..7FH). Address 126 is reserved for commissioning purposes and may not be used to exchange user data. Address 127 is reserved as the broadcast address, which is recognized by all slave devices on a network. The slave responds by placing its own address in the SA field of its response, to let the master know which slave is responding, and it places its assigned master's address in the DA field of its response. Remember that a slave does not issue a response to broadcast messages (address 127).

Telegram header and data frame

SD	LE	LEr	SD	DA	SA	FC	DSAP	SSAP	DU	FCS	ED
1b	1b	1b	1b	1b	1b	1b	1b	1b	var	1b	1b

FIGURE 12.9 Profibus Telegram Header Data and Frame

Profibus DP telegram header abbreviation and frame bytes

SD	1 byte	Start delimiter (used to distinguish telegram format
LE	1 byte	Net data length (DU)+DA+SA+FC+DSAP+SSAP
LEr	1 byte	Length repeated
DA	1 byte	Destination address - where this message goes to
SA	1 byte	Source address- where this message come from. The address of the sending station
FC	1 byte	Function Code (FC= Type/Priority of this message).Used to identify the type of the telegram, such as request , acknowledgement , or response telegrams (FC=13 signal diagnostics data)
DSAP	1 byte	Destination service access point (COM port of the receiver). The destination station uses this to determine which service to be executed.
SSAP	1 byte	Destination service access point (COM port of the sender)
DU	1-32 b (or 1-244b)	Data units/Net data from 1 to 244 bytes
FCS	1 byte	Frame checking sequence (ASIC addition of the bytes within the specified length)
ED	1 byte	End delimiter (always 16H)

FIGURE 12.10 Profibus DP telegram Header abbreviations and Frame Bytes

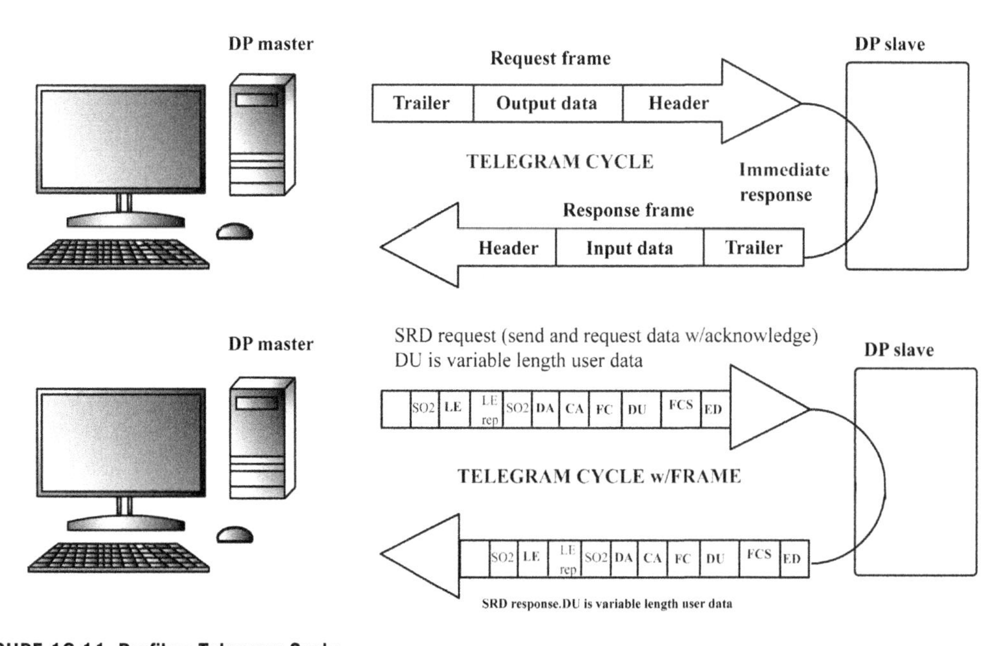

FIGURE 12.11 Profibus Telegram Cycle

Table 12.3 Profibus Telegram formats

Telegram Format	Value	Data Field Length
SD1	10H	0 bytes (no data field)
SD2	68H	1–32 bytes (or up to 244)
SD3	A2H	8 bytes fixed
SD4	DCH	0 bytes (no data field)
SC	E5H	0 bytes (no data field), short acknowledge

Note that the inclusion of a DSAP or SSAP entry in a request or response telegram is identified by setting the highest bit in the address byte of the DA or SA, respectively. This may look like an address greater than 127, but only the lower seven bits of the DA and SA contain the actual address.

12.4.4 FUNCTION CODE OR FRAME CONTROL (FC)

The Function Code (FC) or Frame Control field specifies the type of telegram (request, response, and acknowledgment), type of station (passive or active/slave or master), priority, and telegram acknowledgment (successful or unsuccessful) are given in Tables 12.4 and 12.5.

12.4.4.1 Server access points (SSAP and DSAP)

Data exchanges are handled in the telegram header using SAPs. The SAP determines the data to be transmitted or which function is to be performed. Only telegrams that include data fields use DSAP and SSAP bytes (i.e., SD2 and SD3 telegrams). SRD transmission combines an output message with an input response in a single telegram cycle. The telegram header contains an SSAP and/or DSAP that indicates the service(s) to be executed. One exception is the cyclical data exchange telegram, which is performed with the default SAP (SSAP or DSAP is not provided in its header). Additionally, some telegrams may only provide a DSAP or SSAP, but not both.

Table 12.4 Profibus DP function codes for request telegrams

FC Code	Function (The MSB in FC = 1)
4	SDN low (send data with no acknowledge)
6	SDN high (send data with no acknowledge)
7	Reserved/request diagnostic data
9	Request FDL status with reply
12	SRD low (send and request data with acknowledge)
13	SRD high (send and request data with acknowledge)
14	Request ID with reply
15	Request LSAP status with reply

Table 12.5 Profibus DP function codes for acknowledgement telegrams

FC Code	Function (The MSB in FC = 0)
0	ACK positive
1	ACK negative (FDL/FMA1/2 user error UE, interface error)
2	ACK negative (no memory space for send data (RR))
3	ACK negative (no service activated (RS), SAP not activated)
8	Response FDL/FMA ½ data low and send data OK
9	ACK negative (no response FDL/FMA ½ data and send data OK)
10	Response FDL data high and send data OK
12	Response FDL data low, no resource for send data
13	Response FDL data high resource for send data

The inclusion of a DSAP or SSAP entry in a request or response telegram is identified by setting the highest bit in the address byte of the DA or SA, respectively. Based on the detected SAPs, each station is able to recognize which data has been requested and which response data is to be supplied. PROFIBUS DP uses SAPs 54–62 listed as follows, plus the default SAP (Table 12.6).

SAP55 is optional and may be disabled if the slave does not provide nonvolatile storage memory for the station address. Note that SAPs 56, 57, and 58 are not enabled until the DP slave assumes the data exchange state. SAPs 59, 60, 61, and 62 are always enabled. Note that the DSAP and SSAP entries in a request telegram are also included in the response telegram, where DA + SA + DSAP + SSAP in the response message corresponds to SA + DA + SSAP + DSAP in the request telegram (content position flips).

Table 12.6 Profibus SAP services

SAP	Service
Default SAP = 0	Cyclical data exchange (Write_Read_Data)
SAP 54	Master-to-master (M–M communication)
SAP 55	Change station address (Set_Slave_Add)
SAP 56	Read inputs (Rd_Inp)
SAP 57	Read outputs (Rd_Outp)
SAP 58	Control commands to a DP slave (Global_Control)
SAP 59	Read configuration data (Get Cfg)
SAP 60	Read diagnostic data (Slave_Diagnosis)
SAP 61	Send parameterization data (Set_Prm)
SAP 62	Check configuration data (ÇHk_Cfg)

12.4.4.2 Data unit

The data unit (DU) field contains the data for the station at DA (in case of request data) or the data for the station at SA (in case of response data). DU is generally limited to 32 bytes, but the standard allows for lengths up to 244 bytes.

- Frame Check Sequence (FCS): This field contains the FCS or telegram checksum (00H.. FFH). It is simply the sum of the ASCII bytes of information from DA to DU modulus 256. Checksum = (DA + SA + FC + DU) mod 256. This is simply the bytes added together and divided by FFH (255). This is an integrated function that is normally performed by the PROFIBUS ASIC.
- End Delimiter (ED): This byte identifies the end of a ProfiBus telegram and has a fixed value of 16H.

12.4.5 THE PHYSICAL LAYER

12.4.5.1 RS485

The RS485 is the transmission technology most used in PROFIBUS. Its main characteristics are as follows:

- Twisted pair shielded cable.
- NZR (nonreturn to zero) asynchronous transmission.
- 9.6 kbit/s to 12 Mbit/s configurable Baud rates (when the system is configured, only one transmission rate is selected for all the devices in the bus).
- 32 stations per segment, 127 stations max.
- Distance dependent on the transmission rate (Table 12.7).
  - Expansible distance up to 10 km with the use of repeaters
  - 9-Pin D-Sub connector (Figure 12.12).

There is the need for an active termination on the bus at the beginning and end of each segment (Figure 12.13).

Table 12.7 Transmission rate with distance

Baud rate (kbit/s)	9.6	19.2	93.75	187.5	500	1500	2000
Segment length (m)	1200	1200	1200	1000	400	200	100

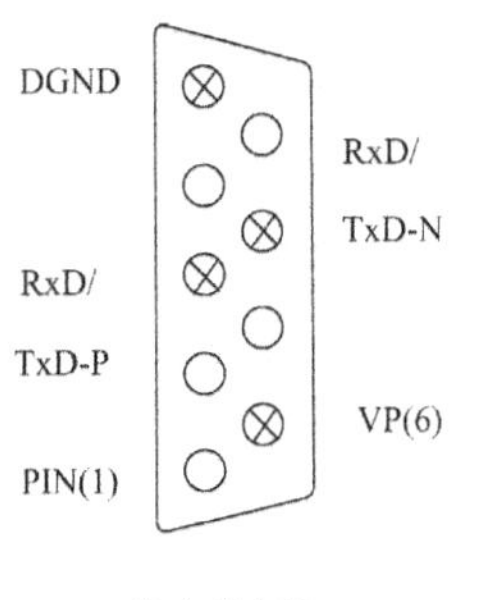

FIGURE 12.12 Profibus 9 pin Sub D

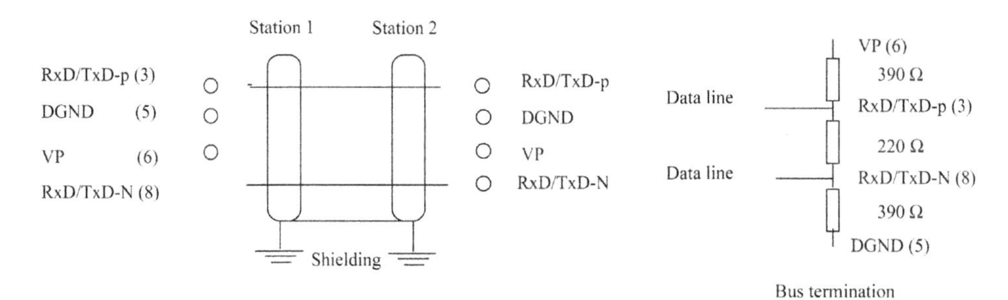

FIGURE 12.13 1) Profibus Pin Out 2) Bus Termination

12.4.5.2 Manchester Coded Bus Powered (MBP)

The Manchester Coded Bus Powered (MBP) was designed specifically to be used in PROFIBUS PA. It permits the transmission of both data and power. The transmission technology is synchronous with Manchester codification in 31.25 kbits/s. It is defined according to the IEC 61158-2 and was elaborated to satisfy the requests by chemical and petrochemical industries – intrinsic safety and powering the field equipment through the bus. The work options and limits in potentially explosive areas were defined as per the Fieldbus Intrinsically Safe Control (FISCO).

The main characteristics of IEC 61158-2 are

- Communication Signal – Codification with voltage modulation
- Transfer rate of 31.25 kbits/s
- Topology supported: bus, tree, and point-to-point
- Power supply options: via bus or external medium
- Intrinsic safety possible
- Number of equipments on the link depends on power consumption – max. 32 (non-Ex)/9 (explosion group IIC)/23 (explosion group IIB)
- Maximum span of 1900 m, expansible up to 10 km with four repeaters
- Spur maximum of 120 m/spur

12.4.5.3 Fiber optic

The fiber optic solution helps in eliminating noise, potential differences, long distances, ring architecture, physical redundancy, and high-speed transmission.

The distance supported on fiber depends on the type of cable used (Table 12.8).

Table 12.8 Fiber Optic distance in Profibus

Fiber Type	Characteristics
Monomode glass fiber	2–3 km mean distance
Multimode glass fiber	Long distance > 15 km
Synthetic fiber	Short distance > 80 km
PCS/HCS fiber	Short distance > 500 m

ACKNOWLEDGMENTS

- http://www.rtaautomation.com/profibus/
- http://www.smar.com/profibus.asp
- http://www.isa.org/Template.cfm?Section=Books1&template=Ecommerce/FileDisplay.cfm&ProductID=6959&file=Chapter1_Profibus.pdf
- http://www.kelburn.net/pdfs/profibus_introduction_698a.pdf

FURTHER READINGS

Profibus Draft Technical Guideline, Profibus-PA User and Installation Guidelines, Version 1.1, September 1996, PROFIBUS Nutzenorganisation e.V, Karlsruhe, Germany, 1994.

Technical Description, PROFIBUS Nutzenorganisation e.V., September 1999.

PROFIBUS, Draft, PROFIBUS Profile, PROFIBUS – PA, Profile for Process Control Devices, Version 3.0, April 1999.

CHAPTER

FOUNDATION FIELDBUS COMMUNICATION 13

13.1 FIELDBUS TECHNOLOGY

13.1.1 INTRODUCTION

In the early 1990s, foundation fieldbus (FF) emerged in the market. Two parallel supplier consortiums, Interoperable Systems Project (ISP) and WorldFIP North America, merged to form the Fieldbus Foundation Organization. The new organization immediately brought critical mass to achieve an internationally acceptable fieldbus protocol. The foundation organized development programs and conducted field trials for end users to drive FF technology in industrial applications.

FF is a communication protocol, which is "all-digital, serial, two-way multidrop communication system that interconnects fieldbus devices with the control system".

FF is fully digital, serial, two-way, multidrop, communication system running at 31.25 Kbits/s, which will be used to connect intelligent field equipment such as sensors, actuators, and controllers. It serves as a local area network (LAN) for the instrumentation used within process plants and facilities with built-in capability to monitor and distribute control applications across the network.

13.1.2 OVERVIEW

An FF system is a distributed system composed of field devices and control/monitoring equipment integrated into the physical environment of a plant or factory. FF devices work together to provide I/O and control for automated processes and operations. FF systems may operate in manufacturing and process control environments. Some environments require intrinsic safety where devices typically operate with limited memory and processing power and with networks that have low bandwidth.

13.1.2.1 Features

The fieldbus retains the desirable features of the 4–20 mA analog system such as given below:

- Single loop integrity
- Standardized physical interface to the wire
- Bus-powered devices on a single wire pair
- Intrinsic safety options

In addition, FF enables:

- Enhanced capabilities owing to full digital communication, reduced wiring, and wire terminations due to multiple devices on one wire

- Increased selection of suppliers due to interoperability
- Reduced loading on control room equipment due to distribution of control and input/output functions to field devices
- Connection to the HSE backbone for larger systems

The advantages of the FF over the legacy systems are:

- The signals are digital, therefore more immune to noise.
- The requirement for damping to filter out noise is eliminated.
- FF automatically detects all connected devices and includes them on a live list.
- Addresses are automatically assigned, eliminating any possibility of duplicate addressing.
- Traditional I/O use 16- or 32-channel cards; these are costly and a weak point. Module failures can generally cause all associated loops to crash. Accidental removal during faultfinding will affect all 16 or 32 loops.
- Minimizing the components reduces the failure probability.
- No requirement to manually configure alarms to detect transmitter failure or broken signal cable. FF builds this automatic safety function.
- FF uses engineering units, not scaled ranges; hence measuring actual process variable, not scaled or a % of 4–20 mA. This eliminates the need for range configuration.
- Conflict in the ranges is not possible.
- Analog to digital conversion is removed, improving accuracy and reliability.
- Dual measurement of parameters is possible from a single instrument.
- Failure prediction is possible to the data available.

13.1.2.2 Benefits

FF enables increased capabilities due to full digital communications, reduced loading on control room equipment due to input and output functions being migrated to field devices, and reduced wiring and terminations due to multiple devices on one wire besides providing vast amount of additional information from each field device that can be utilized for asset management and health maintenance. In fieldbus, multiple process variables, as well as other information, can be transmitted along a single wire pair. This is quite different from the traditional approach of connecting 4–20 mA devices to a DCS system using dedicated pairs of wires for each device. Field devices and segments become a part of DCS. This shall require an integrated approach for configuration, data management, and system architecture approach to field network design. Also, some design activities have to be performed earlier in the project cycle. Complex functions can be achieved with the FF devices, which shall have significant cost designs and which appreciably reduce the commissioning and start up time.

FF can be utilized for majority of the process control applications, including the field instruments connected to the DCS. This shall include control valves, motor-operated valves, transmitters, and local indicators.

13.1.2.3 Fieldbus architecture

The fieldbus architecture is depicted in Figure 13.1. The characteristic of the foundation architecture is to ensure device interoperability with fully specified, standard user layer based on “blocks” and device descriptions (DDs).

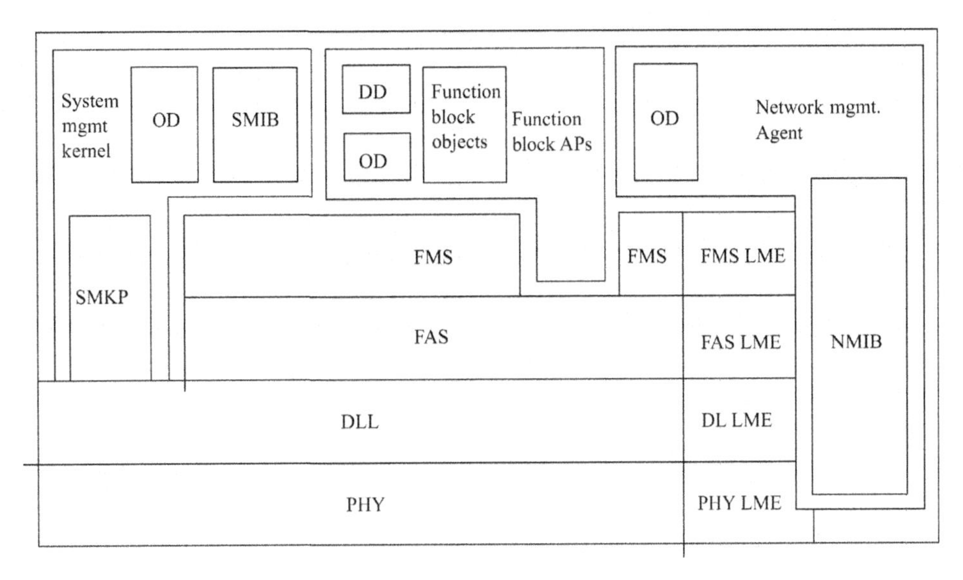

FIGURE 13.1

Fieldbus device architecture

The FF system architecture provides a framework for describing these systems as a collection of physical devices interconnected by a fieldbus network.

The data communicated over the fieldbus is called object description. Object descriptions are collected together in a structure called object dictionary (OD). The object description is identified by its index in the OD. ODs and DDs are independent of the underlying environment, and therefore do not require adaptation.

The FF system architecture defines a specific type of application process, the function block application process (FBAP), to address a variety of functional needs. The FBAP has been designed to support a range of functional models, each addressing a different need.

The FF architecture uses the concept of a virtual field device (VFD). A VFD is used to remotely view local device data in the object dictionary. A device has at least two VFDs:

- Network and system management VFD
- User application VFD

Network management is part of the network and system management application. Network management refers to managing various parameters to carry out fieldbus communication. The VFD used for network management is also used for system management. It provides access to the network management information base (NMIB) and to the system management information base (SMIB).

System management (SM) manages the parameters needed for the construction of a functional control system, rather than communication. SMIB data includes device tag and address information, and schedules function block execution.

The system management kernel protocol (SMKP) communicates directly with the data link layer. The SMKP assigns end user defined names, called tags, and data link layer addresses to devices, as they are added to the fieldbus. It contains an OD that can be configured and interrogated using FMS operating over client/server virtual communication relationships. The use of EDDL allows the development of new devices while still maintaining compatibility. The second is to maintain distributed application time so that function block execution can be synchronized among devices.

Electronic device descriptions (EDDs) created by device description language (EDDL) for a field device support the management of intelligent field devices. EDDs contain the description of all device parameters, parameter attributes, and device functions. EDDs also include a grouping of device parameters and functions for visualization and a description of transferable data records. The DD may be supplied with the device on a disk, or downloaded from the Fieldbus Foundation Web site, and loaded into the host system.

The architecture of a fieldbus device is based on function blocks, which are responsible for performing the tasks required for the current applications, such as data acquisition, feedback and cascade loop control, calculations, and actuation. Every function block contains an algorithm, a database (inputs and outputs), and a user-defined name.

13.1.3 FF NETWORK

13.1.3.1 Fieldbus network classification

The FF has two types of network as depicted in Figure 13.2. One is the FF H1 running on 31.25 Kbits/s, which is used to interconnect the field equipment like sensors, actuators, and I/Os. The other one is high speed Ethernet (HSE) running at 100 Mbits/s, which provides integration of high-speed controllers (PLCs), H1 subsystems (via a linking device), data servers, and workstations.

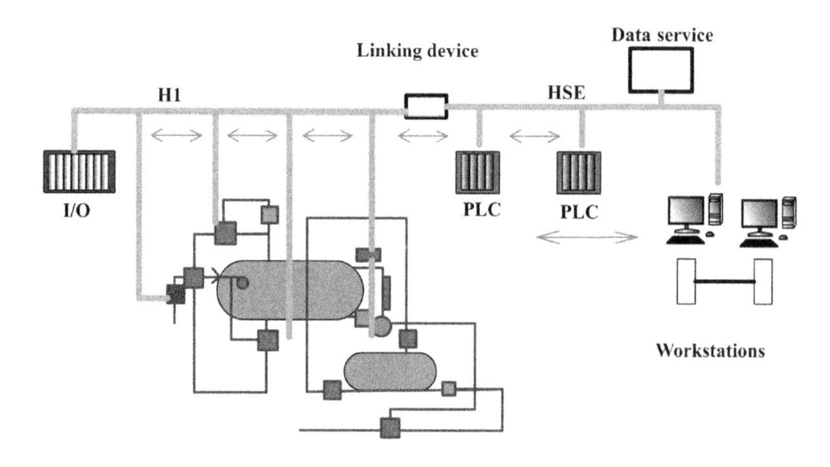

FIGURE 13.2

Fieldbus network classification

13.1.3.2 Fieldbus topology

FF network consists of fieldbus device, spur, trunk, device coupler, and fieldbus power supply, as illustrated in Figure 13.3.

- Link: A link is the logical medium by which H1 fieldbus devices are interconnected. It consists of one or more physical segments interconnected by bus, repeaters, or couplers.
- Segment: A section of a fieldbus that is terminated in its characteristic impedance, meaning a cable and devices installed between a pair of terminators. Repeaters are used to link segments to form a longer fieldbus.
- Trunk: Trunk is the cable between the control room and the junction box in the field. Being the longest cable path on the fieldbus network, it is the main fieldbus communication cable.
- Spur: Spur is a cable between the trunk cable and a fieldbus device, usually connected to the trunk via a device coupler. It is the cable that connects a device to the trunk.
- Device coupler: FF device coupler is located where the trunk is connected to the various device spurs. The couplers have built-in short-circuit protection to minimize the impact of a short circuit at one device affecting the whole segment.
- Fieldbus power supply: Fieldbus requires a special kind of power supply. If an ordinary power supply were to be used to power the fieldbus, the power supply would absorb signals on the cable because it would try to maintain a constant voltage level. For this reason, an ordinary power supply has to be conditioned for fieldbus. Putting an inductor between the power supply and the fieldbus wiring is a way to isolate the fieldbus signal from the low impedance of the bulk supply. The inductor lets the DC power on the wiring, but prevents signals from going into the power supply.

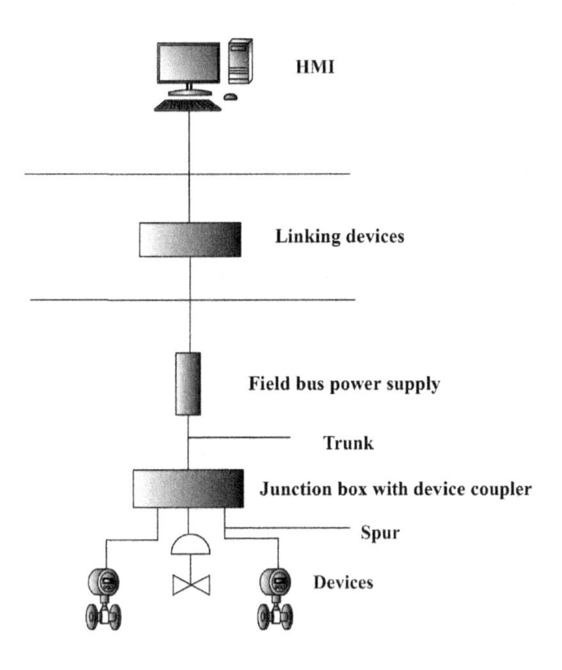

FIGURE 13.3

Fieldbus network overview

- Fieldbus device: In a conventional 4–20 mA DCS, two wires are used to connect to a device, and in this case, the data acquisition and control lies with the controller. With the introduction of FF, data acquisition and control lies in the fieldbus device itself.

13.1.4 FIELDBUS CONNECTION TOPOLOGIES

There are several possible topologies for fieldbus networks. This section illustrates some of the possible topologies and their characteristics.

13.1.4.1 Point-to-point topology

This topology consists of a segment having only two devices. It could be a field device such as a transmitter connected to a host system for monitoring or a slave and host device operating independently such as a transmitter and valve with no connection beyond the two (Figure 13.4).

13.1.4.2 Tree topology

This topology is also known as chicken foot topology. It consists of a single fieldbus segment connected to a common junction box to form a network (Figure 13.5).

13.1.4.3 Spur topology

This topology consists of fieldbus devices connected to a multidrop bus segment through a length of cable called a spur. A spur can vary in length from 1 m to 120 m (Figure 13.6).

13.1.4.4 Daisy chain topology

In this topology, the fieldbus cable is routed from device to device on this segment and is interconnected at the terminals of each fieldbus device (Figure 13.7).

13.1.5 FF TECHNOLOGY

The FF technology is based on OSI (open system interconnect) model of layered communication. Figure 13.8 illustrates the comparison between a fieldbus model with an OSI model of network communications. Layers 3–6 are not present in the fieldbus model, since their services are not needed in a process control application. Moreover, application layer is divided into two sublayers, namely fieldbus message specification (FMS) and fieldbus access sublayer (FAS). An additional layer known as user application layer is present in the fieldbus model, which is used to define control tasks for process plants. The user application layer is not a part of the OSI model.

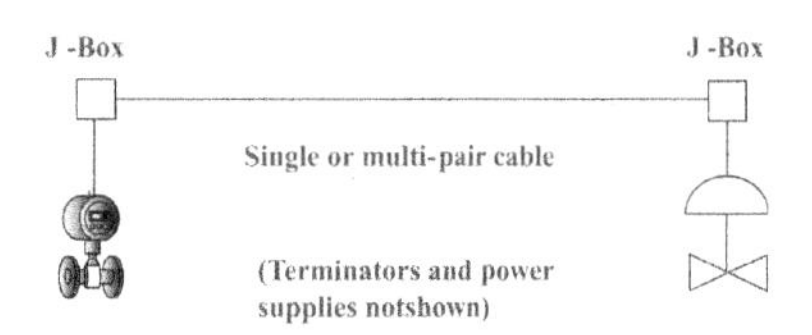

FIGURE 13.4

Point-to-point topology

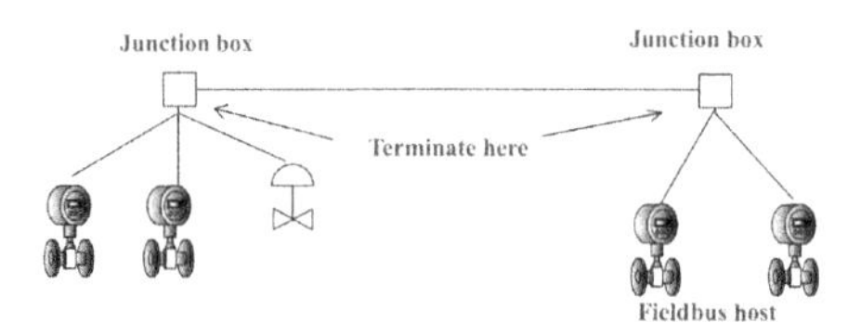

FIGURE 13.5

Tree topology

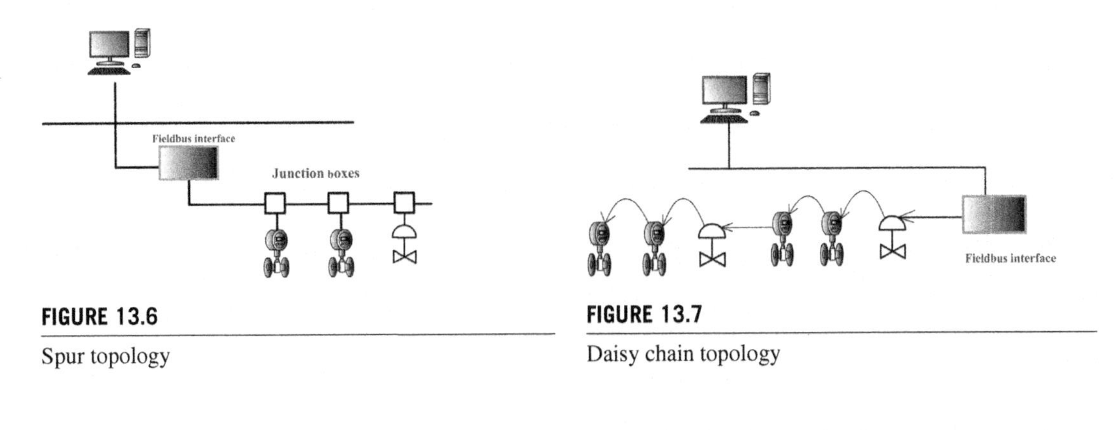

FIGURE 13.6

Spur topology

FIGURE 13.7

Daisy chain topology

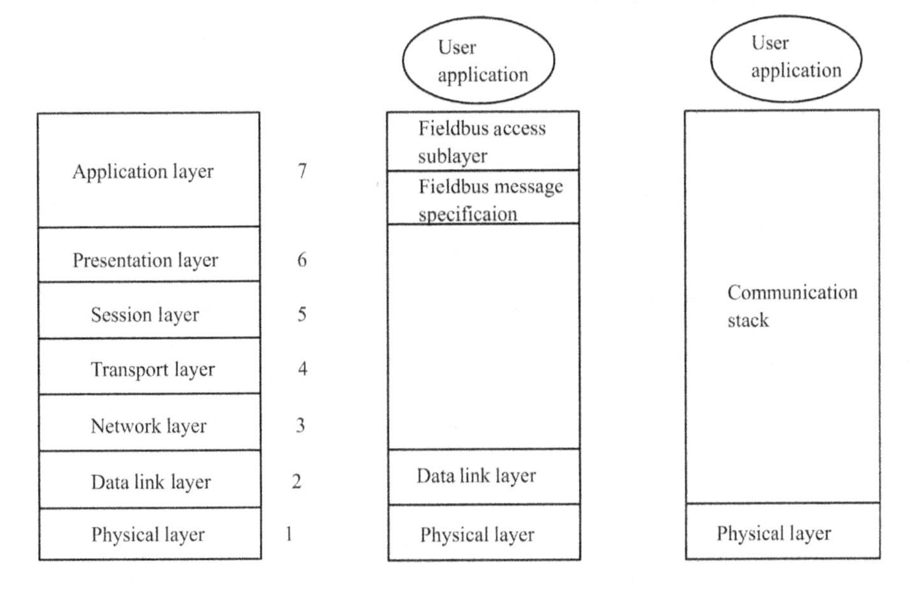

FIGURE 13.8

Communication model

13.1.5.1 Physical layer

The physical layer or the H1 bus interconnects field devices such as sensors, actuators, and I/Os with the host in the control system. Communication signals and current signals are conveyed along the H1 bus at 31.25 Kbps.

The equivalent electrical circuit of physical layer is shown in Figure 13.9. Voltage is applied through impedance conditioner. DC through the impedance is then applied to the devices present in the fieldbus network.

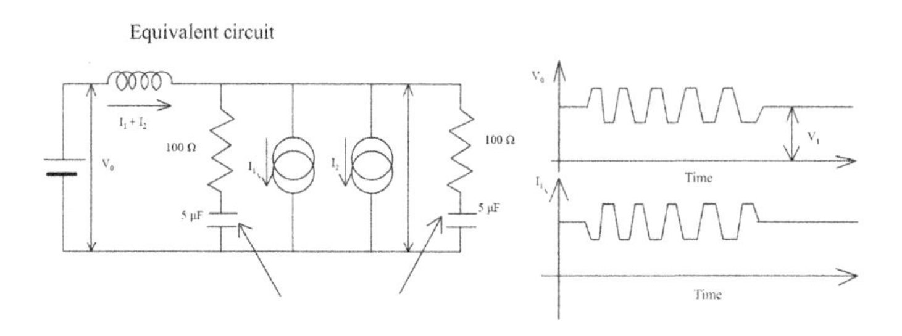

FIGURE 13.9

Electrical equivvalent circuit of fieldbus physical layer

13.1.5.2 Device power requirements

The DC supply voltage across the H1 bus can range from 9 to 32 V. All the fieldbus devices can be powered directly from the H1 bus. The DC supply voltage can range from 9 to 32 VDC. At each end of the cable terminators of 10 Ω, impedances are provided to maintain a balanced transmission line so that high-frequency signals can be transmitted with less distortion. The transmitting device delivers ±10 mA at 31.25 Kbit/s into a 50 Ω equivalent load to create a 1.0 V peak-to-peak voltage modulated on top of the direct current (DC) supply voltage. Loss of power or disturbance to one power supply module shall not result in loss of field device in any circumstances.

The function of the physical layer is to convert the messages received from the communication stacks by adding and removing preambles, start delimiters, and end delimiters into physical signals. The physical signals are then transmitted on the H1 bus. A typical waveform of the physical layer is shown in Figure 13.10.

The start delimiter and end delimiter indicate the start and end of the physical layer signal (fieldbus message). The receiver uses start delimiter to find the beginning of message and accepts the message until end delimiter is used. The function of the preamble at the start is to synchronize the internal clock with incoming fieldbus signal by the receiver. Fieldbus signals are encoded using Manchester biphase-L technique Figure 13.11. Data signal arc combined with the clock signal to create fieldbus signal. Since clock information is embedded into the serial data, the signal is called synchronous serial.

The number of devices possible on fieldbus link depends on many factors such as power consumption of each device, type of cable used, and whether devices are intrinsically safe. Without the use of repeaters, the length of H1 cable can be as long as 1900 m.

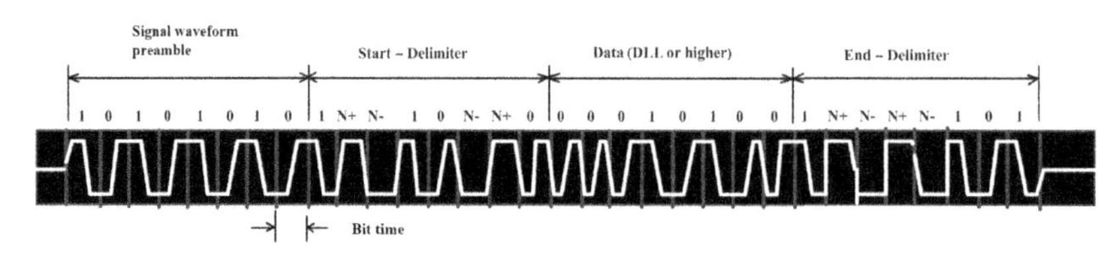

FIGURE 13.10

Fieldbus signal

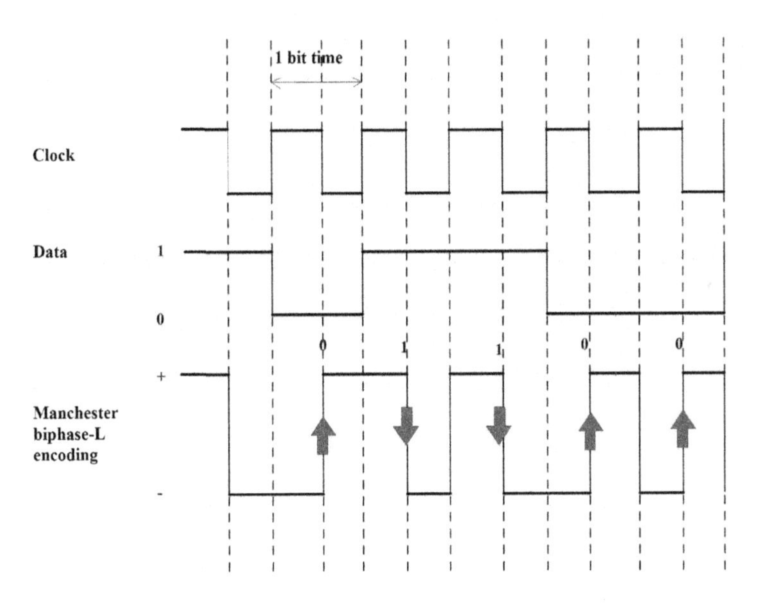

FIGURE 13.11

Manchester biphase-L encoding

13.1.6 DATA LINK LAYER

Data link layer controls the flow of messages onto the fieldbus physical layer. It transfers the data from one device to other devices on the network.

FF devices are classified as given below:

- Basic devices: Basic devices are not capable of functioning as link masters. They do not have link active scheduler (LAS) functionality. A basic device performs all the basic communications required for a field device, except scheduling communication.
- Link master device: Link master devices are capable of functioning as link master, with LAS capability. They are capable of scheduling communication in an H1 network. Any H1 network should have one link master device. It can be DCS or any other device such as an actuator or a transmitter. Once the device configuration is downloaded in the device, the control strategy can work even when the computer is disconnected. The link can have many devices configured as link master, but only one device actively controls the bus at any given time. The bus that controls the bus is called link active scheduler.
- Bridge: Bridge devices connect H1 link together. They are link master devices and must act as link active scheduler.

13.1.6.1 Link active scheduler

All communications on the fieldbus are controlled by a single device called link master. Every host has the capability to act as a link master. The link master has a function known as LAS. LAS decides which device has to transmit data and at what time. This helps in avoiding collision of messages. Following are some of the functions that are performed by LAS.

When it is time for the device to send data on fieldbus layer, LAS issues a compel data (CD) message to the device. On receipt of the CD, the device broadcasts or publishes data on the bus for all other devices on the H1 network. This is the highest priority function that is performed by the LAS. The devices that are configured to read the data published on the bus are known as subscriber devices, and the device that publishes the data is called publisher.

The following are the main functions of the LAS:

- Live list maintenance: Live list contains the list of all the devices that properly respond to pass token (PT). Whenever devices are added or deleted, LAS broadcasts changes to the live list. All link master devices maintain a copy of the live list.
- Data link time synchronization: Time distribution across all the devices is achieved through LAS. All devices are time stamped with the host time by LAS. The LAS sends time distribution messages periodically to all the devices, across the fieldbus network, to synchronize time stamp of all devices with time stamp of LAS.
- Redundancy in LAS: Fieldbus can have many link master devices, which are capable of functioning as LAS. Suppose if current LAS link master device fails, other devices configured as link master take over the control. The link operation does not get failed in any condition.
- Token passing: The LAS sends PT messages to all devices on its network. The device transmits its data when it receives the PT.

13.1.6.2 Types of communication

There are two types of communication in FF.

- Scheduled communication: Scheduled communications are also called cyclic communications. The transfer of regular, cyclic control loop data between devices on the fieldbus network is known as scheduled communication. Function block execution and communication between function blocks – publish/subscribe virtual communication relationship (VCR).
- Unscheduled communication: All the devices on fieldbus network send unscheduled messages between scheduled communications. Unscheduled communication takes place when LAS grants permission by issuing PT. Examples of this type of communication include events and user operation. Unscheduled communication (for event-driven or on-demand communication) – client/server VCR.

The LAS is responsible for scheduled communication that takes place between the various function blocks, which is also known as the publisher/subscriber model of communication. The H1 card contains an LAS algorithm that automatically determines the macrocycle and schedules all connected devices. No manual intervention is required.

During macrocycle time, time other than scheduled communication time is used for nondeterministic bus communications known as unscheduled communication. During unscheduled communication, alarm transmission and set-point changes take place.

There should be sufficient spare time in the macrocycle to satisfy the general sparing conditions of this project. The recommended unscheduled (free asynchronous) time will be 55% of the macro cycle. (i.e., scheduled time shall be maximum 50% of the macrocycle). Of the available unscheduled time, 25% should be unallocated to defined tasks (i.e., function block views and DCI data).

13.1.6.2.1 Device addressing concept

The devices in FF are represented by an address that is an 8-bit primary number. In decimal, address ranges from 0 to 255. The following tabular column shows the distribution of addresses in FF.

Address in Decimal	Address in Hexadecimal	Allocated for
0–15	00–0F	Reserved
16–247	10–F7	Permanent address
248–251	F8–FB	Temporary address
252–255	FC–FF	Visitor address

The devices are addressed by the host system, usually DCS in the network. A maximum of 32 devices can be connected in an H1 network.

13.1.6.2.2 Fieldbus access sublayer (FAS)

The FAS uses the scheduled and unscheduled features of the data link layer to provide communication service for the FMS. VCR describes the types of FAS services.

The VCR is the communication channel between fieldbus devices. All data has to pass through their own VCRs.

There are three types of VCRs:

- Client–server: The client/server VCR type is used for unscheduled, operator-initiated communication between devices on the fieldbus. When a device receives a PT from the LAS, it can send a request message to another device on the fieldbus. The requester is called the Client, and the device that received the request is called the Server. The Server sends the response when it receives a PT from the LAS.
- Publisher–subscriber: The publisher/subscriber VCR type is used by the field devices for cyclic, scheduled publishing of user application function block input and outputs such as process variable (PV) and primary output (OUT) on the fieldbus. When a device receives the CD, the device "publishes" or broadcasts its message to all devices on the fieldbus. Devices that wish to receive the published message are called Subscribers (Figure 13.12).
- Source–sink: The source–sink VCR type is used for unscheduled, operator-initiated, one-to-many communications. It is typically used by fieldbus devices to send alarm notifications to the operator consoles.

13.1.6.3 Fieldbus message specifications

FMS services enable user applications to send messages to each other across the fieldbus using a standard set of message formats.

FMS describes the communication services, message formats, and protocol behavior needed to build messages for the user application.

13.1.6.3.1 FMS services

- Establish and release VCRs and determine the status of a VFD.
- Enable user application to access and change variables associated with an object description.
- Enable user application to report events and manage event processing.
- Remotely upload or download data and programs over the fieldbus.

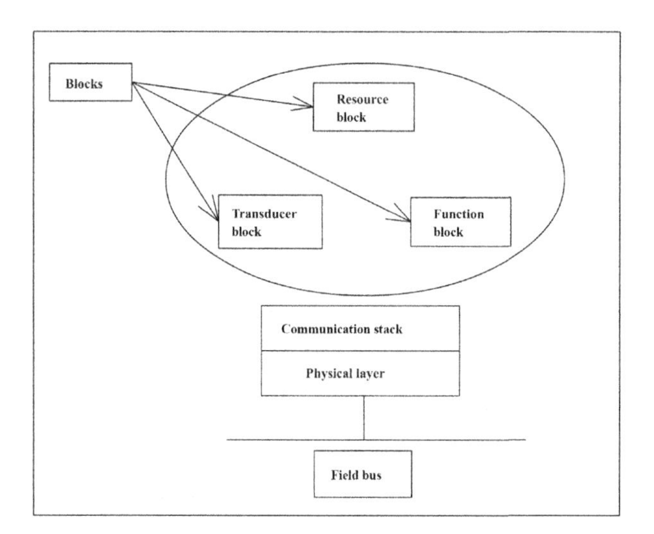

FIGURE 13.12

Scheduled data transfer

13.1.6.3.2 User application layer

The user application layer defines the way of accessing information within fieldbus devices so that such information may be distributed to other devices or nodes in the fieldbus network. This is a fundamental attribute for process-control applications. User application layer defines an FBAP using resource blocks, function blocks, transducer blocks, system management, network management, and DD technology (Figure 13.13).

- Resource block: A resource block shows the contents of a VFD by providing the manufacturer's name, device name, DD, and so on. The resource block controls the overall device hardware and function blocks within the VFD, including hardware status.
- Transducer block: A transducer block converts the signal from physical to digital form. A transducer block is important because it is also used to capture and store all the diagnostic and maintenance-related data for a device.

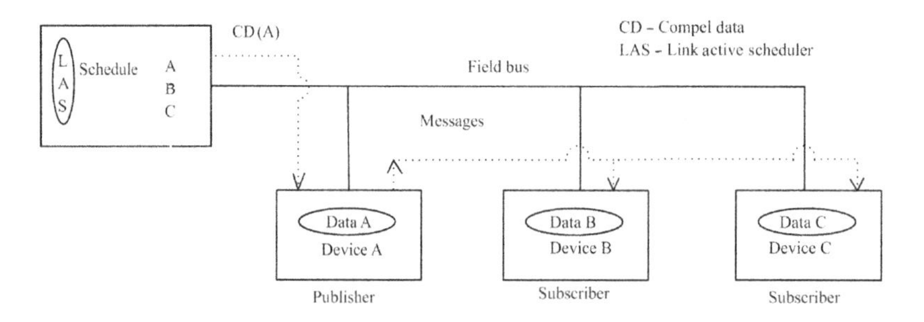

FIGURE 13.13

Fieldbus architecture

- Transducer specifications: The device developers generally define transducer specifications. The transducer specifications establish the base scope of transducer functions. A device may have additional functions, but it must contain the functions specified in the specification to be interoperable within the given specification.
- Function block: The function blocks are responsible for performing the tasks required by the current applications, such as data acquisition, feedback and cascade loop control, calculations, and actuation. Every function block contains an algorithm, a database (inputs and outputs), and a user-defined name. The three function block classes are as follows:
  - Standard block
  - Enhanced block
  - Vendor-specific block

13.1.6.4 Function block application process

The FBAP is composed of a set of function blocks configured to communicate with each other. The outputs from one function block can be linked to the inputs of another function block. Function blocks may be linked within a device, or across the network. The field data are acquired by the function block through the transducer block.

13.1.6.5 Applications of FF technology

The important feature of FF in control system is "control in field application." This is explained in the following examples.

13.1.6.5.1 Example of simple control loop

The analog input (AI) value from device A provides the process value to the PID of the device B, which controls the output depending upon the set point given to the PID block. The processed output is fed into the final control element through the AO block (Figure 13.14).

Device A in this case could be an input device used to measure process parameters such as temperature, pressure, flow, or level transmitter, which provides process value to the PID of Device B from which the output goes to AO block of the same device, which in turn actuates the final control element.

13.1.6.5.2 Example of cascade control loop

Consider the following example (Figure 13.15) of a flow control loop involving a temperature transmitter (device A) and flow transmitter (device B), along with a control valve (device C) that is link master

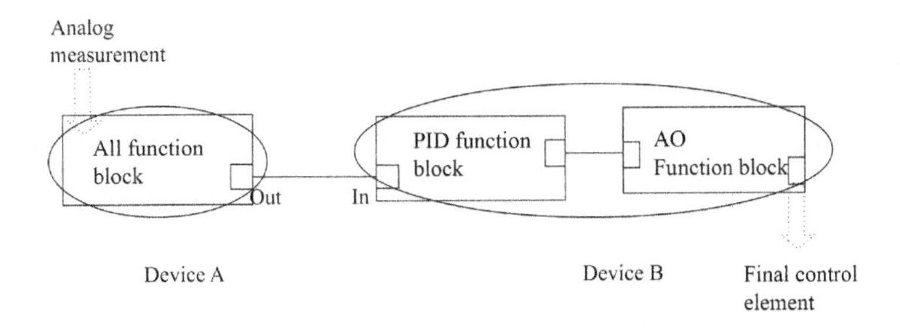

FIGURE 13.14

Example of simple fieldbus control loop

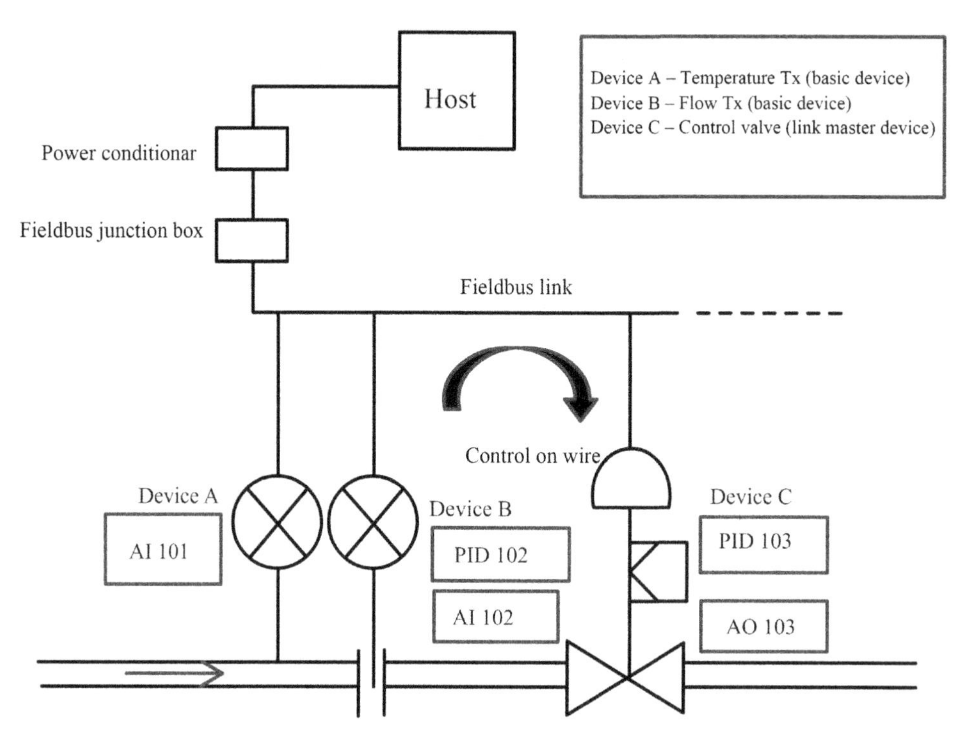

FIGURE 13.15

Application of fieldbus – cascade loop

device. The loop illustrated in the following figure is a cascade loop, where the primary element is temperature and secondary element is flow. The temperature transmitter senses and gives the output to primary PID (device B), which is controlled using the operator set point. The output of primary PID is given as set point to secondary controller, which is nothing but PID (device C) of final control element, which in turn controls the final flow. In the below example, the PIDs (controllers) are residing inside the transmitter and final control valve (Figure 13.16).

Advantages of cascade control loop are as follows:

- Reduced wiring
- Reduced hardware and panels
- Host supports different fieldbus devices irrespective of vendor, that is, seamless interoperability between devices and hosts
- Easy installation and maintenance
- Easy calibration and troubleshooting
- Advance diagnostics
- Predictive maintenance
- Applicable for almost all applications including intrinsic safety

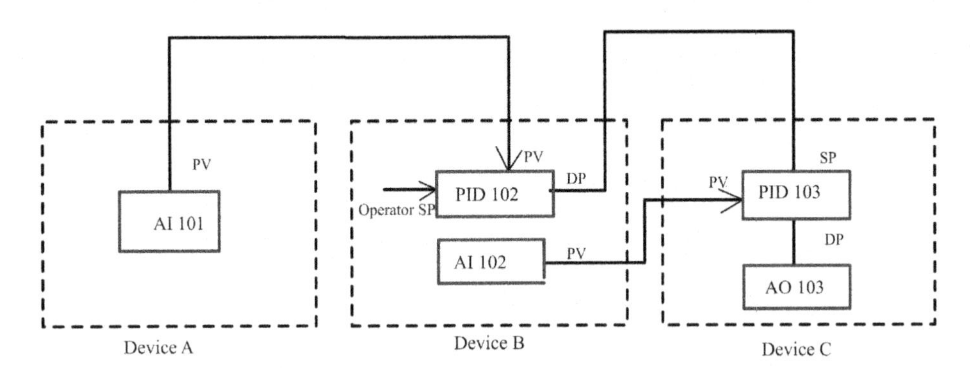

FIGURE 13.16

Loop representation of Figure 13.15

Disadvantages of cascade control loop are the following:

- More expensive
- Needs training and expertise to handle fieldbus devices
- Designing fieldbus segments requires experience and training
- Fieldbus cannot handle faster loops that require fast response, such as antisurge control

FURTHER READING

ISA-50.02, Part 2-1992 Fieldbus Standard for Use in Industrial Control Systems, Part 2 Physical Layer Specifications. Research Triangle Park NC: ISA, ISBN 1-55617-317-2, 1992.

Fieldbus Foundation, 31.25 Kbits/s Intrinsically Safe Systems, Applications Guide AG-163, Fieldbus Foundation, Austin, TX, 1996.

Fieldbus Foundation, Wiring and Installation 31.25 kbit/s, Voltage mode, Wire Medium, Applications Guide AG-140, Fieldbus Foundation, Austin, TX, 1996.

Foundation Specifications. Transducer Block Application Process Part 1, FF-902, revision PS 3.0, April 21, 1998.

Technical Overview, FD-043, Revision 2.0. Fieldbus Foundation, 1998.

Foundation Specifications. Transducer Block Application Process Part 2, FF-903, revision PS 3.0, April 21, 1998.

Foundation Specifications. Function Block Application Process Part 1, FF-890, revision 1.4, June 29, 1999.

Foundation Specifications. Function Block Application Process Part 2, FF-891, revision 1.4, June 29, 1999.

Foundation Specifications. Function Block Application Process Part 3, FF-892, revision 1.4, June 29, 1999.

Verhappen, I., Pereira, A., Foundation Fieldbus: A Pocket Guide. ISBN: 1-55617-775-5 (note – ISA publication). Also available in Portuguese.

Berge, J., Fieldbuses for Process Control: Engineering, Operation and Maintenance. ISBN: 1-55617-760-7 (note – ISA publication). Also available in Chinese.

Fieldbus Technical Overview – FF.org.

Relcom, Inc., Foundation Fieldbus Wiring Design & Installation Guidelines. Download @ http://www.relcominc.com/download/fbguide.pdf.

http://www.pacontrol.com/download/Fieldbus-System-Engineering-Guidelines.pdf.

Fieldbus System Engineering Guide – http://www.Fieldbus.org/images/stories/enduserresources/technicalreferences/documents/system_engineering_guidelines_version_3.pdf.

CHAPTER

WIRELESS COMMUNICATION 14

14.1 INTRODUCTION

Wireless communication in industrial automation is now a reality and no longer an academic topic of discussion. Most industries and manufacturing firms already have some presence of wireless instrumentation in their plants. Wireless instruments are increasingly becoming inevitable to industrial automation just as cell phones are to telecom. There are many benefits of wireless instrumentation and related data acquisition and control system, forcing industries to switch to wireless instrumentation.

14.1.1 BENEFITS OF WIRELESS INSTRUMENTATION

The benefits of wireless instrumentation are as follows.

1. A wireless system simply translates to savings on wiring cost. This becomes a very important factor for choosing a wireless option. An industrial plant occupies a huge area, often up to few square kilometers. The distance between the control room and actual field transmitters sometimes is in kilometers if not few hundred meters. Wireless instrumentation and control system makes the whole investment redundant.
2. Cable lay, connection, cable tray setup, and all other related activities are labor intensive. A huge amount of skilled labor cost is saved by opting for wireless instrumentation. Using wireless field instruments and transmitters saves the data cable cost and power cable cost. These transmitters are self-battery powered and last for years.
3. A wireless system is easy to commission so that the whole system can be commissioned in few days, as opposed to an equivalent wired system that can take months to commission. Project setup, installation, and commissioning time is directly related to revenue and earnings for the company, be it a new plant or an existing plant upgrade. The faster a plant starts production, the faster the company starts earning revenue. In this regard, wireless system installation and commissioning scores heavily over the traditional wired system.
4. Junction box is an important component in the loop of the wired instrument. It is a junction point between the field and the control room. The junction box device can be a 2/3/4 wire transmitter. A multicore cable comes from the control room to the junction box, and then from the junction box, a 2/3/4-core cable goes to the transmitter or to the final control element location. It aggregates many 2-core cables to few multicore cables. A thick multicore cable is then routed to the control room from the junction box. It is easier to place and route fewer number of multicore cables than routing many 2-core cables. There can be many multicore cables that can originate from a single junction box. Depending upon the density of transmitters in the particular area of the plant, a junction box might have connections of few devices to 40/50 devices. Wireless

instrumentation effectively eliminates all project engineering complexities of having different core cables and junction box.

5. The cables coming out from the junction box are placed on the cable tray. All the multicore cables are placed on top of the cable trays and routed till the control room. This calls for different sizes of the cable tray. The cables from different location of the plant come together and accumulate slowly as they progress toward the control room. This requires a lot of planning and documentation to mark which cable is carrying signal from which field device. The cable routing has to follow certain rules, for instance, they cannot be routed together with high-tension cable, the way the cable trays can cross each other, and so on. Again, wireless instrumentation eliminates all these complex schematics.
6. Wireless instrumentation also eliminates the need of control room side terminal panels and marshalling cabinets. Marshalling cabinets are enclosures that contain a multicore cable from the field terminates. These cables terminate in the terminal panel. The terminal panel has connections on both the sides. The other side connection goes to the barrier (Zener barrier) that prevents higher current to go into hazardous environment in the field.
7. Field termination panels are the interface cards put before the input/output (I/O) card. The other side of the barrier connects to the field termination panels. Sometimes, if the loop does not need the barrier, then the wire is directly connected between the terminal panel and the field termination units. Again, all these hardware becomes redundant in wireless connections.

   The actual processing of the signal happens in the I/O cards. The I/O cards vary based on the signal type such as 4–20 mA I/O, 0–5 V, mV signal for temperature, and HART protocol-enabled signals. But the wireless transmitters are intelligent enough to do all the processing and send signals that do not need any of the I/O card to process further. The I/O card is expensive and supports between 8 and 16 device data. One can easily imagine how many I/O cards are required to bring thousands of data point in the control system. Smart wireless devices acting as single I/O totally eliminates this hardware.

   - A huge set of documentation needs to be prepared and maintained for wired instrumentation. Each and every tag or process points must be traced back from the control room to the field location. The loop and line diagrams should specify each and every terminal numbers, junction box, marshalling cabinets, and so on. Moreover, throughout the cable, there must be cross-ferruling that should specify where the other end of the cable is terminated. All these become superfluous for wireless instruments.
8. Increased Availability: Wireless devices operate in mesh network where every device has multiple paths for communication. In wired methods, there is no redundancy available for paths. Breakage/corrosion of wire can result in putting one or more devices out of service depending on the location of the fault.
9. Scalability: Once wireless network is established with a process area, new devices can be added effortlessly without worrying about spare capacity in the junction boxes and adding any new hardware on the distributed control system (DCS) side.
10. Enhanced Safety Compliance: Wireless communication enhances safety as readings from difficult-to-reach locations can be now brought to the control room with reduced need for operator/people to travel to such locations.
11. Increased Energy Efficiency and Decreased Flaring/Hydrocarbon Losses: Wireless technology makes it possible to have devices such as wireless acoustic transmitters. These devices were not

available before; they can effectively detect passing flare valves/fault steam traps. This helps in 24×7 monitoring of critical flare valves/steam traps providing early detection and saving flare/steam losses.

14.2 BASIC CONCEPTS OF INDUSTRIAL WIRELESS COMMUNICATION

In any wireless communication, the data transmission primarily happens between two devices called the transmitter and the receiver. The transmitter sends data and the receiver receives it. Any node in the wireless network performs the role of transmitter and receiver interchangeably. An industrial wireless network primarily has two types of devices – field devices and infrastructure nodes. Field devices are the field sensors equipped with wireless capability, which are responsible for process value measurement. Infrastructure nodes do not measure any field parameters. These are used to collect the signals and dump them over the backbone or can be used as a repeater. Backbone network is the main network infrastructure throughout the plant responsible for transporting data. The backbone can be a normal Institute of Electrical and Electronic Engineers (IEEE) 802.3-based Ethernet infrastructure or an IEEE 802.11a-based wireless mesh network or any other means of communication. Backbone can be considered as the main highway with higher capacity of traffic flow, that is, higher bandwidth. Like all the arterial roads with lesser bandwidth, industrial protocols lead to the highway. They collect the data and put on the main backbone with higher capacity.

Wireless communication depends upon several factors, or rather the communication is controlled by these parameters. The main real estate of wireless communication is the frequency spectrum. The transmitter and the receiver have to be tuned in the same frequency range to communicate. When we tune our FM radio receiver to a particular frequency (e.g., anywhere between 88 and 107 MHz), it receives the signal from the broadcasting or transmitting radio station. Even the cell phone handsets have to be tuned in to a particular frequency to communicate with cell phone towers. But all these frequency bands are used for commercial purposes and thus are called licensed bands. A valid license is required to transmit in these frequency bands. These frequency spectrums fetch billions of dollars for the government. Frequency spectrums need not be purchased for industrial wireless communication. There exist some frequency ranges that are not licensed across the world. License-free industrial, scientific, and medical (ISM) band of 2.4 GHz is mostly used for industrial wireless communication. It is precisely 2402–2482 MHz (in some countries, it is up to 2483.5 MHz).

The range or distance of communication between the transmitter and the receiver is determined by many factors but the primary factor is the transmit power. The output power from the transmitting device defines the range. Generally, it is measured in Watt or mWatt or dBm. The range or distance of communication is directly proportional to the output power. The output power is restricted for ISM band-compliant devices. As it is unlicensed, it should not interfere with another device in the vicinity. Every country has a regulation of permissible output power allowed for ISM-complaint devices. It varies from −5 dBm to a maximum of 23 dBm from country to country. The licensed band devices such as our cell phones, mostly the transreceivers at the cell tower end, can operate in much higher power as there is no other operator in the same frequency range. Therefore, the range becomes higher.

Antennas also play an important role in determining the range. The gain and direction of the antennas can significantly increase the range.

It is important for the receiver to detect and sense this attenuated signal. The greater the receiver side sensitivity, the better the receiver can pick up a signal. The receiver sensitivity, measured in dBm, also determines the range of communication. The sensitivity of the radio chip used in wireless field devices is typically −85 to −98 dBm. The received signal at the receiver end is called received signal strength indicator (RSSI) and represented in dBm. For optimum operation of field devices, the RSSI should be less than −70 dBm. A signal strength of less than −50 and −30 dBm is considered a good and excellent signal strength, respectively.

The distance between the transmitter and the receiver is defined by clear line of sight (LOS) distance. This distance specification can be between 300 and 2000 meters LOS between the transmitter and the receiver, considering different output power, antenna gain, and receiver sensitivity.

Most of the radios used for industrial purpose follow IEEE 802.15.4 standard in MAC and PHY layer. Different standards have modified the upper layers according to their own specification. These radios are of the direct sequence spread spectrum type having 5 MHz band. Only 5 Hz out of 80 MHz (2402–2482 MHz) available in the ISM band is effectively used by these narrowband radios for communication.

IEEE 802.11 standard also specifies the minimum and maximum frequency range of different frequency channels. IEEE 802.11a/b/g (especially b/g radio) also operates in the 2.4 GHz ISM band, but these radios operate in the bandwidth of 22 MHz (Figure 14.1).

Since there is no restriction on the devices used in the license-free band of 2.4 GHz, the industrial wireless standard has to be rugged and made reliable to overcome any possible chances of interference in case of coexistence. Frequency-hopping spread spectrum (FHSS) provides the answer to this problem. Most the industrial standards resort to frequency-hopping techniques to avoid this issue. In the 80 MHz width of ISM, a 5 MHz band radio can have a maximum of 15–16 nonoverlapping distinct frequencies to operate. After every receiver and transmitter communication, the next communication happens in a changed frequency. This technique gives more reliability, security, and ruggedness in the protocol. The frequency hopping normally follows a specific sequence to avoid security threats. Advanced algorithms

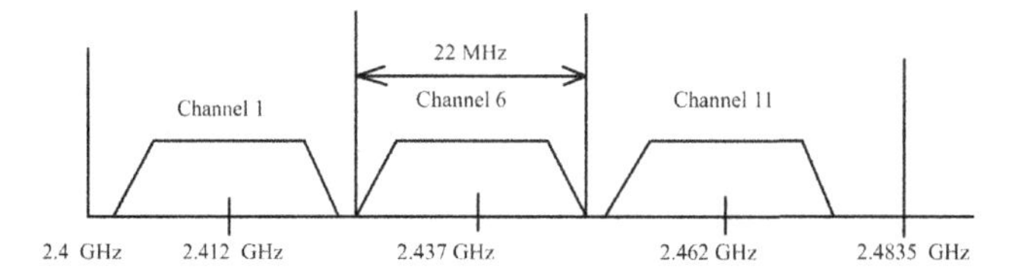

(a) IEEE 802.11b North American Channel selection (nonovarlaping)

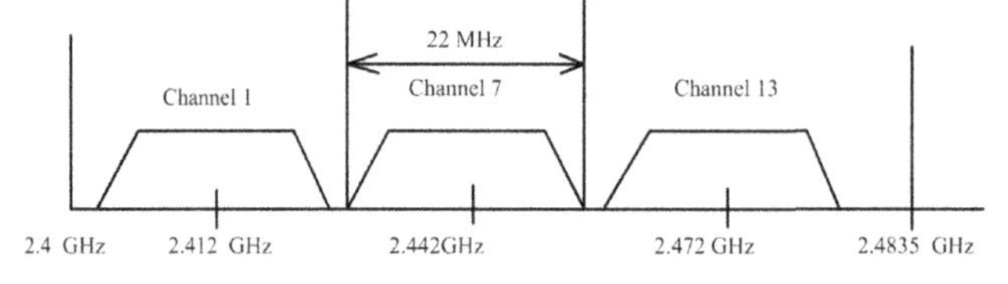

(b) IEEE 802.11b European Channel selection (nonovarlaping)

FIGURE 14.1

(a) IEEE 802.11b North American channel selection (nonoverlapping). (b) IEEE 802.11b European channel selection (nonoverlapping)

are used to identify high packet error rate (PER) in a particular frequency. If packets are missed on continuous basis because of persistent interference in few specific frequencies, the system manager can identify them and blacklist those channels. This is called channel blacklisting. This is followed by the system manager updating all the devices not to use those channels for communication. System manager is the supreme authority sitting on top of all devices, responsible for all resource allocation and management. The system manager allocates and de-allocates resources when a new device joins or a device is taken out from the network in addition to deciding the frequencies to be hopped.

Industrial wireless system depends on time division multiple access (TDMA) for communication. In a typical industrial setup, communication happens between all the field devices or between field devices and infrastructure nodes. A parent node could have multiple child nodes, and the parent node is responsible for routing data packets sent by its child nodes. The receiver of the parent node can listen to only one particular frequency coming from one child. TDMA is used to accommodate multiple nodes. The parent's available time for communication is divided into many tiny slots and these time slots are uniquely addressed. The child nodes reserve these slots for communication. Child 1 reserves slot 1 for its communication, child 2 for slot 2, and so on. When the parent node receives data in slot x, it knows that this data is coming from a device with a particular address. Some slots are kept free and used for emergency communication, such as alarm reporting or retrying of data packets in case first attempt was not successful. One time slot is typically in the range of 10–12 msec. Again, the system manager is responsible for assigning the slots and resources.

Two types of communication happen between the receiver and the transmitter – "Publish–Subscribe" and "Request–Response". In publish–subscribe mode, the transmitter publishes at a scheduled interval and the receiver subscribes to it. This is also known as scheduled or synchronous or latency controlled or deterministic communication. This guarantees that the data reaches the destination at a regular interval. Important data such as main measured process value is transferred using this mechanism. The request–response method of communication is used to transfer the data not required by the control system at regular intervals. This is only initiated when control system or system manager requests for some data. But the response is not guaranteed within a certain time. If there are multiple requests, one could get priority over the other. This type of communication is also known as contention controlled or asynchronous or probabilistic communication.

Time synchronization plays an important role in wireless communication. For example, if the receiver has to receive a packet from a particular addressed transmitter in the designated time slot, then both have to tune to the particular frequency at that instant. When the transmitter sends the packet, the receiver listens to it. The receiver acknowledges the packet and sends the ACK packet to the transmitter. This transmission and reception happens in few milliseconds. All nodes in the wireless network are time synchronized in microsecond accuracy. Time correction from the parent node to its child node happens at frequent intervals.

Time synchronization is especially essential when the devices exercise sleep and wake-up feature for power savings. There are two versions of a wireless transmitter – one without any wire and the other with external power option. Generally, externally powered transmitters use power from any nearby source but transmit the data over a wireless source. But a pure wireless transmitter draws power from internal batteries. The internal radio and sensor circuits are designed in such a way that the batteries last for 2–10 years. A very advanced sleep and wake-up algorithm is generally used to save power when nodes wait for their turn to publish data in the TDMA kind of a design. Reporting rate also determines the battery life. If a device has to publish data at every 5-sec interval, the device wakes up every 5 sec, publishes the sensed field parameter, ensures data delivery in terms of acknowledgment, and calculates few other things and goes to sleep. Typically, data communication needs few milliseconds and the devices are idle rest of the time. The electronic circuit goes to hibernate/sleep mode and thus saves power. This makes time synchronization critical. Unless the device wakes up at the correct time and designated slot (TDMA), it will fail to publish its packet. A device with a slower reporting rate (say 30 or 60 sec, or even higher) has longer battery life than the faster reporting devices because slower reporting rate devices sleep longer, therefore saving energy.

14.2.1 ISM BAND

The ISM radio bands are radio bands (portions of the radio spectrum) internationally reserved for the use of radio frequency (RF) energy for ISM purposes other than communication. ISM band is used by devices and applications such as cordless phone, Walkie-Talkie, Bluetooth devices, WiFi-based devices, laptops, smart phones, microwave ovens, and medical instrumentation. 2.4 GHz is the universally accepted ISM band in almost all the countries. There is another license-free band within the range of 5.4–5.8 GHz.

14.2.2 TRANSMIT POWER AND EIRP

In radio communication systems, equivalent isotropically radiated power (EIRP), or alternatively, effective isotropic radiated power is the amount of power that is emitted by an isotropic antenna (that evenly distributes power in all directions and is a theoretical construct) to produce the peak power density observed in the direction of maximum antenna gain. EIRP can take into account the loss in transmission line and connectors and includes the gain of the antenna. The EIRP is often stated in terms of decibels over a reference power level that is the power emitted by an isotropic radiator with equivalent signal strength. The EIRP allows making comparisons between different emitters regardless of type, size, or form. From the EIRP, and knowledge of a real antenna's gain, it is possible to calculate real power and field strength values. Antenna gain is expressed relative to a (theoretical) isotropic reference antenna (dBi). Radio power is the also called transmit power of the radio chip.

$$\text{EIRP (dBm)} = [\text{Radio power (dBm)}] - [\text{Cable loss (dB)}] + [\text{Antenna gain (dBi)}]$$

This is the equation to convert Watts to dBm. dBm stands for decibel equivalent in milliwatts (Table 14.1).

$$\text{dBm} = 10 * (\log(1000 * P))\ [P = \text{power in Watts}, 1000\,\text{mW} = 1\,\text{Watt}]$$

Table 14.1 Decibel Equivalent in Milliwatts

dBm	Watts
0	.001
15	.032
24	.250
27	.500
30	1.000
36	4.000

14.2.3 RSSI

The RSSI is a measurement of the power present in a received radio signal. RSSI is a generic radio receiver technology metric, which is usually invisible to the user of the device containing the receiver, but is directly known to users of wireless networking of IEEE 802.11 protocol family. RSSI is similar to the signal bars shown in cell phone handsets.

14.2.4 PACKET ERROR RATE

The Packet Error Rate (PER) is the number of incorrectly received data packets divided by the total number of received packets. A packet is declared incorrect if at least 1 bit is erroneous.

14.2.5 THROUGHPUT

In communication networks, such as Ethernet or packet radio, throughput or network throughput is the average rate of successful message delivery over a communication channel. This data may be delivered over a physical or logical link, or pass through a certain network node. The throughput is usually measured in bits per second (bit/s or bps), and sometimes in data packets per second or data packets per time slot. The system throughput or aggregate throughput is the sum of the data rates that are delivered to all terminals in a network.

14.2.6 BANDWIDTH

Bandwidth or digital bandwidth is a measure of rate of data transfer, bit rate, or throughput, measured in bits per second. It tells about the capacity of the network as how much traffic or data can flow through that network.

Bandwidth is the width of the range (or band) of frequencies that an electronic signal uses on a given transmission medium. In this usage, bandwidth is expressed in terms of the difference between the highest-frequency signal component and the lowest-frequency signal component. A higher width of the band translates to a better capacity of data transfer.

14.2.7 TDMA

Time Division Media Access (TDMA) is a channel access method for shared medium networks. It allows several users to share the same frequency channel by dividing the signal into different time slots. The users transmit in rapid succession, one after the other, each using its own time slot. This allows multiple stations to share the same transmission medium (e.g., RF channel) while using only a part of its channel capacity.

14.2.8 FREQUENCY HOPPING

Frequency Hopping Spread Spectrum (FHSS) is a method of transmitting radio signals by rapidly switching a carrier among many frequency channels, using a pseudorandom sequence known to both the transmitter and the receiver. It is utilized as a multiple access method in the frequency-hopping code division multiple access scheme.

A spread–spectrum transmission offers three main advantages over a fixed-frequency transmission:

- Spread–spectrum signals are highly resistant to narrowband interference. The process of recollecting a spread signal spreads out the interfering signal, causing it to recede into the background.
- Spread–spectrum signals are difficult to intercept. An FHSS signal simply appears as an increase in the background noise to a narrowband receiver. A transmission can only be intercepted if the pseudorandom sequence is known.
- Spread–spectrum transmissions can share a frequency band with many types of conventional transmissions with minimal interference. The spread–spectrum signals add minimal noise to the narrow-frequency communications, and vice versa. This results in efficient utilization of the bandwidth.

14.2.9 INTERFERENCE

Interference is a phenomenon that alters, modifies, or disrupts a signal as it travels along a channel between a source and a receiver. The term typically refers to the addition of unwanted signals to a useful signal. Common examples are:

- Electromagnetic interference (EMI)
- Co-channel interference (CCI), also known as crosstalk
- Adjacent-channel interference (ACI)
- Inter-symbol interference (ISI)
- Inter-carrier interference (ICI), caused by Doppler shift in Orthogonal Frequency Division Multiplexing (OFDM) modulation
- Common-mode interference (CMI)
- Conducted interference

14.2.10 ANTENNA TYPES AND GAIN

The purpose of an antenna is to collect and convert electromagnetic waves to electronic signals. Transmission lines then guide these to the receiver front end. It is essential to know the center frequency and operating bandwidth before selecting an antenna. The size of the antenna is directly proportional to the operating frequency. A high operating frequency translates to a small antenna size. Antenna gain is always measured against a known reference such as an isotropic source (dBi) or a half-wave dipole (dBd) (Table 14.2).

Increasing antenna gain by 3 dB generally requires increasing the size by a factor of 2–3 or by reducing the beam width. Vertical omnidirectional antennas and collinear arrays are used for line-of-sight communications with ground-based mobile units. Sectoring can be accomplished by panel antennas. Fixed point-to-point links generally use a yagi or a parabolic dish. Antennas exhibit reciprocity, which means they have the same gain whether used for transmitting or receiving.

14.2.10.1 IEEE 802.11

The 802.11 standards are a group of evolving specifications defined by the IEEE. Commonly referred to as Wi-Fi, the 802.11 standards define a through-the-air interface between a wireless client and a base station access point or between two or more wireless clients. There are many other standards defined by the IEEE, such as the 802.3 Ethernet standard.

Table 14.2 Antenna Type and Its Typical Gain

Antenna Type	Typical Gain (dBd)
Dipole	0
Omni	0
Gain omni	3–12
Mobile whip	−0.6 to +5.5
Corner reflector	4–10
Log periodic	3–8
Horn	5–12
Helix	5–12
Microstrip patch	3–15
Yagi	3–20
Panel	5–20

14.2.10.2 IEEE 802.11b

In 1995, the Federal Communications Commission had allocated several bands of wireless spectrum for use without a license. The FCC stipulated the use of spread–spectrum technology devices. In 1990, the IEEE began exploring a standard. In 1997, the 802.11 standard was ratified and is now obsolete. Then in July 1999, the 802.11b standard was ratified. The 802.11 standard provides a maximum theoretical 11 megabits per second (Mbps) data rate in the 2.4 GHz ISM band.

14.2.10.3 IEEE 802.11g

In 2003, the IEEE ratified the 802.11g standard with a maximum theoretical data rate of 54 Mbps in the 2.4 GHz ISM band. Signal strength weakens due to increased distance, attenuation (signal loss) through obstacles, or high noise in the frequency band; the data rate is automatically adjusted to lower rates (54/48/36/24/12/9/6 Mbps) to maintain the connection.

When both 802.11b and 802.11g clients are connected to an 802.11g router, the 802.11g clients have a lower data rate. Many routers allow mixed 802.11b/g clients or they may be set to either 802.11b or 802.11g clients only.

To illustrate 54 Mbps, if you have Digital Subscriber Line (DSL) or cable modem service, the data rate offered typically falls from 768 Kbps (less than 1 Mbps) to 6 Mbps. Therefore, 802.11g offers an attractive data rate for the majority of users. The 802.11g standard is backward compatible with the 802.11b standard. Today, 802.11g is still the most commonly deployed standard.

14.2.10.4 IEEE 802.11a

Ratification of 802.11a took place in 1999. The 802.11a standard uses the 5 GHz spectrum and has a maximum theoretical 54 Mbps data rate. Like in 802.11g, data rate automatically adjusts to lower rates (54/48/36/24/12/9/6 Mbps) to maintain connection as signal strength weakens due to increased distance, attenuation (signal loss) through obstacles, or high noise in the frequency band. The 5 GHz spectrum has higher attenuation (more signal loss) than lower frequencies, such as 2.4 GHz used in

802.11b/g standards. Penetrating walls provide poorer performance than with 2.4 GHz. Products with 802.11a are typically found in larger corporate networks or with wireless Internet service providers in outdoor backbone networks.

14.2.10.5 IEEE 802.11n

In January 2004, the IEEE 802.11 task group initiated the process to come up with specifications. There have been numerous draft specifications, delays, and lack of agreement among committee members. The proposed amendment has now been pushed back to early 2010. It should be noted that it has been delayed many times already. Thus, 802.11n is only in draft status. Therefore, it is possible that changes could be made to the specifications prior to final ratification.

The goal of 802.11n is to significantly increase the data throughput rate. Along with a number of technical changes, one important change is the addition of multiple-input multiple-output (MIMO) and spatial multiplexing. Multiple antennas are used in MIMO, which use multiple radios and thus more electrical power.

802.11n is scheduled to operate on both 2.4 GHz (802.11b/g) and 5 GHz (802.11a) bands. This could call for required significant site planning when installing 802.11n devices. The 802.11n specifications provide both 20 and 40 MHz channel options versus 20 MHz channels in 802.11a and 802.11b/g standards. By bonding two adjacent 20 MHz channels, 802.11n can provide double the data rate in utilization of 40 MHz channels. However, 40 MHz in the 2.4 GHz band results in interference and is not recommended nor likely which inhibits data throughput in the 2.4 GHz band. It is recommended to use 20 MHz channels in the 2.4 GHz spectrum as 802.11b/g utilizes. For best results of 802.11n, the 5 GHz spectrum is the best option. Deployment of 802.11n takes some planning effort in frequency and channel selection. Some 5 GHz channels must have dynamic frequency selection technology implemented in order to utilize those particular channels.

Another consideration of 802.11n is the significantly increased electrical power demand primarily because of multiple transmitters in comparison to the current 802.11b/g or 802.11a products.

14.3 ISA100 STANDARD

14.3.1 OVERVIEW

The ISA100 standard uses the open system interconnection (OSI) layer description methodology to define protocol suite specifications, in addition to specifications for the functions of security, management, gateway, and provisioning for an industrial wireless network. The protocol layers supported are the physical layer (PhL), data link layer (DL), network layer (NL), transport layer (TL), and the application layer (AL).

Some of the salient features of this standard are as follows:

- Interoperability: To ensure vendor to vendor interoperability among devices compliant with this standard, all mandatory protocol behaviors are described in the body of this standard. It is also summarized as minimum capabilities expected from these devices. Devices complying with these behaviors and capabilities form functional wireless networks and interoperate with each other.
- Quality of service (QoS): This standard supports multiple levels of QoS to support multiple applications with diverse needs within a network. QoS describes parameters such as latency, throughput, and reliability.

- Worldwide applicability: This standard is intended to conform to established regulations in all major world regions (e.g., North America, South America, Asia, Europe, Africa, and Australia); however, it may be usable in every regulatory environment.
  Scalability: The architecture supports wireless systems spanning the physical range from a single, small, isolated network, as in a gas or oil well or a very small machine shop, to integrated systems of many thousands of devices and multiple networks covering a multisquare kilometer plant. There is no technical limit on the number of devices that can participate in a network composed of multiple subnets. A subnet, a group of devices sharing a specific DL configuration, may contain up to 30,000 devices (limitation of subnet addressing space). With multiple subnets, the number of devices in the network can scale linearly.
- Extensibility: The protocols defined by this standard have capabilities such as fields that are reserved for future use and version numbers in headers that allow future revisions to offer additional or enhanced functionality without sacrificing backward compatibility.
- Simple operation: Upon provisioning, a device can automatically join the network. Automatic device joining and network formation enables system configuration with minimal need for personnel with specialized RF skills or tools. Additionally, this standard supports the use of fully redundant and self-healing routing techniques to minimize maintenance.
- Unlicensed operation: This standard is built upon radios compliant with IEEE Standard 802.15.4:2006 operating in the 2.4 GHz ISM band, which is available and unlicensed in most countries worldwide.
- Robustness in the presence of interference and with non-wireless industrial sensor networks: This standard supports channel hopping to provide a level of immunity against interference from other RF devices operating in the same band, as well as robustness to mitigate multipath interference effects. In addition, this standard facilitates coexistence with other RF systems with the use of adaptive channel hopping to detect and avoid using occupied channels within the spectrum. Adaptive channel hopping can also enhance reliability by avoiding the use of channels with consistently poor performance.
- Determinism or contention-free media access: This standard defines TDMA mechanism that allows a device to access the RF medium without having to wait for other devices. By the use of time-synchronized communication using configurable fixed timeslot durations, such as in the range of 10–12 msec, a device is assigned a timeslot and channel unique to that device and the device to which it will communicate. These timeslot durations are configurable on a per-superframe basis. A superframe is a cyclic collection of timeslots. The ability to configure timeslot duration enables:
  - Shorter timeslots to take full advantage of optimized implementations
  - Longer timeslots to accommodate
  - Extended packet wait times
  - Serial acknowledgment from multiple devices (e.g., duocast)
  - Carrier Sense Media Access (CSMA) at the start of a timeslot (e.g., for prioritized access to shared timeslots)
  - Slow hopping periods of extended length

Support is provided for both dedicated timeslots for predictable, regular traffic and shared timeslots for alarms and bursty traffic. In addition, support is provided for publishing, client/server, alert reporting, and bulk transfer traffic.

- Self-organizing networking with support for redundancy: Fully redundant and self-healing routing techniques, such as mesh routing, support end-to-end network reliability in the face of changing RF and environmental conditions. Special characteristics that allow the network to adapt frequencies used (e.g., adaptive hopping) along with mesh routing can automatically mitigate coexistence issues without user intervention.
- Internet protocol (IP)-compatible network layer: This standard network layer uses header formats that are compatible with the Internet Engineering Task Forces 6LoWPAN standard to facilitate potential use of 6LoWPAN networks as backbone. It should be noted that the use by this standard of headers compatible with 6LoWPAN does not imply that the backbone needs to be based on the IP. Furthermore, the use of header formats based on 6LoWPAN and IP does not imply that a network based on this standard is open to Internet hacking; in fact, networks based on this standard will typically not even be connected to the Internet.
- Robust and flexible security: All compliant networks have a security manager to manage and authenticate cryptographic keys. Security utilizes security primitives defined by IEEE Standard 802.15.4:2006 at the Data Link Layer and transport layers, providing message authentication, integrity, and optional privacy. Device authentication is enabled by the use of symmetric keys and unique device IDs, with an option for asymmetric keys. During normal operation, received data authenticity is verifiable through the use of secret symmetric keys known to both the sender and the receiver. During provisioning, authenticity of received device credentials from a new device may be verified by a system manager through the optional use of public keys shared openly by the new device, and a corresponding asymmetric secret private key kept inside the new device. Messages are protected using the default AES 128 block cipher or other locally mandated cryptographic primitives. Device-to-device communication is secured using symmetric keys.
- System management: This standard includes functions to manage communication resources on each individual device, and as system resources that impact end-to-end performance. System management enables policy-based control of the runtime configuration and also monitors and reports on configuration, performance, fault conditions, and operational status. The system management functions take part in activities such as:
  - Device joining and leaving the network
  - Reporting of faults that occur in the network
  - Communication configuration
  - Configuration of clock distribution and the setting of system time
  - Device monitoring
  - Performance monitoring and optimization

System security management works in conjunction with the system management function and optional external security systems to enable secure system operation. All management functions are accessible remotely via the gateway.

- Application process using standard objects: This standard application process is represented as a standard object containing one or more communicating components drawing from a set of standard defined application objects. These objects provide storage and access the data of an application process. Defining standard objects provide an open representation of the capabilities of a distributed application in a definitive manner, thereby enabling independent implementations to interoperate. Objects are defined to enable interaction among field devices and also interoperation with different host systems. The standard objects and services can directly map existing legacy field device communications onto standard objects and application sublayer communication services, thereby providing a means to adapt legacy devices to communicate over the wireless network.

- Tunneling: Tunneling is a mechanism, where the native protocols defined by this standard allow devices to encapsulate foreign PDUs and transport these foreign PDUs through the network to the destination device (typically a gateway). Successful application of tunneling depends upon how well the foreign protocols technical requirements (e.g., timing, latency, and others) are met by the instantiation of the wireless network.

14.3.2 NETWORK ARCHITECTURE

14.3.2.1 Network description

Figure 14.2 depicts the communication areas addressed by this standard, and those areas (shaded) that are not in scope of this standard. In Figure 14.3, circular objects represent field devices (sensors, valves, actuators, etc.) and rectangular objects represent infrastructure devices that communicate to other network devices via an interface to the network infrastructure backbone network. A backbone is a data network (preferably high data rate) that is not defined by this standard. This backbone could be an industrial Ethernet, IEEE 802.11, or any other network within the facility interfacing to the plants network.

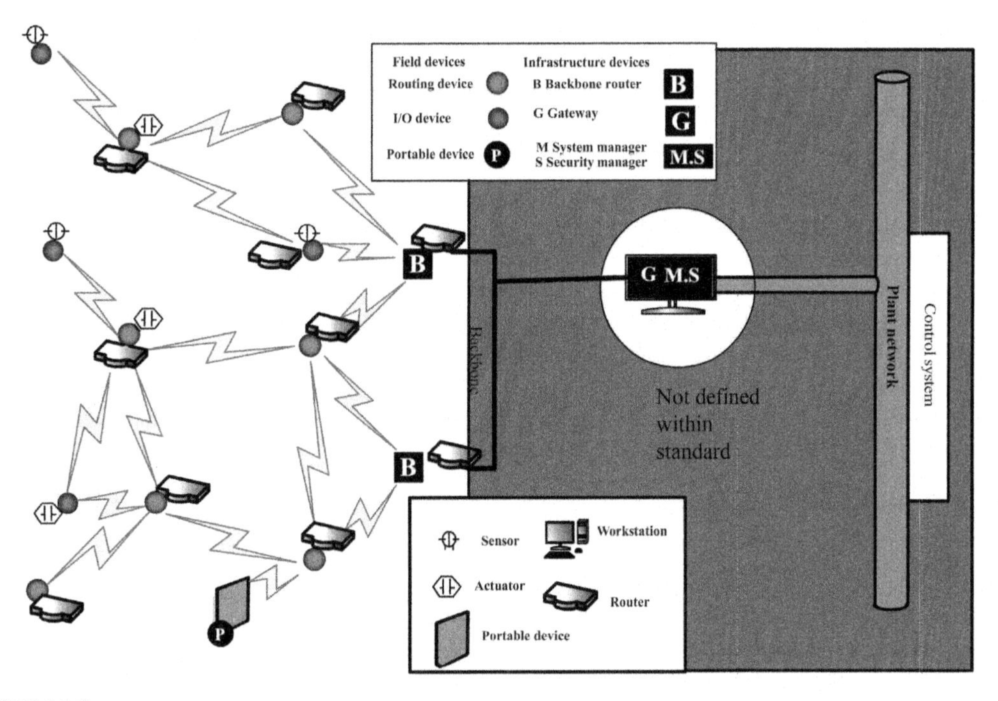

FIGURE 14.2

Standard-compliant network

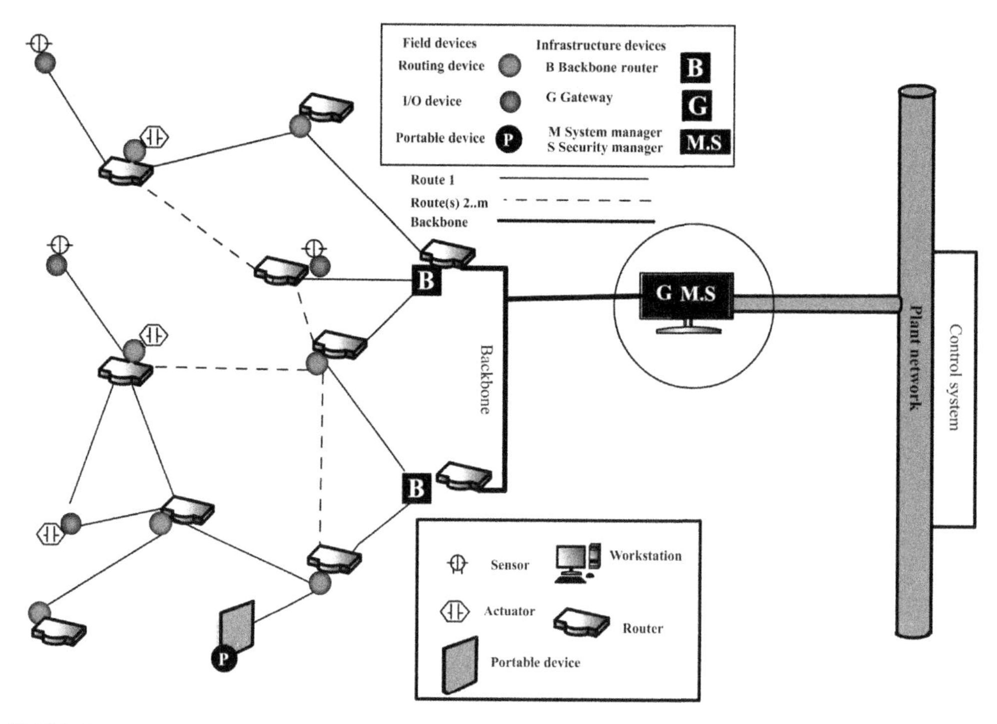

FIGURE 14.3

Backbone network

A field network consists of a collection of field devices that wirelessly communicate using a protocol stack defined by this standard. As shown in Figure 14.3, some field devices may have routing capabilities, enabling them to forward messages from other devices.

A transit network consists of infrastructure devices on a backbone, such as backbone routers, gateways, system managers, and security managers. Since the backbone physical communication medium and its network protocol stack are outside the scope of this standard, it is not specified and may include tunneling compliant packets over external transport or application layers.

Devices that connect two disparate networks must have two interfaces. A backbone router has both a field network interface and a backbone interface. A gateway has at least one network interface (either backbone or field network) and a plant network interface.

14.3.2.2 Network characteristics

Characteristics of a wireless industrial network include:

- Coexistence with other wireless devices in the industrial workspace
- Security, including data integrity, encryption, data authenticity, replay protection, and delay protection
- System management of all communication devices
- Supports application processes using standard objects
- Supports tunneling (transporting other protocols through the wireless network).

14.3.2.3 Coexistence with other RF systems

The system architecture is specifically designed for coexistence with other wireless systems conforming to this standard in addition to other networks operating at 2.4 GHz such as ZigBee™, WirelessHART™, 6LoWPAN protocol networks, IEEE 802.11, Bluetooth™, and radio-frequency identification (RFID) systems.

Operating with very short, time-synchronized communications tends to reduce congestion of RF bands and allows neighboring systems to recover quickly from lost packets. Reduced time of operation on a channel during channel hopping reduces the impact on other radio systems and increases the reliability in the face of interference, since messages can be resent on other, non-interfered channels. Adaptive channel hopping increases coexistence even further by avoiding channels that are occupied. This standard supports the use of clear channel assessment to avoid collisions with other nonsynchronized systems. The architecture supports operations in the event of interference from unintentional radiators, such as microwave ovens, via channel hopping and the use of automatic repeat request (ARQ). ARQ is an error control method for data transmission that uses acknowledgments for successful message reception and delayed retransmission in the case of erroneous reception to achieve reliable data transmission.

14.3.2.4 Coexistence strategies

This standard incorporates multiple strategies that are simultaneously used to optimize coexistence with other users of the 2.4 GHz radio spectrum:

- Leverage infrastructure high data rate communication links: Multi-hop networks repeat the same message multiple times. One fundamental capability of this architecture is to get the data to an infrastructure communication link (preferably high data rate with low error rate) as directly as possible, thus typically reducing the use of this standard's radio channel to one or two Data Link Layer (DL) messages or DL hops per report.
- Time slotted operation: Time slotted operation and scheduled transmissions minimize collisions within the subnet, thus avoiding unnecessary use of the channel for retries.
- Radio-type selection: The radio for the Physical Layer (PhL) is selected based on the ability of overlapping and IEEE Standard 802.11 radios to simultaneously transmit without data loss.
- Low duty cycle: This ensures data transmission for focus applications, thereby minimizing network overhead.
- Staccato transmissions: Expected transmissions are very short (feature of the selected PhL), that is, not bursty, enabling co-located IEEE Standard 802.11 networks to recover quickly in the event of lost packets.
- Time diversity: Many of the focused applications have less stringent latency requirements than other users of the spectrum, providing an opportunity to use time diversity for coexistence. Configurable retry periods potentially spanning hundreds of milliseconds enable the system to coexist with other users of the spectrum that may need to use the band for high priority bursts of activity.
- Channel diversity: The low duty cycle of the radio is spread across 16 channels, further reducing the worst-case opportunity for interference to an expected fraction of 1% under many realistic scenarios.
- Spectrum management: The user may configure superframes within the DL subnet to limit operation to certain radio channels.

- Adaptive channel hopping: The Data Link (DL) can avoid problematic channels on a link-by-link basis, such as channels exhibiting IEEE 802.11 cross-interference or persistent multipath fades.
- Collision avoidance: The DL supports CSMA-CA collision avoidance that, when used, can detect IEEE 802.11 energy and delay its transmission to reduce interference to IEEE 802.11 networks.
- Changing PhL payloads for retransmissions: The probability of an external device mistaking a packet for its own is reduced since retransmissions have different PhL payloads as a result of the DLs inherent security.
- These techniques are not specific to IEEE Standard 802.11 coexistence, and improve coexistence with a wide range of devices sharing the 2.4 GHz band while optimizing first-try success.

14.3.3 SYSTEMS

In the ISA100 standard, a system has an application focus and addresses applications and their needs. Networks, on the contrary, have a communication focus and are devoted to the task of device-to-device communication. Since the focus applications require communication, a network is a part of the system.

14.3.4 DEVICES

A device contains a combination of protocol layers such as Physical Layer (PhL), Data Layer (DL), Network Layer (NL), Transport Layer (TL), and Application Layer (AL), and may include functions such as the system manager, security manager, gateway, and provisioning.

14.3.4.1 Device interoperability

Device interoperability is the ability of devices from multiple vendors to communicate and maintain the complete network. Device interoperability encompasses control over various options, configuration settings, and capabilities of the devices:

- Options: To enable all devices to interoperate regardless of implemented options (those defined within this standard). It is also possible to disable (i.e., not using) options.
- Configuration settings: The system manager is responsible for configuring wireless devices and roles implemented by wireless devices.
- Capabilities: There are minimum capabilities that must be met for devices based upon their role in the system.

14.3.4.2 Profiles

A profile can be described as a vertical slice through the protocol layers. It defines those options in each protocol layer that are mandatory for that profile. It also defines configurations and parameter ranges for each protocol. The profile concept reduces the risk of device interoperability problems between products of different manufacturers.

The ISA100 standard describes two types of profiles:

- Role profile: A role profile refers to the baseline capabilities, including any options, settings, and configurations, that are required of a device to perform that role adequately.
- Radio silence profile: The radio silence profile is the configuration settings required to enact radio silence. Its main purpose is to restrain the radio from transmitting during inappropriate times, such as when transmission is unsafe or when regulations prohibit radio transmissions.

14.3.4.3 Quality of service

The QoS refers to the level of service necessary for the proper operation of an application within a device. This QoS is agreed upon via a contract between the system manager and the requesting device. When the application within a device desires to communicate at a certain QoS level, it sends a request to the system manager notifying it that it wishes to communicate with a specific destination and that it desires a given QoS level. This desired QoS level is indicated by desired contract and message priority, in addition to a certain level of reliability, periodicity, phase, and deadline for periodic messages, and short-term burst rate, long-term burst rate, and maximum number of outstanding requests for client–server messages.

14.3.4.4 Device worldwide applicability

To permit devices to comply with specific regional legislation, devices compliant to this standard shall not implement any characteristics that will preclude their operation in these regions.

14.3.5 DEVICE DESCRIPTION

Devices are the physical embodiment of the behaviors, configuration settings, and capabilities necessary to implement and operate a network in the ISA100 standard. There are many different types of devices depending upon the application, environment, and its function within the network. To fully describe necessary network behavior without defining specific device implementations, this standard defines roles, protocol layers, and field media that devices may embody.

A role defines a collection of functions and capabilities. This standard defines all the roles necessary for the network to operate properly, including system manager, security manager, gateway, backbone router, system time source, provisioning, router, and I/O device. Although all devices conforming to this standard implement at least one role, a device may implement many roles.

The protocol layers describe required behaviors. Not all devices are required to implement all the protocol layers defined in this standard. However, all devices conforming to this standard shall implement the network and transport layers in addition to the DMAP functionality. Every device contains a device management function and a device security management function that coordinate with the system processes to enable secure management of the system resources of a device including their usage.

A field medium is a combination of a PhL and a DL as described in this standard. Although not all devices need to implement a field medium, devices implementing the roles of I/O, routing, or backbone routing must implement at least one field medium defined in this standard.

Figure 14.4 illustrates the distinction between devices (devices supplied by a manufacturer) and the roles they may assume.

Figure 14.4 shows a simple yet complete network compliant with the ISA100 standard. Within this network are several types of devices, including sensors, actuators, routers, a palmtop computer, and a workstation. As shown in the figure, each of these devices may assume different roles within the network. For example:

- The workstation has assumed the roles of gateway, system manager, and security manager.
- Two devices have assumed the role of backbone routers, while seven other devices have assumed the role of routers.

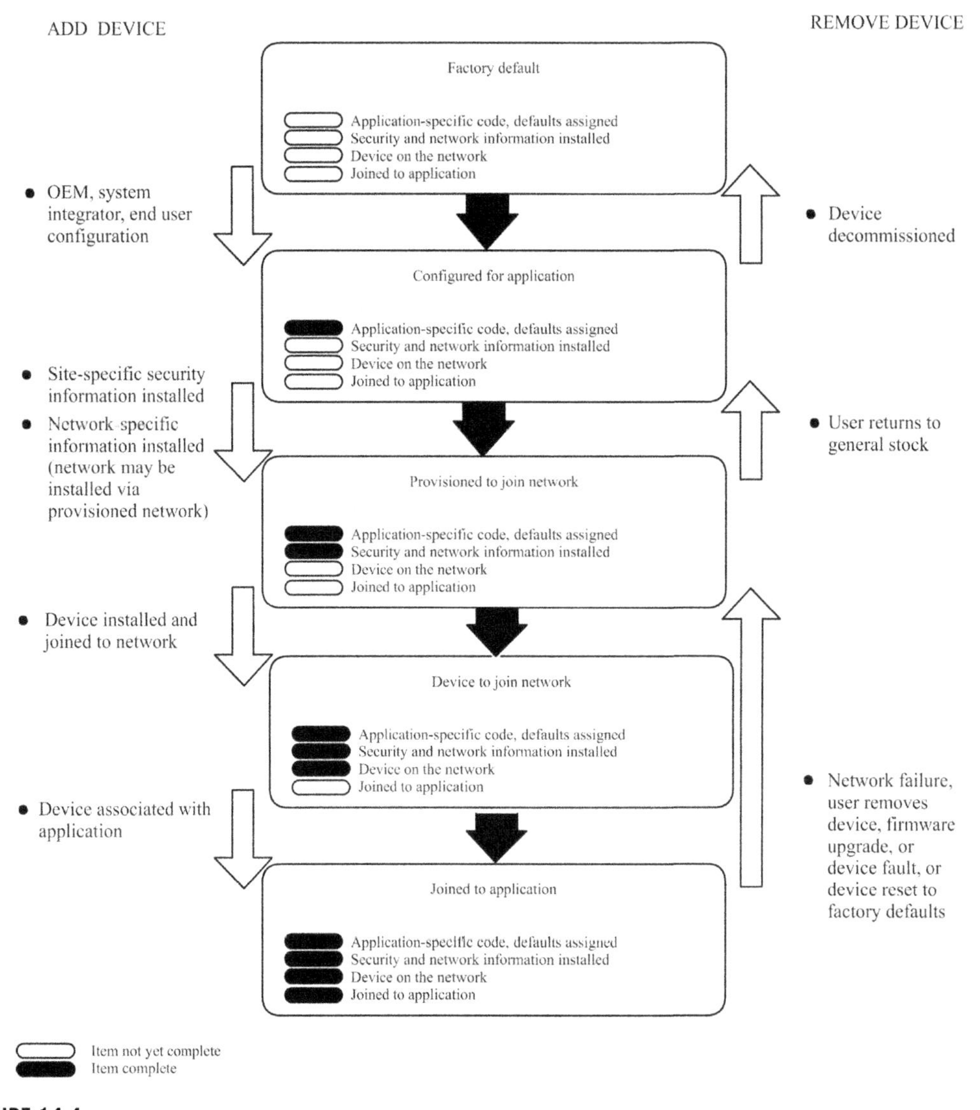

FIGURE 14.4

Physical devices versus roles

- One sensor (at lower left), several actuators and a portable computer have assumed the role of an I/O device.
- The router at the lower left of the figure has assumed a provisioning role, and provisions the new device being introduced.
- Two sensor devices have assumed both the router and I/O roles.

14.3.6 ROLE DEFINITIONS

14.3.6.1 Input/output

A device with the I/O role sources or sinks data and has at least one user application process (UAP) object. A device with only an I/O role has the minimum characteristics required to participate in a network compliant with this standard. The I/O role does not provide a mechanism for message forwarding or routing to any other device. This enables the construction of devices with the least complexity and lowest energy consumption, since they need not expend energy routing other device messages.

14.3.6.2 Router

A device with the role of a router role has routing capability, shall act as a proxy, and have clock propagation capability. These devices can provide range extension for a network and path redundancy and may provide different levels of QoS on a message-by-message basis. The system manager may disable the routing capabilities of the router role to optimize system performance requirements such as message latency or battery consumption.

Devices that implement the router role shall implement the type A field medium.

14.3.6.3 Provisioning

A device with the provisioning role (provisioning device) provisions a device set to factory defaults and implements the device provisioning service object. The provisioning device inserts the required configuration data into a device to allow a device to join a specific network. Devices implementing the PhL are capable of being provisioned using the defined physical interface. This capability can be disabled. Devices that implement the provisioning role implement the type A field medium.

14.3.6.4 Backbone router

A device with the backbone router role has routing capability via the backbone, and acts as a proxy using the backbone. Backbone routers enable external networks to carry native protocol by encapsulating the PDUs for transport. This allows a network described by this standard to use other networks, including longer range or higher performance networks. While the media and protocol suites of backbone networks are not defined in this standard, it is believed that many instantiations of the backbone router are with IP networks. Many of these backbone networks conform to IPv4 as opposed to the newer IPv6. Backbone routers implementing an IP interface support the use of IETF rfc2529 to allow the WISN TLPDUs to be conveyed across an IPv4 backbone.

Devices implementing the backbone router role implement the type A field medium in addition to the backbones network interface unless the device also implements the gateway or system manager or security manager roles, in which case the type A field medium is optional.

14.3.6.5 Gateway

A device with the gateway role implements the high side interface, source and sink data, and has a UAP. The gateway role provides an interface between field devices and the plant network, or directly to an end application on a plant network. More generally, a gateway marks the transition between communications compliant to this standard and other communications and acts as a bridge between this standards application layer and other application layers. There can be multiple gateways in a system.

Devices implementing the gateway role implement either the router role for access to the type A field medium or the backbone router role for access to the backbone medium.

14.3.6.6 *System manager*

A device implementing the system manager role implements the SMAP and sets the time source tree.

The system manager is a specialized function that governs the network, devices, and communications. The system manager performs policy-based control of the network runtime configuration, monitors and reports on communication configuration, performance, and operational status, and provides time-related services.

Devices use a contract to communicate. A contract is an agreement between the system manager and a device in the network that involves the allocation of network resources by the system manager to support a particular communication need of this device. This contract is made between the applications in both devices and the system manager. The system manager assigns a contract ID to the contract, and the application within the device uses the contract for communications. An application may only request the creation, modification, or termination to a contract. It is the sole responsibility of the system manager to create, maintain, modify, and terminate the contract.

Devices implementing the system manager role implement either the router role for access to the type A field medium or the backbone router role for access to the backbone medium.

14.3.6.7 *Security manager*

The system security management is a specialized function that works in conjunction with the system manager and optional external security systems to enable secure system operation. The security manager is logically separable and, in some cases, resides on a separate device in a separate location. Every system compliant with this standard shall have a security manager.

The communication protocol suite between the system manager and the security manager is not defined by this standard.

14.3.6.8 *System time source*

A device implementing the system time source role implements the master time source for the system. A sense of time is an important aspect of the standard; it is used to manage device operation. The system time source provides a sense of time for the entire system.

Devices implementing the system time source role also implement any of the I/O, router, backbone router, system manager, or gateway roles.

14.3.7 FIELD MEDIUM

The ISA100 standard defines one specific field medium, type A. A field media type defines the protocol for the physical and data link layers. Future revisions of this standard may include multiple field media types. The type A field medium consists of the physical layer and data link layer as specified by this standard.

14.3.8 DEVICE ADDRESSING

Each device that implements the type A field medium is assigned a 16-bit DL subnet address for local addressing by the system manager. Each device has a unique 64-bit EUI address.

Each device also has a 128-bit network address that is assigned by the system manager. The system manager may choose to assign the 128-bit address as a logical address to maintain application layer linkage in the event of a device replacement. The 128-bit address may be used by the application to reach a particular device within a system after the join process is complete.

14.3.9 DEVICE PHASES

A device may go through several phases during its operational lifetime. Within each of these phases are multiple states. A notional representation of the phases of the life of a device is illustrated in Figure 14.5.

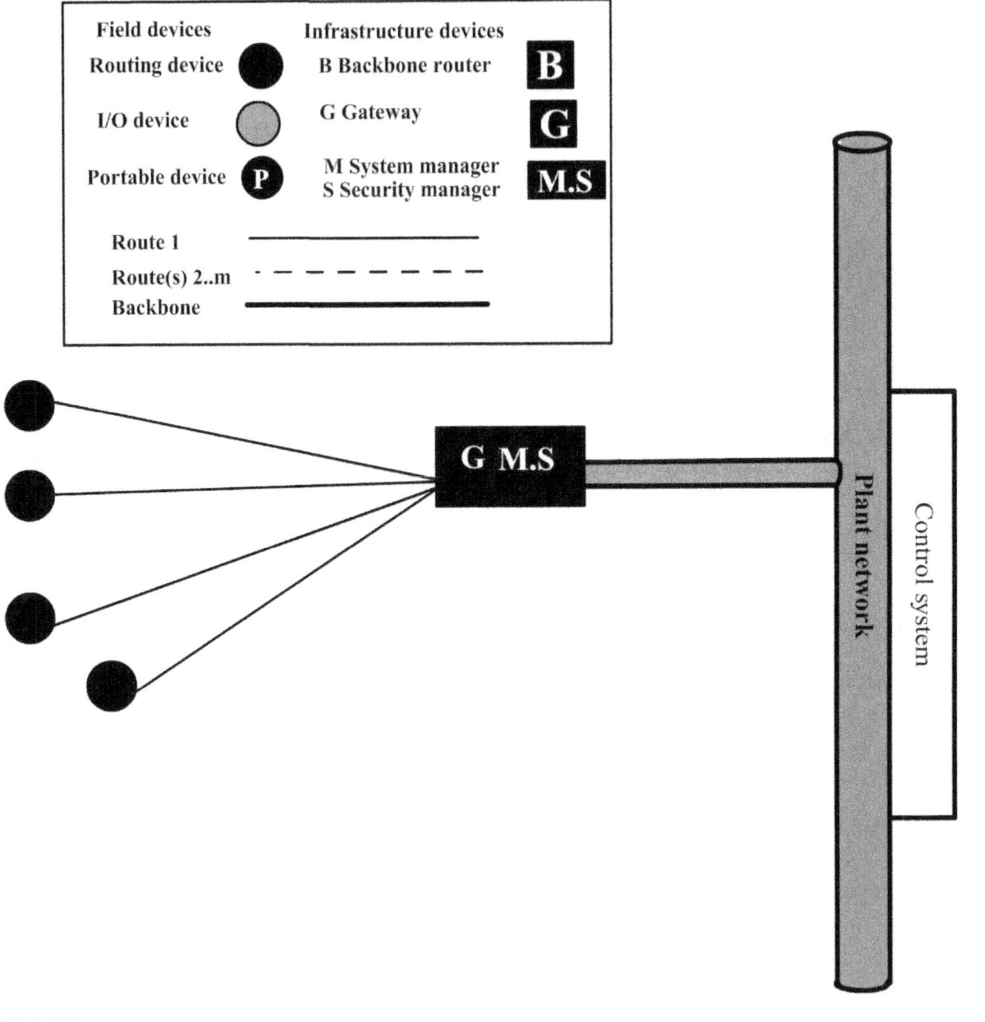

FIGURE 14.5

Notional representation of device phases

A device can pass through these phases several times as it is commissioned and used, then decommissioned and recommissioned for a different application. After joining the network, devices can report their status so that applications know whether a device is accessible and whether it is joined to an application.

14.3.9.1 Factory default

A device is considered nonconfigured if it has not been configured or commissioned with any application- or network-specific information. A nonconfigured device may come from a manufacturer or enter a nonconfigured state as a result of decommissioning.

14.3.9.2 Configured for application

A device is considered configured for an application when it has received its own application-specific programming and when all appropriate defaults have been applied. A device configured for application may come from a manufacturer or may be supplied by a systems integrator or other value-added reseller, already provisioned for the intended application. Over-the-air application program updates can occur, but are handled at the device application layer.

14.3.9.3 Provisioned to join the network

A device is provisioned to join the network when it has obtained the appropriate security credentials and network-specific information. A device typically enters this phase when it has been prepared for installation into an application. Typically, a device does not communicate directly with the security manager; instead, the system manager relays requests to the security manager.

14.3.9.4 Accessible device

A device is considered accessible when it has joined the network and has been authenticated by the system manager. An accessible device can communicate with the system manager.

In this phase, an application object on the device can send or receive information to or from the desired application objects on peer devices.

14.3.9.4.1 Device energy sources

The ISA100 standard does not restrict the types of energy sources a device may use. The standard allows for energy-efficient device behavior to operate for long periods of time (e.g., 5–10 years) using suitable batteries.

The types of energy sources may be grouped into five categories:

- Mains
- Limited battery (e.g., button cell)
- Moderate battery (e.g., lead acid)
- Rechargeable battery
- Environmental or energy scavenging

Devices implementing the roles of I/O or router may be expected to use any category of energy source. The roles of security manager, system manager, gateway, and backbone router are typically performance intensive; therefore, devices implementing these roles are recommended to have larger energy sources such as the mains or moderate battery categories.

The energy source status of devices is critical to proper system management. All devices provide energy supply information to the system manager. This information may be used in making routing decisions.

14.4 NETWORKS

The main focus of a network is device-to-device communication. There are numerous aspects of a network's ability to communicate. These aspects include the atomic network, network topologies, device relationships within a network, protocol suite structure, and the concept of shared time.

14.4.1 MINIMAL NETWORK

A minimal network is a network with the minimum amount of devices implementing the minimum number of roles. The minimum roles for a network are system manager, security manager, provisioning, system time source, and I/O. The system manager and security manager are two separate roles and may reside in the same device or may be split between two physical devices. A single physical device may assume multiple roles. A minimum network shall consist of two devices communicating with each other where one device implements the roles of system manager and security manager, and the roles of provisioning, system time source, and I/O are implemented by either of the devices.

A small representative network of four field devices and one infrastructure device is shown in Figure 14.6. Although such a network is atypical, it represents a small compliant system. In this network, a single physical device has assumed the roles of gateway, system manager, and security manager.

14.4.2 SUPPORTED NETWORK TOPOLOGIES

The following Figures 14.6, 14.7, 14.8 and 14.9 etc provide several examples and illustrate the flexibility of the system architecture. The set of examples is not intended to be exhaustive. These examples are presented here only to provide a better understanding of the system elements.

14.4.2.1 Star topology

This standard supports a simple star topology, as shown in Figure 14.6.

This system configuration can yield the lowest possible latency across the physical layer. It is architecturally very simple, but is limited to the range of a single hop.

14.4.2.2 Hub-and-spoke topology

Expanding the network using the backbone routers allows the user to construct a hub-and-spoke network, as shown in Figure 14.7, wherein devices are clustered around each backbone router, providing access to the high-speed backbone.

In this case, latency is slightly degraded from the simple star topology, but overall throughput can increase, and in larger systems, average latency can decrease because of the multiple data pipes available (one through each backbone router). Although the network can expand further away from the gateway, it is nonetheless limited to single-hop range around a backbone router.

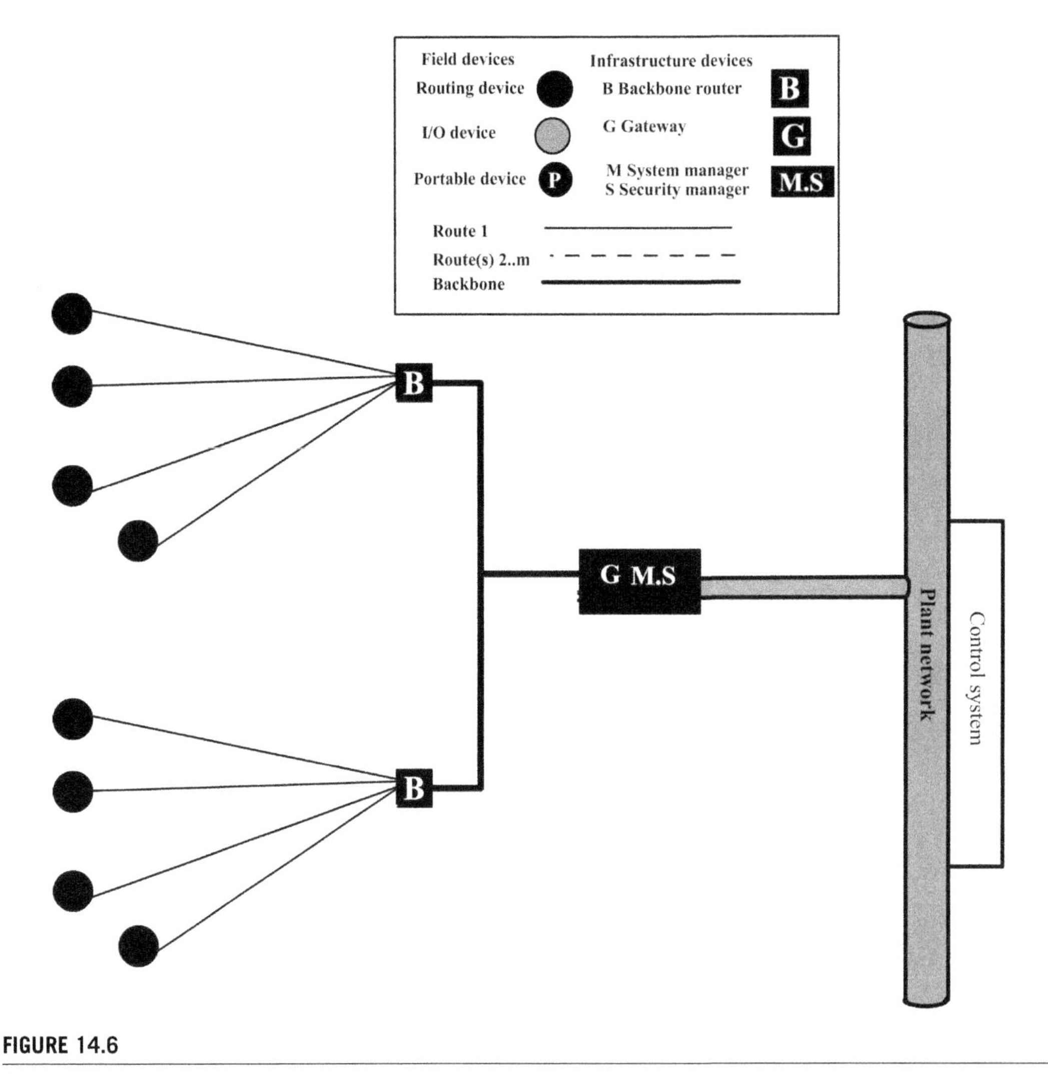

FIGURE 14.6

Simple star topology

14.4.2.3 Mesh topology

This standard supports mesh networking topologies, as illustrated in Figure 14.8.

In some cases, the number of routes a device can support may be limited. The range is extended as multiple hops are supported. Latency is larger, but can be minimized by proper scheduling of transmissions. Throughput is degraded as device resources are used in repeating messages. Reliability may be improved through the use of path diversity.

14.4.2.4 Star–mesh topology

Combining the star topology with the mesh topology is illustrated in Figure 14.9.

This configuration has the advantage of limiting the number of hops in a network.

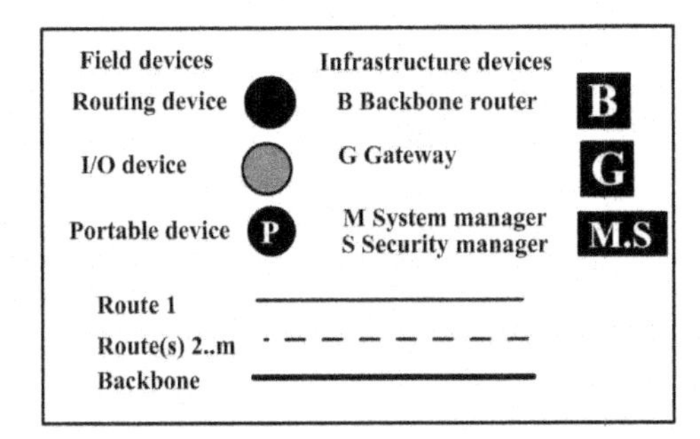

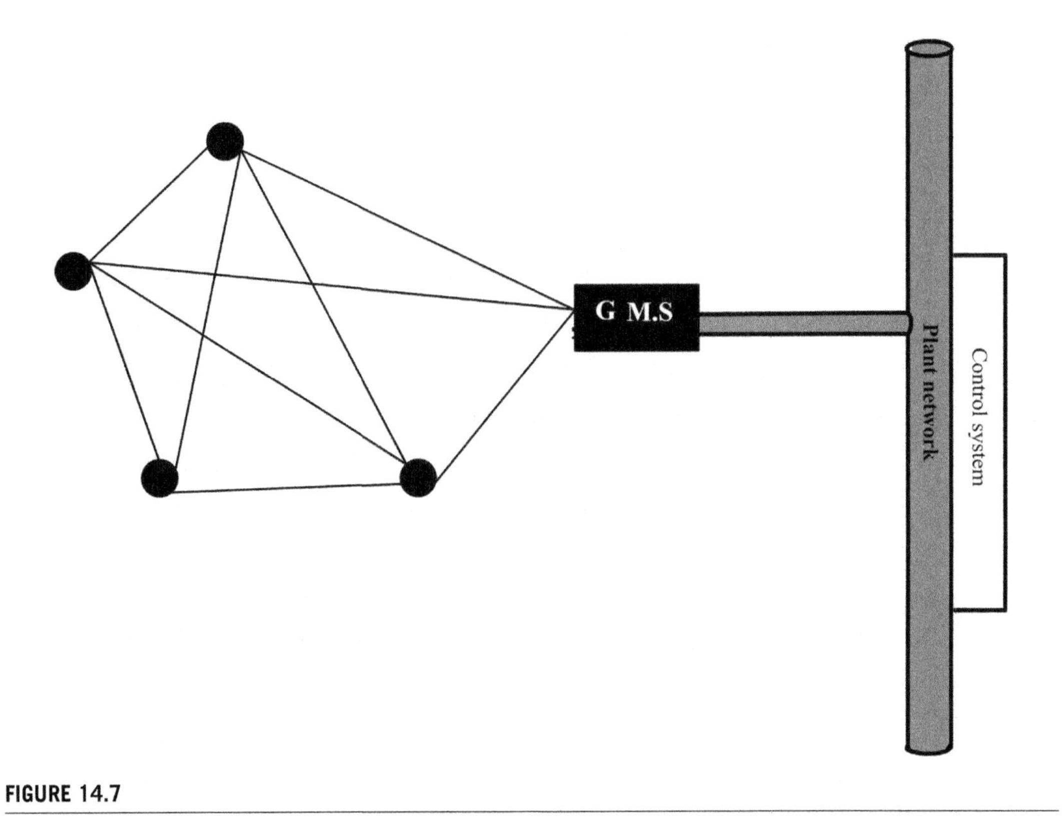

FIGURE 14.7

Simple hub-and-spoke topology

14.4.3 COMBINATIONS OF TOPOLOGIES

The ISA100 standard allows combination of any of the topologies, to enable the construction of a configuration that best satisfies the needs of the application. For example, monitoring systems that span large physical areas within a plant may use the star–mesh topology or a combination of hub-and-spoke and star–mesh topologies, whereas certain control applications where latency is critical may benefit

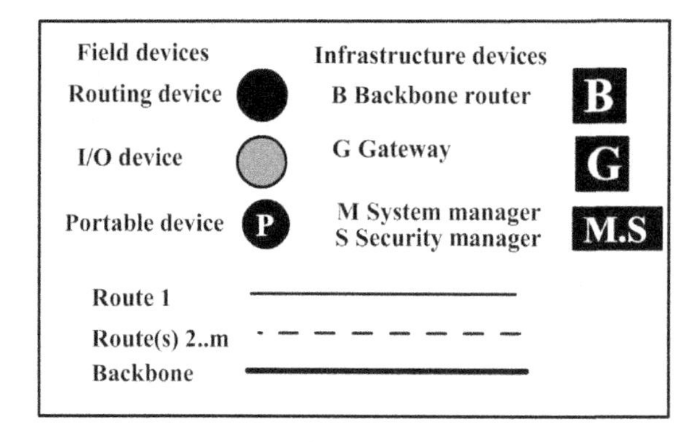

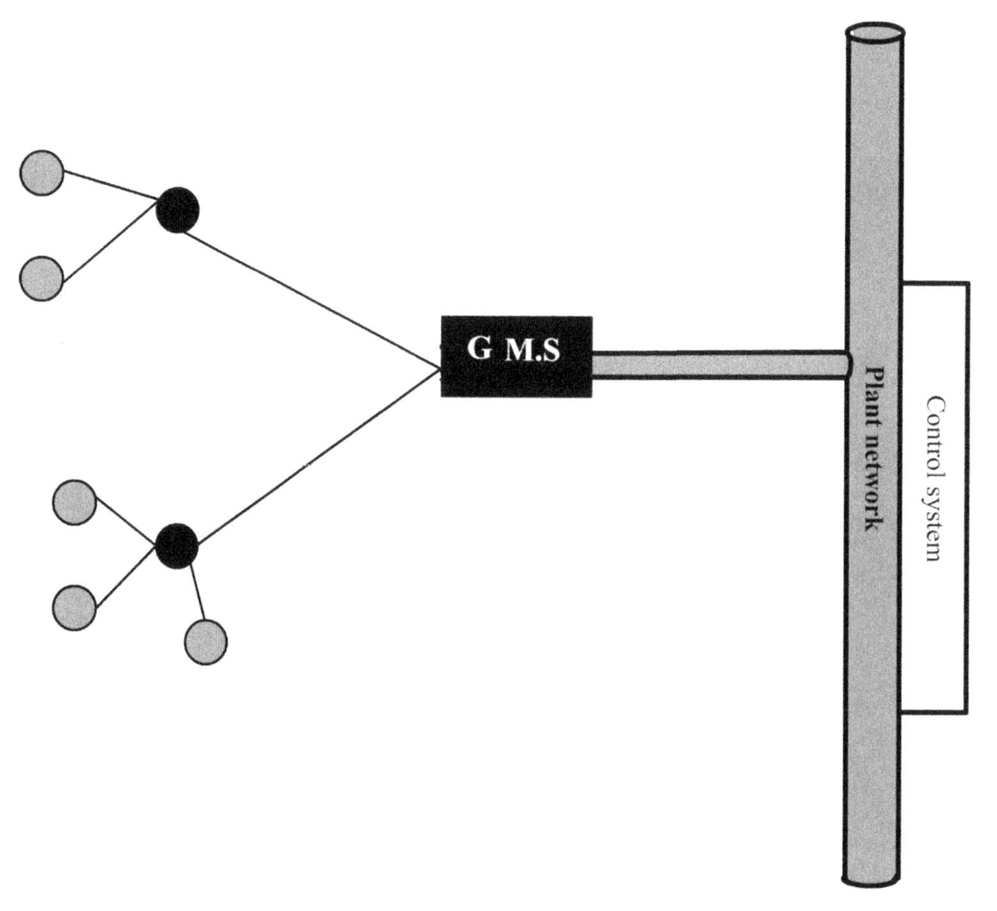

FIGURE 14.8

Mesh topology

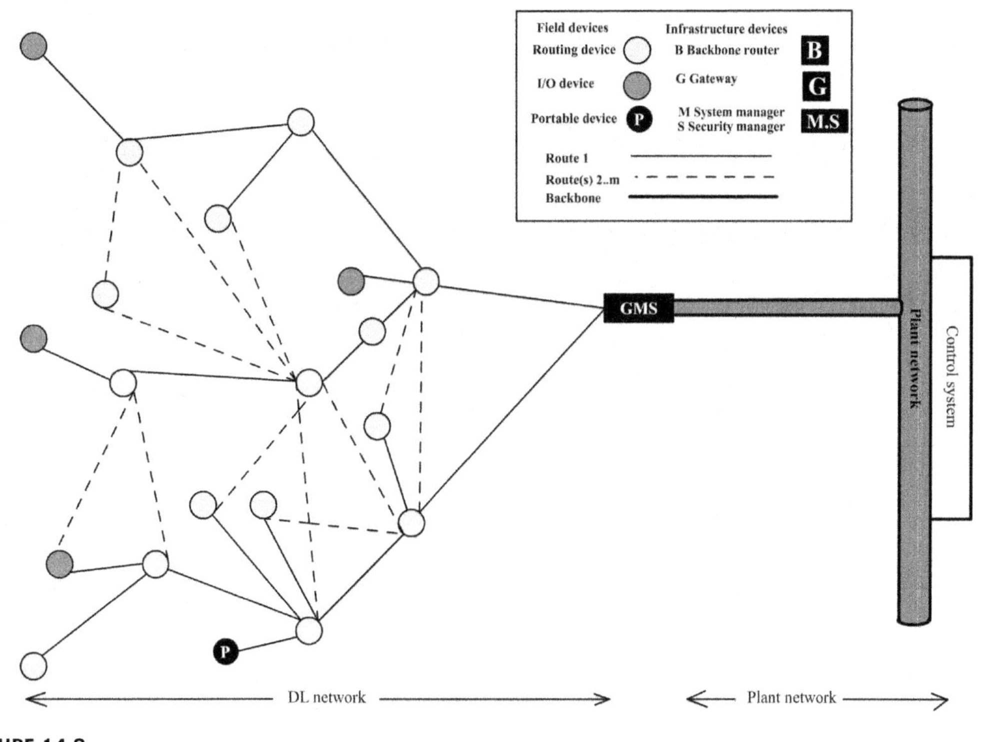

FIGURE 14.9

Simple star–mesh topology

from a pure star or hub-and-spoke topology. The flexibility of the system allows for all of these topologies to operate in harmony, in any combination.

14.5 NETWORK CONFIGURATIONS

The ISA100 standards DL subnet comprises of one or more groups of wireless devices, with a shared system manager and (when applicable) a shared backbone. Although a DL subnet stops at the backbone router, network routing may extend into the backbone and plant network. A complete network includes all related DL subnets and other devices connected via the backbone, such as a gateway, system manager, or security manager. Figures 14.10 and 14.11 illustrate the distinction between a DL subnet and a plant network.

Figure 14.10 illustrates a simple network comprised of a collection of wireless devices called a DL subnet and additional devices that manage the DL subnet and connect it to other networks. In Figure 14.10, the network and the DL subnet are the same.

The DL subnet is comprised of both routing and I/O devices. The solid lines between devices designate the first route established between devices, while dotted lines designate the second route, the third route, and so on. Messages may be routed using any one of the known routes.

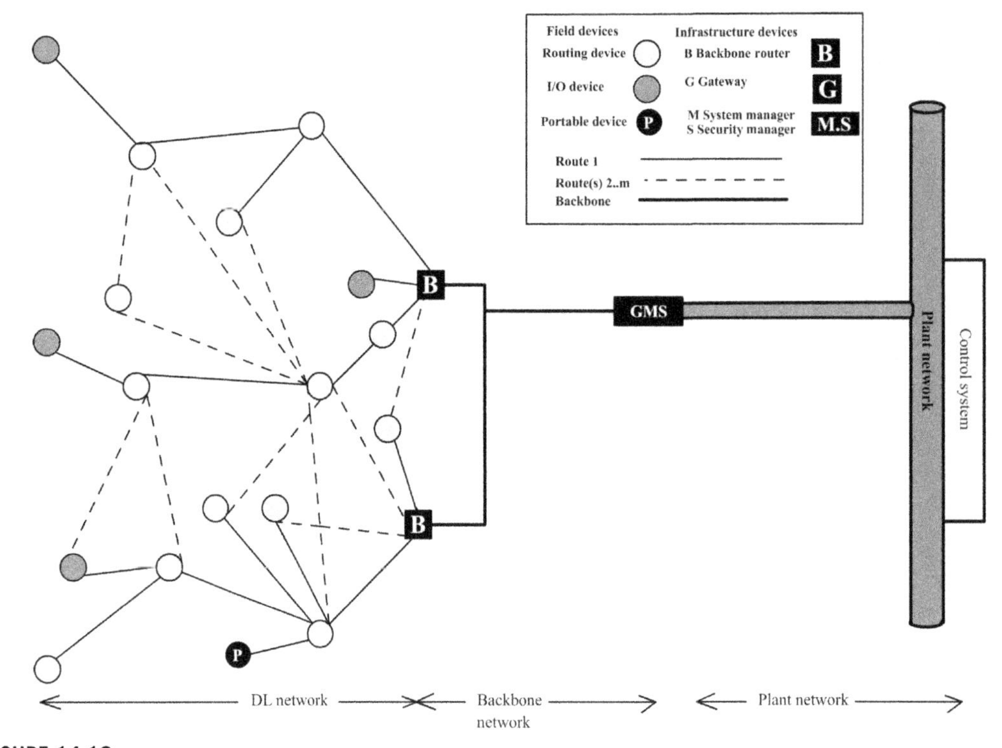

FIGURE 14.10

Network and DL subnet overlap

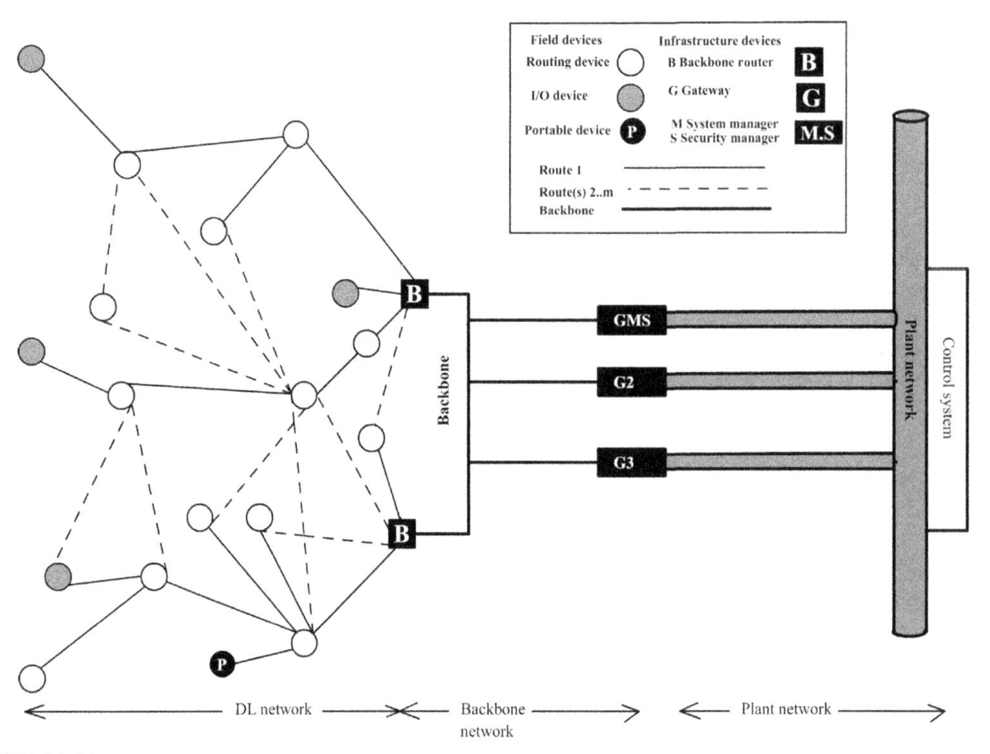

FIGURE 14.11

Network and DL subnet differ

In Figure 14.10 and 14.11, the box labeled G,M,S represents a collection of three separate roles combined into one physical device in this simple network:

- A gateway
- A system manager
- A security manager

In Figure 14.11, the DL subnet includes a collection of field devices up to the backbone routers (boxes labeled B). Backbone routers use connections to a network backbone to reduce the number of hops that messages would otherwise require; this can improve reliability, reduce latency, and extend the coverage of the network.

The network in Figure 14.11 includes the DL subnet, and the backbone and a gateway, system manager, and security manager, which are colocated on the backbone.

14.5.1 MULTIPLE GATEWAYS: REDUNDANCY AND ADDITIONAL FUNCTIONS

Figure 14.12 illustrates a different physical configuration with three gateway devices. One of the gateway devices also implements the system manager and security manager functions.

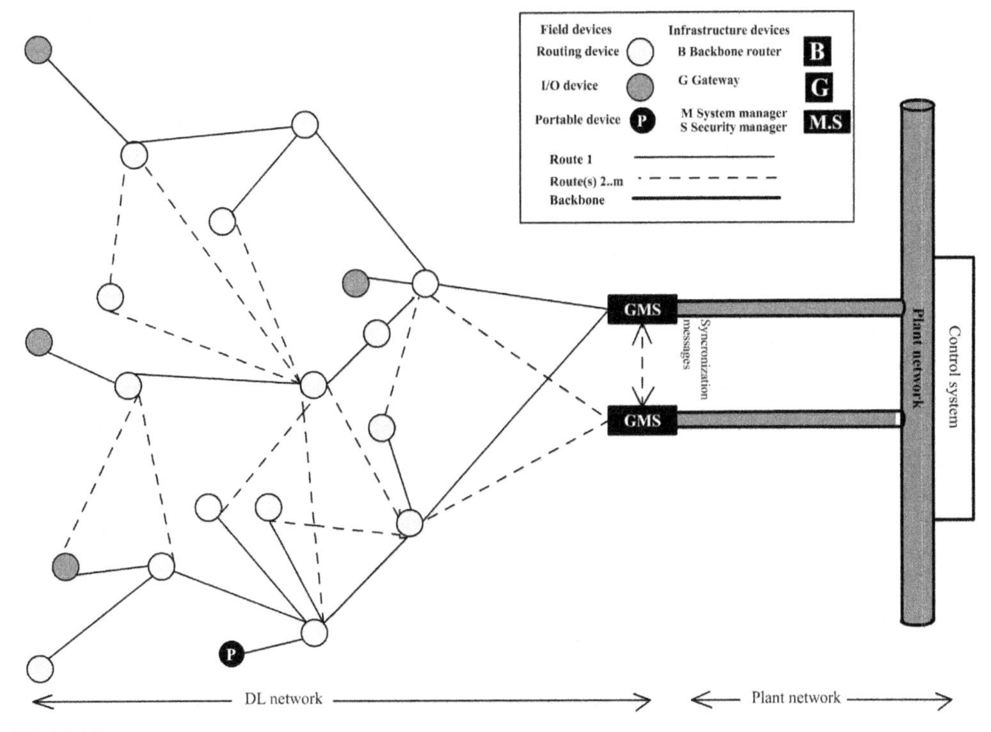

FIGURE 14.12

Network with multiple gateways

The gateway devices may be identical (i.e., mirrored, for redundancy) or unique, for example, with each gateway implementing a software application to handle communications between a particular class of device and a control system attached to the plant network.

14.5.2 MULTIPLE GATEWAYS: DESIGNATING A GATEWAY AS A BACKUP

This standard does not define the functionality of a backup gateway nor the mechanisms for synchronization of backup gateways.

Figure 14.12 is similar to Figure 14.13, but with a second G,M,S device (gateway, system manager, and security manager).

The two G,M,S devices offer identical functionalities and may coordinate their operation via synchronization messages exchanged through a backchannel mechanism not specified by the ISA100 standard. A single G,M,S device may be responsible for all gateway, system manager, and security manager functions, with a second G,M,S device acting as an active standby that remains idle until it is needed. Alternatively, the two G,M,S devices may divide the workload between them until one fails.

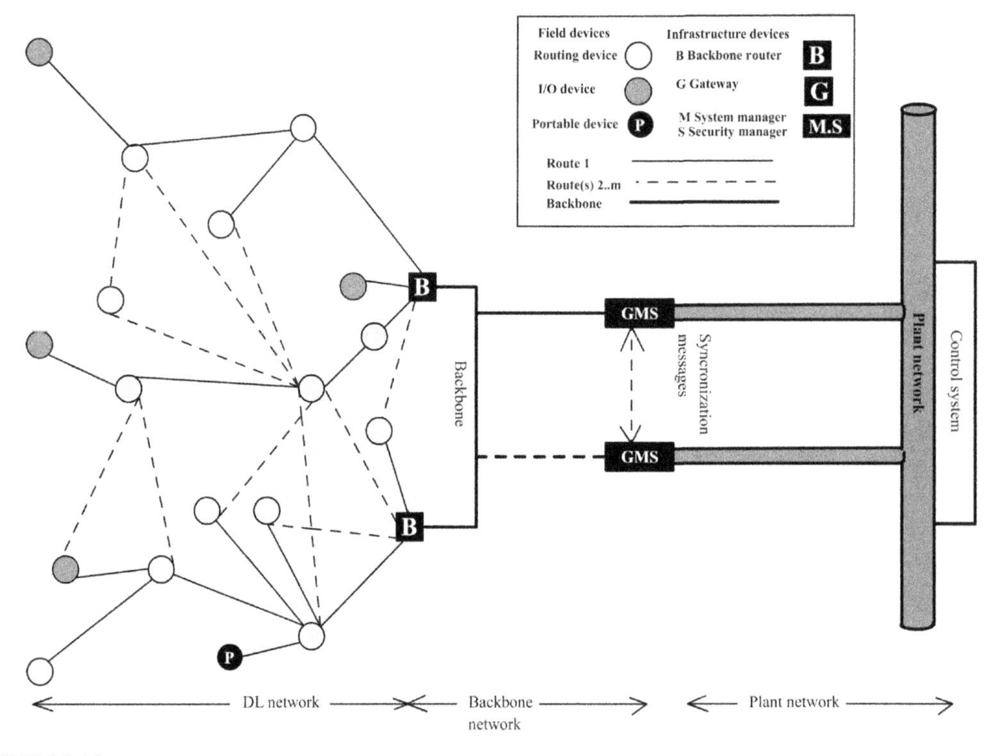

FIGURE 14.13

Basic network with a backup gateway

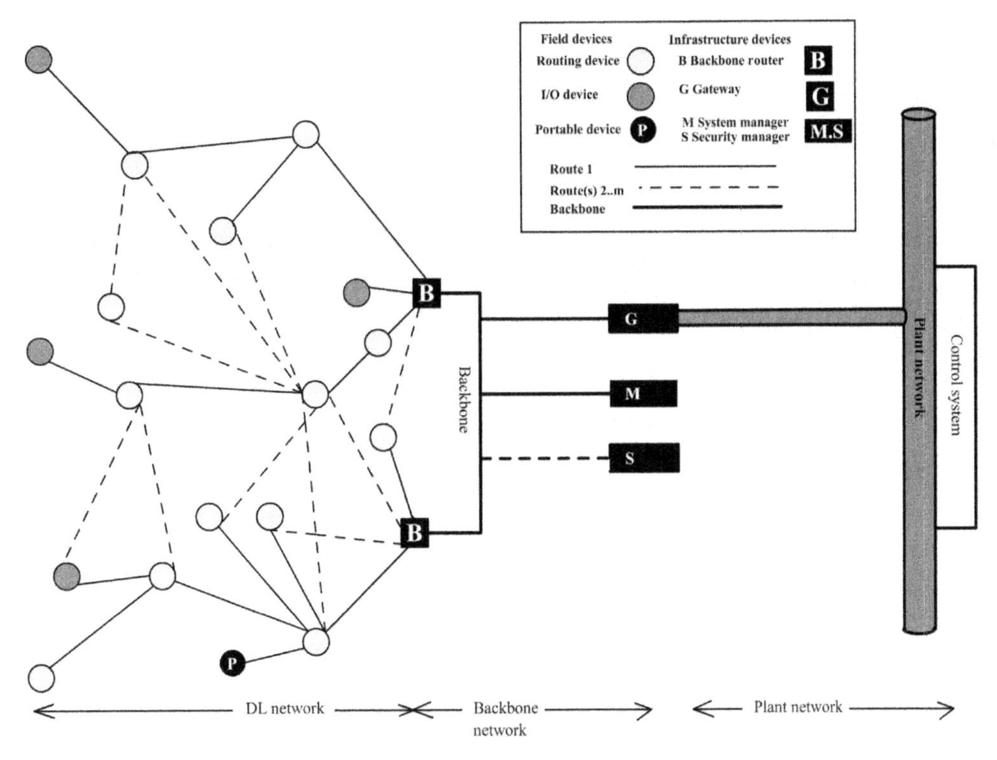

FIGURE 14.14

Network with backbone

14.5.3 ADDING BACKBONE ROUTERS

To the basic network, shown in Figure 14.14, adds the backbone routers (boxes labeled B) that facilitate the expansion of networks compliant with this standard, in terms of both the number of devices and the area the network occupies.

14.6 GATEWAY, SYSTEM MANAGER, AND SECURITY MANAGER

The functional roles fulfilled by the G,M,S device in Figures 14.10–14.12, and 14.14 may be split into multiple physically separated devices as illustrated in Figure 14.15, so that the gateway G, system manager M, and security manager S each operate on a separate device.

The physically separated gateway, system manager, and security manager shown in Figure 14.16 can be implemented only in networks with a network backbone.

14.6.1 DATA FLOW

The following descriptions are intended to provide examples of how data may flow through the system. The set of examples is not intended to be exhaustive.

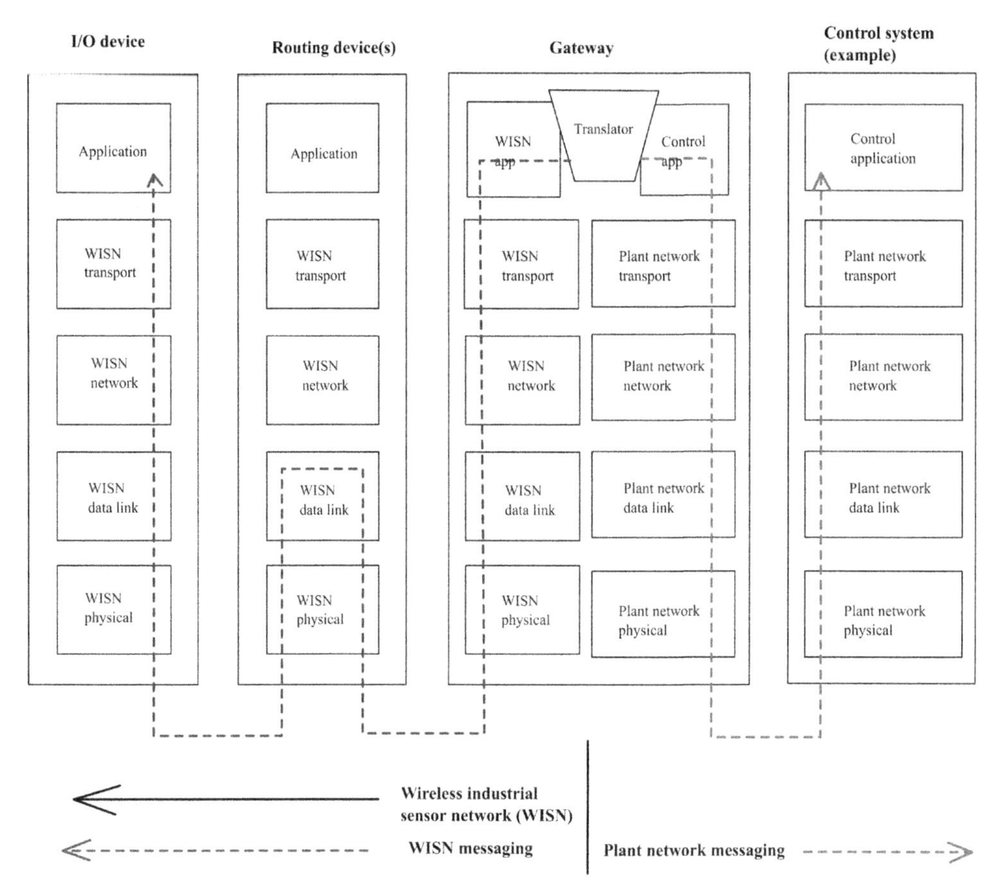

FIGURE 14.15

Network with backbone – device roles

14.6.2 NATIVE COMMUNICATIONS

A device communicates over the network using only ASL-defined services; the payloads are classified as either native or non-native.

14.6.2.1 Basic data flow

The steady-state data flow for a basic network compliant with this standard, such as the one shown in Figure 14.10.

The I/O device is a sensor or actuator device within the DL subnet that contains physical, data link, network, and transport layers as defined by this standard and runs an application that handles the sensor or actuator function.

The router routes messages on behalf of the I/O device. Routing within the DL subnet is performed entirely at the data link layer, not at the network layer. In a real-world network, there is one router for each additional hop between the device and the gateway or backbone router.

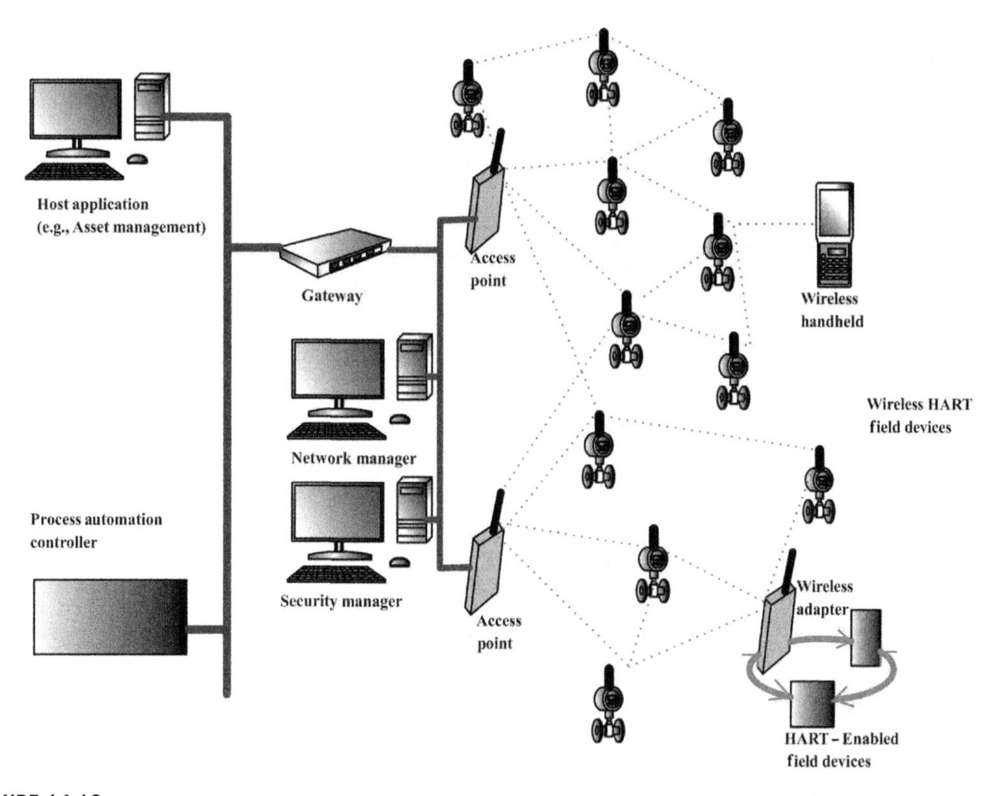

FIGURE 14.16

Physically separated gateway, system manager, and security manager

The gateway translates messages between the DL subnet and the plant network. The application running on the gateway consists of a component that communicates with the application layer of the I/O device, plus a component that communicates with the application layer of the control system, plus any components that facilitate translation between the two, such as a cache.

14.6.3 WIRELESS HART

Wireless HART is a wireless mesh network communications protocol for process automation applications. It adds wireless capabilities to the HART protocol while maintaining compatibility with existing HART devices, commands, and tools (Table 14.3).

The HART protocol implements layers 1, 2, 3, 4, and 7 of the OSI seven-layer protocol model. The HART physical layer is based on the Bell 202 standard, using frequency shift keying to communicate at 1200 bps. The signal frequencies representing bit values of 0 and 1 are 2200 and 1200 Hz, respectively. This signal is superimposed at a low level on the 4–20 mA analog measurement signal without causing any interference with the analog signal. The HART Data link layer defines a master–slave protocol – in normal use, a field device only replies when it is spoken to. There can be two masters, for example, a control system as a primary master and a handheld HART communicator as a secondary master.

Table 14.3 Key Differences Between Wired and Wireless HART

Specification Field	Typical HART Specification	Typical Wireless HART Specification
Output signal	4–20 mA HART	IEC 62591 Wireless HART
Power supply	24V DC loop powered	Intrinsically safe battery
Update rate	1 sec	1 sec to 60 min
Protection type/enclosure	Explosion proof	Intrinsically safe

Timing rules define when each master may initiate a communication transaction. A maximum of 15 or more slave devices can be connected to a single multidrop cable pair.

The network layer provides routing, end-to-end security, and transport services. It manages "sessions" for end-to-end communication with correspondent devices. The data link layer ensures successful propagation of communications from one device to another. The transport layer can be used to ensure successful end–end communication.

The application layer defines the commands, responses, data types, and status reporting supported by the protocol. In the application layer, the public commands of the protocol are divided into four major groups:

- Universal commands: They provide functions that must be implemented in all field devices.
- Common practice commands: They provide functions common to many, but not all field devices.
- Device-specific commands: They provide functions that are unique to a particular field device and are specified by the device manufacturer.
- Device family commands: They provide a set of standardized functions for instruments with particular measurement types, allowing full generic access without using device-specific commands.
- Each wireless HART network includes three main elements:
    - Wireless field devices connected to process or plant equipment. This device could be a device with wireless HART built in or an existing installed HART-cnabled device with a wireless HART adapter attached to it.
    - Gateways enable communication between these devices and host applications connected to a high-speed backbone or other existing plant communications network.
    - A network manager is responsible for configuring the network, scheduling communications between devices, managing message routes, and monitoring network health. The network manager can be integrated into the gateway, host application, or process automation controller.

The network uses IEEE 802.15.4 compatible radios operating in the 2.4 GHz ISM radio band. The radios employ direct-sequence spread–spectrum technology and channel hopping for communication security and reliability, and TDMA synchronized, latency-controlled communications between devices on the network. This technology has been proven in field trials and real-plant installations across a broad range of process control industries.

Each device in the mesh network can serve as a router for messages from other devices. A device toned not directly communicates to a gateway, but just forwards its message to the next closest device. This extends the range of the network and provides redundant communication routes to increase reliability.

The network manager determines the redundant routes based on latency, efficiency, and reliability. To ensure the redundant routes remain open and unobstructed, messages continuously alternate between the redundant paths. Consequently, similar to the Internet, if a message is unable to reach its destination by one path, it is automatically re-routed to follow a known good, redundant path with no loss of data.

The mesh design also makes adding or moving devices easy. As long as a device is within range of others in the network, it can communicate.

For flexibility to meet different application requirements, the wireless HART standard supports multiple messaging modes including one-way publishing of process and control values, spontaneous notification by exception, *ad hoc* request/response, and autosegmented block transfers of large data sets. These capabilities allow communications to be tailored to application requirements, thereby reducing power usage and overhead.

Wireless technology offers opportunities for a wide range of applications – from adding measurements where they were previously out of physical or economic reach, to enabling plant-wide functions such as asset and people tracking, security, and worker productivity. However, the wireless HART specification team recognized that no one technology is right for every application. Their approach was to focus on core process automation functions where no appropriate wireless standard existed.

14.7 APPLICATIONS OF WIRELESS INSTRUMENTATION

There can be several areas of application for wireless instrumentation. Generally, a plant is divided into two areas, in terms of scope of the production. ISBL means *inside battery limit*, which defines the plant production area. Many of the utilities and raw materials used for processing are not always produced inside the plant. There are items such as instrument air, cooling water, and steam that are supplied from OSBL, that is, *outside battery limit*, of the plant. Apart from the utilities, the raw material and produced material comes from OSBL, such as tank firms and storage areas.

Major applications are as follows.

14.7.1 ISBL APPLICATIONS

Major wireless applications for ISBL area include:

- Temperature profiling of columns/reactors
- Skin temperature measurement in boilers/furnaces
- Pressure, differential pressure monitoring
- Inter-unit flow measurement
- Motor bearing/winding temperature monitoring
- Monitoring of important manual valves/bypass valves
- PRVs/flare valves monitoring
- Corrosion monitoring

14.7.2 OSBL APPLICATIONS

Major wireless applications for OSBL area include:

- Vibration monitoring of fin–fan coolers
- Vibration monitoring of cooling towers
- Vibration monitoring of other second/third-tier assets such as pumps/blowers
- Tank farm monitoring
- Effluent Treatment Plants/ Reverse Osmosis (ETP/RO) plants – pressure/temperature/level monitoring
- Sea water intake facility
- Pressure, temperature monitoring
- Steam water analysis system
- Continuous emission monitoring system
- Motor bearing/winding temperature monitoring
- Monitoring of important manual valves/bypass valves
- Corrosion monitoring

14.7.3 STORAGE AND TANK FIRM MONITORING

Every industrial plant needs raw material for production and they use these raw materials to produce their final output. Generally, these raw materials and final outputs are stored in a distant location from the manufacturing location. These storage tanks have different materials that need to be kept in certain conditions. For safe storage of these materials, certain temperature or pressure has to be maintained. The operator might also be interested to know the level in the tanks. So there are enough process parameters that are to be monitored or controlled. Wireless becomes an ideal solution for bringing these process parameters in the control room because of the cost advantage. Moreover, wireless gives flexibility of adding any process point without bothering about extra cable.

14.7.4 MOUNTING ON ROTATING EQUIPMENT

There are many equipment and mechanical instruments that are rotating in nature apart from standard pump, motor, turbine, or compressor. The low-speed rotating equipment such as dryer or kilns are huge in size and some of the measurements need to be done on the rotating body itself. Wireless instrumentation solves the age-old wear-and-tear-related issue of conventional wired device mounted on these equipment.

14.7.5 VIBRATION-PRONE AREA

Process variable measurements are also necessary in vibration-prone areas of reciprocating pumps, centrifuge, pneumatic conveying, and so on. The nature of process is such that these areas or systems experience lot of vibrations and huge pressures that are irregular and fluctuating. This makes instrumentation extremely challenging because conventional wired transmitters with all the terminals and junctions become vulnerable to lose connection. Wireless transmitters are best suited for these purposes.

14.7.6 ACROSS A RIVER

Wireless instrumentation is a boon for data transfer across a barrier where there is a difficulty of cable. Generally, two infrastructure nodes are put across two shores to enable data communication. The sensors data from the other shores can be transported over infrastructure nodes to bridge the gap between the two shores and data can be sent to the control room.

14.7.7 MOVEABLE PLATFORM

Moveable equipment such as a trolley in a rail, cranes, or earth movers are slowly moving toward wireless instruments because of their flexibility and low maintenance. The cables of wired instruments need to be protected with armored shields because of the harsh environment they are deployed. Even protecting shield is not sufficient enough to avoid regular wear and tear. Wireless instrumentation gives the answer for all these problems posed by the harsh environment.

14.7.8 OFFSHORE GAS AND OIL EXPLORATION

Offshore oil and gas platforms are another segment where wireless instruments are getting lot of popularity. The reason behind this choice is slightly different than all others. Offshore platforms have very stringent power usage guidelines. As it is floating in the middle of the ocean, all the power requirements have to be met by its own generation units. Addition of any equipment or instrument is carefully evaluated in terms of its power consumption specifications. Wireless devices clearly come out a winner here, as most of the instruments are battery powered. These batteries are long lasting because of the highly advanced algorithm and sleep features that are implemented in these transmitters. As an example, for a 30-sec reporting rate transmitter, the battery can last up to 10 years for some of the vendors. For 1-sec reporting rate, the battery life can be a minimum of 2 years. Regardless to say that any wiring-related troubleshooting becomes redundant because staff and labor availability is always a question in this environment.

14.7.9 ON BOARD OF A SHIP

Wireless instrumentation is also fast gaining popularity in ships and ocean liners. The popularity is mainly because of its simplicity and less documentation. Wireless becomes ideal for small-scale deployment. It becomes also easy for troubleshooting. Any replacement or new additions become very easy.

14.7.10 MOBILITY WITH HANDHELD STATION

The popular mode of wireless system deployment in larger plants comes with wireless infrastructure development. IEEE 802.11a/b/g/n standards are the popular choice of any wireless back haul or infrastructure. This is an industry-wide common and open standard. 802.11 b/g (also known as WiFi) is used in handhelds, smart phones, laptops, and so on. Typical client-side access has WiFi capability. 802.11a/n is used for main backhaul or mesh throughput. The advanced infrastructure nodes generally have this capability of acting as WiFi access point and also as meshing backhaul along with industrial protocol. A single infrastructure supports both WiFi-enabled devices and industrial transmitters and devices. Mobile operator station is such a device with WiFi connection. Through WiFi access point and using mesh infrastructure, these operator stations can access the main data server of DCS in the control room. The application running in the mobile device gives the same look and feel of the operator stations used for controlling and monitoring inside the control room. This gives lot of flexibility and mobility for the field operators to visualize the running process from anywhere in the field.

14.7.11 VOIP SOLUTION

VoIP phones are the WiFi-enabled phones. Any smart device with WiFi feature can run an application to convert the device into a VoIP phone. This becomes an economical solution to connect thousands of people in a large plant. The field to control room communication can be achieved successfully using these phones. In addition, integrating with an EPBAX system, so that each phone has an extension number and they can communicate to each other seamlessly. All this can be achieved without any additional infrastructure investment.

14.7.12 VIDEO SOLUTION

Live video streaming is becoming an important requirement for process plants. Starting from flare monitoring to critical process area monitoring to perimeter monitoring for security purpose, live video streaming is essential. Wireless communication can also be deployed for this purpose. WiFi-enabled cameras can directly connect to the wireless access point infrastructure and dump the data in the main backhaul. IP-based cameras can be accessed from anywhere in the plant. The main advantage of this solution is the scalability. It eliminates the need to plan cable routing whenever a camera or other accessories need to be added. A WiFi camera can be mounted and operational in few minutes. Only thing that requires attention is the throughput calculation *vis-à-vis* available bandwidth. 802.11a gives a maximum bandwidth of 54 Mbps and 802.11n gives up to 600 Mbps (under certain conditions). But available bandwidth can be much less based on the RSSI, distance, and number of mesh hops present between the camera and the control room.

14.7.13 RFID-BASED SOLUTION

RFID is a technology that uses radio waves to transfer data from an electronic tag, called RFID tag or label, attached to an object, through a reader for the purpose of identifying and tracking the object. RFID tags are small RF tags that can be uniquely identified by a WiFi-based wireless infrastructure. RFID tags are of two types – passive and active. Passive RFID tags do not need any power supply and are ideal for short-range usage. These are very economical (even less than Rs. 10 or $0.2). Active RFID tags are little expensive and come with a battery. These are ideal for long-distance application. RFID tags can be used for unique item identification. In an assembly line-based production system, a part can be easily identified. Even position can be identified through triangulation technique. Location detection can be a very important aspect of a plant operating with hazardous material from the safety point of view. In case of an emergency or an emergency evacuation, people can be located and identified.

14.8 DESIGNING AND ENGINEERING A WIRELESS SYSTEM

A wireless network in an industrial setup must be properly planned and designed to ensure that it works well and becomes fault tolerant. There are series of design steps in arriving at the best performing network. In this section, we explain those design and planning tasks.

14.8.1 NETWORK PLANNING

For network planning, the following points should be considered:

- Decide the best system topology, including time synchronization.
- Determine the optimal number of nodes that has to act as routing nodes or backbone router or infrastructure nodes.
- Determine the number of gateways required if it has larger number of devices. If a gateway supports 100 devices at a required reporting rate and the site has 450 measuring/sensing points, then site needs at least five gateways to support so many devices.
- Determine RF power-level settings according to the location of deployment and country regulations. Nowadays, different tools are available that help in initial estimation and planning. Based on the number of devices or transmitters to be deployed in the field, the tool can suggest the number of routers or backbone routers required in the system considering different system and deployment constraints. The advanced tools can also suggest the locations of these routers.

14.8.2 PHYSICAL LAYOUT

The wireless devices are positioned in a way to minimize obstructions between interconnected devices. The sensing location of a transmitter will be always fixed and it will be as per process requirement. This can never be altered. But the location of the infrastructure node or a backbone router is flexible. Care should be taken to utilize this flexibility to the maximum. The backbone router must be put in a place where it can be in the vicinity of maximum end sensors. That gives better connectivity to the maximum sensors. A typical distance between 300 meters or less is preferred between the sensor transmitter and its immediate router or reporting parent node. This ensures that the signal strength is having enough room for some instantaneous changes or disturbances. A typical field deployment can have −75 dBm to −85 dBm. The intention should be to have a RSSI of −75 dBm and better. But one should avoid going beyond −85 dBm. Maintaining LOS or near LOS improves the wireless link performance. The range varies depending on the environment, LOS, transmit power settings, antenna type, and antenna gain. Height difference and restrictions attenuate the signals further. Sometimes, remote antenna with extension cable will come handy in field deployment, and will work to your advantage to mitigate adverse environmental impacts.

14.8.3 SITE SURVEY

In many cases, site survey of the actual industrial plant might be required to understand the environment and choose right deployment strategy. Site survey is essentially a process through which deployment plan is validated on sample basis. The visibility and expected RSSI between an end sensor and a router is validated by actually putting a device and a router in their prospective locations. Site survey confirms that the theoretical calculation is in accordance with the practical reality of the site. Sometimes, site survey can also help to fine-tune the deployment in finding a better location for the routers or backbone routers. The antenna type and gains can be better selected after seeing the result of the site survey.

14.8.4 NETWORK ID AND SECURITY KEY

Ensure that a unique network ID is selected. In case of a large deployment with multiple gateways, multiple network IDs should be planned. Network ID is a unique identity in the network that binds all wireless devices to a single gateway or the system manager.

Security is an important aspect of any wireless network. Generally, a join security key is must for any industrial wireless network. In some cases, the join key has to be provided by the system installer or user. But in some systems, the system manger generates the security key. Whatever be the way of generating the join key, it must be protected. Restrict access to the Provisioning Device handheld and the system manager.

Use additional security configuration options for the backbone router.

14.8.5 PERFORMANCE

Limit the number of devices connected between the process network and the wireless management application to prevent time delays.

Balance the transmission rate of wireless field devices with the battery life. Performance of a wireless system is measured in two ways. First, reporting rate of devices is an important performance factor. Second, the number of devices supported by the gateway or the system manager is another measure. The faster the reporting rate, the sooner the battery will drain out. The faster the reporting rate, the gateway will become more loaded for supporting same number of devices. Different vendors have different specifications about the reporting rate and number of devices supported by their systems. Typical reporting rate varies from 1 sec to 60 min. It is easy for a router to route the data of a slower reporting device, whereas to route packets of a faster reporting child, the routing parent needs to compromise two aspects. First, the number of devices that it can route is always lower for the faster reporting children. Second, the battery consumption is more for the router in case it has got faster reporting children. The line-powered router does not have an issue of battery life, but it still has the limitation on number of devices it can route based on the reporting rate. Different vendors support different number of devices per gateway, but this specification should always be linked with the reporting rate. For example, vendor X gateway supports: 50 devices @ 4 sec update rate or 100 devices @ 8 sec, whereas another vendor might support 100 devices @ 1 sec or 500 devices at 5 sec.

Try to reduce the number of wireless hops.

14.8.6 SITE-PLANNING CHECKLIST

Use the following checklist for site planning to determine the optimal placement and operating conditions for all:

- Physical obstacles that can be barriers to proper signal path.
- External or internal sources of radio interference.

- Hazardous location certifications for each of the wireless field devices.
- Coverage area required for each device.
- Communication path for each wireless field device.
- Optimal number of routers in the network based on the number of supported wireless field devices.
- Bandwidth and placement restrictions.
- Balance of the transmission rate of wireless field devices with the battery life.
- Locations of wired network access.
- Power access requirements for backbone routers.
- Frequency requirements and channel allocation.
- Transmit power settings.
- Antenna selection.

FURTHER READING

Dick Caro, Wireless Networks for Industrial Automation, ISA.
http://en.hartcomm.org/hcp/tech/wihart/wireless_overview.html.

CHAPTER

OPC COMMUNICATIONS

15

15.1 INTRODUCTION

15.1.1 OPC BACKGROUND

Process industries and commercial buildings meet most of their automation and control requirements using products from various vendors. In the early days of automation, customers were forced to depend on these vendors for even small modifications to the control system as most of these vendors used their own proprietary protocol for communication between their devices and applications. In addition, the use of proprietary protocol made interoperability between various products in a control system a big challenge. A manufacturing plant may have different data sources such as PLCs, DCSs, gauges, database, and RTUs with different network connections such as serial, ethernet, and radio. The process control applications may also use different operating systems such as Windows, UNIX, and DOS. Typically, the vendor fetches data from the data source using an interface that is specific to the connection type and stores the fetched data in a proprietary format. Therefore, when the control system needed modification, the user always had to go to the vendor for doing the required modification. In addition, the use of proprietary protocol and data format called for a custom interface from the application in one system to the device in the other system to be developed if the data from the product of one vendor was needed in the product of another vendor. This often resulted in huge engineering and implementation cost. In effect, the various products that were used for effectively controlling a single manufacturing plant stayed as automation islands due to lack of true interoperability and the customer had to bear the additional cost for enabling data exchange between various systems in the plant (Figure 15.1).

Process industries faced the biggest challenge of enabling interoperability between various automation providers. This challenge was mitigated by a technology standard called OPC, which stands for OLE for process control. OLE is a technology developed by Microsoft that enables the user to create an object with one application (e.g., Microsoft Paint) and then link or embed it in the other application (e.g., Microsoft Word). OLE technology provides a standard way of data access from the data sources in process control systems development regardless of the underlying communication protocols used in the downstream. OPC aims at providing a standard-based infrastructure for exchange of process control data, and by using this OPC technology, the data can be passed from any data source to any OPC complaint application regardless of its vendor.

OPC is based on two components:

- OPC server to read data from the device or to write data to the device.
- OPC client to access data from the OPC server.

A vendor can develop an OPC server to communicate to the data source and expose a standard set of interfaces as specified in the OPC standard specifications. The interfaces exposed by the OPC server

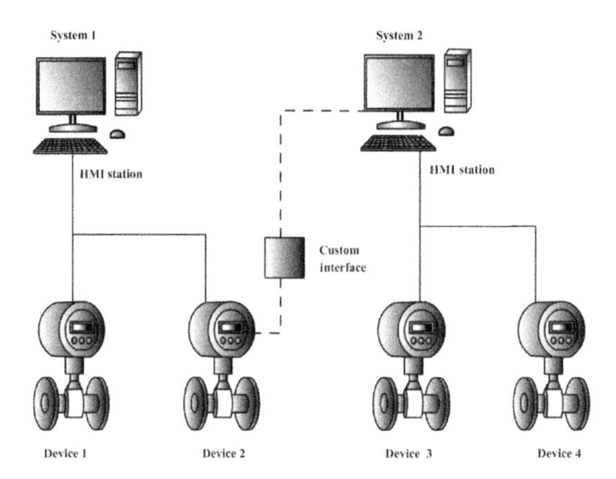

FIGURE 15.1 Automation islands (System 1 and System 2) requiring a custom interface to be developed for data exchange between Device 2 of System 1 and HMI of System 2.

shall be used by any OPC client to access data from the data source. OPC technology therefore provides true interoperability between the products developed by various vendors. This true interoperability provides the customers with a wide range of options to choose from for getting the best-in-class devices and applications without worrying about the communication between them (Figure 15.2).

15.1.2 PROBLEMS SOLVED BY OPC

OPC is based on client/server technology and has been implemented in almost every industry including process manufacturing (such as oil and gas, pulp and paper, utilities, etc.), discrete automation (automotive, packaging, parts, etc.), commercial operations (building automation, HVAC, equipment optimization, etc.), and even municipal operations such as traffic control.

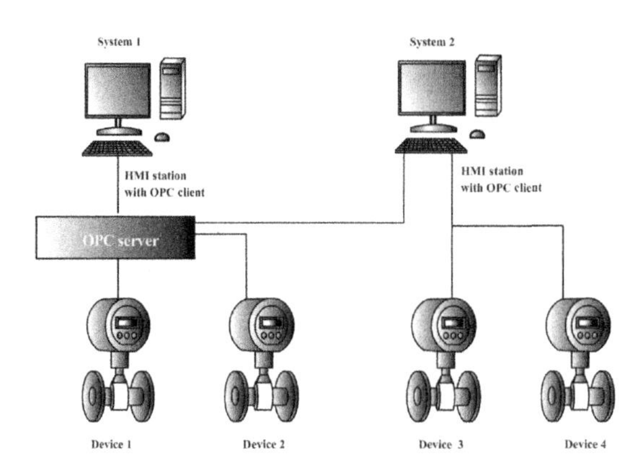

FIGURE 15.2 Automation system with interoperability enabled by OPC technology.

Factors affecting data sharing

Following are the list of factors that have traditionally caused the biggest data sharing issues, which are solved by the use of OPC technology.

Proprietary protocols

Vendors often used proprietary protocols enabling products from a particular product line to communicate among each other, but required custom drivers to communicate with other vendors' products. In addition, different product lines from the same vendor often did not communicate among each other because of different protocols used by each product line, which necessitated the need for additional connectors.

OPC resolves this by making it unnecessary for the data sink to know anything about how the data source communicates or organizes its data.

Custom drivers

Every end-to-end connection required a custom driver to facilitate communications between specific endpoints. For example, if an HMI needed to communicate with a PLC, it required a custom HMI driver written for the specific protocol used by the PLC. If this PLC data was to be historized, the historian required its own driver because the HMI's custom driver could only be used to communicate with the HMI, not the historian. If a custom driver for the specific endpoints was not available, data communications were difficult and expensive to establish. These custom drivers lead to the following problems:

- Duplication of efforts – as every application needs a driver to be developed.
- Inconsistencies between vendors and drivers – hardware features not supported by all driver developers.
- Support for changes in hardware features – a change in hardware features may break some drivers.
- Access conflicts – two packages generally cannot access the same device as they each contain independent drivers.

OPC eliminates the need for custom drivers between each new application and data source. For example, a single standard PLC driver could be shared by both the HMI and the historian via an OPC connector with the added benefit that the OPC connector requires a single connection to the PLC – reducing controller loading. OPC also eliminates all the other problems associated with custom drivers (Figure 15.3).

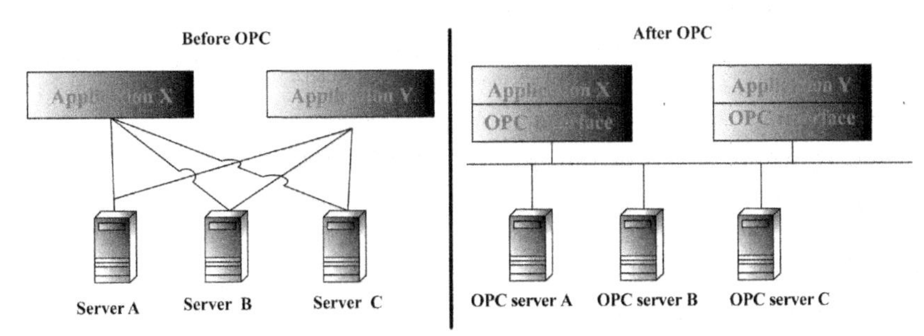

FIGURE 15.3 OPC eliminates the need of custom drivers.

Complex integration

Integration complexity was a challenge arising from the use of custom drivers between every endpoint. Even the integration of a small number of devices and applications involved the use of many drivers. The same HMI running on multiple computers, all communicating with the same device, required multiple installations and configurations of the same driver on each computer. If the HMIs communicated with additional devices, each HMI required its own set of custom drivers for each of the devices. This created a version maintenance nightmare.

Integration is greatly simplified using OPC because once an OPC connector for a particular data source is configured, all OPC-enabled applications can start sharing data with that data source with no concern for additional custom drivers.

Device loading

Each driver establishes its own connection to the device or controller that it is designated to communicate with. A typical installation normally has a large number of custom drivers being used leading to the controller being bombarded by many requests for the same information from every application that it needed to communicate with. In addition, many devices could only accept a limited number of simultaneous connections. If the number of drivers trying to connect to a device exceeded the number of connections it had, further workarounds were needed.

Traffic, and therefore device loading, is greatly reduced by using OPC connectors because a single device-specific OPC connector requires only a single connection to the data source while simultaneously communicating with many applications.

Obsolescence of legacy infrastructure

As vendors release new products, they eventually stop supporting older ones. Often a new version of an HMI may require its own set of device drivers that sometimes may no longer support communications with a device the previous version of the HMI supported.

OPC extends the productive life of legacy systems because once an OPC connector for a legacy system is configured, it allows any OPC-enabled application (most are) to communicate with the legacy system regardless of whether the application natively supports communication with the legacy system or not. Thus, OPC allows the newest applications to continue communicating with the oldest systems.

Enterprise-wide data connectivity

Data-connectivity problems are compounded with the growing need for automation data throughout the enterprise because applications from the corporate side were not designed to communicate with devices and controllers. This can potentially add extra load to the automation infrastructure and raise various automation security concerns.

OPC only needs an OPC connector to make true enterprise-wide automation data sharing possible by enabling approved applications to share data with automation data sources without the need for installing custom drivers.

15.1.3 OPC FOUNDATION AND OPC SPECIFICATIONS

OPC foundation is a nonprofit organization with over 400 vendors of automation systems as its members. It is responsible for establishing and maintaining the OPC specifications that ensures seamless interoperability between various automation products. The OPC foundation holds OPC interoperability sessions around the world in a periodic manner. The automation solution providers send their technical experts to these sessions to connect their OPC products to the OPC products of other vendors. In case of connectivity problems, vendors get together immediately to resolve the issues. These sessions are arranged by the OPC foundation to ensure that the users get the best connectivity experience when they connect multivendor applications together.

The OPC foundation also developed an OPC compliance test suite. The vendors, whose products passed the OPC compliance tests, can submit their test results to the OPC foundation and the OPC foundation permits the vendors to use the OPC compliant logo. OPC foundation also recognizes a select set of independent labs for more thorough testing. OPC compliance logo indicates that the OPC server complies with the particular OPC specifications; it does not mean that all OPC servers are the same. The OPC servers may vary greatly on speed, reliability, and capability.

The OPC foundation maintains many OPC specifications that exist today. These OPC specifications specify the COM interfaces (what the interfaces are), not the implementation (know how of the implementation). In other words, the specifications mention the behavior that the server is supposed to provide to the client applications that use them, not how to develop an OPC server. The most common OPC specifications are as follows:

- OPC data access (OPC DA) – OPC DA provides real-time access to the process data. The client can request the OPC server for the most recent values of flow, pressure, temperature, level, etc. using OPC DA.
- OPC historical data access (OPC HDA) – OPC HDA is used to retrieve and analyze historical data, which is typically stored in a plant historian.
- OPC alarm and events (OPC A&E) – OPC A&E is used for exchange of alarms and events between different vendors. An OPC client can request the OPC server a sequence of events list using OPC A&E.
- OPC data exchange (OPC DX) – OPC DX defines how an OPC server exchanges data with the other OPC servers.
- OPC extensible markup language (OPC XML) – OPC XML encapsulates process control data and thus makes it available across all operating systems.

There are also other specifications such as OPC batch, OPC security, and OPC unified architecture (OPC UA).

The OPC foundation continues to update existing specifications to meet the new and emerging requirements for interoperability. For example, OPC UA was developed in response to OPC's broad adoption across all industries in addition to sharing the data throughout the enterprise. It is important to select the correct specifications for the given application. For example, OPC DA and OPC HDA are different specifications developed for meeting different requirements. Also, there are different versions for each OPC specifications, such as OPC DA release 1, OPC DA release 2, and OPC DA release 3. The user needs to check with the vendor about the OPC specification that their OPC server supports.

The OPC DA, OPC A&E, and OPC HDA are most widely used specifications in the process automation and control industry. The other rarely used OPC specifications are OPC DX, OPC batch, and OPC XML.

OPC data access (OPC DA)

OPC DA provides a standard way to access the real-time data or live data from the process control hardware and software. OPC DA ensures consistent communication between all devices and applications.

OPC server for PLCs, DCS, and other devices provides the data in exactly the same format. Similarly OPC client applications such as HMI, trend applications, and process data archives accept the data in the same format. This enables the process control hardware and software to freely access the data and thereby ensure enterprise-wide interoperability. Many manufacturers adopted OPC DA for real-time data transfer to ensure system scalability (the control system can be expanded easily in future by adding more number of process control application software to fetch data from the device without having to increase the device's capacity to respond to data requests). There are four different releases of OPC DA available (1, 1a, 2, and 3). The users of OPC DA server must use the latest version of server and also ensure that their server is backward compatible.

The OPC DA is independent of the data source. The OPC DA client can be unaware of the type of the device it is communicating with and the protocol used by the device. OPC DA clients send the request to the OPC DA server using the interfaces exposed to the clients by the server. The OPC DA server gets the data from the device and returns it back to the OPC DA client(s). OPC DA server provides access to single-value items, which are called "points." For example, if we consider a simple process controller named "PIC101" in a DCS, this pressure controller will have different parameters such as Setpoint (PIC101.SP), Process Value (PIC101.PV), Output (PIC101.OP), and so on. For the OPC DA server, each parameter is a "point" or an "item"; that is in the example given, PIC101.PV, PIC101.SP, and PIC101.OP are treated as "points" or "items" by the OPC DA server, whereas in the DCS or PLC, PIC101 is treated as a point and PV, SP, OP, and so on are treated as parameters of a point (Figure 15.4).

According to OPC DA specifications, three attributes for each point must be returned by the OPC server to the client requesting the data. The attributes returned are the value of the point, quality of the value, and time stamp at which the value is fetched. For example, if an OPC client application requests

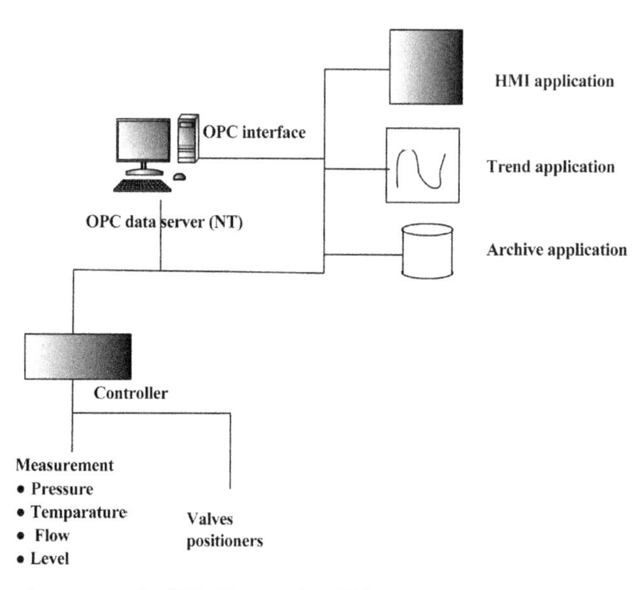

FIGURE 15.4 Real-time data access by OPC clients using OPC DA.

the data for the point PIC101.PV, the OPC server returns the value (e.g., 12.5 PSI), time stamp (e.g., 1-Dec-2011, 14:45:05.552), and the quality (GOOD, BAD, UNCERTAIN). The time stamp usually comes from the device. But if the communication protocol between the OPC server and the device does not allow the time stamp to be passed on to the server, then the OPC server provides its own time stamp to the data returned to the client.

There are various data exchange mechanisms used between OPC server and OPC client in OPC DA, the most common being:

- Synchronous reads or polling
- Asynchronous reads or subscription.

Synchronous reads or polling

In the synchronous reads (or) polling mechanism, the OPC client(s) send request to the OPC server with a list of items that need real-time data (Figure 15.5). The OPC server then responds to the OPC client with the data value, time stamp, and the quality of data for all the items in the list requested by the OPC client. The OPC client then sends the next request for data to the OPC server, which in turn is served by the OPC server. The data exchange mechanism has the following disadvantages:

- The OPC client waits until it gets the response from the OPC server before making the subsequent request for data. If the OPC server does not respond to the request of the client, the client waits or "hangs" until the response is received.
- The OPC server responds to the OPC client with the requested data, even if there is no change in the item's data since the last response. For example, assume that the client made a request

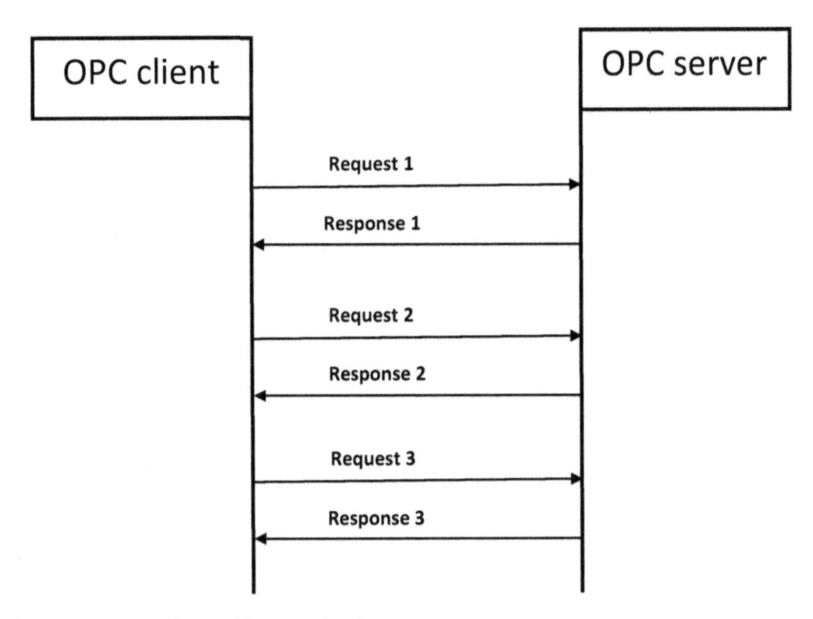

FIGURE 15.5 Synchronous read or polling mechanism.

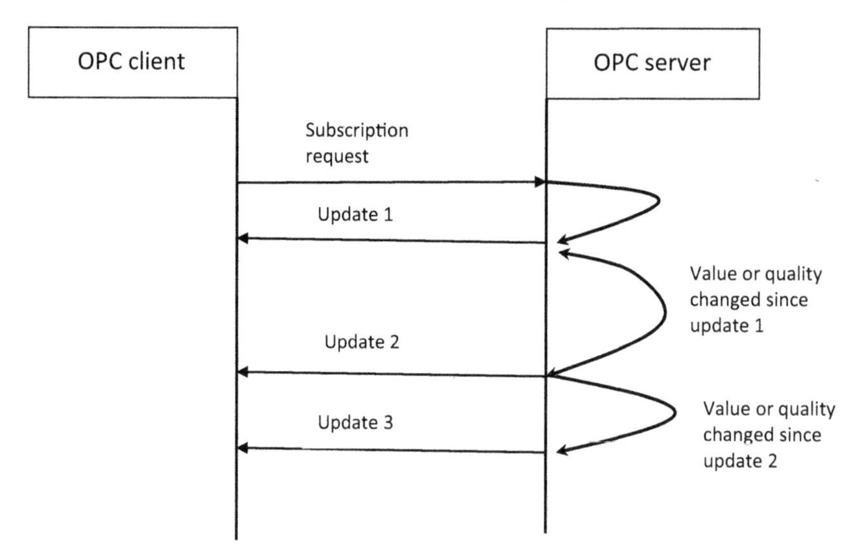

FIGURE 15.6 Asynchronous reads or subscription mechanism.

for the data of item PIC101.PV and got the response as 14 PSI with status as GOOD. When the client subsequently requests for the same data PIC101.PV, the server responds to the request even if the value is still 14 PSI with GOOD status. This unnecessary update (although the data did not change) or back and forth data transfer between the server and client results in a lot of unnecessary data traffic.

Asynchronous reads or subscription

Clients request for subscription of a list of items with a specific subscription period in the asynchronous reads or subscription mechanism (Figure 15.6). The OPC server then creates a "Group" of items, based on the subscription period of the items. The OPC server creates a "Group" for all items that are having the same subscription period. For example, if a client sends a list of 100 items of which 50 items are requested every 1 s, 25 are requested every 2 s, 25 are requested every 10 s, then three groups are created by the OPC server with 50, 25, and 25 items in each group. These groups can be private or global; private groups are created by the OPC server to be bound to a client; global groups are bound to many clients, and the update in this group is published to all associated clients.

When a client makes a subscription request for a "Group" of items, the OPC server responds to the client with the value, quality, and time stamp of all items in the group if the following conditions are met.

1. There is a change in the data value or quality of any of the items in the group since the last update was sent to the client.
2. The subscription period for the group has expired. In other words, even if there is a change in the value or quality of the data, the update is not sent to the client until the subscription period elapses.

This mechanism therefore minimizes the data traffic between the server and the client and also ensures that the client is not overrun with the data at a faster rate than the client can handle. In addition, the client is only notified of the data update and it does not wait for data. Therefore, the client never "hangs." Based on the data exchange mechanisms used by the OPC server to read data from data source or device, there are two types of OPC data read employed at the client(s).

- Device Read: When the OPC client specifically requests for the "Device read" of the required items, the OPC server polls the device for the required data irrespective of whether the data has changed since it was read last or there is no change in data. This type of read increases the traffic between the device and the OPC server.
- Cache Read: By default any read request by the OPC client is taken by the server as the cache read. OPC server has a shared memory allocated which is called as "cache." This cache is updated periodically with the required data value, quality, and time stamp of all the items that are subscribed by various clients connected to the OPC server. The OPC clients are then returned with the requested data based on the data exchange mechanism used between the client and the server.

OPC historical data access (OPC HDA)

Archived process data can be accessed from the historian using the OPC HDA. Plant historians are data storage applications or data archives used for storing the collection of real-time data that are fetched from the data source (DCS, PLC) using OPC DA or the native data access methods (if the plant historian and the data source are from the same vendor, the historian may use native data access method for collecting the real-time data from the data source). This archived data is read from the historian by various applications such as trending application, reporting application, analysis application, ERP, and so on. Different vendors offer various plant historians employ a different architecture for storing the archived data.

For example, a historian from vendor A may use Microsoft SQL Server for storing the archived data and another historian from vendor B may use Oracle database. The OPC foundation established OPC historical data access specifications to enable access of historical data from the historian irrespective of the vendor. This overcame the internal mechanism for storing the established data.

The OPC HDA pulls the historical data from the historian and this pulled data is used by various applications for different purposes such as

- Predicting a possible fault in the process
- Performing root cause analysis in case of a failure in the process
- Performing efficiency calculation of the plant
- Viewing the trend over a week or month
- Performing the financial activities such as billing.

The OPC HDA specifications define how the data must be exchanged between the OPC HDA server and the OPC HDA client. The OPC HDA defines how the historical data is pulled from the historian and also describes how to write the data to the historian. However, the specifications do define how the data is stored in the plant historian. There is no restriction on the data storage format in the historian. Also, OPC HDA does not specify how the OPC HDA server must be implemented; but it explains what interfaces should be exposed by the OPC HDA server so that the OPC HDA client can connect to it and

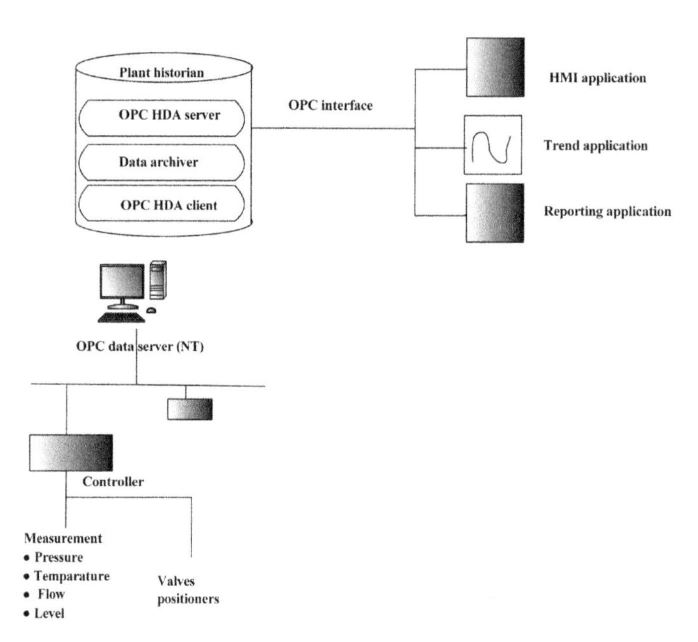

FIGURE 15.7 Historical data access by OPC HDA clients using OPC HDA.

get the required data. OPC HDA has two components: one is the OPC HDA server that accesses data from the historian, and the other is the OPC HDA client that gets the required historical data from the OPC HDA server. Most major plant historians available today have OPC HDA server (Figure 15.7).

The OPC HDA server returns three attributes of the requested item (data value, quality of the data, and time stamp) to the requesting clients, just like OPC DA. OPC HDA returns the stored value or historical value, whereas the OPC DA returns the real-time value or live value. OPC HDA does not have the concept of "Group" creation, whereas OPC DA has "Groups," which are basically a collection of items with the same update rate. Unlike in OPC DA, there is no concept of device read and cache read as the OPC HDA pulls the data that is already read and stored in the historian.

There are two main types of reads in the OPC.

1. Raw Reads: All the stored data is pulled from the OPC HDA server for a particular duration in a raw read. The OPC HDA server does not perform any calculation or conditioning of the data. The OPC HDA client specifies the starting time and end time in its request for data; the OPC HDA server then accesses the historian and pulls all the data stored between the starting time and the end time as specified by the client. There could be huge amount of data returned by the server for a request. The amount of data returned is based on
   a. the time range for which the data is requested and
   b. the interval at which the data is collected and stored in the historian.

   For example, if the data is collected every 1 s and stored in the historian, and if 10 h historical data is requested, it would amount to huge data. As the amount of data returned by the server to the client's request cannot be predicted in advance, the raw reads some time may overload the network when the server returns a huge amount of data for a request from the client. Raw reads are mostly used when the data is to be synched or backed up in which case all the collected data in the given duration needs to be pulled.

2. Processed Reads: In the processed reads, when a client makes a request, the OPC HDA server performs some conditioning or calculations on the data and returns the result of the calculation to the client. When the processed read is made, the OPC HDA server performs data analysis and therefore significantly lesser volume of data is returned to the client by the server therefore greatly reducing network traffic. For example, if the temperature of a reactor is collected every 1 min and stored in the historian over a month's time, the client could request for the average temperature of the reactor every day for the whole month. In this case, the OPC HDA server calculates the average temperature of the day for the whole month and returns the results of the calculation to the client. If the data is not available for the given period, the data is interpolated or estimated by the OPC HDA server. For example, if the flow rate of a process is only recorded for the start of every minute, the OPC HDA server interpolates or estimates the value of flow using the values at Nth minute and $N + 1$th minute when it has to return the value at N minute 30 s. Some of the functions or calculations that the OPC HDA server uses for processing the data are Count, Standard Deviation, Average, Total, Variance, Maximum, Minimum, Time Average, and so on. The OPC foundation has defined rules for implementing these functions to ensure that the results of these functions are the same irrespective of the OPC HDA server used (i.e., when the "Standard Deviation" in the data over a time range is calculated the result obtained should be the same regardless of the OPC HDA server used).

OPC alarm and event (OPC A&E)

OPC A&E specifications were developed for effectively managing alarms and events notifications using a client application. An OPC-compatible alarm and event client can connect to the OPC A&E server and fetch real-time alarms and events. OPC A&E is responsible for passing on the alarm and event notifications from the notification source to the OPC A&E client in the same way as OPC DA is responsible for getting the real-time data from the data source and passing it on to the OPC DA client. Like the other OPC standards, OPC A&E is also vendor independent; that is an A&E client from vendor A can connect to the OPC A&E server of vendor B and subscribe to the alarm and event notifications from the control system. For receiving notifications, the OPC A&E client needs to establish a link to the OPC A&E server using Microsoft's COM or DCOM technology. If the OPC A&E client is running on the same node as the OPC A&E server, then COM is used for the link between them. If OPC A&E is running on a remote computer, the DCOM is used for the link between the OPC A&E server and the client. After establishing the link, the OPC A&E client creates an "Event Subscription" to the server. The OPC A&E client may create multiple event subscriptions for the same connection.

Alarms and events indicate a change in the operating conditions of a plant. The alarms indicate an abnormal operating condition that requires immediate attention, whereas events indicate a change in the operating condition (not necessarily an abnormal condition) that includes the alarm conditions. The alarm condition may be a single-state condition or a multistate condition. The single-state alarm condition has only one state; example for the single-state alarm is the alarm that indicates the controller failure. The multistate alarm has more than one substates; example for the multistate alarm is the flow alarm whose substates are Low, LowLow, High, and HighHigh. However, the events may or may not be associated with a condition. The example of a condition-related event is the alarm event and the event reporting return to normal condition of an alarm. The example of an event not related to any condition is the event that indicates a change in the set point of a control loop. The alarms and events are preconfigured in the system and are generated automatically when the configured state is reached for the control loop or the system. The OPC A&E server automatically captures these alarms and events and makes them available to any OPC client connected to it.

Typically, alarms and events in a process control system are associated with areas. Based on the operator's scope of responsibility, areas are created in the process control system. The alarms and events from the points or tags and the devices that are responsible for controlling a particular area are reported under that specific area. For example, in a PVC manufacturing plant, there could be different areas such as storage area, reactor area, charging area, rundown area, and so on. These areas are created in the control system and all the points and equipment that are responsible for controlling an area are grouped under that area in the control system so that any alarm and event from that area are reported under it. Implementation of areas is optional according to OPC A&E specification. Implementing areas helps the clients to get an area-based filtered list of events.

The OPC A&E specifications describe three types of events:

- Simple Events: Simple events are noncritical events that do not require any action to be taken by the user. They are typically of informational information. Examples of simple events include operator's log-on and log-off to the HMI Station, Start of a batch process, and completion of a batch process. Simple events have a standard set of attributes as follows:
  - Source: This attribute indicates which object has generated the event. For example, if the event is about the change in set point of a control loop (TIC101) then the source of the event will be "TIC101." For system-related events (e.g., HMI Station startup) the source of the event is "System."
  - Time: This attribute indicates the time at which the event is generated.
  - Type: The type of the event; that is, simple, condition, or tracking.
  - Category: This attribute indicates whether the event is a "Process Event" or "System Event" or "Batch Event." Example of a process event is the set point change of a control loop. Example of a system event is the logoff of an operator at a HMI Station. Example of a batch event is the start of a master recipe.
  - Severity: This attribute indicates the severity of the event. The severities could be Urgent, High, Medium, and Low.
  - Message: This attribute is a text that describes the event.
- Tracking Events: A tracking event is similar to the simple event and typically notifies a change in the monitored system such as a configuration change or a user change. It indicates that some specific action was performed on the source by a specific user or other parts of the system. In addition to the standard attributes of the simple event category, the tracking events have one more attribute called "Actor ID" that specifies who or what performed the action on the source. Like simple events, the tracking events also require no specific action from the user.
- Condition Events: Unlike simple and tracking events, the condition events call for an acknowledgment or some sort of responsive action from the user. The condition events are used when the events require a guaranteed delivery and must be acknowledged. The condition events are associated with the detection of some condition in the control system such as pressure in the reactor going higher than the configured value, temperature dropping below the configured value, and so on. Thus, all single-state or multistate process alarms belong to the condition events type.

Condition events have the same set of attributes that the simple events have. In addition to those attributes, they also have the following attributes:

- Condition Name: The name of the condition.
- Subcondition Name: The name of the associated subcondition.
- New State: This indicates the new state of the condition.
- Ack Required: This indicates whether the event requires an acknowledgment or not.
- Actor ID: The identifier of the OPC client or the identifier of the user who acknowledged the event.

Considering the different types of events, there is a scope for filtering the events in order to reduce the traffic over the link between the OPC A&E server and the client. The OPC A&E specification defines a set of common filter criteria that can be applied. The specification also defines the filtering to be done by the server to minimize the traffic over the link between the server and the client. The OPC A&E client can connect to the server and create an event subscription. While creating the subscription, the client can set the filter criteria for each event subscription to get only the filtered list of alarms and events. The filter criteria could be set as follows:

- Based on Severity: Such that the client only receives the high-priority alarms and above.
- Based on Event type: Such that the client only receives the condition events.
- Based on Event category: Such that the client only receives the batch events.
- Based on Area: Such that the client only receives the events from the area "A1."
- Based on Source: Filter criteria for receiving the events from only a specific source can be set.

OPC XML data access (OPC XML DA)

XML, the extensible markup language, aids in describing and exchanging structured information between collaborating applications. XML is a technology that is readily available across all operating systems. OPC XML DA is the OPC foundation's adoption of XML technology for passing plant data across Internet. XML DA-based interfaces simplify the exchange of data among various levels of plant hierarchy (from low-level devices to enterprise system) and to a wide range of platforms. OPC XML DA is developed in a manner that allows the plant data to be passed as a simple object access protocol (SOAP) message over Internet.

SOAP is an XML-based communication protocol that enables information exchange between applications that may run on different platforms. SOAP is language-independent and is easy to get around the firewall. In OPC DA, the applications on different Windows machines communicate using RPC and DCOM. There are two limitations with OPC DA in using it for communication across Internet:

- RPC traffic is typically blocked by all firewalls and proxy servers
- DCOM is typically bound to Microsoft operating systems.

Therefore, SOAP-based OPC DA XML enables communication between applications with different technologies and programming languages running on different operating systems across Internet.

Currently, OPC has not defined any mechanism to differentiate between the nodes that have OPC XML DA servers running or the OPC XML DA servers running on a specific node. The universal description discovery and integration (UDDI) protocol is a widely used standard for web services and is likely to be used by OPC foundation in detecting OPC XML DA servers over network or OPC XML DA servers running on a specific node. Until the UDDI protocol is implemented in OPC XML DA, the OPC XML DA client needs to know the URL of any OPC XML DA server it wants to use.

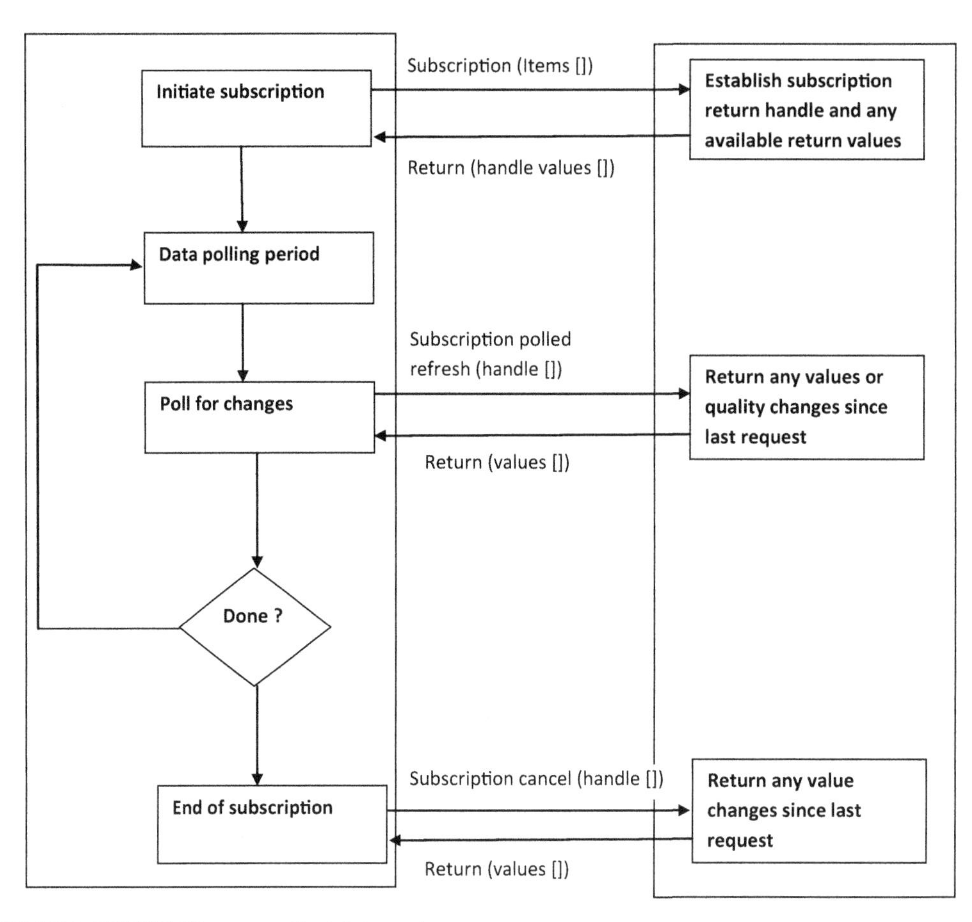

FIGURE 15.8 OPC XML DA server–client data exchange.

The design of OPC XML DA subscription employs a polled-pull style of subscription as illustrated in Figure 15.8. The client application initiates the subscription and agrees to send periodic refresh requests. XML DA supports the following subscription-based services:

- Subscribe: Used to initiate the subscription contract for a list of items for which the data is needed.
- Subscription-polled Refresh: This refresh is made periodically to get the latest changes in the item values.
- Subscription Cancel: This is used for terminating the subscription contract with the server.

OPC batch

A batch process is defined as "a process wherein a finite quantity of output materials is produced by subjecting known quantities of input materials to an ordered set of processing activities over a finite period of time using one or more equipment." A specialized set of control algorithms that execute in sequential manner are required to control a batch process. Unlike continuous process control where the process parameters are continuously monitored and controlled, in batch control, the control actions are taken when certain process conditions are met and then the system waits until the next set of conditions are satisfied after which it performs another set of control actions and so on. Most batch control systems use proprietary interfaces for the distribution and collection of data in a batch process control. It was not possible to augment existing solutions (provided by vendor A) with other improved capabilities (provided by another vendor B) until the OPC batch specifications were released in a plug-and-play environment. To facilitate data exchange, the same infrastructure had to be recreated all over again for the software application to work with the existing solution. The manufacturers and customers wanted to use off-the-shelf open solutions from various vendors for effectively controlling the batch process.

OPC batch specifications provide means to pass batch-related data between system components of various vendors. The OPC batch specifications were developed based on OPC DA specifications and IEC-61512-1 Part-1 Models and Terminology. The "IEC-61512 Part-1 Models and Terminology" is aimed at bringing in uniformity in batch processes and equipment and batch process control. This IEC specifications established process model, physical model, and procedural control model for a batch process. The OPC batch specifications describe the interfaces to be created for transferring batch data related to physical model and procedural control model. Of the data exchanged described below, equipment information (in bullet 2) is the batch data related to physical model and the other data are related to the procedural control model. The data that is exchanged using the OPC batch specifications is as follows:

- Current runtime batch information:
  - Mode and state changes
  - Timing of commands sent to process control devices
  - Timing of the execution of unit procedure events
  - Timing and sequence of allocation, reservation, and release of equipment entities acquired by the unit
  - Status changes in unit equipment, etc.
- Equipment information required to understand the context of the runtime batch information:
  - A string representing the class of the equipment (e.g., reactor, mixer, and so on)
  - A string representing the location or building where the equipment exists
  - A value representing the maximum number of concurrent users of the equipment
  - A value representing the current users of the equipment
  - A string representing the current state of the equipment (e.g., OPEN, CLOSE, RUNNING, STOPPED)
  - A string representing the current mode of the equipment (e.g., MANUAL, AUTOMATIC)
- Historical records of batch execution: The runtime information of the previously completed batches is presented as a historical batch execution record.
- Master Recipe Contents: A recipe refers to the set of information that uniquely defines the requirements for producing a particular product. Master recipe is the plant-specific description of a manufacturing process. The master recipe contents that are exchanged using OPC batch server comprise the data such as
  - Master recipe ID used for a batch
  - Version of the master recipe
  - Product to be produced by the execution of master recipe

- Usage constraints for a process stage (such as "must be succeeded by Unit Operation A" or "must not be run in parallel with Unit Operation B…")
- Scaling rule, such as whether the recipe parameters will be scaled when the batch is scaled.

OPC batch server enables access or communication to a set of data sources. The OPC batch server may communicate to the OPC DA server or it may communicate to a batch application from the same vendor to read the OPC batch-specific data (Figure 15.9).

The OPC batch server may be a stand-alone server or a set of interfaces provided on top of an existing proprietary batch server. The OPC clients connect to the OPC batch server and get the batch data such as the current batch ID, batch execution start time, state of equipments, and so on. The OPC client is a simple display that requires only a few values or a set of complex displays and reports that need a lot of data in different formats.

OPC security

The OPC servers such as DA, HDA, A&E, and so on provide interoperability among the systems from different vendors for the exchange of process data. While unrestricted access to the read of process data is not a cause of concern, write access must be restricted as it may result in incorrect data being written and eventually may lead to severe consequences. Therefore, there is a need to control the write access to protect sensitive plant data from being modified by unauthorized clients. Security must be provided in a consistent manner across all types of servers and vendors to simplify the development process of clients. OPC security must be well integrated with the Windows security and be as much transparent as possible to the client application so that the client need not do something special to connect to the server. The OPC security specifications document provides guidance about how the OPC server must

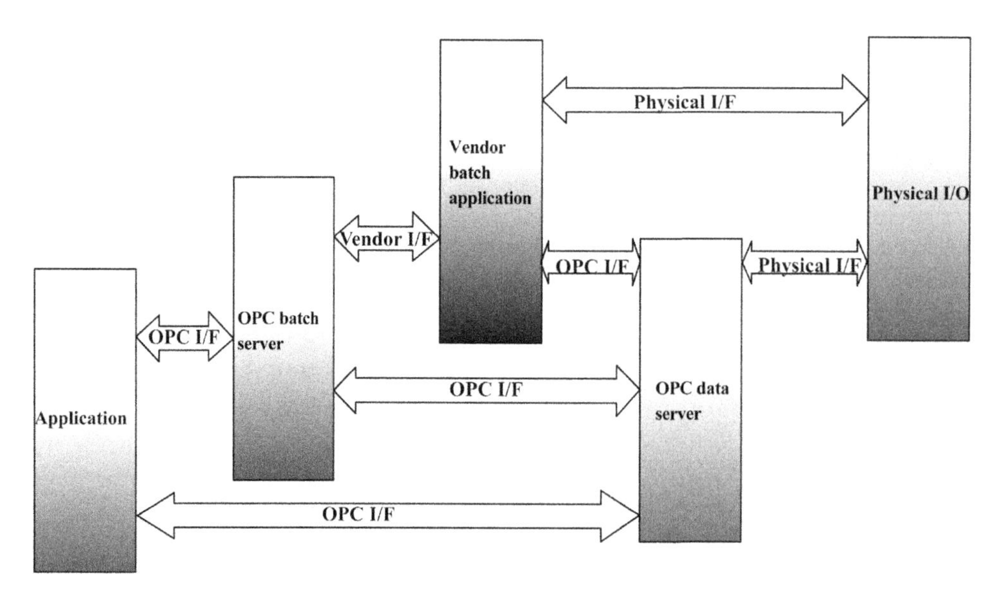

FIGURE 15.9 OPC batch server data access flow diagram.

implement security using operating system facilities. This specification is primarily applicable to Microsoft Windows operating system and other platforms based on the availability of DCOM in such platforms.

OPC security specifications consist of a set of conditions regarding how to secure OPC server applications, OPC client applications (when OPC server performs a callback to the client), and any vendor-specific security objects implemented by the OPC server. The vendor-specific security objects are server interfaces and methods, public groups, and individual or set of data items. The decision regarding the objects to be secured is left to the vendor who implements the OPC server, but the manner in which they are secured must conform to the OPC security specifications.

There are entities called "reference monitors" in the Windows operating systems, which decide on providing access to the associated security objects. In other words, each security object has an associated reference monitor to provide authentication for the clients to access the security object. OPC security specification deals with two reference monitors for OPC secured objects.

- Windows NT reference monitor: The Windows NT reference monitor controls the access to DCOM server and clients. It validates the credentials of the clients requesting connection to DCOM servers and, after successful validation, provides the authorization to connect.
- OPC server: The OPC server is implemented to validate the credentials of the client and make the decision of providing access to its secured objects. They make the authorization decision based on the Windows NT access token or private credentials specific to the vendor. It is recommended that the OPC server implements security validation using NT access token as this enables transparency of the security to the clients. When NT access token is used, the client need not perform any explicit action to generate access certificate. If the NT access token mechanism cannot be used (such as the case where client is running on an operating system that does not support NT access token), the specification defines the ability for the OPC server to implement its own private access certificates using a User Name and Password combination. OPC servers required to support both NT access token and private access certificates must be implemented to support both on a client-by-client basis.

This specification describes the OPC server to implement one of the three levels of security as follows:

- Disabled Security: No security is enforced. The OPC server launch and access permissions are given to everyone. That is, any client can connect to the OPC server and access its data.
- DCOM Security: The access to OPC server is controlled by the Windows NT DCOM security. The OPC server launch and access permissions are controlled by the DCOM object's access control list. If the user account under which the client is running is authenticated by the DCOM object, then only the client will be able to establish the connection and access data.
- OPC Security: The OPC server serves as the reference monitor to control access to its secured objects.

15.1.4 UA – THE NEXT GENERATION OF OPC

OPC UA is the next technological evolution of OPC that has been providing interoperability in process control industry for the past 10 plus years. While OPC specifications for real-time data, alarm & events, and historical data access were defined separately, OPC UA provides a single interface for all types of plant data access. Earlier, three OPC servers were needed for accessing the data, alarms & events, and HDA, but with OPC UA, all three types of information can be accessed using a single interface. Also, devices such as remote terminal unit (RTU) offer all three types of data and OPC UA facilitates access of the data from RTU using a single interface. OPC is Microsoft centric that used DCOM infrastructure for PC–PC communication. Although almost all computers in industrial automation and control systems use Microsoft operating systems, there are also a deep layer of equal number of embedded devices and systems that run real-time operating system (RTOS). The earlier OPC DA always relied on a computer running Microsoft's OS for accessing the data from RTOS devices and opening it to the clients. With OPC UA, the data access from an embedded remote device all the way up to an application running in an enterprise dashboard is made possible. Thus, the OPC UA unifies multiple specifications of the past and redefines the acronym of OPC from OLE for process control to OPen connectivity.

The OPC UA implementations are based on service-oriented architecture (SOA). In SOA, the application's function is presented as a service to the consumer/client applications. The loosely coupled nature of the services (i.e., the interfaces exposed by a service are independent of the implementation of the service) enables the developers to build a client application or another service comprising one or more services without actually knowing the service's underlying implementation. SOA infrastructure is based on three important elements: WebServices description language (WSDL), UDDI, and SOAP. WSDL is an XML-based language used for describing the services offered by a webservice application. UDDI is a platform-independent XML-based registry for registering and locating the web service applications. SOAP is a protocol specification for sending messages between service provider and consumer. A service consumer or client can search the UDDI for the service, get its WSDL, and invoke it using SOAP. The service-based architecture is supported by Microsoft and many other operating systems. This means that the OPC UA is available on more operating systems and embedded systems.

As the system provides more and more open connectivity, the risk of attacks from untargeted malware such as worms circulating on public networks increases. Therefore, it is imperative to provide a tighter security in OPC UA in order to protect the plant from encountering any abnormal situation. The following are the security functions for protecting the systems that use OPC UA for data exchange, as recommended by the OPC UA specifications.

- Authentication: Client and server should prove their identities. It is based on something (e.g., username and password) that the client and server know or have.
- Authorization: The access to read, write, or execute resources must be authorized for only those entities that have need for the access. It shall be something like the server allowing specific actions on specific items for specific users.
- Confidentiality: To protect the data while transmission or storage, data encryption algorithms along with authentication and authorization must be used to provide special secrets and access the secrets.
- Integrity: Receivers must receive the data in the same manner as the transmitters sent it.
- Auditability: Actions performed by the system must be logged to identify the initiators of such actions and to provide confidence to the user that the system actually works as desired.
- Availability: Availability is impaired when the software that needs to run is turned off or the communication mechanism responsible for data transfer is overwhelmed (due to message flooding). Enough protection needs to be provided to ensure high availability of the services.

Therefore, OPC UA is a promising future technology in industrial open connectivity standards and it provides a flexible and secure interface at all levels of a plant, that is, open connectivity from a smart transmitter on a shop floor to the manufacturing execution system – a single interoperable data highway from the shop floor to the top floor.

FURTHER READINGS

OPC Foundation Website http://www.opcfoundation.org/.
OPC Training Institute Website http://www.opcti.com.
Matrikon OPC Website http://www.matrikonopc.com/.
Berge, J., 2002. Field Buses for Process Control: Engineering, Operation, and Maintenance. ISA.
Berge, J., 2005. Software for Automation: Architecture, Integration, and Security. ISA.
http://www.opcfoundation.org/
http://www.opcti.com
http://www.matrikonopc.com/

CHAPTER

ASSET MANAGEMENT SYSTEMS

16

16.1 DEFINITION OF AN ASSET

An asset is an economic resource. Anything tangible or intangible that is capable of being owned or controlled to produce value and that is held to have positive economic value is considered an asset.

Plant is an asset from business or organizational perspective, which is a sum of many smaller assets. The assets range from the process equipment, rotating machinery, electrical equipment, automotive equipment, to instrumentation and control equipment. Hence the industry is offered with various packages of software and hardware to manage and maintain each of the above assets.

16.2 ASSET MANAGEMENT SYSTEM

Modern plants are increasingly deploying asset management systems to improve the operational efficiency. The following are some of the key drivers for the asset management systems.

- Reliability is becoming a competitive business issue and a differentiator for the successful production. Any unplanned plant shutdowns have the significant impact on the production as well as the predictability of the produced products.
- Quality is a survival issue in many/most of the markets for better products. The quality of the produced outcome on the financials of the plant as well as the current and future business.
- Increased production demands on older equipment, which delays the new capital investment while sustaining the current production.
- Manpower reduction requirements to do more with less.
- Cost reductions focus in maintenance and operations.
- Customer requirements are becoming more stringent in various aspects such as product specs, repeatability of the products, and so on.
- Environmental, health, and safety concerns have become a total enterprise issue.
- Value creation is a motivating factor for management decisions.

From plant operations and maintenance perspective, the enterprise asset management systems are available, which help to implement the organizational maintenance strategies, purchase philosophy, auditing and cost control, and so on. In general, an enterprise asset management system manages maintenance scheduling, workflow, inventory and purchase activities of rotating assets (equipment analysis, field advisor), fixed assets (limit monitoring, corrosion monitoring) and automation assets (field device management, host management, network management). In essence, there shall be multiple systems or applications used for the asset monitoring and management, and also an integrated system for managing all these systems and work processes.

Plant asset management can be considered as activities that are provided for sustainable risk over the life cycle of an asset. In general, asset management focuses on lengthening effective life of assets while guaranteeing availability. It reduces capital tied up in goods by eliminating unnecessary assets and delaying capital investment. Asset management is practice of managing assets so that the greatest return is achieved (this concept is particularly useful for productive assets such as plant and equipment). It is the process by which built systems of facilities are monitored and maintained, with the objective of providing the best possible service to users. Asset management system manages plant assets such as boilers, pumps, heat exchangers, turbines, smart field devices, and so on. Asset management system encompasses all activities and measures for maintaining or enhancing the value of a plant. This includes not only plant management, process management, and process optimization, but also and above all, value-maintaining and value-enhancing maintenance and servicing. Instrument asset management system is particularly predestined for asset management at plant level on account of its extensive functionality for the configuration, parameterization, commissioning, diagnostics, maintenance, calibration, and also to indicate the key performance indicators (KPIs) of intelligent field devices and components.

16.3 KEY GOAL OF ASSET MANAGEMENT SYSTEM

The key goals of an asset management system in plant are

- Reduce unmanaged risk
  - Improve health, safety, environmental
  - operability performance
  - Manage complexity and interoperability
  - Sweating of assets
- Achieve sustainability
  - Assure adequate human, financial, and infrastructure resources
  - Lower maintenance cost

An asset management system is a combination of three elements: technology, expertise (diagnostics), and work processes (Figure 16.1). All these above elements play a role for a successful asset management system. The plant needs the adoption of technologies that are ready for asset management functions and needs expertise from systems and maintenance personal on the ability to diagnose, and

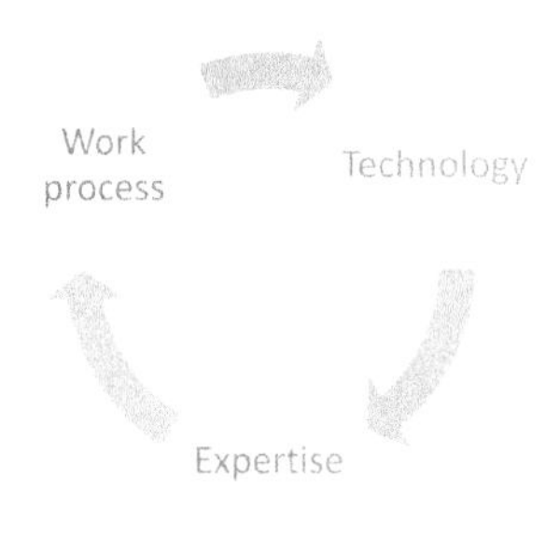

FIGURE 16.1

Elements of asset management system

proper and defined work process that can constantly evolve. In order to achieve the above goals, an asset management system requires a management system with clear accountability of the actions, ability to measure the performance of the assets and personnel managing the assets, ability to create and present the report for the management decisions, and ability to perform audit on the actions.

The plant asset management is a combination of hardware, software, and services used to

1. Monitor: To access the health of plant assets by monitoring asset condition periodically or in real time
2. Predict: To identify potential problems before they affect the process or lead to a catastrophic failure.
3. Analyze: The severity of problems, potential causes, and possible operator actions can be provided to help operators determine the necessary actions.
4. Prevent: Provides solutions to protect equipment against catastrophic damage in case of abnormal conditions.

Different types of assets of a plant at different levels are depicted in Figure 16.2. Within a plant context, there are various departments that are managed based on the organization's design of the individual plant.

At the bottom of the pyramid there are more assets, but of relatively less value. These assets can impact the production, but not complete shutdown. These assets are typically managed from the individual maintenance departments and there are multiple asset management systems available for the same. From an instrumentation and control perspective these were managed by instrument asset management systems (IAMS), which are again integrated with the plant control systems. The upper layer is meant fir for managing pumps, motors, and rotating machinery wherein the number of such equipment might be less compared to the lower level, but the cost of such equipment is high and

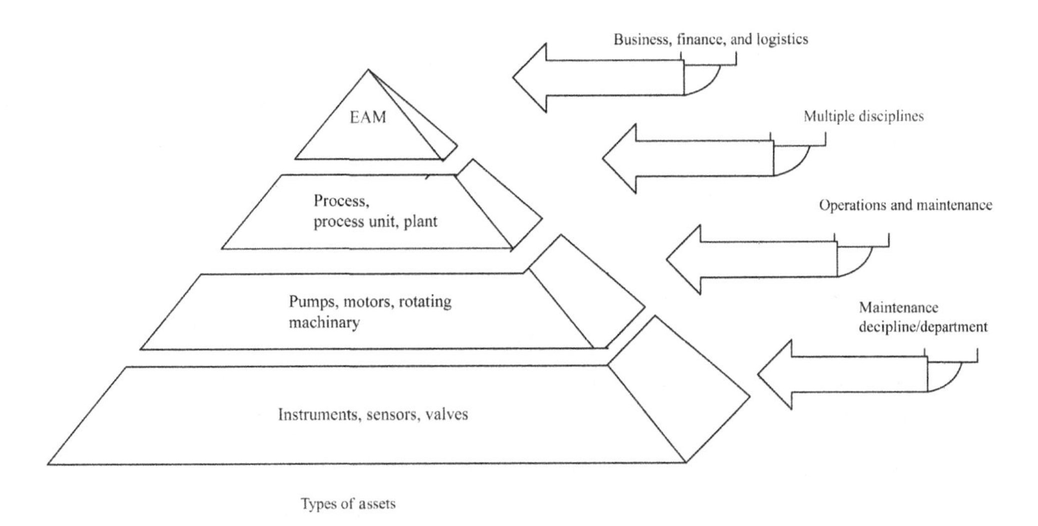

FIGURE 16.2

Elements of asset management system

failure of such equipment has an impact on the production. These equipment were typically managed through the plant asset management system wherein the knowledge of the engineers is translated to fault models to derive the issues and diagnose the problems. The knowledge base helps the operations and maintenance personnel to predict the problems with the equipment early and perform maintenance. The layers above comprise plant equipment and process units, which are small in number and maintenance issues in these equipment has not only production issues, but also in the business operations. For example, boiler, distillation columns, and so on are considered here as process units and maintenance of such equipment is managed through multiple personal from multiple disciplines. The process units are managed through a combination of asset management systems such as instrument asset management, electric asset management systems, and plant asset management systems. At the top of the pyramid, enterprise asset management systems are provided in most of the modern plants. From plant operations and maintenance perspective, the enterprise asset management systems are available, which helps to implement the organizational maintenance strategies, purchase philosophy, and auditing and cost control. The asset management systems, in a nutshell, improve the asset effectiveness. The asset effectiveness is the process of managing the health and performance of an asset. It helps to

- Detect asset issues: an ability to detect the malfunctioning in the device after it is occurred or ability to predict the issues before they actually occur.
- Notify appropriate personnel: ability to inform the appropriate personal on the issue immediately.
- Provide methods to predict and prevent asset issues.
- Help capture and retain of knowledge. The system can provide a custom tool that takes in the inputs on the rich knowledge and becomes a repository of the knowledge.
- Help continuously improve and optimize based on continuous learning on the assets and troubleshooting methodologies.
- Provide information on maintenance performance indicators (KPIs for Maintenance).

In order to achieve the above goals and expectations, the asset management systems are provided with information in the form of models, which act as blue print for the system while detecting the current and future errors. The system will be fed with the fault models, fault symptom models, and tools to quickly locate the documentation and work processes. The following are some of the models required by the asset management systems for an effective functionality.

16.4 FAULT MODELS

A model-based approach is widely used across many manufacturing plants to reduce the amount of energy and time spent on the low-performing equipment. A fault model is a model that discovers how the system handles some common failures that contribute to unsafe situations. Fault models are more abstract than the models of normal behavior. They are easy to construct, comprehend, and customize. They can easily capture interactions between the root cause (faults) and their effects (symptoms) without requiring a detail model to simulate the normal behavior.

In general the models below are supported by asset management system:

- Fault–symptom model
- Calculation model

16.4.1 FAULT–SYMPTOM MODEL

A fault–symptom model is a model used to identify the faults (failure) and the symptoms in an asset. They are generally defined at the type level. The asset management system is a decision–support system, where the problems are identified and repaired by tracking the asset failures and symptoms. Various reliability tools such as RCA, FMEA, and so on are used as inputs for identifying the faults and symptoms.

16.4.2 FAULT TRANSITION STATES

Fault is a state or condition at which a functional failure is identified. These are defined with a conditional expression on the asset's attribute. Faults can move to the following states:

- Faults standing are faults whose state is latched. The state of the fault will not change irrespective of the fault or symptom condition.
- Faults confirmed are faults whose state is satisfied.
- Faults potential are faults whose state is not satisfied but one or more symptom conditions are satisfied.
- Faults eliminated are faults generated that do not agree with the system fault and are forced to be eliminated.

16.5 CALCULATION MODEL

A calculation model in asset management system is a model that defines calculations or logics on the asset's attribute values. The attribute values are updated based on the calculation or logic defined in the application. Asset management systems supports various operators to define the calculation model. This is not limited to physical entity but is also extended to logical entities.

16.5.1 INSTRUMENT ASSET MANAGEMENT SYSTEMS

From an instrumentation and control perspective, all the equipment such as instruments and control equipment can be considered as instrument assets, which need to be maintained and managed to derive the value for the planned duration/life cycle of the equipment or instrument. The equipment such as instruments need to be maintained for proper operation due to the contribution/criticality of the equipment while deriving the value to the production as a whole.

As the ability to self–diagnose device health and integrity improves, available information is too valuable to ignore. For example, standard temperature measurement options offering hot backup redundancy are being expanded into detecting sensor drift and predicting when a temperature sensor will fail. Pressure transmitters now detect plugged impulse lines and inform the operator when an apparently good measurement is, in fact, not valid. Control valve diagnostics and the ability to execute partial valve stroke test or generate valve signatures for online diagnostics allow many valve problems to be easily isolated and remedied without the cost associated with pulling a valve out of service and unnecessarily rebuilding it. All of these developments in device diagnostics help processing facilities practice more preventive and less reactive maintenance. With approximately 50% of the work accomplished in most organizations being reasonably preventable, maintenance, potential cost, and savings from utilizing device diagnostics data are tremendous.

In process control industry, instrument assets range from field devices to hardware/software-based control solutions. In order to maximize the output of an industrial facility, all instrument assets need to be maintained at certain intervals – that is monitored, serviced, refurbished, or replaced. Instrument asset management assists in determining these intervals through continual asset condition monitoring, which predicts time–to–service, detailed diagnostics with guidance of required service actions, and system–supported planning and execution of service tasks. The goal of an instrument asset management solution is proactive rather than reactive maintenance wherever possible.

The last decade has witnessed a lot of improvements in the process industry including large-scale standardization of various integration methods and engineering aspects. Despite the improvements, expectations (demand) from the industry keep increasing, primarily due to various economic situations around the globe. Primary and everlasting expectation is high productivity with relatively less cost. The cost involved for integration of a new plant solution is always seen as critical. Many plants utilize a variety of technologies for different applications, including industrial communication protocols such as Foundation Fieldbus, PROFIBUS, and HART. However, diagnostic information is often represented amongst these networks in different ways. This can include different data structures, different parameter names, and so on. Even within the same protocol, there are areas where each instrument can add additional diagnostic information that is presented in many different formats. Users' need for a common set of instrument asset management tools ensuring important information regarding device status and operating condition gets to the appropriate person within the plant. In turn, the organization proposed a common structure for representing all instrument diagnostics. Plant operators should only see status signals, with detailed information viewable by device specialists. The NAMUR guidelines further recommend categorizing internal diagnostics into four standard status signals, and stipulate configuration should be free, as reactions to a fault in the device may be very different depending upon the user's requirements.

Typically, a plant has field devices and systems from various vendors supporting multiple communication protocols, where each protocol has its own set of parameters for configuration. The users might not be educated or trained enough to work and maximize the value of the information contained within different set of protocol devices. Instrument asset management systems might have been developed even before some of the protocols came into existence and therefore might not even understand niche diagnostic application of the new protocol device. The cost of managing equipment from different device and/or system vendors, inefficiently, leads to very less or no return on the investment.

Instrument asset management systems attempt to address such integration issues resulting in cost saving and effective maintenance of devices in supporting various protocols. Some of the protocols could be vendor specific and not open standard, within a control system, and come with asset management tools with an interface that are easy to deploy and supported by virtually all automation suppliers.

Preventing unplanned shutdowns, reducing downtime, and lowering maintenance costs provide significant financial benefits. Using installed instrument assets to the best of their ability is one way of achieving this. Asset management systems can bring significant operational and financial benefits throughout the plant life cycle and they can be easily used in existing or new plants.

Instrument asset management systems are included as part of a proposal or ordering specifications to ensure complete utilization of assets. Competition for investments comes from both external and internal entities in a global economy. Internal competition leads to budgets being allocated among activities such as: regulatory compliance, safety, quality improvements, energy conservation, and other sustainable initiatives. With capital investments on control systems or process upgrades hard to come by, users are forced to look within their existing operations to optimize operations to positively impact the profitability of the business. Preventing unplanned shutdowns, reducing downtime, and lowering maintenance costs have been shown to provide significant financial and operational benefits. One way to achieve these results is by ensuring that the utilization of all installed instrument assets is maximized to the best of their ability. This is particularly true with regard to smart measurement and control instrumentation. If suppliers and end users can connect instrument asset performance to corporate goals, such as improving compliance, plant and workforce safety, sustainability, and other critical issues, while increasing plant availability, it would secure the interest of capital investments. It is also a good idea to move fixed cost to a variable cost using available technology. For more than 10 years, the majority of instrumentation purchased and installed around the world is smart – meaning that the devices are intelligent and can provide information beyond just the PV (process variable). Likewise, control systems (DCS, PLC, and others) have also become more intelligent, being able to access and use the information from these smart field devices.

Lately, asset management features are available in majority of the distributed control systems (DCS) and smart field devices – often without being utilized by the user. The various standards in the asset management provide standard means to easily access intelligent device information independent of the field communication protocol or control system supplier. Many companies consider the supplier and protocol neutral technologies becoming a universal, life cycle management tool, supported by device and automation system suppliers around the globe. This level of adoption goes beyond international or regional standardization because it is quickly becoming a de facto device standard in both the process and factory automation industries.

The greatest advantage of asset management is that many facilities can start taking advantage of this technology with little investment and very low risk. IAMs are available to help users maximize device intelligence by providing actionable information. Automation professionals are already inundated with data but DTMs provide much more than just data. It displays valuable information that is relevant and timely, improving asset performance.

Instrument asset management system is a vendor-independent application for the configuration, parameterization, commissioning, diagnostics, and maintenance of smart field devices (sensors and actuators) and field components (remote I/Os, multiplexers, control room devices, compact controllers), which in the following sections will be referred to simply as devices. Usually, parameters and functions for all supported devices are displayed in a consistent and uniform fashion independent of their communications interface.

The primary benefits of asset management strategy are increased asset availability and performance and maximized operations and maintenance effectiveness. In the past, asset management tended to focus on expected service life, then on diagnosing failures, and then on predicting and managing maintenance issues. In the current state of affairs, asset management is also finding system bottlenecks and identifying ways to increase system capability.

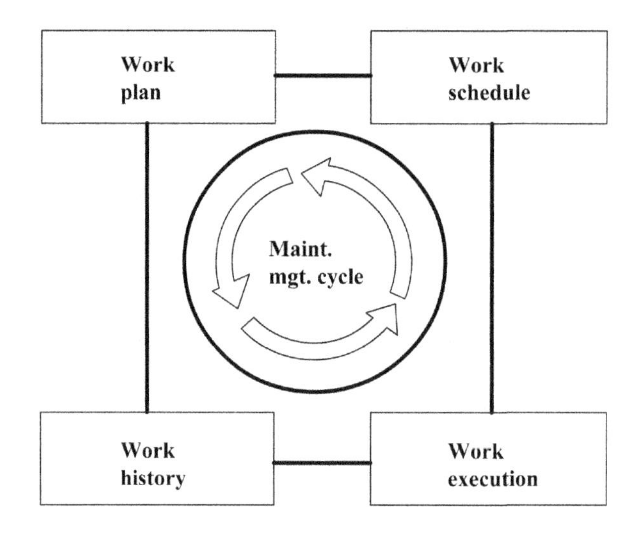

FIGURE 16.3

Maintenance work process

16.6 MAINTAINING WORK PROCESSES

The instrument asset management is a cycle as shown in Figure 16.3, where a work plan is created for the instrument assets. The plan becomes a basis for the schedule, and the schedule becomes the basis for the execution. The execution then becomes the basis for the history of the plant instrument assets and the history of these instruments becomes basis for the plan for the future planning purposes. Based on the history and the plan, the instrument maintenance is further categorized as follows and is coupled with the technology and people.

As discussed earlier, technology, work process, and knowledge are the key elements of the instrument asset management. The maintenance activities of instruments can be categorized as:

- Run to failure: The case in which there is no special maintenance needed and instrument is run until it fails and is replaced with a new item. Generally some items are categorized in this based on the criticality, risk, and cost and time to replace.
- Preventive (scheduled) maintenance: This type of maintenance is required wherein the instrument is maintained at a scheduled interval as defined by the vendor or the plant maintenance strategy or instrument previous history. The schedules for the maintenance of such equipment are predefined. The results are tracked to record the issues, the type of maintenance performed, and next maintenance date.
- Predictive maintenance: Predictive maintenance is an intelligent way to perform the maintenance of the device considering the production schedules, based on the risk and failure history of the instrument type, and sometimes includes some bad actors and root cause analysis of the previous failures. The predictive maintenance has gained lot of acceptance as a practice due to the considerations for the maintenance for less breakdown and judicious strategy.

- Condition-based maintenance: Performing the maintenance of the instrument can be based on the current condition with the predictive diagnostics available from the smart instruments or from the models created using the previous knowledge.
  - Based on real-time diagnostics data – from the various sources of the instrument and also from the instruments in the upstream and downstream of the equipment. The real time plot of the data, alarms, and events in a time series provides valuable insights to an experienced operator on the health of instruments and can be called for maintenance.
  - Multivariable model based equipment/process, which is based on the model created on the functionality of the process equipment. The model reads various parameters on real time and derives the errors using the model. The models can be developed with the help of the equipment manufacturer or based on the in-house knowledge developed over a period of time.
  - Smart device based instrument/valves, which is the current technology of instruments. The current-day technologies provide various diagnostics parameters on the part of instrument. These were the parameters that were not accessible in the early instruments and there were many interoperability issues to access the same. With the advent of standards-based technologies, the parameters are accessible to a wide variety of systems such as DCS, Handhelds, and IAMS.

From the descriptions above, it is clear that instrument assets need maintenance with a clear strategy. Majority of the times there might be no issue in the plant or the device, but the routine maintenance needs to be conducted. It is all about having a risk in a controlled manner. Some studies performed by the industry indicate the trips made by the technicians to the plant or near the instrument may not always conclude the failed instruments. Majority of the times, roughly 63% of the times, there will be either routine checks or no problem. If the instrument asset management systems help to reduce these efforts, then the human resources can be directed for places where there is an issue. These are some of the statistics that justify the value proposition provided by the instrument asset management systems to the end users.

There are standard points of view, and standards such as NAMUR that standardize the way alarms or diagnostics are provided to the users. This way of standardization helps the users to configure the alarms or diagnostics in a specific format and hence focus on key items. NAMUR standard NE107 proposes diagnostic signals/categories that are identified as follows:

Maintenance required: Although the device is still able to provide a valid output signal, it is about to lose functionality or capability due to some external operational condition. Maintenance can be needed short-term or mid-term.

Failure: The instrument provides a non-valid output signal due to a malfunction at the device level.

Check function: The device is temporarily non-valid due to some type of maintenance activity.

Off specification: The device operates out of the specified measurement range. Diagnostics indicate a drift in the measurement, internal problems in the device, or the consequence of some process influence (i.e., cavitations, empty pipe, etc.).

NE107 further recommends the classification and association of a diagnostic event to one of these four levels of diagnosis be configurable by the user. The configuration would depend on the process constraints (e.g., loop criticality) and the role of the addressee, such as an operator, maintenance technician, and so on.

16.7 UNNEEDED TRIPS TO THE FIELD – AVOIDED THROUGH REMOTE DIAGNOSTICS

More than 60% of the time is spent in investigating problems that do not exist. Diagnostics will tell this so it will no longer be necessary to check these reports. Similarly, the sheer volume of device configuration and initial calibration tasks to be completed on a capital project often represents savings potential sufficient to justify the relatively modest incremental costs of smart devices and asset management system. The configuration of every transmitter and valve is checked multiple times before startup, and calibration is verified at least once. These necessary activities can be accomplished more quickly and easily and with fewer errors from a workstation with asset management system than by physically visiting and revisiting each device in the plant, pushing buttons to navigate sometimes extensive menu structures in cryptic text on small local displays. Additionally, all commissioning activities including calibrations and all subsequent changes made through asset management software can be automatically documented without the extra labor and risk of errors associated with manual reporting (Figure 16.4).

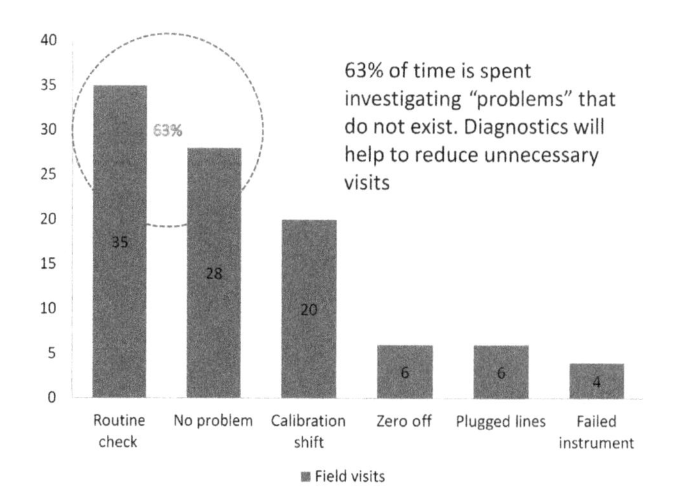

FIGURE 16.4

Field trips

Source: Dow Chemical Company.

16.8 LIFE CYCLE WORK PROCESSES

Instrument asset management systems (IAMS) are required or used in various life cycles of the control systems design. The life cycle activities contribute to the design and configuration of the system from the early life cycle. The following are some of the critical activities performed in the life cycle.

- Front end engineering design: During the initial phases of the project such as FEED, the critical ranking of the equipment in terms of the risk and category of the equipment is performed. Ranking of the instruments will help to prioritize the importance of the equipment for the safe and efficient operations of the plant. The studies performed in these areas indicate that only 5% of the instruments become the critical instruments and rest are in different categories such as essential, important, secondary, and nonessential. Figure 16.5 represents the criticality of the instruments in the form of a pyramid. Subsequently templating the devices with offline parameters and configuration is performed. Sometimes small prototype tests are performed with the devices/instruments selected and critical use cases are tested.
- Design/Factory Acceptance Test (FAT): During the design and FAT, the initial configuration of the instruments is performed. This includes tag names, description of the service, upper and lower range limits, and ranges. For some smart instruments, the alarm configuration and other customization are performed. During the FAT, a representative device is connected to the system and tested. The training is performed at this stage. The training can include inclusion of new devices in the plant, removal, configuration and system maintenance, as well as reporting.

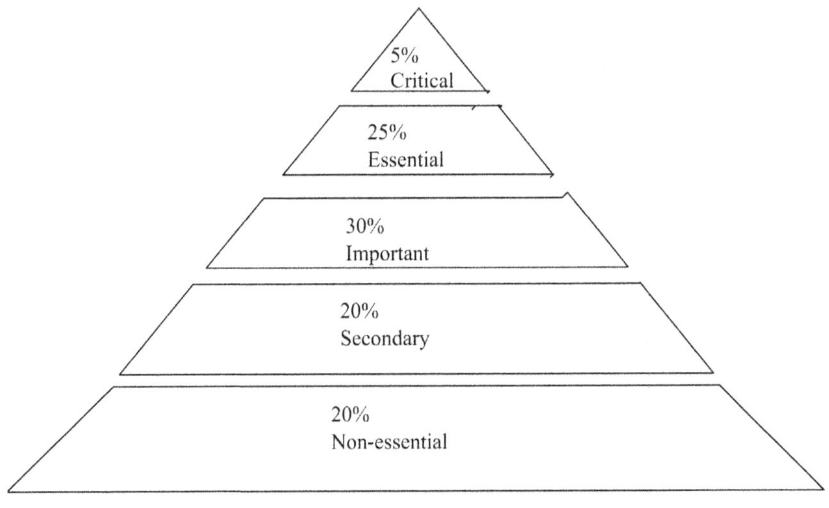

FIGURE 16.5

Instrument criticality pyramid

- Install, commissioning, loop check: The IAMS is used in the installation, commissioning, and loop checks. At this phase of the life cycle IAMS provides good yield in terms of the productivity and cycle time. The initial use of diagnostics can be seen with the loop check, alarm conditions, and other diagnostics.
- Pre startup safety review: The IAMS along with the intelligent devices helps in reducing the cycle time of the pre startup and safety reviews. The field instruments can be simulated to perform certain process conditions while connected to the process and review the safety control performed upon the receipt of such events.
- Routine maintenance: The routine maintenance schedules and calibration schedules and results can be stored in the IAMS database and also many systems offer synchronization of such information from the calibration management systems. The routine documentation of the maintenance can be stored for the future reference and avoid the traditional paper-based work process. The content and the documentation available in the system can be used and accessible for all the software applications above for management information systems.
- Turnaround management: The IAMS helps to measure and improve the turnaround time measurement and hence gives opportunity to improve upon the same for the future. The data from the IAMS becomes the basis for creating the benchmark and improvement actions.

Similar to the maintenance life cycle, operations life cycle is another aspect in which the instrument asset management plays a key role. Figure 16.6 represents the operations life cycle in which there exists operations plan for the production and is the source for deriving the operations schedules. The operations schedules become the basis or the source for the operations of the production. The operations lead to the production history and the same becomes the basis of the operations plans for

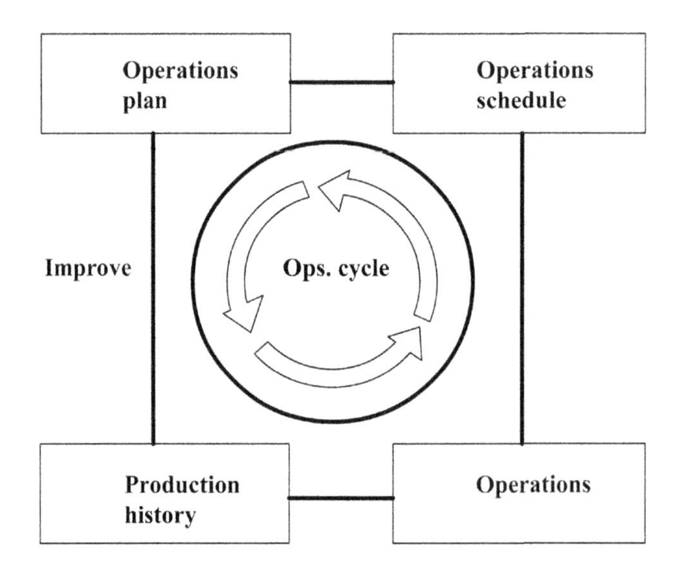

FIGURE 16.6

Operations work process

the subsequent plans with improvisations. The asset management systems can derive the operations plans for the periodic schedules to perform the maintenance without disturbing the operations, and also the production history can provide key insights to the loss of operations and hence to the production due to the failure of the instruments. These data become the basis for the benchmarking for the future improvisations in the maintenance strategy and to the purchase strategy for the instruments or the equipment as well.

16.9 INTELLIGENT FIELD DEVICES – DATA FLOW

Today, the field device revolution is centered on reducing process variable uncertainty and enhancing device functionality and diagnostics while providing more integrated solutions around the desired process measurement. Over the years, plant constructors and operators have consistently pursued two main goals: to lower installation costs and to optimize production conditions. This has led to the widespread use of digital bus technologies for field devices communication in the process and manufacturing industries, as well as development of intelligent automation devices. But, in many cases, the savings potential from fieldbus wiring reductions and digital device communications have already been exhausted. The digital device is becoming a rich source of information and hence eases the maintenance (Figure 16.7).

Intelligent maintenance concepts, on the other hand, still offer tremendous potential for added value. In addition to increasing plant availability, diagnostics-driven maintenance strategies reduce fixed and variable maintenance costs and extend useful asset life by reducing the interval between maintenance events, reducing the cost of failures, and making it easier to plan maintenance and service work. With the advance of intelligent automation components, an extensive amount of data is being generated on all levels of the automation hierarchy and, increasingly, in the field devices themselves (see Figure 16.8). Many of these components provide parameterization options, and some include diagnostic and analytic functions. If modern methods of preventive or condition-based maintenance are to gain more widespread acceptance, the available information must be centrally collated, evaluated, and given to maintenance providers in an understandable form. Asset or lifecycle management systems can only be used consistently and effectively if there is easy access to parameterization, status, and diagnostic data from the field.

Condition-based monitoring (CBM) focuses on optimizing the timing of maintenance. It seeks to avoid unexpected equipment failures on the one hand (too late maintenance) and unnecessary maintenance on the other (too early maintenance). To achieve this goal, individual assets require either embedded intelligence or specific condition monitoring techniques at a higher level. Due to demands for increased availability and uptime, various techniques for monitoring plant assets have been developed. These include using control equipment for monitoring field devices (electro pneumatic positioners monitoring control valves, electrical drives monitoring conveyors, etc.); installing specialized sensors, measuring and diagnostic equipment; and process modeling at a higher system level. Many process industry end users configure their asset management software for predictive maintenance. By using the diagnostic features built into intelligent instrumentation, they track indications of impending failure. Some applications can produce a notice resulting in a work order when repair is required. This is significantly more economical than crisis repair, which is waiting for the failure and repairing the device on an emergency basis.

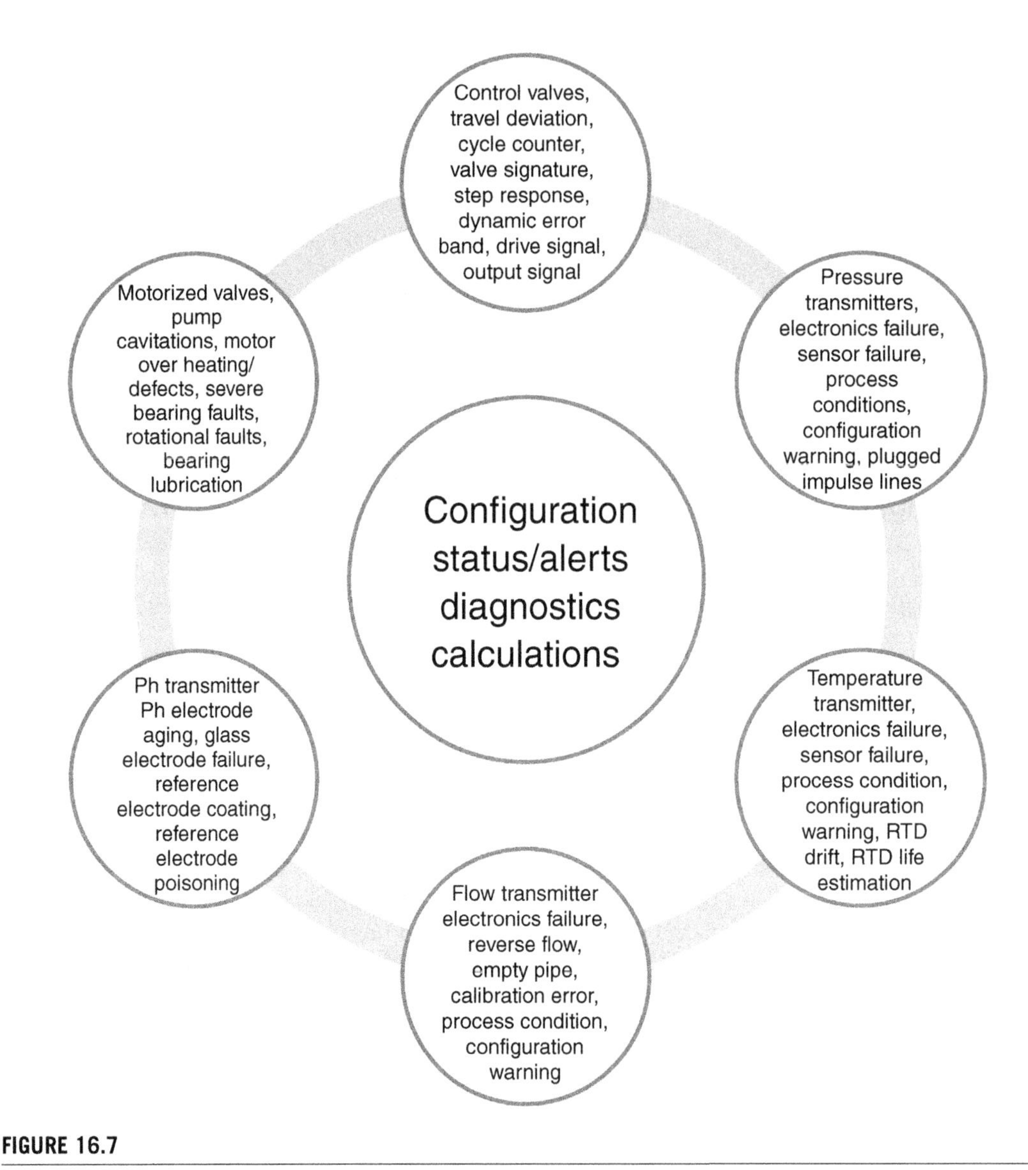

FIGURE 16.7

Asset management functions

16.10 INTEGRATED ASSET MANAGEMENT

An integrated asset management is a combination of systems and tools that drives the asset strategy of the plant for maintenance excellence and operations excellence by monitoring instruments, equipment, and process through various means. The means of the monitoring and diagnostics are augmented by methods such as model-based techniques or intelligent devices and using various standards-based technologies such as DD/EDDL or FDT/DTM or FDI.

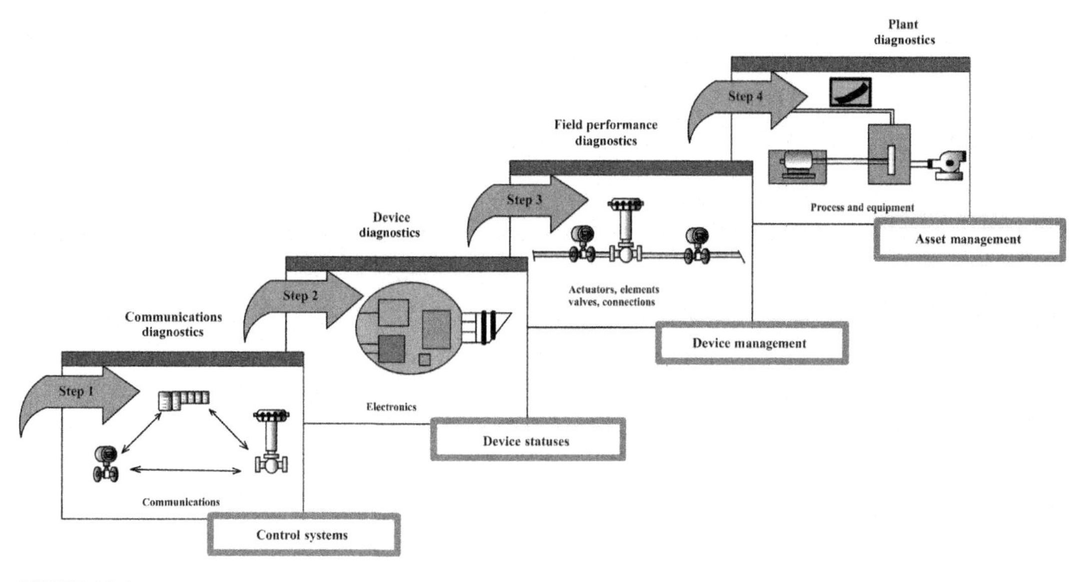

FIGURE 16.8

Data flow in an automation hierarchy.

There are various systems and tools used to achieve the above objectives and some of the tools used in this context are listed below. Tools provide functionality such as:

- Resource planning, tracking, accounting
- Design, calculations, drawing, documentation
- HOSTS:
    - Integrated hosts (DCS)
    - Visitors or bench hosts (handhelds or laptop)
    - Provide configuration management, troubleshooting
    - Human interface

Tools have database that is needed to exchange data (import, export, reformat, reconciliation, change tracking, history, remote access, role-based security, simultaneous access, data comparison).

16.11 USE OF THE TOOLS

The use of the asset management provides the following benefits.

- Most diagnostics go directly to black hole. In the traditional system if not handled properly.
- Diagnostics may need to go to engineer, maintenance, and/or operations.
- Use of priority of diagnostics is determined by applications criticality.
    - Criticality ranking.
    - Alert priority depends on impact severity.
    - Send alerts to operator as well as maintenance as required.

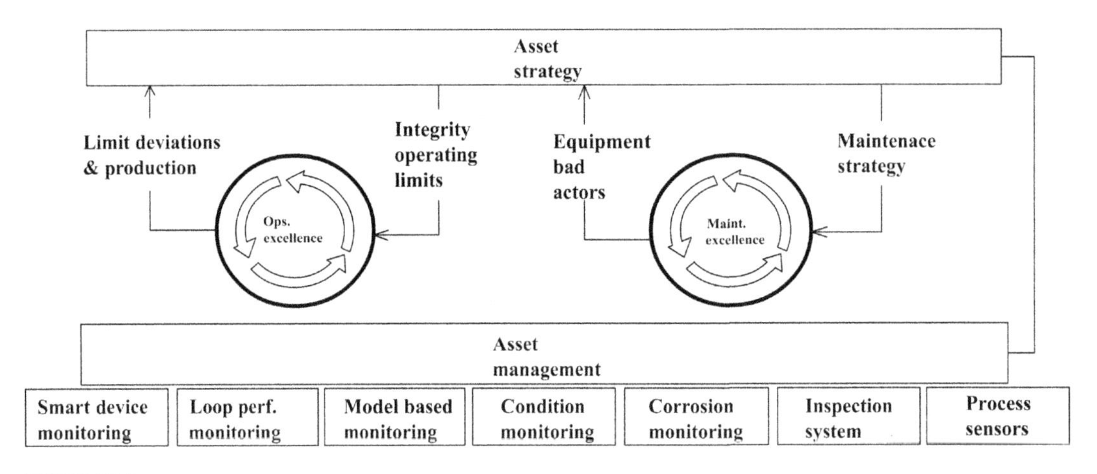

FIGURE 16.9

Asset management strategy

Proposed use of diagnostics depended on GUI.

- Operator view in tree structure.
- Overview to detail in three mouse clicks.
- Cover all system nodes, networks, and modules (Figure 16.9).

16.12 INSTRUMENT ASSET MANAGEMENT SYSTEMS – ARCHITECTURE/SUBSYSTEMS

Figure 16.10 represents general components of typical instrument asset management systems. All the systems constitute a central server for the collection and storage of the data and clients to render the information to the users. The systems are typically supported with various communication technologies either integrated with the DCS or using some gateways.

16.12.1 ASSET MANAGEMENT SYSTEM – SERVER

- Asset management server maintains the record for all device information such as device type, manufacturer, device history, offline configurations, audit trail information, calibration status, and service status.
- Implements centralized management of device information such as the device list, inspection record and schedule, user-created electronic documents, and parts lists.
- This function can generate maintenance alarms for necessary information (phenomena, causes, action) and distribute this maintenance alarm information to personnel who want it.
- Supports different networks such as DCS, safety systems, hardware multiplexers, modem, and FDT networks.
- SAP/ERP interface with auto notification.

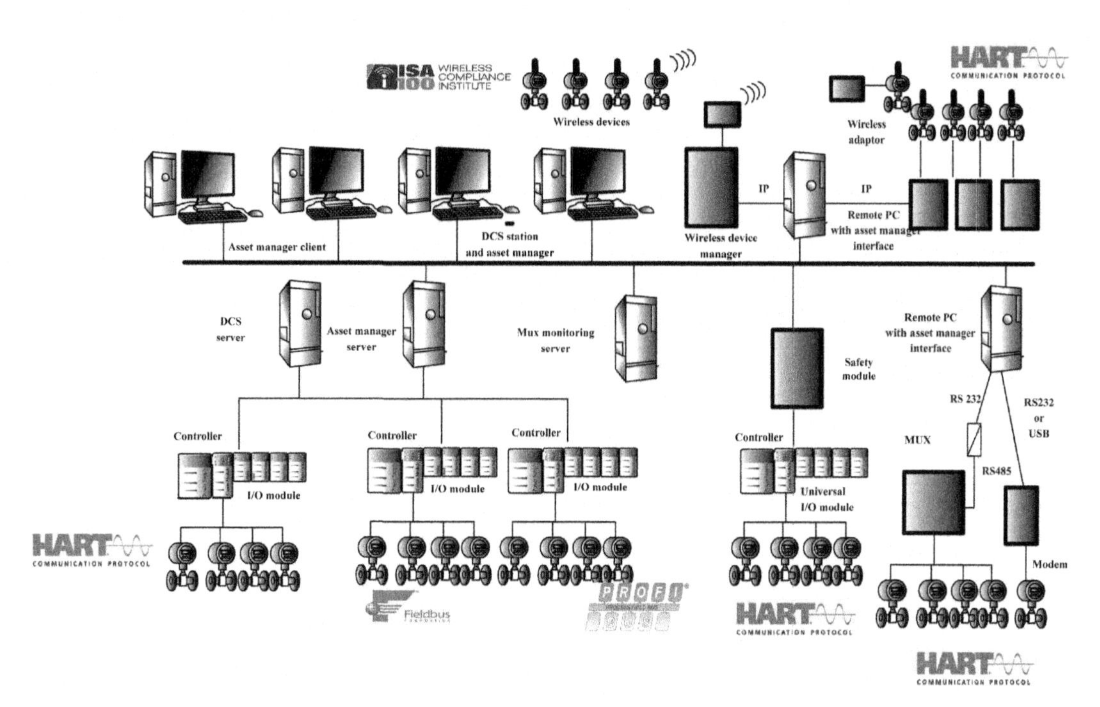

FIGURE 16.10

Architecture of a asset management system

16.12.2 ASSET MANAGEMENT SYSTEM – CLIENT

- User-friendly windows-based operating environment, with explorer-like "Navigator" for selecting a specific device.
- Client provides the user interface and allows for remote connection to the asset management server.
- Performs automatic device recognition, monitors device health status with maintenance alarm functions, monitors communication error, and monitors advanced diagnosis of device, process interface diagnosis, and unit diagnosis.
- A maintenance mark can be assigned to devices for setting and checking maintenance status. A maintenance mark is displayed in device icon; it gives permission for devices to change access level temporarily. It enables to share the situation of operation and maintenance between operators and maintenance personnel.
- The status display icon appears beside the device icon changes depending on communication and device status. This provides an easy–to–understand overview of device health status.
- In addition to maintenance alarm function and standard diagnostic functions, asset management provides add-on components such as advanced diagnosis package especially for valve diagnostics.
- User is able to configure and calibrate the devices supporting different communication protocols such as HART, FF, and PROFIBUS sitting in Control\Calibration room.
- Today's asset management system supports latest technologies such as FDT/DTM (field device tool/device type manager) and EDDL (electronic device description language). These

technologies provide not only more advanced diagnostics but also rich user experience through various graphics.

- Maintains records of audit trials for user actions.

16.13 SMART FIELD DEVICES

Smart field devices provide critical information on their own health, the health of the process, and the health of the equipment around them. This diagnostic functionality is enabled by digital communications such as HART and FOUNDATION Fieldbus protocols. Examples of valuable diagnostic capabilities that exist in some devices today include:

- Control valves: plugging of I/P, travel deviation, insufficient air supply, stuck valve, calibration changes, failed diaphragm, diaphragm leaks, o-ring failures in piston actuator.
- Pressure: electronics failure, sensor failure, process condition, configuration warning, plugged impulse lines.
- Temperature: electronics failure, sensor failure, process condition, configuration warning, RTD drift, RTD life estimation, water ingress, terminal rusting.
- Flow: electronics failure, high process noise, grounding fault, electrode fault, empty pipe, reverse flow, calibration error.
- Analytical: pH electrode aging, glass electrode failure, reference electrode failure, reference electrode coating, reference electrode poisoning.

Devices featuring such diagnostics can be thought of as "data servers" because they do much more than simply measure process variables such as pH, pressure, temperature, level, or flow. These devices provide the basis for abnormal situation management, process optimization, and advanced control. The full value of this capability is only realized when the smart field devices are utilized in combination with asset management system.

16.14 ASSET MANAGEMENT SYSTEM: ROLE-BASED DIAGNOSTICS

Diagnostics are specific to the role of the user. For example, the same diagnostics are used by individual users differently in different context. From a user point of view, the typical roles for the asset management systems are engineers, operators, maintenance technicians, electrical operators, and so on. The asset management systems provide a mechanism to customize the type of the alarms, alerts, and diagnostics. The role and job description are provided below against each role (Figure 16.11).

16.14.1 ENGINEERS

- Configuring and setup of devices during FAT
- Device commissioning during startup
- Assisting maintenance with troubleshooting
- Assisting operations with problem solving

ROLE-BASED DIAGNOSTICS

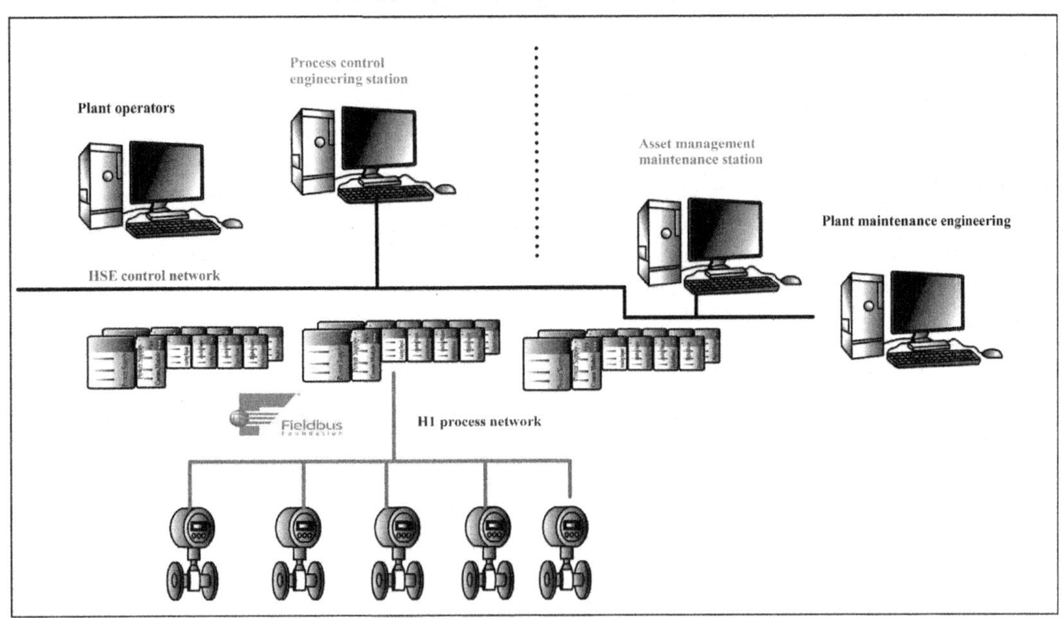

FIGURE 16.11

Role based asset management system

16.14.2 OPERATORS

- Device diagnostics/status checks
- Device information reference for running the process

16.14.3 MAINTENANCE TECHNICIANS

- Performing routine maintenance (especially valves)
- Checking device status (alarms) and troubleshooting
- Device database upload (history) for archive/reference
- Device replacement and needed setup/reconfiguration

16.14.4 MANAGERS

- Create the reports
- Provides the access to the new users/roles

16.14.5 ELECTRICAL ENGINEERS

- Configures the electrical assets
- Receives the alerts

- Schedules the maintenance activities
- Generates the reports

16.15 DEVICE RENDERING TECHNOLOGIES

Information rendering in the IAMS from a rich source of instruments is becoming a key factor for the successful deployment of the solutions. The standards driven by various consortiums are working toward making these technologies more mature over the years. In order to increase the openness of the devices from multiple vendors to common systems and also to standardize the parameters of each device, there are various groups formed to create the standards and technologies. The major technologies used in this context are (1) device description/ electronic device description languages (DD/EDDLs) and (2) field device technology/device type manager (FDT/DTM). The general use case of the user in the context of these new technologies is that the device shall be provided with the DD/EDDL or DTM while purchasing the device in an electronic format. The end user or the engineering team can import these electronic files into the control systems or IASM and then the device works as intended by the vendor. The electronic files are subjected to the versioning and upgrading, and systems are capable to allow the users to upgrade the systems with the new electronic files. These files provide the device interface with rich user interface and wealth of diagnostics information.

The primary purpose of this rendering technologies are as follows.

- Device diagnostics

Device diagnostics can be carried out using a handheld communicator in the field, a laptop in the workshop, or from intelligent device management software as part of asset management solution, either from a dedicated maintenance console or integrated in the operator console. Various device-rendering technologies such as DD/EDDL are used by device manufacturers to control how the device diagnostics is displayed to the technician. These technologies make diagnostics of intelligent devices easier owing to user guidance such as wizards and help, and provide unparalleled consistency of use.

The diagnostics is performed by the device itself, that is, self-diagnostics or self-tests. HART, FOUNDATION fieldbus, PROFIBUS, ISA100, or Wireless HART protocols are used to communicate the result of the self-diagnostics to the handheld communicator or intelligent instrument management software. Each protocol has a different mechanism for reporting diagnostics and therefore perform differently. In a vast majority of cases diagnostics is embedded in the device itself as firmware so that the device is monitored continuously.

- Device integration

Plants have a mix of simple and sophisticated (complex) devices from different manufacturers using different communication protocols. The diagnostics available depends on the type of device:

- Pressure transmitter: sensor failure, plugged impulse line, sensor temperature limits exceeded.
- Temperature transmitter: sensor failure, thermocouple degradation.
- Valve positioner: partial stroke testing (PST), travel histogram, step-response, valve signature, pressure sensor failure, abnormal drive current, travel deviation, reversal count alert, accumulated

travel alert, I/P converter plugged, pneumatic relay leak, valve stuck, actuator leak, low supply pressure, high supply pressure, position feedback sensor failure.

- Flow meter (Coriolis, magnetic, ultrasonic): flow tube stiffness, magnetic field strength, coil resistance, electrode resistance, grounding, noise, empty pipe, flow profile, turbulence, speed of sound.
- pH analyzer: glass electrode impedance, reference electrode impedance.

The diagnostics available vary greatly from one manufacturer to the other. Original device diagnostics (DD) technology from 1992 made it possible to troubleshoot all devices using the same handheld field communicator or laptop software. Only proprietary solutions existed before DD. The original DD technology from 1992 already included "wizards" (aka "methods"), which is a kind of script created by the device manufacturer to guide the technician through more sophisticated test procedures, for instance valve step response test requiring the steps to be predefined and lastly a prompt validation sequence before the self-test actually starts. Wizards thus make diagnostics and troubleshooting easy. Help and conditionals were also part of original DD technology. However, not all devices provided wizards in their DD file and not all intelligent device management software supported wizards. That is, on many systems and for many types of devices, advanced diagnostics in the past was not so easy. However, for some devices, the manufacturer may choose to display test results graphically, such as a valve signature, vibration spectrum, or signal waveform. This was not supported by original DD technology. Therefore, DD did not support the diagnostics of some devices. This includes graphics, menu system, wizards, and conditionals.

- Device troubleshooting made easy

Rendering technologies provides a vast array of possibilities for the device manufacturer to organize the diagnostics display and make the device user-friendly, and also to add graphics such as waveforms for valve signatures and vibration spectrum. Depending on the type of device, the manufacturer uses different methods.

16.15.1 DEVICE DESCRIPTION (DD)

Device description (DD) is an electronic file input prepared complying with device description language (DDL) specifications that describes specific features and functions of a device including details of menus and graphic display features to be used by host applications (including handheld devices) to access all parameters and data in the corresponding device. The DD is written in a standardized programming language known as DDL. A PC-based tool called the "Tokenizer" converts DD source input files into DD output files by replacing keywords and standard strings in the source file with fixed tokens.

16.15.2 DEVICE DESCRIPTION LANGUAGE (DDL)

DDL is a simple structured English language for describing field devices. The DDL helps in defining all the information a host system needs to operate with field devices. It presents the information as a clear, unambiguous, consistent description of field device.

The DDL describes the meaning, or semantics, of user data. In its most basic form, the DDL source is human readable text written by device developers to specify all the information available at the network interface to their device. Once a device description is written, it can be transformed into a compact binary format. Upon arriving at the host application, this binary formatted device description can be decoded and interpreted to provide a complete and totally accurate human text interface to the connected device.

The first step for DD developers is to describe their devices using DDL source language. This language is used to describe the core set of parameter definitions and user group and vendor-specific definitions. Each device function, parameter, or special feature can be described. The resulting DD source file contains an application description of all of the device information accessible. The full benefits of DDL are realized when the host system supports a device description services (DDS) library to decode and interpret the binary format of the data. The binary specifies the parameters contained in the device, and also how the parameters should be presented to a user on a display. Examples of this include how to organize a menu, which text strings should be displayed with an alarm event, what are the steps to take in the calibration procedure, and so on. By using DDS, host system generates displays based on the information contained in the DD binary. The host system can fully and accurately configure a new type of device without any software upgrades.

16.15.2.1 Benefits of DD

The key benefits of using DD for the host integration with the field device are:

- The existing tightly coupled relationship between the release of new field devices and the host applications no longer exists.
- Field device development schedules are not tied to host development or revision schedules.
- The DD can be easily incorporated into a configuration or control system.
- Device suppliers have maximum flexibility in introducing their product and upgrading their installed customer base. The DD binary files can reside in the field devices and therefore be always available. Upgraded DD binaries can be provided in modules that may be inserted into the host. Upgrades may be accomplished through floppy disks or downloaded into the host with PC software.
- Configuration system developers no longer need to be responsible for validation testing of all devices supported in their products. They just have to ensure that they interpret the DDs correctly.

The DDL system architecture consists of a collection of specifications of DDL together with a set of tools that are implemented following these specifications. Specifically the DDL architecture consists of the following components:

- Specifications: A specification of a structured text language, called DDL, is used to describe the meaning of and relationships between device data available via the network. This specifies the syntax of the language used in DDL source files. A specification of a standard encoding of the DDL source file into a binary file format is transported over the network.
- Tools: Tokenizer is a tool for converting the DDL into the binary file format. This tool also checks for proper syntax and conformance to interoperability rules. A tool for extracting information from the DD binary when needed by the applications is referred to as the DDS library (Figure 16.12).

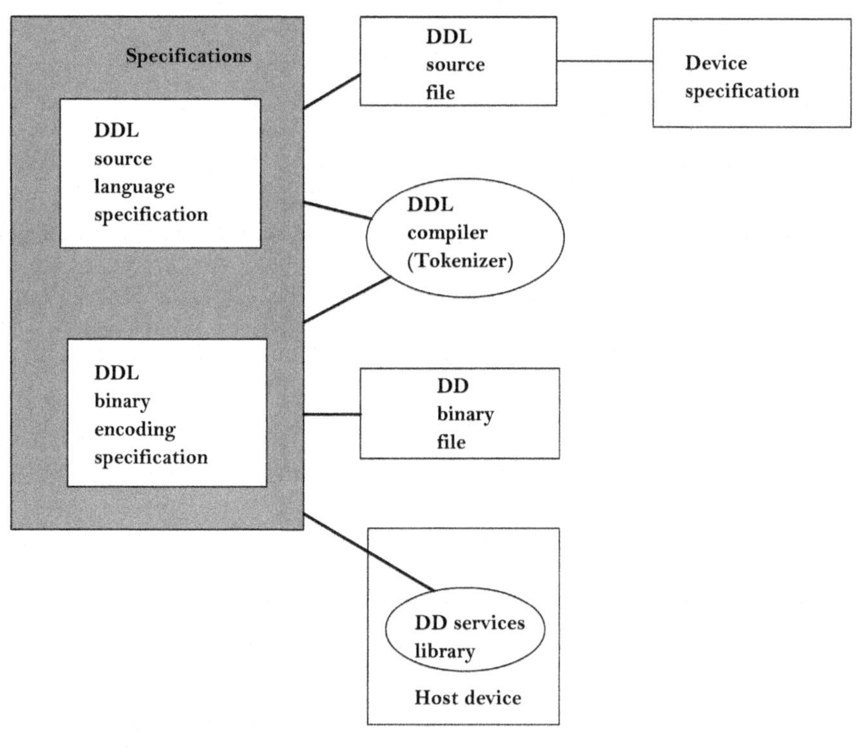

FIGURE 16.12

DD Description workflow

16.16 LIMITATIONS OF DD TECHNOLOGY

- Limitations in the graphical representation of the device at the user interface and complex device signatures and graphical representations cannot be achieved.
- DD can only be represented as a parameter, not as a dash board.
- The solution is more complex.
- Audit trails are not possible in DD.

16.17 ENHANCED DEVICE DESCRIPTION LANGUAGE

16.17.1 INTRODUCTION

Technologies and standards are evolutionary by nature. When discussing digital field automation technology architectures, users state their concerns in a hundred different ways, but in the end, what users seek are assurances that the technology platform they choose provides:

- Freedom of choice in plant floor instrumentation and equipment independent of the host: valves, transmitters, motor starters, remote I/O, and so on.

- Consistency in the engineering of plant floor instrumentation and equipment.
- Flexibility and efficiency in the sharing of plant-floor data throughout the enterprise.
- Ease of maintenance.
- Quantified proof that adopting manufacturers have long-term commitments to expand and improve the underlying technology.

A seldom mentioned, yet extremely important, realization influencing a technology's success is its dependence on other technologies. From a product manufacturer's perspective, the less dependent a technology's architecture is on underlying technologies, the better. Not only does such a design help manage risk, but it also frees up development resources to enhance and improve product features – something end users seek when evaluating similar products from different manufacturers.

16.17.2 OVERVIEW OF EDD AND EDDL

An electronic device description (EDD) contains all device parameter(s) of an automation system component. The technology used to describe an EDD is called EDDL. The EDDL provides a set of scalable basic language elements to handle simple, complex, or modular devices. The EDDL is a descriptive language based on an ASCII format with clear separation between data and program. The EDDL is a structured and interpretative language for describing device properties. Also the interactions between the devices and the EDD runtime environment are incorporated in the EDDL. The EDDL provides a set of language elements for this purpose. It is not necessary to use all of the elements provided by the language. Compatible subsets of EDDL are permitted and may be specified using profiles (e.g. choice of constructs, number of recursions, and selection of options). It is the responsibility of the EDD developers to identify, which profile has been used within each device details.

16.17.3 CONCEPTS OF EDD

The manufacturer of a device or of an automation system component describes the properties of the device by using the EDDL. The resulting EDD contains information such as:

- description of the device parameters
  - description of parameter dependencies
  - logical grouping of the device parameters
  - selection and execution of supported device functions
  - description of the transferred datasets
- EDD is distributed or downloaded but is not present in the device.

EDD supports text strings (common terms, phrases etc.) in more than one language (English, German, French, etc.). Text strings may be stored in separate dictionaries. There may be more than one dictionary for one EDD. An EDD implementation includes sufficient information about the target device for example, manufacturer, device type, revision, and so on. This is used to match a specific EDD to a specific device.

EDDL is enhanced DDL that supports more constructs to enable/enhance the graphical representation for the user. EDDL is a text-based language for describing the digital communication characteristics of intelligent field instrumentation and equipment parameters – device status, diagnostic data, and configuration details – in an operating system and human machine interface (HMI) neutral environment. EDDL-based technologies continue to evolve. EDDL is being enhanced to extend the concept of interoperability to the HMI and diagnostic data. EDDL technologies form the engineering and operating foundation on which all major digital industrial communication protocols – FOUNDATION™, HART®, and Profibus – construct device descriptions. Since EDDL is an open technology with international standard status, it can be easily and effectively applied to any device and any industrial communication protocol. The EDDL technology enables a host system manufacturer to create a single engineering environment that can support any device, from any supplier, using any communications protocol, without the need for custom software drivers for each device type.

The architecture of EDDL technology makes it easy to use multiple communication protocols within the same HMI host application. It eliminates the need for custom tools for configuration, and maintains a common look and feel for operators and maintenance technicians, regardless of the intelligent field instrumentation or equipment being viewed. The EDDL technology enables a host system manufacturer to create a single engineering environment that can support any device, from any supplier, using any communications protocol, without the need for custom software drivers for each device type.

With HMI host systems ranging from large process automation systems, with rich graphic displays, to simple handheld devices with limited display capabilities, EDDL's consistent–look–and–feel philosophy helps to simplify operator and maintenance personnel user training.

Since EDDL technology was designed to eliminate the need for special or proprietary files in host systems, a single EDD file created for a full-featured PAS works equally well in the handheld device.

This important design feature eliminates the expense and burden for intelligent field instrument manufacturers to create, test, register, and maintain different files for different HMI host systems, or a different operating system for a different revision of the same host system. For users, this feature of EDDL technology greatly simplifies changing, adding, upgrading, and evolving HMI host systems.

EDDL technology creates source files containing only content relevant to a specific manufacturer's intelligent field instrument or device. The EDDL technology design philosophy ensures that EDD files remain OS- and HMI-neutral. For manufacturers of intelligent field instrumentation and equipment, this philosophy eliminates the need to expend resources designing and writing software to support operator interfaces – specifying fonts, sizing windows, selecting colors, and so on – for different manufacturers' HMI products and platforms.

16.17.4 ADDRESSING ENGINEERING REQUIREMENTS

EDDL technology addresses the engineering requirements of end users and intelligent field device. Although their reasons differ, the engineering requirements of both groups are similar, each seeking solutions that:

- Ensure operating system independence
- Support multiple industrial communication protocols
- Consistent look and feel across different host vendors
- Avoid special, proprietary, or executable files for different host system platforms
- Support multiple country codes and international languages
- Use standardized testing and validation procedures and tools
- Allow efficient configuration entry
- Use a robust, consistent, easy-to-identify revision control methodology
- Include third-party testing and registration

Besides meeting all of these engineering requirements, EDDL technology also has international standards status such as IEC 61804-2. The EDDL design philosophy provides end users and intelligent field device product manufacturers a consistent and rich engineering environment.

16.18 FDT/DTM

16.18.1 FDT – FIELD DEVICE TOOL

16.18.1.1 Introduction to field device tool

Fundamentals of FDT

FDT technology provides a standardized communication interface between field devices and control or monitoring systems used to configure, operate, maintain, and diagnose intelligent field instrumentation assets for both factory and process automation applications. The versatility of FDT technology makes it suitable for communication protocols deployed in all industry segments, and allow users access to intelligence from a wide variety of process equipment.

FDT does not depend on any communication protocol and the software environment of the host system. FDT technology enables any FDT-enabled device to be accessed from any compliant host using any field communication protocol.

The technology consists of two main components: the frame and the DTM. The frame is either an embedded component of the control system suite or a standalone application, whereas the DTM is a device-specific application that launches within the frame itself. DTMs enable device manufacturers to gain complete control of the attributes displayed for their device in any of the over 15 host frame applications, thus providing the user with access to the full power and capabilities offered in their device. Without FDT technology, device information access may be restricted by the host/control system supplier and may not offer full access to device capability.

With FDT, the "heavy lifting" has been done and integration is seamless between field devices and the frame application, regardless of its manufacturer.

16.18.2 THE FRAME APPLICATIONS

A DTM is displayed or accessed from a frame application, which is a software window that provides the user interface between the device DTM and various applications such as device configuration tools, engineering work stations, operator consoles, or asset management tools. The frame application initializes the DTMs and connects it to the correct communication gateways. A single FDT frame application supports more than 15 of the world's most popular field communication protocols including HART, PROFIBUS, FOUNDATION Fieldbus, Modbus, DeviceNet, Interbus, AS-Interface, PROFINET, IO-Link, CC-Link, and more. FDT supports a mixture of any number of networks and further enables communications to tunnel through any number of networks to reach the end device.

16.19 THE DTM

A DTM is a device-specific software component that contains the application software that defines all the parameters and capabilities included in that device (similar to a device driver for a printer). The DTM provides access and the graphical interfaces needed to easily configure simple or even complex devices. It is particularly useful while commissioning devices, by preventing costly trips to the field and permitting maintenance of devices with sophisticated diagnostic tools. Using a single frame application minimizes technician training and avoids mistakes since all DTMs operate using a similar menu and visualization scheme. Since the device manufacturer is the provider of the DTM, access to the full device capabilities helps in best exploiting the device capabilities. All DTMs work in a common frame application but not all DTMs are created equal. Astute suppliers develop innovative DTMs with broader functionality that can help users improve troubleshooting and maintenance, allowing the manufacturer to further differentiate their products from their competition.

16.20 KEY BENEFITS TO THE USERS

16.20.1 HOST DATA INTEGRATION

Many of today's control and host applications are enabled to access intelligent device information because they include FDT technology. This allows easy access to device information from within the host application. Consider your computer accessing the Internet to access the information on a particular website. This is comparable to a computer (the host system) using the Internet via dial-up or DSL (using a field communication protocol such as HART, FF, or Profibus) opening a browser such as Internet Explorer (a frame application) to access the valuable information located on a website (a DTM accessing the device). The website owner wants the website to be displayed in a certain way, independent of the browser being used by the computer. Likewise, a device supplier wants its device information displayed and presented the same way, independent of the control or asset management system. Similarly, an internet can be accessed independent of the underlying network technologies such as Dial up, DSL, Wireless. Similarly, FDT technology can provide information from the device irrespective of the communication technology and protocols used.

16.20.2 DEVICE INFORMATION ACCESS

The information in intelligent field devices is most often accessed only during device configuration or troubleshooting. The full benefit of this information is only realized when devices are regularly scanned to verify the reliability of the process measurement, device health, or potential process problems. FDT technology offers significant benefits of improved asset management by helping in managing intelligent field assets. FDT technology helps in eliminating the routine manual checking of device health by putting the assets to work and better manage them to increase workforce availability with increased focus on the "real" critical issues. FDT-enabled asset management helps in identifying and effectively scheduling appropriate work orders, to increase reliability of assets while reducing labor and maintenance costs. This is independent of the field communication protocol being used to access the device.

Here are a few of the many benefits that have been realized by users:

- Faster device configuration and commissioning
- Reduced maintenance cost by only repairing devices such as control valves and transmitters that need repair rather than based upon a predetermined maintenance schedule
- Downtime

- Commissioning time
- Change management
- Training and development cost
- Avoid unscheduled shutdowns
- Reduced number of trips into the field – improving safety
- Improved product quality
- Superior scheduling of the operation

In many cases, these benefits can be realized with little investment and minimal risk. FDT technology is likely to be already included with the systems or devices you currently have. If not, new applications can be added without requiring a complete replacement of your existing control system or field devices. Several well-documented case studies that identity significant benefits from using FDT technology are available.

FURTHER READINGS

ANSI/ISA-51.1-1979 (R1993). Process Instrumentation Technology, Research Triangle Park, NC: ISA, ISBN 0-87664-390-4, 1993.

ANSI/ISA–51.1-1979 (R1993). "Process Instrumentation Terminology," Reaffirmed May 26, 1995, ISBN 0-87664-390-4.

FDT Technical Documents from http://www.fdtgroup.org/.

EDDL Brochure and Technical Description on www.eddl.org site.

Berge, J., "Field Buses for Process Control: Engineering, Operation, and Maintenance," ISA, 2002, ISBN 1-55617-760-7.

CHAPTER

CALIBRATION MANAGEMENT SYSTEMS

17

17.1 INTRODUCTION

Calibration refers to the process of checking and accurate adjustment (if necessary) of an instrument response with respect to a device with an established accuracy standard. So the output of an instrument accurately corresponds to its input throughout a specified range. The word standard refers to an authorized basis for a comparison of a unit of measure. It is important to calibrate an instrument for the following reasons:

1. Even the best instruments drift and lose their ability to give accurate measurements. The drift makes calibration necessary.
2. Environment conditions, elapsed time, and type of application can all affect the stability of an instrument.
3. Even instruments of the same manufacturer, type, and range can show varying performance from one unit to another.
4. To maintain the credibility of measurements.
5. To maintain the quality of process instruments at a "good-as-new" level.
6. Safety and environmental regulations.

17.1.1 INSTRUMENT ERRORS

Instrument error can occur due to a variety of factors: drift, environment, electrical supply, addition of components to the output loop, process changes, and so on. Calibration is performed by comparing or applying a known signal to the instrument under test and errors are detected by performing a calibration. An error is the algebraic difference between the indication and the actual value of the measured variable. Typical errors that occur include:

- Span error
- Zero error
- Combined zero and span error
- Linearization error

17.2 NEED FOR CALIBRATION

An instrument needs to be calibrated under the following circumstances:

- When a new instrument is bought.
- When a specified time period is elapsed.

- When an instrument has had a shock.
- Sudden changes in weather.
- Whenever observations appear questionable.

Even though the field devices in industrial plants are calibrated at first to ensure better accuracy, there are some factors that call for recalibration of the devices:

- Accuracy/uncertainty of measurement.
- Risk of a measuring instrument going out of tolerance when in use.
- Tendency to wear and drift.
- Extent and severity of use.
- Environmental conditions.
- Recorded history of maintenance and services.
- Manufacturer's recommendation.

17.3 TRACEABILITY

Traceability is defined by ANSI/NCSL Z540-1-1994 as "the property of a result of a measurement whereby it can be related to appropriate standards, generally national or international standards, through an unbroken chain of comparisons." Traceability is accomplished by ensuring the test standards we use are routinely calibrated by "higher level" reference standards.

Typically, the standards used in the shop are periodically sent out to a standards lab that has more accurate test equipment. The standards from the calibration lab are periodically checked for calibration by "higher level" standards, and so on until eventually the standards are tested against Primary Standards maintained by NIST or other internationally recognized standards (Figure 17.1).

17.4 CALIBRATION STANDARDS

A calibration standard refers to a substance or device used as a reference to compare against an instrument's response.

Types of calibration standard:

1. Primary calibration.
2. Secondary calibration.

17.4.1 PRIMARY CALIBRATION STANDARDS

Primary calibration refers to the calibration of instruments against primary standards. The calibrating process is difficult and proper care should be taken. Calibrating a secondary device is time-consuming but the accuracy is very good. Different types of standards with different types of accuracies and tolerances are used for different purposes. In the USA, the most accurate standards, i.e., those with the smallest tolerances are maintained by a federal agency called National Institute of Standards and Technology (NIST). The standards maintained by NIST are used to calibrate other standards in calibration certification facilities throughout the country. These facilities use their standards in turn to verify the standards of industrial plants.

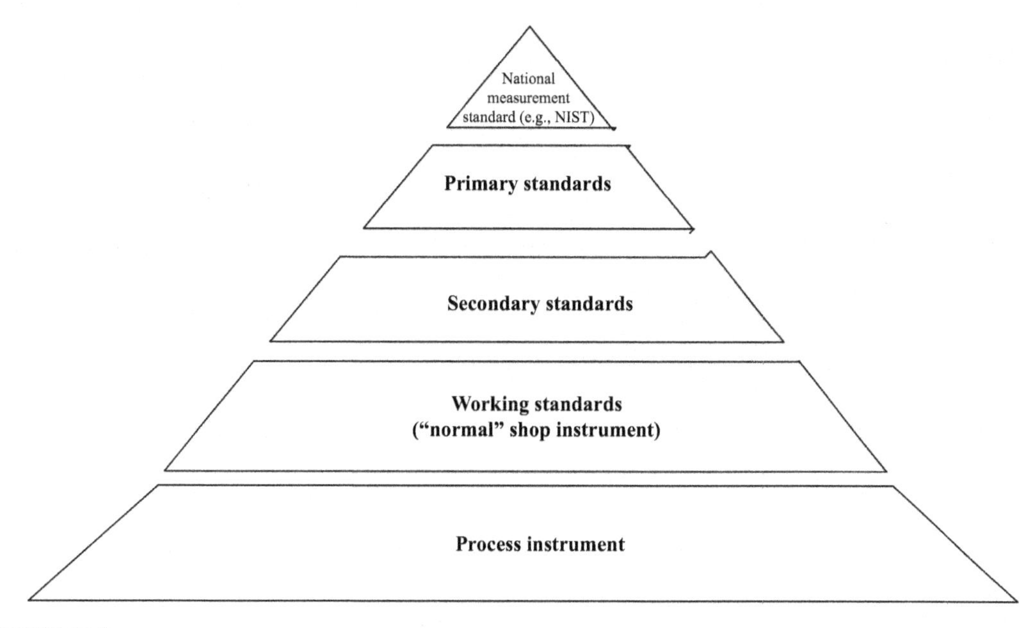

FIGURE 17.1

Traceability flow diagram

The accuracy of an instrument is determined by comparing its response to a value with a known standard. Instruments are calibrated periodically by comparing their readings with special test equipment. If the readings do not match then the instrument is calibrated or adjusted and the tests are repeated.

17.4.2 SECONDARY STANDARDS

The devices routinely used for instrument calibration are called secondary standards. Secondary standards are extremely precise for calibration measurement and controlled instrumentation. Primary standards are based on physical properties of matter and physical laws. For example, in monometers the physical relationship between hydrostatic head and pressure provides the values for testing and adjusting secondary pressure standards. Usually electronic secondary standards are calibrated to primary standards in outside calibration certification facilities.

Experts at NIST provide us a means of tracing measurement accuracy back to primary standards, which are quantities inherently fixed. The vibration frequency of an isolated cesium atom when stimulated by radio energy, for example, is a primary standard used to measure time (forming the basis of the so-called atomic clock). This frequency is fixed in nature and cannot vary. Primary standards therefore serve as absolute references against which we may calibrate instruments.

The machinery necessary to replicate primary standards for practical use are quite expensive and usually delicate. For these primary standards to be useful within the industrial world, we use them to calibrate instruments, which are in turn used to calibrate other instruments, and so on until we arrive at the instrument we intend to calibrate for field service in a process. So long as this "chain" of instruments is calibrated against each other regularly enough to ensure good accuracy at the end-point, we may calibrate our field instruments with confidence. The documented confidence is known as NIST traceability: that the accuracy of the field instrument we calibrate is ultimately ensured by a trail of documentation leading to primary standards maintained by the NIST.

17.5 CALIBRATION CONCEPTS

The Instrument Calibration Series (ISA) defines calibration as "determination of the experimental relationship between the quantity being measured and the output of the device which measures it; where the quantity measured is obtained through a recognized standard of measurement." There are two fundamental operations involved in calibrating an instrument:

- Testing the instrument to determine its performance.
- Adjusting the instrument to perform within specification.

Testing the instrument requires collecting sufficient data to calculate the instrument's operating errors. This is typically accomplished by performing a multiple point test procedure that includes the following steps:

1. Using a process variable simulator that matches the input type of the instrument, set a known input to the instrument.
2. Using an accurate calibrator, read the actual (or reference) value of this input.
3. Read the instrument's interpretation of the value by using an accurate calibrator to measure the instrument output.

This process is repeated for a series of different input values and sufficient data is collected to determine the instrument's accuracy. Depending upon the intended calibration goals and the desired error calculations, the test procedure may require from 5 to 21 input points.

The first test that is conducted on an instrument before any adjustments are made is called the As-Found test. If the accuracy calculations from the As-Found data are not within the specifications for the instrument, then it must be adjusted.

Adjustment refers to the process of manipulating some part of the instrument resulting in its input to output relationship within specification. For conventional instruments, this may be zero and span screws. For HART instruments, this normally requires the use of a communicator to convey specific information to the instrument. The terms "calibrate" and "adjust" are therefore often used synonymously. After adjusting the instrument, a second multiple point test is required to characterize the instrument and verify that it is within specification over the defined operating range. This is called the As-Left test.

It is absolutely essential that the accuracy of the calibration equipment be matched to the instrument being calibrated. Earlier on, a safe rule of thumb stated that the calibrator should be an order of magnitude (10 times) more accurate than the instrument being calibrated. As the accuracy of field instruments increased, the recommendation dropped to a ratio of 4 to 1. Many commonly used calibrators today do not even meet this ratio when compared to the rated accuracy of HART instruments.

17.6 DOCUMENTATION

The development and implementation of a quality calibration program maintains the accuracy and reliability of instrumentation in a facility. A calibration program includes written procedures for performing calibrations in addition to inventorying instruments, determining calibration parameters and intervals, and purchasing appropriate test standards. There are different types of calibration procedures catering to the considerable variation in the level of details contained in calibration procedures. They are:

1. Using calibration procedures from a technical manual: A manufacturer's technical manual is provided for each similar model of instrument. Using calibration procedures from a technical manual have the advantage of the need for very little time/resources to develop procedures and technically accurate and detailed instructions for the specific instrument. But, it does not contain all necessary elements of a calibration procedure and apply to instrument only, not taking into account the application/process.
2. Generic calibration procedures: It can be developed for each instrument type. Generic procedures recommend using the manufacturer's technical manual to perform any necessary adjustments. Generic calibration procedure limits numbers of procedures to a manageable level and it can be a good first step for new facility start-up. Generic calibration procedures have inconsistent methods by different technicians using same procedure and inexperienced technicians need more details.
3. Specific calibration procedures: In most facilities, some calibration procedures have to be very specific to particular instruments. In some cases, specific detailed procedures are required for each instrument. If a generic procedure for a particular pressure instrument fails to adequately address the proper method, a specific procedure should be developed. These specific calibration procedures are performed the same way by all technicians. It may take into account the effect on process and technically accurate and detailed instructions for the specific instrument. Specific calibration procedure has one disadvantage that is it increases resources required to develop, maintain, and track procedures.

17.7 CALIBRATION OF TRANSMITTERS

Transmitters are of two types:

1. Analog transmitters and
2. Digital (intelligent/smart) transmitters.

Before attempting to understand about calibration of transmitters, let us attempt to understand the difference between calibration and reranging.

To calibrate an instrument means to check and adjust (if necessary) its response so the output accurately corresponds to its input throughout a specified range. In order to do this, one must expose the instrument to an actual input stimulus of precisely known (standard) quantity. Ranging an instrument refers to setting the lower and upper range values so it responds with the desired sensitivity to changes in input.

17.7.1 CALIBRATING ANALOG TRANSMITTERS

In analog transmitters the term calibration is nothing but adjusting the zero and span settings. Thus, calibration of analog transmitter includes reranging the instrument. So, the terms calibration and reranging are interchangeable in the context of analog transmitters.

Zero adjustments typically take one or more of the following forms in an instrument:

- Bias force (spring or mass force applied to a mechanism).
- Mechanical offset (adding or subtracting a certain amount of motion).
- Bias voltage (adding or subtracting a certain amount of potential).

Span adjustments typically take one of these forms:

- Fulcrum position for a lever (changing the force or motion multiplication).
- Amplifier gain (multiplying or dividing a voltage signal).
- Spring rate (changing the force per unit distance of stretch).

Figure 17.2 shows the analog pressure transmitter, which consists of three blocks. Low-pass filter is used to reduce the effect of unwanted (noise) signals in the data obtained by the sensor. Amplifier is used to amplify the original signal that comes from low-pass filter or strengthen the weak signal in order to adjust that in a specified range. This is done by adjusting bias (zero) and gain (span) of the amplifier. The output section is the driver circuit, which takes signal from the amplifier and gives the output in the required form and 4–20 mA range.

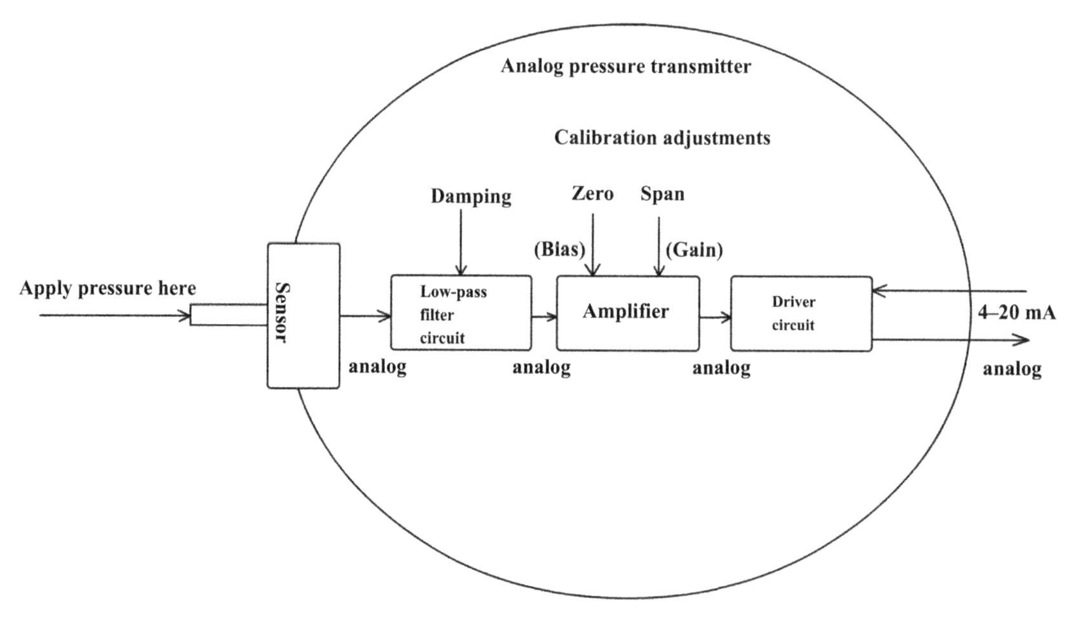

FIGURE 17.2

Block diagram of analog pressure transmitter

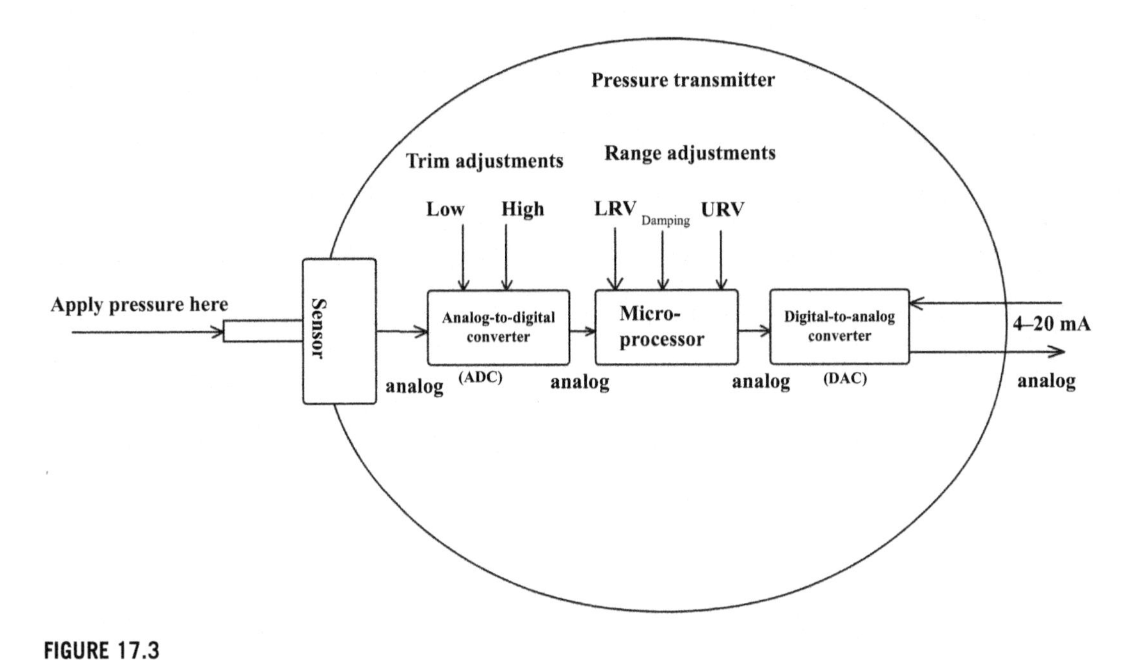

FIGURE 17.3

Simplified block diagram for "smart" pressure transmitter

17.7.2 CALIBRATION OF DIGITAL (SMART/INTELLIGENT) TRANSMITTERS

The advantage of "smart" field instruments containing microprocessors has been a great advance for industrial instrumentation. These devices have built-in diagnostic ability, greater accuracy (due to digital compensation of sensor nonlinearities), and the ability to communicate digitally with host devices for reporting various parameters (Figure 17.3).

The term "calibration" in the context of smart/intelligent transmitters is often misunderstood. In analog transmitters, calibration meant applying a physical input and turning the trim potentiometers to adjust the sensor so that the output current is adjusted according to the desired measurement range. There is a big difference between analog and smart transmitters. In smart transmitters the "calibration" process is divided into three parts:

- Sensor trim (input trim).
- Range setting (reranging).
- Current trim (output trim).

The reason for separating these functions is that the range can be changed without applying a physical input. This is a huge time and cost saver and one of the major reasons for the rapid adoption of smart transmitters. However, it is important not to confuse "sensor trim" with "range setting." Both are part of calibration, but two very different things.

17.7.3 SENSOR TRIM

Over a long span of time, all sensors show some drift depending on the type of sensor, owing to extreme pressure or temperature, vibration, material fatigue, contamination, or other factors. Sensor reading may also be offset due to mounting position.

Sensor trim is used to correct the digital reading as seen in the device local indicator LCD and received over the digital communication. For instance, if pressure is 0 bar but transmitter reading shows 0.03 bar, then the sensor trim is used to adjust it back to 0 bar. Sensor trim can also be used to optimize performance over a smaller range than was originally trimmed in the factory.

Sensor trim requires a physical input to the transmitter. Therefore, the technician must either do sensor trim in the field at the process location, or the transmitter has to be brought back into the workshop to perform sensor trim. This applies to 4–20 mA/HART, Wireless HART, FOUNDATION fieldbus, and PROFIBUS transmitters. Sensor trim in the field is easily achieved using a handheld communicator that is supported by 4–20 mA/HART, Wireless HART, and FOUNDATION fieldbus.

Typically there are three forms of sensor trim:

- Zero sensor trim.
- Lower sensor trim.
- Upper sensor trim.

Zero trim requires the physical input applied to be zero, most often used with pressure transmitters. For best accuracy, sensor trim is performed at two points, close to lower range value and upper range value. This is where lower and upper sensor trim is used. A known physical input is applied to the transmitter to perform the sensor trim; the technician keys in the applied value (on a computer or handheld communicator) communicated to the transmitter, allowing the transmitter to correct itself. The physical input values applied for lower and upper sensor trim, respectively, are stored in the transmitter memory and are referred to as lower sensor trim point and upper sensor trim point, respectively. Sensor trim requires a very accurate input to be applied.

17.7.4 SETTING (RERANGE)

Range setting (reranging) refers to setting the scale for the 4 mA and 20 mA points. It refers to identifying the input at which the transmitter output must be 4 mA, which is the lower range value (LRV) often referred to as "zero" meaning 0%. The input at which it must be 20 mA is the upper range value (URV), sometimes called "full scale" meaning 100%. Note that the term "span" is not the same as URV. Span is the magnitude of difference between URV and LRV. If the sensor trim and output current trim are perfectly done, then transmitter range setting is done without applying input, and therefore can be done remotely from the control room. The range setting is now dependent on turndown ratio (or) range ability, which is defined as the ratio of maximum allowable span to minimum allowable span. For 4–20 mA systems the range is set in both the transmitter and controller. FF and Wireless HART transmitters do not have a 4–20 mA output; therefore, there is no need to set 4 mA and 20 mA range points. For FF and PROFIBUS the range is set in the controller.

These LRV and URV ranges are set in the transmitter by "Direct numeric value" entry and "to applied input". Direct numeric value entering refers to the desired lower and upper range values being simply keyed in from the device software or handheld field communicator, and sent to the transmitter. Range setting to applied input requires a physical input corresponding to the desired range value to be applied to the transmitter. This is often used in level measurement applications. The set PV LRV and set PV URV commands are equivalent to pushing the "Zero" and "Span" buttons found on some transmitters.

17.7.5 TRANSMITTER OUTPUT CURRENT TRIM

This output current trim is often misunderstood because the error location is unclear, i.e., at the sensor input or at the output current. So, it is very important to know when the sensor trim is required and when the current trim is required. It is rare for the output current circuitry of a 4–20 mA transmitter to drift. Current trim is recommended if a handheld communicator is used and it shows any wrong current readings compared to input and output values using standard instruments. Use of range setting to correct current trim or sensor trim is not recommended. However, if the output current is incorrect, use current trim to adjust the output signal. For instance, if the current output is 4.13 mA when it should be 4.00 mA, then current trim is used to adjust it to 4 mA. Current trim is used to match the transmitter current output to the current input of the analog input card channel on the DCS. For instance, the transmitter may be reading 0.00% but the DCS may show 0.13% because of differences in current calibration. Current trim is only applicable to transmitters with 4–20 mA output. So, the current trim is adjusted in the field at the process location by connecting a multimeter to the transmitter test terminals, or the transmitter has to be brought back into the workshop to perform current trim. Current trim in the field is possible using a handheld communicator.

17.8 CALIBRATING A CONVENTIONAL INSTRUMENT

The conventional calibration adjustment involves setting only the zero value and the span value, since there is effectively only one adjustable operation between the input and output (Figure 17.4).

This procedure is often referred to as a Zero and Span Calibration. If the relationship between the input and output range of the instrument is not linear, then we know the transfer function before we can calculate expected outputs for each input value. It is not possible to calculate performance errors without knowing the expected output values.

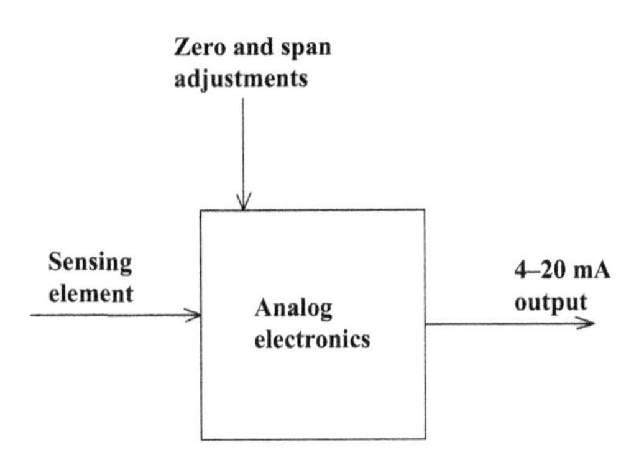

FIGURE 17.4

Conventional transmitter block diagram

17.9 CALIBRATING A HART INSTRUMENT

In conventional instruments there is a clear path between the input and output, i.e., when a pressure sensor is subjected to pressure, it causes some resistance change that gives electrical voltage signal with the help of bridge circuits. However, in HART instruments instead of a purely mechanical or electrical path between the input and the resulting 4–20 mA output signal, a HART transmitter has a microprocessor that manipulates the input data. As shown in Figure 17.5, there are typically three calculation sections involved, and each of these sections may be individually tested and adjusted.

In the first section, the instrument's microprocessor measures some electrical property that is affected by the process variable of interest. The measured value may be millivolts, capacitance, reluctance, inductance, frequency, or some other property. However, before it can be used by the microprocessor, an analog-to-digital (A/D) converter is used to convert it to a digital count.

The output of the first box is a digital representation of the process variable, when read using a communicator. The second section consists of a microprocessor that takes input in digital form and performs some operations on the data and transfers the processed values to output in digital form. This box is strictly a mathematical conversion from the process variable to the equivalent milliamp representation. The range values of the instrument (related to the zero and span values) are used in conjunction with the transfer function to calculate this value. Although a linear transfer function is the most common, pressure transmitters often have a square root option. Other special instruments may implement common mathematical transformations or user-defined break point tables. The output of the second block is a digital representation of the desired instrument output, when the loop current is read using a communicator. Many HART instruments support a command that puts the instrument into a fixed output test mode. This overrides the normal output of the second block and substitutes a specified output value.

The third section is the output section where the calculated output value is converted to a count value that can be loaded into a digital to analog converter. This produces the actual analog electrical signal. Once again the microprocessor must rely on some internal calibration factors to get the output correct. Adjusting these factors is often referred to as a current loop trim or 4–20 mA trim.

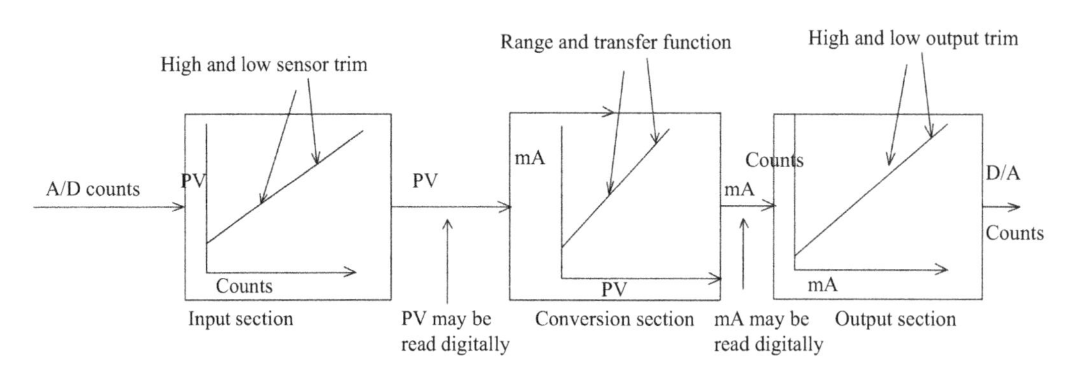

FIGURE 17.5

HART transmitter block diagram

17.9.1 CALIBRATING THE INPUT SECTION

To calibrate the input section, the same basic multiple point test and adjust technique is employed, but with a new definition for output. To run a test, use a calibrator to measure the applied input, but read the associated output (PV) with a communicator. Error calculations are simpler since there is always a linear relationship between the input and output, and both are recorded in the same engineering units. In general, the desired accuracy for this test is the manufacturer's accuracy specification.

If the test does not pass, then follow the manufacturer's recommended procedure for trimming the input section. This may be called a sensor trim and typically involves one or two trim points. Pressure transmitters also often have a zero trim, where the input calculation is adjusted to read exactly zero (not low range). It is important not confuse a trim with any form of reranging or any procedure that involves using zero and span buttons.

17.9.2 CALIBRATING THE OUTPUT SECTION

To run a test, use a communicator to put the transmitter into a fixed current output mode. The input value for the test is the milliampere value that you instruct the transmitter to produce. The output value is obtained using a calibrator to measure the resulting current. This test also implies a linear relationship between the input and output values, and both are recorded in the same engineering units (milliamps). The desired accuracy for this test should also reflect the manufacturer's accuracy specification. If the test does not pass, then follow the manufacturer's recommended procedure for trimming the output section. This may be called a 4–20 mA trim, a current loop trim, or a D/A trim. The trim procedure should require two trim points close to or just outside of 4 and 20 mA. It is important to not confuse this with any form of reranging or any procedure that involves using zero and span buttons.

17.9.3 TESTING OVERALL PERFORMANCE

After calibrating both the Input and Output sections, a HART transmitter should operate correctly. The middle block only involves computations. That is why you can change the range, units, and transfer function without necessarily affecting the calibration. It may be observed that even if the instrument has an unusual transfer function, it only operates in the conversion of the input value to a milliamp output value, and therefore is not involved in the testing or calibration of either the input or output sections. Passing a Zero and Span test like a conventional instrument test does not necessarily indicate that the transmitter is operating correctly. Ranging can be adjusted to meet user requirement if the calibrations of input and output sections are correctly maintained. But the range should be defined by turndown ratio or rangedown. Rangedown refers to the allocation of maximum allowable span to minimum allowable span.

17.10 CALIBRATING FIELDBUS TRANSMITTERS

In fieldbus terminology, "calibration" often refers to the configuration of a transmitter. According to metrology, "calibration" refers to comparing a transmitter to a traceable measurement standard and documenting the results. So, it is not possible to calibrate a fieldbus transmitter using only configurator or configuration software. Also, it is not possible to calibrate a fieldbus transmitter remotely. Fieldbus transmitters are calibrated just like conventional transmitters and need to place a physical input into the transmitter and simultaneously read the transmitter output to see that it is measuring correctly. The input is measured with a traceable calibrator, but also needs to have a way to read the output of the fieldbus transmitter. Reading the digital output is not simple. When fieldbus is up and running, one person in the field measures the transmitter input while another person in the control room reads the output. Naturally these two people need to communicate with each other in order to perform and document the calibration. While fieldbus and process automation systems are idle, we need to find other ways to read the transmitter's output. In some cases a portable fieldbus communicator or a laptop computer with dedicated software and hardware is used. Calibrating a fieldbus transmitter can be tedious, time-consuming, and may require an abundance of resources. There exists no practical way to calibrate fieldbus transmitters.

17.10.1 DOCUMENTING PROCESS FIELDBUS CALIBRATOR

Some suppliers provide documenting process Fieldbus Calibrator, which is a combination of a multifunction process calibrator and a fieldbus configurator. The documenting process calibrator can be used for calibrating Foundation Fieldbus H1 or Profibus PA transmitters. The documenting process Fieldbus Calibrator can calibrate a fieldbus pressure and temperature transmitter, as it can simultaneously generate/measure the transmitter input and also read the digital fieldbus output of the transmitter. The documenting process calibrator can also be used to change the configurations of a fieldbus transmitter. If the fieldbus transmitter fails in calibration, it is used to trim/adjust the fieldbus transmitter to measure correctly.

Being a documenting calibrator, it automatically documents the calibration results of a fieldbus transmitter in its memory, from where the results can be uploaded to calibration management software. This eliminates the time-consuming and error-prone need for manual documenting using traditional methods. The documenting process calibrator is a compact, easy-to-use, and field-compatible calibration solution offering a lot of other functionalities. The documenting calibrator can be used for calibrating Foundation Fieldbus H1 or Profibus PA transmitters.

17.10.1.1 Advantages of the documenting fieldbus calibrator

The most important advantage of the documenting Fieldbus Calibrator is the possibility to calibrate, configure, and trim the Foundation Fieldbus H1 or the Profibus PA transmitters using a single unit. Because it is a combination of a calibrator and a fieldbus configurator, the documenting calibrator is capable of performing traceable calibration on fieldbus transmitters. Fieldbus configurators and configuration software can only be used to read/change configurations. Therefore, calibration needs can be met by one person. It is able to calibrate stand-alone transmitters, or when they are connected to a live fieldbus as with the FF. There is no need for a separate power supply because the documenting calibrator includes an integral power supply for powering up a stand-alone transmitter during calibration. Therefore, the documenting calibrator can also be used during commissioning when the fieldbus and control systems are still idle.

17.11 CALIBRATION MANAGEMENT SYSTEM

Every process plant has some sort of system in place for managing instrument-calibration operations and data. Process measurement devices such as temperature sensors, pressure transducers, and weighing instruments – require regular calibration to ensure they are performing and measuring specified tolerances. However, different companies from a diverse range of industry sectors use very different methods of managing these calibrations. These methods differ greatly in terms of cost, quality, efficiency, and accuracy of data and their level of automation. Calibration management software is one such tool that can be used to support and guide calibration management activities, with documentation being a critical part of this. But in order to understand how software can help process plants better manage their instrument calibrations, it is important to consider the typical calibration management tasks that companies have to undertake. All plant instruments and measurement devices need to be listed, then classified into "critical" and "non-critical" devices. Once this has been agreed, the calibration range and required tolerances need to be identified. Decisions then need to be made regarding the calibration interval for each instrument. The creation and approval of standard operating procedures (SOPs) for each device is then required, followed by the selection of suitable calibration methods and tools for execution of these methods. Finally, the company must identify current calibration status for every instrument across the plant. The execution stage means supervising the assigned calibration tasks. Staff carrying out these activities must follow the appropriate instructions before calibrating the device, including any associated safety procedures. The calibration is then executed according to the plan, although further instructions may need to be followed after calibration. The documentation and storage of calibration results typically involves signing and approving all calibration records that are generated. The next calibration tasks then have to be scheduled, calibration labels need to be created and pasted, then created documents copied and archived. Based on the calibration results, companies then have to analyze the data to see if any corrective action needs to be taken. The effectiveness of calibration needs to be reviewed and calibration intervals checked. These intervals may need to be adjusted based on archived calibration history. If, for example, a sensor drifts out of its specification range, the consequences could be disastrous for the plant, resulting in costly production downtime, a safety problem or leading to batches of inferior quality goods being produced, which may then have to be scrapped. Documentation is a very important part of a calibration management process. ISO 9001:2000 and the FDA both state that calibration records must be maintained and that calibration must be carried out according to written and approved procedures. This means an instrument engineer can spend as much as 50% of his or her time on documentation and paperwork – time that could be better spent on other value-added activities. This paperwork typically involves preparing calibration instructions to help field engineers; making notes of calibration results in the field; and documenting and archiving calibration data. When it comes to the volume of documentation required, different industry sectors have different requirements and regulations. Many plants rely on generic spreadsheets and/or databases for this, while others used a calibration module within an existing computerized maintenance management system (CMMS).

A significant proportion uses a manual, paper-based system. Any type of paper-based calibration system will be prone to human error. Noting down calibration results by hand in the field and then transferring these results into a spreadsheet back at the office may seem archaic, but many firms still do this. Furthermore, analysis of paper-based systems and spreadsheets can be almost impossible, let alone time consuming. Plants can save much time and reduce costs by using calibration management software to analyze historical trends and calibration results. Using software for calibration management enables faster, easier, and more accurate analysis of calibration records and identifying historical trends. Plants can therefore reduce costs and optimize calibration intervals by reducing calibration frequency when this is possible, or by increasing the frequency where necessary. For example, for improved safety, a process plant may find it necessary to increase the frequency of some sensors that are located in a hazardous, potentially explosive area of the manufacturing plant. Just as important, by analyzing the calibration history of a flow meter that is located in a "noncritical" area of the plant, the company may be able to decrease the frequency of calibration, saving time and resources. Rather than rely on the manufacturer's recommendation for calibration intervals, the plant may be able to extend these intervals by looking closely at historical trends provided by calibration management software. Instrument "drift" can be monitored closely over a period of time and then decisions taken confidently with respect to amending the calibration interval. Regardless of industry sector, there seems to be some general challenges that companies face when it comes to calibration management. The number of instruments and the total number of periodic calibrations that these devices require can be several thousand per year. How to plan and keep track of each instrument's calibration procedures means that planning and scheduling is important. Furthermore, every instrument calibration has to be documented and these documents need to be easily accessible for audit purposes.

17.11.1 PAPER-BASED SYSTEMS

These systems typically involve handwritten documents. Typically, this might include engineers using pen and paper to record calibration results while out in the field. On returning to the office, these notes are then tidied up or transferred to another paper document, after which they are archived as paper documents. While using a manual, paper-based system requires little or no investment, it is very labor-intensive and means that historical trend analysis becomes very difficult to carry out. In addition, the calibration data is not easily accessible. The system is time consuming, soaks up a lot of resources and typing errors are commonplace. Dual effort and rekeying of calibration data are also significant costs here.

17.11.2 IN-HOUSE LEGACY SYSTEMS (SPREADSHEETS, DATABASES, ETC.)

Although certainly a step in the right direction, using an in-house legacy system to manage calibrations has its drawbacks. In these systems, calibration data is typically entered manually into a spreadsheet or database. The data is stored in electronic format, but the recording of calibration information is still time-consuming and typing errors are common. Also, the calibration process itself cannot be automated. For example, automatic alarms cannot be set up on instruments that are due for calibration.

17.11.3 CALIBRATION MODULE OF A CMMS

Many plants have already invested in a computerized maintenance management (CMM) system and so continue to use this for calibration management. Plant hierarchy and works orders can be stored in the CMM system, but the calibration cannot be automated because the system is not able to communicate with "smart" calibrators. Furthermore, CMM systems are not designed to manage calibrations and so often only provide the minimum calibration functionality, such as the scheduling of tasks and entry of calibration results. Although instrument data can be stored and managed efficiently in the plant's database, the level of automation is still low. In addition, the CMM system may not meet the regulatory requirements (e.g., FDA) for managing calibration records.

17.12 CALIBRATION SOFTWARE

With specialist calibration management software, users are provided with an easy-to-use Windows Explorer-like interface. The software manages and stores all instrument and calibration data. This includes the planning and scheduling of calibration work; analysis and optimization of calibration frequency; production of reports, certificates and labels; communication with smart calibrators; and easy integration with CMM systems such as SAP and Maximo. The result is a streamlined automated calibration process, which improves quality, plant productivity, and efficiency.

17.13 BENEFITS OF USING CALIBRATION MANAGEMENT SYSTEM

With calibration management system, planning and decision making gets improved. Procedures and calibration strategies can be planned and all calibration assets are managed by the system. Device and calibrator databases can be maintained, while automatic alerts for scheduled calibrations can be set up. The system no longer requires pens and paper. Calibration instructions are created using the system to guide engineers through the calibration process. These instructions can also be downloaded to a technician's handheld calibrator while in the field. Execution is more efficient and errors are eliminated. Using software-based calibration management systems in conjunction with documenting calibrators means that calibration results can be stored in the calibrator's memory, then automatically uploaded back to the calibration software. There is no rekeying of calibration results from a notebook to a database or spreadsheet. Human error is minimized and engineers are freed up to perform more strategic analysis or other important activities. The software generates reports automatically and all calibration data is stored in one database rather than multiple disparate systems. Calibration certificates, reports, and labels can all be printed out on paper or sent in electronic format. Analysis becomes easier too, enabling engineers to optimize calibration intervals using the system's history and trend. Locating records and verifying that the system work is effortless when compared to traditional calibration recordkeeping. Regulatory organizations and standards such as FDA and ISO place demanding requirements on the recording of calibration data (Figure 17.6).

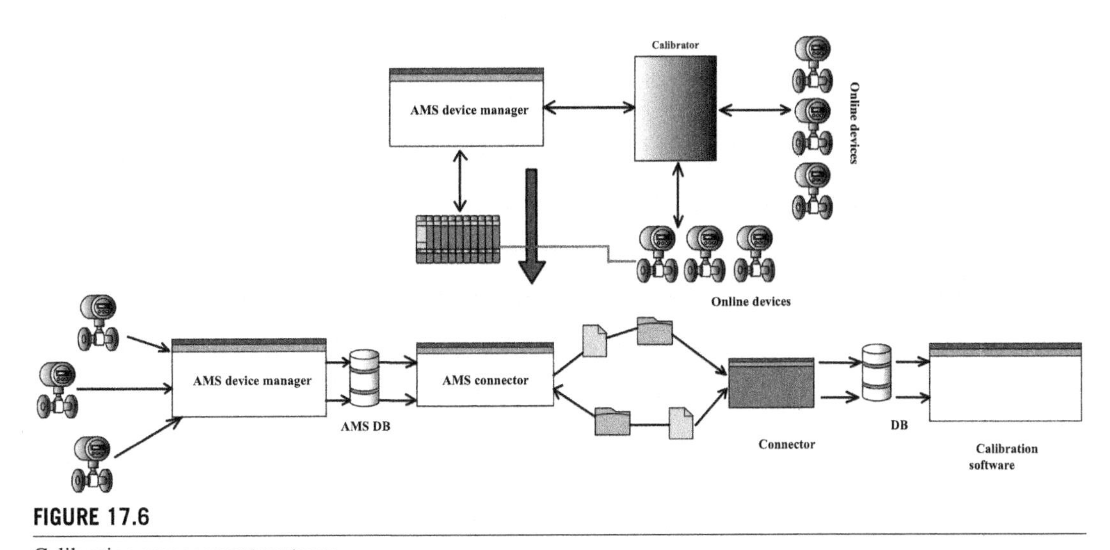

FIGURE 17.6

Calibration management systems

17.14 BUSINESS BENEFITS

For the business, implementing software-based calibration management means overall costs will be reduced. These savings come from the now paperless calibration process, with no manual documentation procedures. Engineers can analyze calibration results to see whether the calibration intervals on plant instruments can be altered. For example, those instruments that perform better than expected may well justify a reduction in their calibration frequency. Plant efficiencies should also improve, as the entire calibration process is now streamlined and automated. Manual procedures are replaced with automated, validated processes, which is particularly beneficial if the company is replacing a lot of labor-intensive calibration activities. Costly production downtime will also be reduced. Even if a plant has already implemented a CMM system, calibration management system can be easily integrated to this system. If the plant instruments are already defined on a database, the calibration management systems can utilize the records available in the CMM system database. The integration will save time, reduce costs, and increase productivity by preventing unnecessary double effort and rekeying of work orders in multiple systems. Integration also enables the plant to automate its calibration management with smart calibrators, which simply is not possible with a standalone CMM system. In summary, every type of process plant, regardless of industry sector, can benefit from implementing specialist calibration management software. Compared to traditional, paper-based systems, in-house built legacy calibration systems or calibration modules with CMM systems, using dedicated calibration management software results in improved quality, increased productivity, and reduced costs of the entire calibration process. Despite these benefits, only one quarter of companies who need to manage instrument calibrations actually use software designed for that purpose.

FURTHER READINGS

Calibrating HART Transmitters HCF_LIT-054, Revision 1.1 Release Date: November 19, 2008.
Lessons in Industrial Instrumentation by Tony R. Kuphaldt Version 1.22 – Last update January 8, 2012.
Calibration: A Technician's Guide. Mike Cable.
Fluke 740 Series Documenting Process Calibrators: Brochure.
Yokogawa Electric Corporation. GS 33Y05Q10-32E. General Specifications of PRM.

CHAPTER

SYSTEM MAINTENANCE

18

18.1 OVERVIEW

The most crucial element for system maintenance is maintenance guide and it is certainly a key to the profitability of companies by ensuring that the control system is maintained so the plant can produce its products.

18.1.1 NEED FOR SYSTEM MAINTENANCE

Maintenance of instrumentation and process control systems from simple gauges to complex distributed control systems (DCSs) is essential for the continuation of the industry. Maintenance personnel should be involved with new installations and upgrading older installations. They should ensure that the system is ergonomically easy to repair and well documented. Training is essential before a new system arrives so the maintenance department can help in installing and checking it out.

Equipment manufacturers usually provide engineering and start-up assistance. Owing to the equipment's complexity, assistance is needed from the original equipment manufacturer. It is essential to be familiar with system requirements and system/instrument capabilities before attempting configuration of control systems and instruments. Instrumentation tells us the process parameters in which we are operating. A simple gage indicates the temperature or pressure; the more complex instrumentation indicates much more about the process. It is important to ensure proper operation of all equipment to make a quality product and to do it safely. The technological advances in the past few years and the advanced technical and specialized equipment call for better-trained maintenance personnel. The types of equipment in control systems cover many disciplines like mechanical, electrical, electronic, computer science, chemical, and environmental engineering. Because of the many different knowledge factors, the individual crafts have to work together. Electrical engineers, mechanical engineers, chemical engineers, and process engineers must understand each other and determine where their responsibilities start and stop.

The field has grown with the application of computers, artificial intelligence, self-tuning, computer-integrated manufacturing (CIM), and so on. Larger companies train pipe fitters to be instrument mechanics in pneumatic plants and electricians to be instrument technicians in electronic plants. Knowledge of the process is essential to designing new systems; therefore, all engineering disciplines must get involved with the instrumentation and control system.

The complexity of control loops and systems requires specialists. The systems concept calls for varied knowledge and the overall concept of control rather than component troubleshooting and replacement. When the control system does not work, the plant does not produce. The control system design can determine the profitability of a company. The production output of the plant can be maximized if it is maintainable and the mechanics, technicians, and engineers are trained. It is important for corrective, preventive, and operational maintenance to be performed by qualified and experienced maintenance personnel.

Many times maintenance experience required owing to the complexity of existing control systems that utilize many fields of expertise. This group is now required to maintain, troubleshoot, and calibrate pneumatic, electrical, electronic, and computerized instruments and systems. The systems approach, which looks at the whole picture to gain an understanding of the process, is the special attribute of I&C maintenance personnel.

Typically, the maintenance supervisors and managers go to instrumentation and control system engineers or maintenance engineers with an I&C background for expert advice on the control system. Instrumentation and control system engineers assist the mechanics and technicians and keep the supervisors and managers informed. They need to be a part of the design and start-up of the control systems. Lot of effort is being spent for training, fault-tolerant systems, redundancy, and new techniques. One simple but essential area that may be neglected is the experience of the past and what it may teach about the present.

Being involved in the problems we encountered and the solutions that were found yesterday helps us make better decisions today. The applications of older systems should be used as the basis for designing newer and generally faster control systems. New problems are encountered in newer systems, but past application experience helps solve the new problems.

Good maintenance saves money. Efficiently working equipment increases the process quality and production. When equipment fails, production normally stops, and many production personnel cannot do their jobs. With good maintenance management, spare parts are available quickly to reduce the mean time to repair (MTTR). When the equipment is repaired properly, the mean time between failures (MTBF) is extended. The proper frequencies of preventive maintenance should provide less down time, and the down time that occurs can be scheduled. We can become pro-active instead of reactive.

18.2 DISTRIBUTED CONTROL SYSTEM MAINTENANCE

A DCS typically controls a control room network connecting multiple control, display, and communication devices. These devices include operator and engineer workstations, continuous and discrete process controllers, field network interfaces, and computing resources, as well as other node types. The DCS architecture enables these devices to be distributed by either function or geographic area. The individual devices are optimized for the specific function they serve. However, this geographical and functional distribution and the quantity and complexity of the individual devices and their interconnections make the development of an effective maintenance program a significant challenge.

18.2.1 MAINTENANCE GOALS AND OBJECTIVES

The successful development of a DCS maintenance program requires a long-term commitment to life cycle planning, budgeting, and training. The planning process begins even before selecting the control system vendor and continues throughout the life of the system. The plan should have the agreement of plant maintenance, operations, and management personnel. Roles and responsibilities must be clearly defined and understood. A quality improvement process should be in place to periodically monitor and evaluate the effectiveness of the maintenance program.

The primary goals of a DCS maintenance program are to:

1. Minimize the number of component failures in the system.
2. Minimize the process impact of failures that do occur and
3. Effectively diagnose the cause of failure and take corrective actions as quickly as possible.

18.2.2 MAINTENANCE STRATEGIES

18.2.2.1 Reactive maintenance

Historically, maintenance has mostly been associated with breakdowns. The philosophy of running equipment until it failed and then sending in an emergency repair crew is a waste of human and material resources. The reactive or "run-to-failure" approach is a costly and difficult strategy to manage. Maintenance costs are high and do not account for production losses, damage to machinery, high inventory, poor morale, and the long range effect on product quality, unit cost, and overall competitive position.

18.2.2.2 Preventative maintenance

Preventative maintenance (PM) planning enables plant and maintenance engineers to implement maintenance activities more effectively by scheduling periodic shutdown for inspection and parts replacement based on equipment-repair histories. Advance preparation is the first step in the transformation towards a proactive maintenance methodology. Since unscheduled maintenance work costs two to three times as much as work planned in advance, this alone can produce significant savings in the maintenance budget. Although preventive maintenance offers clear benefits, there is a downside. Because preventive maintenance activities are performed on a set schedule, there is a risk that functioning equipment may be shut down or replaced unnecessarily, each of which carries its own cost in lost production and wasted capital. Experienced maintenance professionals know that the process of shutdown, inspection, and reassembly itself can introduce a new set of problems. In addition, preventive maintenance cannot entirely eliminate emergency repairs.

18.2.2.3 Predictive maintenance

Operating realities have given rise to the relatively new concept of predictive maintenance. Predictive maintenance is based on accurate measurements of real-time operating conditions. This information is readily available and routinely collected automatically by the DCS. These parameters are used to generate a prediction based on certain criteria, for example, number of days until an alarm condition. Work orders can be scheduled according to demand and required parts ordered "just-in-time" maintenance.

Because it is parameter-driven rather than schedule-bound, predictive maintenance eliminates the expense of unnecessary preventive maintenance on functioning equipment, making resources available for production, or other important areas. Predictive maintenance can also prolong the intervals between preventive maintenance activities, thereby saving on labor and parts. And because problems are identified early, predictive maintenance prevents terrible failure. Predictive maintenance offers the benefit of availability.

18.2.2.4 Proactive maintenance

Proactive maintenance is doing maintenance to avoid failure. These are corrective actions emerging out of failure in one part of the plant and applied across the complex for other equipment of similar type, e.g., alert notes. Apart from these breakdowns, maintenance is done on sudden failure of equipment/plant.

18.2.3 DESIGN FOR MAINTENANCE

The following are the major implications of DCS architecture and design principles on maintenance strategy and tactics:

1. Faults are rare, and little preventive maintenance is required.
2. The faults that do occur are tolerated with little or no degradation of system performance, and they affect only the module in which they occur.
3. For most failures, the system itself determines the correct optimum replaceable unit (ORU) – a board, power supply, peripheral device, etc.
4. The variety of ORUs is held to a minimum, minimizing the spare parts inventory.
5. Test equipment is seldom needed, and no special or complicated test equipment is required.

18.3 MAINTENANCE SOFTWARE

An operator or engineer work station typically has maintenance software that is used to diagnose failures in modules in the control room network that are not diagnosed by the built-in tests and to test the operation of the field network and their process-connected boxes. When an on-process work station is being used for maintenance, one or more of the remaining workstations in the console take over its on-process functions, as illustrated in Figure 18.1.

18.4 MAINTENANCE PROGRAM IMPLEMENTATION AND MANAGEMENT

18.4.1 THE MAINTENANCE PROCESS

Maintenance of a DCS throughout its life cycle can be considered as a sequential series of maintenance cycles. Each maintenance cycle starts with a request for maintenance to be carried out and ends with the issuing of a report giving the outcome of the work done.

Each maintenance cycle requires four key activities:

1. Completing the requested maintenance action.
2. Retesting the system.
3. Recording details of the activities carried out.
4. Noting any resulting changes made to the system configuration.

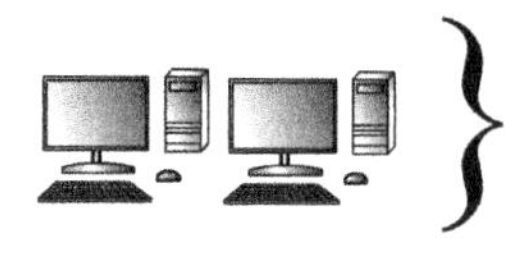

FIGURE 18.1

System maintenance activity

18.4.2 TYPES OF MAINTENANCE ACTIVITIES

There are basically three types of maintenance actions:

1. Preventive maintenance, periodically scheduled routine actions.
2. Corrective maintenance, in response to faults and anomalies.
3. Perfective or adaptive maintenance, brought about by the need to simplify or improve the system or make system changes to reflect changes in the environment.

Preventive maintenance involves all actions aimed at discovering and clearing faults before they affect service. It includes cleaning, routine testing, and periodic replacement of parts that are at the end of their useful life (such as air filters, batteries, etc.). Preventive maintenance is scheduled by the maintenance organization. The periodic replacement of parts demands a rigorous adherence to standard procedures. On completing the maintenance activity, the system should be tested before being put back into service and a report generated to confirm to the operations personnel that all maintenance actions have been successfully completed and tested.

Corrective maintenance involves all actions intended to diagnose and clear faults or anomalous behavior (nonconformance). The diagnosis should identify the cause of, and the circumstances surrounding, the problem. If the corrective action required is the simple replacement of a faulty part without the basic configuration of the software or hardware of the system, an unscheduled part replacement is performed. During part replacement, it is important to check the configuration database to ensure that switch settings and options are correctly set, that the card information for the failed and the replacement card are recorded on a failure report, the successful completion of the operation is reported, and revalidation testing takes place. On completion of this stage the equipment is put back into service.

The use of perfective or adaptive maintenance ensures that any proposed change to a system baseline definition or build level is prepared, accepted, and controlled in accordance with set procedures. Changes to be controlled include corrections, modifications, and enhancements brought about by the need to simplify or improve the system. Plant management compares the estimated cost of modification with the benefit to the business of doing the modification. Some modifications may not be cost effective, but may be essential to the safety or operation of the system. It is important that the modification is planned like any other activity. The change in configuration needs to be approved by plant management who must also agree to the implementation plan.

18.4.3 SAMPLE PREVENTIVE MAINTENANCE ACTIVITIES

One of the main goals of a DCS maintenance program is to minimize the number of failures within a system. The two most important factors in achieving this objective are the quality of the initial installation and an effective PM program.

PM should be part of any DCS maintenance program. The following list may serve as a useful starting point for developing an effective PM program:

1. Check for proper power and grounding on a regular basis.
2. Replace on-board batteries per manufacturer's recommendations.
3. Test switch-over to redundant devices on a regular basis.
4. Keep the installation as clean and dust-free as possible. Change cabinet and air conditioning filters frequently. Keep the DCS equipment room doors and cabinets closed.

5. Check field wiring connections for loose or corroded connections.
6. Install and maintain an air filtration and treatment system in accordance with vendor recommendations. Where corrosive elements are inherent in the process, monitor for their presence in the DCS equipment room. Special inexpensive reactivity coupons are available for this purpose.
7. Periodically review DCS performance with control room operators to bring forward any otherwise unreported incidents or concerns. Coordination with operations is essential in planning for new system installations, major system upgrades, and minor system updates. Periodic operations reviews can also be used to assess the need for additional or refresher training.
8. Finally, consider integrating the DCS preventive maintenance program with your field instrument calibration program. The DCS will function only as well as the date it receives from the field.

PM is a time-based maintenance technique, wherein the instrument is isolated/removed from the service at predetermined intervals/scheduled frequency, to carry out status check/visual inspection/calibration of the instruments and perform adjustments of the instruments and replacement of minor components, if necessary, as a result of such inspection/calibration.

18.4.4 DEVIATION PROCEDURE

18.4.4.1 PM frequency deviation

For fire and gas detectors, PM/calibration is carried out as per the SMP/ESS applicable for a particular type of detector.

PM for instrument systems [e.g., DCS, ESD, machine condition monitoring system (MCMS), PLC, automatic tank gauging system (ATG)] is carried out as per the applicable ESS.

The ESS sheets for instrument systems are available in Tier 4 of the E&M Instrumentation Quality Manual.

Responsibility maintenance chart is as given in Table 18.1.

Records pertaining to PM compliance of QMS instruments of plants are maintained either in hard copy or in SAP system by respective plant instrument section head. Calibration format for QMS instruments is available in Tier-4 of the respective Plant Instrumentation Quality Manual. Retention period of the records for QMS instruments is 3 years.

Procedure for control of trips, alarms, and protective system settings: Equipment and personnel are protected against process upsets, etc., by automatic alarms, trip systems, and other protective devices. These protective systems have been "built in" to the plant during the design stage. Accord-

Table 18.1 An example role and responsibilities

Job Execution	Instrument Technician
Planning, supervision and coordination	Maintenance engineer (I)
Reporting nonconformance and corrective action	Maintenance engineer (I)
Maintaining records	Section head (instrument)
Request for granting concessions	Section head (instrument)
Granting concessions	HOD (plant)

ing to a Reliance Policy, they should always be maintained in good working order. This procedure is applicable to Process and Fire and Gas detection system related trip/alarm settings/defeat. However, for trip and alarm setting changes/defeat related to Rotary equipment, refer to procedure RJN/EMRT/2003.

Protective system: A protective system is a general term that covers the following three types of protective systems.

Trip system: This system detects an abnormal condition and initiates automatic command to stop equipment or plant safely, simultaneously generating an audio-visual alarm.
Alarm system: This system detects an abnormal condition and initiates an audio-visual warning so as to draw attention of the plant operator to the abnormality.
Remotely operated protective systems: A remotely operated protective system is a system by which an operator can take action from a safe location to correct an identified abnormality, e.g., remotely operated emergency trip switches (panel/field) for stoppage/tripping of compressor/heater.

18.4.5 PROCEDURE FOR CHANGE CONTROL

Due to operational requirements, following four types of change to the protective/control systems may arise:

- Defeat of protective systems.
- Modification of protective/control systems.
- Change in the trip/alarm settings.
- Configuration changes of control systems.

18.4.5.1 Defeat of protective systems

As far as possible, defeating of protective systems should be avoided.

In some plants, however, circumstances may arise (e.g., at plant start-up or when maintenance or testing is carried out during continued operation of the plant) when some of the protective systems are to be defeated temporarily. Such defeating of the protective system is carried out in a properly controlled and with due authorization.

To facilitate operation start-up or other operational/maintenance requirements:

- Soft over-ride switches called maintenance over-ride switch (MOS) are provided in DCS.
- Hardwired switches are provided in the operator control console.
- The soft over-ride switches are password protected and these can be activated/deactivated by the operational personnel authorized by the plant production manager.
- Similarly, hard-wired switches on DCS consoles or UCP can be activated by the process personnel.

The following rules are applicable for defeat of a protective system through soft over-ride in ESD/DCS/PLC:

- The request for defeat of a protective system will be made by plant production department in the form given in the procedure.
- The authority and permitted duration for defeating protective system is given in the procedure.

Table 18.2 Change Control Responsibilities

Maintaining of trip defeat register plant	Production in-charge
Authorizing setting changes (other than rotary equipment) plant	Production in-charge (Production manager)
Authorizing defeat of a protective system (other than rotary equipment)	Production in-charge (Production manager) or higher
Authorizing configuration change	Plant manager and HOD CES inst./DCS in-charge.

- The trip defeat details are entered in the trip defeat register, maintained by operations personnel, and should be available at the control room. Format for the trip defeat register is given in the procedure.
- The trip defeats necessitated during the scheduled trip testing jobs (during unit/equipment shutdown) do not call for entry in the form.
- Trip defeat requirements shall be mentioned explicitly in the work permit issued to carry out such PM jobs. All such trip defeats shall be normalized before closure of the work permit.

18.4.5.2 Modification of protective control systems

Where a new trip/alarm system, or major modification to an existing protective/control system is required, procedure for plant change order as specified in procedure no. HSE210 Management of plant changes shall apply. Changes in the trip and alarm settings must be authorized by the plant production manager on the trip/alarm setting change request for all instruments. For trip and alarm setting changes related to rotary equipment, refer to procedure RJN/EMRT/2003.

18.4.5.3 Configuration change in control systems

These changes provide additional facilities to the operator or maintenance personnel and will not call for change in hardware, P&ID or cause and effect matrix. Few examples of such configuration changes are configuration of first-out alarms/additional diagnostic facilities/totalizer in DCS/enabling or disabling existing facilities in the respective control system. Configuration change request must be approved by HOD (plant) and DCS in-charge. This change request should be used only for configuration changes. If change in hardware, P&ID, or cause and effect is required, separate PCO must be raised.

18.4.5.4 Responsibility

Table 18.2.

18.4.6 RECORDS

Trip and alarm schedule should contain the fields of tag number and description of alarm and trip settings as minimum requirement and is to be updated and maintained in soft copy (PC/SAP)/website or hard copy in the respective plants. The cause and effect diagram issued by Bechtel/Existing graphics in DCS/Logics in DCS/PLC/ESD is updated and signed off by respective plant production manager/instrument manager. The updated documents are sent to FDC along with PCO for updation of records. For packages where vendor has not provided cause and effect diagram, the logic chart/diagrams are to be validated and signed off by respective plant production manager and plant instrument.

18.4.7 PERSONAL COMPUTER MAINTENANCE

The development of low-cost software and greater capability at ever-decreasing hardware costs is leading to the application of personal computer systems in the process industries at a rapid pace. Systems have evolved far beyond typical transaction-driven, commercially oriented software applications. Many applications now place these systems in online, real-time control applications. The development of lightweight, portable laptop units could potentially lead to these units becoming a valuable maintenance tool for the instrument technician. Current applications include their substitution for manufacturer-specific PLC programming units.

Like PLCs, most desktop systems share common functionality and architecture. A typical system consists of a central processing unit (CPU), a keyboard, monitor, and a printer. CPUs are typically packaged with a hard disk drive and one or two floppy disk drives. Keyboards are typically 101-key enhanced units providing 10–12 special-purpose function keys. Monitors are available in a wide variety of sizes, resolutions, and color capabilities. Laser and ink-jet printers are the predominant choice for most systems.

The internal layouts for most desktop (or tower design) central processors are functionally similar. The typical unit contains a power supply, a mounting rack for the disk drives, and a motherboard that contains the microprocessor and system control chips, a socket for a coprocessor or overdrive chip, the base memory, and a bus structure with connectors for several expansion boards. Units are typically furnished with the disk controller and a parallel/serial interface card installed. The expansion slots are used for additional parallel/serial interfaces, extended memory, bus mouse, and special application interfaces. Repair by replacement is recommended for failed power supplies, disk controllers and drives, expansion slot interface boards, and chips provided with sockets. Outright replacement or repair by a factory authorized service center is recommended for monitors.

Since keyboards involve moving parts, they frequently require minor maintenance. The low cost of these units encourages some users to replace rather than repair them.

Repair by replacement is the only viable maintenance option for a failed floppy or a hard disk drive. Preventive maintenance for these units extends their useful service life. Minute amounts of dust, lint, or magnetic oxide particles can cause a number of problems with floppy disk drives. Head cleaning kits are available to keep the drive heads free of such contaminants.

Preventive maintenance can significantly extend the life of hard disk drives. Alignment drift and repeated start-up of the read-write heads over data areas eventually leads to deterioration of the drive's low-level format. Utility software allows the user to refresh the low-level format before failures occur. Such software also verifies the operation of the disk controller, disk caching, and the main memory.

A low-level scrub of the disk surface can take several hours, so it is best to run this type of software overnight. A "park" utility provided with the software can be used to ensure that the read-write heads do not begin operation over an area that contains data. If an apparent failure is experienced, try cleaning and reseating the ribbon cable that connects the drive and the controller board. Many apparent drive failures are due to faulty connections. In case of repeated failures, replace the cable and make sure the unit is not exposed to vibration.

18.5 SOFTWARE AND NETWORK MAINTENANCE

18.5.1 OVERVIEW

Most control modernization projects today take advantage of the new and expanding capabilities of microprocessor-based digital control systems (DiDCs), digital DCSs, also called DCSs and PLCs. Because these systems are capable of performing both regulatory and supervisory controls, plant personnel have to learn to maintain computer-controlled systems. Because these systems are more software-oriented than hardware-oriented, maintenance of these systems requires knowledge of software configuration and programming.

Programming generally refers to the development of custom code in which the engineer builds a program, utilizing a computer programming language. The language used is commonly called ladder logic. Configuration, on the other hand, implies the engineer is working with a DCS in which the functional operation is described by a linking of preprogrammed tasks that have been specifically developed for industrial control applications.

The application of computers in the industrial control industries stemmed from the need to advance control capabilities. This need was defined by economic, quality, and safety justifications for better control and automation and the benefit of less expensive installation. The application of computers also translated to the expansion of the maintenance skills required to keep these systems running. The skills must include the ability to understand and work with the system's software configuration and program.

18.5.2 OPERATING SYSTEM SOFTWARE

At the base of the software hierarchy is the computer's operating system. The operating system is the software that allows the user to manage the computer and its resources. There are many operating systems, most of which were developed by the hardware manufacturers. The most widely used operating system is Microsoft's Windows. Another operating system, UNIX, has been a favorite with software developers for a number of years. A UNIX variation, Linux is also gaining popularity.

18.6 COMPUTER OPERATING ENVIRONMENT

The electronic components of digital computers are designed to be operated by electrical current that meets certain specifications and under certain conditions of temperature, humidity, and cleanliness. Failure to respect these conditions may cause the components to fail or affect their function uniformity.

General-purpose computers are intended for use in homes and offices, where the environment threats to electronic components are minimal. Their design, therefore, provides minimal protection against adverse conditions. One way to increase the level of protection is to change the design of the computer to incorporate more polished components.

However, this approach entails a substantial increase in cost, especially since solving some environmental problems may create others. For example, sealing up the outer case of the computer to exclude water, dirt, and harmful vapors will also cut off air circulation, resulting in excessively high temperatures inside unless large heat sinks and cooling systems are added.

Therefore, many industrial computer installations enclose unmodified general-purpose computers in a protective room where an environment suitable for the computers can be maintained. However, the latest DCSs are designed for use in most industrial environments.

PLCs were designed from the outset to function on the plant floor, and their environment requirements are, therefore, less stringent. This does not entirely eliminate the need for protection for PLCs. They are vulnerable to the same environment threats as general-purpose computers, only somewhat less so.

18.6.1 DUST AND DIRT

Dust and dirt quickly form a blanket over all accessible surfaces in a computer. The layer of dirt may prevent physical contacts from closing (as in keyboards) or may block ventilation and other openings. If the dust is conductive, it may cause short circuits; if insulated, it may lead to heat build-up and component failures. Minimizing the number of openings through which dust can enter is an effective solution for dust infiltration. This approach is taken in many PLCs and industrially hardened computers, although it may create heat dissipation problems. The other approach is to filter the air, either at the inlet to the computer room or as it goes in through particular openings in the computer case. This method works well if filter media are appropriate to the contaminants to be guarded against and are changed regularly.

18.6.2 CHEMICAL VAPORS

Corrosive chemical vapors can influence contacts and conductive traces inside a computer, eventually breaking their electrical continuity. They can also damage data recording surfaces on disks. The chemical vapors issue is tackled by either making the computer or the control room tight against the vapors in question or installing appropriate filters. It may be necessary to set up the computer room at a distance from sources of chemical vapors to ensure they remain at a sufficiently low concentration.

18.6.3 WATER

Water can get into a computer in two ways: by direct splash (installation near process equipment using water or accidental spillage) and by condensation. Obviously, water can cause catastrophic short-circuiting inside a computer, and a chronic water problem can also lead to corrosion and failure of components. The operating specifications of most computers include a humidity range within which the computer is designed to function. Designing an entire computer to be watertight is difficult because of heat dissipation requirements, but watertight keyboards or key pads are often recommended on the plant floor and connected to a computer nearby in a protected room. Keeping beverage cups and other sources of nonprocess liquids at a safe distance from computer is a sensible preventive measure.

18.6.4 CONDENSATION

Condensation is usually the result of low temperature or extremely high humidity in the ambient air. If the problem is local, a heater may be used to raise the temperature, which lowers the relative humidity for a given air moisture content per unit volume. In an environmentally controlled room, a humidifier or dehumidifier may also be used. An air conditioning system can control both temperature and humidity.

18.6.5 VIBRATION

The vibration that often accompanies operating process equipment can break contacts and solder joints inside computers, causing malfunctions and data loss in disk drives. The simplest remedy is to isolate the computer from the vibration source, either by moving it away from the vibrating equipment or by mounting it on a shock-absorbent base.

18.6.6 ELECTRICAL DISTURBANCES

Electronic computer components require a relatively unvarying supply of low-voltage, low-current electrical power. Standard distribution voltages (110–120 V or greater) must be stepped down to supply these components, usually by a step-down transformer (often called the computer power supply), which outputs the voltage used on the system bus (typically 5 V). Further voltage reductions may be required for specific components.

18.6.7 VOLTAGE AND CURRENT TRANSIENTS

A high-voltage or excessive power condition, if severe or prolonged, can simply burn out electronic components. Less severe transients may be interpreted as a spurious change of state, causing erratic program function or corruption of data. With too low an initial voltage, the computer can simply fail to function because the voltage is insufficient to raise electronic devices above their threshold values. A severe voltage drop while the computer is in operation may trigger an unplanned restart (warm boot). The most serious condition is fluctuating low voltages, which may cause repeated partial restarts, leaving memory and data in unpredictable states and possibly hanging up the computer altogether. The effects of a complete power loss are actually less drastic, except that the contents of volatile RAM are lost and the contents of a file may be corrupted if a write operation is in progress when the power loss occurs.

18.6.8 INTERFERENCE

Noise in the form of radio frequency interference (RFI) or electromagnetic interference (EMI) can be harder to track down and correct. RFI arises most often from lengths of cable acting as antennae or from mobile communication sources in the plant such as walkie-talkies and lift truck radios. EMI can come from any piece of equipment that generates an electrical or magnetic field (most commonly motors). It is important to check for possible temporary sources of interference; arc welding is a notorious culprit. Computer components such as cables, long communication lines, or disk drive motors may themselves generate interference. Normally, computers are required to meet government specifications for RFI and EMI emissions so that they do not interfere with the operation of other electronic equipment.

The effect of electrical interference is similar to that of fast transients from the power line: the generation of spurious signals that the computer mistakes for data. These may cause erratic operation or, in extreme cases, total computer failure.

18.6.9 POWER FAILURE

If continued operation of a computer is critically important, an uninterruptible power supply (UPS) should be added to the incoming power line. During normal operation, the line power charges a battery, which, in turn, feeds the power supply and the computer. In case of a blackout, the power stored in the battery can maintain the computer in operation for a limited time, typically one to several hours. Depending on the need, for example, time to copy the contents of memory to disk and shut down the computer in an orderly fashion requires around twenty minutes to half an hour; otherwise the cost must be balanced against the importance of the process and the normal length of power outages. Care must be taken to ensure that the wattage rating of the UPS should be sufficient to support the memory, display, and disk drives (typically the biggest power consumer).

18.6.10 STATIC ELECTRICITY

A static discharge is typically low current that can reach extremely high voltages that can damage computer equipment. Components such as memory chips and integrated circuit boards are typically shipped in static-proof bags to protect them from accidental discharge during shipment. Once these components are installed, they are protected by the computer's grounding subsystem. During installation, however, they are vulnerable, and installation personnel should take appropriate precautions, such as wearing ground straps and static-free clothing and using antistatic mats. Maintaining adequate humidity levels helps control static electricity.

18.6.11 HEAT

Electronic components are designed to work within an operating temperature range that is stated in their specifications. Outside this range, their characteristics change beyond design limits, their behavior becomes erratic, and their life is shortened. In extreme cases, it could lead to a fire hazard. The usual method of cooling computer components is by convection to the surrounding air, often assisted by a fan that draws air over the components. Other devices such as monitors (screens) have no fan and cool themselves only by convection. For this method to work, the surrounding air has to be substantially cooler than the components, which is one reason computer rooms are often air-conditioned. Computers have vents to draw cooling air in and out, which must be left unblocked even if they seem to provide a nice shelf for manuals, drawings, and bag lunches. If convection alone cannot maintain the components within their specified operating range, components with a higher temperature tolerance must be substituted.

18.6.12 NETWORK OPERATIONS

Networks enable computing systems to reach out to the wider world to obtain and provide data. However, this is prone to risks and vulnerabilities. File corruption, intruders, viruses, and unauthorized use of computing resources are only some of the threats a network must deal with. In addition, networks are large and complex systems, which require ongoing trouble-shooting and maintenance.

18.6.13 SECURITY

As corporate networks become more and more connected to the outside world, they provide companies not only with a way to get information out to their employees, customers, and suppliers, but an open door that allows others to get in through this open door. Competitors can steal proprietary information, corrupt vital databases, and cripple computer facilities, seriously damaging a company's operations, competitive position, and reputation. This is in addition to the existing challenge of protecting internal networks against unauthorized access and use.

Majority of the control systems available in current days fall under the following category:

- Distributed Control Systems (DCS)
- Emergency Shutdown Systems (ESD)
- Programmable Logic Controllers (PLC)
- Machine Control Monitoring Systems (MCMS).

Microprocessor-based control system/controllers are protected with passwords in order to ensure the controlled environment for changing/modifying the existing programs.

18.6.13.1 Password protection for DCS

DCS is a highly reliable system where false configuration can cause severe loss of system integrity and plant availability with consequent loss of production. Therefore, access must be controlled.

The system security for DCS is divided into three categories as follows:

- Environment passwords and access classes for local operation: Password protection and access class configuration for process graphics, to access system configurator and system administration through environment for local operation.
- Environment password and access class for remote operation: Password protection and access class configuration for process graphics, to access system configurator and system administration through environment for remote operation (X-window).
- Root password protection for remote "login" to workstation.

Various access levels for local operations are given in Table 18.3.

18.6.13.2 Password protection Safety Systems

Following philosophy has been adopted to ensure proper system security to avoid unauthorized access and also to prevent accidental changes by users in system configurations (Figure 18.2):

- All users are divided into five levels.
- Particular user rights are assigned for each level.
- Following user levels have been defined and a unique password has been assigned for each levels.
  - LEVEL 1 User – Plant instrument engineers.
  - LEVEL 2 Plant system engineer.
  - LEVEL 3 System administrator.
  - LEVEL 4 System administrator.
  - LEVEL 5 System administrator.

18.6.14 ISOLATION

For critically important operations, the safest solution may be physical isolation, i.e. disconnecting a particular computer from the network and/or physically preventing certain operations. However, this approach makes data acquisition, maintenance, and troubleshooting more difficult.

Table 18.3 Access levels and privileges

Environment	Access Class	Privileges	User	Password Access
Field_Op_Env	1	View displays	All field operators	NA
Ctrl_Rm_Op_Env	2	View displays, change controller parameters (set point, output, auto-manual, etc.), view, acknowledge and clear alarms. View alarm history.	Panel operator, shift engineer, SS	Plant manager/SS
Supervisor_Env	3	All Ctrl_Rm_Op_Env access, change alarm limits, tune controllers	Shift engineer, SS, instrument engineer	Plant manager/SS/ shift engineer
Maint_Engr_Env	4	All Supervisor_Env access, make configuration changes, access system management and health display (SMDH)	Instrument manager, plant DCS engineer	Plant manager/ plant instrument manager/DCS HOD
Soft_Engr_Env	5	All Maint_Engr_Env access, access software management to write programs, modify files, etc. Change passwords and change environment menus.	DCS core group, plant instrument manager	DCS HOD

	Level 1	Level 2	Level 3	Level 4	Level 5
Functionality	General user	Plant system engineer	System administrator	System administrator	System administrator
Logic monitoring					
IO disable					
System monitoring					
Dictionary editor/config					
Ladder editor					
Download change (increment)					
Download all					
Password change for different levels					

FIGURE 18.2

Role based functionality

18.6.15 FILE SECURITY

The possibility of the same file or device being accessed for use by two people can cause problems. In the case of disk files, a common solution is to "lock out" one user until the other has finished. This locking may be done at the volume, file, or record level. Volume locking ensures that no other user can access any file on the same disk volume until the first user has finished. File locking prevents other users from accessing the file that the first user is accessing, while record locking prevents access only to the record the first user is working on. Clearly, the smaller the unit that is locked, the less the user's free access to data is compromised. Managing the locking and unlocking, however, becomes more complicated.

With several users, it becomes more important to control which files each user can use and how. It may not be desirable for everyone to have access to payroll or personal files. On the other hand, it may be desirable for them to be able to look at, say, process data histories or download them to their spread sheets without being able to change the master copy. Most operating systems include some capability for controlling access to files (setting a read-only bit in the operating system file description, for example). More sophisticated systems can restrict access by individuals, groups, or job functions and allow administrators to decide whether each user may look at the file, make a copy of it, write to it, or delete it.

18.6.16 FIREWALLS

A firewall scrutinizes all transmissions going into or out of a corporate network and blocks or quarantines any questionable ones. The most common kind of firewall is a filtering router, which looks at data packets as they go by and pass or block them depending on the address or information in the packet header. This is complicated by the existence of many different messaging protocols, including FTP (used for Internet file transfers), domain name system (DNS), and Unix windowing protocol (X11). Your firewall may leave you unable to use one or more of these. Sometimes, files that are attached to electronic mail messages often get corrupted or disappear as they pass through firewalls.

Firewalls slow down traffic into or out of your network because of the extra processing they do on each and every data packet. If these delays become too severe or the other restrictions on traffic too burdensome, users may search for alternative means such as modem connections that bypass the firewall. This, of course, defeats the original security objectives.

18.6.17 VIRUSES

A computer virus is a malicious piece of executable code that is intentionally imported into a computer or a network and usually damages some aspect of the computer's or the network's operation. The virus may damage files, directories, other software, system configuration information, any one or all of the above. It may also make unauthorized use of information such as password files or email address books in order to gain access to other computers. In extreme cases, a computer system may need to be shut down and reconfigured from scratch. One of the major tasks of firewalls and other network security measures is to protect against viruses.

Fortunately, a virus can usually be recognized by characteristic patterns in its code or its behavior. Commercial software is available that detects and neutralizes incoming viruses and can be regularly updated as new viruses are developed. Virus detection has often depended on the fact that a virus had to be an executable file with an *.exe, *.vbs, *.pif, or *.com extension. More recently, however, macro viruses have taken advantage of the ability of certain programs (mainly Microsoft word and excel) to include small sets of executable instructions (macro viruses) in a document or spread sheet. A virus masked as one of these instructions sets as opposed to a separate file is much more difficult to detect and remove.

One of the most effective types of virus protection is user education. In a plant environment, the consequences of a virus infection are not only personal inconvenience and possible data loss, but also potential damage to the computer systems that run an operating plant.

18.7 NETWORK MAINTENANCE

Like all large complex systems, local area networks (LANs) have their share of problems, and it is vital that someone be able to solve them quickly. Much of the software and data as well services as printers and electronic mail are accessible only via the network. When an industrial process is involved, the efficient operation of the process or the accuracy of the process information may be compromised if the network goes down. In most of the process control and factory automation systems installed in the last few years, the automation network is vital to the plant's operation. While a good control system design allows a plant to operate through a brief network outage, during prolonged network outages continued operation must not be left unsupervised by human operators. Clear installation and troubleshooting instructions and good vendor support should be high priorities.

Simple overloading is a common cause of network breakdowns or poor performance. Storage gets crowded and fragmented, buffers overflow or frequent collisions, and repeated transmissions clog the network. The best way to overcome this is by selecting the right network, installing it with guidance from the supplier, and planning adequate capacity for present needs, with provision for an easy expansion path for future needs.

18.7.1 OPEN SYSTEMS

Early computer installations were self-contained. If computers were networked at all, it was with other equipment from the same manufacturer, via proprietary interfaces. In the plant, isolated, single-purpose computers monitored or controlled "dumb" equipment. The option of communicating with other devices or computers was not an issue. Early networks encompassing a process area or department were usually purchased from a single supplier and were operated separate from other plant computer systems. The smooth functioning of the network was the responsibility of the supplier, and again connectivity with outside devices or computer systems was not a requirement.

However, networks soon grew to attain plant or corporate scale. Inexpensive computer hardware (PCs) meant a computer on every desk and in every corner of the plant, from various manufacturers, each with its own inventory of add-on hardware and software and PLCs; smart instruments became more and more like specialized computers with their own interfacing needs.

This led to incompatibilities among different manufacturer's hardware and software, which started to become troublesome. Companies wanted to take advantage of new technology, yet could not be sure that it would work with their present systems. They could not meet their expanding and changing needs within existing proprietary systems, but also could not afford to have a custom solution built from scratch or spend the time and money needed to resolve interfacing and communication problems.

The concept of open systems was introduced to solve these problems by defining a system architecture that could seamlessly incorporate products from different manufacturers. The concept was initially used by Sun Microsystems in the mid-1980s as a marketing strategy to differentiate Sun from its mainframe and workstation competitors, which allegedly sold "closed" systems. The strategy proved so successful that other computer manufacturers quickly moved in the same direction. Large computer users in business, government, and defense quickly realized how much money they could save and flexibility they could gain by moving to open systems, so they started to specify interface standards to which their computer suppliers had to adhere.

"An open system is a collection of interacting software, hardware, and human components designed to satisfy stated needs, with interface specifications of its components that are fully defined, available to the public, maintained according to group consensus in which the implementations of the components conform to the interface specifications."

18.7.2 INDUSTRIAL NETWORKS

The primary function of an industrial local area network (ILAN) is to serve as a data highway that connects sensors, controllers, computers, and other electronic devices in an industrial plant. ILANs also support the movement of data that allows human operators and plant engineers both to query and change the operation of these plant devices and to retrieve data for use in business or engineering applications. High-level ILANs are high-speed networks because of the data transfer rates required. They must also provide a high degree of reliability and equipment redundancy. Lower-level networks, with a minimum of extra features, are also available that are specially designed for the efficient movement of real-time sensor and actuator data.

18.7.3 CURRENT MAINTENANCE TECHNOLOGY

To meet global competition, industrial management is now forced to continuously reassess manufacturing processes to meet the objectives of round-the-clock production, high-quality products, and high levels of manufacturing flexibility. Additional major requirements include ensuring plant safety and environmental protection. These goals must be achieved while reducing manufacturing costs. Each of these objectives is directly affected by the maintenance strategies applied.

The evolution of the blend of skilled personnel and the application of maintenance technology has moved significantly beyond on-site analysis and correction of process system errors toward the practical application of the concepts inherent in the science of maintenance, including maintainability as measured by MTTR and, most importantly, reliability associated with MTBF. The latter is associated with predictive maintenance and root cause analysis, i.e., identifying potential problems before failure and if possible, establishing root cause to ensure complete elimination. Beyond the use of these as maintenance technologies is the movement toward applying the same techniques to enhance process performance.

Maintenance technology has made significant advancements to meet these objectives in the last decade. Advances in secured remote support capability enables more effective use of immediate diagnostic support and predictive analysis, pro-active self-monitoring and alarm analysis with dial-back capability, and remote assistance (the vendor has the ability to see and use the customer's unique system displays) including web-enabled information distribution. The application of these technologies facilitates the partnership of responsible plant personnel with correct consultative resources regardless of subject and resource location.

In addition to maintaining system availability, the same technologies can be applied to enhance actual process performance and optimization. This is accomplished using the remote connection with virtual work-groups (experts connected electronically). The process system is the gateway to more extensive plant-wide evaluation. Finally, these tools could then be adopted to allow plant-wide asset availability management.

Given the objectives stated, an integrated support strategy must be developed. Providing timely and effective support is always a major challenge, since industrial customers are often located in remote or isolated areas. This requires the design of a very structured support system that allows for the rapid deployment of support resources to:

1. Improve response time and quality of interactions.
2. Reduce travel to customer sites.
3. Cooperate and share knowledge in a highly cohesive manner.
4. Apply the most qualified resource, regardless of geographic location through virtual workgroups.
5. Ensure project schedules are met through timely assistance.

18.7.4 STRUCTURED GLOBAL CONNECTION

Progressive vendors have developed state-of-the-art technology to remotely connect the experts to customers located around the globe. The use of TCP/IP network strategy takes advantage of the off-the-shelf routers and sophisticated firewall technologies. This innovative approach provides the ability to create a virtual support organization that securely links multiple performance centers together. This enables delivery of uninterrupted, consistent quality service to all customers around the world.

The flexibility of this technology enables the very best people located anywhere in the world working in maintenance support, advanced application, or economic performance consultation to be connected in seconds. Working in virtual workgroups, efficiency is gained and costs controlled by not having to fly people from one location to another. The result is an enabling vehicle allowing the leveraging of global expertise to access remote client systems anywhere in the world through secure network connection.

Security is a major key to success. Customers require safe, secure, and reliable operation of their process control systems during the remote session.

Some recommended security guidelines are:

1. Customer authorization is required for all connections. All remote log-on sessions follow strict procedures to ensure no actions are taken that might affect the ongoing operation of the connected system.
2. The vendor's remote center is a locked facility accessible only to authorized personnel who remotely access the customer's system.
3. The remote support engineers should undergo specialized training in remote analysis techniques. In addition, they should be knowledgeable in general plant operations to help ensure plant safety procedures are followed. Advanced application experts can use the same techniques. This greatly expands the breadth of support capability that can be applied.

4. All communications between the customer system and remote center pass through a secure, virtual private network (VPN) environment, using the highest exportable encryption algorithms.
5. Access to the server database containing sensitive information such as passwords and system configurations is restricted to workstations validated as members of the domain structure. The database is password-protected and all passwords meet strict criteria.
6. Each remote center operates with a firewall configured to allow only predefined networks access to customer sites. This firewall filters traffic to allow only designated protocols and applications to be communicated among its interfaces.
7. Video displays are recorded and all keystrokes are logged and archived. A record of all actions taken can be provided upon request.
8. Once a log-on session is mutually agreed upon between vendor and site personnel, a call-back connection is initiated as follows:
   a. The remote center phones the customer's on-site system to give the correct password authorization.
   b. The connectivity software loaded on the customer's system receives the call, breaks the connection, and reconnects to the vendor's remote center system.
   c. Only then does the flow of data actually begin between the customer's system and the vendor's remote center system.

18.7.5 REMOTE DIAGNOSIS

Once the connection is established, software-embedded diagnostic tools resident in a dedicated maintenance module can move maintenance effectiveness to a higher level.

When users encounter a problem, they first try to resolve it on their own. If they fail to fix the problem, they call the vendor of the process control system and try to solve the problem by discussing it over the telephone. The use of remote monitoring, diagnostics, and maintenance of the process control system is a faster and better way of accomplishing the same process. Typically, a fully featured process control system consists of multiple-bus structures and many components to support the distributed system architecture. Various computers, controllers, and system stations or other modules with specific responsibilities communicate with each other via different types of networks.

Ensuring harmonious operation of all these modules work is "timing logic" that controls both the overall system's own operations, such as sending messages, accessing hard drives, and the control strategies. Network-related problems can be diagnosed by allowing information packets to be filtered and captured for analysis. As a result, the entire process from failure identification to problem correction takes on a new perspective. At the heart of remote support are high-level technical specialists who provide real-time analysis of developing problems. In effect, remote support actually gives plant personnel greater control over how and when equipment service is scheduled and performed. For instance, when a plant personnel first encounters a problem, the remote support function acts as an online diagnostic tool to provide immediate analysis. It logs information relevant to the problem and retains that information for problem resolution. Remote support personnel at the support facility can immediately use this information and initiate appropriate tests. Once the problem is identified and severity assessed, a resolution plan can be developed with scheduled activities. These activities may include actual remote resolution assistance.

18.7.6 PREDICTIVE ANALYSIS

Remote diagnostics can be enhanced to actually predict and circumvent a potential problem before it interferes with the system's availability. This capability gives plant management the option of deferring a repair action until it can be scheduled at an appropriate time. This is a powerful concept that enhances the reliability of the system.

The need to administer a system takes on a special significance if the equipment is to work at peak efficiency throughout its product lifecycle. Over the years, the individuals working on and maintaining the equipment may also change. Further, as the size of a network-based system grows over time, it becomes increasingly important to monitor the performance of all the pieces of equipment. This is because degradation in the performance of one component can affect other components on the network.

With predictive analysis, however, users can monitor the overall system and gradually build up a performance profile that can serve as a relative indicator of its "health." For example, each system component may be monitored to ensure that it is running at the correct revision level and to check that the error levels are within acceptable limits. Each bus is analyzed to ensure that it is operating within specifications, and various critical files are diagnosed to detect corrupted, duplicated, or misplaced data.

The key capacity measurements that are tracked include idle time and memory availability of the CPU that runs the various applications, and the work load of the control processor. Such measurements can quickly become complex, since a single process controller can consist of several different subsystems. Also, the devices that are most likely to cause problems are identified and monitored closely to keep the overall system's performance from degrading. In fact, plant personnel can remain in control of the total system only when they are able to factually pinpoint the source of the trouble. The ability to identify and eliminate potential problems obviously enhances plant safety and reduces environmental upsets.

18.7.7 EVENT-ACTIVATED RESPONSE

The reliability of a process control system can be increased by continuous self-monitoring for designated alarms. If a critical system event triggers an alarm, an automatic event-activated response function simultaneously notifies designated customer operations personnel and the vendor remote support center. Once connected, the system uploads all the data related to the alarm condition for analysis by remote support engineers.

Process systems and their associated problems are too complex to handle individually and manually. The task of predicting and managing failures becomes more manageable when the system itself begins the diagnostic process. Alert notification begins with vendor-supplied software that continuously listens to the system and initiates an "event-activated response" when a process abnormality appears.

In the event of such problems, the plant's technical personnel are alerted by the system's alarm flashing on their workstations. A message describing the alarm type usually accompanies the alarm. At the same time, a program in the remote-monitoring module initiates remote connection to the vendor's technical service center and uploads the particular alarm sequence. At the vendor site, service engineers begin analyzing the problem based on the incoming data. In addition, an engineer could call the plant site to confirm that there is a problem and initiate a process-monitoring session. All the information generated is stored for future reference.

This capability can also extend to performance monitoring. This response is initiated when certain types of process alarms are encountered. All information connected with the system's performance is recorded, followed by data analysis to reveal the source of the problem.

18.7.8 KNOWLEDGE-BASED WEB ACCESS

A library of support documentation can be invaluable because it enables the user to answer questions quickly to determine if an issue has been previously resolved. The knowledge base website delivers up-to-the-minute technical reference information, news, and online communication 24/7.

Topic categories may include software correction text files, helpful hints, product specification sheets, and special notices, among others. One asset of a technical knowledge base website may be access to downloadable software. Another function enabled by website technology is the ability to "push" pertinent news bulletins to registered website users. Keyword search methodology facilitates users' abilities to locate extensive technical support information on vendor products. The website also provides an efficient means to communicate with technical specialists and company staff.

If a product issue is experienced repeatedly or there is a need to formally request an additional feature or enhancement to a product, a customer request mechanism can be instituted as the preferred channel of communication and documentation, through the website. The integrity of website documentation must be well maintained by technical personnel. The day-to-day activities of support personnel can generate much of the technical information posted.

18.7.9 ONLINE INTERACTIVE ASSISTANCE

User assistance is a critical requirement for customers seeking help in all aspects of equipment utilization software, hard drive and memory allocations, how to configure programs, initial setup and usage, and system and application engineering assistance. This new remote service capability revolutionizes customer support in this increasingly important area. The ability to view exactly what the user sees locally is a far superior support technique than blind telephone discussion. The sharing of remote display screens allows both user and vendor personnel to easily guide each other through complex multi-step procedures, leading to faster problem resolution.

18.7.10 REMOTE PERFORMANCE SUPPORT

In today's competitive global marketplace, the appropriate utilization of sophisticated advanced process control technology can often provide a significant economic advantage. Lifecycle benefits can be maximized through advanced remote application support capabilities that enable geographically distributed, virtual work teams to effectively support customers' advanced control applications.

Until lately, the concentration has been on using support technology to ensure the integration of system availability and accuracy of defined functions. Only after this is accomplished can the enhanced utilization of the system application be assured.

The same tools are again applied to the next level of remote performance support. After economic studies are completed and baselines established, remote monitoring is critical to maintain and ensure the on-going benefits.

Remote location of most of the plant facilities poses a major challenge for vendors to provide effective support in a timely manner. This requires vendors to design a very structured support system using technology that rapidly deploys resources to:

1. Improve response time and quality of interactions.
2. Reduce travel to customer sites.
3. Cooperate and share knowledge in a highly cohesive manner.
4. Apply the most qualified resource regardless of geographic location through virtual workgroups.

FURTHER READINGS

Goettsche, L.D. (Ed.), 2005. Maintenance of Instruments and Systems. Second ed. ISA, Practical Guides for Measurement and Control Series.

Lipták, B.G., Editor-in-Chief. 2003. Instrument Engineer's Handbook: Process Measurement and Analysis, vol. I, fourth ed. ISA & CRC Press.

CHAPTER

ADVANCED PROCESS CONTROL SYSTEMS

19

19.1 INTRODUCTION AND NEED FOR ADVANCED PROCESS CONTROL (APC)

Control designs in the process industry are almost exclusively based on PID controllers these days. Even though they are simple to implement and easy to integrate into a control system, these feedback control strategies have their inherent drawbacks of dead time (lag in system response to changes in set point). Although feed forward control avoids the slowness of feedback control, the effects of disturbances should be perfectly predicted for them to be effective. However, most of the complex processes have many variables that are to be regulated and have multiple control loops. These multiple control loops often interact, causing process instability. Thus, the traditional feedback and feed forward control strategies quickly reach their limits when more complexity is involved. Advanced Process Control or APC opens up new opportunities at this juncture.

The area where APC applications operate is described in Figure 19.1.

With APC, even complex situations can be mathematically described with process parameters or variables and then used for automatic and flexible plant operations. APC provides process management that can significantly reduce the consumption of energy and raw material, consistently maintain high quality standards, and contribute to more flexible production.

19.2 HISTORY OF PROCESS CONTROL

A brief look into the history of process control will show the significance of APC.

By mid 1950s, process sensors, pneumatic transmission of process data, pneumatic controllers, and valve actuators had become highly developed forms of automated control. Through these technologies, significant savings in operations were achieved. During this period, process analyzers for on-stream analysis became available providing operators with more specific and timely information than just process flows, temperatures, pressures, and levels. The classical control theory began to be developed by academic institutes and the major control companies.

From the late 1950s to the early 1960s, electronic instrumentation involving electronic transmission and control became more prevalent. The instrument industry held heated debates on the relative merits of electronics versus pneumatics. Some standardization was achieved while instrument vendors championed their own systems. Also, data logging was introduced.

The appearance of low-cost digital mini-computers in the early 1960s brought about a significant milestone – the use of digital computers to control refineries and chemical plants. Joined development efforts between major users and instrument and computer companies were undertaken and computer control systems software began to evolve rapidly.

Optimization	Local LP/QP optimization	Multi unit LP/QP optimization	Rigorous model-based optimization
Model-based control	Smith predictor, IMC, etc.	Multivariable control	
Advanced control	Feed forward control	Dynamic decoupling	Online analyzers, constraint control
Regulatory control	Single PID	Cascade control	

Optimizations

Different technologies at different levels

FIGURE 19.1 Different technologies at different levels

By mid 1960s to mid 1970s, electronic analog instrumentation became prevalent due to its ability to meet industry requirements and provide reliable computer backup and interface to the process. But, the shift from analog to digital was already underway. Visionaries saw that Distributed Control System (DCS) could place multiple loops under one computer chip and implement operator communications through an electronic data highway rather than dedicated panel instruments.

While the hardware as well as operating system software for implementing process control was of primary importance to the success of projects, it was applications and process control strategies that reaped the major economic benefits. Process Control practitioners in the refinery and chemical industry recognized that the control concepts of the period generally outplaced the practical implementation and that much of the theory was difficult to apply to real process. The most prevalent reasons for lack of practical implementation were:

- Lack of understanding of the process
- Strong interactions among variables
- Process nonlinearity
- Few accurate mathematical models.

However, as practitioners in the industry gained experience with computer control of commercial plants, the incentives were recognized and realized. These incentives derived from:

- Improved regulatory and advanced control
- Better understanding of process dynamics and unit operations
- Operating closer to constraints
- Finding the most profitable operating conditions.

With feed forward and interactive controls, the relationships between different variables had become much more complex and far less apparent.

With the advent of much more powerful microprocessors in the late 1970s and early 1980s, the architecture of the microcomputers at the bottom performing repetitive, simple operations and the host computer at the top doing the complex calculations requiring a lot of number crunching power became the backbone of most subsequent computer control installations. By the late 1980s, multivariable model-based predictive controllers enabled control in a single program. With predictive capability, a controller can make the moves necessary to precent any constraint violation before it occurs, rather than reacting after the fact as PID controllers are forced to do. Increasingly, plant testing of process response dynamics to develop and improve the multivariable control models became a strong priority.

By the mid 1990s, essentially every refinery was using LP (Linear Programming) or other simulation models for off-line business optimization and to provide volumetric signals or guidance for raw material selection, operating throughputs and intensities, and desired product slates. By 1995, the technology was working better due to the emergence of the larger model-based multivariable predictive controllers that used the actual plant test data to simulate plant dynamics. Many companies were switching to these multivariable controllers and deactivating some of their single-loop advanced controllers. These large model-based controllers could just manage the dynamic complexities and interactions much better and with some synergy. These predictive controllers were installed on an outboard computer to the DCS and deactivated many loops in the DCS.

19.3 ADVANCED PROCESS CONTROL

Advanced Process Control (APC) is a broad term composed of different kinds of process control tools for solving multivariable control problems or discrete control problems. APC draws its elements from many disciplines ranging from Control Engineering, Signal Processing, Statistics, Decision Theory, and Artificial Intelligence.

Few different process control tools involved in APC are:

- Model Predictive Control (MPC)
- Statistical Process Control (SPC)
- Run2Run (R2R)
- Fault Detection and Classification (FDC).

Some of these are briefed in the sections below.

19.3.1 MODEL PREDICTIVE CONTROL

Model Predictive Control or MPC is an advanced method of process control that has been in use in the process industries such as chemical plants and oil refineries since the 1980s and has proved itself. Model Predictive Controllers rely on the dynamic models of the process, most often linear empirical models obtained by system identification.

MPC possesses many attributes that make it a successful approach to industrial control design:

- Simplicity: The basic ideas of MPC do not require complex mathematics and are "intuitive."
- Richness: All of the basic MPC components can be tailored to the details of the problem in hand.
- Practicality: It is often the resolution of problems such as satisfying control or output constraints, which determines the utility of a controller.
- Demonstrability: It works! – as shown by many real applications in industry where MPC is routinely and profitability employed.

While MPC is suitable for almost any kind of problem, it displays its main strength when applied to problems with:

- A large number of manipulated and controller variables.
- Constraints imposed on both the manipulated and the controlled variables.
- Changing control objectives and/or equipment (sensor/actuator) failure.
- Time delays.

MPC models predict the change in the dependent variables of the modeled system that will be caused by changes in the independent variables. In a chemical process, independent variables that can be adjusted by the controller are often either the set points of regulatory PID controllers (pressure, flow, temperature etc.) or the final control element (valves, dampers, etc.). Independent variables that cannot be adjusted by the controller are used as disturbances. Dependent variables in these processes are other measurements that represent either control objectives or process constraints. MPC uses the current plant measurements, the current dynamic state of the process, the MPC models, and the process variable targets and limits, to calculate future changes in the independent variables. These changes are calculated to hold the dependent variables close to target while honoring constraints on both independent and dependent variables. The MPC typically sends out only the first change in each independent variable to be implemented and repeats the calculation when the next change is required.

19.3.1.1 Linear MPC

While many real processes are not linear, they can often be considered to be approximately linear over a small operating range. Linear MPC approaches are used in the majority of applications with the feedback mechanism of the MPC compensating for prediction errors due to structural mismatch between the model and the process.

Theory behind MPC

MPC is based on iterative, finite horizon optimization of a plant model. At time t, the current plant state is sampled and a cost-minimizing control strategy is computed (via a numerical minimization algorithm) for a relatively short time horizon in future: $[t + T]$. Specifically, an online or on-the-fly calculation is used to explore state trajectories that emanate from the current state and find a cost-minimizing control strategy until time $t + T$. Only the first step of the control strategy is implemented, then the plant state is sampled again, and the calculations are repeated starting from the now current state, yielding a new control and new predicted state path. The prediction horizon keeps being shifted forward, and for this reason, MPC is also called *Receding Horizon Control*. Although this approach is not optimal, in practice it has given very good results.

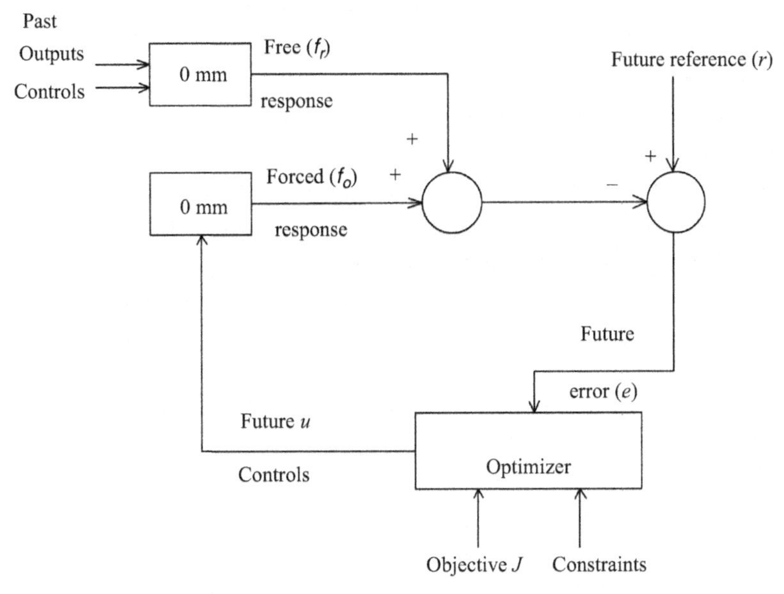

FIGURE 19.2 Principle of MPC Controller

Principles of MPC

MPC is a multivariable control algorithm that uses:

- An internal dynamic model of the process
- A history of past control moves
- An optimization constant J over the receding prediction horizon.

A conceptual diagram illustrating the principles of an MPC controller described above is shown in Figure 19.2.

The heart of the controller is a model M(θ), parameterized by a set θ, which is used to predict the future behavior of the plant. The prediction has two main components: the free response (f_r), being the expected behavior of the output assuming zero future control actions, and the forced response (f_o), being the additional component of the output response due to the "candidate" set of future controls (u). For a linear system, the total prediction can be calculated as $f_o + f_r$.

The reference sequence (r) is the target values the output should attain. The future system errors can then be calculated as $e = r - (f_o + f_r)$, where f_o, f_r, and r are vectors of the appropriate dimensions.

An optimizer having a user-defined objective function $J(e,u)$ is used to calculate the best set of future control actions by minimizing the objective function, $J(e,u)$. The optimization is subject to constraints on the manipulated variables (MVs) and controller variables (CVs).

What makes MPC a closed-loop control law is the use of the receding horizon approach. This implies that only the first of the set of control actions, u, is transmitted to the plant, after which the complete optimization and prediction procedure is repeated using the current plant output.

19.3.1.2 Nonlinear MPC

Nonlinear Model Predictive Control (NMPC) is a variant of MPC that is characterized by the use of nonlinear system models in the prediction. As in linear MPC, NMPC required iterative solution of optimal control problems on a finite prediction horizon. While these problems are convex in linear MPC, in NMPC they are not convex anymore. This poses challenges for both NMPC stability theory and numerical solution.

19.3.2 STATISTICAL PROCESS CONTROL (SPC)

Statistical Process Control (SPC) is the application of statistical methods to the monitoring of a process to ensure that it operates at its full potential to produce conforming product. Under SPC, a process behaves predictably to produce as much conforming product as possible with the least possible waste. While SPC has been applied most frequently to controlling manufacturing lines, it applies equally well to any process with a measurable output. Key tools in SPC are control charts, a focus on continuous improvement, and designed experiments.

19.3.2.1 Concept

No matter how tightly controlled and well run a process is, variations exist in the quality of the resulting product. It is important to evaluate these variations to determine if the resulting product characteristics are within acceptable quality limits.

The causes of variations can be separated into two distinct classes. Some variations are inherent in the process itself and can be called normal (sometimes also referred to as common or chance). Other variations can be attributed to "special causes" that are outside of the process. Possible "special causes" could be improperly calibrated instruments, insufficient training, outdated reagents, etc. As these "special causes" of variations are detected, procedures can be developed and implemented to ensure that these do not continue. SPC allows us to detect when these few special causes of variation are present. Once removed, the process is said to be stable, which means that its resulting variation can be expected to stay within a known set of limits, at least until another special cause of variation is introduced.

SPC techniques use random sampling and statistical analysis instead of continuous monitoring to determine, with almost complete confidence, whether the variation is due to the process itself or due to special causes. When SPC techniques are used, the frequency and timing of measurement testing are usually defined by a statistician.

19.3.2.2 Advantages

Much of the power of SPC lies in the ability to examine a process and the sources of variation in that process using tools that give weight to objective analysis over subjective opinions and that allow the strength of each source to be determined numerically. Variations in the process that may affect the quality of the end product or service can be detected and corrected, thus reducing waste as well as the likelihood that problems will be passed on to the customer. With its emphasis on early detection and prevention of problems, SPC has a distinct advantage over other quality methods, such as inspection, that apply resources to detecting and correcting problems after they have occurred.

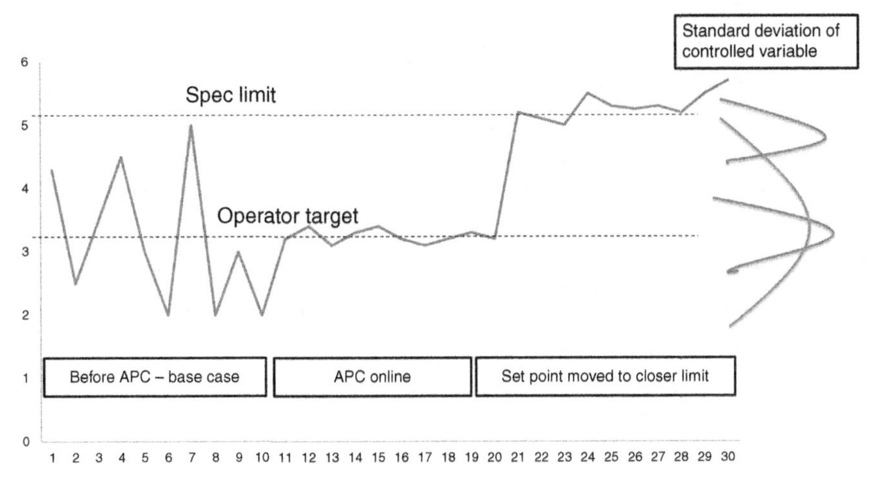

FIGURE 19.3 Sample benefits of APC

19.4 ADVANTAGES OF APC

Typical benefits of APC are improved bottom line results including:

- Efficiency gains
  - Increased yield
  - Increased throughput
  - Reduction in energy and raw material per unit of product
  - Decreased operating costs
- Quality gains
  - Consistent product quality
- Agility gains
  - Increased operating flexibility
  - Improved process stability

By implementing APC, benefits ranging from 2% to 6% of operating costs have been quoted. These benefits are clearly enormous and are achieved by reducing process variability, hence allowing plants to be operating to their designed capacity (Figure 19.3).

19.5 ARCHITECTURE AND TECHNOLOGIES

Many leading companies recognize that APC applications can produce significant improvement in control of complex processes, particularly those with long dead/lag times, interacting loops, highly constrained operations, or inverse response. Historically, APC implementation required very specialized skill-sets and experienced resources to implement and maintain – limiting use of the technology to only very large refineries or petrochemical plants that could justify such an expense. New embedded APC tools offered by some automation suppliers are starting to change this situation. Ease-of-use features designed into these tools aim to make APC blocks almost as easy to use as a PID loop.

A worldwide survey conducted by Qin and Badgewell showed that, of the roughly 2200 installations surveyed, over 82% of all APC applications were implemented in the refining and petrochemical industries and the majority of these applications were in large facilities of the major refineries.

Traditional APC technology is usually implemented in a supervisory architecture similar to that depicted in Figure 19.4.

In this environment, the APC applications are executed in a separate computer interfaced by some means to a DCS. The APC application will calculate the moves required, which are sent to the DCS that

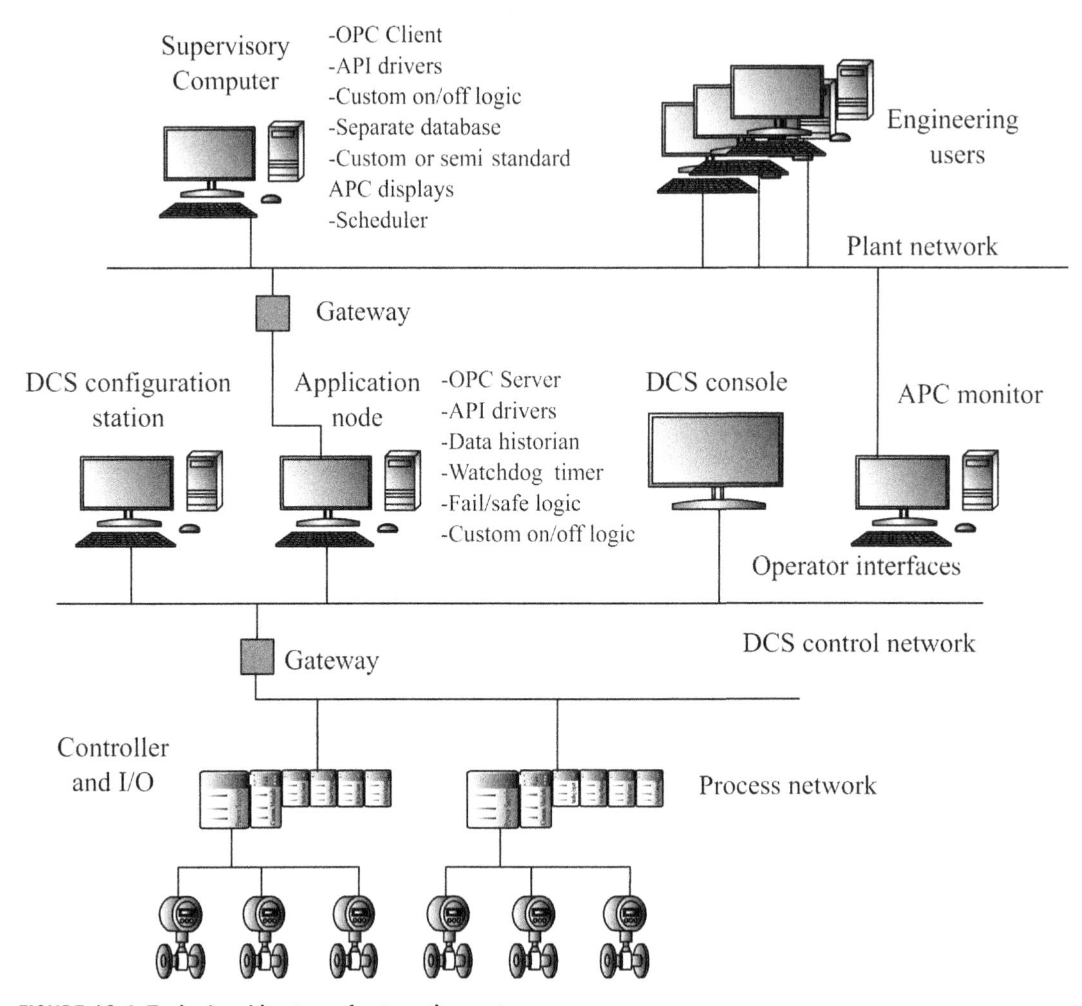

FIGURE 19.4 Typical architecture of automation system

performs the basic control functions through the regulatory controllers and field IOs. In many cases, the DCS and APC systems are interfaced via the industry standard OLE for Process Control (OPC) protocol.

The supervisory system usually has its own user interface, DCS drivers, database, scheduler, and tag synchronization issues. Usually, at least some level of custom programming is required in the DCS to provide the operator on/off functions, fail-safe logic, and watchdog timer functions. Step-testing the process often required 24-h engineering coverage for days or weeks at a time. Furthermore, APC applications historically required very experienced consultants with specialized skills to implement and maintain. As a result, only the largest process with the biggest potential benefits could afford to implement these technologies.

Unfortunately, a significant portion of APC applications implemented over the years have been turned off within a few years after commissioning. The major reasons for this step are:

- Regulatory control problems – The basic regulatory loops must work well before an APC application has any chance of success. Malfunctioning valves, poor tuning, and controllers in manual can cause APC performance to deteriorate.
- Process changes – Any change to the process that affects the controller design or significantly changes the dynamics or gain of the process models will require additional work to update the APC application.
- New constraints or limits – Process or equipment limits that were not considered in the original control design must be incorporated into APC strategy.
- Different control objectives – Sometimes the process operation objectives change from the original design due to changes in economics, feeds, constraints, or operating conditions.
- Controller requires restepping process – Any time process dynamics change significantly, the process needs to be restepped and the models refit to reduce model errors. This can be an expensive and time-consuming process.
- Applications not maintained – Applications need to be continually revised to stay up with the latest operating systems and software versions. Once the applications get too far out of date, it becomes prohibitive to upgrade them without significant investment.
- Lack of operator training – If the operators do not understand what the APC controller is doing, it will get turned off. It is important that operators are properly trained on APC technology and advanced control strategies to ensure uptime is maintained.
- Budget constraints – Many a times, software maintenance is not budgeted like other equipments in the plant. Also, the cost to hire APC experts to redesign, reconfigure, step-test, model, update documentation, and recommission an existing APC application can be almost as much as the original engineering services.

With the advent of new embedded APC tools, many of these problems go away. Embedded APC functions eliminate the need of a separate supervisory system and all the extra databases and programming that go along with it. The new tools are just part of the automation architecture – like a PID block – completely removing a whole layer of complexity in systems, software, databases, and interfaces.

Under the new architecture, APC functions can be distributed and executed on multiple controllers or application stations running on the native control system bus as in the architecture shown in Figure 19.5.

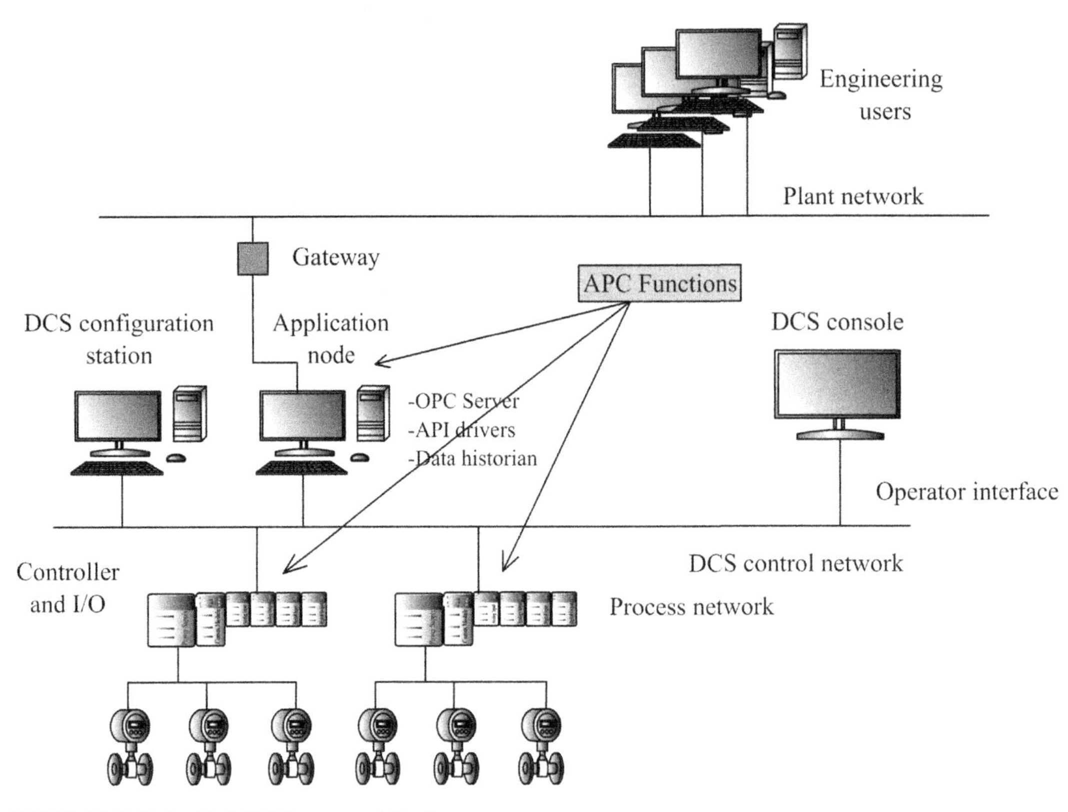

FIGURE 19.5 Embedded APC in new architecture

As a result, the effort to implement and maintain these applications is dramatically reduced. With these new systems, there are:

- No extra databases to maintain on the supervisory system.
- No database synchronization issues as points are added, changed, or recalibrated.
- No watchdog times required to confirm that the APC application is still working.
- No controller fail/shed logic design to automatically handle failure of an APC application.
- No interface configuration or programming to communicate between the DCS and the supervisory computer.
- No separate operator interface monitors or custom graphics for the APC functions.

A few vendors offer embedded APC tools that can run entirely in automation system controllers, in a high-speed, robust, and redundant environment. This architecture opens the technology to a whole new class of control problems, including those with very high-speed dynamics or applications that need to output directly to a valve instead of a PID controller set point.

FURTHER READINGS

http://en.wikipedia.org/

http://www.pacontrol.com/AdvancedCtrl.html

http://upetd.up.ac.za/thesis

http://www.emersonprocessxperts.com/articles/HydrocarbonProcessing

https://www.honeywellprocess.com

Blevins, T., McMillan, G.K., Wojsznis, W.K., Brown, M., 2003. Advanced Control Unleashed – Plant Performance Management for Optimum Benefits. ISA.

Kane, L.A. (Ed.), 1999. Advanced Process Control and Information Systems for the Process Industries. Gulf Publishing.

CHAPTER

TRAINING SYSTEM

20

20.1 INTRODUCTION TO PROCESS MODELING

Mathematical representation of any physical system or process is called model. A model is formulated based on performing mass, energy, and pressure balance on the system or process. Apart from the mass, energy, and pressure balance, thermodynamic calculations are needed to predict the properties and conditions of fluids in the system or process. Simulation is the process of solving the mathematical model.

The model equations can be classified as steady state, dynamic, lumped, distributed, stochastic, and empirical modeling. Stochastic modeling is not mostly used in training simulators/systems environment.

Process modeling and an understanding of simulation are prerequisite for understanding training simulators/systems. Therefore, let us attempt to understand these topics before we discuss training simulators.

20.1.1 STEADY-STATE OR DYNAMIC MODELING

A steady-state model does not account for the element of time, while a dynamic model does. Steady-state models are typically represented by algebraic equations, whereas dynamic models are represented by differential equations.

20.1.2 LUMPED OR DISTRIBUTED MODELING

A distributed model accounts for variation of physical properties with respect to space (for example, concentration gradient in three dimensions x, y, and z) and time. Such a system defined in mathematical equation will give partial deferential equation.

While lumped model assumes the properties are equal in space, it only considers variation with respect to time. A lumped model is typically represented by an ordinary differential equation (ODE), whereas a distributed model is represented by a partial differential equation (PDE).

20.1.2.1 Governing equation for modeling

Conservation of mass and energy of the system is the basis of modeling.

For steady-state model

$$\text{Rate of input} - \text{Rate of output} = 0 \tag{20.1}$$

For dynamic model

$$\text{Rate of input} - \text{Rate of output} = \text{Rate of accumulation} \tag{20.2}$$

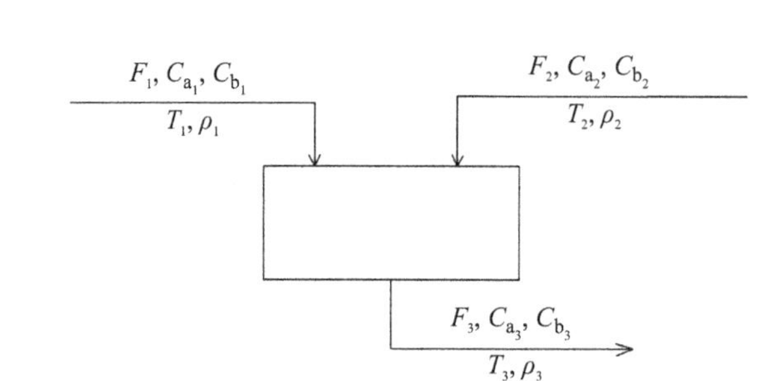

FIGURE 20.1

Input and output of process

20.1.3 STEADY-STATE MODELING

Steady-state systems are typically represented by algebraic equation, as illustrated by the following simple example. If two different streams at different flow rate (F, m^3/h), temperature (T, °C), and concentration (C, moles/m^3) enter a mixing tank, they are mixed thoroughly to get uniform temperature and concentration outlet stream (Figure 20.1).

Mass balance:

$$F_1\rho_1 + F_2\rho_2 = F_3\rho_3$$

Component balance:

$$F_1.C_{a_1}.\text{MW}_a + F_2.C_{a_2}.\text{MW}_a = F_3.C_{a_3}.\text{MW}_a$$

$$F_1.C_{b_1}.\text{MW}_b + F_2.C_{b_2}.\text{MW}_b = F_3.C_{a_3}.\text{MW}_b$$

Energy balance:

$$F_1\rho_1.C_{p_1}(T_1 - T_R) + F_2\rho_2.C_{p_2}(T_2 - T_R) + Q = F_3\rho_3.C_{p_3}(T_3 - T_R)$$

where

F_1, F_2, and F_3	inlet and outlet streams, m^3/h
T_1, T_2, and T_3	temperature of streams, °C
ρ	density of streams, kg/m^3
C_a and C_b	concentration components A and B kgmoles/m^3
MW	molecular weight of components
C_p	heat capacity, kJ/kg
T_R	°C

20.1.4 DEGREE OF FREEDOM

Mixing tank is considered as a system in steady-state modeling. It is represented by four algebraic equations, with 16 variables. Solving a set of equations is called simulation. Therefore, steady-state simulation will involve the solving of the above four algebraic equations to simulate the system.

Degree of freedom is one of the basic fundamentals behind simulation of any system. The degree of freedom should be zero for any system to be simulated.

Degree of freedom (DOF) = Number of variables − Number of equations

For example, the calculation of number of variables present in a sample model is as follows:

For inlet streams F_1, C_{a_1}, C_{b_1}, T_1, and ρ	$5 \times 2 = 10$
Similarly, for outlet stream	= 5
Energy input	= 1
Total	= 16

In this case DOF is (16 − 4 = 12). To make the DOF zero we have to specify 12 variables to simulate the system.

20.1.5 DYNAMIC MODELING AND SIMULATION

Dynamic modeling is accounted for by time elements. Therefore, simply put, dynamic modeling is the modeling of process with respect to time. Steady-state model equation gives the output condition for a given input condition; it does not tell us how the particular output variable is responding for given input over time before reaching steady state. This kind of transient response can be studied using dynamic modeling. This is represented by ODE with time derivative. The following simple example illustrates the dynamic model equation for a tank with isothermal condition (constant temperature) (Figure 20.2).

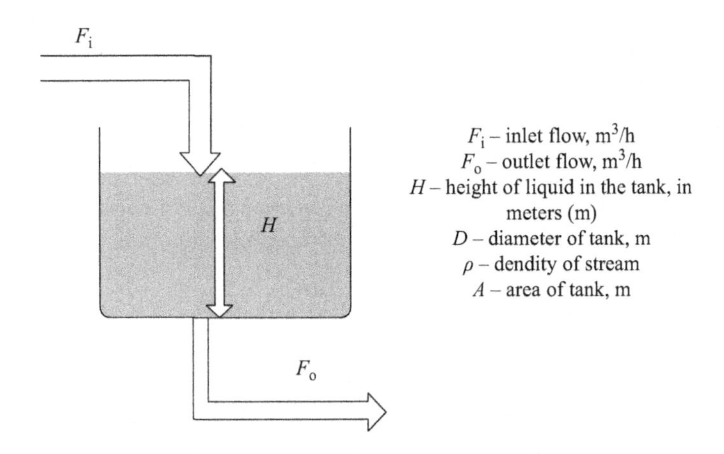

FIGURE 20.2

Process for simple modelling

Mass balance:

$$F_1\rho_1 + F_2\rho_2 = F_3\rho_3$$

Rate of mass accumulation = Inlet flow rate − Outlet flow rate

$$\frac{\rho \mathrm{d}V}{\mathrm{d}t} = F_\mathrm{i}\rho_\mathrm{i} + F_\mathrm{o}\rho_\mathrm{o}$$

$$\rho A \frac{\mathrm{d}H}{\mathrm{d}t} = F_\mathrm{i}\rho_\mathrm{i} + F_\mathrm{o}\rho_\mathrm{o} \tag{20.3}$$

If we assume a constant outlet flow rate in this system, then a small change in the inlet flow rate affects the height of liquid in the tank. Solving the above equation helps us in finding the variation in height of liquid in the tank over time.

In the same example if we remove the assumption that F_o is constant and varies with H then,

$$F_\mathrm{o} = f(H)$$

Substituting the above equation in Equation (20.3) and solving the same gives us the transient response of H and F_o before reaching the steady state.

$$A\rho \frac{\mathrm{d}H}{\mathrm{d}t} = F_\mathrm{i}\rho_\mathrm{i} + f(H)\rho_\mathrm{o}$$

20.1.6 APPLICATION OF MODELING AND SIMULATION

Modeling is being used extensively in research and development, design, and plant operation. Unlike earlier times, the use of computers has made it easy to solve modeling (simulation).

1. Determining reaction kinetics and parameter from pilot plant study – by steady-state simulation.
2. Control studies and optimization of operating condition – by dynamic simulations.
3. Sizing of equipment, scale-up and scale-down studies of production plant – by steady-state simulation.
4. Simulating startup, shutdown, and emergency situation to analyze the control strategy and designing safety system – by dynamic simulations.
5. To study the interactions of various equipment – by steady-state and dynamic simulation.
6. Operator training to handle normal plant operation, emergency situation, and troubleshooting.

20.2 TRAINING SYSTEMS

Training simulator systems is a tool that imitates any system (process or control) digitally to train the human resources. Dynamic modeling and simulation play a major role for training simulator. Robustness and accuracy of dynamic modeling (model equations) make training simulator more realistic.

The training simulator aims to simulate the real system behavior with less complicated equation with limited assumptions. One of the major assumptions in training simulator is that any system is considered as lumped (hold-up of any system is considered as single entity with uniform condition such as temperature, pressure, and composition). This eliminates distributed modeling and leads to partial derivative equation, which is difficult to solve.

The following are some of the benefits of training simulator.

1. Improves the safety of any process plant.
2. Improves the operator efficiency.
3. Reduces the off spec or non-conformances that improve the yield of a process.
4. Operator/Engineer can be trained on startup and shutdown procedure without loss of material.
5. Control strategies are validated for some extreme conditions of process plant or unit operation to validate the process control.
6. Operators can be trained to handle emergency situation of process plant to ensure recovery of plant to normal condition with minimal loss of material.
7. Operators/engineers can be trained before commissioning the plant.
8. If the model is accurate, engineering studies can be conducted for some important unit operations such as reactor, column, and so on.
9. Good training simulator model can be used for advanced process control implementation (research).

20.3 COMPONENTS OF TRAINING SIMULATORS SYSTEM

The following are the basic components of training simulator architecture

- Process model
- Control model
- Connectivity of process model and control model.

20.3.1 PROCESS MODEL

All simulators use a modular building block approach to create a mathematical model of the operational process. The modules are coded to create computer subroutines, which represent the basic operating features of the physical equipment and regulatory control functions. Sometimes an individual piece of equipment may require several basic modules to represent its operation. The mathematical model is created by selecting the modules necessary to represent the process and connecting the associated streams in accordance with the process piping, control configuration, and logic diagrams. The process model replicates the dynamic behavior of the actual plant.

20.3.2 CONTROL MODEL

A control model refers to the execution of all distributed control system (DCS)/PLC strategies/logics. Just like in a real plant, control models are needed to monitor and control the simulated process for

simulated dynamic process model. The control model can run on the DCS control simulator or it could run as control emulation within training simulator.

20.3.3 CONTROL SIMULATION AND CONTROL EMULATION

Simulation and emulation terminologies can mean different things in different contexts. We are discussing these terminologies in the context of training simulator.

Control simulation: Embedded (real) controllers are used for collecting and controlling the plant parameters (transmitters and control valves) through the input and output channels of DCS modules. Instead of embedded controllers, the DCS vendors perform the similar operation on computers using simulated controllers and I/Os. This is called control simulation.

20.3.4 NEED FOR CONTROL SIMULATOR

The preparation of control strategies/logic begins before the actual plant is commissioned. Once the actual plant is ready, the DCS engineer validates if the control strategies/logic are working as per input and output from/to real I/Os. This could take considerable amount of time and delay the plant startup time.

To overcome this problem, lately, DCS vendors came up with their own control simulators. These simulators run on computers instead of embedded hardware and they have simulated I/Os. This enables the DCS engineer to build the strategies using the simulated controllers and simulated I/Os, and validate them using simulated values.

Migrating the strategies/logic from simulators to real controller is accomplished at the click of a few buttons, thereby drastically reducing engineering effort. Similarly, all the logic is tested without real I/Os. Control emulation: Instead of running control model in DCS/PLC simulators, replication of similar algorithm in the training simulator's own dynamic simulator is called emulation.

20.3.5 NEED FOR CONTROL EMULATION

In the absence of control simulators in the mid-1980s and 1990s, training simulator vendors had to emulate the control logics in their own dynamic simulators for control models.

Emulation refers to the working of control strategies/logics the way they work in DCS. For example, a PID controller working in a particular DCS must work in a similar way in dynamic simulator (this includes the dynamic responses).

This is a tedious task and is almost like rewriting another DCS control algorithm for a dynamic simulator, requiring huge time. However, the benefit of emulation was both control and process model run in a single system and the customers need not procure additional computers and licenses from DCS vendors for training simulator. However, emulated system may not mimic real control system look and feel.

As of now, most of the training simulator solutions are delivered using control simulations and not control emulations because all DCS vendors have their own control simulators. In addition, the graphic-user interface (GUI) in training simulator is similar to that of real DCS and the training simulator operators get same look and feel as real systems. Also, it is preferred since the operator gets fully familiarized with real control system GUI.

20.4 ARCHITECTURE OF A TYPICAL TRAINING SIMULATORS

20.4.1 TRAINING SIMULATORS FEATURES

- Training simulator access level (engineer, instructor, and trainee)
- Time manager (run, stop, speedup, slowdown)
- State save and restore
- Event capture
- Trend
- Alarms
- Malfunctions
- Training and evaluation
- Playback of captured events
- GUI
- Training exercises
- Interface with control simulators (Figure 20.3)

Training simulator features might be different across different training simulators. However, the following information is applicable for most of training simulator.

There are three access levels (engineer, instructor, and trainee) in training simulator:

- Training simulator engineer: Training simulator engineer is responsible for training simulator maintenance. The engineer may be involved in the creation or modification of training simulator simulation models, testing and tuning of simulation model performance, and the incorporation of new process parameters that reflect changes in the plant.

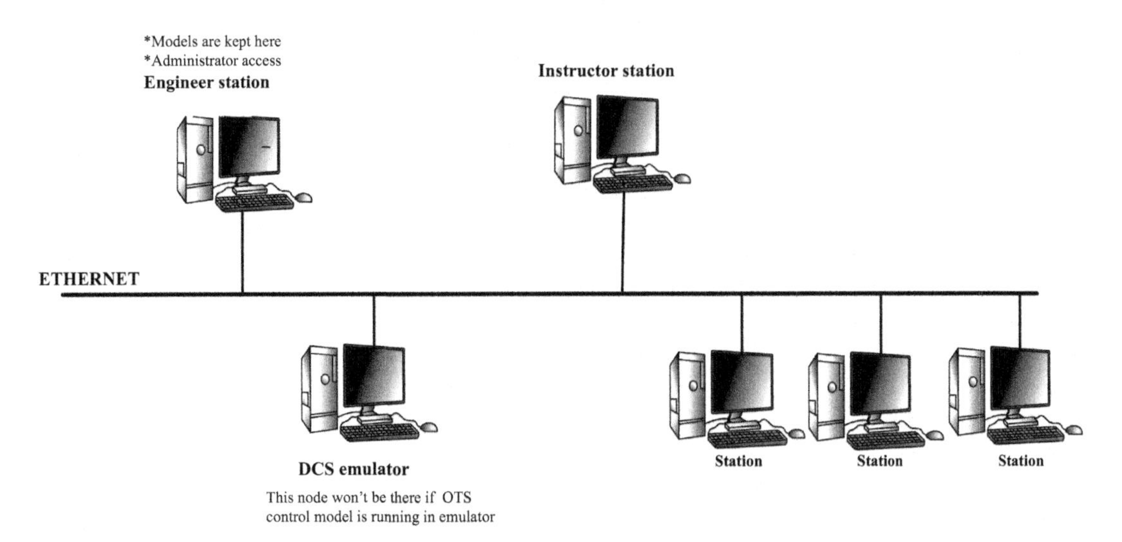

FIGURE 20.3

Architecture of a training system

- Training simulator instructor: The instructor uses training simulator to teach process operation procedures to trainees. The training process involves the development of trainee exercises. These training exercises are based on a process event or series of events that occur during the course of the simulation. The instructor uses the training simulator user interface to administer these training exercises, to prepare testing exercises, and to evaluate a trainee's progress. During a training exercise, an instructor can control the simulation through changing the simulation time, introducing malfunctions, or manipulating control points. The trainee's responses are monitored and evaluated by the instructor.
- Training simulator trainee: The trainee uses training simulator to learn process operation and control procedures. A trainee may be an operator, an engineer, or any individual that requires instruction with DCS console process operation. The trainee will respond to process requirements through manipulation of control devices, process equipment, and field-based systems.

20.4.2 TIME MANAGER

Training simulator is a dynamic modeling based tool that dynamically imitates real plant behavior with respect to time. It enables viewing the response of system in real time or future of model for a given change in the condition of the system. Therefore, the time element plays a major role in the training simulator.

Time manager enables the user to control the model simulation time. It is possible to freeze the model calculation, which is equivalent to stopping the real system at a particular condition, or state training simulator operator can stop the plant simulation and start at any time from the state where they stopped.

Simulation can be speeded up to see the behavior as soon as possible instead of waiting in real time to see the response. Similarly, user can slow down the simulation and see the response to understand some spontaneous critical response, which is very difficult to observe in real time.

20.4.3 STATE SAVE AND RESTORE

In a real system it is not possible to restore the state of a plant directly; operator must start from cold condition and they can be trained on particular state of the process. In the case of a training simulator, it is possible to save the state of particular simulation condition and restore whenever needed, and we can train on same state repeatedly without having to start the plant from cold condition. This leads to time saving and ease of training.

Therefore, any state in the simulation can be saved and restored at any condition and the training can be proceeded without any time delay unlike a real plant.

20.4.4 TRENDS

The trending feature of the training Simulator enables the tracking and monitoring of any parameter, just like in a real-time DCS. Therefore, a set of interrelated parameters can be grouped together for the trend to be plotted continuously. A trainee/operator can view important system parameter response at a single place and take necessary action.

20.4.5 ALARMS AND EVENTS

Training simulator system supports alarm reporting and logging simulation, operator action, and other events. Alarm reporting will be mainly used by trainee. It serves the same purpose of alarms provided in real control systems and the trainer also uses the alarms reported during the training session to evaluate the performance of the trainee. Similarly, events are used to analyze when a simulation is started, stopped, state is saved, when malfunction is triggered, compare the actions taken by an operator against standard operating procedures to evaluate a trainee, and also to create scenarios based on the occurred events.

20.4.6 MALFUNCTIONING, TRAINING, AND EVALUATION

The primary purpose of training simulator is to train the operator enough to operate a real plant. Since training simulator is a virtual plant, running the simulator for long durations does not lead to inconsistent behavior or failure of equipment, as in a real plant, as there is no wear and tear in the simulated system.

So, real-time equipment failure and emergency scenario is modeled in addition to the normal modeling and enabled intentionally while training for access to the operator or trainee. Based on the action taken, the trainee is assessed. There is a separate malfunction section for each unit operation and some evaluation methods to evaluate trainee performance.

20.4.7 TRAINEE EXERCISES

A standard exercise can be created for playing readily to repeat the same set of scenario for a particular process model training for a large number of participants. This leads to savings in trainer time along with reducing the dependency of skilled operator or trainer.

20.4.8 PLAYBACK OF EVENTS

This feature enables automatic playback of a set of predefined events at a specified time. Playback is used to repeat the operator actions automatically to explain the wrong steps executed by a trainee during a training session. Also, some of the repetitive actions can be configured as scenario and played back whenever needed.

20.4.9 GRAPHIC-USER INTERFACE

The training simulator has a GUI, which mimics a real DCS panel graphic environment, making the training simulator more user-friendly. Different training simulator may use different building-built or third-party graphic tool to support good user interface. The GUI supports process flow diagrams and displays some important parameters value enabling the user to navigate easily from one section to other section and have overall view of the entire process plant. It also provides a color indication for alarm and guidance to trainees as well.

FURTHER READINGS

ARC Advisory Group. ARC Insight #2005-21MPH, April 2005.

Blevins, T., et al., 2003. Advanced Control Unleashed: Plant Performance Management for Optimum Benefit. ISA, pp. 383–385; 389.

Dale, E., 1954. Audiovisual Methods in Teaching. Revised ed. Dryden Press. Lieb, S. "Principles of Adult Learning." VISION, 1991.

CHAPTER

ALARM MANAGEMENT SYSTEMS

21

21.1 INTRODUCTION

Alarm management refers to the effective design, implementation, operation, and maintenance of industrial manufacturing/process plant alarms. Alarm management is necessary in a process plant environment controlled by an operator using a control system, such as a DCS, or a Programmable Logic Controller (PLC). Alarms are triggered when the process value deviates from normal operating conditions, i.e., on the occurrence of abnormal operating situations. An abnormal situation occurs when a disturbance in a process causes plant operations to deviate from its normal operating state, calling for human intervention. The inability to diagnose and control abnormal situations has billions of dollar impact on the economy. Alarm management basically aims at preventing, or at least minimizing, physical and economic loss through operator intervention in response to the condition that was alarmed.

Alarm management refers to the processes and practices for determining, documenting, designing, monitoring, and maintaining alarm messages from process automation and safety systems. In reality, alarm management does not always achieve this because they are improperly designed, poorly documented, changed without adequate review, or fail to provide enough information to the operator. An effective alarm system is a key part of a safe and reliable process. Incorrectly designed and poorly functioning alarm systems can have serious consequences and lead to ineffective alarm; which in turn leads to alarm flooding, high number of standing alarms, inadequate prioritization of alarms and improper or no alarm action.

Advanced alarm management system is a combination of applications that help to make effective alarm systems, conforming to industry recommendations and standards such as Engineering Equipment & Materials Users' Association (EEMUA) publication 191, International society of Automation (ISA)-18.2, and Abnormal Situation Management (ASM) Consortium best practices.

Advanced alarm management system offers the following benefits:

- Improves operator effectiveness, protecting plant uptime and safety, and reducing losses
- Minimizes the number and impact of abnormal situations
- Reduces time and effort to develop, deploy, and maintain alarm systems
- Identifies alarm system problems and performance, as well as operator workload

Furthermore, an ineffective alarm management leads to conditions such as:

- No alarm at all in spite of occurrences of the event
- Untimely response to an alarm due to poor prioritization
- No show of the alarm due to inhibited or disabled alarms
- A flood of alarms and missing events due to poor configuration
- Flood of alarms leading to a missing sequence.

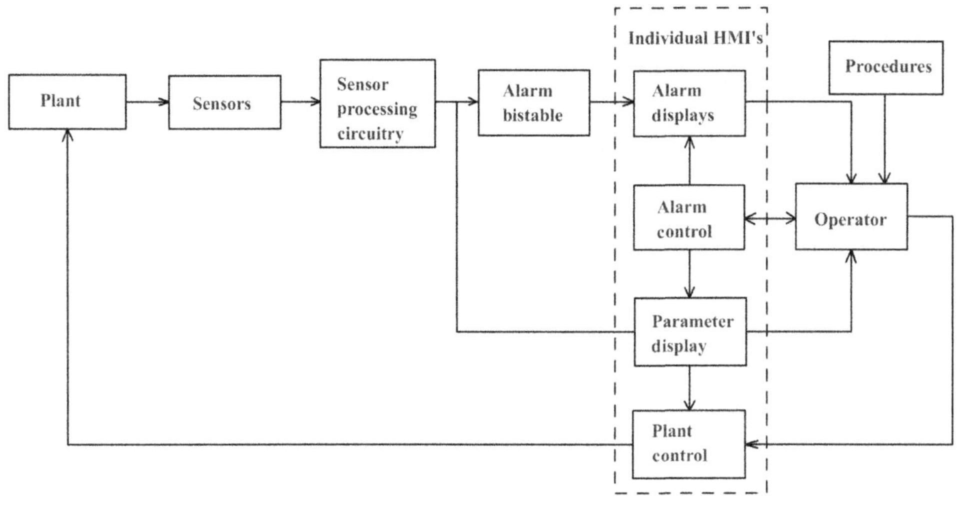

FIGURE 21.1

Conventional alarm system

Ambiguous or missing instructions to the alarms are some of the harmful effects seen in the operations and many a times, the alarm management systems are designed to address these issues.

21.2 CONVENTIONAL AND ADVANCED ALARM SYSTEMS

21.2.1 CONVENTIONAL ALARM SYSTEM

Sensors such as resistance temperature detectors (RTDs), bellows, pressure detectors, etc., are used to measure various plant parameters such as temperature and pressure. The output signal of the sensors is processed electronically and sent to various circuits that serve as controls, displays, and alarms. Figure 21.1 shows the inputs to a parameter display and to an alarm bistable. Each alarm circuit for a parameter has a setpoint and actuates the alarm display. The control room operators then make judgments about the plant state and the actions to take based upon the parameter displays and procedures. The operators would also review other information sources (e.g., access other displays, contact plant personnel) and make adjustments to the plant systems and components through plant controls. These adjustments would affect plant processes and the results would be detected by the sensors and transmitted back to the alarm displays of HMI.

21.2.2 ADVANCED ALARM SYSTEM

Advanced alarm management helps to protect plant uptime and safety by reducing the losses caused by ineffective alarming. Reducing the number of alarms that require operator intervention improves the effectiveness of operators. In addition following best practices reduces time and effort to develop, deploy, and maintain an alarm system:

Figure 21.2 presents a block diagram for an advanced alarm system. The plant, the sensors, and the signal processing circuitry are similar to that in a conventional alarm system. However, the advanced

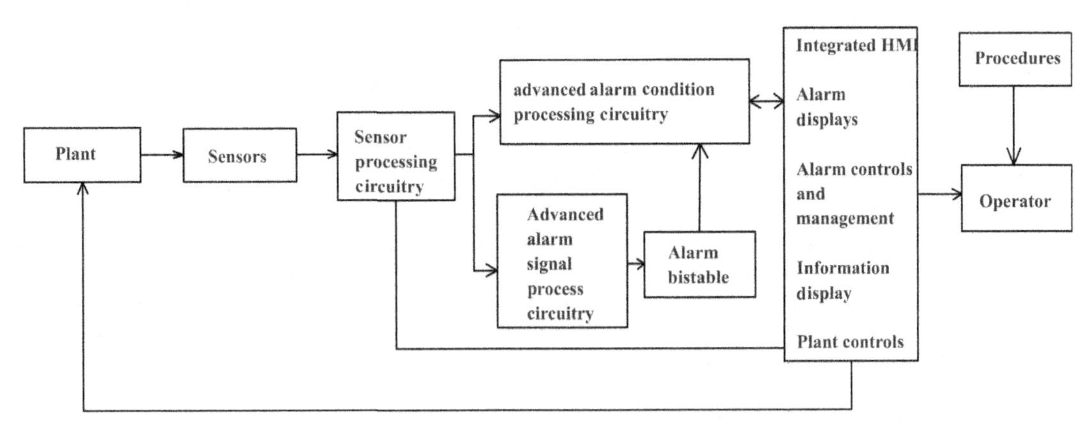

FIGURE 21.2

Advanced alarm system

alarm system typically contains more extensive information processing capabilities. The outputs from the advanced alarm system are typically input to some integrated HMI that may use visual displays or other display devices to present alarm information to the operators. The operators would use their procedures and the HMI to assess the situation, plan responses, and take any necessary action to control the plant.

Examples of alarm types:

- PV high limit
- PV low limit
- De-energized on PV high limit
- De-energized on PV low limit
- Fault diagnosis and FAIL output.

An effective configuration of an alarm should meet the following goals.

- An alarm should indicate the need for the operator intervention.
- An alarm should indicate when the control system can no longer cope up with the situation.
- An alarm should indicate a timely operator intervention.

21.2.3 ALARM DEFINITION

Alarm definition refers to specifying process parameters monitored and displayed by the alarm system and set-points used to define alarm conditions. The following are important considerations for alarm definition.

21.2.3.1 Alarm conditions

- Challenges to critical safety functions
- Deviations in key plant parameters
- Conditions representing hazardous to personnel
- Challenges to equipment having safety function
- Deviations from technical specifications

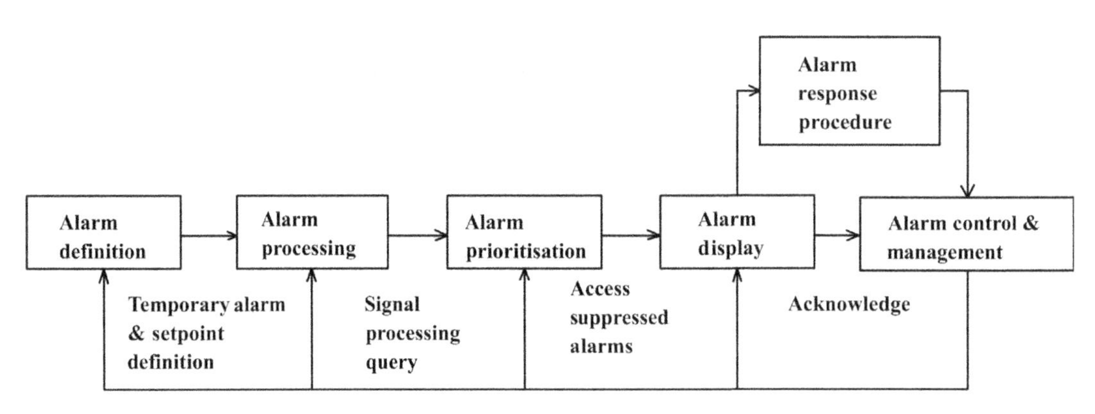

FIGURE 21.3

Alarm system functional elements

- Deviations from emergency procedure deviation points
- Safety considerations related to the plant mode
- Abnormal situation occurrence

21.2.3.2 Alarm signal processing

Alarm signal processing refers to the process by which signals from plant sensors are automatically evaluated to determine whether any of the monitored parameters have exceeded their set-points, and to determine whether these deviations represent true alarm conditions. Alarm signal processing includes techniques for analyzing normal signal drift and noise signals. They are used to eliminate signals from parameters that momentarily exceed the set-point limits but do not indicate a true alarm. Figure 21.3 illustrates the incorporation of signal processing into the circuitry of an advanced alarm system.

21.2.3.3 Alarm prioritization

Alarm prioritization refers to determining the relative importance of alarm conditions to the operating crew. This is achieved by applying alarm condition processing in advanced alarm systems. Alarm prioritization shall be based on safety risk, environmental risk, and financial risks if the same is not addressed.

21.2.3.4 Alarm information display

Alarm information is organized and displayed to control room personnel on the *information display*. To support different functions of the alarm system, multiple visual display formats may be required, such as a combination of separate displays (e.g., alarm tiles) and integrated displays (i.e., alarms integrated into the process displays). Therefore, the display format of alarm information and the degree to which it is presented separately or is integrated with other process information are important safety considerations.

21.2.3.5 Alarm controls

The types of devices used to operate the computer-based communication system, including input devices and conventional controls, must be identified. In conventional plants, dedicated, hardwired controls and pushbuttons are used to support the alarm silence, acknowledge, reset, and test functions. In advanced control rooms, operator interaction with the alarm system is achieved through interfaces that are also used for other purposes. This means that the operator may use the same input or control devices to interact with the alarm system and other control or displays.

21.2.3.6 Alarm response procedures

Alarm response procedures provide more detailed information about the type of alarm condition in the alarm message, typically – the source of the alarm, set-point, causes, automatic actions, and operator actions. Automatic response procedures are especially important to operators when an unfamiliar alarm is activated or when an alarm seems inconsistent with the operator's understanding of the plant's state.

The following characteristics of advanced response procedures are important:

- Information content (descriptions of the alarms, operator actions, and support material)
- Format (the way in which information is arranged)
- Location (the accessibility of the ARPs to control room personnel)
- Methods of user interaction with ARPs

21.2.4 LIFE CYCLE OF ALARM MANAGEMENT

Understanding the lifecycle of alarm management in the process industries provides framework for customers to manage every aspect of the alarm system. The existing alarm management practices are represented by the life cycle model. This model covers design and maintenance activities from philosophy to management of change, and is useful in identifying the requirements and roles for implementation (Figure 21.4).

21.2.4.1 Philosophy

Development of an alarm philosophy marks the starting point in the alarm lifecycle. The alarm philosophy provides guidance to the other lifecycle stages. It includes key definitions like alarm definition, which by itself is a critical element to alarm management. It takes into account the alarm handling capabilities of the control system and other site-specific considerations. The philosophy ensures planning and documentation of the processes for other lifecycle stages. The alarm philosophy captures information for each alarm, such as the basic control system information:

- Tag – the tag refers to the tag number of the alarm in the database.
- Alarm type – the alarm type describes the alarm as high, low, or a discrete state.
- Description – the description is for the tag, from the same tag database.
- Units/states – the units are the engineering units for an analog type value, and the states are the discrete states of a digital value.
- Setting/alarm state – the setting is the analog alarm limit or the discrete state that generates the alarm.

Some additional information is necessary to document an alarm for procedures and training:

- Consequence of deviation
- Corrective action

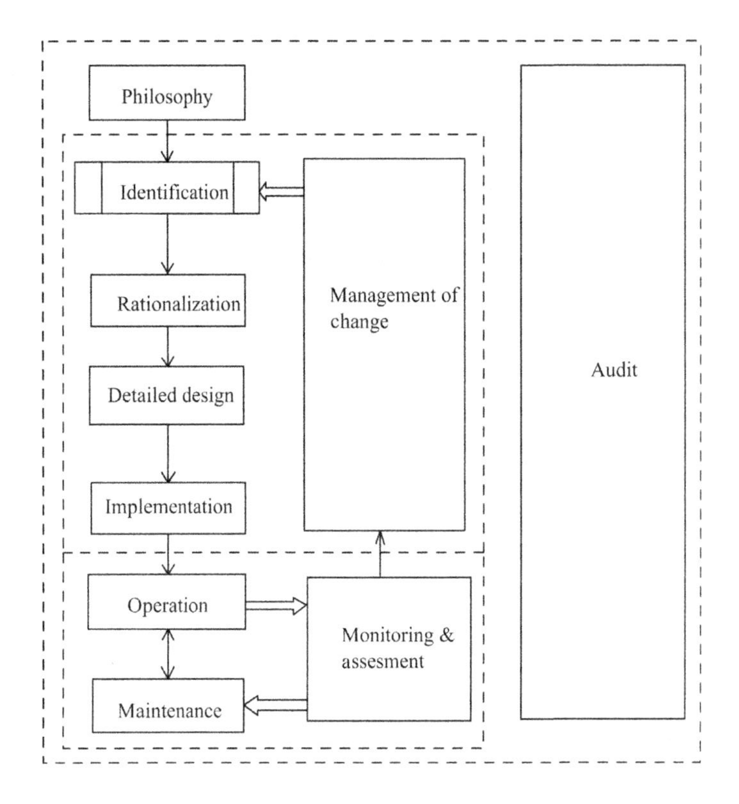

FIGURE 21.4

Alarm management lifecycle

- Time for response
- Consequence category
- Basis

This information is required to train operators to respond to alarms – specifically what action is necessary, and how fast it must be completed before the consequence. Documenting the basis for the alarm allows re-evaluation of the consequences, especially with process changes.

- Priority
- Retention period
- Report requirements
- Notification requirements.

This information specifies properties of the alarm. The priority in the operator interface is a critical way to designate the importance of the alarm. The alarm record may need to be kept for a certain period of time, included in certain reports, or the alarm may be set to trigger e-mail, pager, or voice mail messages. These functions are defined in the philosophy, and the rationalization identifies individual alarms that require these functions (Table 21.1).

Table 21.1 Example of alarm philosophy

	Alarm 1	Alarm 2	Alarm 3	Alarm 4
Tag	LIG502	LIG502	PIG502	PIG502
Alarm type	LL	HH	LL	HH
Description	T502 level	T502 level	T502 pressure	T502 pressure
Units/states	%	%	INWC	INWC
Setting/alarm state	10	90	1	10
Consequence of deviation	Cavitate pump	Overflow tank	Air intrusion	Excess venting
Corrective action	Stop pump	Close inlet valve	Stop pump	Close inlet valve
Response time	2 min	2 min	10 min	10 min
Consequence category	Equipment	Safety	Safety	Environmental
Basis	Pump cavitation at 2%	Tank overflow at 107%	Vacuum breaker setting	Conservation vent setting
Priority	Low	Emergency	High	High
Retention period	1 year	5 years	5 years	5 years
Report requirements	Pump report	Safety report	Safety report	Environmental report
Notification requirements	None	None	None	Environmental coordinator

21.2.4.2 Identification

The identification stage of the alarm lifecycle includes activities such as P&ID reviews, process hazard reviews, layer of protection analysis, and environmental permits that identify potential alarms. ISA-18.2 does not prescribe requirements for alarm identification methods. These methods are already well documented. To ensure that the results are useful as an input to the alarm rationalization stage, it is helpful to document the cause, potential consequence, and the time to respond for each identified alarm.

21.2.4.3 Rationalization

The second part is the alarm rationalization. This systematic process optimizes the alarm database for safe and effective operation. It usually results in a reduction of the number of alarms, alarm prioritization, validation of alarm parameters, evaluation of alarm organization, and the number of alarms assigned to an operator, and finally alarm presentation. In the rationalization stage, each potential alarm is tested against the criteria documented in the alarm philosophy to justify that it meets the requirements of being an alarm. The consequence, response time, and operator action are documented. Alarm priority should be set based on the severity of the consequences and the time to respond. Classification identifies groups of alarms with similar characteristics (e.g., environmental or safety) and common requirements for training, testing, documentation, or data retention. The results of the rationalization are documented in a Master Alarm Database (Figures 21.5 and 21.6) (Table 21.2).

21.2.4.4 Detailed design

In the detailed design stage of the alarm lifecycle, an alarm is designed to meet the requirements documented in the alarm philosophy and rationalization. Poor design and configuration practices are a major

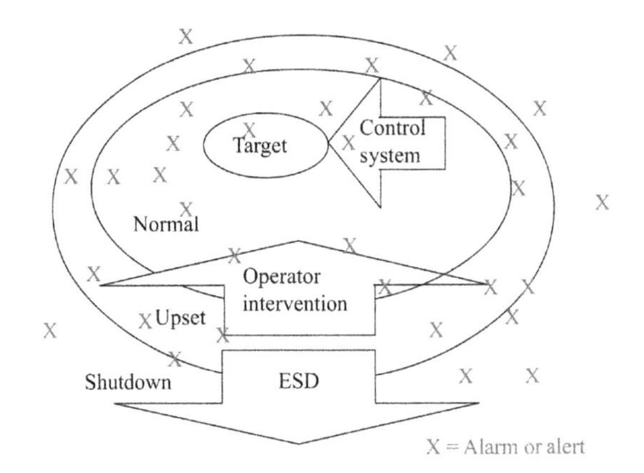

FIGURE 21.5

Alarms and boundaries without philosophy and rationalization

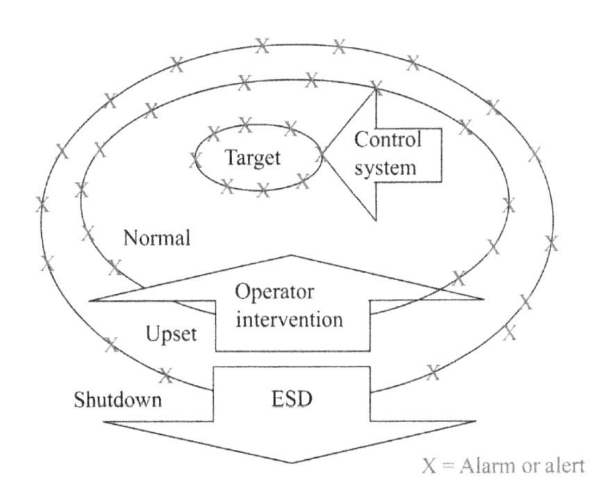

FIGURE 21.6

Alarms and boundaries with philosophy

Table 21.2 Example of alarm priority features

Priority	Color	Tone	Shape
Emergency	Red	Tone 1	Red triangle, point up
High	Yellow	Tone 2	Yellow diamond
Low	Orange	Tone 3	Orange triangle, point right

cause of alarm management issues. Alarm design includes – basic alarm design, setting parameters like the alarm deadband or off-delay time, advanced alarm design like using process or equipment state to automatically suppress alarms, and HMI design, displaying the alarm to the operator to effectively detect, diagnose, and respond to it. The detailed design phase involves configuring the system using the information contained in the Master Alarm Database (such as alarm limit and priority).

Basic alarm design: on-line tuning

Deadbands and delay timers are determined by measuring the noise and trajectory of the PV preceding the alarm.

- For the process conditions where the measurement slowly approach an alarm trip point, requiring a relatively large deadband, as the noise will cause the signal alarm and return repeatedly.
- Process conditions where the measurement rapidly approach an alarm trip point requiring a relatively smaller deadband as the signal travels cleanly through the alarm point and the noise causes the signal alarm and return repeatedly.

Offline tuning: setting the value of an OFF-delay timer

The optimum value can be estimated by analyzing the histogram of time between alarms and alarm duration.

Advanced alarm design

Enhanced and advanced alarm methods are additional alarm management techniques other than those normally employed in control systems. They generally provide added functionality over the basic alarm system design and may be particularly useful to guide operator action during plant upsets or other multiple alarm situations. This includes:

- Information linking
- Logic based alarming
- External notifications
- Batch considerations
- Alarm attribute enforcement

21.2.4.5 Implementation

The implementation stage of the alarm lifecycle involves putting the alarms into operation. It includes training, testing, and commissioning of the alarms. Testing and training are ongoing activities, particularly as new instrumentation and alarms are added to the system over time or process design changes are made.

21.2.4.6 Operation

In the operation stage of the alarm lifecycle, an alarm notifies the operator of the presence of an abnormal situation. Key activities in this stage include exercising the tools the operator may use to deal with alarms, including shelving (suppression in DeltaV terminology) and mechanisms for operator access to information fleshed out during rationalization such as an alarm's cause, potential consequence, corrective action, and the time to respond.

21.2.4.7 Maintenance

The process of rendering an alarm out-of-service transitions the alarm from the operation stage to the maintenance stage. In the maintenance stage the alarm is nonfunctional. The standard describes

Table 21.3 Average alarm rates

Long-Term Average Alarm Rate in Steady Operation	Acceptability
More than 1 per minute	Very likely to be unacceptable
One per 2 min	Likely to be over-demanding
One per 5 min	Manageable
Less than one per 10 min	Very likely to be acceptable

the recommended elements of the procedure to remove an alarm from service and return it to service. The state of out-of-service is not a function of the process equipment, but describes an administrative process of suppressing (bypassing) an alarm using a permit system.

21.2.4.8 Monitoring and assessment

The monitoring and assessment stage of the alarm lifecycle encompasses data gathered from the operation and maintenance stages. Assessment involves comparison of the alarm system performance against the stated performance goals in the philosophy. The rate alarms are presented to the operator in this stage. To provide adequate time to respond effectively, an operator should be presented with a maximum of one to two alarms every ten minutes. A key activity during this stage is identifying "nuisance" alarms – which are alarms that annunciate excessively, unnecessarily, or do not return to normal after the correct response is taken (e.g., chattering, fleeting, or stale alarms) (Tables 21.3 and 21.4).

21.2.9.9 Management of change

The management of change stage of the alarm lifecycle involves authorization for all changes to the alarm system, including the addition of alarms, changes to alarms, and the deletion or removal of alarms. Once the change is approved, the modified alarm is treated as identified and processed through the stages of rationalization, detailed design, and implementation again. Documentation like the Master Alarm Database is updated and the operators are trained on all changes since they must take the actions.

21.2.4.10 Audit

The audit stage of the alarm lifecycle is primarily focused on the periodic review of the work processes and performance of the alarm system. The audit phase aims to maintain the integrity of the alarm

Table 21.4 Alarm rates following an upset

Number of Alarms Displayed in 10 min Following a Major Plant Upset	Acceptability
More than 100	Definitely excessive and very likely to lead to the operator abandoning use of the system
20–100	Hard to cope with
Under 10	Should be manageable – but may be difficult if several of the alarms require a complex operator response

system throughout its lifecycle and identify areas of improvement. The alarm philosophy document may need to be modified to reflect changes resulting from the audit process.

21.2.5 ALARM SYSTEM PROBLEMS

The main problems in alarm management are nuisance alarms, stale alarms, alarm floods, and clarity of the alarm to the operator. The processes defined in the alarm philosophy, implemented with operational discipline, can address these problems.

21.2.5.1 Nuisance alarms

Nuisance alarms indicate an abnormal condition when none exists, or when no change in process condition has occurred. Nuisance alarms desensitize the operator, reducing the response to all alarms. Instrument problems or alarms set within the normal operating range often cause nuisance alarms. Measurement of the alarm frequency by tag is used to detect nuisance alarms at a threshold defined in the alarm philosophy, for example, 10 alarms per day. Once detected, nuisance alarms should be investigated and corrected as soon as possible. Typical alarm reports indicate a very small percentage of tags are responsible for the majority of alarms. Without monitoring and prompt follow-up, nuisance alarms can quickly deteriorate the performance of an alarm system to the point where tens of thousands of alarms are recorded per day.

21.2.5.2 Stale alarms

Stale alarms remain in the alarm state even when no abnormal condition exists or no operator action is required. Stale alarms form a baseline of alarms that require no action and train the operator to ignore certain alarms. Alarm configuration problems or alarms set within the normal operating range cause stale alarms. Measurement of the time in alarm by tag is used to detect stale alarms at a threshold defined in the alarm philosophy, for example, 24 hours. Without monitoring and follow-up, the number of stale alarms slowly increases, decreasing the effectiveness of the alarm system.

21.2.5.3 Alarm floods

Alarm floods refer to a temporary high rate of alarms, usually associated with an event like a process upset. Alarm floods overwhelm the operator, masking the important alarms and reducing the operators ability to respond to the upset appropriately. Configuring multiple alarms for a single event is one of the main causes of alarm flood. Alarm floods are detected by measuring the rate of alarms in a given time interval with a threshold defined in the alarm philosophy, for example, 10 alarms per 10 min. Alarm floods are one of the more difficult problems to solve, but a problem closely linked with plant disasters. Monitoring can detect and report alarm floods, but reducing floods takes detailed process understanding and good alarm practices. Duplicate alarms are reduced by rationalization. Advanced alarming techniques can reduce the number of alarms during an upset. Figure 21.7 shows the alarm floods caused by a process, which is controlled by a distributed control system.

21.2.5.4 Alarm clarity

Clarity of alarms relates to configuring the alarms and training the operator to respond to the alarm. Alarm documentation generated during rationalization serves as an input for training. Alarm clarity problems are a difficult thing to measure. Operator training can provide the opportunity to identify

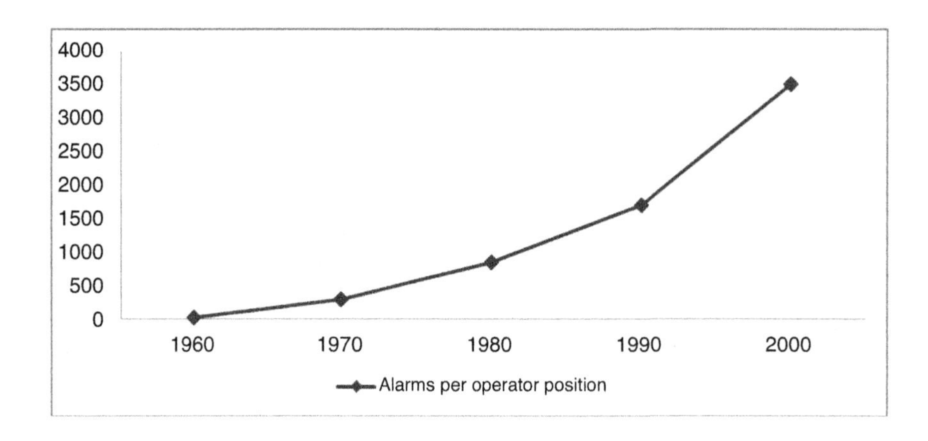

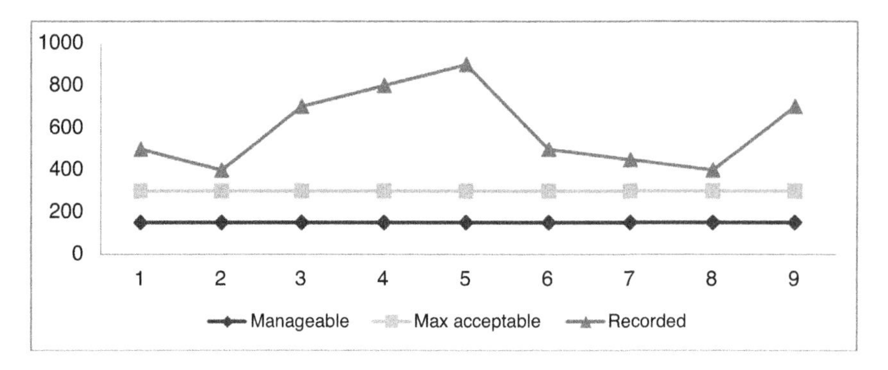

FIGURE 21.7

DCS alarm system problem

clarity problems. Sometimes the problems can be resolved with changes to the basic alarm configuration. Advanced alarm techniques are employed to produce fewer alarms with clear meaning.

Improper configuration leads to abnormal situations. Figure 21.8 illustrates the causes for abnormal situations. It shows the failure rate corresponding to different errors.

21.2.6 ADVANCED ALARM PROCESSING TECHNIQUES

There are a wide variety of alarm processing techniques. Advanced alarm processing systems often employ combinations of them. Each processing technique changes the resulting information provided to the operator. There are four classes of alarm processing techniques:

- Nuisance alarm processing
- Redundant alarm processing
- Significance processing
- Alarm generation processing.

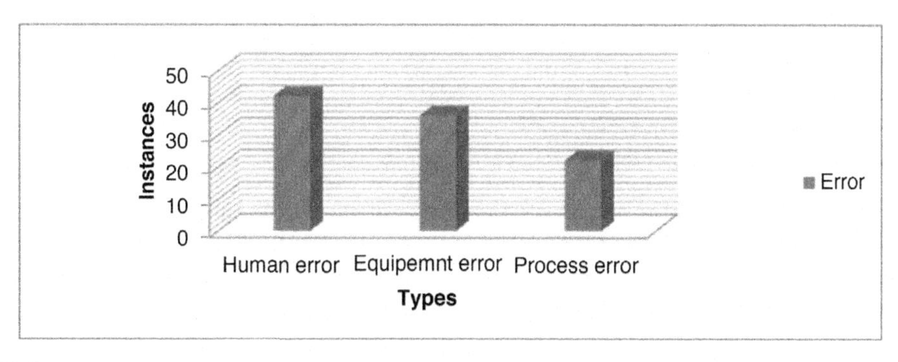

FIGURE 21.8

Causes of abnormal situations

21.2.6.1 Nuisance alarm processing

Nuisance alarm processing employs techniques to eliminate alarms having no operational safety importance. For example, mode-dependent processing eliminates alarms that are irrelevant to the current mode of the plant; for example, the signal for a low-pressure condition may be eliminated during modes when this condition is expected, such as start-up and cold shutdown, but be maintained during modes such as normal operation when this condition is not expected.

21.2.6.2 Redundant alarm processing

Redundant alarm processing analyzes if alarm conditions are true or valid but are considered to be less important because the information they provide is redundant with other alarms and theoretically supplies no new or unique information. For example, in causal relationship processing, only causes are alarmed and consequence alarms are eliminated or their priority is lowered. However, such techniques may minimize information that the operator uses to confirm that the situation represented by the true alarm has occurred, for situation assessment, and for decision making. Therefore, processing methods may qualitatively affect the information given to the operating crew in addition to quantitatively reducing alarms.

21.2.6.3 Significance processing

This class of processing includes techniques that analyze alarm conditions that are true or valid but are considered to be less significant than other alarm conditions. For example, in an anticipated transient without scram, alarms associated with minor disturbances on the secondary side of the plant could be eliminated or their priority lowered.

21.2.6.4 Alarm generation processing

This class of alarm processing evaluates the existing alarm conditions and then generates alarm messages that:

- Give the operator higher level or aggregate information
- Notify the operation when unexpected alarm condition occurs
- Notify the operation when expected alarm conditions do not occur.

These processing techniques generate new (e.g., higher-level) alarm conditions. Alarm systems should reduce errors, which often reflect the overloaded operator's incomplete processing of information. Alarm generation features may help mitigate these problems by calling the operator's attention to plant conditions that are likely to be missed.

In general the following can be considered as good characteristics of a good alarm.

- Relevant: The alarm should be relevant, which means that it should not be spurious alarms or an alarm of less value to the operator.
- Unique: The generated alarm should be unique and only once. There shall not be multiple alarms of the same type or multiple alarms triggered by the same event.
- Timely: The alarm should be timely in terms of the presentation to the operator, such as neither too early nor too late for any actions.
- Prioritized: Alarm should clearly indicate the importance of the event upon its occurrences for the proper action. Too many prioritizations also dilute the attention. Hence, a properly engineered system represents a correct priority.
- Understandable: Having a message that is clear and easy to understand for the operator.
- Diagnostic: Identifying the problem that has occurred.
- Advisory: Indicative of the action to be taken.
- Focusing: Drawing attention to the most important issues.

FURTHER READINGS

ANSI/ISA-18.01-1979 (R2004), Annunciator Sequences and Specifications, p. 9.
ISA-RP77.60.02-2000, Fossil Fuel Power Plant Human-Machine Interface: Alarms, p. 9.
ANSI/ISA-84.00.01-2004-Part 2 (IEC 61511-2 Mod) – Functional Safety: Safety Instrumented.
2004 Part 1 (IEC 61511-1 Mod) – Informative. Sections 9.4.2 and 9.4.3.
Alarm Systems: A Guide to Design, Management and Procurement. EEMUA, p. 105.
Alarm Systems: A Guide to Design, Management and Procurement. EEMUA, p. 107.
Systems for the Process Industry Sector – Part 2 Guidelines for the Application of ANSI/ISA-84.00.01.

CHAPTER

DATABASE SYSTEMS

22

22.1 HISTORIAN DATABASE

22.1.1 INTRODUCTION TO HISTORIAN DATABASE

The plant historian or plant history database is responsible for capturing data from various automation subsystems in the plant and consolidating and presenting them in a usable and analytical form.

The fundamental challenge in a refinery/plant where multiple products are manufactured in geographically distributed establishments relates to maintaining the quality of the products. Maintaining uniform quality across the products irrespective of the manufacturing location is the ultimate goal of all organizations. Alongside, there is a pressure to not only sustain but also continuously improve quality, for which the need for correct metrics/measures is of great importance.

In such a situation, there is a need for an automated system that provides the data for monitoring quality. Using an automated system for this translates to lower scope for human data entry errors or manipulations. Due to the above reasons database systems are essential and are used in many automation topologies.

Following are some of the primary requirements of such system and their importance.

- Huge data storage capacity: Storing huge data is one of the key factors because data from multiple sites/multiple plants in the same site needs to be handled appropriately. Hence, the data ranges from medium to very large on the basis of capturing frequency. For example, if the process variable values of a specific section in a plant are captured at the rate of 1 s, the resulting data may be of the size of 1 GB per day; if the same is captured at 2 s, then it apparently comes down to 0.5 GB approximately. Therefore, it is essential to optimize data capture according to the available physical memory, to avoid possibility of data corruption.
- Data storage: Storing the data for long periods is important for decision making. When data dynamics are captured over a period, it enables people to easily find a pattern of how the system is behaving and conclude about the performance on the basis of it.
- Powerful statistical tools: Powerful statistical tools are vital for analyzing various dependencies of the parameters. Identifying dependent parameter behaviors, for instance, helps in understanding the corelation between two parameters. This helps. Many control system trends have good trending capacities but poor statistical analysis capabilities. The need for tools that provide good statistical analysis plays a very important role here.
- Flexible data retention methods: Flexible data retention and data mining techniques are important for data storage over very long periods to use for further decision making. Data is not retained in the system for long periods like a period of 1 year, but is archived in media. When needed, the media can be plugged in to access the data; this makes decision making more reliable, since it is made on observations over long periods of data.

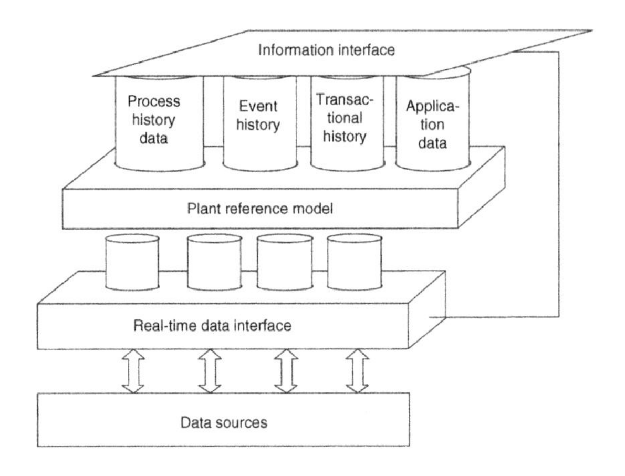

FIGURE 22.1

Structure of database systems

The historian is a application that is used for monitoring and analyzing rather than controlling any process in the plant. It helps in quality control and decision making (Figure 22.1).

Data sources provide information to the real-time data interface (RDI). The data sources are the following levels in a distributed control system.

- Level 0: This level comprises all the field instruments; for example, flow/temperature/pressure and so forth are sensed at this level.
- Level 1: This level comprises of the controllers (embedded) controllers that communicate with field devices and contain control strategies to respond to the inputs from field devices.
- Level 2: This level comprises supervisory platforms for viewing the overall controllers' performance and plant performance. It contains interfaces for configuring and supervising the control strategy.
- Level 3: This is the level where database systems resides and collects the tag information from data sources from the distributed control systems that lie on the levels below the plant historian.
- Level 4: There is a plant historian shadow server at level 4 that gives consolidated information from various historian servers that reside at level 3 of various sites. Level 4 provides a very high-level overview of multiple systems unlike the details that a level 3 system has. Therefore, levels 3 and 4 can be considered as the levels hosting plant historian.

22.1.2 INTRODUCTION TO REAL-TIME DATA INTERFACES

Having understood the levels and data sources of Historian, let us understand what makes a historian (Figure 22.2):

- Real time data interface
- Database
- Desktop applications
- Advanced historian

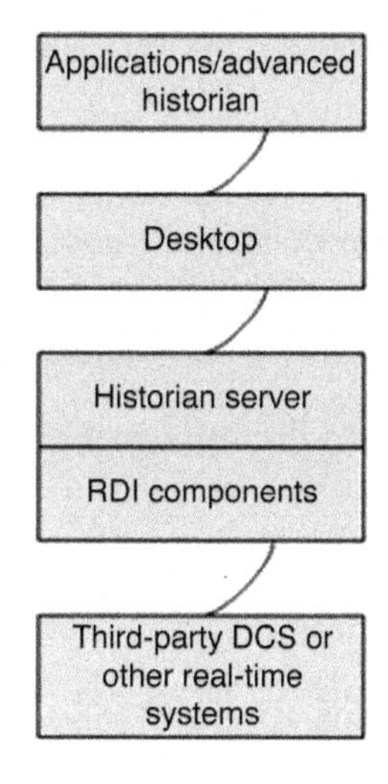

FIGURE 22.2

Database system data flow

RDI, also known as real-time data interface, is responsible for obtaining data from the plant onto a historian. The data retrieved from RDI is stored in a database for further processing. Therefore, the heart of the entire system is constituted by the database. SQL/oracle is one of the popular database choices. The database is designed with incorporating the following vital features:

1. Stores plant configuration data and plant reference for all components. This is a key feature for associating process data and business data, which is useful for engineers/supervisors/operators and executives/plant management, respectively.
2. Since database is central to the entire historian, even if other components of the system fail, the database is in a position to be plugged in to a new system for retrieval of the old data. Therefore, robustness and data retrieval is the key of the database.
3. It is compatible with other application sitting along and above the levels where the historian resides.

Desktop is the primary interface for users to interact with historian. It helps in converting the data into a usable format. Common desktop tools are Windows tools such as word/excel/outlook and the like. In many cases, the capabilities of excel are exploited as an add-on to retrieve and give away data to historian.

Applications are advanced features of the historian, for example, event journal collection, data synchronization, and data transfer from relational database to historian and vice versa.

Advanced historian supports shadow historian to collect data from multiple sites and has a feature of virtual tags for advanced calculations.

22.1.3 REAL TIME DATA INTERFACE (RDI)

RDIs are classified into three types:

Type A RDI is owned by the manufacturer of the historian and offers better compatibility with the control system.
Type B RDIs are based on the popular technologies used for communication across multiple control system, for example, a company X historian can obtain data from a third-party control system through object linking and embedding process control (OPC); therefore, although there are various control systems from different vendors, they can communicate to a single historian using common protocols such as OPC/Modbus and the like. This also eliminates the incompatibility issues across various vendors.
Type C RDIs are provided to the user by the historian manufacturer on special request unless the technology is available otherwise.

22.1.3.1 Database

On a high level, there are two types of databases: a relational database and a historian database; a historian database and a relational database work hand in hand with each other. The relational database to historian database has the capability of plugging in to a new system in case of failure of the historian system.

Historian collects, stores, and manages plant historical data, including real-time and relational data. The historian integrates real-time data with an industry-standard relational database. The relational database stores the plant model and operating events and transactions.

The following are the chief functions of a plant data historian:

- Interfaces with third-party DCS and other real-time systems.
- Collects, integrates, and maintains long-term history of real-time continuous and discrete measurements.
- Provides data capture and historization, as well as intelligent data retrieval required to support operations' analysis and layered plant information applications.
- Stores configuration and reference data, also called the fixed plant data book.

Reference model tools stores reference and configuration information about the plant. It includes historian information and integrated plant operations reference information used by the plant business applications. The integrated tools unifies both historian and plant operations information.

Some of the major types of information contained in the reference model tool include

- equipment names, attributes, and groupings
- product names, attributes, and groupings
- equipment operating modes
- stream definitions and lineup configurations
- quality test property definition and sample points
- company and report headings

22.1.3.2 Desktop

Desktop provides interactive access to the data and information stored in the database system. The desktop consists of a family of tools and application user interfaces that provide easy access to data from desktop PCs.

Developers can use tools, such as the Microsoft Excel companion add-on, to access and use stored data from the database system and other sources. Desktop users have access to data for analysis, trending, reporting, incident replay; and process upset diagnostics, data configuration, and error cause identification. The desktop also includes Visual OLE (Object Linking Embedded) Automation and ActiveX components to enable application development that support access to current and historical data.

22.1.3.3 Applications

The event journal collection and storage mechanism allows events from the journal files to be recorded in a relational database. The following event message types are collected:

- Process alarms
- SOE (Sequence of Events) events
- Operator messages
- Process changes
- System status
- Status notification
- System maintenance
- System error messages

There can be individual relational tables, one for each of the message types listed above. Each event message is stored in one row in the associated relational table. Data is retrieved at an interval configurable from 1 to 60 min. The event journal collector supports automated size control by limiting the number of days during which data is stored in the journal tables.

Reference data synchronization: References data in historian may be optionally synchronized with the control systems. The tag description, high and low extreme values for each tag, when applicable, may be synchronized. Relational to historian is a facility within historian that provides data values from relational into historian history. This provides a convenient mechanism for users to insert results from third-party systems or other data providers into the history database.

Historian to relational: This facility within the historian provides regularly scheduled transfers of data from historian to relational tables in a relational database. This allows historian summary data placed into a relational database to be accessible by a wide range of tools. If any factors prevent data from being extracted at the scheduled time, this facility automatically synchronizes the next time when that extraction is possible.

22.1.3.4 Advanced historian

Some advanced historians have a unique "shadowed" architecture that allows data to be collected from a variety of locations and control systems, and be consolidated into one central location.

This is achieved through one configuration and management tool, rather than one per system. Advanced historian "virtual tags" provide the ability to do tag-based calculations "on the fly". These calculations appear to the user more like collected values, but do not take up valuable storage space. Advanced historian provides support for both the OPC data access and historical data access standards, providing superior openness and the ability to integrate to a wide variety of applications supporting this open standard.

22.1.4 TECHNOLOGY OF HISTORIAN

A historian acquires the data from RDIs; tags are defined for acquiring the data from the control systems. Tag is related to a control strategy that controls the output or input of a particular parameter in the plant. For example, consider a loop in the plant controlling the flow in a specific area; the output of the loop can be one tag, whereas the input of the loop can be another tag.

The popular methods for addressing the parameters in the control systems are tag based and address based. Address-based definitions are more complex in nature especially when communicating across multiple systems because of the interoperability issues, whereas tag can be treated as a generic way of calling the data across multiple systems. It is comparable to addressing a known person by name or a number; apparently referring by name is always easier than referring by number.

While defining a tag, it is represented by its name, tag descriptions, engineering units, and flags to represent the active/inactive state of the tag, allowing data storage and data editing capabilities. Once the tag starts acquiring data, they are plotted for numeric data that is represented as a continuous series of linear segments of variable duration, each with an associated confidence factor of 0 to 100. The value of a process variable or tag is defined for any time as the point on the linear segment corresponding to that time. This representation as has the following advantages over traditional constant-interval averaging methods.

- Data retrieval independent of data collection: Applications may query data for times independent of point scan rates or storage intervals. Accurate measurements for operation transactions are determined regardless of the transaction times or duration. For example, if tag data is collected every 2 min, the values for this tag may be requested at 30-s intervals so that it may be compared to a tag that is retrieved every 30 s. The values returned for the 2-min tag are interpolated.
- Optimized data compression: Data compression is optimized such that data enough to reconstruct the linear segments, to within specified tolerances, is stored. The effective storage sample rate of the tag is automatically varied according to how dynamically the data changes. If data compression for a tag is not enabled, the raw collected data is stored.
- Effective missing data handling: If data for a tag is temporarily unavailable, its value is automatically estimated as a linear interpolation between known data points.
- Accurate data reductions: Highly accurate data reduction calculations are performed, such as averages, min, max, totalization, and regression.
- Data confidence factors: Historian provides accurate composite confidence factor estimations. Time-weighted averages of confidence factors are used in data reduction calculations. A data reduction containing a missing data interpolation has a confidence factor weighted according to the relative amount of missing data. For example, a 25% missing data duration has a 75% confidence factor.

22.1.5 DATA PROCESSING AND COMPRESSION

22.1.5.1 Automatic engineering unit conversions

Historian provides standard engineering unit conversions between absolute values, rates, and accelerations. For example, a flow measured in barrels/day can be requested as gallons/minute or cubic meters/hour. Data conversion between imperial, US, and metric units is also supported. Information can be requested for a measurement in the units required without having to build conversion functions into various user tools and applications.

The engineering unit conversions are also user configurable. Users can add new engineering unit conversions and modify existing engineering unit conversions. In addition, new base units can also be added to the historian system. All engineering unit conversions are derived from a base unit. For example, the base unit for length is meters. All other engineering units of length type are derived from the meter base unit. New base units can be added to the historian system. Other engineering units of the same type are then configured as a derivation of that base unit.

22.1.5.2 Time-weighted data reductions

The historian enables time-weighted data reductions, which include min, max, delta, mean, linear regression, moving average, and standard deviation. Averages are calculated on a time-weighted basis rather than on a sample or arithmetic basis. These reductions return both a reduction value and a composite confidence factor based on the reliability of the source data. The confidence value assigned to the returned reduction value is the least confident component used in the calculation.

Time-weighted reduction calculations avoid the intrinsic error associated with simple arithmetic calculations. Time-weighted averages can be thought of as the average of the area under the curve. Simple arithmetic calculation considers sample points that are picked at random rather than a time-weighted average, which is more accurate.

22.1.5.3 Data compression

Historians support data elimination compression technique. Since data is represented as a series of virtual linear segments between points, intermediate values can be eliminated if they are estimated to within an error tolerance specified for the tag. Decompressed data can be retrieved on demand.

22.1.5.4 Data reliability factors

Historian maintains a reliability factor for each measurement value, which reflects the integrity of the value. Measurements that are bad, missing, and unavailable, have gross errors, or are outside the instrument calibration range are assigned zero reliability. Historian calculates an overall weighted reliability for accumulations involving multiple values that may each have different reliability. Time-weighted averages of confidence factors are used in data reduction calculations.

22.1.5.5 Tag calculations

Historian provides a system of easily managed calculated tags whose data values are derived via a mathematical and/or logical procedure. Calculations are based on collected process variables or other calculations. Each calculation is specified using its source tag names TAG1 × (TAG2 + TAG3). These virtual tags can be used to:

- Fill in missing instrumentation.
- Provide redundant measurements and actual versus theoretical comparisons.
- Perform continuous mass/volume conversions.
- Track unit and furnace efficiencies or other complex calculations.

Punctuation	Identifies
&, |	Logical and, or
<, <=, =.!=, >=, >	Logical comparison
+, −	Addition, subtraction
*, /	Multiplication, division
^	Raise to power
!, −, +	Negation, negative, positive

FIGURE 22.3

Logical operations with identities

Tag values are calculated when a data request is made for the tag. The valid data time range for a virtual tag is bound only by the available range for other tags referenced in its procedure. The following functions are supported within calculations (Figure 22.3):

- Use of if-then-else, if-else, and endif statements for conditional execution.
- User-specified reductions for selected time periods such as AVG, DELTA, MAX, MIN, REGSLOPE, and STDEV.
- Nested function calls to reference standard calculations such as meter corrections, mass conversions, vessel level to quantity conversions, and so on.
- Built-in function references such as TIME, LOOKUP, ABS, EXP, LOG10, SIN, and so on.
- Character-string data processing such as if property = "RED' then...
- Support for expression operators.

22.1.6 DATA STORAGE AND RETRIEVAL

22.1.6.1 Multifile/multilayer long-term archive system

The historian archive system is designed to provide maximum flexibility using a multiple-file archive system that permits overlapping time ranges between archives. This system has the following advantages:

- Data from one archive can be resampled, for example, on a 10-min basis, and stored in another archive to reduce on-line data storage space requirements. Depending on site requirements, a number of archives with various storage resolutions could be employed.
- Archives with overlapping date ranges are layered so that higher-resolution data takes priority over resampled lower-resolution data.
- A backup archive can be restored at any time and connected to the system. Any programs querying data from the appropriate date range automatically access the recovered archive.

22.1.6.2 Importing historical data to the history database

Once tag definitions are loaded into the historian, the tag's corresponding history data can be recovered from a different historian. As the data is recovered, the compression rules defined for each tag are applied against the history.

22.1.6.3 History recovery

History module recovery is supported for the RDI/DCS interfaces that support it. Recovery functions automatically initiate the recovery of missed data since the last successful collection period for all tags configured for recovery on the DCS. The ability to recover history and the length of history that can be recovered depends on the history module storage available on the DCS.

22.1.6.4 Off-line history archive

The historian creates an archive that may be removed, backed up to tape or another medium, and later restored without shutting down the system.

22.1.6.5 Securing and auditing

Historian has extensive security capabilities. Read and write access to current and history data can be restricted to individual users or groups of users. Historian also provides full audit trail capabilities for auditing changes to history data and tag configuration data.

Access to audit information is available through standard reports. Historian security functions are optional. When enabled, security functions require a username and password to log on and use the system.

Historian security can be divided into two functional areas:

1. The need to manage the historian system.
2. The requirement to access historian data.

The first is the responsibility of the historian administrator. The second may be a general requirement with users being restricted to specific subsets of the data based on roles. Roles are assigned functional levels of permission. Users can be assigned more than one role definition and inherit the highest level of permission from each role.

22.1.6.6 Historian system management

System management functions are performed by the historian administrator using menus and forms not generally available to application users. Management functions include the following:

- Installing client application
  - Customize application menu captions
- Database attachment (attach tables)
- Role/Function definition
  - Create new roles
  - Drop roles
  - Assign function security to roles
  - Assign menu access to roles
- User enrollment
  - Create new users
  - Grant users to roles
  - Revoke users from roles
  - Activate and deactivate existing users

FURTHER READINGS

Date, C.J., 1999. An Introduction to Database Systems, Seventh ed. Addison Wesley Longman.

Gray, J., October 1999. Evolution of Data Management. IEEE Computer, 38–46.

Harrington, J.L., 2002. Relational Database Design Clearly Explained, Second ed. Morgan Kaufmann.

Litwin, P., 2003. Fundamentals of Relational Database Design. http://r937.com/relational.html.

Stankovic, J.A., Son, S.H., Hansson, J., June 1999. Misconceptions About Real-Time Data Bases. IEEE Comput., 29–36.

Strong, D.M., Lee, Y.W., Wang, R.Y., 1997. Data Quality in Context. Commun. ACM 40 (5), 103–110.

Wang, R.Y., Storey, V.C., Firth, C.P., 1995. A Framework for Analysis of Data Quality Research. IEEE Trans. Knowledge Data Eng. 7 (4), 623–640.

CHAPTER

MANUFACTURING EXECUTION SYSTEMS

23

23.1 INTRODUCTION

In today's complex manufacturing environment, it is tough to imagine managing without computer systems and software. A manufacturing execution system (MES) consists of the computer systems used to manage production and daily operations within a process facility, above the level of automatic control. MES delivers information that enables the optimization of production activities from order launch to finished goods. An MES guides, initiates, responds to, and reports on plant activities as they occur, based on accurate current data. The resulting rapid response to changing conditions, coupled with a focus on reducing non-value-added activities, drives effective plant operations and processes. The ISA-95 standard defines MES to include links to work orders, receipt of goods, shipping, quality control, maintenance, and scheduling. This is in addition to supporting functions such as management of security configuration, information, documentation, and regulatory compliance. An MES largely works at the information system, or business system level. The user community is diverse, and includes management, engineering, and operations staff. An MES applies to all types of process industries such as refining, oil and gas, chemicals, pharmaceuticals, mining, minerals, pulping, paper, and power.

An MES often links to, and overlaps with, enterprise resource planning (ERP) systems. ERP systems provide global supply, demand, and order tracking systems, but for the most part lack real-time information and support for the work processes within a plant. MES links automated control systems with ERP systems. MES makes use of common functions such as data management, visualization, and production reporting. The major functional areas within an MES may be classified as follows:

- Production planning: This involves creating feasible and optimized operating plans, equipment assignment plans, and recipes. It also determines how to handle production orders and assesses the current capabilities of the manufacturing facility.
- Production execution: This entails execution of production plans safely, reliably and efficiently, including links to the automation and control systems.
- Operations management: This functional area determines safe operating limits, monitors process conditions, communicates operating plans, reports on shift activities, and manages operating tasks and procedures.
- Production management: This functional area measures the actual production, reconciles production for internal and fiscal reporting, applies quality information to production assessment, and reports actual production performance.

Today's manufacturers are under tremendous pressure to develop high-quality products quickly and cost-effectively. In such an environment, adopting an MES is one way to operate profitably that makes shop floor information available to the rest of a company, allowing them to respond more rapidly to changing requirements and conditions. MES provides a lot of benefits to the manufacturing industry including:

- Reduces scrap and waste: The setup tends to be quick and consistent, enabling immediate problem identification so that the process can be stopped, limiting the number of bad parts and wasted material.
- Captures costs more precisely: With an MES, labor, scrap, downtime, tooling, and other costs can be captured directly from the shop floor as they occur. This makes the information more reliable and actionable for pricing new work and renegotiating unprofitable business.
- Increases uptime: Uptime is the time when the machines are running and used for producing goods as planned. It is really hard to make money if the machines are not running. A modern MES often includes integrated scheduling and maintenance. A job or part is not scheduled unless the source inventory is available, the machine is properly maintained, and the right tooling is ready.
- Reduces inventory: In earlier days, industries often used to maintain a spare inventory to cater to sudden future needs. This was because there was no proper means of keeping track of inventory regularly. With MES, as inventory records are constantly updated with new production, scrap, nonconforming material, and the like, purchasing, shipping, and scheduling people know what material is really on hand and when to schedule the next procurement of materials.
- Improves customer satisfaction: As MES keeps proper track of inventory and machine conditions, it becomes easier for the manufacturers to find materials and plan the production to meet an emergency requirements from customers.

23.1.1 EVOLUTION OF MES

Before attempting to understand MES, it is critical to understand the automation and computerization of the manufacturing process and how various developments led to the evolution of MES. The history of computerized automation of the manufacturing industry can be classified as follows:

1. Accounting systems and simple inventory management systems;
2. Manufacturing resource planning (MRP) systems that emphasized a material planning approach;
3. Niche systems for job shop, repetitive process, and so forth;
4. Functional extensions and new information technology as embodied in MES.

Since accounting systems were so well defined and established, they were one of the first business applications to be automated. In fact, the first manufacturing applications occurred as a by-product of the computerization of accounting, which often included components for inventory accounting. The need for software specifically designed for manufacturing, rather than financially oriented information used to solve operations problems, led to the eventual development of MRP and a variety of software packages.

MRP concepts, or closed-loop manufacturing systems, reached their zenith during the late 1970s and early 1980s. Essentially, they are designed to integrate all the operational functions for a manufacturing organization from engineering through production, and replace a reactionary management culture with top–down planning disciplines. These systems were and still are successful for companies that are really determined to change. For the most part, discrete manufacturers with deep bills of material (BOM) realized the great benefit of MRP, simultaneously reducing inventories while improving customer service. As MRP technology matured, other functions were added to provide routing definition, shop floor control, and capacity planning. These functions provide expanded cost definition, generate production order paperwork, and track job status. However, without finite capacity scheduling and real-time feedback, the generated dispatch lists typically are not very useful.

MES rectified the lack of real-time feedback in MRP systems as MES applications provide companies with the tracking capability to monitor floor activities with greater resolution (hour/minute), and integrated with finite capacity scheduling to provide fast reaction to changes. It also continually analyzes activities in order to be responsive to events as they occur on the shop floor. MES produces information to support the material requirement and the necessary constrained resources for executing a manufacturing plan. Support activities are part of the critical manufacturing path and coordinated so that the entire organization follows one coherent, workable plan. The planning resolution (hour/minute) of an MES is more precise than a stand-alone MRP system. MES compliments MRPII applications and extends their capabilities by incorporating an execution-driven approach. Together these solutions provide companies with more realistic schedules that compress cycle time, reduce work in process, and improve the value added time while maximizing return on assets

23.1.2 OVERVIEW OF MES

Figure 23.1 describes the architecture of a typical control system across all levels (sensor-level network, control-level network, and plant/factory network). As illustrated, the MES resides in a plant information network and has access to the control network.

MES is designed to fulfill the needs of a broad manufacturing enterprise, by coupling front office accounting with the factory supervisory control systems and products. In addition to linking, MES also closely ties the outputs of these three layers of information systems: (1) those residing in the planning functions, such as MRP, (2) execution functions, such as supervisory control software or quality control, and (3) control systems that create the data utilized, so that the enterprise has full access to the separate databases of information that exist within the organization. MES's functions such as scheduling, resource allocation, process management, quality management, and operation analysis, all operate to "translate" the real-time data occurring on the factory floor into information that is useful from a process control/management standpoint. This further ripples into other adjunct processes such as labor, equipment, and materials management; product tracking; and supportive systems such as quality and documentation.

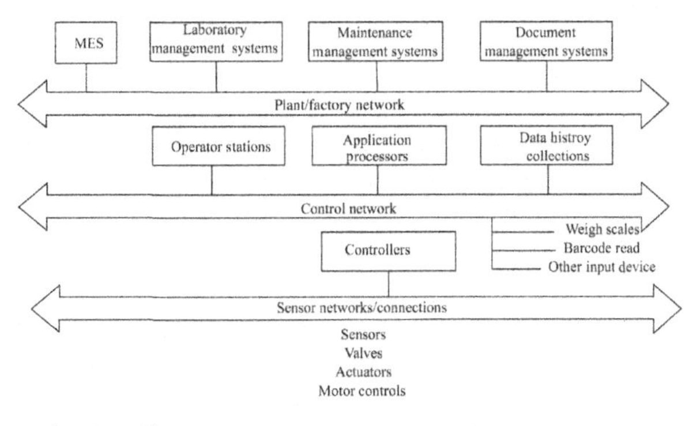

FIGURE 23.1 MES in Layered automation

The control systems' focus is on the process or operation itself – sequencing and manipulating the process to assure that tolerances are kept within defined limits, that material flow is maintained, and that all of the people, equipment, and resources involved within the process are fully utilized. Interaction between the control and MES layers is dynamic; control information/status can be accessed as it is created, satisfying statistical process control (SPC) and statistical quality control (SQC) needs; or it can be batched for later observation, meeting labor tracking or maintenance needs. Likewise, the control layer can query the MES (execution) layer for status changes, recipe updates, or other operating instructions. This results in supporting the functionalities of the MES layer" resource allocation, operations scheduling, production dispatching, and the supply of real-time data from the outputs of the control layer.

23.1.2.1 Data flow between MES and control layer

Figure 23.2 describes the operations, interactions, and data flow between the execution layer (MES) and the control layer.

The execution layer is represented by functions that are in charge of the overall flow of the product and/or process, such as those found in supervisory control and data acquisition (SCADA) systems. A SCADA system controls the execution of a specific recipe at a specification operation that the execution

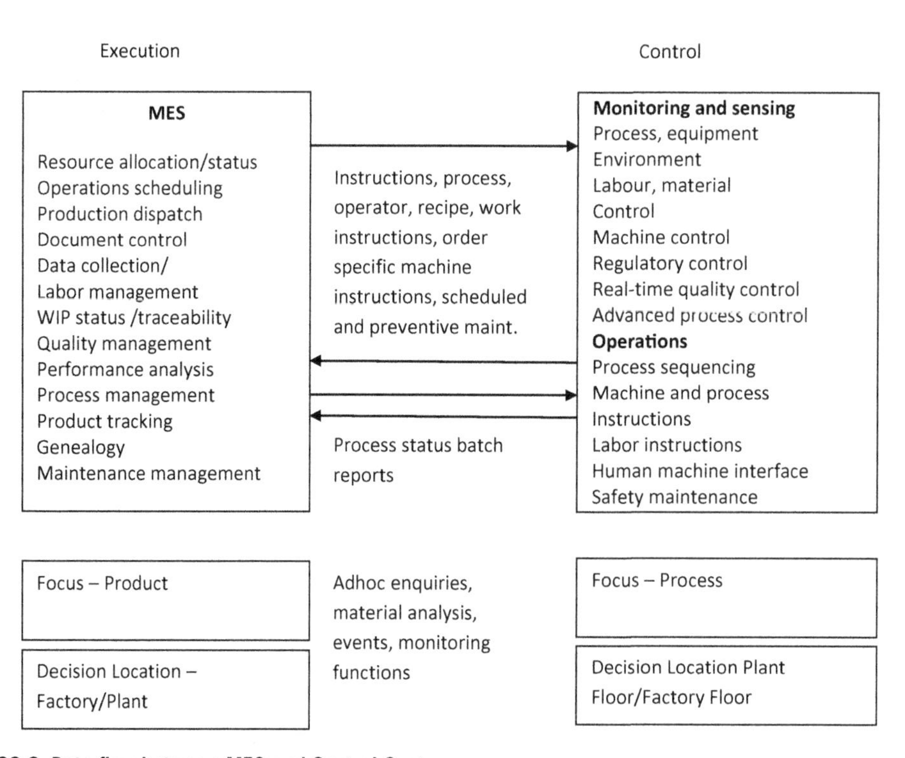

FIGURE 23.2 Data flow between MES and Control Systems

system passed to it. Central repositories of data are collected from various locations within the factory, rather than being localized within a particular area. The specific performance of a piece of equipment, or an operator, may not be as important to the execution layer as it would be to a control operator. The results from these discrete operations are blended into the process control data, and then those results are communicated to the execution systems.

Then the execution layer downloads instructions (for a specific recipe) to the control layer. These instructions supply direction to the people and machines required to carry out the manufacturing operation. The controls people and machines are responsible for monitoring and controlling their own operations, to assure that the outputs are in compliance with the requirements set from the execution layer. This monitoring and control can include separate software and hardware products for quality, process control, data acquisition, safety, and maintenance. Drill-down inquiries or status indicators from the execution layer can spontaneously access information created on an as-needed basis for process control. Bilateral inquiries can emerge from either layer; these inquiries can be used to measure progress-to-plan; to communicate unscheduled changes; or to announce alarms, events, or changes that have occurred. The controls function is responsible for using all of the equipment on the factory floor: hardware, software, and people, in a manner consistent with the goal of producing a product or process that falls within the parameters set forth by the corporation. MES serves as a two-way window into the manufacturing process, integrating, and facilitating key information flow, and commands between controls and business planning systems for total resource and enterprise management.

23.1.2.2 Enterprise data flow

Figure 23.3 depicts the data flow between various systems in a manufacturing enterprise.

In the data flow diagram, from the left you can see the MRP system. They normally do not work in "real-time," but rather in a batching mode; that is the 100× time factor. The MRP system notes product usages, customer orders, and materials requirements, and sends requests to the execution (MES) layer

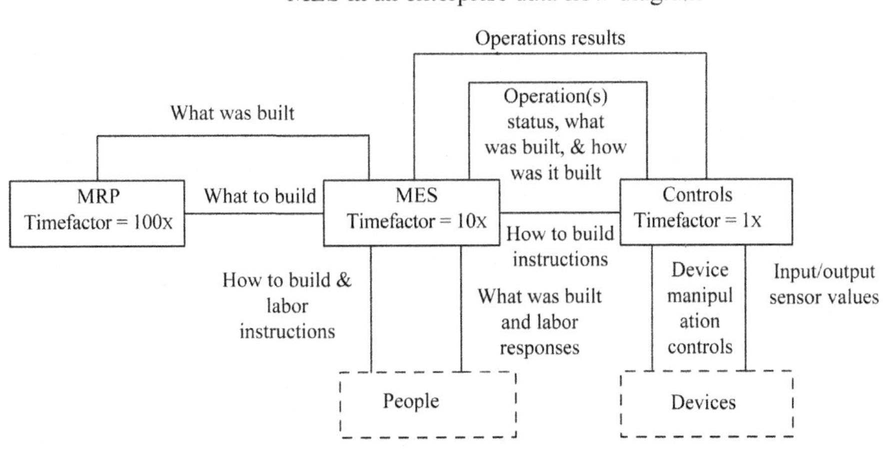

FIGURE 23.3 MES enterprise dataflow

to build more products to fulfill these needs. The MES systems are responsible for product manufacture, and all operations associated with the creation of those products. The MES layer stores recipes that detail the "how to build" instructions for the control layer. These recipes are used for both labor and physical devices. As the MES systems work in short time spans, but normally not in the speeds associated with controls, the time factor is 10×. Once the instructions, programs, documents, software, and other manufacturing requirements for support systems are transmitted, the controls layer is then responsible for carrying out the process. Controls work in real time; their time factor of 1× means that there are always operations occurring to refine, or correct, the process to maintain desired tolerances (or outputs). The dotted line boxes (physical devices and people) carry out the finite instructions for the process output.

23.1.3 MES FUNCTIONALITIES

MES has an important role in a manufacturing industry, in that it employs the planning techniques that improve communication both inside and outside the manufacturing enterprise. Manufacturers that implemented MES have made major improvements in coordination with cross-functional activities to identify potential customer's shipment and vendor receipt problems. Let us now attempt to understand the various functionalities offered by MES.

23.1.3.1 Supply chain planning and execution

Supply chain planning and execution includes the functions to organize, manage, and optimize the supply chain. Many of these functions are performed by headquarters of the manufacturing industries. Typically, there is a handoff from the planning performed at the enterprise level in this area and the planning performed at the manufacturing facility level in manufacturing planning and execution. Supply chain management includes supply and demand planning, order fulfillment, distribution and transportation, and inventory management.

- Supply and demand planning defines and coordinates all aspects of the supply chain, including raw materials and finished products. Demand planning includes capability to promise (CTP), market allocation, constrained planned, and unconstrained planning. Safety stock is the quantity of materials acquired held to satisfy fluctuations in demand and processing. The users typically are business staff, often at headquarters. This area is focused on planning at the enterprise level, which often spans multiple manufacturing facilities. In contrast, production planning in the manufacturing planning and execution category is focused on planning within a manufacturing facility.
- Order fulfillment function involves managing sales orders including taking orders, transaction management, tracking deliveries, billing, and financial reports.
- Distribution and transportation function involves managing shipments, receipts, and transportation logistics. It manages different types of transportation (truck, rail, ship, pipeline, and the like.), and the details unique to each. The users typically are business staff at headquarters and plant sites. Inventory management tracks inventory of raw materials, intermediates, and finished products. It confirms receipt of goods and product shipments, tracks both physical inventory and inventory that is available to promise, and tracks work in progress (WIP) and inventory valuation. The users typically include business staff at a facility, operators responsible for inventory, receipts, and shipments. While physical inventory is important to all industries, the details of how it is determined vary considerably

23.1.3.2 Manufacturing planning and execution

Manufacturing planning and execution includes the functions to plan, execute, and manage production within a single manufacturing facility. These functions typically relate to running the facility safely and efficiently, starting with production orders and ending with products ready for shipment, taking into account the unique aspects of the manufacturing facility. This area includes production planning, production execution, operations management, and production management.

- Production planning consists of creating feasible and optimized daily/weekly operating plans, equipment assignment plans, and recipes, considering available material, capacity, and production resources and tools. It determines the optimal way to handle production orders, and assesses the current capabilities of the manufacturing facility. Production planning is focused on a manufacturing facility, assuming production orders are provided by supply chain modules. It also encompasses recipe and specification management. The typical users are business planning staff at a facility, including planners and schedulers; engineers; operations supervisors. Note that allocation of production to customers when operating at capacity is assumed to be part of the supply chain module. The planning aspect of production planning is focused on planning production, whereas the planning aspect of operations management is focused on the operational aspects of meeting the production plans.
- Production execution involves the execution of production plans safely, reliably, and efficiently. This area is the link to automation and control systems. The users typically include console operators, process engineers, and control engineers.
- Operations management supports console and field operators, and the operating department. It defines safe and appropriate operating limits for process variables, ensures that process conditions are within limits, and enables follow-up on root causes. Operations management communicates current and planned conditions from one shift to another shift to the management. The users typically come from the operations department, and include supervisors, console operators, field operators, process engineers, and engineering specialists. Note that operations planning is focused on planning to meet production plans that come from the production planning area. Work instructions include daily or standing instructions from specialists.
- Production management measures actual production, taking into account, rates, properties, and genealogy. It determines reconciled production figures for internal and fiscal reporting, and determines allocations. The users typically are business staff at a facility, as well as engineers. Note that the yield and loss accounting function determines actual production and production losses, and is often closely linked with inventory management functions. Production allocation applies to joint ventures when production is allocated to multiple entities. While, quality control entails the use of quality information as part of production management, quality management is concerned with entering, storing, and maintaining quality data irrespective of how the data are used. Energy management includes all aspects of predicting, monitoring, and improving energy consumption, including financial aspects.

23.1.3.3 Management and support

Management and support includes decision support, data mining, and analysis tools, which are generally used after the situation of plant condition occurred, to analyze what happened and how to improve. This area also includes assuring product quality, documenting quality compliance for audits, and regulatory compliance and reporting. Functional areas include production reporting and analysis, quality management, and compliance management.

- Production reporting and analysis provides accurate and timely information about profitability, reliability, and other performance metrics. It supports what-if analysis and troubleshooting. Manufacturing analytics includes SPC and overall equipment effectiveness (OEE) analyses. The users typically are business staff, engineers, and management.
- Quality management assures product quality, including managing the processes for defining, measuring, and reporting product quality. It enables continuous process improvements, and manages audits (inspections or close examinations) meant to ensure that products and business processes meet specified criteria. Quality management includes laboratory functions. The users typically include laboratory staff, engineers, and business staff.
- Compliance management ensures compliance with regulatory and statutory requirements. It includes product safety, dealing with hazardous materials, occupational health and safety, carbon emissions, environmental reporting, and other reporting required by government agencies. The users typically include business staff and management. Compliance management includes most regulatory reporting.

23.1.3.4 Asset management

The primary objective of asset management is achieving optimal asset conditions at the lowest cost and best asset performance, taking into account capital expenditures, operating costs, asset utilization, maintenance costs, and supporting process performance. Asset management manages equipment and other fixed assets for optimal overall process performance combined with the minimum costs for assets. It includes costs of acquiring and disposing of assets, maintenance, procedures for change management, shutdown and turnaround planning, and procedures for servicing assets. Maintenance staff, equipment engineers, and, in some cases, operators are some of the users of the asset management function.

23.1.3.5 Common functions – infrastructure

This functional area provides common tools and infrastructure needed to collect and store data from around a manufacturing facility; provides access to all types of information such as reports, key performance indicators, and calculation logic; organizes workflow, collaboration, and notifications throughout the facility; and manages documents. The users include engineers, operators, and many others at a manufacturing facility, and sometimes engineers and business staff at other locations.

23.1.4 MES DEVELOPMENT

Let us attempt to understand the guiding principles in deploying an MES and the key requirements to be met by an MES. The following goals are to be taken into consideration while deploying an MES:

- Simplicity (keep it simple): It is essential for the systems to be easy to use, easy to install, easy to configure, easy to customize, and easy to administer. Even the most sophisticated technologies benefit from simple user interfaces, easy to use administration tools, and so on.
- Minimize servers: The best way to minimize costs for everyone is to minimize the number of requisite server grade computers, especially for small projects. This is because server grade computers are costlier than the client or workstation grade computers.

- Loosely coupled solutions: Loosely coupled solutions enable customers to upgrade or add new components to their solution without requiring the upgrading of the entire system. Loosely coupled solutions provide release independence for applications, databases, and other solutions.
- Industry standards such as ISA S95, B2MML, PRODML, and others are used to define interfaces among modules, where practical. The use of industry standards helps deliver loosely coupled solutions.
- Build for differing environments: different plants have deployment needs that address different network topologies (Internet, WAN, process control networks, firewalls, demilitarized zone (DMZ), and the like.), different performance, transport, and security requirements.

23.1.4.1 Visualization

Visualization refers to the presentation of data in a way that helps end users to easily understand what the data mean, to put it in context, and to understand what they can do with the information. Visualization techniques especially apply to casual and summary interactions, often involve data from multiple sources, and often involve graphics or pictures. In some cases, improved visualization simply means a simpler, easier to use user interface. In other cases, improved visualization involves graphics to organize and present information.

Not all components need to run in all environments, but at least some MES components run in each of these environments. A common framework is defined for each of the five environments, with a common look and feel so that MES components work smoothly in at least one or more of the five standard environments. The following are the some of the most important visualization environments for MES:

- Rich client: Users require a highly interactive environment for both operational and configuration-based tasks.
- Web/enterprise portal: Thin client solution provides basis for informational desktop where data from the enterprise or across the MES application suites can be brought together. Usage pattern in this environment is more informational than high user interaction.
- Office applications: Similar requirements to web/enterprise portal, office applications are seen as an informational tool where data that MES solutions create/manage can be displayed or further analyzed (e.g., KPIs in Microsoft Excel, schedules in Microsoft Outlook)
- Human–machine interface (HMI) station: Users require seamless integration of some MES solutions with the operational environment of human-machine interface station (e.g., integration of operations logbook capabilities in HMI station displays)
- Handheld devices: In MES, handheld devices are used more for information and simple acknowledgement tasks, and leverage devices like android or IOS devices.
- Common look and feel: Applications that look similar, behave similar, and present similar interaction patterns enable users easily gain proficiency and be productive in their assigned responsibilities.

23.1.4.2 Models

Automation systems have traditionally offered rich integration with the data managed by a specific application; however, integration of that same data with other solutions, third-party applications, and customer environments has not been easy or straightforward. This results in duplication of models for organizing information, and new applications do not take advantage of what has already been configured. The MES deployment places a central role on the data that is generated and utilized by the solutions. To that end, two different views are defined for the MES data: (1) an entity model, and (2) a data model.

The entity model enables the applications to interact with information using strongly typed object definitions that are based on industry standards, promote abstraction of data access mechanisms, provide compile time validation that speeds development, and improve quality.

The data model, while abstracted through the entity model, provides the mechanism through which third-party applications interact with the information managed and maintained by MES solutions. Stored procedures and views are defined that abstract the physical data model, and enable MES to secure and manage the integrity of the data.

The following are some of the key features of these models:

- Defined information (entity) model: With the definition of an entity model, a strongly typed object model can be employed that enables solution providers, application developers, and project customization to more easily create new extensions for MES.
- Defined data model: A published/defined data model enables integration of MES with third-party solutions that are not able to easily support higher levels of integration such as the information model or workflows.
- Documentation: The object model and data model are documented, the documentation is kept up to date, and the documentation is published. Failure to adequately document the object model and data model could lead to failure of the component.
- Defined support for industry standards: Compliance with industry standards (ISA S95, MIMOSA, and the like) is an entry requirement.
- Support for extensible data/information models: The ability to easily extend or integrate with the MES data model hinges on the definition of an appropriate basis model that is designed for application extension. This mechanism enables data reuse and configuration once.

23.1.4.3 Security

Secure by default, secure in deployment, and secure in operation are the core tenets of MES. Security is a broad topic covering many areas and is addressed in all aspects of the MES development ensuring that MES solution is completely secure from possible virus attacks on the open network. Integration with Microsoft's Active Directory for providing authentication and the base authorization needs of MES solutions is considered during MES development to provide secure access to MES. A key need in secure solution is the ability to deliver stand-alone solutions that can be deployed for a single user or to a single machine that follows the core security tenet above without large infrastructure cost or configuration. Another important point to address is that not all personnel related configuration is necessarily security configuration and can be performed by authorized users such as the designation/assignment of job responsibilities or workflow escalation. While storage in such technologies as Microsoft's Active Directory may be utilized, the configuration by non-IT personnel is also required.

Key requirements in deploying a totally secure MES system are as follows:

- Integration with Microsoft Active Directory: Security configuration is generally allotted to IT departments; integrating MES security configuration and needs with Microsoft's Active Directory enables plant IT departments to utilize the standard security configuration tools. If the MES

supports only Windows, then active directory becomes a requirement. If not, then use of active directory is useful, but optional.
- Support for distributed environments: MES solutions are deployed in enterprise environments where a single domain trust is not common. The ability to connect and work with data across nontrusted domains is needed.
- Support for alternate authentication models (e.g., X.509 Certificates): Certificate-based technologies provide a mechanism for establishing enterprise and extranet identity when deployed in distributed environments.
- Secure deployment for single user/machine installations: While many MES solutions need to integrate with multiple systems or be utilized by multiple users, there are some MES solutions that can be effectively utilized by a single user on a single machine. Requiring a large server infrastructure or deployment to accommodate the needs of these single-user scenarios is a barrier to adoption.
- Nonsecurity related user configuration: Solutions built on MES take advantage of user profile information not necessarily related to security authorization decisions (e.g., equipment certifications) and needs to be configurable without requiring system administrator/IT level privileges.
- Secure data: Increasing amounts of data are being generated and stored in and by MES solutions, ensuring that it is safe, unchanged, and accessible to only those that require it. This is a fundamental requirement.
- Locked-down environments: Customers are operating in locked-down security environments and coupled with advanced in secure deployment by base operating systems, MES solutions need to be properly designed to operate in nonprivileged locked-down environments.
- Role- and scope-based security: Role-based access control mechanisms provide a mechanism to identify the responsibilities that users of the system are allowed to perform. Additional restrictions based on an area of responsibility allow further refinements in compartmentalization of responsibilities.

23.1.4.4 Integration

Integration is another key aspect of MES not only covering integration with other MES components but also with the existing applications, as well as third-party solutions. The sum of the whole is greater than the sum of the parts. MES components achieve integration through common definitions for core information elements, and through the use of standards such as ISA S95, B2MML, PRODML, and similar recognized standards. The requirements to be met here are as follows:

- Integration with other applications: Integration with ERP, third-party, and legacy systems are key aspects of MES.
- Interoperability in a heterogeneous technology environment: Customers have varying technology environments (e.g., .NET, Windows, SAP xMII, SAP NetWeaver, IBM WebSphere, and so on) into which MES needs to be deployed and integrated.

23.1.4.5 Workflow

Workflow describes the business processes that applications deliver. These have traditionally been coded into the application: however, plant users increasingly require the ability to customize these workflows to their environments and their unique business processes. The MES supports user-interface workflow, intra-application workflow, and inter-application workflow using a mixture of workflow technologies that enable workflows to be easily extended, customized, and monitored. The following workflow requirements need to be met while developing an MES:

- Workflow monitoring: This requires having the ability to see where an active workflow is and to enable troubleshooting, task monitoring, or workflow override.
- Workflow customization: Customers are successful because of the work practices that they institutionalize within their organization and require that the solutions they purchase and deploy to support those work processes, or be customized to support them rather than changing their work practices to accommodate the quirks of a particular solution.

23.1.4.6 Reporting

Reporting is a common need across all applications. Reporting lends itself to making use of third-party vendors that specialize in reporting, especially as customers often have requirements to use specific third-party reporting systems. Reporting in MES is a common application that can be used (or not) by all other applications. It embeds a third-party reporting system, such as Business Objects' Crystal Reports11, Microsoft's SQL Server Reporting Services, or the like. This "reporting application" is designed such that it could be replaced or revised to make use of a different third-party reporting system, should the need arise. The MES offers the following reporting services:

- Supports commercial reporting tools: plant users have preferences for Microsoft's SQL Server Reporting Server, Business Object's Crystal Reports, Cognos, and others. The key requirement is to support one of these third-party technologies. A secondary, less important, requirement is to provide a choice.
- Provides documented data models: The creation of custom reports requires both the ability to access the data that MES solutions are using/storing and documentation on how it is modeled/structured.

23.1.4.7 Data management and data access

The MES provides a solution that requires at most one database server. Most of the products offered by several industrial automation solution providers frequently require both SQL server and Oracle, which hikes the cost and maintenance. The system works entirely on a single database manager. Some customers prefer SQL server or Oracle, while others do not have a preference. Many MES components require or benefit from a process historian. The MES system defines a data entity model that includes data access methods for information stored in the database manager (SQL server or Oracle) and the historian. MES applications interact with the underlying database manager and historian only through the defined data access methods, thereby providing isolation between applications and data storage. This enables the physical data storage to change without requiring the applications to change.

The data management component handles information from any combination of the MES infrastructure, standard MES products, and custom applications. Applications are able to extend the data storage and data access layers to include their own data. An MES system offers the following benefits:

- Requires at most one database server: Customers require the ability to deploy and support at most one database server for MES solutions.
- Data management strategy and guidance: The system provides a home for all types of data, including data storage for the MES infrastructure, standard products, and custom applications, which might be used in any combination.

- Defined data access methods: The system provides access to all managed data through defined methods, such as OLE for process control unified architecture (OPC UA), Web services, or abstracted database queries.
- Data access guidance and examples: Provides guidance and examples on how to efficiently access data from disparate systems, project data through services, and integrate with user interfaces or other systems.
- Decouple applications from database and other applications: Data storage and data access methods loosely couple applications to one another and to the underlying database manager so that customers can change database managers or upgrade the system without having to upgrade all applications.
- Data migration: Customers can upgrade from one version of the MES system to another, and have all data migrated without loss of information and without manual intervention.

23.1.4.8 Configuration and deployment

Configuration and deployment today is expensive, time-consuming, and often requires significant experience to do correctly. MES reduces deployment costs in several ways. The system speeds configuration efforts and reduces errors through the combination of configuration wizards, and bulk import techniques. This enables customers and less experienced engineers to perform the configuration.

MES systems can be large, and often are first installed and configured in staging areas, and later moved into a production environment. The system simplifies deployment from staging areas to production using an approach similar to bulk import configuration. The system helps in packaging and exporting all configuration information (data and files) in one system, such as a staging system, and then importing this information into another system, such as production system. This includes not only data contained/managed by MES components but also that which is managed by other systems that MES is built upon (e.g., active directory roles). The following features make the configuration and deployment of MES an easy process:

- Guided configuration: Configuring advanced applications is complex. Stepwise guidance through configuration wizards enables configuration to be done correctly with minimal errors.
- Bulk configuration: This enables configuration to be performed more efficiently through the use of bulk import techniques that in turn enables the use of tools like Microsoft Excel to quickly and efficiently identify and modify data required for configuration. Bulk configuration supports both full and partial data loading.
- Support to rollout from staging to production: Customers often do initial and ongoing system configuration in staging areas, and then move the tested systems into production. It lowers the cost of ownership and administration when it is easy to move configuration information from test to production

23.1.4.9 System management

System management refers to all aspects relating to the management, monitoring, and administration of MES. Systems today suffer issues ranging from inadequate health information, to inability to easily manage or diagnose application issues. A compounding factor is that IT administrators usually support multiple systems, and are able to figure out what to do with minimal training. The goal of improved system management tools is to simplify administration and reduce administration costs. In addition, good system management tools help to enable remote and outsourced system administration.

The MES provides centralized environments where application's health, performance, and workflow can be monitored, where application diagnostics can be collected, viewed, and packaged, and where applications can be managed for common tasks such as starting, stopping, password/user configuration changes, and launching other more specific application configuration. These centralized environments are capable of addressing single-node installations as well as multiple server-based installations from a single local or remote location. The following features enable effective management of MES:

- Performance monitoring: Understanding and tracking the performance of the system and of the applications enables customers to better deploy, manage, and plan for expansion. It also serves as an invaluable tool for diagnostics when the system is not performing to expectations or promises.
- Health/reliability monitoring: Similar to performance monitoring, health/ reliability monitoring is monitoring and capturing the overall health and robustness of the MES system.
- Workflow management/monitoring: With workflows spanning applications, systems, mechanisms for monitoring their progress, as well as modifying or altering their current state is needed.
- Application management: Beyond the configuration tasks associated with deploying MES solutions, applications themselves have elements of management (e.g., feature enablement) that are generally more administrative in nature than the engineering or configuration work for a specific plant/site.
- Integrated management, monitoring, and diagnostics cockpit: An integrated environment for managing, monitoring, diagnosing, and administrating MES solutions that pulls together the health, performance, and workflow information into a single environment facilitates management and configuration as well as enables improved diagnostics through integrated logging/trace information.
- Network view: Remote access and a network-wide view is needed to effectively manage a large system.

23.1.4.10 Customization

A major limitation with some of the existing MES solutions is the ability to tailor standard product to meet the needs of different industries or to be tailored to the needs of any customer without reengineering. MES enables customization of out-of-the-box integrated solutions for customers and operations. The ability to easily customize and extend MES solutions provides lower project delivery costs, improved profitability of projects, and shorter time for return on investment. The MES is highly customizable:

- Customize the look and feel of a solution: Both customers and different industries have the need to tailor a standard solution to their specific needs. This can be as simple as setting a new color scheme or changing to industry-specific jargon to more complex reorganization of the user interface elements.
- Integration of third-party, standard product, and custom solution into a comprehensive solution: Building new solutions or customizing existing solutions through the use of other standard components, custom developed components, or third-party solutions/applications enables comprehensive solutions to be built and tailored to a customer's specific needs.
- Modify the workflow of a solution: Customers often want an MES to fit their work processes, rather than require wholesale changes to their work processes to use the software. (There are cases

where customers want new work processes, but that is a different subject.) An MES offers enough options to allow the system to adapt to load workflow needs.

23.1.5 SUMMARY

MES can be the key implementing cornerstone in achieving an ERP/MES/control integrated enterprise of tomorrow. MES association (MESA) which is a trade association consisting of MES vendors conducted a survey among the users of MES in order to determine the real, demonstrated benefits of MES.

- Sixty-six percent (66%) of the manufacturers responding reported a reduction in manufacturing time of 45% or greater.
- Sixty-six percent (66%) of the manufacturers responding reported a reduction in entry time of 75% or better.
- Sixty-three percent (63%) of the manufacturers responding reported a reduction in paperwork between shifts of 50% or better.
- Sixty-three percent (63%) of the manufacturers responding reported reduction in lead time of 35% or better.

The results of the survey proved that MES greatly helps in reducing the cycle time significantly in a manufacturing process. MES leap over the gap between front office and factory floor, and empower production employees with a down-to-earth, easy-to-use system that supplements the office-oriented system. MES, in conjunction with current planning systems, is easy to install, and employees are quick to learn its benefits without starting from scratch. MES makes sure that the game plan devised in the office is a winner on the floor.

FURTHER READINGS

MES Functionalities and MRP to MES Data Flow Possibilities. White Paper Number 2. MESA International, 1994.

ANSI/ISA-95.00.01-2000. Enterprise/Control System Integration – Part 1 Models and Terminology.

ANSI/ISA-95.00.02-2000. Enterprise/Control System Integration – Part 2 Object Model Attributes.

ANSI/ISA-95.00.03-2005. Enterprise/Control System Integration – Part 3 Models of Manufacturing Operations Management.

Cox, III, James F., Blackstone, Jr., J.H., 1998. APICS Dictionary, Ninth ed. APICS.

White papers published by MESA

Williams, T.J., 1992. The Purdue Enterprise Reference Architecture: A Technical Guide for CIM Planning and Implementation. ISA.

www.mesa.org

www.honeywell.com

CHAPTER

CYBER SECURITY IN INDUSTRIAL AUTOMATION

24

24.1 PLANT CONTROL NETWORK

Plant Control Network is the network where all the components related to industrial measurement and control connect and exchange information to automate the process or production. For example, industries such as refinery, petrochemical, and power have their own measurement and control devices and production processes, which can be automated using systems like distributed control system (DCS) or programmable logic controller (PLC). All the measurement and control devices, the controllers, which make control decisions based on the measured inputs and the logic/algorithms built in them, human machine interface (HMI), or operator stations (OS), are connected to network known as plant control network (PCN). PCN also contains some other nodes that are not directly used in measurement and control, but to effectively manage and maintain the core control system, devices and process. These nodes are grouped together and named as manufacturing operation management system. Figure 24.1 shows basic components of Plant Control Network.

Plant Control Network, sometimes, is also referred to as OT (operational technology) network, whereas the corporate or business network is known as IT (information technology) network.

24.1.1 IMPORTANCE OF PHYSICAL SECURITY

In ensuring efficient and effective plant operations, having proper physical security measures in place is just as critical as the flow of raw materials, efficient process control, and competent workforce. The chief aim of cyber security, the core concept of physical security, is to detect and prevent an intrusion. Even though it is a different kind of intrusion, consequences of failing to protect can be disastrous. For this reason, it is clear that physical security cannot be placed on the low priority.

Therefore, plants must make sure that they do not neglect having an effective physical security system in place amid the cyber security. The common approach remains "defense-in-depth." This is similar to the "layers-of-protection" philosophy used when designing integrated safety systems, i.e., ensuring multiple measures are deployed on top of each other so if one layer is penetrated, another one is there to further safeguard. With this system, no layer or measure is responsible for being a catch-all.

But recent trends in manufacturing (process safety incidents, increasingly tighter budgets, the aging workforce, regulations, and increased demand) are forcing plants to dig a little deeper with the defense-in-depth concept. Specifically, defense-in-depth means more today than just adding layers of protection, it means ensuring that those layers are folded into the core plant controls along with other subsystems.

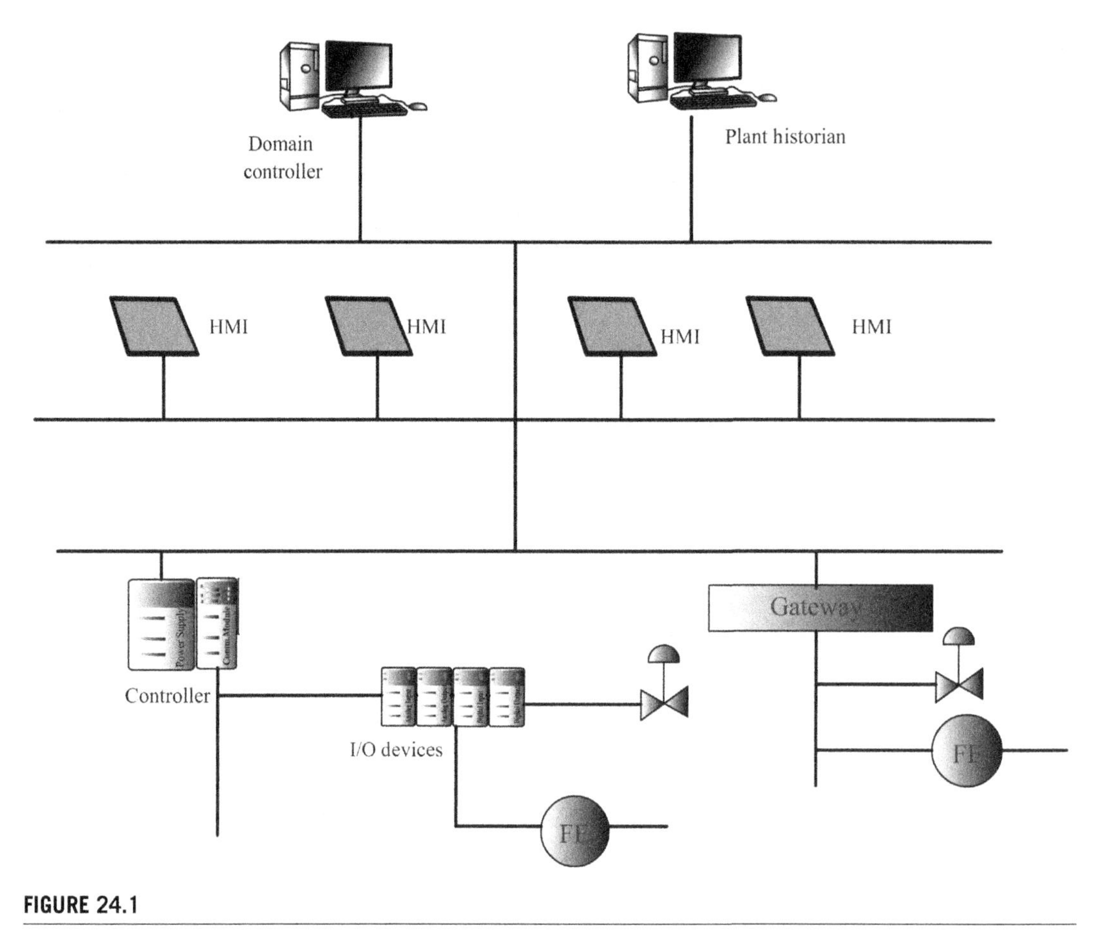

FIGURE 24.1

Industrial network

24.1.2 THE CONTROL-IN-DEPTH APPROACH

Plants are always focused on protecting their people, assets, and the environment. At its core, physical security helps achieve this objective by keeping the wrong people on the right side of the fence to prevent incidents of vandalism, theft, and other malicious acts.

Since protecting people, assets, and profitability demands a holistic point of view, visualizing what physical security at a plant looks like can be difficult as the concept is fairly abstract. The most obvious ways to protect facilities are physical: fences, barricades, and guards. But these safeguards provide only one layer of security, and physical security is more than a barbed-wire fence around a facility. It involves tracking and managing people and assets, complying with federal and industry regulations, monitoring premises through video recordings, implementing access control, and establishing perimeters. There is no single approach to best implement or strengthen a physical security plan, but there are blueprints from successful implementations that can provide a guide in terms of the steps to follow and considerations to take.

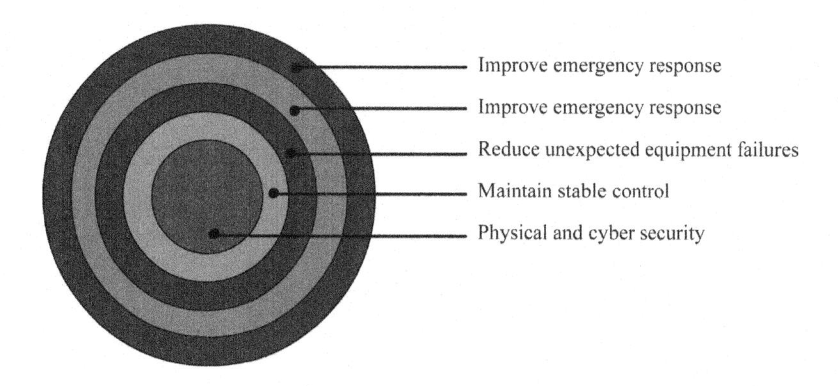

FIGURE 24.2

Defense in depth

An integrated approach must provide protection from their process control room to the perimeter, protecting all the assets of a plant. It should include components such as video surveillance, access control, perimeter intrusion detection, and command centers that work together to provide an extensive method of protection for plants.

An integrated physical security solution allows plant managers and operators to ensure that their top safety priorities such as stable control of the plant, are secured while concentrating on other components that factor in to the safe, secure operation of a plant. Therefore, the design of an integrated system must be clearly aligned with existing safety and process control systems. This approach enables certain layers of protection to deter incidents in the first place, while others can provide detection and alerting, and associated guidance.

Layers can either be automated or require human interaction. Some layers offer easily quantifiable risk-reduction benefits but require pre-identification of risks. Others are less tangible and offer softer benefits (Figure 24.2).

The core of a layered architecture is a well-designed and implemented process design that embodies the business, safety, and production considerations necessary for effective operations. The process must be controlled by a secure process control network that extends across the entire plant and business networks. Managing the plant's assets ensures functioning of the process design as intended, while protecting the plant from incidents with an early indication of failing assets.

With the correct work practices and technology in place, in the event of an abnormal situation, disrupting safe operations, an emergency response plan can be executed, controlled, and monitored to minimize the impact of the incident.

24.1.2.1 *Example of plant security*

When operating under the defense-in-depth plant security concept, it is necessary to have technologies that can integrate with each other to reinforce the layered protection necessary in securing a plant. Video surveillance technology plays a key role here. The installation of a digital video system necessitates the ability to integrate video with the control room.

Connecting the closed-circuit television system with a plant's control room provides the operators with widespread access and monitoring capabilities that touch all corners of the facility. Operators can view, record and archive the functions from a single location.

Table 24.1 Most common weaknesses identified on installed Control Systems

Rank	Site Assessment	Incident Response	Gap Areas
1	Credentials management	Network design weaknesses	Lack of formal documentation
2	Weak firewall rules	Weak firewall rules	Audit and accountability (event monitoring)
3	Network design weaknesses	Audit and accountability (event monitoring)	Permissions, privileges, and access controls

By integrating digital video into one platform, monitoring a facility – remotely located hundreds of miles away – can now be done from a central control room. This is especially critical today, considering that many new sites are smaller and more widely dispersed compared to the massive facilities of a few decades ago. A distributed video architecture (DVA) allows security personnel to view video from multiple digital video systems on one virtual platform, across a single campus or across the country.

DCS systems are particularly beneficial for multi-site and critical infrastructure applications because it provides central monitoring and control of field equipment on the network. DCS systems in a distributed architecture are scalable, making the extension of a system much more cost-effective.

DCS systems also reduce impact to plants when deploying traditional surveillance measures. The availability of alarms or alerts 24/7 radically improves the speed and accuracy by which operators respond to abnormal situations. Vital plant events are responded to appropriately, and possibly, often averting costly plant shutdowns. Deployment costs are also reduced, with one system able to integrate disparate systems such as security, fire and gas, and control. In addition to reducing operating costs, Video surveillance also reinforces the oversight and response ability for plant operators.

Control Systems differs from other computer systems because of legacy-inherited cyber-security weaknesses or vulnerabilities (Table 24.1).

Control Systems vendors and owners can learn and apply many common computer-security concepts and practices to secure and protect their systems. Security should be designed and implemented by qualified security and ICS experts, who can verify that the solutions are effective and can make sure that the solutions do not impair the system's reliability and timing requirements.

Given the nature of the vulnerabilities found in Control Systems, asset owners cannot always directly fix them. Thus, as asset owners wait for vendor patches and fixes, the design and implementation of defense-in-depth security strategies that aid in protecting the Control Systems from attack is part of an effective, proactive security program. Such a program is a necessity because attack strategies are constantly evolving to compensate for increasing defense mechanisms.

Designing security into the system and using secure coding and best practices regarding security can also minimize damage from attacks by insiders, social engineers, or anyone else with access behind the PCN perimeter (Table 24.2).

1. Vulnerabilities

   Control Systems vulnerabilities are presented according to the categories that describe a general problem observed in multiple security assessments.

Table 24.2 Vendor mitigations

Vendor Mitigations	Assets Owners Mitigations
Educate /train developers in secure coding – lack of input validation, authentication & integrity checks	Redesign network layouts to take full advantage of firewalls, VPNs, etc.
Expeditiously test and provide security patches to affected customers	Implement a layered network topology (critical communications in secure and reliable layer)
Implement and test strong authentication and encryption mechanisms	Restrict physical access to the ICS network and devices
Increase the robustness of network parsing code	Deploy timely security patches after testing
Develop network traffic firewall and IDS rule sets – create custom protocol parses for common IDS	Customize IDSs for the ICS hosts and networks
Conduct third-party security source code audits	Restrict ICS user privileges (i.e., establishing roles based access control)
Redesign network protocols with security	Develop a password management plan (strong password)
Develop advanced cyber test suites for security issue in product lines	Change all default passwords
Develop encryption and/or cryptographic hashes to ICS data storage and communications	

These three general categories are grouped by:

a. Vulnerabilities inherent in the Solutions
b. Vulnerabilities caused during the installation, configuration, and maintenance
c. The lack of adequate protection because of poor network design or configuration.

CSET self-assessment tool. The primary goal of the Control System Security Program CSSP cyber-security assessments is to improve the security of the critical infrastructure by delivering to each industry partner a report of all security problems found during the assessment along with associated recommendations for improving the security of their product or infrastructure (as appropriate).

The CSSP has performed assessments on a large variety of systems, and for each assessment, CSSP tailors the assessment plan and methodology to provide maximum value to the customer owning the system. System configurations also vary considerably depending on Control Systems functionality negotiated objectives, and whether the assessment was conducted in the laboratory or onsite. In all cases, the architecture and boundaries for the system under test are carefully determined. Assessment targets are developed individually for each assessment based on the system configuration and assessment focus, in order to address the concerns of the partners. Although a common approach is used for all assessments, the details of each assessment vary; the fact that a vulnerability was not listed on a particular system report does not imply that it did not exist on that system (Table 24.3).

24.1.3 FINDING THE STARTING POINT

An inside-out approach by plant managers is essential to effectively secure a facility. It is a good idea to begin with securing the heart of their plants (the process control network) and gradually build layers of protection that extends all the way to the property perimeter.

Table 24.3 Common security weaknesses identified

Category	Common Vulnerability
Improper input validation	Buffer overflow
	Command injection: • Operating system (OS) command injection • SQL injection
	Cross-site scripting
	Path traversal
Permission, privileges, access control	Improper access control (authorization) Incorrect default permissions
Improper authentication	Channel accessible by no endpoint (man-in-the-middle)
Insufficient verification of data authenticity	Cross-site request forgery
	Missing support for integrity check
	Download of code without integrity check
Indicator of poor code quality	NULL point dereference
ICS software security configuration, and maintenance (development)	Poor patch management • Unpatched or old versions of third party applications incorporated into ICS software
	Improper security configurations • Security functions / options not used during development • Information exposure through debug information
Credentials management	Insufficiently protected credentials • Plaintext storage of a password • Unprotected transport of credentials
	Use of hard-coded credentials
	Weak password policies • Use of default username and password

There are several thorough steps that must be taken to achieve these layers of protection and ensure effective integration, such as:

- Site vulnerability assessment
- Understanding available security systems
- Determining mitigation steps
- System implementation
- Reassessment

A site vulnerability assessment determines possible holes in a plant's overall security system and prioritizes improvement opportunities. The assessment includes the study of the impact of a security breach and its repercussions on security personnel and process operators, and also examines the gaps that exist in the plant's physical security application.

Once the site vulnerability is assessed, a thorough understanding of the latest security technology is necessary to determine threat mitigation steps and how to fill the observed security gaps. Physical security subsets such as – access control, visitor management, video surveillance, and perimeter and intrusion control must be considered for effective plant security. A sound understanding of the latest technologies and their advancements is essential to having an effective, holistic approach to security.

Vulnerabilities must be categorized and prioritized, before identifying mitigation steps. Mitigation steps are unique to each site, but practices such as strengthening the situational awareness of process operators and security personnel can help integrate process control and security systems. It is useful to build more awareness of nonsecurity incidents to security personnel and of security incidents to process personnel; this type of compromise and awareness ensures a more effective, holistic approach to industrial security.

Most basic protection involves right password protection. Common password protection methods are as follows:

- **Passwords:**
- Each user shall have a separate password required for login to the system.
- Management and administration of passwords shall be done from a central location within the system. If a user updates his password on any system, every system connected shall have access to the updated password. Separate passwords for individual workstations on the system is not permitted.
- The system shall be capable of enforcing password policies for administration of user passwords.

  The following policies shall be capable of being configured as a minimum:

- Password aging – the system shall be capable of configuring and enforcing a maximum password age. Users shall be required to change their password within the password-aging period. Users shall be notified during login when the current password is about to expire. Users who do not change their password within the password-aging period shall be locked out of the system.
- Password complexity – The system shall be capable of configuring and enforcing the policies for password construction. As a minimum, passwords shall be required to meet a minimum length requirement.
- Password uniqueness – The system shall be capable of configuring and enforcing a minimum number of unique passwords to be used prior to a password being re-used. This prohibits the user from entering the same password.

System implementation creates an integrated architecture that enables plant operations staff to improve collaboration and responsiveness to reduce security risks. The capability of process control system to integrate security and process systems, and its ability to tie third-party systems together, makes it essential when making integration a reality. Without a standard, it is difficult to achieve effective communication that allows a site to be more aware of a security issue, and awareness allows for increased responsiveness (Figure 24.3).

When fully executed and online, an integrated approach to plant security helps to improve their business performance and peace of mind. It includes independent, yet interrelated layers of protection to deter, prevent, detect, and mitigate potential threats, while increasing the flexibility, scalability, and cost-effectiveness of the entire system. Implementing technology-driven solutions may provide some relief to the pending safety pressures. For a satisfactory solution to the physical security problem, a site must consider independent yet interrelated layers of protection to deter, mitigate, and prevent potential threats.

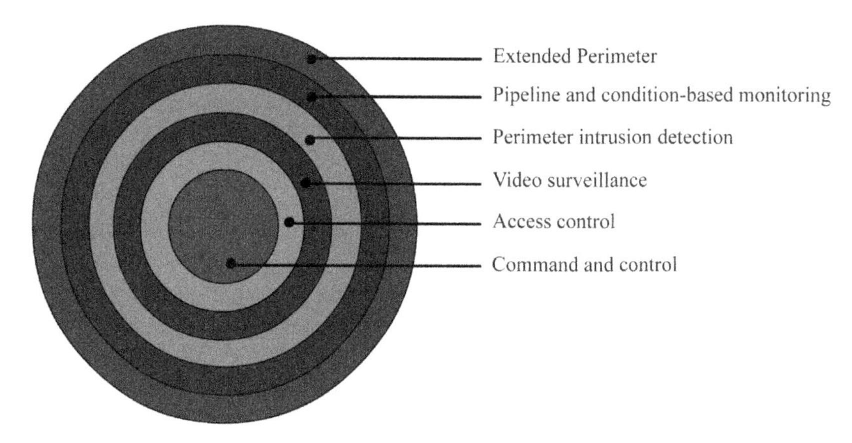

FIGURE 24.3

Defense in depth for cyber security

Supplier shall submit a detail specification specially covering system and network security features available in the offered system. The specification shall include as minimum the following requirements.

24.1.4 SYSTEM HARDENING

- Removal of unnecessary services and programs
- Host intrusion detection systems
- Changes to file system and OS permissions
- Hardware configuration
- Heartbeat signals
- Installing OSs, applications, and third-party software

24.1.5 PERIMETER PROTECTION

- Firewalls
- Network intrusion detection system
- Canaries

24.1.6 ACCOUNT MANAGEMENT

- Disabling, removing, or modifying well-known or guest accounts
- Session management
- Password/authentication policy and management
- Account auditing and logging
- Role-based access control for control system applications
- Single sign-on
- Separation agreement

24.1.7 PROGRAMMING PRACTICES

- Programming for security

24.1.8 FLAW REMEDIATION

- Notification and documentation from vendor
- Problem reporting

24.1.9 MALWARE DETECTION AND PROTECTION

- Malware detection and protection

24.1.10 HOST NAME RESOLUTION

- Network addressing and name resolution

24.1.11 END DEVICES

- Intelligent electronic devices
- Remote terminal units
- Programmable logic controllers
- Sensors, actuators, and meters

24.1.12 REMOTE ACCESS

- Dial-up
- Dedicated line modems
- TCP/IP
- Web-based interfaces
- Secure virtual private networks
- Serial communications security

24.1.13 PHYSICAL SECURITY

- Physical access of cyber components
- Physical perimeter access
- Manual override control
- Intraperimeter communications

24.1.14 NETWORK PARTITIONING

- Network devices
- Network architecture

24.2 CYBER ATTACKS

Cyber security, until recently was plagued with immaturity, reactive technical solutions, and a lack of security sophistication that promoted critical outbreaks, such as Code Red, Nimda, Blaster, Sasser, SQL Slammer, Conficker, and myDoom among others. The security community has evolved and grown smarter about security, safe computing, and system hardening but so have our adversaries. The current decade is rising up to be the exponential jumping off point. The adversaries are rapidly leveraging productized malware toolkits that allow them to develop more malware than in all prior years combined, and they have matured from the previous decade to release the most insidious and persistent cyberthreats ever known.

The Google hacks announced in January 2010, and the WikiLeaks document disclosures of 2010 have highlighted the fact that it is nearly impossible to prevent external and internal threats. Miscreants continue to infiltrate networks and ex-filtrate sensitive and proprietary data upon which the world's economies run. When a new attack emerges, security vendors are obligated to share findings to protect those not yet impacted and to repair those who have been.

24.2.1 ANATOMY OF AN ATTACK

The Night Dragon attacks work by methodical and progressive intrusions into the targeted infrastructure. The following basic activities were performed by the Night Dragon operation (Figure 24.4):

- Company extranet web servers are compromised through SQL-injection techniques, enabling remote command execution.
- Commonly available hacker tools are uploaded on compromised web servers, allowing attackers to pivot into the company's intranet and giving them access to sensitive desktops and servers internally.
- Using password cracking and pass-the-hash tools, attackers gain additional usernames and passwords, enabling them to obtain further authenticated access to sensitive internal desktops and servers.
- Using the company's compromised web servers as command and control (C&C) servers, the attackers discovered that they needed only to disable Microsoft Internet Explorer (IE) proxy settings to enable direct communication from infected machines to the Internet.
- Using the remote administration tool (RAT) malware, they proceeded to connect to other machines (targeting executives) and infiltrating email archives and other sensitive documents.

24.2.1.1 Details of the attack

Attackers using several locations leveraged servers on purchased hosted services and compromised servers to wage attacks against global oil, gas, and petrochemical companies, and individuals and executives to acquire proprietary and highly confidential information. The attackers used a variety of hacker tools, including privately developed and customized tools that provided complete remote administration capabilities to the attacker. Remote Administration Tools (RATs) provide functions similar to Citrix or Microsoft Windows Terminal Services, allowing a remote individual to completely control the affected system.

To deploy these tools, attackers first compromised perimeter security controls, through SQL-injection exploits of extranet web servers, and targeted spear-phishing attacks of mobile worker laptops, and

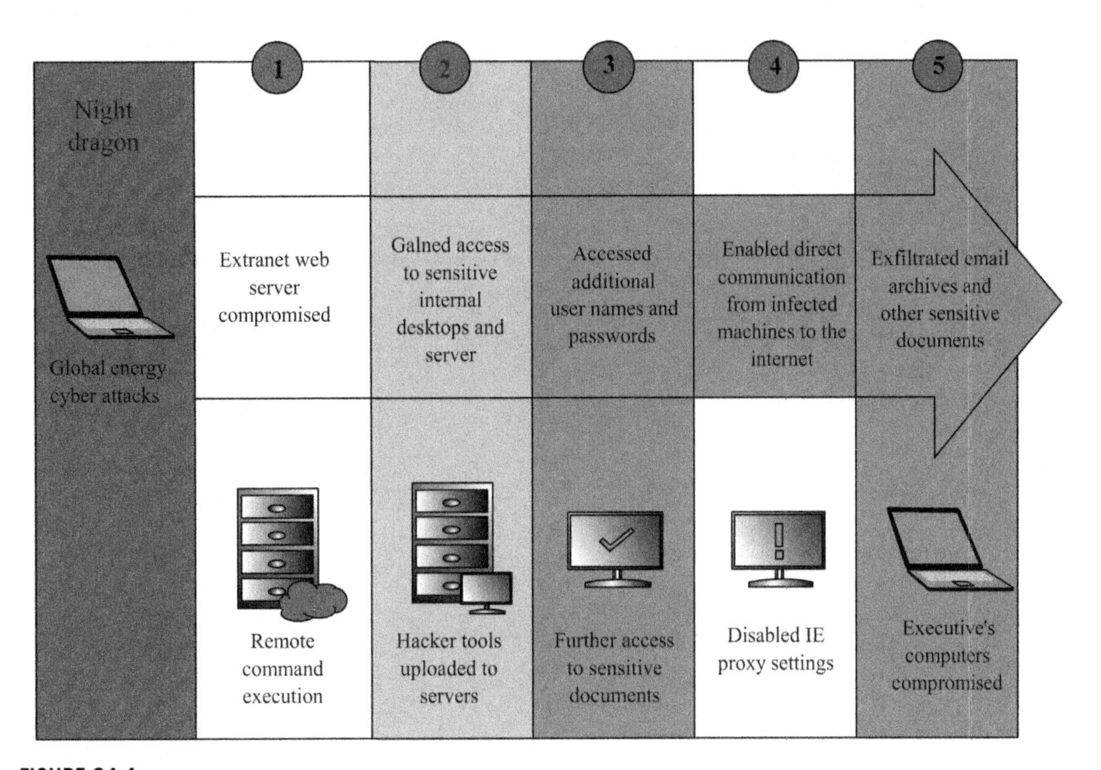

FIGURE 24.4

Activities of accessing

compromising corporate VPN accounts to penetrate the targeted company's defensive architectures (DMZs and firewalls) and conduct reconnaissance of their networked computers (Figure 24.5).

Many hacker websites offer these tools for download, including links to reduh, WebShell, ASPX-Spy, and many others, plus exploits and zero-day malware.

Once the initial system was compromised, the attackers compromised local administrator accounts and active directory administrator (and administrative users) accounts. The attackers often used common Windows utilities, such as SysInternals tools and other publicly available software, including hacking tools developed in China and widely available on Chinese underground hacker websites, to establish "backdoors" through reverse proxies and planted Trojans that allowed the attackers to bypass network and host security policies and settings. Desktop antivirus and antispyware tools were also disabled in some instances, this is a common technique of targeted attacks.

ISA SP99 recommends "defense-in-depth" concept to secure industrial network from probable outside attacks. Control traffic, which is time critical and important for deterministic task completion, must not get disturbed because of noncritical network traffic flow like system health monitoring in the same network. Segregation of control network and prioritization of control traffic over other traffics is one important recommendation. Use of DMZ (de-militarized zone) and firewall in between

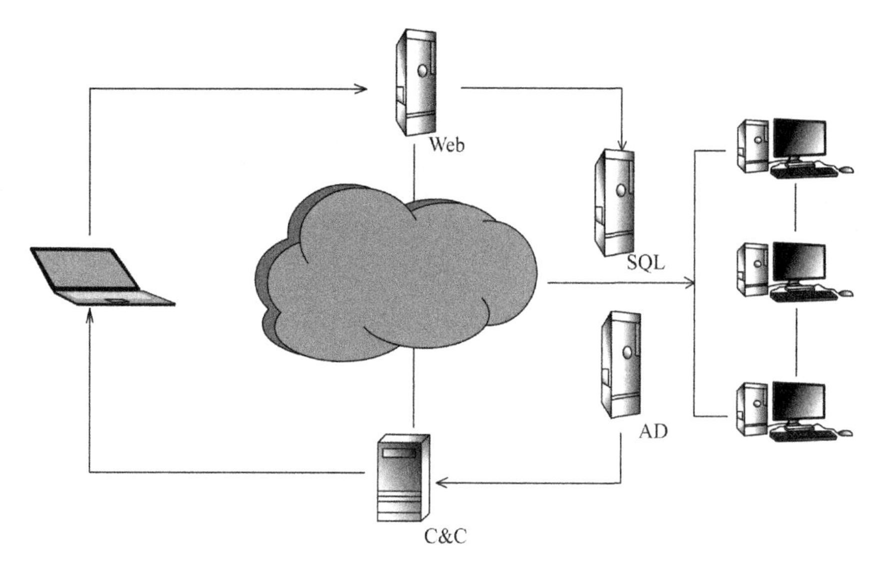

FIGURE 24.5

SQL Injection

corporate network and industrial control network for data transfer instead of giving direct access from corporate network to industrial control network is another must to have to secure any industrial solutions. These recommendations help to isolate network traffics and avoid attacks like denial of service as well.

Weak firewall rules, opening up multiple ports through firewall, access control only based on ports without specifying IP address, etc. are other examples of network design weaknesses. Defense-in-depth increases the hurdles to attack a system and hence decreases the chance that attackers will be able to cross all hurdles. By this, it increases the chance that the attackers will give up before realizing the goals of attacking the control system network.

Following diagram shows an example of defense-in-depth network implementation (Figure 24.6).

In this example, Level 0 network consists of end measurement and control devices, which directly measures the process variables and control the process. Level 1 network contains all the controllers, which have the logics/control algorithms to take control actions based on the measured variables. In Level 1 network control traffic should be given highest priority over any other traffic. Level 2 network is the place where all OS or HMIs, server, etc. gets connected. Level 3 networks mainly contain applications used to optimize plant operations (advanced control) like field device maintenance application, domain controller, process historian, etc.

Level 0, Level 1, and Level 3 combined is called process control network (PCN).

Level 3.5 is the other name of DMZ, which is used to get process control network data on business layer (Level 4) applications.

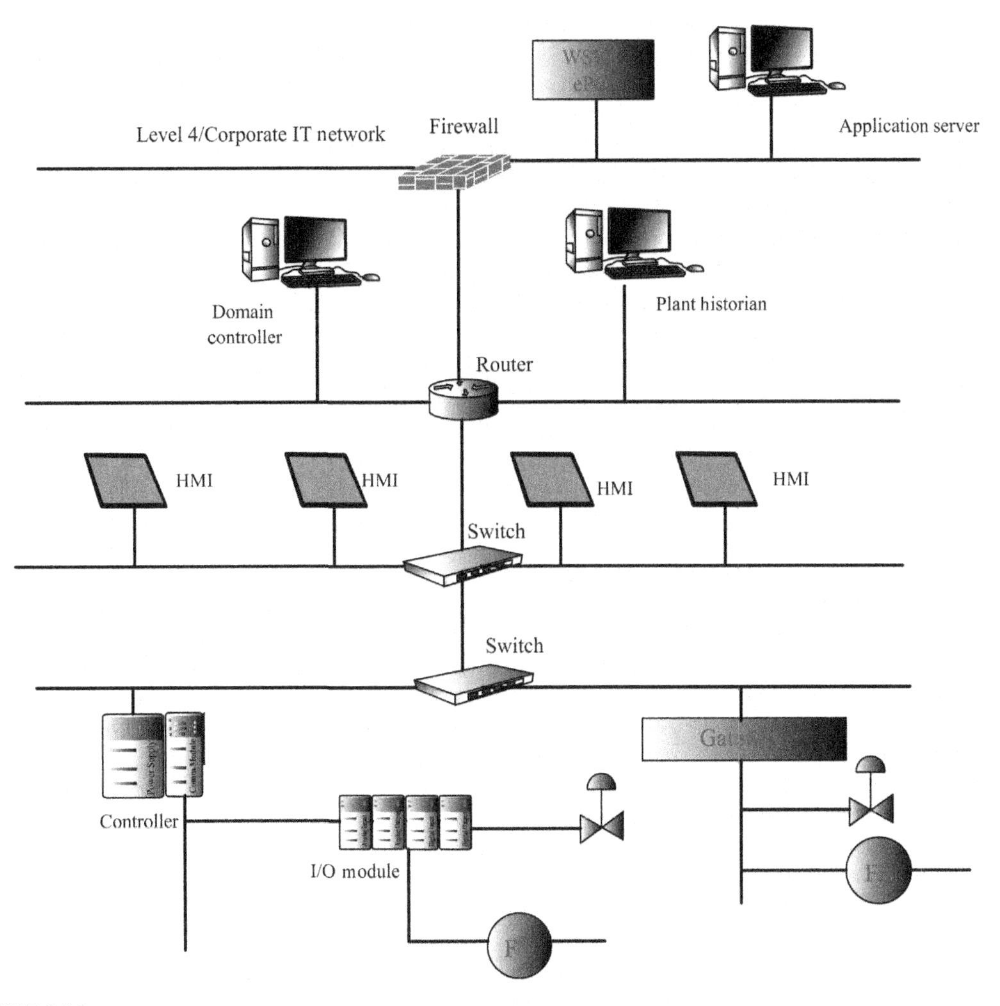

FIGURE 24.6

Defense-in-depth example

24.3 UNDERSTANDING COMMON PCS VULNERABILITIES

A major difference in securing PCS and a typical computer system is in the ICS components that do not use standard IT hardware or software. Custom PCS hardware and software have not been scrutinized like common computer products, and refresh rates are typically much lower.

PCS security objectives are typically prioritized as:

1. Availability
2. Integrity
3. Confidentiality

24.4 COMMON PCS SOFTWARE SECURITY WEAKNESSES

24.4.1 IMPROPER INPUT VALIDATION

- Buffer Overflow: Input validation is used to ensure that the content provided to an applications does not grant an attacker access to unintended functionality or privilege escalation. Buffer overflow vulnerabilities are result of programmer error.
- Lack of Bounds Checking: Lack of input validation for values that are expected to be in a certain range, such array index values, can cause unexpected behavior.
- Command Injection: This allows the execution of arbitrary commands and code by the attacker. If a malicious user injects a character (such as semi-colon) that delimits the end of one command and beginning of another, it may be possible to insert an entirely new function.
- SQL Injection: This injection has become a common issue with database-driven websites. The flaw is easily detected and easily exploited, and as such, any site or software package with even a minimal user base is likely to be subject to an attempted attack of this kind.
- Cross-Site Scripting: Cross-site scripting vulnerabilities allow attackers to inject code into the web pages generated by the vulnerable web application. Attack code is executed on the client with the privileges of the web server.
- Improper Limitation of a Pathname to a Restricted Directory (Path Traversal): Directory traversals are commonly associated with web applications, but all types of applications can have this class of vulnerability.

24.4.2 POOR CODE QUALITY

- Use of Potentially Dangerous Functions: Application calls a potentially dangerous function that could introduce vulnerability if used incorrectly.
- NULL Pointer Dereference: A NULL pointer dereference occurs when the application dereferences a pointer that it expects to be valid, but is NULL, typically causing a crash or exit.

24.4.3 PERMISSIONS, PRIVILEGES, AND ACCESS CONTROLS

Permissions, privileges, and other security features are used to perform access controls on computer systems. Missing or weak access controls can be exploited by attackers to gain unauthorized access to PCS functions.

- Improper access control (authorization)
- Execution with unnecessary privileges

24.4.4 IMPROPER AUTHENTICATION

Much vulnerability identified in PCS due to the software failing to sufficiently verify a claim to have a given identity.

- Authentication bypass issues
- Missing authentication for critical function
- Client-side enforcement of server-side security

24.4.5 IMPROPER NETWORK DESIGN

Lack of network segmentation is the most widely seen network design weaknesses in industrial control domain. ISA SP99 recommends "defense-in-depth" concept to secure industrial network from probable outside attacks that is explained in previous sections.

As we have seen the weaknesses till now, there are certain best practices that can be followed in order to minimize the risk arising due to these vulnerabilities, which are as follows:

1. Vulnerability in the product can be mitigated through product development best practices, few of which are as follows:
   - Input validation to handle vulnerabilities such as buffer overflow, SQL injection, cross-site scripting, path traversal, lack of bounds checking, and so on.
   - Limit use of potentially dangerous function calls: Unsafe C/C++ function call is dangerous as this might end up into buffer overflow. Most of the unsafe function calls are seen in communication processing codes. Following secure coding practice is recommended to all system vendors.
   - Authenticated and encrypted communications between different nodes on industrial control network: This is important to reduce the man-in-the-middle attacks.
   - No download of code without integrity checking: Use of cryptographic digital signatures and building automation code to verify the signature before download is recommended to secure PCSs as this method provides data integrity.
   - Credentials should not be hard-coded: System vendors should come out of hard-coded credentials and move toward secure authentication.
2. Vulnerabilities caused during the installation, configuration, and maintenance of the PCS can be minimized by following certain standard operational procedures a sample of which is provided here:
   - Periodic update of security hot fixes and antivirus patches: This is important to minimize attacks through known vulnerabilities. Once the vulnerability is open to all and patch is also available to address that, if the same is not applied, the system remains vulnerable for that window. This kind of window is easy target for attackers and hence hot fix and patch application at regular interval is critical.
   - Strict account management (includes creation, activation, group access, disabling, periodic reviewing of all the user accounts): If credentials are passed as clear text, attackers can capture that and use to increase privilege and gain access to critical resources of PCS. Account lockout policy, enabling password complexity, forcing users to change password at regular interval, removal of unnecessary privileges from the groups that are not required to carry out the defined tasks, etc. are important to make the industrial network and the system safe.
   - Any security hot fixes and antivirus updates must be tested on off-process system to make sure no collateral damages before deploying those on industrial on-process nodes. PCS is complex in nature and many such systems were built long ago when security concepts were not present. Directly applying new security hot fixes and patches can impact the working of few applications or interaction of one application with other which can result into unpredictable system behavior, application, or service crashes, performance degradation, etc.
   - Periodic backup of all control system nodes so that the same can be recovered quickly in case of any disaster: As taking backup is important, transferring the backups to a secured location is also important. All the system-level data must be backed up and stored at a secured location at regular interval.

- Proper maintenance of security documentations: First step to make PCS safe is to acknowledge the risks and document the detailed plan of addressing those risks. This should be aligned with organization's business case. Support from senior management is must to make implementation of security recommendations a success, and for that documentation of the plan, assigning proper business priority, etc. are important and should be kept up to date.
- Security audits and assessments is another recommended practice to keep PCS secured. Management of change (MOC) with audit trail is one critical aspect of this. Event loggings must be turned on and reviewed at regular interval. Network logging is important to monitor as this can give accurate information if any unwanted access/attack is being tried.

3. The lack of adequate protection due to poor network design and configuration can be minimized by following a set of recommended best practices that are given in subsequent sections, for explanation and understanding on these levels it is recommended to refer defense-in-depth section:
- Level 1 traffic should never reach to Level 3 network: Level 1 traffic mainly consists of peer-to-peer messages between controllers, IO data between controller and gateways. These are sensitive data and can be exploited if goes outside. As corporate network has visibility till Level 3, it is recommended to avoid any such traffic coming to Level 3 network.
- Level 1 critical peer-to-peer traffic must get higher priority over any other network traffic: Peer-to-peer traffic between controllers is used for industrial control purpose. For deterministic data transfer, this traffic should be given higher priority over other traffic such as supervisory message exchange between Level 1 and Level 2. Nowadays all switches are smart and have options to prioritize the traffic at switch level.
- Broadcast and multicast traffic should be limited at Level 1: Broadcast and multicast traffic are mainly used for diagnostic and health monitoring purposes. If this traffic increases in Level 1, control operation might get impacted.
- Protection against broadcast and multicast storm should be implemented at both Level 1 and Level 2: Storms might get generated in industrial network because of wrong network connections (say loop formation), or because of faulty hardware. This can bring down the whole industrial network and hence resulting into complete loss of control and loss of view. To avoid such catastrophe, storm limit must be set at all switches to make sure the problematic node (which generates the storm) is isolated automatically from the network if storm traffic crosses certain predefined limit. This would ensure that other parts of industrial network remain unaffected because of one problematic node.
- Router between Level 2 and Level 3 should have proper ACLs configured to limit the connection to Level 2 from above: All Level 3 nodes do not need access to Level 2 nodes and vice versa. To minimize the traffic at PCN (Level 1 and Level 2) level, it is recommended to use proper access control lists (ACL) at router.
- Level 3 network should not be directly accessed from Level 4: Level 4 network is in general exposed to outside world and hence has maximum probability of cyber attacks. If Level 4 nodes have direct access to Level 3, it becomes easier for the attacker to target Level 3 and hence the complete industrial network. To limit such attacks, it is not recommended to provide direct connectivity to Level 3 from Level 4.
- Any Level 4 application that needs process control data must come to DMZ/Level 3.5. Level 3.5 should access data from Level 3 and serve those to Level 4 applications: Same as above

point, DMZ is created to increase one more defense layer in the "defense-in-depth" concept and also to isolate the access levels.

- PCSs must provide all the required port details that should be opened through ACLs for specific application needs: ICS providers must provide list of ports that are used for different applications/features it supports. This would make sure that only "needed" ports are open through ACLs based on the end user's feature needs and connectivity.
- Unused switch ports (at Levels 1, 2, and 3) should be protected as any unwanted node can get connected to those free ports and create problem in the industrial network: Although proper isolation is done using DMZ and other layers, ACLs, and so on still any outside computer can get connected to free port on Level 1 or Level 2 network and can spread malware or hack the complete industrial control network. In this case, after following all recommended isolation also network could not be protected. To avoid such issues it is recommended to lock all unused switch/router ports in industrial control network.
- Firewall should limit access to the different lower network levels based on necessary communications: This point is similar to router ACLs. Firewall ACLs should also be reviewed at periodic interval to make sure only "required" ports and IPs are opened through the ACLs.
- Firewall rules should be strong enough to restrict any unwanted communication to industrial control network: By default firewall should be configured as "deny all" and very few ports should be open. Allowing DCOM through firewall is an example of "weak firewall configuration" as DCOM uses wide range of ports and almost all ports get opened if DCOM is allowed through firewall. In such cases it is always recommended to use OPC tunneller and only open one port that tunneller uses.
- Intrusion detection system (IDS) and intrusion prevention system (IPS) should be in industrial control network to monitor any attacks:
  - Signature-based IDS – this monitors the network traffic, compares with the signature of known malicious attacks/threats and alerts the system or network administrator. This is similar to the way antivirus software works and needs periodic updates.
  - Anomaly-based IDS – this first monitors the traffic and baseline the network signature. And then it starts comparing the current network traffic with the baseline one to detect any threats and alerts the system or network administrator.
  - IPS – this is also known as IDPS (intrusion detection and prevention system). This not only does the intrusion detection, it also tries to prevent/block the detected intrusions such as dropping the malicious packet, blocking traffic from the suspected node, and so on.

24.4.5.1 Use of remote administration tools

Remote administration tools (RATs) are commonly used administrative tools that enable hackers (and administrators) to manage victims' computers (or managed systems) and completely control their use and function. RAT features often include screen and webcam spying, keystroke logging, mouse control, file/registry, and process management, and, of course, remote command shell capability.

Once the attackers gain complete control of the targeted internal system, they dump account hashes with gsecdump and use the Cain & Abel tool to crack the hashes to leverage them in targeting ever more sensitive infrastructures. Files of interest focused on operational oil and gas field production systems and financial documents related to field exploration and bidding that are later copied from the compromised hosts or via extranet servers. In some cases, the files were copied to and downloaded from company web servers by the attackers. In certain cases, the attackers collected data from SCADA systems.

24.4.5.2 Detecting an attack

The methods and tools used in these attacks are relatively unsophisticated, as they simply appear to be standard host administration techniques, using standard administrative credentials. This is largely why they are able to evade detection by standard security software and network policies. Since the initial compromises, however, many individual unique signatures have been identified for the Trojan and associated tools by security vendors; yet only through recent analysis and the discovery of common artifacts and evidence correlation has been established to determine that a dedicated effort has been ongoing for at least 2 years or more.

The following artifacts can help to determine whether a company or System has been compromised:

- Host files and/or registry keys
- Antivirus alerts
- Network communications

The Trojan components are manually copied or delivered through administrative utilities to remote systems. They do not include any worm or self-replicating features, nor can the Trojan "infect" other computers. Removing the Trojan components is simply a matter of deleting the related files and registry settings.

24.4.5.3 Antivirus alerts

Antivirus patterns are defined according to samples submitted by clients or analysts as they are discovered. Some Trojans exhibit characteristics of other types of malware, such as worms or viruses that have the ability to infect other systems. RATs do not typically include such features, and because they are defined with unique configurations for custom purposes, they commonly change faster than unique samples can be identified.

An antivirus pattern that is generic enough to detect regardless of configuration changes can only be defined when an entire toolkit is found. The package necessarily includes application server, the generator utility for creating droppers, related droppers, and backdoors – and a sufficient number of each to correlate the toolkit. Several unique patterns have been developed from submitted samples.

24.5 STANDARDS

Cyber security standards have been created recently because sensitive information is now frequently stored on computers that are attached to the Internet. Also many manual tasks are computerized now; therefore there is a need for information assurance (IA) and security. Cyber security is important in order to guard against identity theft. Businesses also have a need for cyber security because they need to protect their trade secrets, proprietary information, and personally identifiable information (PII) of their customers or employees. The government also has the need to secure its information. One of the most widely used security standards today is ISO/IEC 27002 which started in 1995. This standard consists of two basic parts. BS 7799 part 1 and BS 7799 part 2 both of which were created by British Standards

Institute (BSI). Recently this standard has become ISO 27001. The National Institute of Standards and Technology (NIST) has released several special publications addressing cyber security. Three of these special papers are very relevant to cyber security: the 800-12 titled "Computer Security Handbook," 800-14 titled "Generally Accepted Principles and Practices for Securing Information Technology," and the 800-26 titled "Security Self-Assessment Guide for Information Technology Systems." The International Society of Automation (ISA) developed cyber security standards for industrial automation control systems (IACS) that are broadly applicable across manufacturing industries. The series of ISA industrial cyber security standards are known as ISA-99 and are being expanded to address new areas of concern.

24.5.1 ISA SECURITY COMPLIANCE INSTITUTE

The ISA Security Compliance Institute (ISCI) has developed compliance test specifications for ISA99 and other control system security standards. They have also created an ANSI-accredited certification program called ISASecure for the certification of industrial automation devices such as PLCs, DCSs, and safety instrumented systems (SIS). These devices provide automated control of industrial processes such as those found in the oil & gas, chemical, electric utility, manufacturing, food & beverage, and water/wastewater processing industries. Government bodies and private industry is increasingly concerned regarding the risk of these systems being compromised by "evildoers" such as hackers, disgruntled employees, organized criminals, terrorist organizations, or even state-sponsored groups. The recent ICS malware known as Stuxnet has heightened concerns about the vulnerability of these systems.

Stuxnet has clearly been a wakeup call for the industry and companies need to ensure cyber security to protect their operations.

24.5.2 CASE STUDY

Let us consider a process plant which has two areas namely "production" and "utility." Both these plant areas are geographically isolated and connected through optical fiber link. Because of this, two dedicated control systems are controlling these two areas. There are needs of data transfer between these two control systems. Also few Level 4 plant network OPC clients (advanced applications) need some data from these two control systems.

So, basic end user needs are:

- Data transfer between the two areas of the plant
- L4 applications need access to control system data via OPC

The network architecture that serves both the needs mentioned above is shown in Figure 24.7.

24.5.2.1 Automation systems interconnectivity

- Production area has one control system with controllers, IO modules, server with OPC interface, etc.
- Similarly utility area has another control system with controllers, IO modules, server with OPC interface, etc.
- Level 2 network of both the control systems are connected to firewall.
- Third leg of the firewall is connected to Level 4 network.
- Firewall rules are configured to achieve the following tasks:
    - OPC data transfer between Level 2 control system and Level 4 OPC client
    - Data transfer between two control systems

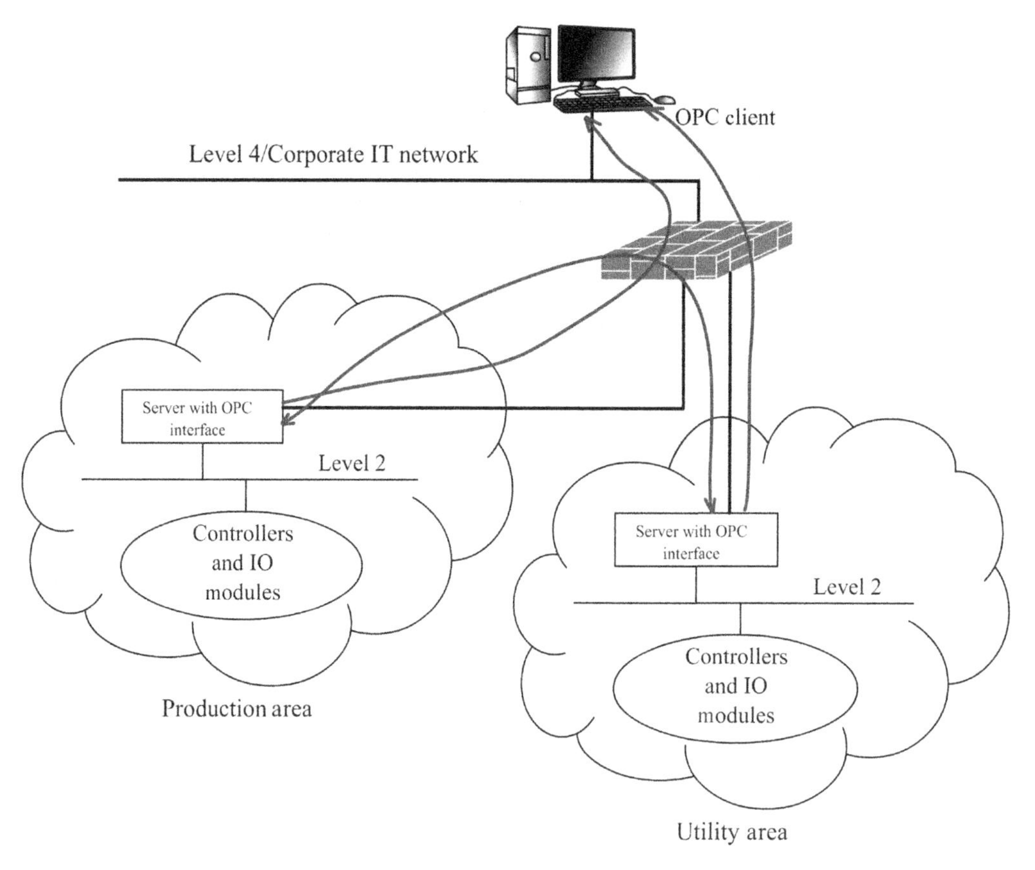

FIGURE 24.7

Plant automation systems interconnections

24.5.2.2 Security guideline violations

Above solution violates many of the industrial automation security guidelines. The recommendation is to stop here and try to list down the security guideline violations. Write down your findings here before proceeding to subsequent sections.

24.5.2.3 Findings

1. Level 2 networks are directly connected to firewall. There is no Level 3 network in between Level 2 and Level 4. So, defense-in-depth recommendation is violated here.
2. Control data transfer between the two plant areas would happen through the Level 4 firewall which means L1 control data is going through Level 4 firewall. This is again violation of security guideline.
3. For Level 4 OPC client to get data from Level 2 OPC server, almost all the ports (Port 135 and Port 1024 – 65535) need to be opened at firewall. Which means the security is completely compromised. If all the ports are open as mentioned earlier, the firewall in between becomes useless.

24.5.2.4 Recommended network connectivity

Figure 24.8.

Following are the changes made in system interconnectivity to make the industrial automation system secured:

1. Router is introduced between the firewall and Level 2 network to create Level 3 network in between.
2. Control data transfer between the two plant areas will happen through Level 3 network and hence the control data is not going through Level 4 firewall.
3. OPC server with Tunneller is added at Level 3 network, which collects the data from Level 2 servers and acts as data source for Level 4 OPC client. The tunneller uses one particular

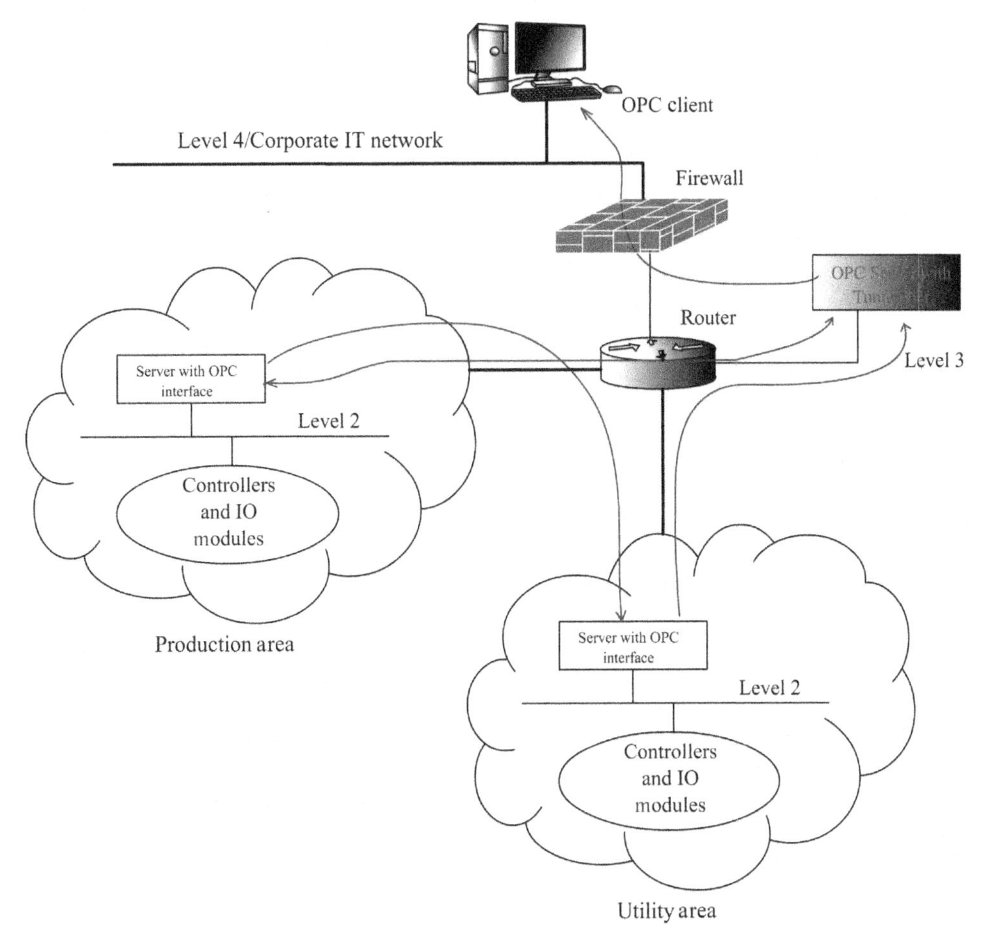

FIGURE 24.8

Recommended plant automation systems interconnections

user-defined TCP port to transfer the data across firewall and hence no need to open wide range of ports that DCOM demands to work across firewall.

4. Level 4 OPC client is not accessing the Level 2 server directly. Now Level 3 OPC server is acting as data source for Level 4 applications. Thus, another layer of isolation is added (defense-in-depth).

FURTHER READINGS

ANSI/ISA-TR99.00.01-2004 – Security Technologies for Manufacturing and Control Systems.

ANSI/ISA-TR99.00.02-2004 – Integrating Electronic Security into the Manufacturing and Control Systems Environment.

Cyber Telecom: Security Surveying Federal Cyber Security Work.

The Standard of Good Practice (SoGP).

Department of Homeland Security, A Comparison of Cyber Security Standards Developed by the Oil and Gas Segment.

International Society of Automation (ISA) as an ISA Fellow.

CHAPTER

MOBILE AND VIDEO SYSTEMS 25

25.1 INTRODUCTION

Mobile Technologies and mobility based applications are increasingly used in many field of engineering. Further, the process manufacturing industry is beginning to adopt and deploy mobile computing applications with an emphasis on operational excellence and operator-driven reliability (ODR) programs. In many cases, executive management is placing emphasis on reliability improvement initiatives that require plant operators to uncover hidden opportunities to improve efficiencies, and reduce operating costs.

Wireless technology provides opportunities to unlock new value in process automations and plants at economically viable levels previously unheard of. The possibilities of enabling significant productivity and efficiency gains are endless.

Below are the major driving factors for mobile applications in industrial automation and control industry:

- Support maintenance excellence initiatives
- Enable ODR programs
- Operations, quality, safety, maintenance, and reliability

In industrial automation and control world the devices that are going to be used for mobility applications has to comply with hazardous area standards to be used inside the plant. The key platforms that are being used in this area are palmtops or small industrial grade tablet PC's as well as handheld devices with wireless connectivity.

Wireless technologies empower the outside operator to make decisions in the field where decisions must be made in order to succeed in the case of incident avoidance. Finally, the fact that today's wireless computing products can connect the field and the associated data or observations with the rest of the plant's computing infrastructure.

Figure 25.1 shows typical plant with WLAN infrastructure as well as wireless sensor networks along with mobility devices such as mobile process station and mobile handheld devices coexisting with each other being part of the same wireless infrastructure.

All the applications are providing data back into host applications such as DCS, asset management applications and host of other applications.

With the specifics of this chapter we will see mobile devices such as mobile process consoles and mobile handheld devices in detail in further sections along with what needs and benefits are driving the mobility in industrial automation and control field.

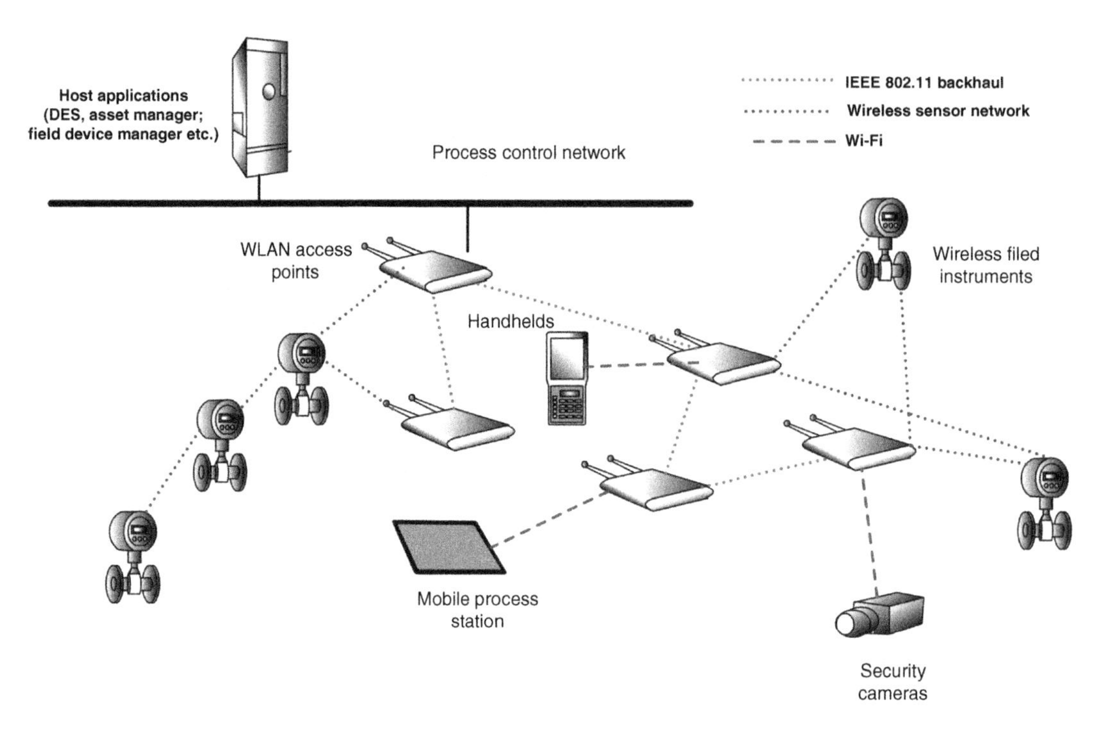

FIGURE 25.1 One network supports all applications

25.2 MOBILE PROCESS MONITORING CONSOLE

Mobile console as the name suggests extends access to critical process information, historical data, graphics, maintenance data, and other key functions into the field. Wireless technology expands the scope of process automation beyond the control room so that remote users can see the impact of their activities on the plant's operation in real time.

For any mobility application, wireless infrastructure backbone is very important and key thing. Once the user has a right wireless backbone network in place, the possibility of mobility-based applications is very high.

25.3 KEY BENEFITS OF WIRELESS PROCESS MOBILE CONSOLE

25.3.1 IMPROVE OPERATIONS

Mobile console provides plant staff with access to process information that they can capture and distribute across the operations. A collaborative decision support environment between the field and the control room helps improve daily operations, emergency response during incidents and improves production uptime. It also unifies the workspace between the board operator and the field operator – a benefit not offered by traditional field-centric distributed control systems.

25.3.2 IMPROVE MAINTENANCE

Mobile console provides maintenance staff with access to real-time maintenance information. They can capture and distribute data on plant asset management applications.

Mobile process monitoring console use available wireless LAN infrastructure inside the plant for connectivity back into the plants distributed control system. These devices act as Wi-Fi client to the access points using 802.11 b/g communications.

25.4 HANDHELD MOBILE DEVICE SOLUTIONS

Wireless field handhelds are being widely accepted as a merging technology for improving operational efficiency and reliability as well as in reducing plant incidents. Working toward the goal of flawless operation, these handheld mobility devices helps to realize a reduction in production downtime/ slowdowns and improved decision support (Figure 25.2).

Below are the some of the needs that are driving mobility–based applications:

- Cost reductions
- Worker and task management
- Inventory
- Greater data collection production information
- Operator rounds
- Standards compliance
- Managing assets deployment
- Auditing, maintenance, repair, and calibration

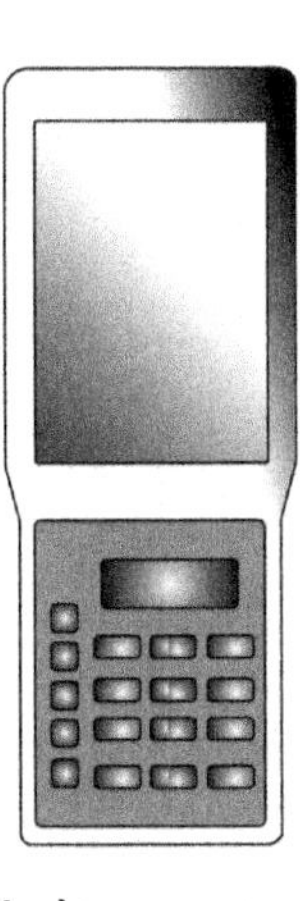

FIGURE 25.2 Typical mobile wireless handheld device

These devices in general support Wi-Fi based wireless communication (802.11 b/g) and can connect with any access point infrastructure nodes, which are available in the field. Additionally these devices also support GPS-based location support.

25.5 SOME OF THE MAJOR BENEFITS OF FIELD-BASED MOBILITY SOLUTIONS

- Remove paper-based data records with handheld digitization.
- Extend equipment life and avoid secondary damage by catching impending failure earlier.
- Ensure field operators adhere to best practices when performing their tasks.
- Train new field operators quickly and ensure they achieve the required competency level.
- Appropriate actions can be taken when critical limits are exceeded for an asset.
- Ensure the sharing of data across teams concerned with asset health.
  - Maintenance, operations, and reliability

Mobile computers being wireless and to enhance the security features, these applications are generally integrated into the level 3/4 network along with firewall protection. The whole purpose of this approach is to keep the process network traffic (generally level 1/2) away from application-centric traffic (Figure 25.3).

25.6 MOBILE DEVICE BASED SOLUTIONS

As the smart mobile phones are gaining access in people's life, we do see the situation not far away in terms of usage of mobile-based application for better connectivity and maintenance scenarios being done using smart phone based applications in industrial environment. Following are the major mobile operating systems, which are popular and being used by vendors for porting the applications.

Following are the major development platforms:

- Android
- iOS
- Windows mobile

Below are various types of mobile applications:

- Native mobile application
- Mobile web application
- Hybrid application

Many automation vendors are exploring the options around usage of above platforms for bringing applications to smart phones. This is one of the major trends of future mobility, which is still in initial phase and will grow as acceptance of mobile-based applications increases inside the process industry.

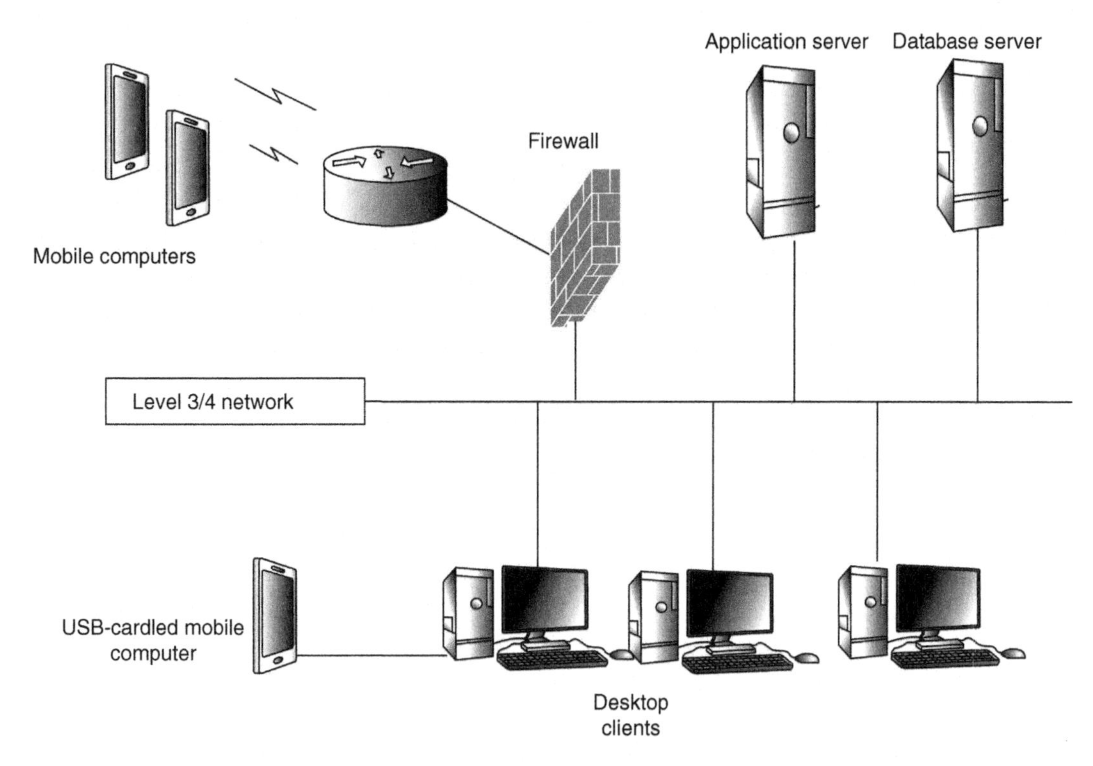

FIGURE 25.3 Typical system architecture for mobility applications

25.6.1 VIDEO SOLUTIONS

Video solutions in plants are intended to monitor critical, high risk, and inaccessible areas. These areas include high-temperature processes such as kilns, boilers, etc. These solutions involve a psychological aspect, as it applies in all industries, i.e., an experienced personnel can gauge the hiccups in the plant by seeing the way process is getting operated, however, since he cannot have access to all areas and mimic displays are not an absolute substitute for viewing the process exactly the way it is running, video solutions are the best method to monitor.

When video solutions have originated, they are always connected through wired networks, as the time progressed the video systems have taken the path of wireless and became more versatile in terms of viewing the systems which means that, apart from video cameras mobile stations, etc. have become part of the technology.

The position of video networks in a DCS is for the purpose of supervision. However, since there is no control component involved, it is placed one level above supervisory networks, in a network that is dedicated for video Ethernet network and brought into the supervisory network for monitoring it manually. Therefore, the location of video systems in DCS, which is wired is as follows:

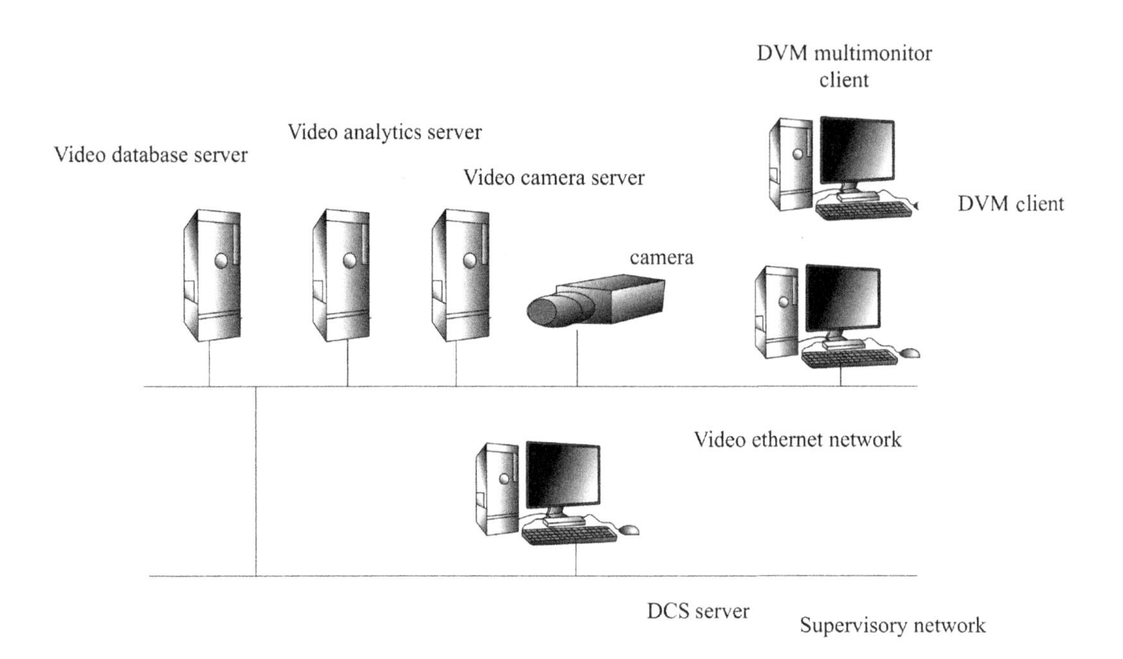

25.6.2 VIDEO SYSTEM ARCHITECTURE

It is designed to have a flexibility of integrating it to a DCS HMI (human machine interface) or on any that is on the WAN, this gives the flexibility of operators in plant to monitor the systems or for the managers who can view it over WAN.

A video system is connected to a separate switch by isolating it from the rest of the process. The intention is since the video transmission uses high bandwidth it should not choke the other traffic available on the network. Hence the best practice is to dedicate a switch for the video system, apply certain filters to transmit traffic that is needed for DCS view thereby reduce the load on overall network.

Components of VIDEO SYSTEM:

1. Cameras
2. Video streamers
3. Video system database
4. Video system analytics
5. Video system camera server
6. DCS server
7. Video system/client

Cameras: Modes of the camera and supported formats are not covered in depth because they are out of the scope of this book, camera can be NTSC or PAL CC TV camera with COAX/Ethernet output, NTSC and PAL are the formats supported for video capturing and streaming. Cameras with PTZ support also can be used; PTZ stands for Pan, Tilt, and zoom, which are used for movement of the camera horizontally,

Resolution	Compression	Image size (kb)	Max frame rate (fps)	Bandwidth per fps (kbps)	1-hour recording disk space per fps(mb)
(160 x 120)	Maximum	1.3	21	10.27	4.51
	High	1.4	21	11.60	5.10
	Medium	2.1	21	16.89	7.42
	Low	3.0	21	23.36	10.27
	Minimum	5.1	21	40.77	17.85

FIGURE 25.4 Sample information for selecting a streamer is as given (all dependent parameters information are also included)

vertically, and zooming into the image, respectively. Others can be fixed cameras with built in streamers. Camera streaming is a very important feature that is required for a video systems.

- Camera streaming converts analog input-to-digital format, streams digital output to TCP/IP network
- Receives command from VIDEO SYSTEM server and executes PTZ commands accordingly.
- Streamer-based motion detection.

Streamers: Choosing a streamer is dependent upon the complexity of the image. Normally an outdoor image is considered to be a complex image because of the varying colors in the picture. A medium complexity image is the one kept in corridors, which covers longer distances with contrasting lighting but the lighting is fixed. Lower complexity is the one, which monitors smaller areas. Factors that should be considered for selecting a streamer is given in Figure 25.4.

25.6.3 VIDEO SYSTEM DATABASE

Normally database server and camera server can be on the same machine if the system is not too large; however, it is generally preferred to have separate servers for database and camera. Such an architecture not only improves the performance of video analytics, but it also ensures that peaks in video analytics processing load (which occur when a lot of objects are being detected/tracked) do not compromise your system's recording or viewing functions.

This server is responsible for:

- storing the configuration and real time information of the video system
- distributing state information to the clients
- accepting requests from other video system clients to view video.

25.6.3.1 Redundancy support in database server

Redundancy is a configuration that allows a backup database server to automatically continue the role of the master (referred to as preferred) database server in the event of a hardware or software failure on the master database server.

The preferred database server is the database server video system clients will connect to if available. The backup database server is fully operational and ready to take over the role of database server if the preferred database server fails. The backup database server is constantly synchronizing with the preferred database server.

25.6.3.2 Considerations

Video system database server redundancy uses SQL server merge replication to keep the SQL database synchronized. Therefore, after a failover it is possible that the data on the backup database server might not be consistent with that on the preferred database server. When the preferred database server is working again, the data can be synchronized between the two servers so they are restored to their correct state.

It is possible to install a camera server on a redundant server. However, it is not a recommended configuration for large systems. When the primary server is restarted after a failover, it automatically resynchronizes with the backup server, provided it has not been down for long. If it has been down for more duration (this longevity varies from vendor to vendor), servers must be manually synchronized.

The date and time on all servers must be synchronized to ensure that all dates and/or times associated with events in the database are consistent between servers.

25.7 VIDEO SYSTEM ANALYTICS

Video analytics involves using specialized algorithms to detect movement in the video and, optionally, to track and classify moving objects. A system recording and executing these analytics is known as video system analytics server. These algorithms attempt to replicate the way in which security personnel analyze and react when they detect motion in a video.

Popular video system algorithms, each of which has their own features:

- Motion detection algorithms
- Object tracking algorithm
- Object tracking and classification algorithm

25.7.1 MOTION DETECTION ALGORITHMS

The motion detection algorithms only detect motion – in essence, they detect changes in the video between adjacent frames.

The motion detection algorithms run on either the server (server-based motion detection) or the streamer (streamer-based motion detection).

Server-based motion detection imposes an extremely high load on the server because the server must decompress the incoming video and perform motion detection on each frame.

Streamer-based motion detection does not impose any load on the server. (Note, however, that not all streamers support streamer-based motion detection.)

The server-based motion detection algorithms are:

- Standard algorithm: designed for indoor use or in scenes where the background is completely stationary
- Premium algorithm: designed for scenes where there is some background movement (such as trees), it can filter out this movement to prevent false alarms

- Object tracking algorithm: The object tracking algorithm attempts to track an object as it moves about, after it has detected the initial movement.
- Object tracking and classification algorithm: The object tracking and detection algorithm extends the capabilities of the object tracking algorithm by classifying objects as "conveyor," "boiler," and so on.

25.8 REGIONS OF INTEREST

A region of interest (ROI) is a region within the field of view that you want the algorithm to monitor. In practice, ROIs allow you to exclude regions that you do not want to monitor, such as fork lifts, trees, and sides of buildings. You define ROIs when configuring an algorithm by drawing one or more rectangles or polygons over the video image.

25.9 MINIMUM OBJECT SIZE

The MOS is the smallest object that the algorithm is instructed to detect. You typically use the MOS to:

- Prevent false motion detection caused by small objects, such as tools.
- Prevent detection of people, when you only want to detect process you define the MOS when configuring an algorithm by drawing a rectangle over the video image. (Note that the position of the MOS is not relevant.)

25.10 VIDEO SYSTEM CAMERA SERVER

Video Systems camera server is used for just storing the recordings from the cameras and then communicate with analytics server to provide the data for further analysis, it works hand in hand with analytics server and database server, actually analytics server is like a camera server with analytics capability. Only thing owes the higher processing capability that an analytics server need, VIDEO SYSTEM camera server is separated. If you change servers, video stored on the old server is not lost because the VIDEO SYSTEM database stores the location of every piece of captured video.

Software and hardware needed for camera server are same as that of the database server.

25.11 DCS

Detailed description of DCS is provided in another section of this book. DCS server is used to configure the video system configuration on its HMI, hence an operator can view the plant details in HMI as mentioned earlier. video system integration with DCS is so tight that any alarm in the DCS calls for an attention from the cameras of the respective areas, this is how it works. While defining the video system cameras assets are defined for them as per the location they belong to in the plant when an alarm or events comes up it comes with the asset as well, hence the respective camera is called up.

25.12 OPERATOR CONSOLE

This is the location where DCS operators can view the plant process as the output from the camera is hooked up with HMI of the server; the luxury of the data accessibility comes on the basis of the security features configured, which are as follows. Security is configured in two ways, one is operator based and another one is station based.

Operator-based security has access levels ranging from operator->supervisor->engineer->manager as the level goes authentication changes from view only to configure every parameter.

Whereas console-based security works as follows, these are multiple clients connecting to server with a station address and some of these stations may need more privileges or less privileges, for example, critical locations need to have restricted authenticity for safety purposes hence stations themselves are defined with limited access. Hence when it comes to clients, based on above security levels only the features of camera can be accessed or denied.

25.13 VIDEO SYSTEM CLIENT

Video system client works similar to DCS console only difference is it will access data from video system servers on a different network. It has multimonitor client support as well; a maximum number of monitors that are supported here are four. And each monitor can have 4 locations monitored within them; hence a user draws benefit of monitoring 16 locations at one shot.

FURTHER READINGS

Barsamian, A., 1999. Inline Blending Minimizes Tankage, Inventory, Giveaway, Boosts Client Satisfaction and Bottom Line. World Refining, Nov–Dec, 1999, p. 46 (Catalog of Blending Benefits).

Barsamian, A., 2001. Genetic Algorithm Optimizers for Blending Applications. ISA 2001 Proceedings.

Index

A

D

E

G

H

I

J

K

L

P

T

X

Z